Advances in brazing

Related titles:

Welding and joining of aerospace materials
(ISBN 978-1-84569-532-3)

Microjoining and nanojoining
(ISBN 978-1-84569-179-0)

Welding and joining of magnesium alloys
(ISBN 978-1-84569-692-4)

Details of these books and a complete list of titles from Woodhead Publishing can be obtained by:

- visiting our web site at www.woodheadpublishing.com
- contacting Customer Services (e-mail: sales@woodheadpublishing.com; fax: +44 (0) 1223 832819; tel.: +44 (0) 1223 499140 ext. 130; address: Woodhead Publishing Limited, 80, High Street, Sawston, Cambridge CB22 3HJ, UK)
- in North America, contacting our US office (e-mail: usmarketing@woodheadpublishing.com; tel.: (215) 928 9112; address: Woodhead Publishing, 1518 Walnut Street, Suite 1100, Philadelphia, PA 19102-3406, USA)

If you would like e-versions of our content, please visit our online platform: www.woodheadpublishingonline.com. Please recommend it to your librarian so that everyone in your institution can benefit from the wealth of content on the site.

We are always happy to receive suggestions for new books from potential editors. To enquire about contributing to our Materials series, please send your name, contact address and details of the topic/s you are interested in to francis.dodds@woodheadpublishing.com. We look forward to hearing from you.

The Woodhead team responsible for publishing this book:

Commissioning Editor: Jessica Rowley
Publications Coordinator: Lucy Beg
Project Editor: Sarah Lynch
Editorial and Production Manager: Mary Campbell
Production Editor: Mandy Kingsmill
Freelance Project Manager: Annette Wiseman
Copyeditor: Sue Clements
Proof reader: Jonathan Webley
Cover designer: Terry Callanan

Advances in brazing

Science, technology and applications

Edited by
Dušan P. Sekulić

WP
WOODHEAD PUBLISHING

Oxford Cambridge Philadelphia New Delhi

Published by Woodhead Publishing Limited,
80 High Street, Sawston, Cambridge CB22 3HJ, UK
www.woodheadpublishing.com
www.woodheadpublishingonline.com

Woodhead Publishing, 1518 Walnut Street, Suite 1100, Philadelphia,
PA 19102-3406, USA

Woodhead Publishing India Private Limited, G-2, Vardaan House,
7/28 Ansari Road, Daryaganj, New Delhi – 110002, India
www.woodheadpublishingindia.com

First published 2013, Woodhead Publishing Limited

British Library Cataloguing in Publication Data
A catalogue record for this book is available from the British Library.

Library of Congress Control Number: 2012951830

ISBN 978-0-85709-423-0 (print)
ISBN 978-0-85709-650-0 (online)

The publisher's policy is to use permanent paper from mills that operate a sustainable forestry policy, and which has been manufactured from pulp which is processed using acid-free and elemental chlorine-free practices. Furthermore, the publisher ensures that the text paper and cover board used have met acceptable environmental accreditation standards.

Typeset by RefineCatch Limited, Bungay, Suffolk
Printed and bound in the UK by the MPG Books Group

Contents

Contributor contact details

(* = main contact)

Editor and Chapter 3

Professor Dušan P. Sekulić
Department of Mechanical Engineering
Institute for Sustainable Manufacturing
414F CRMS Building
College of Engineering
University of Kentucky
Lexington
Kentucky 40506-0108
USA

E-mail: sekulicd@engr.uky.edu

Chapter 1

Dr Nicolas Eustathopoulos, Professor Fiqiri Hodaj* and Dr Olga Kozlova
SIMaP – UMR CNRS 5266
Grenoble INP – UJF
BP 75, F-38402
Saint Martin d'Hères
France

E-mail: fiqiri.hodaj@simap.grenoble-inp.fr

Chapter 2

Dr Yury Flom
NASA Goddard Space Flight Center
Code 541 Greenbelt
Maryland 20771
USA

E-mail: yury.a.flom@nasa.gov

Chapter 4

Dr Victor F. Khorunov* and Dr S. V. Maksymova
E. O. Paton Electric Welding Institute
11 Bozhenko str.
Kiev
03680
Ukraine

E-mail: khorunov@paton.kiev.ua; maksymova@paton.kiev.ua

Chapter 5

Dr A. Rabinkin
Brazing and Joining Consultant, LLC
Metglas Inc.
Hitachi Metals
78 Brooklawn Dr.
Morris Plains
New Jersey 07950
USA

E-mail: arabinkin@gmail.com; anatol.rabinkin@metglas.com

Chapter 6

Dr A. Rabinkin*
Brazing and Joining Consultant, LLC
Metglas Inc.
Hitachi Metals
78 Brooklawn Dr.
Morris Plains
New Jersey 07950
USA

E-mail: arabinkin@gmail.com; anatol.rabinkin@metglas.com

Dr Alexander E. Shapiro
Titanium Brazing, Inc.
5977 Coventry Cross Lane
Hilliard
Ohio 43026
USA

E-mail: ashapiro@titanium-brazing.com; ashapirox@gmail.com

Dr M. Boretius
Listemann AG
LI-9492 Eschen
Wirtschaftspark 34
Liechtenstein

E-mail: m.boretius@listemann.com

Chapters 7 and 8

Dr He Peng
State Key Laboratory of Advanced Welding Production Technology
Harbin Institute of Technology
Harbin 150001
China

E-mail: hithepeng@hit.edu.cn

Chapter 9

Dr Victor F. Khorunov* and Dr O. M. Sabadash
E. O. Paton Electric Welding Institute
11 Bozhenko str.
Kiev
03680
Ukraine

E-mail: khorunov@paton.kiev.ua; sabadash@paton.kiev.ua

Chapter 10

Dr Hui Zhao*
Creative Thermal Solutions
Urbana
Illinois 61802
USA

E-mail: hui.zhao@creativethermalsolutions.com

Dr Ralph Woods
Pleasanton
California
USA

E-mail: raawjan@comcast.net

Chapter 11

Dr Rajiv Asthana*
Department of Engineering & Technology
University of Wisconsin-Stout
Menomonie
Wisconsin 54751
USA

E-mail: asthanar@uwstout.edu

Dr M. Singh
Ohio Aerospace Institute
NASA Glenn Research Center
Cleveland
Ohio 44135
USA

E-mail: mrityunjay.singh-1@nasa.gov

Chapter 12

Dipl.-Ing. Susann Hausner* and Professor Dr.-Ing. habil. Bernhard Wielage
Chemnitz University of Technology
Institute of Materials Science and Engineering
Erfenschlager Str. 73
09125 Chemnitz
Germany

E-mail: susann.hausner@mb.tu-chemnitz.de

Chapter 13

Dr Toshi-Take Ikeshoji
Tokyo Institute of Technology
Japan

E-mail: Ikeshoji.t.aa@m.titech.ac.jp

Chapter 14

Professor Wolfgang Tillmann*
Head of the Institute of Materials Engineering
Faculty of Mechanical Engineering
Technical University of Dortmund
Leonhard-Euler-Str.2
44227 Dortmund
Germany

E-mail: wolfgang.tillmann@udo.edu

Dr Ahmed Elrefaey and Dipl.-Ing. Lukas Wojarski
Institute of Materials Engineering
Faculty of Mechanical Engineering
Technical University of Dortmund
Leonhard-Euler-Str.2
44227 Dortmund
Germany

E-mail: ahmed.elrefaey@uni-edu.de; lukas.wojarski@udo.edu

Chapter 15

Dr Harald Krappitz
Innobraze GmbH
Fritz-Müller-Strasse 97
D-73730 Esslingen
Germany

E-mail: harald.krappitz@innobraze.de

Chapter 16

C. A. Walker
Metallurgy and Materials Joining Department
Sandia National Laboratories
PO Box 8500 MS-0959
Albuquerque
New Mexico 87185-0959
USA

E-mail: cawalke@sandia.gov

Chapter 17

M. Salvo*, V. Casalegno, S. Rizzo, F. Smeacetto, A. Ventrella and M. Ferraris
Department of Applied Science and Technology
Politecnico di Torino
Corso Duca degli Abruzzi 24
10129 Torino
Italy

E-mail: milena.salvo@polito.it

Chapter 18

Dr-Ing. Ina Hoyer* and Professor Dr-Ing. habil. Bernhard Wielage
Chemnitz University of Technology
Institute of Materials Science and Engineering
Erfenschlager Str. 73
09125 Chemnitz
Germany

E-mail: ina.hoyer@mb.tu-chemnitz.de

Chapter 19

Dr D. K. Hawksworth
Sapa Group
109 Thomas Street
PO Box 69533
Oakville
Ontario
L6J 7R4
Canada

E-mail: doug.hawksworth@aol.com; doug.hawksworth@sapagroup.com

Preface

Brazing is a technology for bonding similar or dissimilar substrate materials by using temperatures above 450°C to effect the phase change in an added filler or clad, preferably without markedly impacting the integrity of the substrate(s). By convention, a related process – soldering – is defined as a similar bonding that occurs at temperatures below 450°C. In fact, the physical and chemical processes involved may be quite similar for both kinds of bonding. The significant difference is in the selection of materials; the different materials will lead to somewhat different phenomena during the heating–bonding–cooling sequence of the process.

Brazing began not as a science but as an art, that is, as a body of knowledge and skills developed through trial and error. As such, brazing belongs to one of the oldest technological advances of humankind and traces its origins back to the dawn of civilization. More than 3000 years before the Common Era, as documented on many archeological sites of the Sumerian civilizations, people had understood something of the art of bonding metals. The related skills spread subsequently to Egypt, and later to China and elsewhere. With the advent of modern technological practices, implementations of the associated procedures have been perfected to fascinating levels. This was possible to a large extent thanks to sophisticated experimental techniques, such as scanning electron microscopy, atomic force microscopy, X-ray diffraction analysis, etc., assisted by the development of numerical tools for simulation, such as phase-field methods implemented on supercomputers. Still, the science of brazing is under continuing development, and its rightful connection to the art waits to be fully established.

This book intends to assist professionals in the field of brazing not only to become better informed about aspirations of science to fully understand brazing phenomena, but also to gain an insight into the technology development and its implementation through applications.

Technical literature in the field of brazing, although not without several highly respected and broadly distributed monographs and handbooks devoted to the subject (focusing mainly on the art and paying scant attention to fundamental issues involving advanced phenomena), lacks an in-depth exposition of a variety of technological issues, from metallurgical behavior of participating material systems to process optimization and brazed structures' mechanical, thermal and/or corrosion behavior. This book is intended to address this gap by offering a collection of in-depth presentations devoted to a number of state-of-the-art brazing topics. Nineteen teams, consisting of multiple or single authors, have accepted the Editor's invitation to prepare chapters in selected areas; however, the authors were free to approach the topics as they wished.

In addition to serving the primary purpose mentioned, i.e. offering to the brazing community a sourcebook about the state of the art of the discipline, an additional motivation has shaped its content. The invited authors are from a number of countries, representing some of the most respected world institutions in the field, including national laboratories, agencies, academic organizations, research centers from industrial corporations, and other experts in the field. The R&D efforts and interests, the needs of industry, the technological advances and interactions between industry and institutions of learning still differ significantly with respect to the research/technology development emphasis. For example, teams of authors from Asia, Europe, and/or the Americas may offer a markedly diverse set of sources for further study, hence providing to the professionals from other centers a unique opportunity to get a broader and deeper insight into the relevant information. Furthermore, researchers from different centers inevitably approach the same and/or similar topics of current interest with a somewhat different emphasis. This diverse approach to a variety of topics is, by definition, hardly possible to experience in single-author monographs or even handbooks, which usually advocate for a certain view and provide a standardized approach.

This book is somewhat artificially split into three broad sections: (i) science/fundamentals, (ii) materials/technology and (iii) applications. Clearly, each chapter includes all three aspects and it would have been unproductive to structure the content in a rigid framework. So, the book's structure should be interpreted quite flexibly.

Part I starts with a chapter on wetting prepared by Dr Eustathopoulos, Prof. Hodaj and Dr Kozlova of the Materials and Processes Science and Engineering Laboratory, Grenoble Institute of Technology, France. Fundamental equations of wetting and adhesion, including an interpretation of wetting on different types of solids (metals, ionocovalent ceramics, carbon based ceramics, etc.), are included. The thermodynamics of the process of wetting and spreading and capillary infiltration in brazing processes of non-reactive and reactive systems are discussed. In the second chapter, Dr Flom of the NASA Goddard Space Flight Center, USA, presents an approach to failure assessment of brazed structures. An assessment of the engineering tools of Finite Element Analysis based on well-established theories of Tresca, von Mises, and Highest Principal Stress is provided. Subsequently, the interaction equations and Coulomb–Mohr failure criterion are used to estimate margins of safety of brazed joints. In the third chapter of Part I, Prof. Sekulic from the University of Kentucky, Lexington, KY, USA discusses modeling in brazing. Approaches to modeling thermal cycles of a brazing process, modeling of the thermomechanical behavior of brazed structures, and modeling of micro-scale phenomena (including solid state diffusion, melting of clad/filler and subsequent spreading of capillary driven flows, and solidification during brazing) have been summarized.

Part II opens with the team from the E. O. Paton Electric Welding Institute in Kiev, Ukraine, led by Academician Khorunov and Dr Maksymova, writing about brazing superalloys with boron-free filler metals and γ-TiAl intermetallic alloys with copper-free filler metal. This is the first in a series of chapters devoted to brazing of various material systems, the most voluminous part of this book. High nickel-based and TiAl-based intermetallics are good examples of heat-resistant materials. The discussion of structural transformations occurring in alloys of Ni–Cr–Zr systems is also offered. Melting ranges and the ranges of concentrations have been listed for Ti–Zr–Fe and Ti–Zr–Mn systems, including the structural features and mechanical properties. The use of active filler alloys for brazing ceramic cutting materials is addressed as well. Dr Rabinkin Metglas Inc of Hitachi Metals, USA, offers a comprehensive review of high temperature brazing with a special emphasis on filler metals and processing. An outline of existing classes of brazing filler metals, including new systems of Ni/Fe/Cr-based eutectic alloys including phosphorus, has been provided. The Ni-based alloys with Ge or Zr/Hf/Cr additions are discussed as well. The same chapter offers a detailed discussion of microstructures, properties and optimal brazing conditions for brazing steels and alloys. Dr Rabinkin leads another team, including Drs Shapiro of Titanium Brazing Inc., Coventry Cross Lane, Ohio, USA, and Dr Boretius of Listemann AG, from Liechtenstein, in writing a chapter on brazing of diamonds and cubic boron nitrides. The emphasis on the nature of interactions across the diamond and metal interface is addressed first, followed by diamond graphitization issues during annealing and brazing, as well as wetting of diamonds and CBNs by metals. The chapter concludes with a discussion of brazing techniques. Two chapters prepared by Prof. He Peng of the Harbin Institute of Technology, China, address two different topics: (i) brazing of oxide, carbide, nitride and composite ceramics, and (ii) brazing of nickel aluminides, iron aluminides and Ti–Al intermetallics. The former topic belongs to the collection of chapters in this section devoted to the important field of brazing ceramics. Ceramics are difficult to bond, and brazing is considered as one of the few technologies that can perform the task effectively. The chapter also offers insight into the difficulties of ceramics brazing for each of the ceramic families considered. The second chapter by the same author addresses bonding of intermetallic materials, such as Ni–Al, Fe–Al and Ti–Al. Physical properties of each family are reviewed, followed by a description of brazing methods. The subsequent two chapters, prepared by Academician Khorunov and Dr Sabadash of the E. O. Paton Electric Welding Institute, Kiev, Ukraine, Dr Zhao of Creative Thermal Solutions, Urbana, USA, and Dr Woods, USA, are devoted to aluminum brazing. The first of the two chapters is on aluminum and aluminum to steel brazing, in particular assisted by a reactive flux. The work of Drs Zhao and Woods offers a detailed review of a number of aspects of controlled atmosphere brazing of aluminum performed in the presence of a potassium fluoroaluminate flux. Active metal brazing of advanced ceramic-matrix composites to metallic systems is addressed in a chapter written by Prof. Asthana of the University of Wisconsin-Stout and Dr Singh of the Ohio Aerospace Institute, NASA Glenn Research Center, Cleveland, OH, USA. Advanced ceramic-matrix composites have demonstrated a great potential for demanding applications, but bonding of assemblies including these materials is challenging. Related topics such as wetting and molten metal infiltration are discussed. Subsequently, brazing practices for SiC–SiC, C–SiC, C–C, and ZrB_2-based ultra-high temperature ceramics and oxide, nitride and silicate-based composites are reviewed. Details of interface microstructure, mechanical and physical properties are also discussed. The last chapter on the topic of brazing ceramics to metals in this section is prepared by Dr Hausner and Prof. Wielage of the Institute of Materials Science and Engineering, Chemnitz University of Technology, Germany. As emphasized by the authors, this chapter presents the state of the art in brazing of metal–ceramic compounds. For selected material system combinations, more detail about microstructures and mechanical behavior is provided. The chapter offers information on ceramic properties, standard ceramic brazing processes, brazing of metallized ceramics, active brazing, and mechanical properties of brazed metal–ceramic joints, as well as a discussion of brazing techniques including several examples. The last chapter in this section is devoted exclusively to carbon–carbon (C/C) composites brazing to metals. Dr Ikeshoji of the Tokyo Institute of Technology, Japan, first addresses the properties of C/C composites, followed by a discussion of brazing filler alloys, including filler metal specifications and recommendations. The chapter concludes with a discussion of brazing procedures.

Part III discusses applications of brazing, beginning with a chapter written by a team from the Institute of Materials Engineering of the Technical University in Dortmund, Germany, consisting of Dr Elrefaey and L. Wojarski, and led by Prof. Tillmann. The authors discuss brazing of cemented carbides (tungsten carbide and a metallic binder) and high performance ceramics to substrates such as structural steel, as applied to cutting materials for machining. Suitable brazing alloys for bonding cutting materials made of such systems are discussed in depth. An interesting application of brazing processes is discussed in the chapter by Dr Krappitz, from Innobraze, Esslingen, Germany: coating by brazing. Functional coatings are applied to substrates of different components of assemblies exposed to severe operating conditions. The applications include surface restoration, wear protection, and protection against corrosion and oxidation. Metal–nonmetal brazing for electrical, packaging and structural applications is discussed by C. A. Walker of Sandia National Laboratories, Albuquerque, New Mexico, USA. As stressed by the author, 'Although metal–nonmetal brazing technology is a time-tested and proven joining method used for fabrication of high-voltage, high-current devices and insulators for the electrical power industry, it can offer a host of opportunities for today's materials engineer or designer seeking to utilize unique material properties of recently engineered materials . . .' The chapter starts by discussing design and continues by listing functional requirements, brazing system selection (including both braze and mating substrates), and fixturing. A discussion of metallization schemes follows. The chapter concludes with a review of brazing methods and testing. The use of glasses and glass-ceramics as sealants in solid oxide fuel cells and as joining materials for SiC-based materials is considered by the team from the Materials Science and Chemical Engineering Department of the Politecnico di Torino, Italy. The team is led by Prof. Salvo, and includes as co-authors Dr V. Casalegno, S. Rizzo, Prof. Smeacetto, A. Ventrella and Prof. Ferraris. The first application is a planar solid oxide fuel cell that requires thermomechanically and thermochemically stable sealant materials for long-term exposure to oxidizing, wet-reducing environments at up to 800°C. The second set of applications relates to glass-ceramics use in joining of SiC-based materials in nuclear (fusion and fission) and high-temperature applications. An important application of brazing involves Ni-based filler metals for brazing of metallic materials intended for human use (such as in contact with drinking water or for the food industry). Prof. Wielage and Dr Hoyer, Chemnitz University of Technology, Germany, provide a detailed review of an investigation of the possibility for Ni-based material to be included in the list of harmless materials. The current status of the associated regulations and standardization has been reviewed. Brazing with aluminum brazing sheets is applied to a vast number of contemporary products in the automotive, power, chemical and aerospace industries. In most cases these applications require mass-production by traditional flux brazing. Dr Hawksworth, Sapa, Sweden/Canada, discusses an application of a novel aluminum brazing sheet that does not require additional fluxing. The chapter addresses flux and fluxless brazing materials/techniques, and discusses the use of the novel material. Information about background chemistry and metallurgy influencing fluxless brazing is reviewed. Subsequently, the new composite filler metal–flux material is discussed.

Handling the vast diversity and complexity of the topics assembled in this volume could not be imagined without the involvement of experts in the field willing to provide input and review the material's technical soundness. All contributions have been so reviewed. I am grateful to a number of individuals who volunteered to provide reviews and whose points were incorporated into the final versions of edited contributions. These individuals include, listed alphabetically, but are not restricted to, the following experts: Prof. R. Asthana, University of Wisconsin, USA; Dr P. Garcia, Solvay Fluor, Hanover, Germany; Prof. H. Karaca, Department of Mechanical Engineering, University of Kentucky, Lexington, KY, USA; Dr L. Orman, Solvay Fluor, Hanover, Germany; Dr A. Salazar, Institute of

Research for Technology Development, University of Kentucky, Lexington, KY, USA; Dr A. Shapiro, Titanium Brazing, Ohio, USA; Dr H.-W. Swiderski, Solvay Fluor, Hanover, Germany; and W. Liu, Institute for Sustainable Manufacturing, College of Engineering, University of Kentucky, Lexington, USA.

Appreciation is extended to my students, too numerous to mention by name, whose dedicated work in my laboratory at the College of Engineering, University of Kentucky, Lexington Campus, has served as a constant inspiration in this work.

Dušan P. Sekulić

Part I

Fundamentals of brazing

1

The wetting process in brazing

N. EUSTATHOPOULOS, F. HODAJ and
O. KOZLOVA, SIMaP, France

DOI: 10.1533/9780857096500.1.3

Abstract: The fundamental equations of wetting and adhesion are presented and the contact angles of non-reactive liquid metals and molten oxides on different types of solids are interpreted. The main features of reactive wetting are briefly described and illustrated. The two types of brazing used in practice (capillary brazing and sandwich brazing) are defined and the thermodynamics and kinetics of capillary infiltration are presented. The different configurations expected to occur in 'sandwich brazing' at varying intrinsic contact angles are described and illustrated. Three examples of brazing of metals and ceramics in non-reactive and reactive systems are discussed to show how wetting can affect brazability and the properties of brazed joints.

Key words: contact angle, liquid metals and oxides, brazability, infiltration kinetics, non-reactive and reactive brazing.

1.1 Introduction

It is widely accepted that joining by brazing is possible if the liquid braze wets the solids to be joined. The most frequently used criterion to quantify this condition is a contact angle lower than 90°. One of the aims of this chapter is to go beyond this criterion and provide a more detailed description of the relationship between *wettability* and *brazability*. This relationship is presented and illustrated in Section 1.3.

The technique most widely used for high-temperature wetting experiments is the sessile drop technique. Sessile drop experiments are quite easy to perform and are not expensive (the braze mass and the substrate size involved in an experiment are small, typically 100 mg and $15 \times 15 \times 1$ mm respectively). Moreover, well-documented procedures are now available describing how to conduct these experiments in order to obtain significant information, avoiding artefacts and erroneous deductions (Eustathopoulos *et al.*, 1999, 2005). Finally, after cooling, this type of specimen can be easily used for characterizing braze/solid reactivity by optical microscopy, scanning electron microscopy or microprobe analysis.

Sessile drop experiments can provide very useful information on a number of choices concerning brazing: method of brazing, type and composition of the braze, brazing time, furnace atmosphere, etc. These potentialities of wetting experiments are highlighted in Section 1.4, where three different cases of brazing of metals and ceramics are presented and discussed. However, there are some major differences between sessile drop experiments and brazing tests. For example, size effects can be observed in brazing (but not in sessile drop experiments), because brazing involves particularly low values of braze volume per unit area of solid/braze interface. Furthermore, when two dissimilar materials A and B are brazed, the information obtained from two individual wetting experiments (braze/solid A and braze/solid B) is essential but not sufficient, because the interactions between the liquid braze and one of the solids can affect the interactions between the braze and the second solid. These phenomena are also illustrated and explained in Section 1.4.

In Section 1.2, after a brief review of the fundamental equations of wetting and adhesion, the contact angles of non-reactive liquid metals and molten oxides on different types of solids (metals, ionocovalent ceramics, carbon-based materials, etc.) are given and interpreted by comparing the strength of solid/liquid interactions with the strength of interactions in the bulk liquid. The main features of reactive wetting are also briefly described and illustrated.

1.2 Wetting of solids by liquid metals and oxides

1.2.1 Non-reactive wetting

The wetting of solids by non-reactive liquids is described by the classical Young (Eq. 1.1a) (Young, 1805) and Young-Dupré (Eq. 1.1b) (Dupré, 1869) equations:

$$\cos\theta = \frac{\sigma_{SV} - \sigma_{SL}}{\sigma_{LV}} \qquad [1.1a]$$

$$\cos\theta = \frac{W_a}{\sigma_{LV}} - 1 \qquad [1.1b]$$

According to Eq. 1.1b, the intrinsic contact angle θ in a non-reactive liquid/solid system (Fig. 1.1) results from two types of competing forces: (i) adhesion forces that develop between the liquid and the solid phases, expressed by the quantity of work of adhesion W_a, which promote wetting, and (ii) cohesion forces of the liquid taken into account by the surface energy of the liquid σ_{LV} acting in the opposite direction (the cohesion energy of the liquid is equal to $2\sigma_{LV}$). W_a is related to the surface energies of the solid (S)/liquid (L)/vapour (V) system by $W_a = \sigma_{SV} + \sigma_{LV} - \sigma_{SL}$ (Dupré, 1869).

Liquid metals are high surface energy liquids. Their σ_{LV} is close to 0.5 $J.m^{-2}$ for low melting point (m.p.) metals such as Pb, Sn and In, about 1 $J.m^{-2}$ for moderate m.p. metals Ag, Au, Cu and close to 2 $J.m^{-2}$ for high m.p. metals such as Fe, Ni, Pt or Mo (Eustathopoulos *et al.*, 1999). These values are one to two orders of magnitude greater than the surface energies of room temperature liquids in which bonding is achieved by weak, intermolecular interactions ('physical interactions') – see Table 1.1. Then, according to Eq. 1.1b, good wetting (i.e. a contact angle close to zero) of a liquid metal on a solid substrate

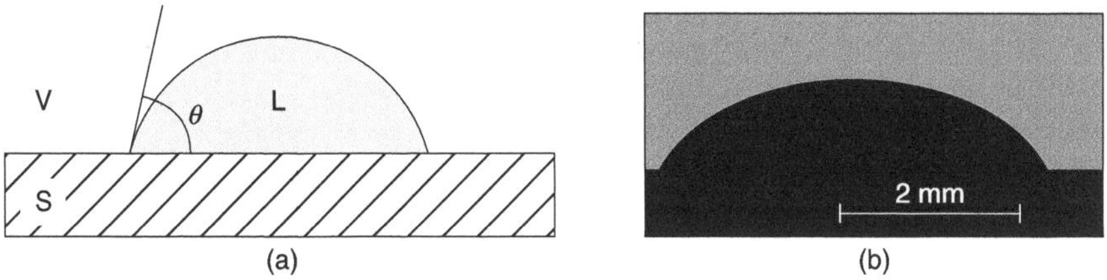

1.1 (a) Definition of contact angle θ in a solid (S)/liquid (L)/vapour (V) system. (b) CuAg drop on a steel surface at 900°C in high vacuum (Kozlova, 2008).

Table 1.1 Surface energy of some typical low temperature liquids at 20°C (Johnson and Dettre, 1993) compared to the surface energy of liquid metals at their melting point (given in parenthesis) (Eustathopoulos *et al.*, 1999)

Liquid	σ_{LV} (mJ.m^{-2})
Pentane	16
Propanol	24
Benzene	29
Water	73
Mercury	500 (−39)
Gold	1138 (1063)
Iron	1850 (1536)
Tungsten	2310 (3380)

can be observed if the work of adhesion is close to the cohesion energy of liquid $2\sigma_{LV}$, which is possible only if the interfacial bond is strong, i.e. chemical in nature (metallic). This condition is fulfilled for liquid metals on solid metals regardless of the miscibility between the liquid and the solid, because in this type of system the interfacial bond is metallic, i.e. strong. Liquid metals also wet semi-conductors such as Si, Ge or SiC (Table 1.2) because these solids, which are covalent in the bulk, have a metallic character near the surface. Finally, liquid metals also wet ceramics such as carbides, nitrides or borides of transition metals because a significant part of the cohesion of these materials is provided by metallic bonds. Among the solids that are not wetted by liquid metals are the different forms of carbon, the ionocovalent oxides and the predominantly covalent ceramics with a high band gap like BN or AlN. In these non-wetting systems adhesion is provided by weak, van der Waals interactions.

It should be emphasized that even solids wettable by liquid metals can lead to non-wetting contact angles due to the presence on their surface of skins acting as wetting barriers (see Section 1.4). For example, most ceramics with a partially metallic character (such as borides or nitrides of transition metals) are

Table 1.2 Wetting of different types of solids by non-reactive liquid metals. The contact angle values are from the review paper (Dezellus and Eustathopoulos, 2010) except for CuSi/SiC (Rado and Eustathopoulos, 2004)

Type of substrate	θ	Contact angle examples
Solid metals	$\theta \ll 90°$	Pb/Fe: 40°
Semi-conductors	$\theta \ll 90°$	Sn/Ge: 40° CuSi/SiC: 30–45°
Ceramics with a partially metallic character	$\theta \ll 90°$	AgCu/Ti_3SiC_2: 10° Au/ZrB_2: 25°
Carbon materials	$\theta \gg 90°$	Au/C: 119–135°
Ionocovalent ceramics	$\theta \gg 90°$	Cu/Al_2O_3, Cu/SiO_2:120–140° Au/BN: 135–150°

easily oxidized solids and their wettability can be hindered by the presence of oxide films.

Molten oxides do not wet carbon-based solids (vitreous carbon, graphite), the contact angles being around 130–140°, which corresponds to W_a values close to 0.1 J.m^{-2} (Table 1.3). These values are typical of systems in which adhesion is provided by van der Waals interactions (Eustathopoulos *et al.*, 1999).

Oxidic glasses wet ($\theta < 90°$) metallic solids such as Pt, Au or Cu (Table 1.3), leading to W_a values in the range 0.3–0.5 J.m^{-2}, which are close to the values obtained for liquid metal / solid oxide systems. Therefore, the different behaviour observed in molten oxide / solid metal systems (wetting) and in molten metal/ solid oxide systems (non-wetting) is not due to different W_a values but to different values of σ_{LV}. For instance, σ_{LV} is 1.30 J.m^{-2} for liquid Cu and only 0.30–0.40 J.m^{-2} for oxidic glasses. Thus, despite the low value of W_a, non-reactive glasses may wet metal surfaces because of their low surface energies (Eustathopoulos *et al.*, 1999; Pech *et al.*, 2004). Enhanced wetting of a metal by a molten oxide can be expected if the metal surface has been oxidized, promoting the development of oxide–oxide interactions across the interfaces. This is illustrated in Table 1.3 by results obtained with the same molten oxide on oxidized and non-oxidized iridium.

Table 1.3 Contact angles of molten oxides on different types of solids. Soda-lime is a common glass consisting mainly of a mixture of SiO_2, CaO and Na_2O

Type of substrate	θ	Contact angle examples
Solid metals	50–80°	Soda-lime/Pt: 75° (Pech *et al.*, 2004) $2Bi_2O_3x3GeO_2$/Ir: 70° (Duffar *et al.*, 2010)
Carbon	$\gg 90°$	Soda-lime/C_v: 135° (Pech *et al.*, 2004)
Solid oxides	0–30°	$2Bi_2O_3x3GeO_2$/IrO_2: 0° (Duffar *et al.*, 2010)

C_v, vitreous carbon.

In non-reactive systems *the spreading rate*, expressed by the velocity U of the triple line, is controlled by the viscous flow and described by a power function U versus the instantaneous contact angle value θ (de Gennes, 1985):

$$U \sim \theta^n \qquad [1.2a]$$

For a spherical-shaped drop with a base radius R, the spreading rate U is *by definition* equal to dR/dt while the angle θ can be easily expressed as a function of drop volume and of R. Then the integration of Eq. 1.2a leads to:

$$R^{3n+1} \sim t \qquad [1.2b]$$

Two different models have been developed, leading to different values of the exponent n. In the first model, viscous dissipation is assumed to occur in the drop bulk ($n = 3$) (de Gennes, 1985), while in the second the dissipation occurs mainly in a region of nanometre size close to the triple line ($n = 2$) (Blake, 1993). Regardless of the value of n, Eqs 1.2a and 1.2b indicate that the wetting rate is very high for short times (i.e. for high θ values) and decreases strongly for longer times, in agreement with the experimental results (Fig. 1.2; Pech *et al.*, 2004). Molten metals and semi-conductors are very low viscosity liquids, their viscosity close to the melting point being a few mPa.s. For this type of liquid *the spreading time*, t_{spr}, that is, the time needed for millimetre-sized droplets to attain capillary equilibrium, is around 10 ms (Naidich *et al.*, 1992; Protsenko *et al.*, 2010; Saiz and Tomsia, 2004). The viscosity of molten glasses is often several orders of magnitude greater than that of molten metals. As a consequence, the spreading time t_{spr} for oxides can attain several seconds or hundreds of seconds (Fig. 1.2) and even more.

1.2.2 Reactive wetting

Wetting of a ceramic by a liquid metal can be assisted by the formation at the interface of continuous layers of a new compound wetted better than the initial substrate by the liquid metal. The thermodynamics and kinetics of reactive wetting, reviewed in Dezellus and Eustathopoulos (2010) and Eustathopoulos (2005), can be summarized as follows (Fig. 1.3):

- The contact angle in a reactive system varies between two characteristic contact angles: the initial contact angle θ_0, which is the contact angle on the unreacted substrate S ($\theta_0 = \theta_S$), and the final contact angle θ_F, which is the contact angle on the reaction product P ($\theta_F = \theta_P$). This is true not only when the reaction product is better wetted than the initial substrate but also when the opposite situation occurs.
- The spreading time t_{spr} in metal/metal and metal/ceramic reactive systems is in the range 10–10^4 s, i.e. several orders of magnitude higher than $t_{spr} \approx 10^{-2}$ s observed for non-reactive metals. Therefore in a given reactive system we have $t(\theta_F) \gg t(\theta_0)$, which implies that the spreading rate of the reactive stage

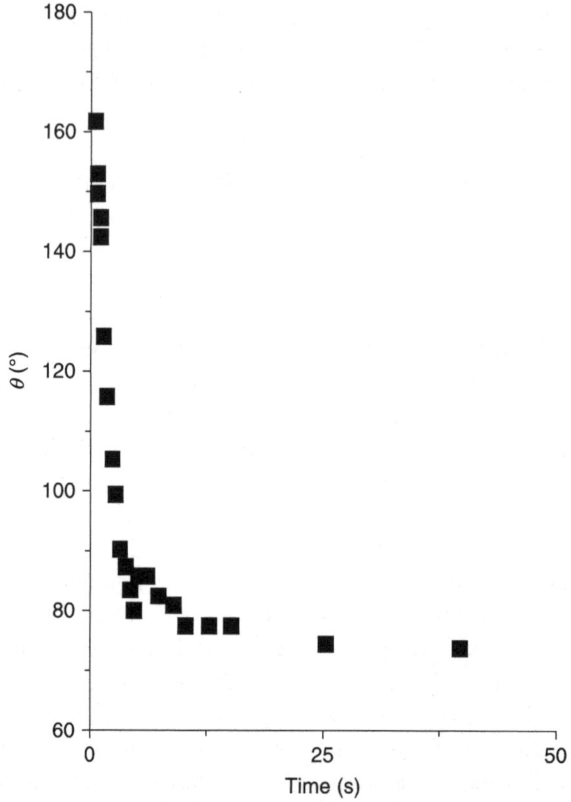

1.2 Wetting kinetics of a drop of soda-lime glass (13.4 wt% Na_2O, 10.9 wt% CaO, 1.6 wt% Al_2O_3, 1.4 wt% MgO, SiO_2 bal.) on steel at 1200°C in He ($\eta = 10^2$ Pa.s). According to Pech *et al.*, 2004.

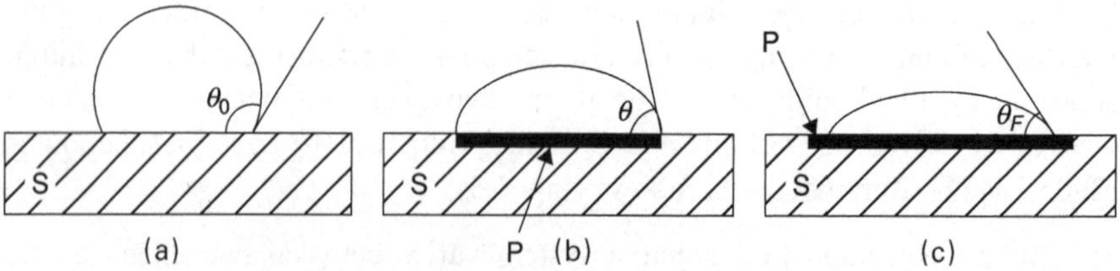

1.3 Schematic representation of the 'Reaction Product Control' model. The initial contact angle θ_0 is the contact angle on the unreacted ceramic substrate surface S (a). After a transient stage, a quasi-state configuration is established at the triple line where the advance of the liquid is hindered by the presence of a non-wettable substrate in front of the triple line (b). Thus, the only way to move ahead is by lateral growth of the wettable reaction product layer P until the macroscopic contact angle equals the equilibrium contact angle θ_P of the liquid on the reaction product, which corresponds to the final contact angle θ_F ($\theta_P = \theta_F$) (c). According to Eustathopoulos, 2005.

is not limited by the viscous flow in the liquid, but by the interfacial reaction close to the triple line.

- The rate of the interfacial reaction at the triple line may, in turn, be controlled by the slower of two successive phenomena that intervene in the reaction process: diffusive transport of reacting species to (or from) the triple line and local reaction kinetics at the triple line. In the case of control by the *local reaction at the triple line*, the rate of reaction and hence the triple line velocity are constant with time in a large domain of θ. An example of linear wetting is shown in Fig. 1.4 for molten Al–Si on vitreous carbon (Calderon *et al.*, 2010). In the case of rapid reactions and for alloys diluted in the reactive solute, diffusion from the liquid bulk to the triple line can be rate-limiting, resulting in non-linear spreading kinetics ($R \sim t^{1/4}$). An example is droplets of copper containing 1–2 at.%Cr on vitreous carbon, a system in which wetting is promoted by the metal-like chromium carbide formed at the interface (Hodaj *et al.*, 2007).

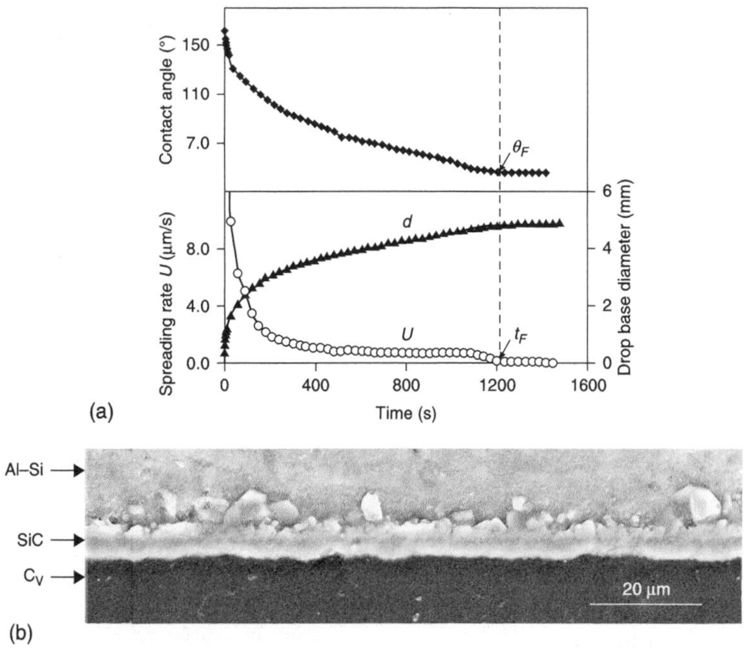

1.4 (a) Contact angle θ, drop base diameter *d* and spreading rate *U* versus time for an Al-25at% Si drop on vitreous carbon at 1000°C in high vacuum. (b) Cross section of the reaction layer of an Al-25at% Si drop on vitreous carbon C_V at 1000°C. The holding time is 60 min. According to Calderon *et al.*, 2010.

1.3 Wetting versus brazing: general considerations

Brazing can be performed using either the 'sandwich' (Fig. 1.5(a)) or the 'capillary' (Fig. 1.5(b)) configuration. In the first case, the braze is placed directly between the two parts to be joined. In the second case, the braze is placed on a surface outside the gap such that the gap is filled by infiltration of the liquid braze from one edge of the gap to the other. This method can be used for complex geometries but it needs good wetting and high infiltration rates.

1.3.1 Capillary brazing

Consider the formation of a joint by non-reactive brazing of two flat parallel plates of unit width made of the same material, in a vertical position (Fig. 1.6). The thickness of the gap is e and the infiltration takes place from the bottom to the top of the gap (upward infiltration).

The variation in surface energy of the system when the liquid front rises from $z = 0$ to z is $2z(\sigma_{SL} - \sigma_{SV})$, which, according to Young's equation (Eq. 1.1a), is equal to $-2z\sigma_{LV}\cos\theta$, θ being the equilibrium contact angle. The work necessary to overcome gravity forces when z changes from zero to z is $ze\rho g\frac{z}{2}$, where ρ is the liquid density and g gravity. The variation in total free energy ΔF is then:

$$\Delta F = F(z) - F(z=0) = -2z\sigma_{LV}\cos\theta + ze\rho g\frac{z}{2} \qquad [1.3]$$

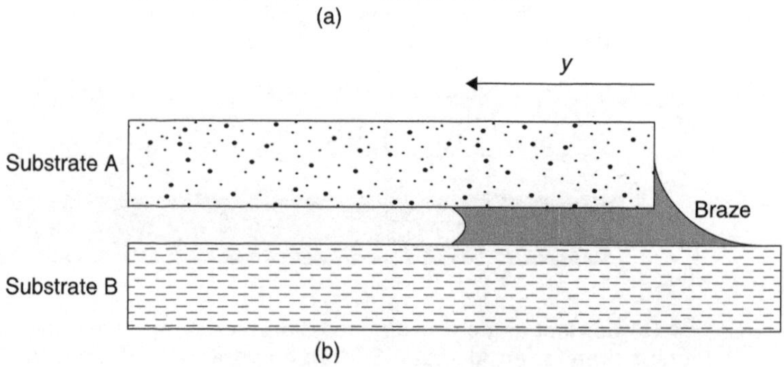

1.5 Schematic presentation of brazing: (a) sandwich configuration, (b) capillary configuration; *y* is the infiltration distance given by equation 1.7

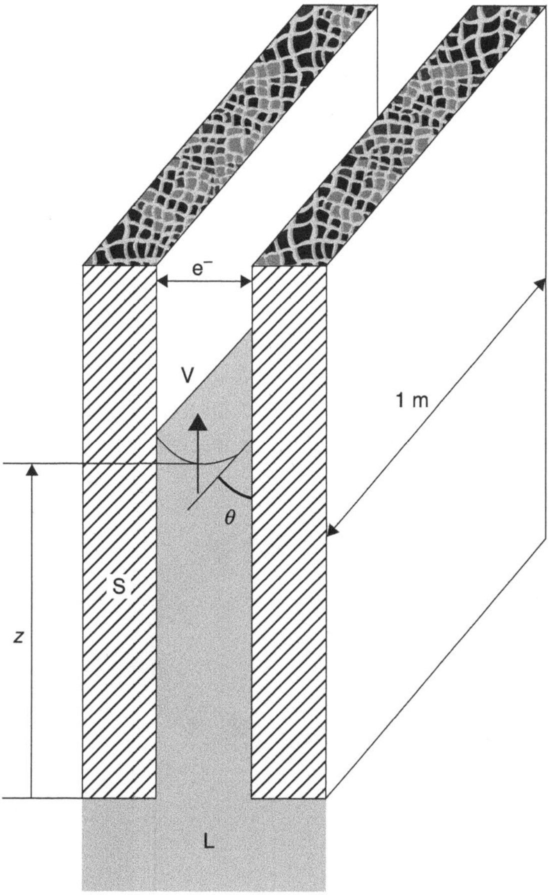

1.6 Non-reactive capillary infiltration. The filling is done from the bottom to the top of the joint. The arrow indicates the direction in which the capillary pressure is exerted. Letters S, L and V indicate solid, liquid and vapor phases respectively; e is the thickness of the gap.

Setting $\frac{d(\Delta F)}{dz} = 0$, the following equation is obtained, giving the equilibrium rise z_{eq} in the capillary:

$$\rho g z_{eq} - \frac{2\sigma_{LV}\cos\theta}{e} = 0 \qquad [1.4a]$$

In Eq. 1.4a the first term is the hydrostatic pressure P_h and the second term the capillary pressure P_c:

$$P_c = \frac{2\sigma_{LV}\cos\theta}{e} \qquad [1.4b]$$

By taking typical values for a metallic braze, $\sigma_{LV} = 1$ J.m^{-2}, and $\rho = 5 \times 10^3$ kg.m^{-3}, we find, for $e = 20$ µm and $\theta = 45°$, $z_{eq} = 2$ m. The values of z involved in brazing are usually much lower than z_{eq}, meaning that the gravity forces are negligible compared with the capillary forces. Therefore the total pressure P_t applied to the infiltrating liquid is nearly equal to the capillary pressure (Eq. 1.4b). According to this expression, gap infiltration is possible if the equilibrium contact angle θ is lower than a threshold contact angle $\theta^* = 90°$. The same criterion applies for other geometries, the only condition being that the gap area perpendicular to the infiltration front must be constant. An example is a gap formed between two coaxial cylindrical-shaped solids (Fig. 1.7).

When the condition of constant gap area is not fulfilled, θ^* is no longer equal to 90°. For instance, in the V-shaped horizontal gap of Fig. 1.8, consisting of two plates of unit width forming an angle φ, the infiltration does not correspond just to the replacement of an area of solid/vapour surface by an equal area of

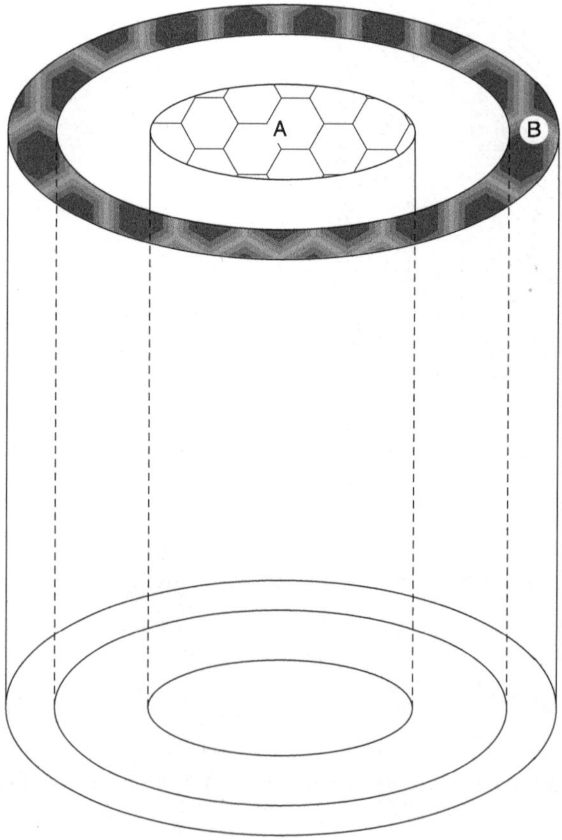

1.7 Gap of constant area formed between two coaxial cylindrical-shaped solids: rod A inserted in tube B.

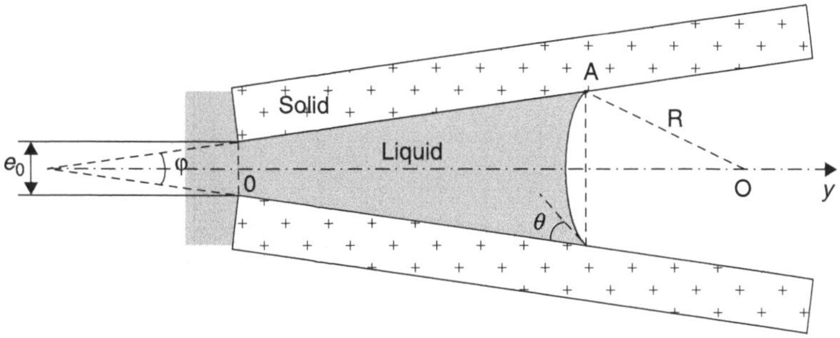

1.8 Schematic representation of liquid infiltration in a V-shaped capillary gap of angle φ. e_0 is the gap thickness at the entrance of the gap where $y = 0$.

solid/liquid interface, but also to an increase in the liquid/vapour surface area. As a consequence, in this case, the capillary pressure P_c becomes a function of y. Taking into account that for the geometry of Fig. 1.8 one of the curvatures of the liquid/vapour surface is zero (this is the curvature in a direction perpendicular to the figure), P_c is given by:

$$P_c = \frac{\sigma_{LV}}{|R|} = \frac{2\sigma_{LV}\cos(\theta + \varphi/2)}{(e_0 + 2y.tg(\varphi/2))} \qquad [1.5]$$

where R is the radius of curvature in the plan of the figure with $|R| = \text{AO}$. According to this equation the infiltration is spontaneous (i.e. $P_c > 0$) if the equilibrium contact angle θ is lower than a threshold value θ^*:

$$\varphi\theta < \theta^* = 90° - \varphi/2 \qquad [1.6]$$

If $\varphi = 0$ (i.e. parallel plates), the condition 1.6 becomes $\theta < \theta^* = 90°$, as expected.

In the preceding section, the capillary brazing of two solids of the same nature has been considered. When brazing concerns two different materials A and B (for example a metallic substrate A with a ceramic substrate B), the wetting of the liquid braze on both substrates must be taken into account (Fig. 1.9). In this case, it is easily shown that, from a thermodynamic point of view, spontaneous infiltration occurs if $\theta_A + \theta_B < 180°$. Taking a typical value $\theta_A = 30°$ for the contact angle of a metallic braze on a metallic solid, this means $\theta_B < 150°$. The contact angles of non-reactive liquid metals on ionocovalent ceramics or on carbon-based solids are in the range 120–150° (Table 1.2), i.e. close to but lower than 150°, thus indicating that spontaneous infiltration would occur. However, for non-wetting liquids with $\theta_B \gg 90°$, *on real solids*, that are solids with rough surfaces, a true solid ceramic/liquid interface cannot be formed at any point on the ceramic surface, because the liquid cannot penetrate the surface asperities present on real materials. For this 'composite wetting', the contact angle θ_B is higher (and usually

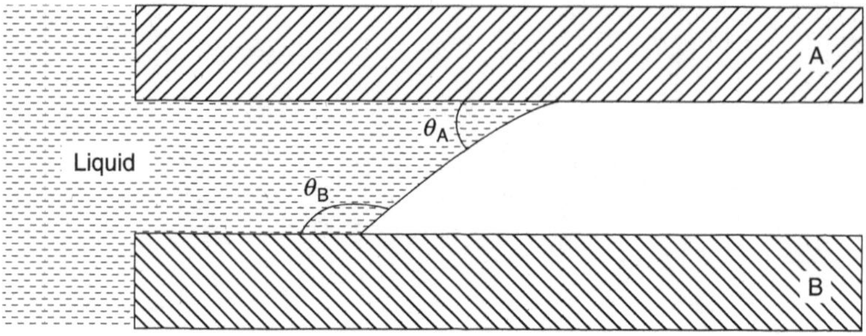

1.9 Infiltration of a liquid in a capillary gap between two parallel plates of dissimilar materials.

much higher (Eustathopoulos *et al.*, 1999)) than the intrinsic contact angles given in Table 1.2, thus excluding spontaneous infiltration in practice.

In order to perform capillary brazing in a given application, the time taken by the liquid to infiltrate a gap completely must be within the brazing times used in practice, i.e. 5–30 minutes. The flow of a liquid in a capillary gap consisting of two parallel and horizontal flat plates separated by a distance *e* is laminar. The fluid velocity at a distance *y* from the gap entrance $U = dy/dt$ is proportional to the pressure gradient P_c/y and to the inverse of the viscous resistance, a quantity which varies as η/e^2, where η is the dynamic viscosity: $U = e\sigma_{LV}\cos\theta/(6\eta y)$ (Haon *et al.*, 2008). The integration of the expression of *U*, first with respect to distance *y* and then with respect to time *t*, leads to the classical Washburn's parabolic infiltration law (Washburn, 1921):

$$y^2 = e\frac{\sigma_{LV}\cos\theta}{3\eta}t \qquad [1.7]$$

For more complex gap geometries, Eq. 1.7 remains basically valid if the gap area perpendicular to the infiltration front is constant. The only influence of the geometry is on the value of the numerical constant. If the gap is vertical and the hydrostatic pressure is not negligible in comparison with the capillary pressure, the infiltration rate must be calculated taking into account the total pressure P_t, which leads to a more complex, non-explicit function $y(t)$ (Eustathopoulos *et al.*, 1999; Milner, 1958).

Figure 1.10 presents the variation in infiltration distance *y* as a function of time for $\theta = 45°$ taking for $\sigma_{LV}/\eta = 250$ m.s^{-1}, a value that is typical for liquid metals and alloys. This figure shows that an infiltration distance $y = 100$ mm is obtained with infiltration times $t \sim 2$ and ~ 20 s for $e = 100$ and 10 μm respectively, meaning that the infiltration process is very rapid. Even for $e = 1$ μm, a gap thickness corresponding to the space between two plates in apparent contact, where *e* is determined by the surface roughness of the plates, the infiltration time is only a few minutes, which is compatible with the brazing times used in practical

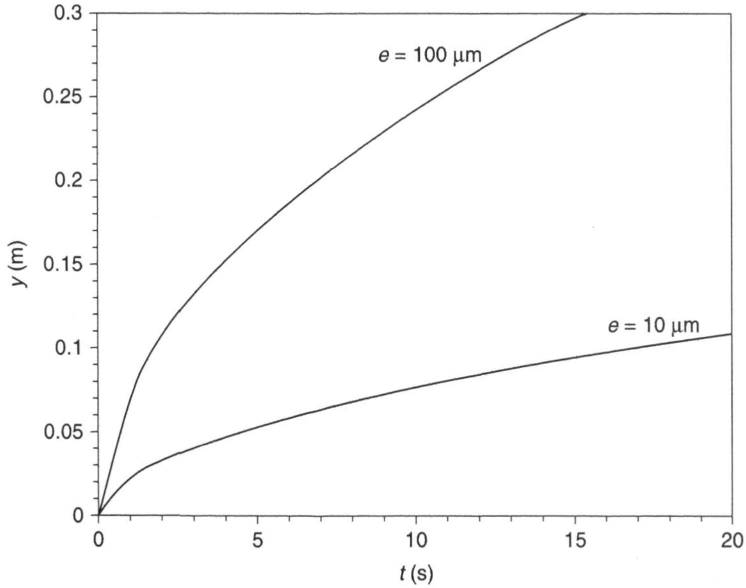

1.10 Variation of the infiltration distance y as a function of time in a horizontal capillary gap formed by two parallel flat plates for two values of thickness e of the joint. $\theta = 45°$, $\sigma_{LV}/\eta = 250$ m.s^{-1}.

situations. Note, however, that in the case of molten oxides, which are much more viscous than liquid metals, the infiltration time can be several orders of magnitude higher than the previously calculated infiltration times. Another limitation of capillary brazing concerns reactive brazing, where the infiltration process is no longer governed by the viscous flow of the liquid in the gap but by the interfacial reaction. In this type of infiltration the average infiltration rate, equal to the wetting rate, is only a few μm.s^{-1} (see an example in Fig. 1.4(a)), therefore excluding the use of capillary brazing in most practical applications.

Figure 1.11 gives an example of the capillary brazing of two pieces of steel obtained with a non-reactive CuAg eutectic braze at 860°C. This shows that the joint is completely filled throughout its length.

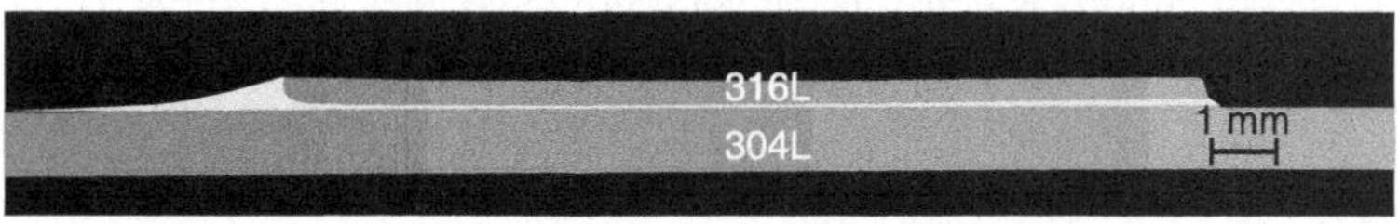

1.11 Back-scattered electron microscopy of a cross section of a joint between two stainless steels obtained by capillary brazing using a CuAg eutectic braze. The braze was placed initially on the left side of the joint. $T = 860°C$, $t = 20$ min. The compositions of steels 316L and 304L are given in Table 1.4 (Kozlova, 2008).

1.3.2 Sandwich brazing

For the sandwich brazing technique (see Fig. 1.5(a)), four configurations can be observed depending on the value of the intrinsic contact angle θ of the liquid alloy on the solid parts (Kozlova, 2008):

$\theta \gg 90°$

On real solid surfaces, i.e. solids with a certain roughness at micron scale, a liquid with intrinsic contact angle $\theta \gg 90°$ (typically 130–150°) contacts the solid only at a limited number of points, thus leading to composite wetting (see Fig. 1.12(a), left) (Eustathopoulos *et al.*, 1999). In this case, even a small stress generated during cooling can lead to detachment of the solidified braze from the two parts (no bonding – see Fig. 1.12(a), right).

$\theta > 90°$

Below a certain contact angle, the value of which depends on the roughness parameters of the surface and on the intrinsic contact angle (Eustathopoulos *et al.*, 1999), the configuration corresponding to composite wetting is no longer stable and the contact between the braze and the solid is intimate at any point of the interface (see Fig. 1.12(b), left). However, a tendency to dewetting exists, as shown schematically in Fig. 1.12(b), right. Dewetting, which can be initiated at defects such as porosities and bubbles, is detrimental to the mechanical performance of the joint.

$45° < \theta \leq 90°$

The solids and the braze meet at intimate interfaces and total filling of the gap is observed. If the external pressure P_{ext} applied to the liquid is negligible compared with the capillary pressure, the liquid will stay inside the gap (see Fig. 1.12(c) and 1.13).

The liquid can leave the gap only if the external pressure is greater than a threshold value P_c^* corresponding to the position of the meniscus defined in Fig. 1.12(c) (in this position the angle formed by the meniscus on the lateral surface of solid A is equal to the equilibrium contact angle on A, θ_A). P_c^* can be easily calculated as a function of parameters σ_{LV}, θ_A, θ_B and e_0 (Kozlova, 2008):

$$P_c^* = \frac{\sigma_{LV}(\sin\theta_A - \cos\theta_B)}{e_0} \tag{1.8}$$

Equation 1.8 applied to the case of two plates made of the same material gives:

$$P_c^* = \frac{\sigma_{LV}(\sin\theta - \cos\theta)}{e_0} \tag{1.9}$$

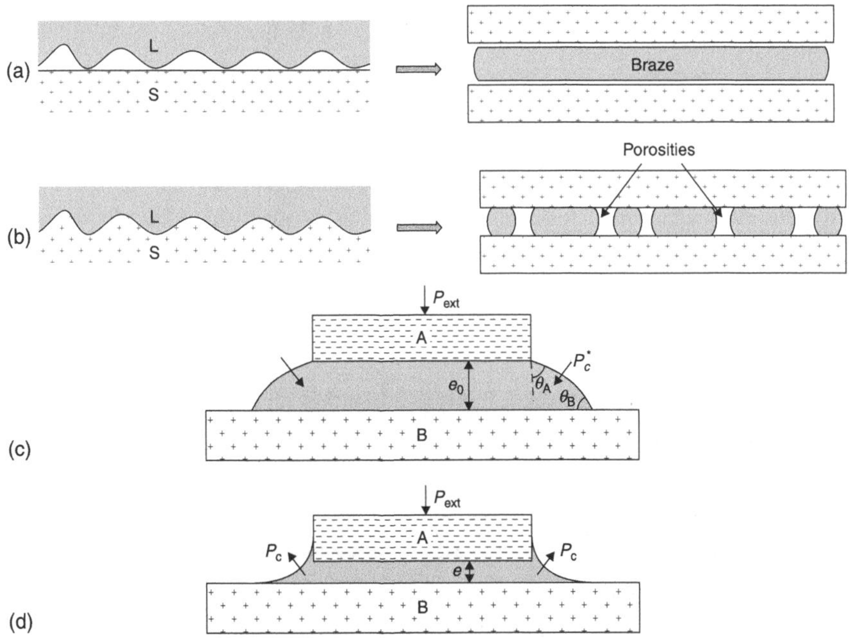

1.12 (a) $\theta \gg 90°$: At the microscopic scale, the liquid contacts the rough surface of the solid only on a few points (left). After brazing, non-bonding of solid parts is observed at room temperature (right). (b) $\theta > 90°$: At the microscopic scale, an intimate contact exists at any point of the interface (left). Dewetting with formation of porosities can be observed at room temperature (right). (c) $45° < \theta \leq 90°$: The liquid is retained inside the gap except if P_{ext} exceeds the threshold capillary pressure P_c^* given by Eq. 1.8. (d) $\theta < 45°$: A meniscus is formed spontaneously around the gap regardless of the value of P_{ext}. The final thickness e is lower than e_0.

1.13 Sandwich brazing of two stainless steels by CuAg eutectic at 820°C ($t = 20$ min, high vacuum). The liquid stays inside the gap. The compositions of steels 316L and 304L are given in Table 1.4 (Kozlova, 2008).

It can be easily verified that for $\theta > 45°$ P_c^* takes positive values.

A numerical application with $\sigma_{LV} = 1$ J.m^{-2}, $e_0 = 100$ μm and $\theta = 60°$ gives $P_c^* =$ 3600 Pa. The minimum value of P_{ext} is the pressure exerted by the weight of the upper solid. Taking $\rho = 5 \times 10^3$ kg.m^{-3} and a thickness for the upper plate $h =$ 10 mm, P_{ext} is 500 Pa. This value is much lower than the threshold pressure P_c^*, which means that, in this case, the liquid cannot leave the gap without exerting an additional external pressure.

$\theta < 45°$

As in the previous case, the interfaces between the solid parts and the braze are intimate everywhere and total filling of the gap is observed with spontaneous formation of a meniscus outside the gap. Indeed, given that P_c^* is negative in this case, both external and capillary pressures act in the same direction, thus leading to a decrease in the final thickness of the joint (see Figs 1.12(d) and 1.14).

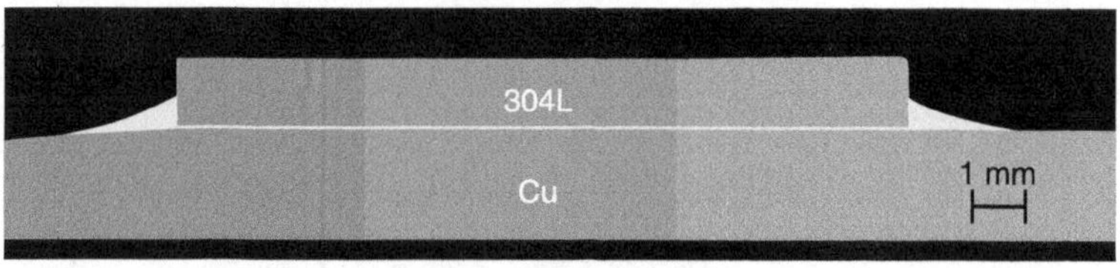

1.14 Sandwich brazing in high vacuum of a steel plate to copper (T = 860°C, 20 min). The joint thickness is much lower (by a factor of 5) than the initial thickness of the sheet of CuAg braze. The composition of steel 304L is given in Table 1.4 (Kozlova, 2008).

1.4 Brazing of metals and ceramics by non-reactive and reactive alloys

1.4.1 Wetting of stainless steels by CuAg eutectic

Stainless steels (ss) are covered by tenacious oxide layers, making them highly resistant to oxidation. These passive films are also a barrier to wetting and brazing by liquid metals and alloys (Eustathopoulos *et al.*, 1999). However, above a certain temperature, heating in high vacuum leads to good wetting (contact angle $\theta \ll 90°$), thus enabling these materials to be joined by brazing (McGurran and Nicholas, 1985; Ambrose and Nicholas, 1996). The wetting temperature is an important technical parameter, since an excessively high brazing temperature can affect the microstructure and properties of a metallic partner of the steel or of the steel itself. For instance, brazing of a ss to copper by CuAg at $T > 900°C$ leads to excessive dissolution of the Cu part into the liquid braze (Kozlova, 2008). A comprehensive study of wetting of ss by CuAg eutectic and CuAg based alloys was performed by Kozlova *et al.* (2008) by varying the temperature, the level of vacuum in the furnace and the type of steel (Table 1.4).

Figures 1.15 and 1.16 present the results of sessile drop experiments performed by the sessile drop technique in two different high vacuum furnaces (Fig. 1.15) and for four different steels (Fig. 1.16). In all cases, the initial contact angle θ_0

Table 1.4 Carbon and titanium content in stainless steels studied in Kozlova (2008). Other elements (in wt.%): Cr: 16–20; Ni: 8–14; Mn ≈ 2; Mo ≈ 0–2; Si ≈ 0.75–1; P ≈ 0.04; S ≈ 0.03; Fe = balance

Steel	C (wt.%)	Ti (wt.%)
316L	0.03	–
316	0.08	–
304L	0.03	–
321	0.08	0.40
316Ti	0.09	0.45

observed at melting is much higher than 90° and changes hardly at all for the first 15 minutes. At a certain temperature, T^0, θ starts to decrease rapidly towards a contact angle θ_M well below 90° defining the change in the slope of the contact angle versus time curve. Both T^0 and θ_M values depend on the vacuum level in the furnace and on steel composition. Thereafter, the decrease in θ continues at a much slower rate down to a nearly steady angle θ_F observed after 30–40 min of holding at 860°C. The contact angles in the range 120–140° observed at melting are typical of noble metal/ionocovalent oxide systems and are considered to be those on oxidized steels. Conversely, the wetting contact angles $\theta_M \approx$ 30–60° are typical of non-reactive liquid metal/solid metal systems (Table 1.2), meaning that the transition from θ_0 to θ_M is controlled by steel deoxidation. Finally, the spreading

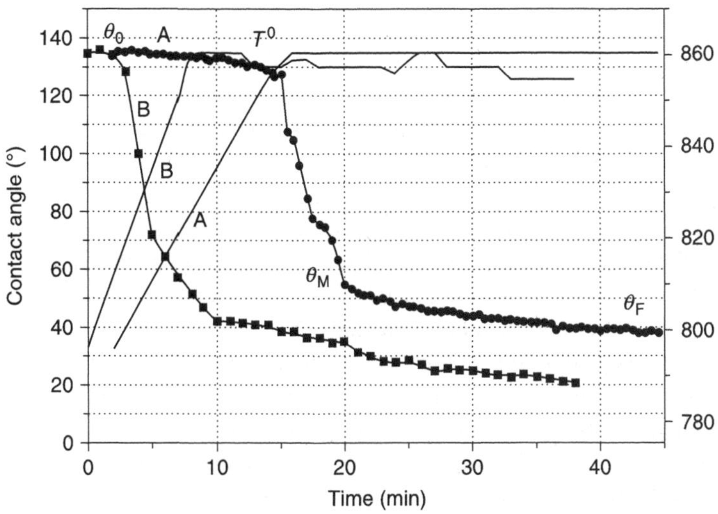

1.15 Comparison of wetting curves obtained by the sessile drop method in alumina ($P = 9.7 \times 10^{-5}$ Pa, curves A) and metallic ($P = 3 \times 10^{-5}$ Pa, curves B) furnaces. CuAg eutectic on a 321 stainless steel (Kozlova *et al.*, 2008).

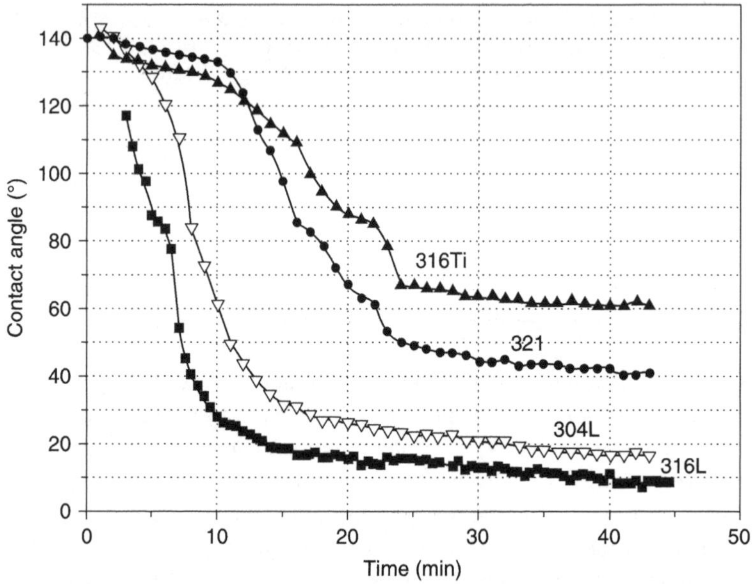

1.16 Wetting curves obtained by the sessile drop technique in an alumina furnace ($P = 9.7\times10^{-5}$ Pa) for CuAg eutectic with four different stainless steels during temperature rise to 860°C (Kozlova *et al.*, 2008).

stage from θ_M to θ_F is related to dissolution occurring at the liquid alloy/steel interface (Kozlova *et al.*, 2008).

As discussed in Kozlova *et al.* (2008), steel deoxidation in high vacuum occurs by reduction of chromium oxide in the passive film by residual carbon (ss contains several hundreds of ppm of carbon, Table 1.4) that can reach the steel surface by rapid grain boundary diffusion:

$$3[C] + Cr_2O_3(S) \rightarrow 3CO(G) + 2[Cr] \qquad [1.10]$$

where the symbol used for C and Cr denotes that these elements are in solution in the steel. A ss surface in contact with air at room temperature is instantaneously covered by a continuous, nanometric thick oxide film. When the steel is heated in a furnace, thickening of the oxide film occurs, because the chromium in the steel acts as a getter of the oxygen present as an impurity in the furnace atmosphere. This phenomenon predominates up to a certain temperature T^0, where carbon diffusion becomes effective, thus enabling steel deoxidation through reaction 1.10. This analysis may explain why a rather limited increase in the total pressure in the furnace (from 3×10^{-5} to 9.7×10^{-5} Pa), implying an equivalent increase in the partial pressure of residual oxygen, causes a very significant shift (by 50°C) towards the higher temperatures of the beginning of spreading (Fig. 1.15). The Ti contained in the steel (321 and 316Ti steels, but not 304 and 316L steels, contain a significant amount of Ti; see Table 1.4) leads to a strong decrease, by nearly two orders of magnitude, in carbon activity due to the formation of titanium carbide (Kozlova *et al.* 2008). This decrease produces a significant increase in the temperature T^0 and a 30–40° increase in the final contact angle (Fig. 1.16).

The above results clearly indicate that the two main factors determining *the brazability* of ss in high vacuum are the quality of the furnace vacuum and the content of carbide-forming alloying elements in the steel.

1.4.2 Wetting of SiC by molten metals and molten silicides

When a small piece of a high melting point metal such as nickel is placed on silicon carbide, the shape of the drop formed on melting is disturbed due to a strong reaction between nickel and silicon carbide (Fig. 1.17). This reaction, evidenced by the deep crater formed under the droplet, consists of SiC dissolution into nickel while the excess carbon forms graphite precipitates close to the interface (Rado *et al.*, 1999):

$$SiC(S) \rightarrow [Si] + C_{gr}(S) \qquad [1.11]$$

Given this exacerbated reactivity, especially the brittle graphite particles that are detrimental to mechanical properties, the use of Ni (or any other high melting point metal) for silicon carbide brazing is not possible.

In order to decrease the reactivity, one solution would be to use moderate melting point metals such as silver or gold. In this way the reactivity with SiC at temperatures around 1000°C is indeed considerably reduced. However, at the same time non-wetting is observed, as the contact angles are much higher than

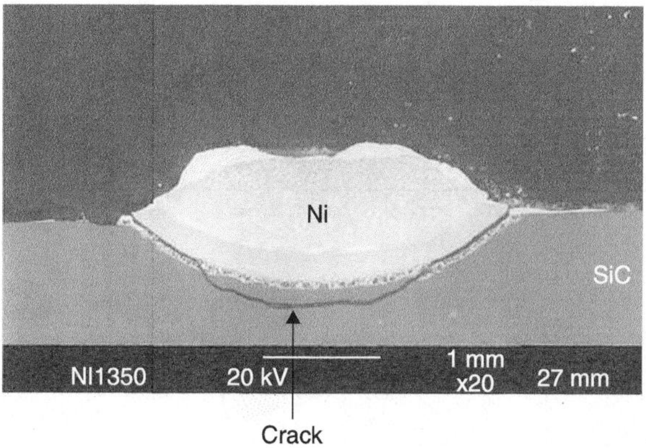

1.17 Cross section in a Ni/SiC specimen showing the strong reactivity between Ni and SiC. The holding time was 15 min at 1350°C. (Due to SiC dissolution into Ni, complete melting occurs at a temperature lower than the melting temperature of pure Ni (1452°C).) The black line is a crack formed in cooling. According to Rado *et al.*, 1999.

90° (Naidich, 1981). These results disagree with the results contained in Table 1.2, indicating that semi-conducting solids such as SiC are wettable by liquid metals. This apparent contradiction was solved by Rado (Rado, 1997; Rado *et al.*, 1999), who showed that the non-wetting of SiC by liquid metals observed up to 1100°C does not reflect the intrinsic properties of silicon carbide surfaces. This is due to the presence, on the SiC surface, of a tenacious silica skin forming instantaneously at room temperature in air and subsequently growing during the temperature rise in the furnace due to the oxygen impurity in the furnace atmosphere. This film acts as a barrier to wetting, just like the passive layers present on ss.

The skin can be removed *in situ* using a metal–silicon alloy instead of the pure metal. Indeed, silicon can react with the silica layer, forming the volatile silicon monoxide:

$$SiO_2(S) + [Si] \rightarrow 2SiO(g) \qquad [1.12]$$

This *in situ* cleaning of the SiC surface is illustrated by the results presented in Fig. 1.18 for a model system consisting of a Cu–Si alloy on preoxidized SiC (Dezellus *et al.*, 2002). Wetting in this system occurs in two stages. In the first stage the contact angle decreases rapidly towards 100°, which is the contact angle of the alloy on oxidized SiC, i.e. on silica. At this point the contact angle remains constant for a certain time needed for the alloy to deoxidize the substrate

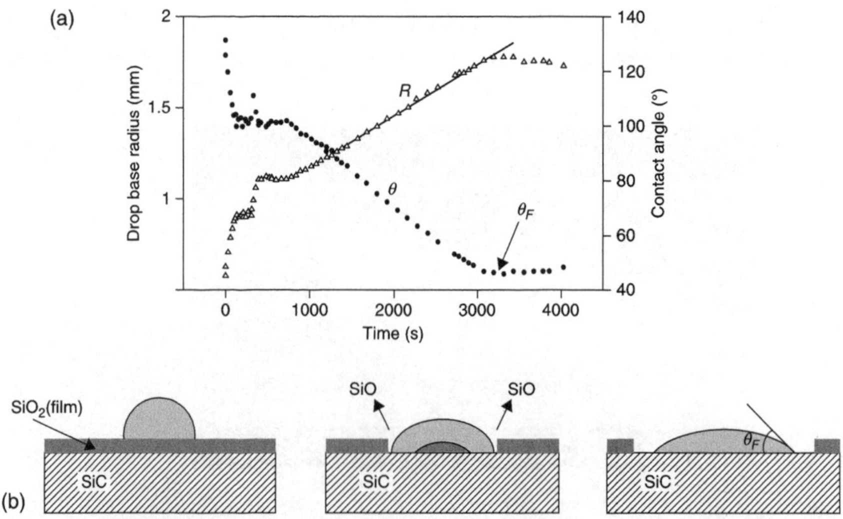

1.18 (a) Contact angle and drop base radius, *R*, versus time for a Cu–40at%Si droplet on preoxidized SiC (Dezellus *et al.*, 2002). (b) Schematic representation of the transition from the initial contact angle on the oxidized SiC to the final contact angle on a clean SiC surface.

around the triple line, thus coming into direct contact with silicon carbide. The contact angle then decreases again and wetting continues until the contact angle becomes equal to the equilibrium contact angle of the alloy on SiC.

This mechanism is effective for any metal–silicon alloy. For a given system, the silicon concentration is chosen such that the formation of graphite precipitates is suppressed. In this way a non-reactive nickel silicide / silicon carbide interface is obtained (Fig. 1.19) together with excellent wetting (the contact angle in this system is around 20°) (Rado *et al.*, 1999).

The removal of the oxide film is crucial in brazing. Thus, when two pieces of SiC are brazed with a metal–Si alloy in the sandwich configuration, deoxidation of SiC is not very effective due to silicon monoxide entrapment. This leads to the presence in the gap of large voids that are detrimental to the mechanical properties of the assembly (Fig. 1.20(a)). Conversely, in the capillary configuration, the evacuation of the volatile SiO occurs easily at the infiltration front, thus leading to complete deoxidation of SiC and complete filling of the gap (Fig. 1.20(b)) (Koltsov, 2005; Koltsov *et al.*, 2008).

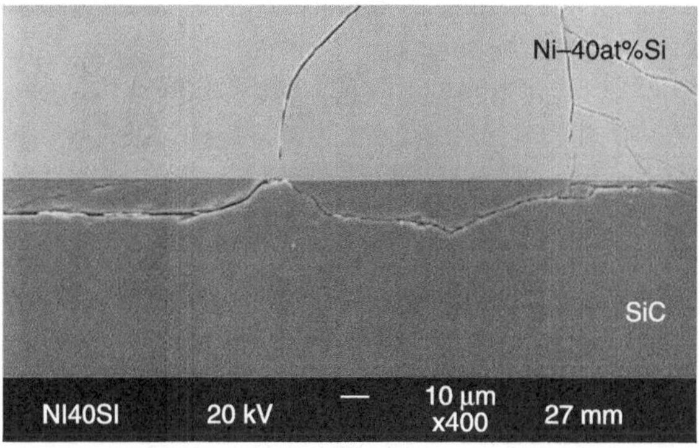

1.19 Cross section in a Ni–40at%Si/SiC specimen showing the absence of reactivity between nickel silicide and SiC (20 min at 1350°C) (Rado *et al.*, 1999).

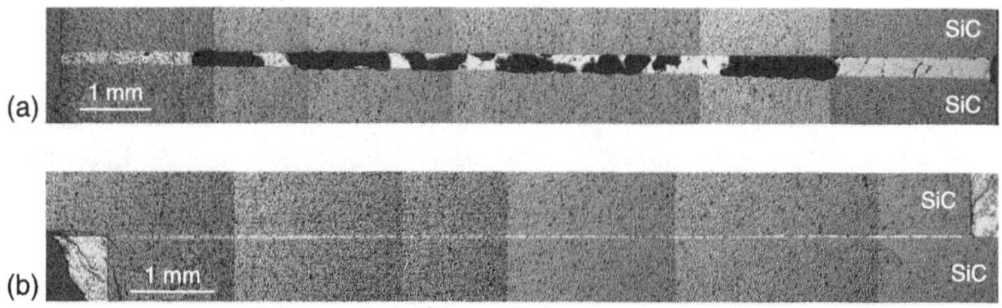

1.20 Brazing of SiC by a Pr silicide in high vacuum. Sandwich (a) versus capillary (b) brazing (15 min at 1250°C) (Koltsov, 2005; Koltsov *et al.*, 2008).

The accurate knowledge of basic phenomena occurring at metal/silicon carbide interfaces, thanks to the previously cited studies, has led to the development of a series of patented high melting point metal–silicon brazes with several applications in materials technology (Gasse, 1996).

1.4.3 Wetting of alumina by reactive CuAgTi alloys

Brazing of alumina to itself or to metallic partners is often performed using CuAg alloys containing a few per cent of Ti, an element that promotes wetting and adhesion. The effect of Ti on wetting is illustrated by the sessile drop experiment, depicted in Fig. 1.21, in which a CuAg–1.75wt% Ti alloy was processed *in situ*, by placing a small quantity of Ti on top of a piece of AgCu on an alumina substrate (Voytovych *et al.*, 2006).

The large, non-wetting, contact angle observed at melting, which is typical of a non-reactive noble metal on an oxide, is the contact angle of CuAg alloy on alumina. This angle remains constant for several minutes, the time needed for Ti to dissolve in the liquid and reach the metal/oxide interface by diffusion. Afterwards, the contact angle decreases rapidly towards a value close to 10°, which corresponds to excellent wetting.

The micrograph in Fig. 1.22 presents the interface formed between the alloy and alumina at the end of the wetting experiment. The reactive interface is

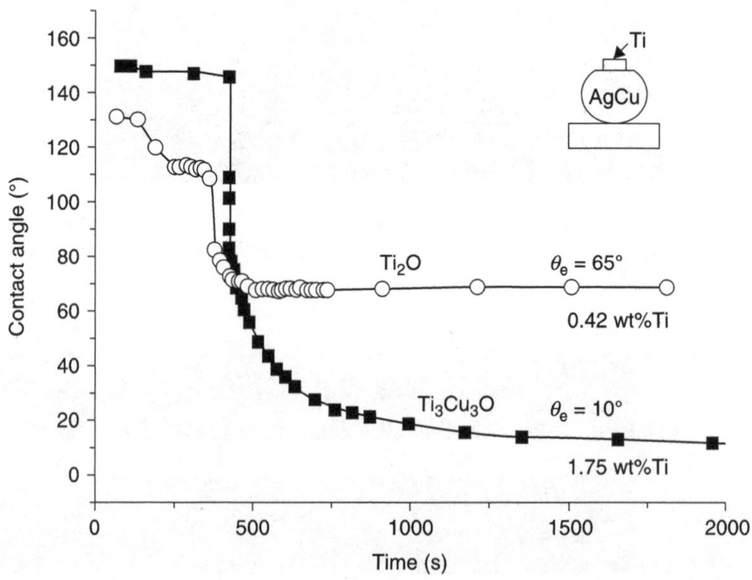

1.21 Contact angle as a function of time for two different composition CuAg–Ti droplets processed *in situ* on monocrystalline alumina in He (plateau temperature 900°C). According to Voytovych *et al.*, 2006.

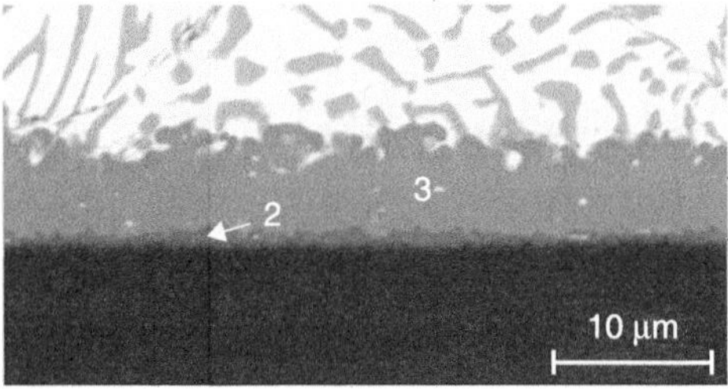

1.22 Cross section in a CuAg–1.75wt%Ti/alumina specimen, 900°C, 30 min, He atmosphere. The number 2 indicates the submicronic Ti_2O layer and the number 3 the Cu_3Ti_3O layer. According to Voytovych *et al.*, 2006.

composed of two layers. Layer II, formed in contact with the liquid alloy, was identified as the Cu_3Ti_3O compound. In the case of brazed joints, this M_6O type compound, first evidenced in Kelkar and Carim (1993), also contains a few per cent of aluminium resulting from the decomposition of alumina. A thin, submicronic layer (layer I) is also formed in contact with alumina and identified as the Ti_2O suboxide. As argued in Voytovych *et al.* (2006), the final contact angle observed by the alloy is on the Cu_3Ti_3O compound, which is the solid contacted by the liquid alloy. The low final contact angle observed in this system is precisely due to the metallic character of this compound. By decreasing the Ti concentration in the alloy from 1.75 wt% to 0.42 wt%, only the Ti suboxide was formed at the interface and, as a consequence, a very significant increase in the equilibrium contact angle was observed. Even higher contact angles can be observed with further reduction in Ti concentration or, more generally, of titanium *activity* in the alloy, which is in reality the relevant thermodynamic quantity involved in the interfacial reactivity. Such a reduction in Ti activity was produced by Kritsalis *et al.* (1994), where the metal matrix of the alloy was changed from CuAg to NiPd. Indeed, due to the very strong Ni–Ti and Pd–Ti interactions, reflected by the very negative values of partial enthalpy of mixing of titanium in nickel (-187 kJ.mole^{-1}) and in palladium (-255 kJ.mole^{-1}), the activity of Ti in NiPd is, for the same Ti concentration, several orders of magnitude lower than in the CuAg eutectic. This results in the formation of high oxidation-level Ti oxides such as Ti_3O_5 or Ti_5O_9 at the interface, the corresponding contact angles being close to and even higher than 90° (see Fig. 1.23). When these oxides are formed, the stress generated during cooling leads to interfacial failure, indicating a weak mechanical interface: a very low titanium activity is detrimental to both wetting and mechanical strength (Kritsalis *et al.*, 1994).

The average spreading rate of CuAgTi alloys on alumina is about 5 μm.s^{-1} (Kozlova *et al.*, 2011). Therefore, only sandwich brazing can be used for reactive

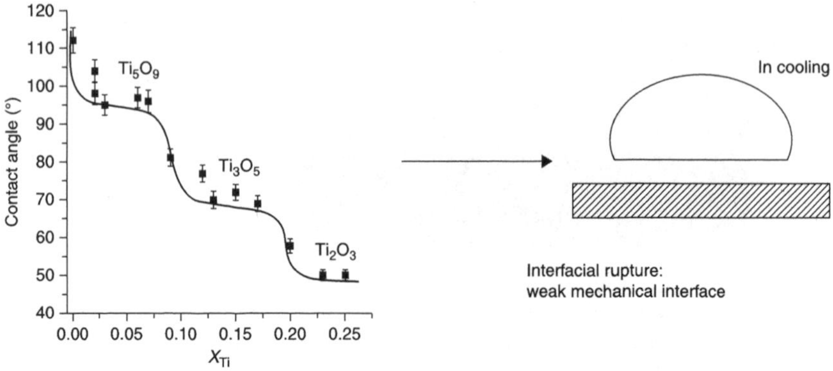

1.23 Equilibrium contact angles versus Ti mole fraction in NiPd alloy at $T \approx 1300°C$. The three plateaus correspond to three different Ti oxides. In all cases stress generated in cooling led to interfacial rupture. According to Kritsalis *et al.*, 1994.

brazing of alumina. There are two main differences between brazing and wetting experiments in reactive systems:

- The volume of the liquid alloy per unit area of interface (and thus the quantity of Ti available for the reaction) is up to two orders of magnitude lower in brazing than in sessile drop experiments involving droplets of several millimetres. Therefore size effects can occur in brazing.
- When alumina (or another ceramic) is brazed to a metallic partner Me, the interactions between Me and the brazing alloy can affect the joint microstructure and the interfacial chemistry. This is because the diffusion time in a liquid gap of 100 microns is only a few seconds, which is much lower than the usual brazing times (several minutes).

Size effects are obviously more pronounced for alloys diluted in Ti. In the brazing experiment depicted in Fig. 1.24(b), the initial Ti concentration was high enough to form layer II (Cu_3Ti_3O) at the very start of brazing. However, due to the significant decrease in Ti concentration caused by the interfacial reactions during brazing (but not during the sessile drop experiment), layer II is not stable for longer periods. To avoid this complication, it is preferable to braze with alloys containing 2–3 wt% of Ti. Note that Ti concentrations higher than 3 wt% must be avoided, as they can lead to the formation in the joint bulk of Cu–Ti intermetallics that lower the mechanical performance of the assembly (Kozlova, 2008; Kozlova *et al.*, 2010).

When one of the alumina pieces is replaced by a metallic piece, for instance by a copper–nickel plate, the dissolution of Ni and Cu from the plate in the liquid braze leads to a dramatic change in joint microstructure and also a net change in reactivity at the alumina/braze interface (see Fig. 1.25). This effect on interfacial

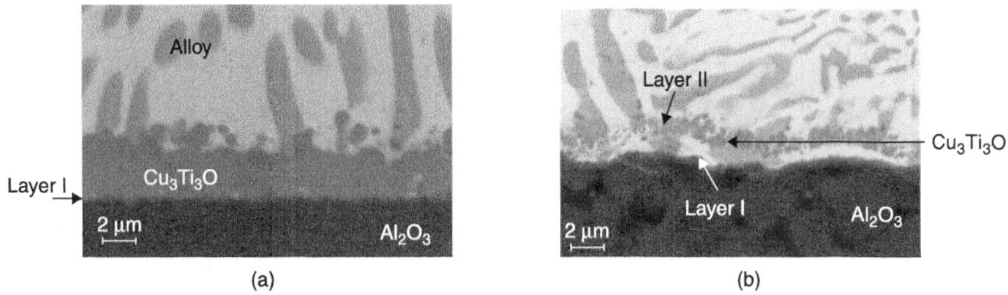

1.24 Cross sections in two CuAg–0.80 wt%Ti/alumina specimens obtained in wetting (a) and brazing (b) experiments ($T \approx 850°C$, 15 min). According to Kozlova, 2008 and Kozlova *et al.*, 2010.

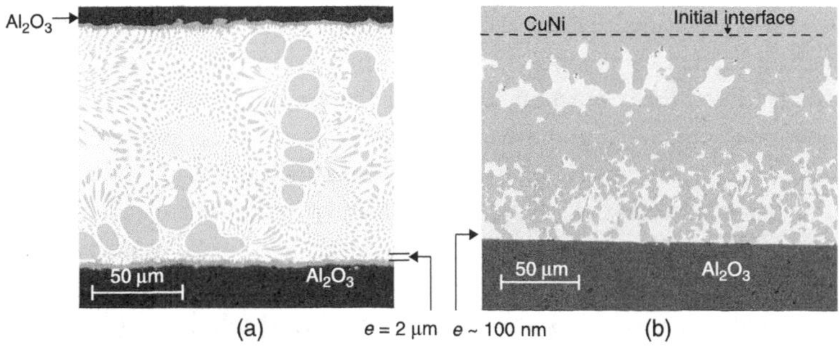

1.25 Cross sections of alumina to alumina (a) and of alumina to CuNi (b) assemblies brazed by a CuAgTi alloy (900°C, 15 min). *e* denotes the thickness of the interfacial reaction product. According to Valette *et al.*, 2005.

reactivity is due mainly to the very strong Ni–Ti interactions, which are responsible for the strong decrease observed in the thickness of the interfacial reaction layer (from ~ 2 μm to ~ 200 nm). Moreover, instead of the Cu_3Ti_3O compound formed at the interfaces in the alumina to alumina joint, high oxidation-level Ti oxides are formed in the case of the CuNi/alumina joint, resulting in the mechanical weakening of the interface and of the actual joint (Valette *et al.*, 2005).

From the above results, it appears that when ceramics are brazed to metallic parts, allowance must be made not just for the well-known problem of thermo-mechanical compatibility between these two types of solids but also for the interactions between the metallic part and the braze that can affect wetting and interfacial chemistry as well as the joint composition and microstructure.

1.5 Conclusion

As a general rule, liquid metals and alloys wet metallic solids well (contact angles much lower than 90°), the only condition being to proceed with clean, non-oxidized surfaces. Conversely, most low and moderate melting point metals, such as Pb, Sn, Ag, Cu . . ., do not wet ionocovalent oxides (alumina, silica, magnesia, zirconia . . .) or carbon-based solids, and adhere weakly to them. Non-wetting is also observed with non-oxide ceramics, such as SiC, AlN or Si_3N_4, that are usually covered by tenacious oxide layers acting as barriers to wetting and brazing. Wetting and adhesion can be promoted by certain alloying elements forming at the interface, by reaction with the ceramic, metal-like reaction products that are wettable by the liquid metals. Typical examples are Ti in $CuAg/Al_2O_3$ and Cr in Cu/C_{gr}.

Brazing can be performed either by infiltration of the liquid braze into the gap (*capillary brazing*) or by melting the braze foil or powder placed directly between the solids to be joined (*sandwich brazing*). Capillary brazing is possible only with non-reactive brazes that wet the solid parts well (contact angles well below 90°). Infiltration is driven by the capillary pressure and its rate is limited by the viscous flow. It is described by the classical Washburn equation in which the infiltration distance increases parabolically with time. Sandwich brazing is, in principle, possible even for contact angles higher than but close to 90°. Nevertheless, it is shown that contact angles lower than 45° are suitable for obtaining reliable joints that can be easily controlled by simple visual inspection.

In the last section of this chapter, in order to illustrate the relation between *wettability* and *brazability*, three particular cases of joining by brazing, both reactive and non-reactive, were described and discussed: (i) homogeneous brazing of *stainless steels* by non-reactive (or weakly reactive) CuAg eutectic alloys, (ii) homogeneous brazing of *silicon carbide* by metal–silicon containing alloys and (iii) heterogeneous brazing of *alumina to copper and copper–nickel* using reactive CuAgTi alloys. With these examples it is also shown how the information obtained with simple sessile drop experiments can help in defining the type of braze and the process parameters to be used in a particular system.

1.6 References

Ambrose JC and Nicholas MG (1996), 'Wetting and spreading of nickel-phosphorus brazes: Detailed real time observations of spreading on iron-chromium substrates', *Mater Sci Technol*, 12, 72–80.

Blake TD (1993), 'Dynamic contact angles and wetting kinetics', in: *Wettability*, Ed. JC Berg, New York, Marcel Dekker Inc., ISBN 0-8247-9046-4, 251–309.

Calderon NR, Voytovych R, Narciso J and Eustathopoulos N (2010), 'Wetting dynamics versus interfacial reactivity of AlSi alloys on carbon', *J Mater Sci*, 45, 2150–2156.

de Gennes PG (1985), 'Wetting: statics and dynamics', *Rev Modern Phys*, 57, 827–863.

Dezellus O and Eustathopoulos N (2010), 'Fundamental issues of reactive wetting by liquid metals', *J Mater Sci*, 45, 4256–4264.

Dezellus O, Hodaj F, Rado C, Barbier JN and Eustathopoulos N (2002), 'Spreading of Cu-Si alloys on oxidized SiC in vacuum: experimental results and modelling', *Acta Mater*, 50, 979–991.

Duffar T, Bochu O and Dusserre P (2010), 'Effect of oxygen on the molten BGO/Ir wetting and sticking', *J Mater Sci*, 45, 2140–2143.

Dupré, A (1869), *Théorie Mécanique de la Chaleur*, Gauthier-Villars, Paris.
Eustathopoulos N (2005), 'Progress in understanding and modeling reactive wetting of metals on ceramics', *Current Opinion in Solid State and Materials Science*, 9, 152–160.
Eustathopoulos N, Drevet B and Nicholas N (1999), *Wettability at High Temperatures*, Pergamon Materials Series, Volume 3.
Eustathopoulos N, Sobczak N, Passerone A and Nogi K (2005), 'Measurement of contact angle and work of adhesion at high temperature', *J Mater Sci*, 40, 2271–2280.
Gasse A (1996), *Rôle des interfaces dans le brasage non réactif du SiC par les silicures de Co et de Cu*. PhD Thesis, Polytechnique Institute of Grenoble, France.
Haon C, Saqure H, Daniel M, Drevet B, Camel D, *et al.* (2008), 'The role of capillarity in the gravitational molding of metallic glass pieces', *Mater Sci Eng*, 495A, 215–221.
Hodaj F, Dezellus O, Barbier JN, Mortensen A and Eustathopoulos N (2007), 'Diffusion-limited reactive wetting: effect of interfacial reaction behind the advancing triple line', *J Mater Sci*, 42, 8071–8082.
Johnson RE and Dettre RH (1993), *Wettability*, Surfactant Science Series, Volume 49, Ed: JC Berg, New York, 1–74.
Kelkar GP and Carim AH (1993), 'Synthesis, properties, and ternary phase stability of M6X compounds in the Ti-Cu-O system', *J Am Cer Soc*, 76, 1815–1820.
Koltsov A (2005), *Physico-chimie du brasage de AlN: Mouillage et reactivité*, PhD Thesis, Polytechnique Institute of Grenoble, France.
Koltsov A, Hodaj F and Eustathopoulos N (2008), 'Brazing of AlN to SiC by a Pr silicide: Physicochemical aspects', *Mat Sci Eng A*, 495, 259–264.
Kozlova O (2008), *Brasage réactif Cu/acier inoxydable et Cu/alumina*, PhD Thesis, Polytechnique Institute of Grenoble, France.
Kozlova O, Voyvotych R, Devismes M-F and Eustathopoulos N (2008), 'Wetting and brazing of stainless steels by copper-silver eutectic', *Mat Sci Eng A*, 495, 96–101.
Kozlova O, Braccini M, Voytovych R, Eustathopoulos N, Martinetti P *et al.* (2010), 'Brazing copper to alumina using reactive CuAgTi alloys', *Acta Mater*, 58, 1252–1260.
Kozlova O, Voytovych R and Eustathopoulos N (2011), 'Initial stages of wetting of alumina by reactive CuAgTi alloys', *Scripta Mat*, 65, 13–16.
Kritsalis P, Drevet B, Valignat N and Eustathopoulos N (1994), 'Wetting transitions in reactive metal-oxide systems', *Scripta Mater*, 30, 1127–1132.
McGurran B and Nicholas MG (1985), 'A study of factors which affect wetting when brazing stainless steels to copper', *Brazing & Soldering*, 8, 43–48.
Milner DR (1958), 'A Survey of the Scientific Principles Related to Wetting and Spreading', *Brit Weld J*, 5, 90–105.
Naidich YV (1981), *Progress in Surface and Membrane Science*, vol. 14, ed. DA Cadenhead and JF Danielli, Academic Press, New York, 353–484.
Naidich YV, Sabuga W and Perevertailo V (1992), 'Temperature effect on kinetics of spreading and wetting in systems with different interactions at the contacting phases', *Adgeziya Raspl Pajka Mater*, 27, 23–34, in Russian.
Pech J, Braccini M, Mortensen A and Eustathopoulos E (2004), 'Wetting, interfacial interactions and sticking in glass/steel systems', *Mater Sci Eng A*, 384, 117–128.
Protsenko P, Garandet JP, Voytovych R and Eustathopoulos N (2010), 'Thermodynamics and kinetics of dissolutive wetting of Si by liquid Cu', *Acta Mater*, 28, 6565–6574.
Rado C (1997), *Contribution à l'étude du mouillage et de l'adhésion thermodynamique des métaux et alliages liquids sur le carbure de silicium*, PhD Thesis, Polytechnique Institute of Grenoble, France.

Rado C and Eustathopoulos N (2004), 'The role of surface chemistry on spreading kinetics of molten silicides on silicon carbide', *Interface Science*, 12, 85–92.
Rado C, Kalogeropoulou S and Eustathopoulos N (1999), 'Wetting and bonding of Ni-Si alloys on silicon carbide', *Acta Mater*, 47, 461–473.
Saiz E and Tomsia AP (2004), 'Atomic dynamics and Marangoni films during liquid-metal spreading', *Nat Mater*, 3, 903–909.
Valette C, Devismes M-F, Voytovych R and Eustathopoulos E (2005), 'Interfacial reactions in alumina/CuAgTi braze/CuNi system', *Scripta Mater*, 52, 1–6.
Voytovych R, Robaut F and Eustathopoulos N (2006), 'The relation between wetting and interfacial chemistry in the CuAgTi/alumina system', *Acta Mater*, 54, 2205–2214.
Washburn EW (1921), 'The dynamics of capillary flow', *Phys Rev*, 17, 273–283.
Young T (1805), 'An Essay on the Cohesion of Fluids', *Philosophical Transactions of the Royal Society of London*, 95, 65–87.

2

Strength and margins of brazed joints

Y. FLOM, NASA Goddard Space Flight Center, USA

DOI: 10.1533/9780857096500.1.31

Abstract: Despite the great advances in analytical methods available to structural engineers, designers of brazed structures have great difficulties in determining load-carrying capabilities of the brazed assemblies and predicting their failures. In this chapter we will review why such common engineering tools as finite element analysis (FEA) as well as many well-established theories (Tresca, von Mises, Highest Principal Stress, etc.) do not work well for brazed joints. This chapter will show how the classic approach of using interaction equations and the lesser-known Coulomb–Mohr failure criterion can be employed to estimate margins of safety (MS) in brazed joints.

Key words: brazed joints, margins of safety, failure assessment diagram, interaction equations, Coulomb–Mohr failure criterion.

2.1 Introduction

It is hard to overestimate the importance of brazing in modern manufacturing processes. Sophisticated designs of structures and mechanisms used in the aerospace, aircraft, automotive, power and medical industries quite often expect various brazed joints to perform under complicated multiaxial loading conditions.

Despite great advances in brazing technology and applications, reliability of brazed joints remains one of the least developed fields of structural analysis. Mechanical, welded or adhesively bonded joints in metallic and composite structures are routinely assessed for their load-carrying capabilities in accordance with widely accepted engineering analysis techniques and failure criteria (Blodgett, 1963; Bruhn, 1973; *Astronautic Structures Manual*, 1975; Hart-Smith, 1973; Shigley and Mischke, 1989; Tong and Steven, 1999). A search for information on the engineering practice of estimating or predicting load-carrying capability of brazed joints, however, produces almost no results.

Consequently, there is a need for a simple engineering methodology that would enable designers and structural analysts to evaluate strength margins of the brazed joints exposed to combined shear and normal stresses.

This chapter reviews the challenges of using common failure criteria in predicting failure of brazed joints and offers an alternative approach based on developing failure assessment diagrams (FADs). The first step in constructing FADs is to identify or develop brazed joint failure criteria. It would be highly desirable for such criteria to satisfy the following conditions:

- Criteria should be applicable to any brazed joint geometry.
- They should be fairly conservative to account for the many uncertainties associated with the brazing process, properties and analysis of the joints.
- They should be sufficiently simple and easy to use so that designers and structural engineers find them helpful for practical applications.
- They should be based on clearly defined properties of the brazed joints that can be determined in a fairly straightforward manner.

For now, this chapter is limited only to static loading of the brazed assemblies. As our knowledge expands, future revisions and/or updates of this chapter may address the brazed joints subjected to dynamic loading.

2.2 Applicability of common failure criteria to analysis of brazed joints

2.2.1 Maximum normal stress

This criterion predicts failure when the largest normal principal stress reaches the uniaxial tensile strength of the material (Dieter, 1976; Dowling, 1993). This criterion is commonly used in predicting failure of brittle materials. It is perhaps the simplest failure criterion in use, and can be expressed as:

$$\sigma_1 = \sigma_o \qquad [2.1]$$

where σ_1 is the maximum normal or, according to convention, the first principal stress and σ_o is the yield tensile strength obtained from a tensile test of the standard tensile test specimen (ASTM E8/EBM-09, 2009). It has to be pointed out that in most structural applications yielding is considered to be a form of failure. σ_o is calculated as P/A, where P is the yield load observed during the tensile test and A is the initial cross-sectional area of the test specimen. The engineering community defines yield load as a load causing 0.2% strain in the uniaxial tension test. As one can see, σ_o is an *average* stress – an important point in our upcoming discussion. σ_o is commonly used in structural design as a mechanical property of the material (MMPDS-05, 2010) or tensile yield strength allowable. The value of σ_1 is calculated, typically, using finite element analysis (FEA) performed on the entire structure or its component.

An attempt to use this criterion for failure assessment of brazed joints leads to several complications. First of all, the 'uniaxial tensile strength of the material' is not defined for brazed joints. For homogeneous metallic material, uniaxial tensile strength is a mechanical property of that material, determined, as mentioned above, from the standard tensile test, such as, for example, described in (ASTM, E8/EBM-09, 2009). The uniaxial tensile test of the butt-brazed tensile specimen determines the tensile strength of the *joint*, not the strength of a specific material – brazing filler metal or adjacent base metal. Obviously, properties of the filler metal and the base metal do contribute to the overall strength of the brazed joint.

It is a well-established fact that the tensile strength of butt-brazed joints exceeds, by far, the tensile strength of the filler metal tested in bulk form (Brazing Handbook, 2007; Rosen and Kassner, 1993). The second problem with this criterion is when it is applied to ductile and lap shear brazed joints. Such joints undergo relatively large plastic deformation prior to failure, which is quite different from the brittle behavior. Consequently, if selected, the maximum normal stress criterion may be applicable only to uniaxially loaded butt-brazed joints, where a filler metal constraint results in a highly triaxial stress state causing the braze joint to behave in a quasi-brittle manner.

2.2.2 Maximum shear and octahedral stress

These two criteria are very similar. Maximum shear stress or the Tresca criterion predicts failure, manifested by yielding, when the maximum shear stress τ_{max} on any plane reaches a certain critical value τ_o, as expressed in Eq. 2.2:

$$\tau_{max} = \tau_o \qquad [2.2]$$

where, again, τ_o is the shear yield strength of the material, i.e. a material mechanical property. Determination of τ_o is not as straightforward as tensile allowable. In fact, only a thin wall tube subjected to a pure torsion renders a direct measurement of τ_o and such tests and their results are not readily available. A more common approach is to perform a uniaxial tension test and calculate τ_o from σ_o using the relationship between maximum shear and principal normal stresses. Equation 2.2 can be written as:

$$\tau_{max} = \frac{\sigma_1 - \sigma_3}{2} = \tau_o \qquad [2.3]$$

where σ_1 and σ_3 are the first (maximum) and third (minimum) normal principal stresses. In uniaxial tension $\sigma_3 = 0$ and $\sigma_1 = \sigma_o$. Consequently,

$$\tau_{max} = \frac{\sigma_o}{2} = \tau_o \quad \text{or} \quad \tau_o = \frac{\sigma_o}{2} \qquad [2.4]$$

As we can see, according to the maximum shear stress theory, the maximum value of τ_o is 0.5 of σ_o.

The maximum octahedral (von Mises) or maximum distortion energy criterion predicts failure when shear stress τ_h in the octahedral plane reaches a critical value τ_{ho} (Dowling, 1993), or:

$$\tau_h = \tau_{ho} \qquad [2.5]$$

where τ_{ho} is also a material property that now represents a critical value of shear stress on the octahedral plane that causes yielding. The shear stress on the octahedral plane can be expressed in terms of principal stresses as (Dowling, 1993):

$$\tau_h = \frac{1}{3}\sqrt{(\sigma_1 - \sigma_2)^2 + (\sigma_2 - \sigma_3)^2 + (\sigma_1 - \sigma_3)^2} \qquad [2.6]$$

Again, applying the von Mises criterion to the uniaxial tension test, $\sigma_1 = \sigma_o$ and $\sigma_2 = \sigma_3 = 0$, we obtain:

$$\tau_h = \frac{1}{3}\sqrt{(\sigma_o)^2 + (\sigma_o)^2} = \tau_{ho} \qquad [2.7]$$

From Eq. 2.7 the value of τ_{ho} can be also obtained from the uniaxial tension test as:

$$\tau_{ho} = \frac{\sqrt{2}}{3}\sigma_o \qquad [2.8]$$

Consequently, in accordance with the von Mises criterion, $\tau_{ho} = 0.47 \cdot \sigma_o$ compared with 0.5 in Tresca theory.

There are several problems with using these criteria for predicting failure in brazed joints. Both Tresca and von Mises criteria are essentially yielding criteria that are typically used to predict an onset of yielding in homogeneous or isotropic ductile metals (Dieter, 1976; Dowling, 1993). Brazed joints, however, are quite far from isotropic. Physical and mechanical properties within the brazed joints undergo significant changes over very short distances as we traverse the brazed joint from one adjacent side of the base metal into another. The situation becomes even more complicated when drastically dissimilar materials form a brazed joint. Furthermore, the concept of brazed joint yielding is not well defined. Tensile tests of the butt-brazed specimens as well as the lap shear pull tests show practically no difference in stress–strain behavior between the brazed and solid specimens up to the point of failure of the brazed joints (Flom, 2011; Spingarn *et al.*, 1983; Flom and Wang, 2004), as shown in Fig. 2.1–2.3.

Recall that successful failure criteria should be applicable to any brazed joint geometry. When a butt-brazed joint is tested in uniaxial tension, a mechanical constraint provided by the base metal develops a triaxial tensile stress state within the braze layer. Even in such a ductile filler metal as pure silver, the level of constraint is so high that hydrostatic stress is very close to the axial one, which means that the values of principal stresses are very similar (Rosen and Kassner, 1993). In the pure hydrostatic stress state (all principal stresses are equal), shear and von Mises stresses are zero and failure occurs without plastic deformation. Similarly, in a butt-brazed joint under uniaxial tension, shear and von Mises stresses are quite low. Consequently, these stresses would not be good criteria in brazed joints with a high mechanical constraint and would result in underprediction of failure.

Let us consider lap shear brazed joints in which both base and filler metals are ductile. It would appear that shear or distortion energy criteria would be much better suited to joints of this type, which typically undergo large plastic deformation prior to failure. However, due to non-uniform distribution of shear stresses within the lap shear joints, the values of the shear and von Mises stresses could be quite

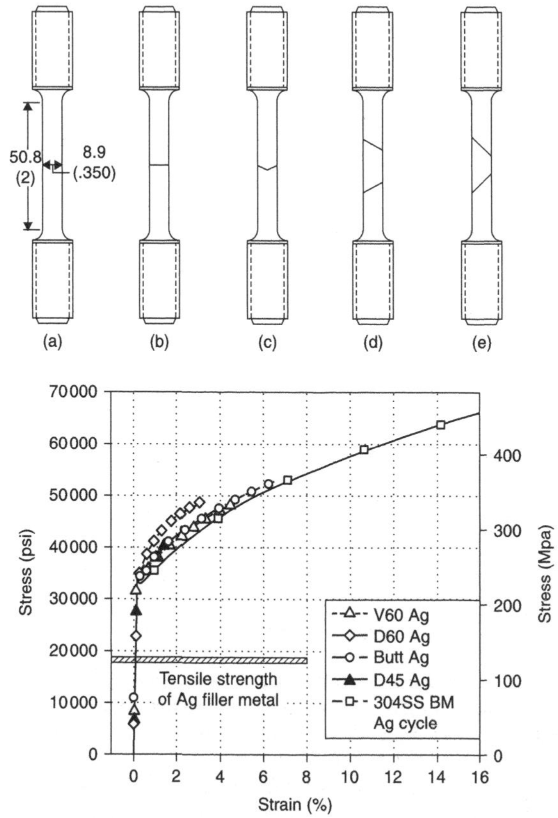

2.1 (a) to (e) Various types of silver-brazed 304SS joints: (a) base metal blanks; (b) butt joints; (c) V60; (d) D60; (e) 45°. (f) Stress–strain curves of Ag brazed specimens and 304SS base metal after Ag cycle. Note that strain hardening rates are very similar to the base metal. Reprinted from Flom (2011).

high, particularly at the joint ends (Flom and Wang, 2004), as shown in Fig. 2.4. Experimental results indicate that such values could exceed the strength of the filler metal by two or three times. If one attempts to correlate the highest von Mises stress values with the event of failure in lap shear joints, such values are going to be much greater than those observed in butt-brazed specimens, as shown in Fig. 2.5.

Consequently, von Mises stress is not a very good choice for a failure criterion, due to its great variation with brazed joint geometry. In addition, a major difficulty in estimating von Mises stress in a brazed joint is a reliance on our knowledge of the elastic modulus and the yield strength properties of the braze layer within the brazed joint. Without such knowledge, the use of finite element analysis in

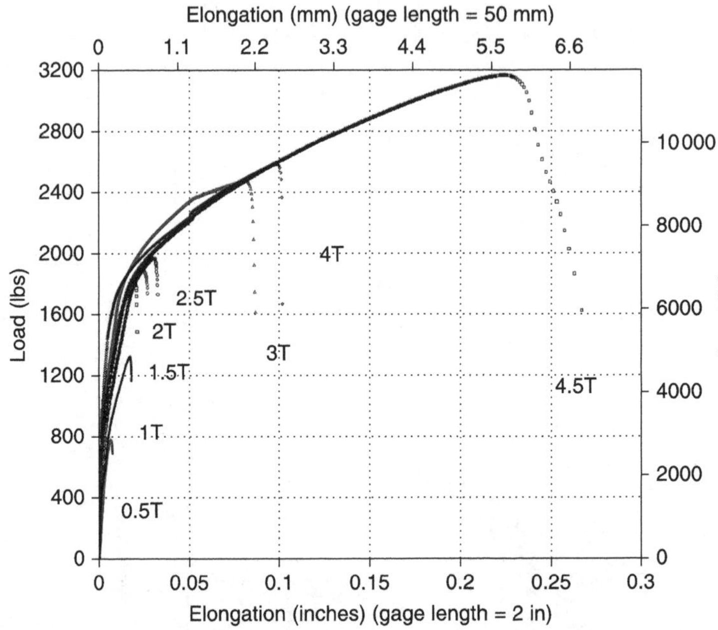

2.2 Tensile test of 347SS lap shear specimens brazed with Ag filler metal. Load vs. elongation results are plotted for various overlap lengths ranging from 0.5*T* to 4.5*T*, where *T* is the thickness of the base metal. Note that the strain hardening rate is essentially the same for different overlaps, which indicates that plastic deformation within the brazed joint is largely obscured by the plastic deformation of the base metal (Flom and Wang, 2004). Reprinted with permission from *Welding Journal.*

calculating von Mises stress within the filler metal layer is very limited, if not impossible.

2.2.3 Interaction equations

Interaction equations were introduced to predict failures in structures subjected to combined loading conditions (Shanley and Ryder, 1937). These equations incorporate both maximum shear and normal stresses and are expressed in terms of the stress ratios. In their simplest generic form interaction equations can be written as (Peery and Azar, 1982):

$$R_\sigma^m + R_\tau^n = 1 \quad [2.9]$$

where R_σ and R_τ are normal and shear stress ratios, respectively, and the exponents *m, n* are determined experimentally. Typically, experimental results are plotted as

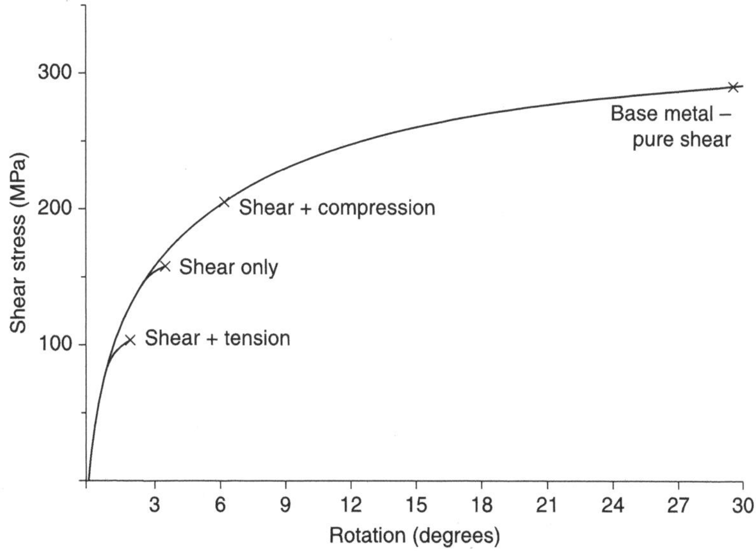

2.3 Incoloy 800 specimens brazed with Ni-based AWS BNi-8 filler metal and torsion tested at 650°C. Note that the behavior of the brazed joints is very similar to the base metal. Credit to Sandia National Laboratory, Spingarn *et al.* (1983).

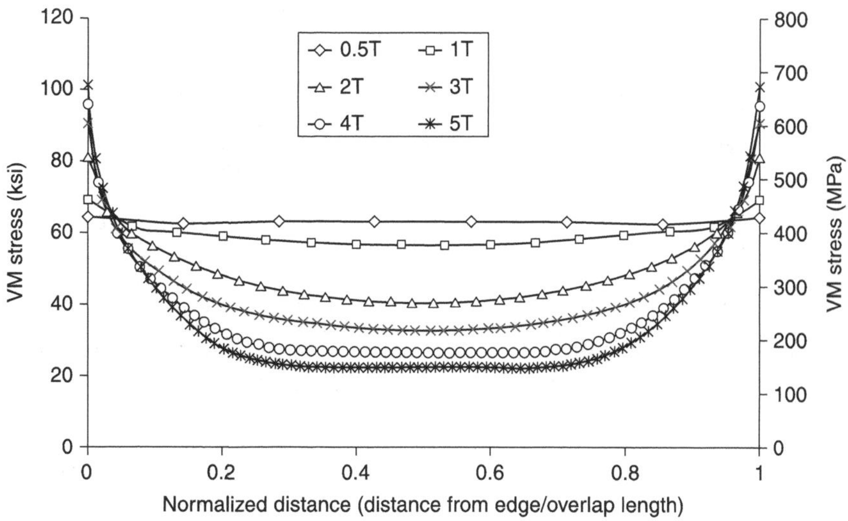

2.4 Distribution of von Mises (effective) (VM) stress within the 347SS/Ag lap shear brazed joints tested to failure. The stress is plotted as a function of the distance from the joint edge for specimens having overlap lengths ranging from 0.5*T* to 5*T*, where *T* is the thickness of the base metal. Note that stresses in many locations, particularly near the joint edge, exceed, by far, the ultimate strength of the bulk silver, which is about 35 ksi (250 MPa) (Flom and Wang, 2004). Reprinted with permission from *Welding Journal*.

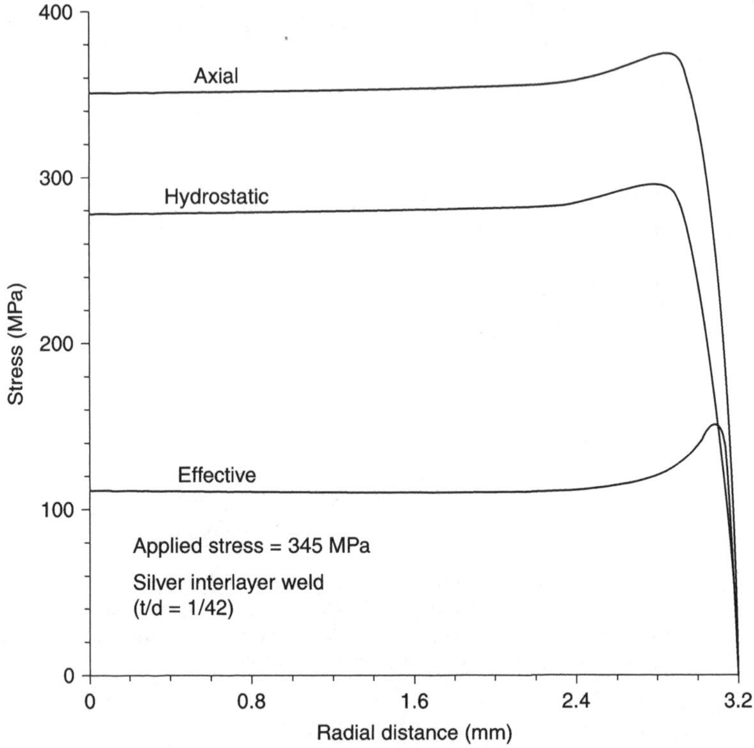

2.5 Stress distribution in a silver-brazed butt joint determined using FEA. Aspect ratio here is 1/42 or approximately 0.024. Most structural brazed joints have much smaller aspect ratios, of the order of 0.008. In such joints, the axial stresses are even closer to hydrostatic stresses than shown on this plot (Rosen and Kassner, 1993). Reprinted with permission of ASM International. All rights reserved. www.asminternational.com.

shown in Fig. 2.6. Stress ratios are determined by calculating maximum normal and maximum shear at some specific point in the structure and dividing them by their respective tensile and shear allowables, such as:

$$R_\sigma = \frac{\sigma_1}{\sigma_o} \quad \text{and} \quad R_\tau = \frac{\tau_{max}}{\tau_o} \qquad [2.10]$$

Over the years, interaction equations have evolved into comprehensive and quite effective relationships verified experimentally for different structural shapes and loading conditions, such as tension, compression, bending, shear and torsion (Blodgett, 1963; *Engineering Stress Memo Manual*, 2008). Before FEA became a standard tool in structural analysis, these interaction equations were used very successfully to predict failures in astronautic and aircraft metallic structures

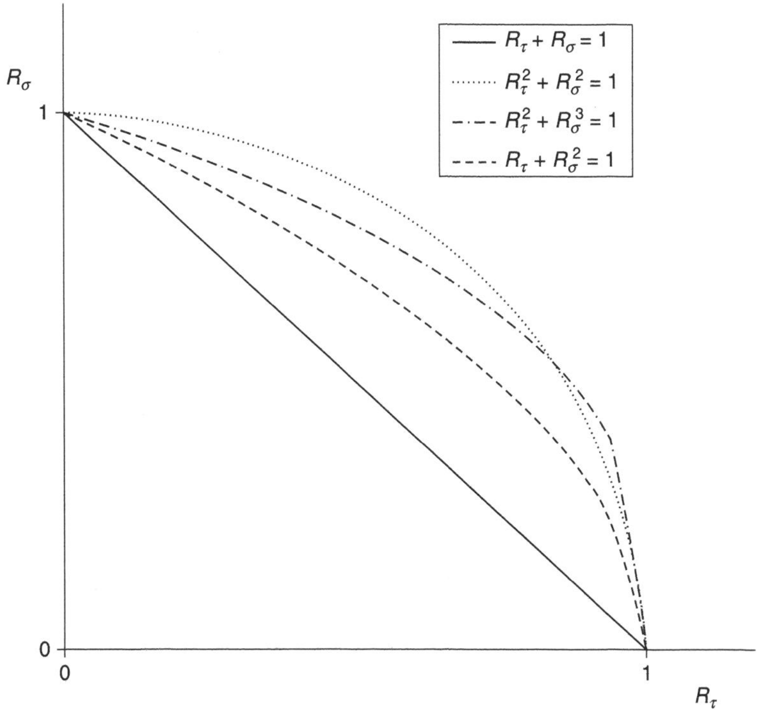

2.6 Plots of several interaction equations. Note that the straight line represents the most conservative case.

Table 2.1 Some well-known interaction equations

Loading combinations	Interaction equations	Margins of safety (MS)
Normal and bending stresses	$R_\sigma + R_b = 1$	$\frac{1}{R_\sigma + R_b} - 1$
Normal and shear stresses	$R_\sigma^2 + R_\tau^2 = 1$	$\frac{1}{\sqrt{R_\sigma^2 + R_\tau^2}} - 1$
Bending, torsion and compressions	$R_b^2 + R_\tau^2 = (1 - R_c)^2$	$\frac{1}{R_c + \sqrt{R_\sigma^2 + R_\tau^2}} - 1$
Bending and torsion	$R_b + R_\tau = 1$	$\frac{1}{R_b + R_\tau} - 1$

(*Astronautic Structures Manual*, 1975; Bruhn, 1973; Sarafin, 1998). Some examples of such equations are shown in Table 2.1 (*Astronautic Structures Manual*, 1975).

A fairly comprehensive compilation of interaction equations is provided in the *Engineering Stress Memo Manual* (2008). In addition to their great practical value, interaction equations have an interesting feature that may be quite useful in a discussion of brazed joints. There is no requirement that stresses used in interaction equations have to act on the same planes. Interaction equations are well suited for structural analysis of homogeneous metallic materials. It would be interesting to investigate their suitability for predicting failures in brazed joints. Preliminary results indicate that interaction equations can be used for a conservative, lower bound estimate of the failure conditions in several base metal/filler metal combinations (Flom, 2011; Flom *et al.*, 2011).

2.2.4 Coulomb–Mohr failure criterion

This criterion states that fracture takes place in a given plane when a critical combination of normal and shear stresses has occurred (Dowling, 1993). This is a very interesting criterion and definitely worth considering when discussing brazed joints that fail within the braze plane. The Coulomb–Mohr criterion has a very simple form that assumes a linear relationship between normal and shear stresses:

$$\tau + \mu\sigma = c \qquad [2.11]$$

In this expression, μ and c are material-specific parameters. Later in this chapter we will spend more time discussing this criterion. A modified form of the Coulomb–Mohr criterion, proposed by Christensen (2004), shows better correlation with experimental results for homogeneous materials. It offers a more general form of failure condition and considers a combined effect of hydrostatic (dilatation) and distortion (von Mises) components of stress. However, its practical use in analysis of brazed joints is rather limited, since it requires a detailed FE analysis of the brazed joints, which, in turn, relies on a knowledge of the mechanical properties of the braze layer.

2.2.5 Fitness-For-Service (FFS) approach

This approach was first introduced in the welding industry and is now widely accepted in analysis of critical welded structures containing discontinuities (Dowling and Townley, 1975; Webster and Bannister, 2000; Gordon, 1993; API, 2007). The brand new or existing welded structure is evaluated on the basis of whether it can safely operate under given loading and environmental conditions. Using a certain amount of mechanical testing and analytical techniques, a safe-to-operate zone is established for a particular weldment containing flaws. FADs are constructed to define such safe zones. In order to construct FADs, the welding industry has adopted a specific failure criterion based on static strength and fracture mechanics characteristics of the weld joint containing flaws. An example of such a FAD for a weld joint is shown in Fig. 2.7.

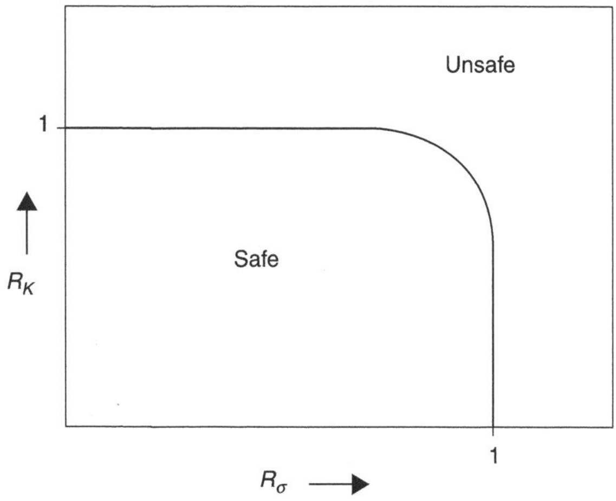

2.7 An example of a FAD used in the welding industry. The vertical axis represents resistance to brittle fracture and the horizontal axis represents resistance to plastic collapse (Gordon, 1993). Reprinted with permission of ASM International. All rights reserved. www.asminternational.com.

FADs are plotted in terms of the fracture toughness and plastic collapse stress ratios (Gordon, 1993). The vertical axis represents ratios of $R_K = K_1/K_{mat}$ and the horizontal axis represents ratios of $R\sigma = \sigma/\sigma_{pc}$, where K_1 is the stress intensity at the existing flaw due to the applied load, K_{mat} is the fracture toughness of the material, σ is the applied stress and σ_{pc} is the plastic collapse stress for a given welded component. An attempt to apply this fracture toughness and plastic collapse-based FAD to brazed joints is described by Leinenbach *et al.* (2007).

There are several problems with using a fracture toughness and plastic collapse stress-based criterion for predicting failures in brazed joints. First, the concept of brazed joint fracture toughness is not well defined. It is not clear what fracture toughness of the material means in the case of a brazed joint consisting of two adjoining base metals (similar or dissimilar) and a thin layer of the filler metal, metallurgically different from the base metals. Consider the following argument. Recall that, for a crack to grow, the energy available for crack extension should exceed the crack resistance, which is the energy required to create new crack surfaces. Each material has its own surface energy – it is a physical property. If a crack propagates through the filler metal, the energy required to create new crack surfaces is different from the energy required for creation of new crack surfaces in the base metal. If a crack chooses to propagate through the base metal/filler metal interface, the surface energy again will be different. And, finally, if the crack meanders through the base, the filler and interface regions of the brazed joint, and the energy of the newly created surfaces will be even harder to define. If the fracture toughness of the brazed joint does not have a well-established definition, measurement of the brazed joint fracture toughness is even less clear, particularly when it comes to a discussion of the validity of the fracture toughness testing. Similar problems arise when attempting to define and/or measure the plastic collapse stress of brazed joints. One can see now how problematic, particularly in the practical sense, it would be to try to incorporate fracture toughness and plastic collapse stress into failure criteria for brazed joints.

2.3 Alternative approach for developing failure assessment diagrams (FADs)

Let us consider whether the Coulomb–Mohr criterion can be used to predict failures in brazed joints while satisfying as many of the conditions listed in the Introduction as possible. To begin, we are going to treat the brazed joint as a system, rather than trying to study separate regions of the joint which are influenced by complex metallurgical reactions of the brazing process and, as a consequence, have significantly different material properties. The latter has been the more traditional approach taken by many investigators attempting to study the mechanical properties of brazed joints (Flom and Wang, 2004; Rosen and Kassner, 1990; Tolle *et al.*, 1995; Wen-Chun Jiang *et al.*, 2008). As was briefly mentioned earlier in this chapter, mechanical properties of the braze layer are not readily available or capable of being determined through conventional methods. An evaluation of mechanical properties of brazes is a rather challenging endeavor. A very thin cast layer of filler metal consists most often of multiple phases and is affected by the braze gap size and dilution with the base metal as well as by possible formation of intermetallic compounds located at the filler metal/base metal interfaces. Even the properties of a braze layer consisting of a single-phase pure metal may be significantly different from mechanical properties of identical wrought pure metal tested in the bulk form (Rosen and Kassner, 1990), as shown, for example, in Fig. 2.8. The presence of additional phases and eutectics makes the situation even worse. For example, it is a well-known fact that joints brazed with Ni-based filler metals, such as AWS BNi2, could develop a variety of microstructures and properties depending on joint geometry and the brazing cycle (*Brazing Handbook*, 2007; Lugscheider and Partz, 1983). Consequently, trying to predict mechanical properties of the braze layer, either by testing some bulk form of the filler metal or by using some software specialized in calculating the material properties of a given alloy (Wen-Chun Jiang *et al.*, 2008), may produce unreliable results. Other attempts to measure properties of the braze layer were based on testing miniature test specimens containing braze (Leinenbach *et al.*, 2007). This approach may also present some problems, since the aspect ratios of such brazed joints are significantly higher than in typical structural joints. Typical ratios of the brazed gaps to the joint size (diameter or width) are less than 0.005. Thus, the level of constraint within the braze layer and triaxiality of the stress state are very

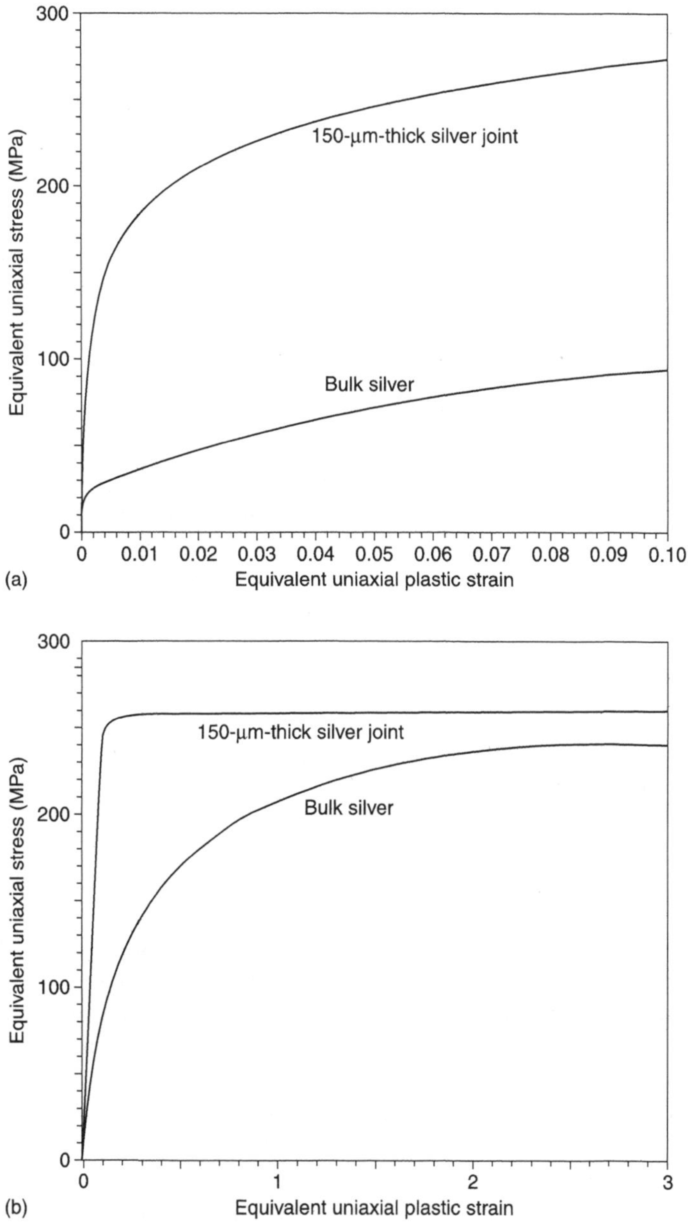

2.8 Equivalent uniaxial stress vs. equivalent uniaxial strain plot for silver-brazed maraging steel butt-brazed specimens tested in torsion. An expanded view of the initial region of the test, up to 0.1 strain, is shown in (a). The entire test is captured in (b). Note a significant difference in strain hardening rate between the bulk silver and a silver layer in the brazed joint. Reprinted with permission from Rosen and Kassner, 1990. Copyright 1990, American Vacuum Society.

different from those in actual brazed joints, which, in turn, may result in unrealistic ductility and stress–strain response of such miniature specimens. Consequently, rather than trying to predict the material properties of the braze layer, we will focus on the properties of the brazed joint as a whole, or as a system. Also, this will be more in line with the FFS approach.

Now, as we prepare to apply the Coulomb–Mohr criterion to the brazed joints, we can think of the constants μ and c (see Eq. 2.11) as representing the properties of the system, not the material properties of the braze layer. Since our goal is to develop a braze joint failure criterion applicable to any brazed joint geometry, it is constructive to start with the two most fundamental ones: a lap and a butt-brazed joint.

When the brazed joint is subjected to pure shear loading conditions, normal stress acting on the braze plane is zero. Lap shear brazed joints tested under uniaxial load do not meet, strictly speaking, the conditions of pure shear. Consequently, the results are reported in terms of an average shear stress. In this case, there is a non-uniform distribution of shear stresses within the lap and the peel effects at the joint ends are ignored (AWS, 2008). Therefore, in a practical sense, the presence of normal stresses in a standard lap shear test specimen is also ignored. If we let $\sigma = 0$, the Coulomb-Mohr expression for the lap joint can be written as:

$$\tau = c \qquad [2.12]$$

Now constant c can be interpreted as simply the average shear strength of the lap joint, as written below:

$$\tau = c = \tau_o \qquad [2.13]$$

where τ_o is the shear strength or allowable of the lap shear brazed joint. It is interesting to note that for lap shear brazed joints, which represent the most ductile geometric configuration, Eq. 2.13 takes the form of the Tresca criterion, discussed in Section 2.2.2. There are various types of lap shear brazed joint used in the brazing industry to test specimens. Several of them are shown in Fig. 2.9. Some of these joints may be a little closer to pure shear than the others, and the question is which lap shear type is more suitable for measuring shear allowable. A detailed discussion on this subject is provided by Peaslee (1976). It turns out that a single lap shear specimen is quite adequate for measuring average shear strength of brazed joints, and any attempts to develop braze specimen or test fixtures to eliminate or minimize the peel effects as well as the bending of the specimen during testing are of little or no consequence. Apparently, the behavior of the lap shear-type joints is predominantly controlled by shear, which causes extensive plastic deformation, and the influence of normal components of stresses is typically obscured by the experimental scatter when testing any type of brazed specimens.

On the other hand, when a butt-brazed joint is tested in tension under uniaxial loading conditions, the shear stress within the brazed plane is essentially zero and

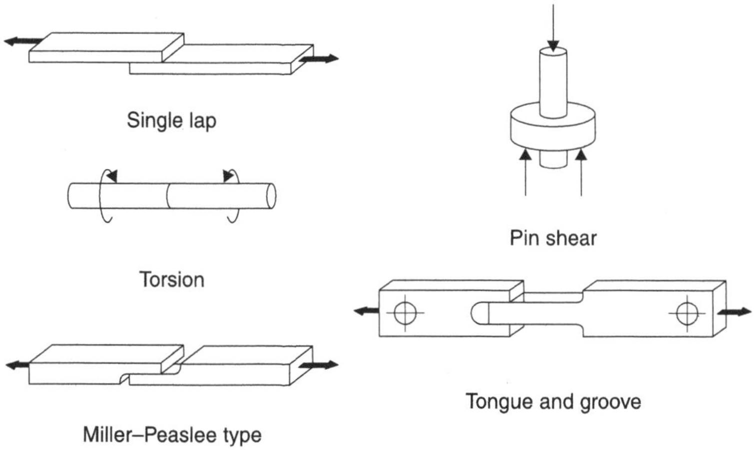

2.9 Several types of brazed shear specimens. The single lap joint has been used in various studies to determine shear strength of brazed joints. The effect of eccentricity of the test specimen on average shear strength is much less than the experimental scatter and variability of the brazing process itself (*Brazing Handbook*, 2007). Consequently, the single lap test specimen, described by Peaslee (1976), remains the standard test specimen for measuring the shear strength of brazed joints.

can be ignored in the practical sense. Therefore, assuming $\tau = 0$ and substituting $c = \tau_o$, the Coulomb–Mohr criterion for the butt-brazed joint can be written as:

$$\mu \cdot \sigma = \tau_o \quad \text{or} \quad \sigma = \frac{\tau_o}{\mu} \qquad [2.14]$$

When testing standard butt-brazed test specimens to failure, we obtain their ultimate tensile strength (AWS C3.2, 2008) or tensile allowable σ_o. It is important to keep in mind that the value of σ_o is constant for a specific base metal/filler metal combination as long as it is determined from testing standard butt-brazed test specimens. This is no different from testing the material properties of any metallic materials. Consequently, a butt-brazed joint, or any other brazed joint in which the braze plane is subjected to predominantly normal stresses, is going to fail when the maximum normal stress $\sigma = \sigma_o$. For such a condition, Eq. 2.14 can be expressed as:

$$\sigma = \frac{\tau_o}{\mu} = \sigma_o \quad \text{or} \quad \mu = \frac{\tau_o}{\sigma_o} \qquad [2.15]$$

Now we can define constant μ as the ratio of the brazed joint shear and tensile allowables. It is important to remember that τ_o and σ_o are not the properties of a specific material, but the properties of a collection of materials, forming the brazed joints or systems subjected to their two most extreme conditions: 1) the butt-brazed joint under uniaxial load, which results in the highest degree of constraint, and 2) the lap shear joint also subjected to uniaxial load, which, in turn, results in the most ductile behavior. Substituting μ and c in Eq. 2.15 and dividing by τ_o, the Coulomb–Mohr criterion for brazed joints can be written as:

$$\tau + \frac{\tau_o}{\sigma_o} \cdot \sigma = \tau_o \quad \text{or} \quad \frac{\tau}{\tau_o} + \frac{\sigma}{\sigma_o} = 1 \qquad [2.16]$$

As one can see, for the brazed joints dominated by shear stresses, i.e. $\sigma = 0$, Eq. 2.16 transforms simply into the failure criterion of the lap shear joint, such as $\tau = \tau_o$. Likewise, for the brazed joints dominated by normal stresses ($\tau = 0$), such as, for example, butt-brazed joints, Eq. 2.16 becomes simply the failure criterion of the butt-brazed joint subjected to uniaxial tensile load: $\sigma = \sigma_o$.

The same equation, using Eq. 2.10 and 2.11, can be written in terms of stress ratios R_τ and R_σ:

$$R_\sigma + R_\tau = 1 \qquad [2.17]$$

As one can see, Eq. 2.17 is identical to interaction Eq. 2.9 when exponents m and n are equal to 1. In other words, Eq. 2.17 represents perhaps one of the most conservative forms of interaction equations. Graphically, this expression can be plotted as a straight line, as shown in Fig. 2.6.

A number of brazed joint systems subjected to multiaxial loading were tested to see whether Eq. 2.17 can be used to conservatively predict failure of various brazed joint geometries (Spingarn *et al.*, 1983; Flom, 2011; Flom *et al.*, 2009, 2011). The base/filler metal combinations used in these studies are listed in Table 2.2.

For each base metal/filler metal family of brazed joints, standard lap shear and butt-brazed tensile specimens were tested to determine tensile σ_o and shear τ_o strengths (allowables) of the respective brazed joints. In addition to standard braze test specimens, more complex (verification) test specimens, designed to create combined tensile and shear stresses in the brazed joints, were fabricated and tested using identical braze processes and test temperatures. For a detailed description of the verification specimens, the readers are referred to the references listed in Table 2.2. For convenience, all experimental results, expressed in terms of the stress ratios, are plotted on the same graph, shown in Fig. 2.10. As one can

Table 2.2 Base and filler metal combinations tested in previous studies

Base metal	Filler metal	Test temperature	Source/year
Incoloy 800	AWS BNi-8	650°C	Spingarn *et al.*, 1983
Albemet 162	AWS BAISi-4	RT	Flom *et al.*, 2009
304 Stainless steel	AWS BAg8	RT	Flom, 2011
304 Stainless steel	Pure silver	RT	Flom, 2011
Ti-6Al-4V	Al 1100	RT	Flom *et al.*, 2011

see, FAD based on Eq. 2.17 represents a very conservative lower bound estimate of failure condition in the brazed joints. It is a well-known fact that brazing process variables can greatly affect the strength of the brazed joints (*Brazing Handbook*, 2007). Our ability to account for all process variables and their influence on properties of the brazed joints is rather limited and accompanied with considerable uncertainty. Therefore, in an attempt to address such uncertainties and increase the level of conservatism, the following procedure was used for data analysis and graphing the plots shown in Fig. 2.10:

- First, using global model FEA of the brazed specimens, maximum normal and maximum shear stresses were calculated in the finite elements located in the brazed plane of the brazed joints. It was assumed that maximum normal and maximum shear stresses all acted on the braze plane.
- Second, it was assumed that maximum normal and maximum shear stresses were located in the same finite element within the braze plane even though their actual locations could be different.

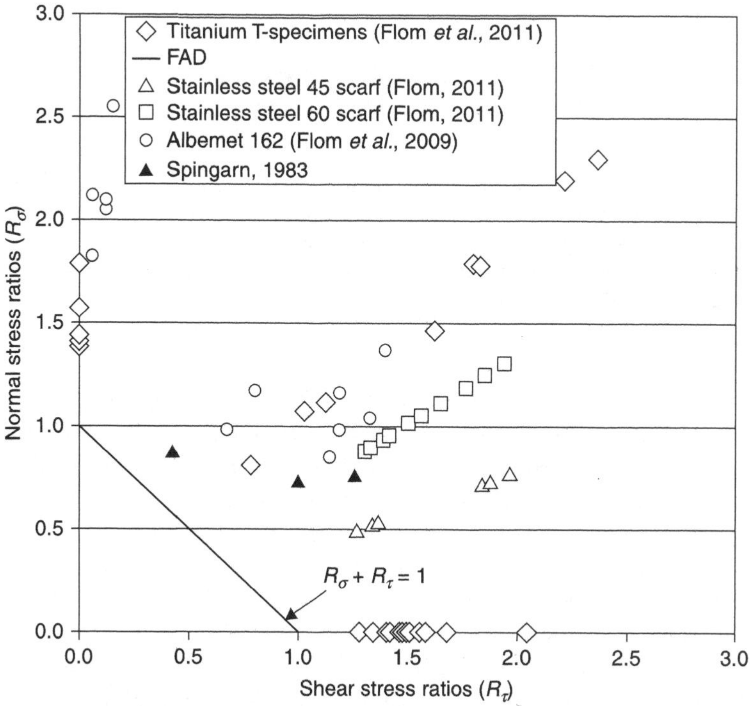

2.10 Test results from previous studies plotted as stress ratios. As one can see, the FAD line defined by $R_\sigma + R_\tau = 1$ can be used quite conservatively for the lower bound estimate of stress combinations that may cause brazed joint failure. Titanium lap shear and butt-brazed specimens (Flom *et al.*, 2011) are also plotted along shear and normal stress axes, respectively.

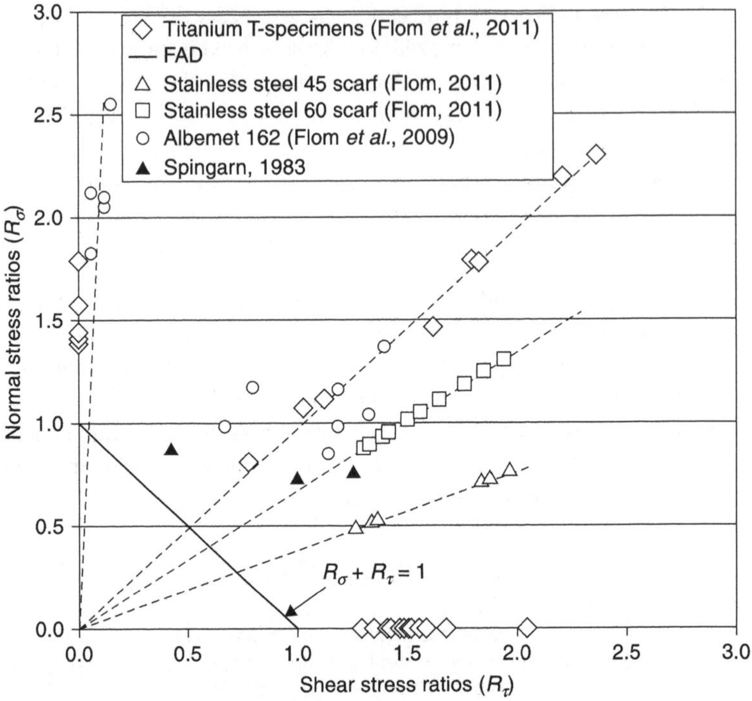

2.11 The same results as in Fig. 2.10, shown here with the addition of the trend lines (dashed). Each trend line is associated with a specific braze joint geometry. We can also call them braze joint lines. The position of a data point on such a line indicates a particular combination of stresses.

Two interesting observations can be made when examining the experimental results plotted in Fig. 2.10. One is that all data points representing specific joint geometries are aligned along the same trend lines, as emphasized in Fig. 2.11. The second observation is that the stress ratios representing failure of the brazed joints fall on the same trend lines regardless of what combination of base and/or filler metal is used to fabricate the brazed joints, or even their test temperature. These observations can be explained with the help of the Mohr circle construction, using scarf joints as an example, as shown in Fig. 2.12. However, this explanation is not sufficient to rationalize the behavior of the T-type titanium specimens also showing a similar trend (see Fig. 2.10 and 2.11). At the time of writing, no rigorous explanation of the observed trend line for T-specimens has been developed. A well-pronounced data dispersion or scatter observed on the graphs is, most likely, related to the inherent scatter in test results associated with testing of the brazed specimens, even the standard ones (AWS C3.1, 1963). For example, the results for lap shear and butt-brazed specimens, presented in Fig. 2.10 as the data points

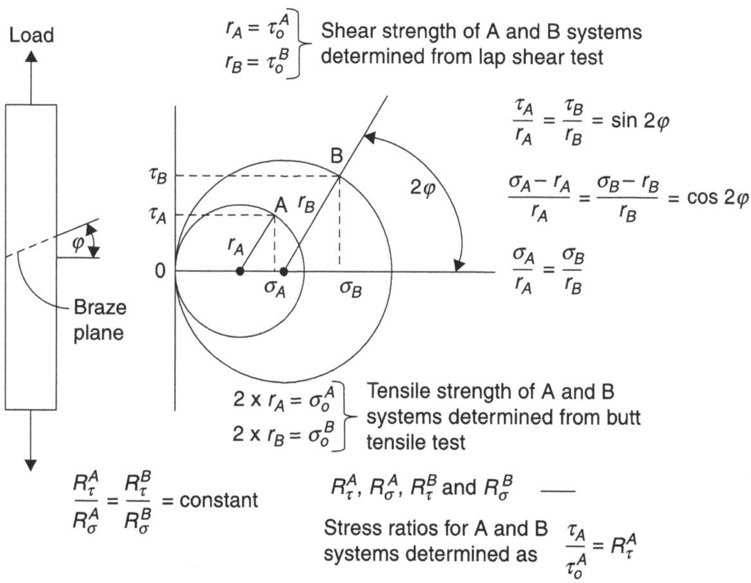

2.12 Mohr circle construction representing the stress state on the braze plane of the scarf joint subjected to uniaxial tension. The smaller circle represents a scarf joint fabricated from the base/filler metal combination, denoted as the A system. The large circle represents a scarf joint with the identical angle, but fabricated from the B system, which is stronger than A. Following the nomenclature above and using simple geometric observations, one can see that shear and normal stress ratios for scarf joint A form the same proportion as the stress ratios for joint B. This explains the experimental observation that stress ratios in the identical scarf brazed joints fabricated from different base/filler metals fall on the same trend line.

located along each axis, also show considerable scatter. The scatter is most likely a result of the variability in internal quality of the brazed joints. Consequently, due to propagation of scatter or error, it is quite reasonable to expect even larger scatter in testing more complicated brazed specimens. The observed trend lines could be an important tool during preliminary design of brazed structures. They can help in estimating brazed joints' margins of safety and predicting their failures. An example of using such a tool will be given later in this chapter.

These findings lead to the following simple engineering methodology of developing and using Coulomb–Mohr-based FADs:

1. Determine tensile and shear allowables by testing standard brazed test specimens (AWS C3.2, 2008) and construct the FAD line.
2. Using FEA, determine maximum normal and maximum shear stresses acting on the braze plane in the actual structural brazed joint subjected to a small

arbitrary load. This load should act on the joint in the same manner as the design load. These stresses will define the coordinates of the point needed to construct the trend line.

3. Connect the origin with the point determined in Step 2 and construct a trend, or, as we call it, a braze joint line (see Fig. 2.13).
4. An intercept of the FAD line and the braze joint line corresponds to a zero safety margin condition or a very conservative, lower bound failure condition.
5. Fabricate a small number (two or three) of identical or 'realistic' specimens representing actual brazed joint geometry and test them to failure under identical loading conditions. This step will help to determine the actual values of maximum normal and maximum shear stresses causing failure. Combination of these failure stresses will define the failure point that can be plotted to validate the brazed joint line. If the actual level of stresses causing failure is of no consequence, this step can be omitted.

Now we can demonstrate how beneficial the brazed joint lines can be in designing a brazed assembly. Suppose such a structure is to be fabricated from a material that is expensive and difficult to work with, such as beryllium, for example. Unless one can find a reliable source for tensile and shear allowables that have

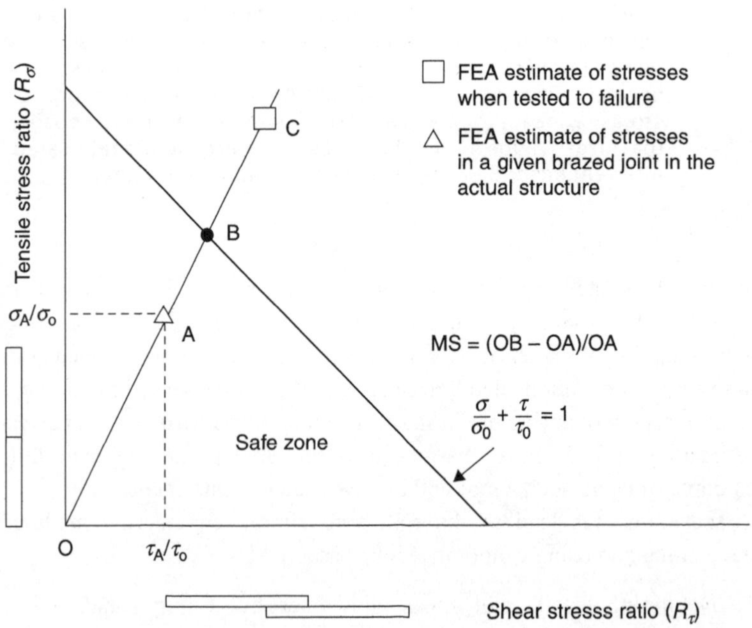

2.13 This graph illustrates the process of constructing the braze joint line and determining margins of safety of the brazed joint. The safe zone represents safe combinations of stresses.

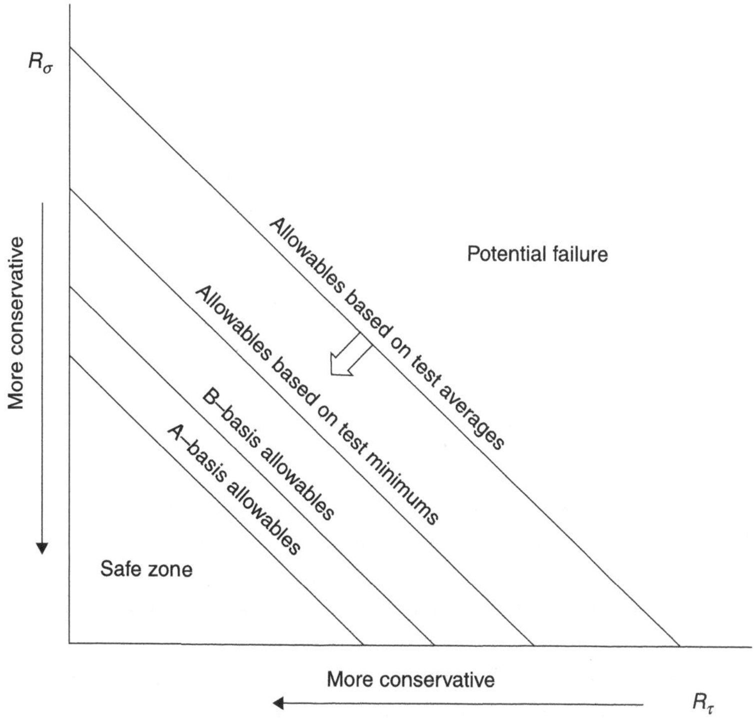

2.14 Tensile σ_o and shear τ_o allowables can be determined using various levels of statistical requirements, as illustrated on this graph. More conservative or stringent allowables result in a smaller safe zone. The graph shows relative position of FAD depending on the relative value of the allowables. For example, if the test average values of σ_o and τ_o are twice as high as their 'B'-basis counterparts, it is easy to see that the line $R_\tau + R_\sigma = 1$, representing test averages, changes to $R_\tau + R_\sigma = 0.5$, representing 'B'-basis results.

already been established, it would be necessary to test a certain number of standard lap shear and butt-brazed specimens. Depending on how rigorous and conservative the requirements are, it would be up to the program to determine the quantity of test specimens, so, for instance, either A-basis or B-basis (MMPDS-05, 2010) allowables could be established. The more critical the requirements are, the smaller is the zone representing safe combinations of stresses (safe zone), as shown in Fig. 2.14. It is very important to mention that the standard brazed test specimens should be manufactured using the same brazing procedure and by the same vendor selected for brazing of the actual brazed structure. When Step 1 of the procedure outlined above is completed and the FAD line is constructed, the designer needs to define the braze joint line or trend line specific to the joint geometry considered for the brazed structure. In order to define the trend line, all one needs to do is to perform simplified FEA of the imaginary brazed joint fabricated from *any* well-characterized, more conventional base metal/filler metal combination; for example, titanium base metal brazed with 1100 Al alloy filler metal. Assuming that our observations of the brazed joints' trend lines (see Fig. 2.10 and 2.11) hold true and are validated in future studies, a ratio of maximum normal and maximum shear stresses should define the point to construct the braze joint line, as described in Steps 2 and 3. After that, margins of safety (MS) can be estimated as (see Fig. 2.14):

$$(OB/OA) - 1 \qquad [2.18]$$

This geometric procedure, when expressed mathematically, takes the form:

$$\frac{1}{R_\sigma + R_b} \qquad [2.19]$$

Equation 2.19 is identical to some of the expressions listed in the last column in Table 2.1. By changing the brazed joint geometry and/or design (for example, changing the angle in the scarf joint), one can actually see how such a change affects the MS of the brazed joint. Then, the brazed joint design can be modified to achieve the desired MS.

Obviously, more base metal/filler metal combinations need to be tested before this simple engineering procedure can be accepted in the brazing industry. It is our opinion, however, that the preliminary results described in this chapter are very encouraging and may generate enough interest in the industry to warrant additional studies of other brazed joint systems.

A word of caution when developing brazed joint shear strength allowables. It is a well-established fact that average shear strength of the brazed lap joints drops as the overlap length increases (*Brazing Handbook*, 2007; AWS C3.2, 2008). Thus, when testing lap shear specimens having short overlap lengths (less than $2T$, where T is the thickness of the base metal), one may get a rather high values of shear strength. This is a typical behavior of joints brazed with ductile filler metals. In practice, however, when the lap shear joint geometry-based structures are fabricated, the practical overlap lengths are kept at $4T$ or even higher in order to achieve the full load-carrying capability of the brazed joints. Full load-carrying capability or full strength of the brazed joint is considered to be achieved when the joint strength becomes equal to the strength of the base metal. Such a condition is observed in lap shear joints when their overlap lengths are $4T$ or greater (*Brazing Handbook*, 2007). Therefore, testing specimens with short overlap lengths may lead to artificially high values of shear strength allowables. Consequently, a series of lap shear specimens covering a range of overlap lengths, typically between $1T$ and $5T$, should be tested in order to establish the lowest, and not the highest, average shear strength or allowable, as described in AWS C3.2 (2008).

2.4 Conclusion

- Construction of FADs based on the Coulomb–Mohr failure criterion provides a simple engineering tool for conservative estimate of margins of safety in braze joints.
- More work is required to further validate and better understand the trend line behavior of brazed joints.

2.5 Acknowledgements

The author would like to acknowledge the help of his co-workers at NASA Goddard Space Center, who are too many to name, in the preparation of this manuscript.

2.6 References

ASTM, E8/E8M-09 (2009), *Standard Test methods for Tension Testing of Metallic Materials.*

Astronautic Structures Manual (1975), NASA TM X-73305, MSFC Vol. 1, pt 1.

AWS C3.1-63 (1963), *Establishment of a Standard Test for Brazed Joints,* A Committee Report, American Welding Society (AWS).

AWS C3.2M/C3.2:2008 (2008), *Standard Method for Evaluating the Strength of Brazed Joints*, American Welding Society (AWS).

Blodgett, O.W. (1963), *Design of Weldments*, The James F. Lincoln Arc Welding Foundation, Cleveland, Ohio.

Brazing Handbook (2007), 5th ed., American Welding Society.

Bruhn, E.F. (1973), *Analysis and Design of Flight Vehicle Structures*, Jacobs Publishing.

Christensen, R.M. (2004), A Two-property Yield, Failure (fracture) Criterion for Homogeneous, Isotropic Materials, *J. Eng. Materials and Technology*, 126: 45–52.

Dieter, G.E. (1976), *Mechanical Metallurgy*, 2nd ed., McGraw-Hill.

Dowling, N.E. (1993), *Mechanical Behavior of Materials, Engineering Methods for Deformation, Fracture, and Fatigue*, Prentice Hall.

Dowling, A.R. and Townley, C.H. (1975), The Effect of Defects on Structural Failure: A Two-criteria Approach, *Int. J. Pres. Ves. and Piping*, 3, Applied Science Publishers Ltd, England.

Engineering Stress Memo Manual (2008), Lockheed Martin Co.

API (2007), *Fitness-for-service Engineering Assessment Procedure*, American Petroleum Institute, API 579-1/ASME FFS-1.

Flom, Y. (2011), *Failure Assessment Diagram for Brazed 304 Stainless Steel Joints*, NASA/TM-2011-215876.

Flom, Y. and Wang, L. (2004), Flaw Tolerance in Lap Shear Brazed Joints – Part 1, *Welding Journal*, 83(1): 32-s–38-s.

Flom, Y., Wang, L., Powell, M.M., Soffa, M.A. and Rommel, M.L. (2009), Evaluating Margins of Safety in Brazed Joints, *Welding Journal*, 88(10): 31–37.

Flom, Y., Jones, J.S., Powell, M.M. and Puckett, D.F. (2011), *Failure Assessment Diagrams for Titanium Brazed Joints*, NASA/TM-2011-215882.

Gordon, J.R. (1993), *Fitness-for-Service Assessment of Welded Structures*, ASM Handbook, Vol. 6, 10th ed., pp. 1108–1116.

Hart-Smith, L.J. (1973), *Adhesive-Bonded Double-Lap Joints*, NASA CR-112235.
Leinenbach, C., Lehmann, H., Schindler, H.J., Dűbendorf and Schweiz (2007), *Mechanisches Verhalten und Fehlerempfindlichkeit von Hartlötverbindungen*, Carl Hanser Verlag, Műnchen, MP Materials Testing, 49.
Lugscheider, E.F. and Partz, K.D. (1983), High Temperature Brazing of Stainless Steel with Nickel-based Filler Metals, BNi-2, BNi-5 and BNi-7, *Welding Journal*, 62(6): 160-s–164-s.
MMPDS-05 (2010), *Metallic Materials Properties Development and Standardization (MMPDS)*, prepared by Battelle Memorial Institute under contract with Federal Aviation Administration.
Peaslee, R.L. (1976), The Brazing Test Specimen – Which One?, *Welding Journal*, 55(10): 850–858.
Peery, D.J. and Azar, J.J. (1982), *Aircraft Structures*, 2nd ed., McGraw-Hill.
Rosen, R.S. and Kassner, M.E. (1990), Diffusion Welding of Silver Interlayers Coated onto Base Metals by Planar-magnetron Sputtering, *J. Vac. Sci. Technol. A*, 8(1).
Rosen, R.S. and Kassner, M.E. (1993), *Mechanical Properties of Soft-Interlayer Solid State Welds*, ASM Handbook, 10th ed., Vol. 6, pp.165–172.
Sarafin, T.P. (1998), *Spacecraft Structures and Mechanisms, From Concept to Launch*, Kluwer Academic Publishers.
Shanley, F.R. and Ryder, E.I. (1937), Stress Ratios, *Aviat. Mag.*, June.
Shigley, J.E. and Mischke, C.R. (1989), *Mechanical Engineering Design*, McGraw-Hill.
Spingarn, J.R., Kawahara, W.A. and Napolitano, L.M. Jr (1983), *Shear Strength of a Nickel-Based Brazement at 923 K*, Sandia Report SAND82-8037.
Tolle, M.C., Kassner, M.E., Cerri, E. and Rosen, R.S. (1995), Mechanical Behavior and Microstructure of Au-Ni Brazes, *Metallurgical and Materials Transactions A*, 26A: 941–948.
Tong, L. and Steven, G.P. (1999), *Analysis and Design of Structural Bonded Joints*, Kluwer Academic Publishers.
Webster, S. and Bannister, A. (2000), Structural Integrity Assessment Procedure for Europe-of the SINTAP Programme Overview, *Engineering Fracture Mechanics*, 67: 481–514, Elsevier Science Ltd.
Wen-Chun Jiang, Jian-Ming Gong, Hu Chen and S.T. Tu (2008), Finite Element Analysis of the Effect of Brazed Residual Stress on Creep for Stainless Steel Plate-Fin Structure, *ASME, Journal of Pressure Vessel Technology*, 130.

3

Modeling of the sequence of phenomena in brazing

D. P. SEKULIĆ, University of Kentucky, USA

DOI: 10.1533/9780857096500.1.55

Abstract: The phenomena involved in a brazing process constitute a set of very complex spatially distributed, transient physical and chemical processes, influenced by multiple parameters. This has been difficult to model reliably. This chapter offers a high level review of three broad areas of modeling: (i) the heating/cooling process during brazing, (ii) the thermomechanical behavior of brazed assemblies, and (iii) micro-scale phenomena, taking place in the joint zone and primarily related to transport phenomena. This discussion illustrates the importance of modeling brazing processes and its increasing contribution to the long-lasting success of the art.

Key words: brazing, phenomena, modeling.

3.1 Introduction

Materials bonding by brazing have a very long history, and the *art* continues to evolve into domains of more advanced applications, covering a wider range of material selections and satisfying more intricate designs under ever more stringent processing conditions (Shapiro and Sekulic, 2008). Undoubtedly, brazing is one of the oldest and most highly developed manufacturing processes. However, under pressure from the need to accommodate new application requirements, multiple challenges have presented themselves. Therefore, very detailed insight into the sequence of phenomena associated with the elements of the brazing process becomes mandatory for successful application of the technology. Consequently, developing a *science* of brazing is becoming much more important and increasingly more sophisticated.

Historically, the *art* of brazing has followed well-educated guesses and has frequently used a 'cook and look' empirical approach (Davies and Roberts, 2000). Exhausting trial and error procedures have been standard, with a very slow introduction of predictive, analytical and numerical modeling tools.

So, it must be clear that this field of inquiry and its associated technological applications require science-based predictions. To a large extent, detailed and very elaborate understanding of materials processing and its metallurgical aspects has been acquired over the years. However, modeling and science-based predictions (as opposed to empirical trial-and-error inquiries) have often been confined to applications of commercially available numerical tools, suited for engineering analysis of the outcomes of brazing processes, but not necessarily intended to uncover/describe an evolution of the fundamental phenomena involved. These efforts have usually been aimed at predicting overall outcomes of technological processes.

Modeling attempts in the context of the scope implied above have had variable levels of success, depending on the depth of knowledge of interrelated segments of the domains of inquiry, often featuring fuzzy boundaries between disciplines, and different levels of rigor and complexity.

The first modeling direction to be discussed is primarily intended to answer questions involving the brazed structure response on a *macro*-scale level. It often refers to the use of numerical modeling tools for either predicting process outcomes and/or establishing a mechanical/thermal (thermomechanical) state of an assembly after a brazing process. A typical example would be an attempt to model the residual stress status of a brazed assembly. Directly related efforts follow a need to define manufacturing *process parameter fields*, such as spatial distributions of the temperature of the whole assembly, including fixturing and/or processing environment; that is, they are aimed at understanding the heat transfer history of the structure or its components, and identifying the consequences inflicted by the thermal treatment involved. This includes efforts to optimize a brazing net-shape manufacturing process as a whole, in an attempt to mitigate issues imposed by non-uniformity of physical fields (e.g. temperature), but without venturing deeply into the intricacies of associated physical phenomena and micro/nano-scale processes. Finally, mostly on a *micro* level, although one would not exclude from consideration the entirety of a brazed assembly and its interactions within the engineering setting, a modeling effort may focus on individual phenomena, more or less decoupled from each other. For example, the heating rate to which a material in the solid state is exposed up to the solidus/liquidus temperature causes structural/phase changes (say in the solid state, and subsequently causing/affecting melting, surface tension-driven flow, solidification, etc.), so the impact of these changes at smaller physical length scales should be interpreted carefully. More specifically, for example, a phase change of the filler metal – say either during melting or during solidification – would be more or less distant from an equilibrium steady state. The diffusion that takes place in either solid or liquid states leads to various structural changes. The issue is how one can predict the solid phase growth and microstructure formation during solidification. Liquid metal spreading, surface tension-driven molten medium (filler, clad, flux) flow, etc., impose limitations on the brazing process outcomes, e.g. joint formation. Consequently, the question is how to take these into account and verify that the clearances are capillary for a given design. A reliable prediction of the dynamics of microstructure growth has largely been unresolved for complex material systems. Moreover, one should understand that in reality many distinct aspects of material/structure behavior at different length scales cannot evolve in a decoupled way, so an adequate approach ultimately must consider multiple relevant aspects for the particular problem.

The macro-scale aspects of modeling are much more broadly implemented and frequently utilized in industry, possibly due to more confidence in well-developed numerical tools addressing/predicting macro level consequences relatively reliably. Moreover, these aspects are important for control of a process at hand in a real-time domain. Micro-, and increasingly nano-, scale domain phenomena are often at a very fundamental level of understanding, and the complex interactions manifested within any of these parts of the process are more often studied empirically rather than predicted numerically or analytically.

One may not need to venture deeply into either of the discussed aspects to appreciate the fact that success in generalization and predictability of a brazed process outcome depends on fundamental knowledge of the process intricacies, in addition to acquiring empirical evidence. In particular, this is the case with solving many state-of-the-art technological problems.

In this chapter, we intend to cover the three identified modeling domains through a review of related bibliographic sources. The emphasis of our review will not be on the topics in depth, but on a broad illustrative set of typical case studies.

3.2 Modeling brazing systems

Figure 3.1 represents a schematic of a materials processing system used to execute a brazing task. For the sake of clarity, let us consider a system consisting of only

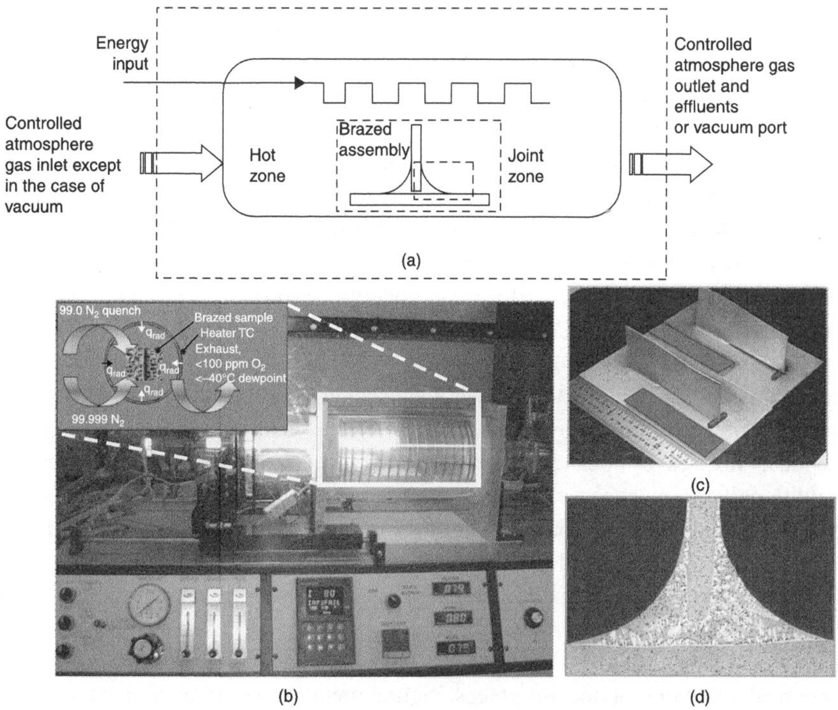

3.1 A brazing furnace system. (a) The hot zone, schematic, (b) controlled atmosphere brazing (CAB) furnace, (c) the assembly (aluminum brazing sheet and non-clad aluminum mating surfaces), (d) joint zone.

a few main components: (i) a furnace (hot zone) subsystem (Fig. 3.1(a) and (b)); (ii) a brazed assembly subsystem (Fig. 3.1(c)); including (iii) a joint zone (Fig. 3.1(d)) where transport and phase change phenomena relevant to joint integrity take place.

The focus is on modeling processes within each of these subsystems, as well as modeling the system as a whole. Certainly, a vast number of different brazing methods and material systems have been available, and a comprehensive insight into a generally applicable model would not be possible. We will, instead, illustrate an approach to modeling some of the phenomena of interest, so that each of them may be appreciated in both breadth and depth. Readers are advised to explore available sources for a detailed insight.

Let us assume that in the facility of Fig. 3.1(a) and (b) two or more aluminum mating surfaces of an assembly (or multiple assemblies, Fig. 3.1(c)) have to be bonded. The bonding process will be performed by brazing under a protective atmosphere following a prescribed temperature history designed so that the filler clad metal forms a metallurgical bond upon melting and solidification along the interfaces of the mating surfaces (Fig. 3.1(d)). It is clear that the assembly must be exposed to a more or less complex heating/cooling cycle. It should be executed so that spatial temperature fields are established by (1) radiation between the heating elements and the assembly within the hot zone cavity, (2) convection between the protective gas (say high purity nitrogen) and the assembly surfaces, and (3) conduction within the assembly, the associated fixture, and the hot zone elements in contact with the fixture. Certainly, the heating may be implemented using different approaches (localized laser beam, induction heating, furnace heating, etc.), but, for the sake of simplicity, we will address only a radiation–convection furnace brazing. The process would be inherently transient, and would lead to a non-uniform temperature field in the assembly at any point in time. The art of brazing strives to achieve as uniform a temperature as possible in the local bonding zone between the mating surfaces and the filler metal, and to hold it within the spatial domain of interest sufficiently long so that the filler can change phase from solid to liquid, as well as to achieve a subsequent cooling of the assembly and solidification in such a way as to reduce material thermal distortion caused by residual stresses established in particular if dissimilar materials are present. Additionally, in most cases (particularly if a vacuum is not used), a flux must be present on the mating surfaces to facilitate destruction of the oxide layer and to achieve good liquid metal wetting. The filler/clad material changes its phase (melting and subsequent solidification) depending on the temperature history imposed. In the meantime, the surface tension-driven metal flow fills capillary gaps between the mating surfaces and is associated with solid state diffusion, chemical reactions at the interfaces, liquid metal penetration along the grain boundaries, dissolution of substrates into the liquid phase, and subsequent formation of microstructures dictated by the system's state and its evolution conditions. This, although a straightforward and well-developed technology perfected over many years into the state-of-the-art controlled atmosphere brazing (CAB), constitutes a succession of a complex sequence of materials processing steps, intermingled with heat and mass transfer phenomena and structural changes of multiple materials, which are very difficult to describe precisely in theoretical modeling. Still, after introducing a number of assumptions, and establishing the well-controlled physical fields and their state uncertainties, one may, with more or less success, predict an outcome of the process. Let us briefly discuss a few such modeling efforts.

3.2.1 Hot zone modeling

One good example of modeling at the overall brazing system level is the attempt to simulate the operation of a brazing furnace. Thermal modeling of such a system is not trivial. As emphasized in Hosking *et al.* (2000), the model must capture, in general, the impact of all three heat transfer mechanisms: (1) convective heat transfer (if a vacuum is not used), (2) a complex thermal radiation exchange in the enclosure containing the brazed assembly and the corresponding fixture (or multiple assemblies if brazed simultaneously), and (3) the thermal conduction through all 3-D (or at least a 2-D) components of assemblies under markedly transient conditions – as well as (4) phase change phenomena. Figure 3.2 presents

3.2 High-temperature furnace and FEM mesh (Hosking, 2003; Burchett, 2003).

a batch furnace and illustrates its interior finite element discretization, including the brazed assemblies. It must be clear that a complex, conjugate heat transfer and fluid flow problem in furnace modeling cannot be reliably simulated unless detailed numerical modeling is executed. In this example, Hosking *et al.* (2000), a series of multiple experimental and numerical tests were executed in a hydrogen brazing furnace with simple test geometry pieces consisting of Mo, alumina and Kovar™ test bars. Single (Mo) and multiple (alumina and Kovar™) test pieces were used following a 50Au–50Cu brazing cycle with, say, a ramp-up till the first hold at approximately 800°C and the peak dwell between 950°C and 970°C. Measured data include temperatures of the part(s) and the furnace. Modeling of the hydrogen flow and heat transfer was performed for a 2-D axisymmetric fluid flow using Galerkin finite-element code (GOMA: Schunk *et al.*, 1996, 1997) and a non-linear, heat conduction algorithm COYOTE (Gartling *et al.*, 1994, 2010), a 3-D transient finite-element code implemented with parallel computing. Ultimately, the furnace model takes into account radiation heat transfer coupling between the load and the furnace cavity, including auxiliary internal supporting structures. Advection by hydrogen and external losses were not modeled. Figure 3.2 illustrates the finite element mesh of loaded test bars (the mock-up for a molybdenum work rack loaded with both small and large monolithic test bars; Hosking *et al.*, 2000).

Simulation of the temperature history of a Mo test piece in the furnace features initial lagging of the predicted temperature vs. measured values, but at higher temperatures the temperature difference has reversed in sign and ultimately converged to zero at the peak. During quenching, at lower temperatures in particular, the measured values decreased faster than numerically predicted (Hosking *et al.*, 2000). Studies of the impact of the loading configuration and material selection (using alumina and Fe–29Ni–17Co wt%) demonstrate, as expected, that larger thermal capacity and lower thermal diffusivity materials are more sluggish in transient temperature changes, with larger deviations of predictions at lower temperatures during cooling if experimental conditions during quench are uncontrolled (Hosking *et al.*, 2000). It was also clear that the hot zone support structures interfere with the temperature history of the test material, causing undesired temperature gradients in work pieces. Similarly, temperature gradients are identified within the heating zone. Finally, thermal emissivity differences between the part and the hot zone walls/structures leads to sizable temperature history changes.

Dempsey *et al.* (2003) used the developed FE modeling tools for determining temperature responses in the furnace environment of Hosking *et al.* (2000), and the statistical design of experiments (Box *et al.*, 1978) to map the furnace brazing process space during brazing of high voltage components. Brazing was facilitated with copper and 50Au–50Cu filler metals, including modeling of fixtures made of Kovar or molybdenum. A dramatic reduction in the number of experimental runs needed has been obtained as a key outcome of the study.

3.2.2 Continuous brazing furnace modeling

The examples discussed so far involve a batch furnace with transient changes imposed by the *in situ* spatially distributed stationary heaters. However, in a continuous brazing furnace, a series of brazed assemblies travels through the multizonal sequence of chambers/mufflers. Under such conditions, a similar temperature history can be imposed on a brazed assembly, but the heating elements of different zones are at different temperatures, so that ramp-up and cool-down processes are achieved by guiding an assembly on a conveyor through the furnace at a pre-specified speed. In Fig. 3.3, such a setting is presented, showing only three heating and a couple of cooling zones (clearly the thermal degreasing and fluxing stages are not shown).

Figure 3.4 presents a data set representing temperature differences across a compact heat exchanger core that was brazed in the continuous CAB furnace of Fig. 3.3. It is obvious that, although the 12 heating zones actually used were fine-tuned so that local temperature differences remained within only approximately 5° at the peak, during heating and cooling the local differences may increase up to 35 K (see the large diagonal, locations 4, 5, 1 and 8, Fig. 3.4). Reducing these differences to below 5 K is a demanding task and requires excellent understanding of the transient temperature changes of a complex assembly during brazing. One may model the assembly behavior by using, for example, Computer Fluid Dynamics (CFD) codes with structured grids, say CFD 2000 (1999) or Ansys Fluent (2011). The level of detail in numerical simulation depends on the smallest length scale $O(10^{-4}$ m) and overall assembly length scale $O(10^{0}$ m), which may lead to an excessively large model size of about 10^{10} to 10^{12} nodes, possibly surpassing the capability even of some supercomputers. The problem, though, may be resolved by considering a model of the structure as consisting of a number of spatial zones with different but spatially uniform properties at different scales. According to Sekulic *et al.* (2003), a complex modeling problem may be simplified through the hypothesis that the transient conjugate radiation–conduction (convection) behavior of a

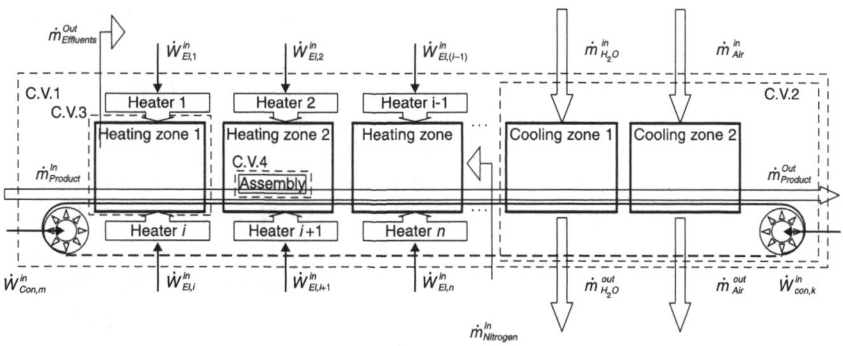

3.3 A continuous brazing furnace (Sekulic, 2011).

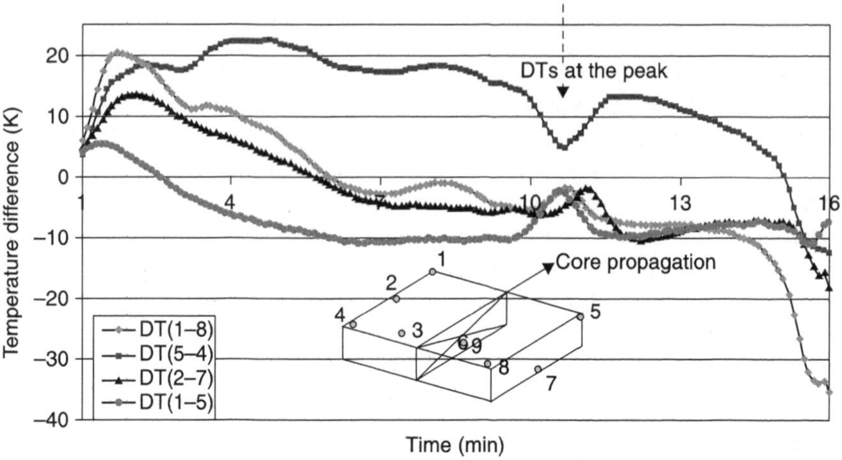

3.4 Temperature difference distribution within a CAB aluminum porous assembly during brazing in a continuous furnace (Sekulic, 2009; Sekulic *et al.*, 2003).

composite heat transfer structure can be modeled in time (within the imposed, predetermined margin of error) by using a multizonal solid-state model, if each zone is characterized by its effective thermophysical properties. How this potentially radical simplification can be implemented is illustrated in a simple manner (see Plate I in the color section between pages 268 and 269).

The worst case scenario would be to consider the core of the assembly and the header areas together as having averaged, lumped thermophysical properties (see Plate I(a) and (c)). Far better prediction of the brazing conditions in the critical zones of the complex assembly may be obtained if each of these zones is modeled as having spatially different properties but lumped within the zone (Plate I(b)). Following the same logic, further refinement at smaller spatial scales may be implemented selectively.

Slightly better insight into the intricacies of temperature distribution variability within a non-homogeneous assembly during brazing can be obtained from Plate II in the color section between pages 268 and 269, representing a sequence of time instants to which an aluminum sample was exposed in a typical ramp-up heating (Sekulic *et al.*, 2001).

Note the presence of large temperature gradients in the header area within the joint zone (Plate II). The most critical segment of the temperature field evolution is at the brazing temperature peak (which has been modeled reasonably well, but not as well as one might wish for the current state of the art), featuring less than 10 K deviation from actual measured data (for local temperature differences between larger mass tube assembly parts vs. smaller mass fin material assembly parts).

Radiation-only simulation of a CAB furnace has been performed by Ratts (2000). This study reported a prediction of temperature differences across a brazed heat exchanger (approximately 0.5 by approximately 0.5 m and 0.4 m in thickness) ranging from 10 to 20 K. The modeling results are very sensitive to the selection of thermal emissivity of the assembly material, i.e. this selection may sizably impact the temperature response of the brazed object. Although a relatively small number of nodes in a finite difference calculation routine were used, good agreement between predictions and experimental data was reported.

3.3 Finite element analysis of residual stresses in brazed structures

Let us start the discussion of this topic with a review of modeling using numerical simulation of residual stresses formed during the cool-down phase of a brazing process. Most often this is facilitated by using multidimensional, non-linear thermomechanical analysis of the stress–strain behavior of the brazed assemblies. When a system encounters a problem, typically one may identify brazing of dissimilar materials characterized by a sizable mismatch of thermomechanical properties. This problem is the easiest to identify when brazing ceramics to metals. In what follows, a series of studies devoted to the listed topics will be briefly summarized.

The multiaxial stress state of a brazed assembly can rarely be simulated by analytical elastic material models. In some cases such an approach may be used as a first approximation of the behavior of simple geometric configurations. For example, a biaxial stress state of a simplified plate-like configuration (three elastic infinite plates bonded together), excluding the end and edge effects, can be modeled by an analytical elastic model based on elementary bending theory (i.e. with the metal in tension and the ceramic in compression at the interface) (Iancu *et al.*, 1990). A similar analytical model has been demonstrated for brazing different metallic materials. For example, brazing copper, austenitic steel and nickel-based alloy to a ferritic steel using AgCu braze may be modeled with an acceptable level of certainty (Bing *et al.*, 1994). However, more complex configurations – and in particular when inclusion of elastoplastic behavior with temperature-dependent properties is a must – require thorough numerical modeling (Williamson *et al.*, 1993). In this study, residual stresses were investigated at graded and non-graded interfaces for Al_2O_3–Ni as a material system. The graded zone has been considered as a series of composite interlayers, each layer having different material properties. In this model both Ni and Al_2O_3–Ni composites are assumed to be elastic–plastic, including the von Mises yield condition and isotropic hardening, while pure alumina was considered as elastic. The thermal expansion coefficients for both Ni and Al_2O_3 were assumed to decrease linearly along the cooling path.

The number of design parameters relevant to dissimilar material brazing is extensive and includes geometric shape and dimensions, brazing temperature history, material properties, pressure loading and capillary clearances. This list, though, is open-ended. Proper design would greatly benefit from modeling the impact that some of these factors may have on brazing outcomes. Such analysis may be performed through numerical simulation. For example, Gong *et al.* (2009) illustrate this point by modeling stainless steel (AISI SS304) fin-to-plate brazing for a heat exchanger assembly using a boron–nickel–chromium filler metal (BNi-2). Finite element modeling was straightforwardly performed using a mesh consisting of quadrilateral elements. Modeling was executed by assuming that the coefficient of thermal expansion, Poisson's ratio, Young's modulus and yield strength are temperature dependent. All materials were assumed to be isotropic and elastoplastic. The residual stresses modeled include longitudinal, transverse and shear stresses. Such a study has the capacity to provide a detailed view of the impact of various design parameters.

A good understanding of the residual stress field is necessary to minimize tensile stresses in the ceramic part appearing due to the thermal mismatch of metal–ceramic components of the brazed assembly during cooling from the brazing temperature down to room temperature. The finite element modeling-based calculations reveal the residual stress field most closely if temperature-dependent elastic–plastic properties of the materials are considered. Such modeling may also be based on a plasticity model characterized by the assumption that material behaves as though having an invariant strain rate (Pintschovius *et al.*, 1994). In that study, brazing of steel to ceramic, Si_3N_4 and ZrO_2 with Y_2O_3 was accomplished by using AgCuTi braze. The numerical model is structured so that each numerical iteration step features increments of the total strain as being composed of elastic, thermal and plastic strain contributions. These contributions were modeled using Hooke's law, a temperature-dependent thermal expansion coefficient and yield law in a plastic potential representation. The ceramic material was assumed to be elastic, while steel and braze materials are considered as isotropic and ideally elastic–plastic (a von Mises yield function and the Prandtl–Reuss yield law). A lumped capacitance temperature field was assumed. The comparison between modeling and experiments indicates that the residual stresses would be seriously overestimated if plastic deformation in the steel part during cooling to room temperature was not taken into account. Moreover, modeling gives stress overestimation, which can be explained by questioning the selection of material properties.

The residual stress status developed during a brazing cycle of nodular cast iron (with spherical graphite nodules in a matrix of pearlite and ferrite) bonded to stabilized zirconium (partially stabilized with magnesia, containing cubic, tetragonal and monoclinic polymorphs) with a silver Ag–Cu–Ti alloy in a vacuum (the specimen had a cylindrical, axisymmetric geometry) was modeled successfully by Wang *et al.* (1996). Brazing was executed in a vacuum at 850°C for 20 minutes and with a cooling rate of 10 K/min (Hammond *et al.*, 1988). Modeling postulates an isotropic thermal expansion for both mating parts, while iron and the braze materials were assumed to be elastic–plastic (von Mises yield strength and isotropic hardening). Young's modulus, Poisson's ratio, the coefficient of thermal expansion, yield strength and tensile strength for iron and the filler were considered again as temperature dependent, while zirconia was considered as elastic. Modeling predictions of residual strain distributions were corroborated with neutron diffraction data. The best agreement was established for the cast iron. The modeling failed the validity check with respect to predicted large stress concentrations, which is attributed to limitations involving spatial resolution. The same holds for the details of the strain–stress field within the filler metal.

Active brazing of sintered Si_3N_4 with addition of Y_2O_3 to stainless steel (SS316) and with copper and/or molybdenum as interlayers and Ag–Cu–Ti alloy as braze at 850°C in vacuum leads to thermally induced residual stresses. These stresses were modeled by 3-D finite element elastoplastic simulation using commercial software (Kim *et al.*, 2001). The maximum residual stress occurs within the Si_3N_4 material near the interface with the Cu interlayer. The modeling offers an insight into the optimal magnitude of the interlayer to minimize the residual stresses, as well as to accommodate dissolution of Cu into the braze. The use of a Mo interlayer leads to additional residual stresses within that layer due to a sizable mismatch in the coefficients of thermal expansion for Mo and steel. As a consequence, fracture may occur either in Si_3N_4 or in the Mo interlayer, depending on the Mo interlayer thickness. These conclusions agree reasonably well with experimental data (Kim *et al.*, 2001).

Let us re-emphasize that a combination of ceramics and metals is used to exploit the advantages of both materials (i.e. resistance to corrosion and wear vs. stability of strength, respectively) but causes a series of design problems – brittleness of the ceramics, large disparities in thermal expansion (as already considered) – leading ultimately to a need for sophisticated bonding techniques. The influence of the thermal expansion mismatch requires design consideration, along with the need for simulation of the bonded assembly behavior after reaching the peak brazing temperature. As emphasized earlier, the most often used tool of analysis is the numerical finite element method. It is important to note that this method has always been executed with a series of simplifying assumptions, so that the resulting prediction of residual stresses would inevitably feature a certain margin of error. For example, bonding Al_2O_3 to steel (SS 304) with Ni or Ni–Cr foils in combination with Cu and Ti interlayers using transient liquid phase brazing was analyzed to explore the validity of the prediction of bond strength trends as well as stress distribution within the joint zone (Zhang *et al.*, 2002). The consequences of the imposed stresses were manifest in the appearance of fractures either along the filler/metal interface, across both filler and ceramic, or the ceramics only. Such fractures can be reasonably well predicted, regardless of an apparent disparity between the numerically predicted stress–strain fields and the actual stress state established by experimental means. In yet another study (Heikinheimo and Saarenheimo, 2000), a numerical simulation of thermal residual stresses in a brazed cylindrical alumina sample with a Ti (ASTM Grade 2) joint was explored. The filler metal brazing foil was 59Ag–27.25Cu–12.5In–Ti, and brazing was performed at 750°C for 15 minutes followed by 100 minutes cooling down to room temperature, while modeling was performed under these and modified conditions with a doubly shorter cooling time and, optionally, a cooling time three times longer. The commercial FE code ABAQUS was used (ABAQUS, 1998). Creep and plastic deformations were mutually independent. The von Mises yield function and isotropic strain hardening were considered for elastic–plastic behavior. It was concluded that the effect of creep has to be taken into consideration for reliable prediction of thermal residual stresses.

The accuracy of a simulation of metal-to-ceramic brazing depends on the braze material model. An example of an approach to modeling the viscoplastic behavior of braze alloys is presented by Neilsen and Stephens (2000). The braze alloys considered were 82Au–15.1Ni–0.7Mo–2.1V and 62.2Ag–36.2Cu–1.6Ti for brazing an alumina ceramic tube to a Kovar disk. The alumina was modeled as a linear elastic material with given values of Young's modulus and Poisson's ratio, while the metal was considered to be a temperature-dependent elastic–plastic material. Braze material is modeled as viscoplastic (Miller, 1976; Bammann, 1990; Freed and Walker, 1993), with the hyperbolic sine kinetic equation for the inelastic strain rate dependent on temperature, which accounts for kinematic hardening. This modeling was able to offer an insight into the effects of braze alloy variations on the stress state generated during brazing (significantly higher stresses generated in the ceramic when the temperature increases and when the higher strength braze alloy was used). Neilsen *et al.* (2003) used the same model in a study of a eutectic silver–copper (72Ag–28Cu in wt%) braze. The ability of the model to capture the behavior for uniaxial compression, as well as both primary and secondary creep, was well documented. A simulation of the ceramic to metal brazing with the eutectic silver/copper filler braze for ceramic/metal bonding was performed as well (Neilsen *et al.*, 2003). Mating surfaces were alumina and iron–nickel–cobalt rods (both modeled as elastic materials). A comparative simulation was performed with active filler metal (63Ag–35.25Cu–1.75Ti). It was demonstrated that the active braze generates a lower peak stress (active braze creeps more readily at temperatures above 450°C), but also generates a higher residual stress (it has a higher strength than the eutectic silver–copper alloy at temperatures below 450° C) (Neilsen *et al.*, 2003).

Carbon steel, low alloy CrMo steel and NiCrMo tool steel can be bonded to cemented carbide [WC-(Ti, Ta, Nb)C, Co] using Cu supported by a Ni mesh. Such brazing may be performed in hydrogen at 1100°C, but the cooling process would cause the appearance of residual stresses. If cemented carbide is assumed to be a temperature-independent linear elastic material, while steels are considered as being temperature-dependent elastic–plastic materials, and Cu as having a temperature-dependent thermal expansion coefficient (elastic–plastic material as well), the behavior of such an assembly may be simulated numerically using finite element modeling. It may be confirmed that such predicted stress fields (Pintschovius *et al.*, 1999) are in general agreement with available empirical evidence (neutron analysis (Bing *et al.*, 1996) and X-ray stress analysis). It is indicative to note that for this assembly the relatively high residual stress values were determined only in the neighborhood of the steel and cemented carbide interface. It is clear that the residual stress state is impacted by a larger thermal shrinkage of steel vs. cemented carbide, as well as elastic–plastic deformations (Pintschovius *et al.*, 1999). The topography of the assembly, 3-D FEM mesh and residual stresses maps in all three directions of an orthogonal coordinate system are presented in Plate III in the color section between pages 268 and 269 (Pintschovius, 2011; Pintschovius *et al.*, 1999). Designation CK 45 refers to the carbon steel, while the bottom set of the maps corresponds to the tool steel, both brazed to the WC-based sintered alloy with copper.

Figure 3.5 shows the schematic of the assembly and indications of the distortion across a principal plane. The residual stresses diminish rapidly away from the interfaces between the mating surfaces (domain between 0 and 5×10^{-3} m).

One of the most important factors responsible for the stress state in a ceramic–metal system is the elastoplastic response of the resolidified filler. Consequently, as we discussed earlier, a brazing filler cannot be modeled as an elastic participant in bonds between dissimilar metals, if good modeling of the bond is needed. In several modeling attempts referenced above, the ceramic has been considered as elastic, but it must be assumed that the metal or filler, or both, are elastic–plastic to obtain adequate results. Hence, the modeling of the mechanical behavior of brazed ceramic–metal assemblies needs the true thermomechanical properties and

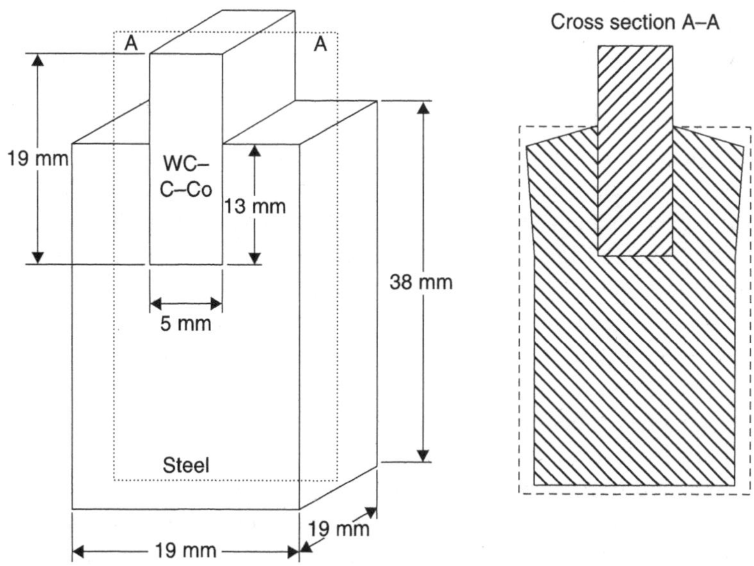

3.5 A brazed assembly and a sketch showing the distortions; redrawn and modified from Pintschovius *et al.* (1999).

must assume proper elastoplastic characteristics of the participating media (Galli *et al.*, 2009). In that study, Si_3N_4/TiN–steel joints were brazed with AgCuTi filler alloy in vacuum and analyzed experimentally and numerically. The study illustrated an acceptable agreement between experimentally determined principal stresses in the ceramic component and those obtained by 3-D finite element modeling, while assuming that both steel and ceramic stay in the elastic range (but with temperature-dependent mechanical properties) while the brazing alloy was considered to be elastoplastic (von Mises yield criterion).

A modeling framework for an evaluation of the strength characteristics of multiple brazed layers (ductile interlayers) in a system representing a ceramic-to-metal joint was studied numerically and experimentally by Park *et al.* (2002). As an example, Si_3N_4 was brazed to Inconel 718 in a multilayer assembly. The finite element method was complemented by an order of magnitude scaling analysis in order to devise analytical relationships between the strain and properties. Modeling was aimed at an analysis of the residual stresses and establishing criteria for design of the sequence of interlayers ahead of brazing. A relatively simple geometry of an axisymmetric set of discs was brazed together at 880°C in a vacuum with thin foils of Ni and W as interlayers in various combinations, and Ag–Cu–Ti braze filler. An interlayer with low yield stress reduces the strain energy in the base materials. However, there is a trade-off between this relief and plastic deformation in the interlayer. The heating ramp rate was 8°C/min and the slow cooling rate was 2°C/min. In this study, both mating surfaces and interlayers were modeled as elastoplastic materials with temperature-dependent properties. Finite element modeling indicates that inclusion of multiple interlayers can be beneficial for a reduction of strain energy in the ceramic part.

An analysis of the impact of multiple layers, i.e. Cu, Kovar, Mo and W interlayers, on thermal stresses of brazed joints involving Si_3N_4 to steel, executed by performing both FEM calculations and experimental brazing with an active Ag–Cu–Ti filler in a vacuum (Zhou *et al.*, 1991), indicates that joining involving low yield strength and high ductility interlayers features smaller thermal stresses and higher strength than using low thermal expansion and high yield strength interlayers. Modeling was performed assuming that the interlayer properties such as elastic modulus, yield strength, strain hardening rate and thermal expansion are temperature dependent during the cool-down after brazing, and that plastic flow takes place within the steel mating substrate, interlayer and filler metal. It was established that an adequate selection of an interlayer (high ductility and low yield strength) has a dominant influence. Similarly, the thickness of the interlayer impacts the level of thermal stress.

An interesting aspect of the impact of non-homogeneous mechanical properties within the joint area is related to the fact that different phases within the formed filler metal fillet may have significantly different properties. It is therefore necessary to consider not only macro tensions between the base metals and filler but also tensions within the microstructures of the joint area (Wielage *et al.*,

2000). The mechanical properties of the different phases must be known. With this objective, Wielage *et al.* (2000) studied the bonding of NiCo steel, mild steel, silicon nitride alumina and zirconia using Ag–Cu–Ti and Ag–Cu–In–Ti fillers. Multiple layers of different phases along the interfaces were characterized experimentally. In addition, reinforcement materials may be added, forming additional phases in surrounding areas (for example in the reaction zone between the alumina, Ag–Cu–Ti filler metal and graphite fibers). The properties of individual phases may be determined with various levels of uncertainty using techniques such as nano and/or micro-indentation.

3.4 Micro-scale brazing phenomena modeling

The most challenging, and undoubtedly the most intricate, aspect of modeling brazing phenomena – necessary for a good understanding of the brazing process – occurs at the micro-scale. This is a consequence of the fact that the least understood physical and chemical phenomena take place at micro- and/or nano-scale levels. These domains constitute a vast field of basic and applied science investigations, in various domains of materials science, physics and chemistry. Clearly, it would be impossible to provide a comprehensive review of this important field of inquiry in a short review chapter such as this, so we instead discuss a few selected topics considered over the years in the author's laboratory.

We will briefly discuss several modeling features: (1) joint formation, (2) melting of the clad/filler and residue formation, (3) spreading of molten metal and (4) solidification. Each of these phenomena has been studied extensively by many researchers in the field and there is a large body of relevant literature.

3.4.1 Joint formation modeling

Let us reiterate that a typical brazing cycle starts with heating a material in the solid state. The solid-state phase is exposed to significant changes, including chemical reactions in the presence or absence of flux, but most importantly in the presence of diffusion processes (for example, migration of elements across the interface between, say, clad and the material core and vice versa, triple line movements over solid substrates during wetting and associated phenomena, solid grain phase boundary migration, etc.). For example, after the onset of the destruction of the oxide layer, and upon reaching the melting point of the clad/filler, molten metal is produced, driven by surface tension into the capillary gaps, hence forming the joint. Concurrently and subsequently, complex reaction mechanisms at the interfaces and within the base and filler domains often lead to the formation of intermetallics and associated chemical reaction products.

A typical joint is presented in Fig. 3.6. In this example, a bond has formed between the molybdenum and molybdenum–rhenium mating surfaces; the filler

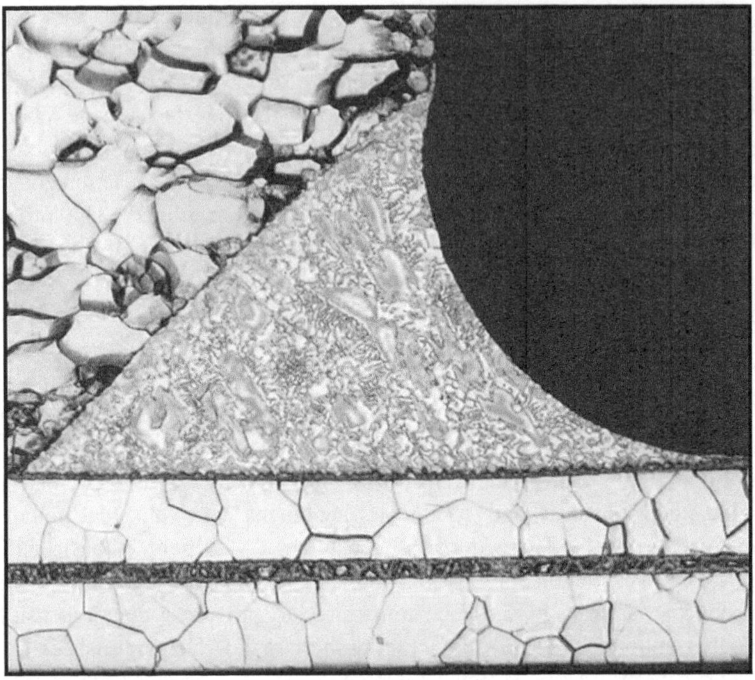

3.6 Mo–Ni fillet between Mo and Mo–Re mating surfaces (Busbaher *et al.*, 2010).

metal is made of Mo–Ni microparticles doped with 3% Mo–Ni nanoparticles (<100 nm). Brazing was performed in a H_2–N_2 atmosphere at 1350°C (Busbaher *et al.*, 2010). The fillet's shape topography was established as an equilibrium free surface at the onset of liquid metal solidification. This typical 'elastica curve' of the free surface is formed because the molten metal establishes an equilibrium shape that corresponds to the minimum potential energy of the free surface (Sekulic, 2001; Zellmer *et al.*, 2001).

Modeling of the shape requires (Sekulic, 2001; Zellmer *et al.*, 2001):

$$\min(E_p) = \min(E_s + E_g + E_w) \quad [3.1]$$

subject to

$$\int_V \mathrm{d}V = V_{joint} \quad [3.2]$$

In Eq. 3.1, the E's represent the total potential energy, free surface molten metal energy, gravitational energy and wetting surface tension (molten metal–solid) energy, respectively. In Eq. 3.2, V_{joint} in an engineering application would be the specified joint domain size, responsible for the necessary mechanical and thermal integrity. It is easy to conclude that a design task may require this volume (i.e. the mass of the fillet, including its topography and properties) to be specified, in addition to a full characterization of the mechanical integrity properties of the resolidified phases and the fillet as a whole. With this input, one may optimize the joint shape (Eq. 3.1) so that this constraint is satisfied. A detailed analysis of such numerical modeling is presented in Sekulic (2001) and Zellmer *et al.* (2001).

Determining the actual mass/volume of the fillet, under specified brazing conditions, but without empirical input, is a difficult (virtually impossible) task. Namely, not all available filler metal would end up in the joint zone. Let us consider this problem with an example involving a typical fin-tube brazed joint of a compact aluminum heat exchanger traditionally manufactured using a state-of-the-art CAB process. Contemporary designs, as a rule, use brazing sheets with a clad layer as the source of the molten metal. Upon melting, the liquid formed is driven by surface tension to fill the clearances between the fin surfaces and the tubes. For the known clad ratio, one may expect that the thus known mass of the molten filler produced would easily be associated in equal proportion with each of the joints formed – if the geometry of the mating surfaces is known. This simple answer, however, is not right. Even before melting, and in particular upon melting, silicon from the clad diffuses into the core and reduces its content within the clad at the interface. Hence, a lower Si content in that clad domain leads to the formation of a residue *in situ* so that only a portion of the molten clad/filler will actually start moving, driven by surface tension, and ultimately reach the joint zone. This mass, if known, would be proportional to the volume of the constraint of Eq. 3.2, and any subsequent modeling and/or prediction of the fillet size would then be possible. Based on this consideration, a prediction of the fillet topography of aluminum brazed joints can be performed (Sekulic *et al.*, 2004).

The molten clad mass of the joint area, based on the mass conservation principle, for the brazing sheet is

$$m_{\text{joint}} = m_{\text{clad}} - m_{\text{residue}} = m_{\text{clad}} - (m_{\text{depleted}} + m_{\text{solid solution}} + \Sigma\, m_i) \qquad [3.3]$$

In Eq. 3.3, m_{clad} represents the mass of clad before the brazing sheet is exposed to heating, and m_{residue} is the mass of metal that did not change phase from solid to liquid upon heating to the peak. Non-melting is due to (i) Si depletion of the interface portion of clad in contact with the Si lean base as a consequence of Si diffusion (Gao *et al.*, 2002), m_{depleted}, (ii) the solid solution portion of clad upon melting formed *in situ*, $m_{\text{solid solution}}$, and (iii) the mass trapped *in situ* due to auxiliary phenomena, such as viscosity during the surface tension-driven flow phase. Clearly, one can model the depletion process as well as solid solution formation *in situ* by invoking a Fickian diffusion of Si into the core and assuming, or not, the validity of the equilibrium phase diagram (Sekulic *et al.*, 2004).

Figure 3.7 illustrates the results of such a model of joint formation. A set of non-orthogonal mating surfaces is bonded and modeled under different assumptions involving the participating clad. A validity of either equilibrium (the

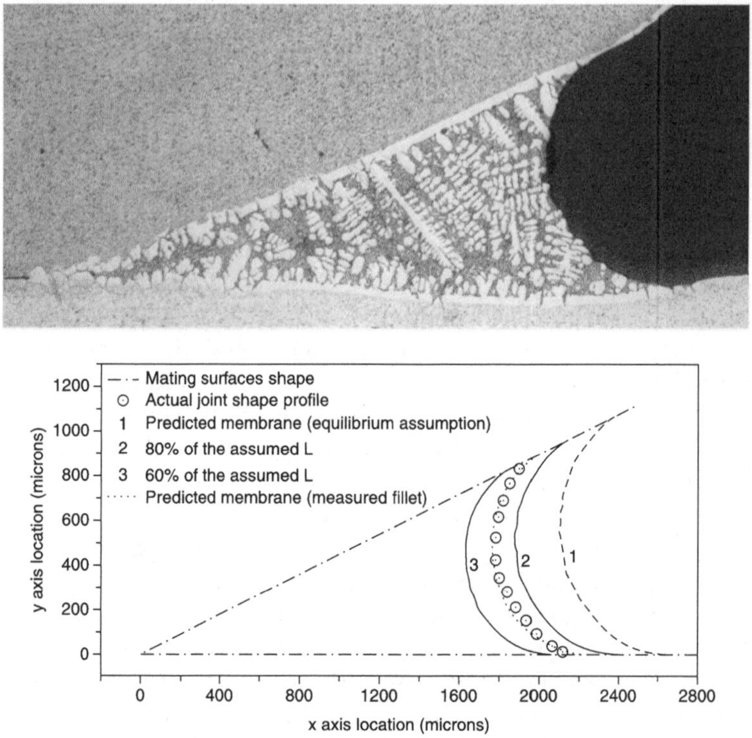

3.7 Modeling of the fillet size/shape. L is the length of the assumed clad layer participating in the surface tension-driven flow (Sekulic *et al.*, 2004).

lever rule) conditions or non-equilibrium melting conditions (Sekulic *et al.*, 2004; Zhao and Sekulic, 2006) was postulated. A heuristic model of diffusion-controlled melting of a clad layer can be formulated assuming the clad layer to represent a matrix of grains of specified size having initial Si content corresponding to the state prior to thermal processing. This content of Si evolves along a diffusion-controlled Fickian profile. An increase of clad temperature is followed by phase transformation and the formation of liquid with separation of solid solution within the mushy state. The liquid produced is driven away by surface tension. This process progresses until a solidus concentration is reached within the remaining solid state of the grain, forming the residue. This *simplified* scenario allows the corresponding Si diffusion differential equation to be formulated with corresponding initial and boundary conditions (Zhao and Sekulic, 2006). Figure 3.8 shows a comparison of the residue predicted in this way for different initial clad layer thicknesses.

The importance of including diffusion in models describing brazing processes is clearly visible in attempts to model Transient Liquid Phase (TLP) bonding. For

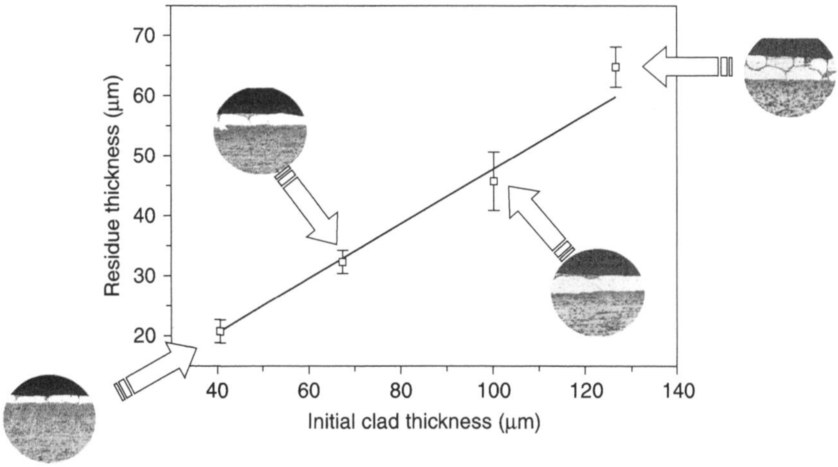

3.8 Comparison of modeled and experimental thickness data for the formation of residue (a heuristic model of diffusion controlled melting of a clad layer during brazing with an AA3003/AA4343 brazing sheet), peak brazing temperature 605°C and dwell at the peak 2 min in ultra high purity N_2 (Zhao and Sekulic, 2006).

example, extensive volume and grain boundary precipitation when brazing nickel superalloys (Inconel 625 and 718, SS 410) with BNi-2 filler metals (Arafin *et al.*, 2006) requires the diffusion of solute atoms into the base alloy during base metal dissolution and liquid homogenization to be taken into account. According to Arafin *et al.* (2006), the isothermal solidification times were successfully predicted.

3.4.2 Modeling of surface tension-driven flows

The surface tension-driven molten metal phase of a brazing process associated with wetting of the substrate has been modeled extensively, as presented in Chapter 1 of this book. Hence, we will not consider this aspect of the brazing process in detail. It suffices to say that the spreading of a molten clad or filler metal during brazing represents, as a rule, a reactive wetting and spreading over rough surfaces, with a variable topography, both spatially and temporally. Two main settings of significant relevance for brazing are most often considered: (i) spreading of a sessile drop of a liquid metal on a smooth substrate, modeled as a rule with a Tanner's-type law (a correlation between the triple line movement and time; Tanner, 1979), and (ii) spreading through/over a capillary and/or rough surface substrate modeled as a rule with a Washburn-type power law (Washburn, 1921): in general, a power law relationship of the type $r^n \sim t$, where n may take values different from 2 (which would be typical for non-reactive wetting), r

denotes the triple line location and t is time. For example, $n \gg 1$ may fit very well a reactive wetting of Al+Si filler over an aluminum substrate in the form of a sessile drop (Zhao and Sekulic, 2008). On the other hand, the spreading of AgCuTi-based filler over TiAl features $n \sim 1$ (Li *et al.*, 2011) regardless of the roughness (i.e. reactions and intermetallics formation at an interface have a strong impact). Moreover, if spreading of liquid metal occurs with little or no interface reaction, the kinetics of the triple line would follow $n = 2$, as for solder spreading over Sn–Cu intermetallics (Liu and Sekulic, 2011). In many cases, modeling of the molten metal spreading during brazing may be presented in the form of a power law.

Modeling of the kinetics of liquid metal spreading during brazing, such as high-temperature brazing involving spreading of liquid silver and SiO_2–CaO–Al_2O_3–TiO_2 glasses on molybdenum in near non-reactive spreading, has been extensively studied (Saiz and Tomsia, 2006). Low-temperature systems have been modeled using the Hoffman–Voinov–Tanner law (Kistler, 1993). High-temperature spreading has been discussed by a number of authors; see, for example, Eustathopoulos *et al.* (1999). Modeling of the spreading kinetics of reactive brazing alloys on ceramic substrates has also been considered using semi-empirical models (Meier *et al.*, 1998) and thermodynamics (Kritsalis *et al.*, 1991; Delannay *et al.*, 1987). The rate of wetting kinetics may be constrained by the rate of interfacial product formation. An interesting problem for high-temperature brazing in metal–ceramic systems is the presence of solid solubility and the appearance of micro-topographic modifications at the triple line locations, causing the appearance of so-called 'ridging' (Saiz *et al.*, 1998).

The wetting of copper–titanium alloys (Ti between 1 and 20 wt%) over polycrystalline alumina substrates was analyzed using semi-empirical modeling of the spreading kinetics by assuming that the rate of spreading is determined by the rate of change of the amount of interface reaction products (Meier *et al.*, 1998). The rate of change of reaction products formed was expressed through modeling of the fraction of interface coverage by the reaction product. It is difficult to establish whether (i) the magnitude of the Gibbs free energy of the reaction phase formed at the interface between the solid ceramic and liquid metal or (ii) the electron structure of the reaction products is the most important for spreading (Kritsalis *et al.*, 1991). The possibility of the existence of a continuous reactive phase layer beyond the solid–liquid interface has also been explored (Landry *et al.*, 1997). Modeling of the dissolution at the interface between the solid substrate and the liquid metal was conducted by establishing coupling between the dissolution process and the rate of spreading (Warren *et al.*, 1998). In spite of significant advances in understanding wetting mechanisms of molten metals over solid surfaces, a comprehensive numerical modeling of reactive wetting may be for some cases beyond the range of the liquid systems considered. Non-reactive wetting is modeled relatively well (Braun *et al.*, 1995).

3.4.3 Modeling joint solidification

Finally, let us briefly address modeling of the final stage of the brazing process, braze solidification. This process has also been studied extensively, but very little has been offered in terms of reliable modeling. To appreciate its complexity, let us consider the formation of alpha phase dendrites in a resolidified undercooled Al–Si braze typical of aluminum brazing (Sekulic *et al.*, 2005a, 2005b). In Fig. 3.9 a typical population of alpha phase dendrites within resolidified Al–Si, near the eutectic hypoeutectic alloy braze can be identified. The population of the dendrites differs depending on initial Si concentration and solidification conditions, in particular the level of subcooling.

An attempt to model dendrite formation (Fig. 3.9) was based on a series of assumptions: (i) the molten metal is assumed to be a chemically inert, binary, non-eutectic alloy (solidification into the solid solution only) with undercooling present; (ii) the system is considered to be non-isothermal, two phase under constant pressure; (iii) convection is neglected; (iv) all properties are constant; (v) solute diffusion is non-Fickian in the liquid but is neglected during quench in the solid. The full set of governing equations includes the energy equation, mass balance, Fick's law (taking into account a relaxation of the solute diffusion flux) and evolution of the solid volume (Sekulic *et al.*, 2005a). Computational modeling was executed to resolve a significant difference between physical length scales of

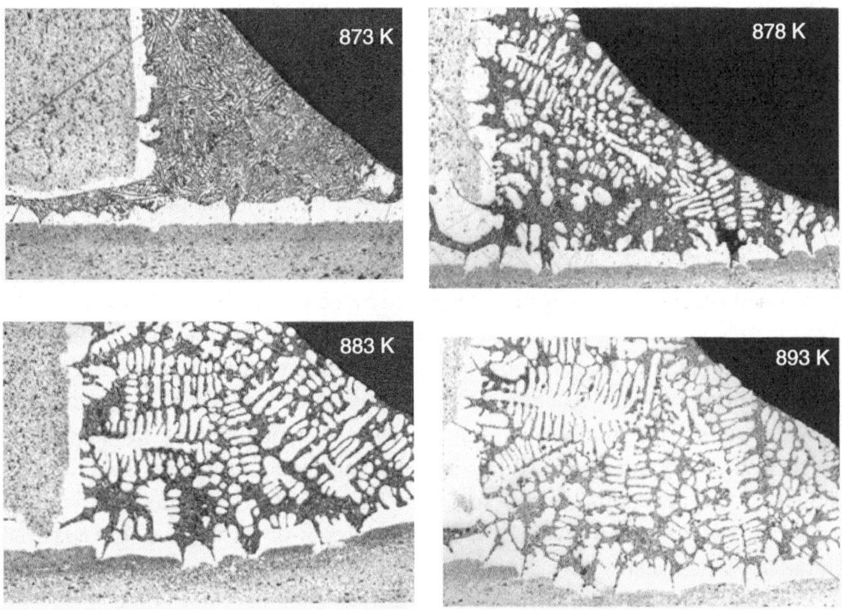

3.9 Population of alpha phase dendrites within resolidified AA4343 between AA3003 mating surfaces (Sekulic *et al.*, 2005a).

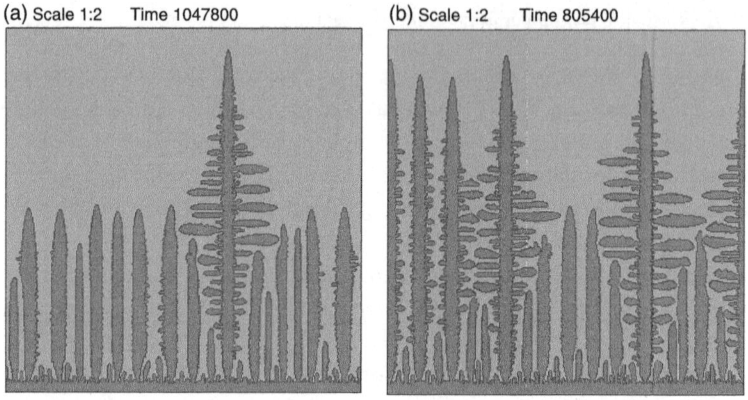

3.10 Simulation of a solidification microstructure, Al+Si solid solution formation from a melt. Quench is from 873 K (a) and 883 K (b) (Sekulic *et al.*, 2005b).

heat and mass transfer in a non-isothermal model (Galenko and Krivilyov, 2000; Sekulic *et al.*, 2005b). Figure 3.10 illustrates the simulation of solidification microstructures. A long computational time and sizable computer memory were necessary (a node of a HP N-4000 supercomputer having 8 PA-RISC 8500 at 440 MHz with 8 GB RAM of memory).

Comparison of Figs 3.9 and 3.10 indicates a smaller population of dendrites are formed during the quench from a lower temperature level vs. a higher dendrite population if the quench is performed from a higher temperature level. It can be concluded that one of the main governing parameters driving the pattern evolution in the solidification process during aluminum brazing is the undercooling level for the binary alloy composition during brazing, in addition to the parameters associated with the state of the melt at the onset of solidification vs. its location within the phase diagram.

Simulation of the microstructure evolution during solidification may be performed successfully by using phase field modeling (Apel and Böttger, 2012). This interesting simulation study was performed on a Ag–Cu–O alloy under a reactive air brazing. The simulation addresses microstructure evolution in the filler but disregards any reaction at the substrate interfaces (both the ceramic and the steel sides were modeled with no flux boundary conditions). Modeling was performed with a parametrically variable, but in a given simulation fixed, contact angle (20° for good, 90° for poor and 160° for no wetting conditions). The onset of the simulated sequence was at 1423 K and the cool-down rate was 0.17 K/s. The initial nucleation states were generated at random. A two-phase melt of Ag8Cu+O after nucleation was simulated in time. During cooling, CuO is the first solid phase to appear (the nucleation undercooling is an input parameter) at 1232 K. The complex phase transformation sequence modeled includes a two-phase ionic liquid state, followed by monotectic and eutectic reactions, leading to the solid two-phase structure composed of fcc-Ag (solid solution) and CuO. This model demonstrated (supported by experimental data) how nucleation undercooling for the CuO phase and a limited oxygen interchange alter the phase formation (Apel and Böttger, 2012).

Numerical optimization techniques have also been used for modeling tasks without requiring a deep insight into the physical aspects of the process. An illustration of a semi-empirical computer-aided alloy-designing optimization approach was presented in a study of a suitable composition of insert metals for TLP bonding of high-aluminum Ni-base superalloys (Nishimoto and Saida, 2000). Identification of the key influential factors (e.g. melting point, hardness, critical interlayer width and the void ratio in the bonded interlayer) was followed by multi-objective, multidimensional mathematical programming to optimize the performance using a limited database. Based on the modeling performed, it appears that Ni–3.5%Cr–3.5%B–3%Ti insert metal is superior to MBF-80.

3.5 Conclusions

As indicated in the Introduction, the main objective of this chapter is to provide a review of a selection of various modeling efforts associated with brazing phenomena. The most frequent attempts are to model the thermomechanical and thermal behaviour of brazed assemblies as well as the hot zone (including the assembly) during ramp-up and cool-down phases of the brazing cycle. Micro-scale processes, involving complex transport and phase change phenomena, are more difficult to model, and in this domain the challenges for further advances will persist.

3.6 References

ABAQUS Theory Manual (1998) V. 5.8, Hibbit Karson Sorensen, Inc. RI.

Ansys (2011) http://www.ansys.com/Products/Simulation+Technology/Fluid+Dynamics/ANSYS+FLUENT, accessed October 2011.

Apel, M. and Böttger, B. (2012) Phase field simulations of microstructure formation in an Ag-CU-O brazing filler under reactive air brazing conditions, IBSC 2012, *Proceedings of the 5th International Brazing and Soldering Conference*, 22–25 April 2012 (R. Gourley and C. Walker, Eds), ASM International, pp. 458–464.

Arafin, M.A., Medrai, M., Turner, D.P. and Bocher, P. (2006) *Thermodynamic modeling and experimental investigation of brazed joints used in aerospace industry*, IBSC 2006, ASM and AWS, ASM International, Materials Park, OH, pp. 189–196.

Bammann, D.J. (1990) Modeling temperature and strain rate dependent large deformations of metals, *Appl. Mech. Rev.*, Vol. 43, No. 5s, pp. 312–319.

Bing, K., Eigenmann, B., Scholtes, B. and Macherauch, E. (1994) Brazing Residual Stresses in Components of Different Metallic Materials, *Materials Science and Engineering*, Vol. A174, pp. 95–101.

Bing, K., Pintschovius, L., Eigenmann, B. and Macherauch, E. (1996) Combination of X-Ray and neutron diffraction, hole drilling and FE-calculations for the determination of the residual stress state in brazed joints of steel and cemented carbide, *Proc. of the 4th European Conf. on Residual Stresses* (S. Denis, Ed.), ECRS 4: 4th European Conf. on Residual Stresses, Cluny, France, 4–6 June 1996. Proc. Vol.2, S.627–636.

Box, G., Hunter, W. and Hunter, S. (1978) *Statistics for Experimenters*, Wiley & Sons, NY.

Braun, R.J., Murray, B.T., Boettinger, W.J. and McFadden, G.B. (1995) Lubrication theory for reaction spreading of a thin drop, *Phys. Fluids*, Vol. 7, pp. 1797–1810.

Burchett, S.N. (2003) *Computational Modeling of Joining Processes*, SECA SOFC Seal Meeting, 8–9 July 2003, http://www.netl.doe.gov/publications/proceedings/03/seca-seal/Burchett.pdf, accessed October 2011.

Busbaher, D., Liu, W. and Sekulic, D.P. (2010) High Temperature Brazing using Nano-Particles Doped Filler Metal for Dispenser Cathode Application, *IEEE Int. Vacuum Electronics Conference*, 11 IVEC 2010, Monterey, CA, 18–20 May 2010, pp. 151–152.

CFD 2000 (1999) Adaptive Research (A Division of Simunet Corporation), CFD 2000 v. 3.3 – User's Guide Supplement, Alhambra, CA, http://www.adaptive-research.com/cfd2000_software.htm, accessed July 2012.

Davies, D.R. and Roberts, P.M. (2000) *Best practice brazing by process auditing: The European experience*, Advanced Brazing and Soldering Technologies (IBSC 2000), ASM & AWS, Miami, FL and Materials Park, OH, pp. 367–388.

Delannay, F., Froyen, L. and Deruyttere, A. (1987) Review: The wetting of solids by molten metals and its relation to the preparation of metal-matrix composites, *J. Mater. Sci*, Vol. 22, pp. 1–16.

Dempsey, J.F., Dykhuizen, R.C., Crowder, S.V., Dudley, E.C. and Hosking, F.M. (2003) *Braze process characterization using thermal analysis modeling and statistical design of experiments*, IBSC 2003, ASM & AWS, Miami, FL and Materials Park, OH, CD Edition, Paper 10.7.

Eustathopoulos, N., Nicholas, M.G. and Drevet, B. (1999) *Wettability at high temperatures*, Pergamon, New York.

Freed, A.D. and Walker, K.P. (1993) Viscoplasticity with creep and plasticity bounds, *Int. J. Plasticity*, Vol. 9, No. 2, pp. 213–242.

Galenko, P.K. and Krivilyov, M.D. (2000) Modeling of crystal pattern formation in isothermal undercooled alloys, *Modelling Simul. Mater. Sci. Eng.*, Vol. 8, pp. 81–94.

Galli, M., Botsis, J., Jamczaak-Rusch, J., Maier, G. and Welzel, U. (2009) Characterization of the residual stresses and strength of ceramic-metal braze joints, *Journal of Engineering Materials and Technology*, ASME, Vol. 131, pp. 021004-1-8.

Gao, F., Zhao, H., Sekulic, D.P., Qian, Y. and Walker, L. (2002) Solid state Si diffusion and joint formation involving aluminum brazing sheet, *Materials Science and Engineering*, A337, pp. 228–235.

Gartling, D.K. and Hogan, R.E. (1994) A Finite Element Computer Program for Nonlinear Heat Conduction Problems, Part I (Theoretical Background) and Part II (User's Manual), Sandia Report, Part I, SAND94-1173, Part II SAND94-1179.

Gartling, D.K., Hogan, R.E. and Glass, M.W. (2010) COYOTE – A Finite Element Computer Program for Nonlinear Heat Conduction Problems Part I – Theoretical Background; SAND2009-4926.

Gong, J., Jiang, W., Fan, Q., Chen, H. and Tu, S.T. (2009) Finite element modeling of brazed residual stress and its influence factor analysis for stainless steel plate-fin structure, *Journal of Materials Processing Technology*, Vol. 209, pp. 1635–1643.

Hammond, J.P., David, S.A. and Santella, M.L. (1988) Brazing ceramic oxides to metals at low temperatures, *Welding Journal*, Research Supplement, Vol. 67, 10, pp. 227s–232s.

Heikinheimo, L.S.K. and Saarenheimo, A. (2000) Numerical simulation of thermal residual stresses in a brazed alumina to titanium joint, *International Brazing and Soldering Conference Proceedings*, IBSC 2000, American Welding Society and ASM International, Miami, FL, Materials Park, OH, USA.

Hosking, F.M. (2003) *Sandia Brazing Research and Modelling Capabilities*, SECA SOFC Seal Meeting, 8–9 July 2003, http://www.netl.doe.gov/publications/proceedings/03/seca-seal/Hosking.pdf, accessed October 2011.

Hosking, F.M., Gianoulakis, S.E., Givler, R.C. and Schunk, R.P. (2000) Thermal & Fluid Flow Brazing Simulation, in *Advanced Brazing and Soldering Technologies*, IBSC 2000, American Welding Society and ASM International, Miami, FL, and Materials Park, OH, pp. 389–397.

Iancu, O.T., Muntz, D., Eigenmann, B., Scholtes, B. and Macherauh, E. (1990) Residual Stress State of Brazed Ceramic/Metal Compounds, Determined by Analytical Methods and X-ray Residual Stress Measurements, *Journal of American Ceramic Society*, Vol. 73, 5, pp. 1144–1149.

Kim, T.-W., Chang, H.W.I.-S. and Park, S.-W. (2001) Re-distribution of thermal residual stress in a brazed Si_3N_4/stainless steel joint using laminated interlayers, *Journal of Materials Science Letters*, Vol. 20, pp. 973–976.

Kistler, S.F. (1993) Hydrodynamics of wetting, in *Wettability* (J.C. Berg, Ed.), Marcel Dekker, NY, pp. 311–429.

Kritsalis, P., Coudurier, L. and Eustathopoulos, N. (1991) Contribution to the study of reactive wetting in the CuTi/Al2O3 system, *J. Mater. Sci.*, Vol. 26, pp. 3400–3408.

Landry, K., Rado, C., Voitovich, R. and Eustathopoulos, N. (1997) Mechanism of reactive wetting: The question of triple line configuration, *Acta Mater.*, Vol. 46, pp. 3079–3085.

Laurent, V., Chatain, D. and Eustathopoulos, N. (1987) Wettability of SiC by aluminium and Al-Si alloys, *J. Mater. Sci.*, Vol. 22, pp. 244–250.

Li, Y., Liu, W., Sekulic, D.P. and He, P. (2012) Reactive wetting of AgCuTi filler metal on the TiAl-based alloy substrate, *Applied Surface Science*, Vol. 250, pp. 343–348.

Liu, W. and Sekulic, D.P. (2011) Capillary driven molten metal flow over topographically complex substrates, *Langmuir*, Vol. 27, pp. 6720–6730.

Meier, A., Chidambaram, P.R. and Edwards, G.R. (1998) Modeling of the spreading kinetics of reactive brazing alloys on ceramic substrates: Copper-titanium alloys on polycrystalline alumina, *Acta Mater.*, Vol. 46, 12, pp. 4453–4467.

Miller, A. (1976) An inelastic constitutive model for monotonic, cyclic, and creep deformation, Part 1 – Equations Development and Analytical Procedures, *J. Engr. Mater. Tech*, Vol. 98, 2, pp. 97–105.

Neilsen, M.K. and Stephens, J.J. (2000) *Residual stress in metal-to-ceramic braze joints: Advanced braze alloy constitutive model*, IBSC 2000, American Welding Society and ASM International, Miami, FL, and Materials Park, OH, pp. 411–418.

Neilsen, M.K., Stephens, J.J. and Gieske, J.H. (2003) *A viscoplastic model for the eutectic silver-copper alloy*, IBSC 2003, American Welding Society and ASM, Miami, FL and Materials Park, OH, CD Edition, Paper 9.5, p. 9.

Nishimoto, K. and Saida, K. (2000) *Computer-aided alloy design of insert metal for transient liquid phase bonding*, IBSC 2000, American Welding Society and ASM International, Miami, FL and Materials Park, OH, pp. 398–405.

Park, J.-W., Mendez, P.F. and Eagar, T.W. (2002) Strain energy distribution in ceramic-to-metal joints, *Acta Materialia*, Vol. 50, pp. 883–899.

Pintschovius, L. (2011) Personal communication, Karlsruhe Institute of Technology.

Pintschovius, L., Pyka, N., Kussmaul, R., Munz, D., Eigenmann, B., *et al.* (1994) Experimental and theoretical investigation of the residual stress distribution in brazed ceramic-steel components, *Materials Science and Engineering*, Vol. A177, pp. 55–61.

Pintschovius, L., Schreieck, B. and Eigenmann, B. (1999) Neutron, X-ray, and finite element stress analysis on brazed components of steel and cemented carbide, *Textures and Microstructures*, Vol. 33, pp. 263–278.

Ratts, E. (2000) A Study of the Radiant Heat Exchange within a Controlled-Atmosphere Furnace, *Heat Transfer Engineering*, Vol. 21, pp. 55–64.

Saiz, E. and Tomsia, A.P. (2006) *Kinetics of High Temperature Spreading*, IBSC 2006, ASM International, Materials Park, OH, pp. 203–206.

Saiz, E., Tomsia, A.P. and Cannon, R.M. (1998) Ridging effects on wetting and spreading of liquids on solids, *Acta Mater.*, Vol. 46, 7, pp. 2349–2361.

Schunk, P.R., Sackinger, P.A., Rao, R.R., Chen, K.S. and Cairncross, R.A. (1996) *GOMA – A Full-Newton Finite Element Program for Free and Moving Boundary Problems with Coupled Fluid/Solid Momentum, Energy, Mass, and Chemical Species Transport: Users Guide*, Sandia Technical Report SAND95-2937, January 1996.

Schunk, P.R., Sackinger, P.A., Rao, R.R., Chen, K.S., Cairncross, R.A., *et al.* (1997) *GOMA 2.0 User's Guide*, Sandia Technical Report SAND97-2404.

Sekulic, D.P. (2001) Molten aluminum equilibrium membrane formed during controlled atmosphere brazing, *Int. J. of Eng. Science A*, Vol. 39, pp. 229–241.

Sekulic, D.P. (2009) An Entropy Generation Metric for Non-energy Systems Assessments, *Energy*, Vol. 34, pp. 587–592.

Sekulic, D.P. (2011) Energy and Exergy: Does one need both concepts for the study of resources use? Chapter 2 in *Thermodynamics and the Destruction of resources* (B. Bakshi, T. Gutowski and D.P. Sekulic, Eds), 2011, Cambridge University Press, Cambridge, UK, New York, USA, pp. 45–86.

Sekulic, D.P., Salazar, A.J., Gao, F., Rosen, J.S. and Hutchins, H.F. (2001) Transient Behavior of Compact Heat Transfer Surface during Brazing, *Experimental Heat Transfer, Fluid Mechanics and Thermodynamics*, Vol. 1 (G.P. Celate, P. Di Marko, A. Goulas and A. Mariani, Eds), Edizioni ETS, Pisa, pp. 803–808.

Sekulic, D.P., Salazar, A.J., Gao, F., Rosen, J.S. and Hutchins, H.F. (2003) Local Transient Behavior of a Compact Heat Exchanger Core During Brazing. Equivalent Zonal (EZ) Approach, *Int. J. of Heat Exchangers*, Vol. 4, pp. 91–108.

Sekulic, D.P., Gao, F., Zhao, H., Zellmer, B. and Qian, Y.Y. (2004) Prediction of the fillet mass and topology of aluminum brazed joints, *Welding Journal – Research Supplement*, Vol. 83, 3, pp. 102s–110s.

Sekulic, D.P., Galenko, P.K., Krivilyov, M.D., Walker, L. and Gao, F. (2005a) Dendritic growth in Al-Si alloys during brazing. Part 1: Experimental Evidence and Kinetics, *Int. J. of Heat Mass Transfer*, Vol. 48, pp. 2372–2384.

Sekulic, D.P., Galenko, P.K., Krivilyov, M.D., Walker, L. and Gao, F. (2005b) Dendritic growth in Al-Si alloys during brazing. Part 2: Computational Modeling, *Int. J. of Heat Mass Transfer*, Vol. 48, pp. 2385–2396.

Shapiro, A. and Sekulic, D.P. (2008) A New Approach to Quantitative Evaluation of a Design for Brazed Structures, *Welding Journal, Research Supplement*, Vol. 87, pp. s-1–s-10.

Tanner, S.H. (1979) The spreading of silicone oil drops on horizontal surfaces, *J. Phys. D: Appl. Phys.*, Vol. 12, pp. 1473–1484.

Wang, X.L., Hubbard, C.R., Spooner, S., David, S.A., Rabin, B.H., *et al.* (1996) Mapping of the residual stress distribution in a brazed zirconia-iron joint, *Materials Science and Engineering*, A211, pp. 45–53.

Warren, J.A., Boettinger, W.J. and Roosen, A.R. (1998) Modeling reactive wetting, *Acta Mater.*, Vol. 46, pp. 3247–3264.

Washburn, E.W. (1921) The dynamics of capillary flow, *Phys. Rev.*, Vol. 17, pp. 273–283.

Wielage, B., Klose, H., Martinez, L. and Schüler, H. (2000) *Determination of mechanical characteristic values for brazed ceramic-metal joints*, IBSC 2000, American Welding Society and ASM International, Miami, FL and Materials Park, OH, pp. 406–410.

Williamson, R.L., Rabin, B.H. and Drake, J.T. (1993) Finite Element Analysis of Thermal Residual Stresses at Graded Ceramic-metal interfaces. Part I. Model Description and Geometrical Effects, *J. Appl. Phys.*, Vol. 74, 2, pp. 1310–1320.

Zellmer, B.P., Nigro, N. and Sekulic, D.P. (2001) Numerical modeling and experimental verification of the formation of 2D and 3D brazed joints, *Modelling Simul. Mater. Sci. Eng.*, Vol. 9, pp. 339–355.

Zhang, J.X., Chandel, R.S., Chen, Y.Z. and Seow, H.P. (2002) Effect of residual stress on the strength of an alumina-steel joint by partial transient liquid phase (PTLP) brazing, *Journal of Materials Processing Technology*, Vol. 122, pp. 220–225.

Zhao, H. and Sekulic, D.P. (2006) Diffusion-controlled melting and re-solidification of metal micro layers on a reactive substrate, *Heat and Mass Transfer*, Vol. 42, pp. 464–469.

Zhao, H. and Sekulic, D.P. (2008) Wetting of a hypo-eutectic Al-Si system, *Materials Letters*, Vol. 62, pp. 2241–2244.

Zhou, Y., Bao, F.H., Ren, J.L. and North, T.H. (1991) Interlayer selection and thermal stresses in brazed Si_3N_4–steel joints, *Materials Science and Technology*, Vol. 7, pp. 863–867.

Part II

Materials used in brazing

4
Brazing of superalloys and the intermetallic alloy (γ-TiAl)

V. F. KHORUNOV and S. V. MAKSYMOVA,
E. O. Paton Electric Welding Institute, Ukraine

DOI: 10.1533/9780857096500.2.85

Abstract: This chapter investigates the problem of brazing heat-resistant materials – high nickel-base and γ-TiAl-base intermetallic alloys. Diffusion brazing with filler metals of the Ni–Cr–B and Ni–Cr–B–Si systems is inefficient for nickel alloys, while composite filler metals based on these systems yield satisfactory results. When used as an alternative to boron- and silicon-containing filler metals, alloys of the Ni–Cr–Zr systems do not form hard and brittle phases at the interface in the joints. Filler metals based on the Ti–Zr system are best suited for joining γ-TiAl-base intermetallic alloys. Ti–Zr–Fe and Ti–Zr–Mn alloys, which show promise as brazing filler metals are selected, and structures and mechanical properties of the brazed joints are presented.

Key words: superalloy, high-strength nickel alloy, vacuum brazing, brazing filler metal, Ni–Cr–Zr system, intermetallic alloy, seam structure, seam mechanical properties.

4.1 Introduction

The chapter consists of two sections. The first section is dedicated to the problem of brazing high heat-resistant nickel alloys (superalloys). It gives a concise description of the peculiarities of the alloys, problems arising in their welding and brazing, specific properties of brazing filler metals used to implement the brazing process, and their advantages and drawbacks, as well as key features of the vacuum brazing process performed using different types of brazing filler metals. It is noted that the overwhelming majority of the brazing filler metals are based on utilisation of boron and silicon (collectively or separately) as depressants, which from the standpoint of the authors is not always reasonable.

Consideration is given to a different system of formulation of filler metals for brazing superalloys, which is based on employing elements of groups IV and V of the periodic system as depressants. Results of investigations of the Ni–Cr–Zr system for nickel-rich alloys, the positions of monovariant eutectics and a ternary eutectic, and phase compositions of the alloys over the entire concentration range are given. Data on structure and properties of the joints produced by vacuum brazing using the suggested brazing filler metals are presented. Experience of applying these systems of the brazing filler metals, including the composite ones, to arc brazing is described.

The second section considers structural peculiarities of γ-TiAl-base intermetallic alloys produced by different methods. The choice of the alloying system for development of filler metals for brazing of these alloys is substantiated. Liquidus surfaces for the Ti–Zr–Fe and Ti–Zr–Mn systems are plotted, and ranges of concentrations of the alloys showing promise for use as brazing filler metals are identified. Morphological peculiarities in the formation of the brazed seams produced by vacuum brazing using known and new brazing filler metals are investigated. Results of strength tests of the brazed joints at room and higher temperatures, as well as data on their long-term strength, are given. It is shown that brazing filler metals of the Ti–Zr–Fe and Ti–Zr–Mn systems have a high potential for providing brazed joints with a structure and properties close to those of the base metal.

4.2 Brazing of superalloys on a nickel base

Of the many commercial heat-resistant materials, the alloys used to manufacture the parts of the hot section of gas turbine engines, and turbine blades in particular, attract the most interest. In the literature they are referred to as superalloys. The characteristics of these alloys are described in monographs.[1–6] At present, in the majority of cases these are high nickel (and partially cobalt) alloys. They are characterised by solid solution, carbide and intermetallic strengthening (due to precipitation of the γ'-phase), which imparts good properties at high temperatures, and by the presence of resistant oxides (of chromium, aluminium, titanium, etc.) on their surface. The emergence of dispersion-strengthened, single crystal and eutectic alloys, as well as alloying with rhenium, opens up good possibilities for upgrading gas turbine engines. The most typical alloys developed in the USSR or the CIS are shown in Table 4.1.

4.2.1 Weldability of superalloys

It is a known fact that blades and other parts may contain defects originating from casting or service, which need to be repaired. The low ductility of nickel alloys with a high content of the γ'-phase may be a serious problem when welding them. Weldability of these alloys is usually related to the total content of aluminium and titanium, which exert the main effect on the amount of the γ'-phase. The diagram shown in Fig. 4.1 divides the alloys into weldable and unweldable. This diagram differs from the currently familiar version[7] in that it additionally shows data on nickel-base superalloys developed in the USSR and CIS countries. It can be seen from this diagram that the majority of modern superalloys are unweldable. Therefore, as noted by Xiaowei Wu *et al.*,[8] for these alloys welding can be applied only to repair minor defects, when only local heat input can limit strains. So, the key method for joining these alloys is brazing.

Table 4.1 Some high-strength alloys with a nickel base developed in the USSR and CIS

Alloys	Co	Cr	C	Fe	Al	Nb	Ti	W	Mo	Other elements
ZhS32	8–10.5	4.3–5.6	0.12–0.18		5.6–6.3	1.4–1.8		7.8–9.5	0.8–1.4	4Ta, 3.5–4.5Re
ZhS26	8–10	4.3–5.6	0.13–0.18		5.5–6.2	1.4–1.8	0.8–1.2	10.9–12.5	0.8–1.4	0.8–1.2V
ZhS36	8–9	3.5–4.5	0.015	1	5.5–6.2	0.7–1.5	0.7–1.5	11–12.5	1–2.2	0.1Ta, 1.6–2.3Re
ZhS6U	9–10.5	8–9.5	0.13–0.2	1	5.1–6	0.6–1.2	2–2.9	9.5–11	1.2–2.4	
ZhS6F	6–10.5	4–7	0.12–0.19		5.1–5.6	1.2–1.7	0.8–1.5	11–13	0.8–1.5	
TsNK7	8–9.5	14–15	0.08–0.1	1	3.4–4.5		3.4–4.4	6.2–7.5	0.2–0.6	0.3Mn, 0.3 Si
ChS70-VI	11	15.7	0.1		2.7	0.2	4.7	5.5	1.5	
EP 539		17–19	0.05		3–4.0		2–3.0	2.5–4	5–7,0	
ChS104	11.2	21		0.5	2.5	0.25	3.5	3.5	0.6	0.3Mn, 0.3 Si 0.06–0.1B
VZhL-12U	12.5	8.5–10.5	0.14–0.2	–	5.0–5.5	0.5–1.0	4.2–4.7	1.0–1.8	2.7–3.4	–

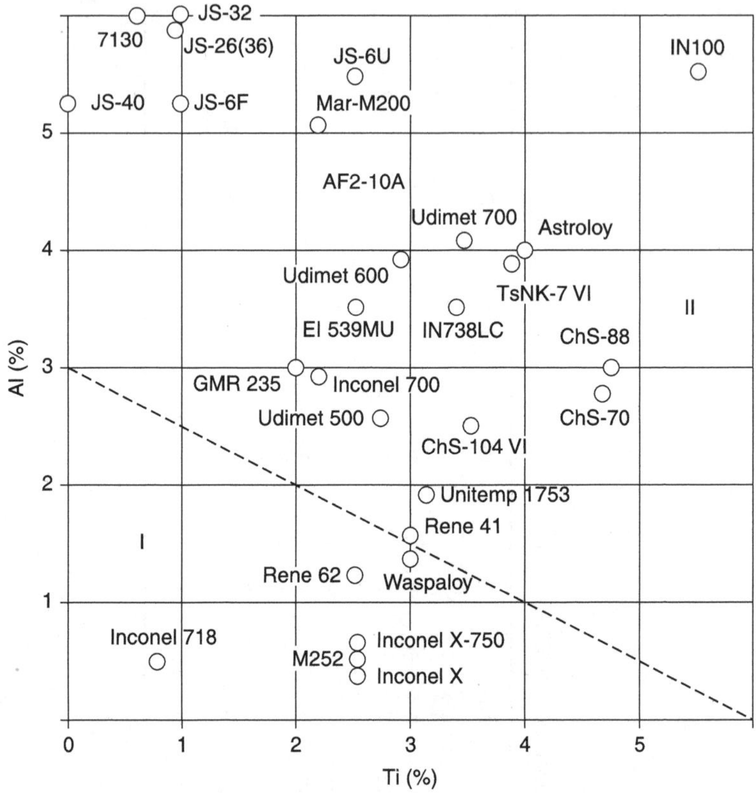

4.1 Evaluation of weldability of high nickel alloys depending on the aluminium and titanium proportions.

4.2.2 Conventional brazing techniques

Brazing of superalloys is also a challenge. When deciding on a brazing filler metal and technology, one should bear in mind that it is necessary to provide, first, wetting of the surface covered with resistant oxides (of chromium, aluminium, titanium, etc.), and, second, properties of the brazed joints close to those of the base metal.

In worldwide practice the first problem is usually addressed by using brazing filler metals of the traditional Ni–Cr–B system or the Ni–Cr–Si–B system, and by performing brazing in vacuum or a neutral atmosphere.[8–17] It is well known that brazing filler metals of the above systems were developed for brazing stainless steels,[18,19] with no regard for the structural peculiarities and strength characteristics of superalloys. Therefore, using them under common brazing conditions fails to give positive results: a brittle and hard silicon-rich eutectic forms at the seam centre, while boron actively diffuses into the base metal to form boride phases at the grain boundaries. In this connection, extensive investigations were carried out to study the application of so-called diffusion brazing. The primary objective of this method is to provide a microstructure and mechanical properties of the brazed joints identical to those of the base metal using a long isothermal holding time. This method was proposed long ago,[16] but is still under investigation.[8–13]

It has been established as a result of the investigations that there are factors limiting the use of this method for repair purposes. First, this as a long-term temperature effect on a workpiece, which may cause functional properties of the base metal to deteriorate, and, second, it is economically inexpedient. Therefore, investigators are naturally striving to find methods for reducing the isothermal holding time. To a certain degree, this can be achieved through decreasing the thickness of the brazing filler metal and pressing the mating surfaces.

4.2.3 Innovative brazing techniques for superalloys

Amorphous alloys

Expectations were raised in this area with the development of amorphous alloys of the Ni–Cr–Si–B system. In 1981, Allied Signal Corporation (USA) took out a patent for a 'homogeneous plastic brazing foil' 0.025–0.06 mm thick. Within the scope of this patent Allied Signal Corporation offered a number of alloys, which can be produced in the amorphous state. Properties and examples of application of these materials are considered in detail in references.[17,19] Table 4.2 shows alloys developed in the USA and USSR (and in Russia).

As seen from the investigation results obtained by different authors,[8–13] the amorphous foils used for brazing fail to substantially reduce the isothermal holding time and provide an acceptable structure and properties for the joints. For example, when a single-crystal superalloy was brazed at 1230°C using an amorphous foil of the Ni–Cr–B system with 8 h holding time, the resulting seam metal under fracture stresses had a much lower ductility than the base metal. The authors explained this by precipitation of boron atoms at sub-boundaries of the base metal.[12]

Composite filler metals

A radical reduction of the brazing time can be achieved by using so-called composite filler metals.[20–26]

Composite filler metals consist of a filler metal proper and a filling agent, the melting point of which is higher than the brazing temperature. Dissolving in the filler metal, the filling agent increases the viscosity of the melt and keeps the latter in wide (non-capillary) gaps, and its particles act as additional solidification centres.

Different investigators used high-temperature vacuum brazing to repair components of gas turbine engines using the composite filler metals, which gives

Table 4.2 Amorphous nickel-base filler metals

Filler metal	Chemical composition, wt.% (Ni – rest)						T_s/T_l, T_b
	Cr	Si	B	Fe	C	Other	(°C)
MBF 10/10A Metglas	13–15	4–5	2.75–3.5	4–5	<0.06	–	970/1040
MBF 15/15A Metglas	12–14	4–5	2.5–3.2	3.5–5	<0.03	–	–
MBF 30/30A Metglas	–	4–5	2.75–3.5	≤0.5	0.06	–	980/1040 foil
MBF 80/80A Metglas	14.5–16	–	3.17–4.2	–	<0.06	–	T_b = 1175
STEMET 1301	7	4.6	3.1	3	–	–	T_l = 980, T_b = 1010–1177
STEMET 1311 MIFI-AMETO	0.4	4	4	5	–	16 Co	T_l = 985, T_b = 1020–1050
MBF-50 Metglas	19	7.3	1.5	–	–	–	
MBF51 Metglas	15	7.25	1.4	–	–	–	1030/1126
VPr 11 VIAM	14–16	4–5	1.8–3.2	3–5	0.3–1.0	0.1–1 Al	980/1050
VPr 24 VIAM	6–7	2.5–3.0	0.2–0.3	–	–	1 Ti; 10–12 Nb; 8.5–9.5 W; 1.6–2.0 Mo; 4–5 Al	1150/1190 T_b = 1200–1220
VPr 27 VIAM (Ni–Cr–Al–B)	–	–	–	–	–	–	1030/1080 T_b = 1150–1200
VPr 42 VIAM (Ni–Cr–Si–B)	–	–	–	–	–	–	1050/1070

a chemical composition and physical–mechanical properties of the brazed seams close to those of the base metal. In particular, the technological peculiarities of isothermal vacuum brazing of superalloys with different alloy levels (Table 4.1) are considered in several papers.[21–25] These studies involved a wide range of investigation methods to comprehensively analyse the results of brazing using composite filler metals, in which boron, or collectively boron and silicon, served as a depressant. Brazing filler metals of the first type were compositions consisting of 40% filler metal (Ni–9Co–14Cr–3.5Al–2.5B) and 60% filling agent (Rene-142 alloy powder). Filler metals of the second type additionally comprised 15–25% of a silicon-containing eutectic-composition filler metal (Ni–12Si), due to the decreased content of boron-containing filler metal. The structure and phase composition of the joints in the as-brazed state and after heat treatment were investigated, and the mechanical properties at room and higher temperatures, as well as long-term strength values, were determined. Brazing was performed at a temperature of 1210–1230°C with holding for 15–20 min, which was followed by a two-stage heat treatment in vacuum.

The seams formed in the case of the boron-containing filler metals were characterised by a wide scatter of microhardness values, corresponding to a wide variety of structural components in the seam metal. Boron diffused into the base metal to a depth of 300 μm.

Long-term strength tests of specimens of the brazed joints revealed substantial structural changes, such as melting of eutectic components of the seam metal at high temperature and active diffusion of boron into the base metal, for the seams formed using the boron-containing filler metals consisting of 40% #1 + 60% Rene-142. After 55 min of testing, boron had penetrated to a depth of 1–2 mm on each side.

The filler metal consisting of 20% (Ni–9Co–14Cr–4Al–2.5B) + 20% (Ni–12Si) + 60% Rene-142 formed seams, the structure of which experienced dispersion strengthening due to the γ'-phase, in gaps from 100 to 800 μm wide on alloy ZhS6U. Approximately identical volume contents (54–59%) of the strengthening phase was fixed in all zones of the brazed joints, including the seam metal, diffusion zone and base metal. The depth of diffusion of boron into the base metal did not exceed 70 μm. Moreover, investigations showed refinement of the carbide phase precipitates in the seam metal, as well as precipitation of mostly the Ni_3B phase instead of the CrB phase.

When using the silicon-containing filler metals, the structure of the seam metal and diffusion zone did not experience any significant changes during the long-term strength tests, the seam size did not change, and the depth of diffusion of silicon into the base metal after holding at a temperature of 900°C for 16 h or longer was no more than 50 μm. The strength values achieved at room temperature were close to those of the base metal (Table 4.3). The differences in strength data

Table 4.3 Strength of joints on nickel superalloys brazed with composite filler metals (T_{test} = 20°C)

Base material	Filler metal	Tensile strength σ_B (MPa)
ChS70-VI	#1+60%Rene-142	732.5
	40%#1+20%HC12+60%Rene-142	778.0
VZhL-12U	#1+60%Rene-142	721.0
	40%#1+20%HC12+60%Rene-142	873.0
ZhS26VI	#1+60%Rene-142	692.0
	40%#1+20%HC12+60%Rene-142	718.5
ZhS26NK	#1+60%Rene-142	766.0
	40%#1+20%HC12+60%Rene-142	875.0

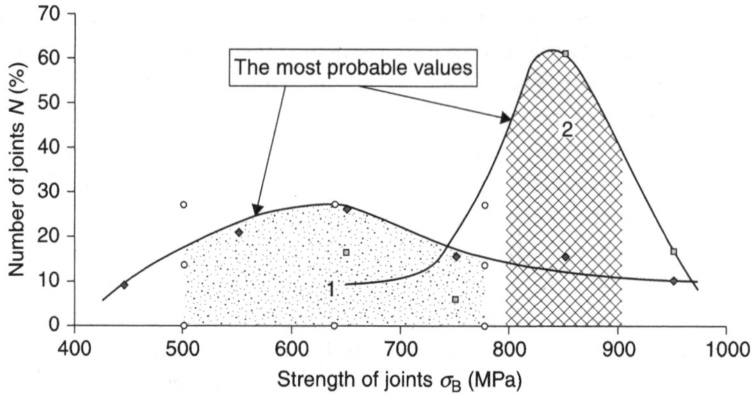

4.2 Statistical curves of strength of joints on alloy VZhL-12U brazed using boron-containing filler metals: 1, without silicon; 2, with silicon.

between the boron-containing filler metals and filler metals containing both boron and silicon are demonstrated in Fig. 4.2.

Analysis of experience around the world shows that the use of filler metals containing both boron and silicon involves a number of serious difficulties, which are associated mainly with diffusion of boron along the grain boundaries of the base metal, and which can be overcome by employing special techniques that allow joints with properties close to those of the base metal to be produced by the vacuum brazing process. However, it should be recognised that this area uses ideas that have been known for a long time, and whose implementation can be improved by advances in investigation methods and methods of production of metal alloys in different structural states.

Work on development of alternative systems of brazing filler metals has been carried out for a number of years. Elements of groups IV and V of the periodic table, such as Zr, Hf, Ti, V and Ni, are used as depressants in a new generation of filler metals. An example is brazing filler metal VPr 24 (Table 4.2), in which niobium, in addition to silicon, is used as a depressant. The role of boron is minimised. The best-studied filler metals are those based on eutectic alloys of the Ni–Cr–Zr system.[27–29] Some consideration has also been given to alloys of the Ni–Cr–Hf system.[27] The structure of the Ni–Cr–Zr ternary system in a region of nickel-rich alloys was examined in detail. The region of four-phase peritectic ($Ni_7Zr_2L=Ni+Ni_5Zr$) transformation, as well as positions of monovariant eutectics ($Ni+Ni_5Zr$ and $Ni+Ni_7Zr_2$) and a ternary eutectic in the diagram, were determined (1220°C). Phase composition, hardness and melting point of various high nickel alloys were studied. These data are given in detail in Khorunov *et al.*[29] For the development of brazing filler metals, eutectic alloys containing 8.8 at.% Zr are of greatest interest. Figure 4.3 shows variations in melting temperatures

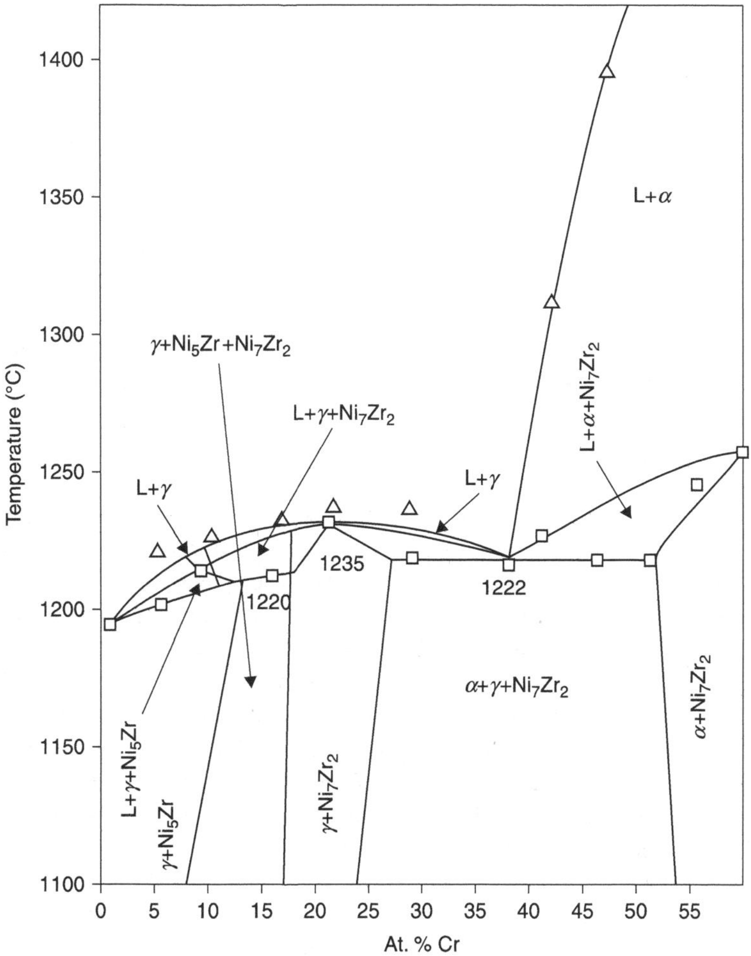

4.3 Polythermal section of constitutional diagram of Ni–Cr–Zr along 8.8 at.% Zr isoconcentrate.

and structural compositions of such alloys depending on the chromium content. It should be noted that alloying of binary eutectic Ni+Ni_5Zr with chromium leads to an insignificant increase in solidus temperature. At the same time, the microhardness of the eutectic component in the alloys with a constant content of zirconium (8.8 at.%) substantially changes. For instance, the microhardness of the eutectic within the binary γ-Ni+Ni_5Zr region is 5300–5700 MPa. But with the appearance of low-hardness intermetallic Ni_7Zr_2 in the structure of the alloys the value of microhardness becomes much lower (2000–3000 MPa). This makes it possible to provide the brazed seams with a hardness close to that of the base metal by adjusting the composition of a brazing filler metal. That is, in this case we can eliminate the key drawbacks of boron- and silicon-containing filler metals – high brittleness and hardness. Moreover, the liquidus temperature of the alloys hardly changes with an increase in the chromium content, which is important for selection of compositions of filler metals.

The system under consideration was found to have a number of alloys with an acceptable melting point for brazing of superalloys,[27,28] allowing them to be used as brazing filler metals. It should be noted that filler metals of the system considered have a high content of zirconium, which is an adhesion-active element, and that in vacuum they provide good wetting of the surfaces of high nickel alloys, despite the fact that the latter have a high content of Cr, Al, Ti and other elements forming strong oxide films on the metal surface. The brazing filler metals can be used under the conditions for capillary and non-capillary brazing.

Capillary brazing

Experiments on capillary brazing of nickel superalloys were conducted using three filler metals of the Ni–Cr–Zr system with different chromium contents. Spreading of the filler metals over the surfaces being brazed, their erosion activity, structure and properties of the brazed joints at room and higher temperatures were investigated. The results were compared with those obtained by using traditional filler metal VPr 27 of the Ni–Cr–Al–B system.

It was established that the properties of the filler metals are greatly affected by the chromium content. For example, the area of spreading of the filler metals decreases with an increase in the concentration of chromium (Fig. 4.4). The solid solution structure can be formed in gaps 20–30 μm in size only when using a brazing filler metal with low chromium content (Ni–4.7Cr–14.7Zr). The seams made using filler metals with high chromium content (15 and 24.4%) have the eutectic structure. An increase in the holding time at the brazing temperature can lead to erosion of the base metal along the grain boundaries. This can be explained by the fact that zirconium atoms are big, compared with boron atoms, and solubility in nickel is low.[30,31] These factors lead to the conclusion that the application of diffusion brazing using the considered filler metals, at least with high chromium content, is inexpedient.

To investigate the properties of the brazed joints, the joints were made on alloy JS6U brazed with the above filler metals of the Ni–Cr–Zr system at a temperature of 1220°C and holding time of 15 min, as well as with filler metal VPr 27 at a temperature of 1160°C and the same holding time. This was followed by homogenising annealing at 900°C for 4 h. The gap for brazing was 0.2 mm. A distribution of the microhardness in a cross-section of the joints is shown in Fig. 4.5. When using filler metal VPr 27, the microhardness of the eutectic was 8600–8700 MPa, which was twice as high as that of the base metal (4300–4400 MPa). In addition, the microhardness of the diffusion zone was also high (6500–6600 MPa). In joints brazed with filler metals of the Ni–Cr–Zr system

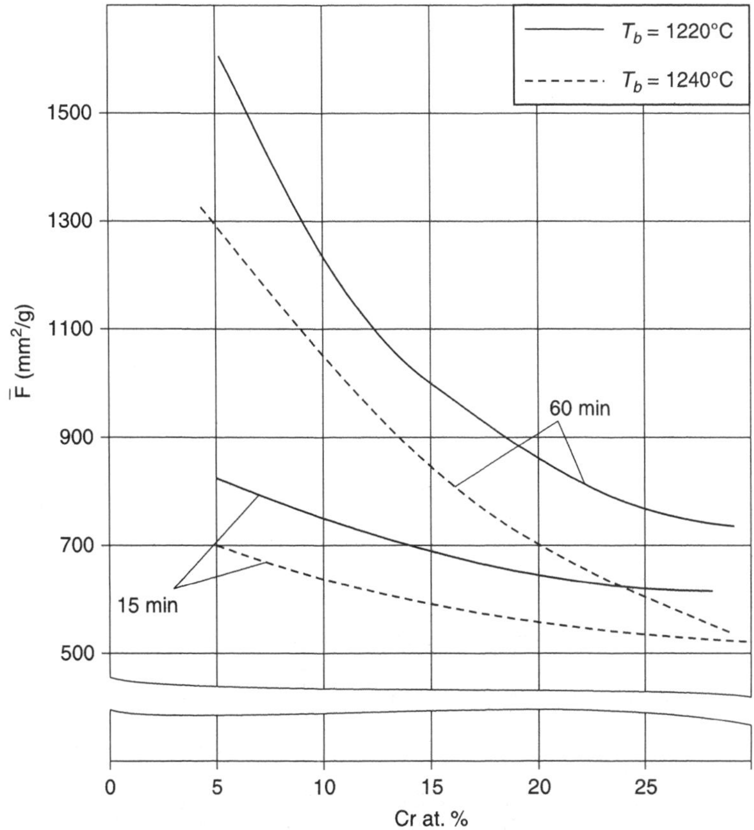

4.4 Spreading of filler metals (F̄) of the Ni–Cr–Zr system against the concentration of chromium in the melt.

the distribution of the microhardness was of an absolutely different character (Fig. 4.5). Depending on the composition of the filler metal, the microhardness of the eutectic solidified along the seam axis can be either at the same level as that of the base metal (filler metal Ni–4.7Cr–14.2Zr) or even lower than that of alloy JS6U (filler metals Ni–15Cr–14.7Zr and Ni–24.4Cr–14.6Zr). The microhardness of the diffusion zone in the brazed joint was at the same level as that of the base metal.

Tensile strength values of the joints brazed with different filler metals of the Ni–Cr–Zr system were close to each other (Table 4.4), despite the above-mentioned difference in structure of the seams. At the same time, considerable differences were revealed by scanning electron microscopy of fracture surfaces. The seam with the solid solution structure was characterised by tough fracture (Fig. 4.6(a)), whereas the joints with the eutectic structure comprised regions with brittle intercrystalline fracture (Fig. 4.6(b), (c)). The best results in tensile tests of

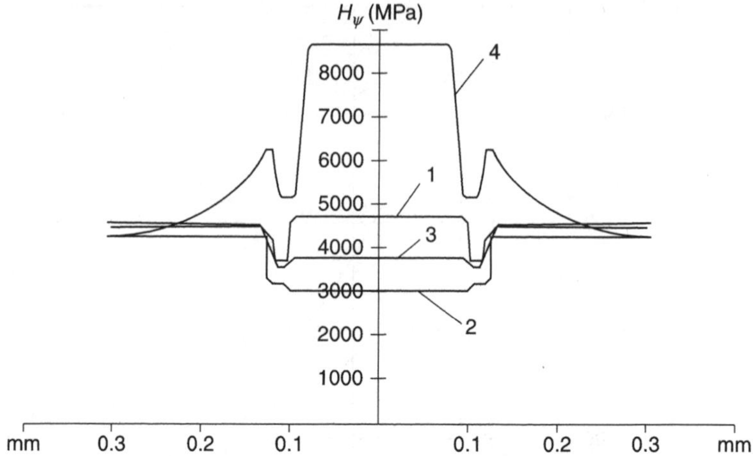

4.5 Distribution of microhardness in cross-section of alloy ZhS6U brazed with filler metals (wt.%): 1, Ni–4.7Cr–14.2Zr; 2, Ni–15Cr–14.7Zr; 3, Ni–24.4Cr–14.6Zr; 4, VPr 27.

Table 4.4 Tensile strength of butt joints on alloy ZhS6u

Grade or composition of alloy	Tensile strength σ_t^{20} of butt joints (MPa)	Tensile strength σ_t^{900} of butt joints (MPa)
Ni–4.7Cr–14.2Zr	860. . .932/896	482. . .524/500
Ni–15.0Cr–14.7Zr	888. . .936/912	360. . .440/390
Ni–24.4Cr–14.6Zr	859. . .909/884	545. . .750/676
VPr 27	850. . .880/865	Unstable results

the butt specimens at a test temperature of 900°C were exhibited by the joints brazed with filler metal Ni–24.4Cr–14.6Zr (Table 4.4).

Capillary brazing is applied mainly for fabrication of new products, e.g. sectional blades, for replacement of part of a defective blade by a specially made insert, etc. However, when used for repair, its capabilities are limited. The most common defects are cracks, porosity, etc. In all these cases it is necessary to deposit a substantial amount of filler metal. So, capillary brazing is unsuitable here. There are good grounds to consider that brazing with composite filler metals is more efficient. However, this area has not been adequately investigated as yet.

Alternative heating methods

Vacuum heating can provide satisfactory results using the Ni–Cr–Zr system filler metals. However, this process is expensive and complicated. In this

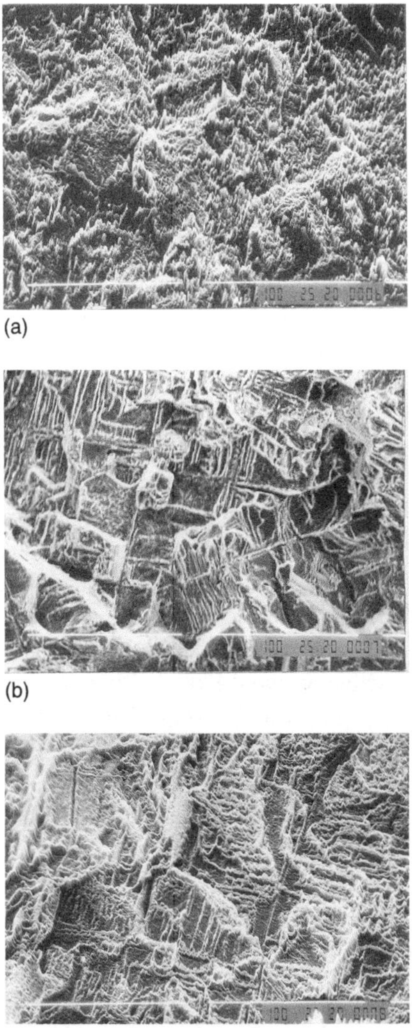

4.6 Character of fracture of specimens brazed using filler metals of the Ni–Cr–Zr system with different chromium content (wt.%): (a) 4.7; (b) 15.0; (c) 24.4.

connection, intensive research is now being carried out to find technologies with alternative heating methods, and successful results in this area are reported (arc discharge with hollow cathode, light beam, etc.). The most attractive methods are tungsten-electrode arc (TIG) and microplasma heating methods. Brazing was performed with and without preheating, using cast and composite filler metals. Alloy ZhS6U was used as a base metal.

It is common knowledge that alloys of this type are unweldable. Investigations of welding by setting the weld pool on the base metal and then using the Ni–Cr–Zr system filler metal showed, as expected, the presence of hot cracks in the fusion zone. They propagated along the grain boundaries in the base metal (Fig. 4.7(a)). However, when the arc brazing process was performed in such a way that the base metal was not melted, but only wetted with the molten filler metal, the sensitivity to hot cracking was reduced, and all the cracks were filled with the filler metal (Fig. 4.7(b)). Most probably, the cracks formed at a temperature above the temperature of solidification of the seam metal, and a more fluid component of the filler metal penetrated along the grain boundaries in the base metal.

Further decrease of the thermal effect on the fusion zone metal can be achieved by using powder-containing composite filler metals (mixtures of a low-melting point component, i.e. alloys of the Ni–Cr–Zr system, and a refractory component, i.e. high superalloy ZhS 26U). Naturally, the important role in this case is played by the ratio of the low-melting point component to the refractory component of the composite filler metal. On one hand, the higher the weight content of the refractory component, the higher are the properties of the deposited metal. But, on the other hand, the weight content of the low-melting point component should not be too low, because it should provide good wetting of the base metal and be capable of healing hot cracks, if any. With the optimal low-melting point component to refractory component ratio it is possible to avoid hot cracking and provide the deposited metal with a homogeneous fine structure (Fig. 4.7(c)) favourable for subsequent heat treatment.

X-ray spectral microanalysis shows that the seam consists of an alloy with a composition that can be classed with heat-resistant ones (Table 4.5). It should be emphasised that the alloy has a high weight content of aluminium (5.5%) and titanium (2.9%). The content of zirconium is 1.9%. A solid solution of a variable concentration (dendrites and metal in dendrite spacings) with a low content of zirconium (about 0.2 wt.%) predominates in the seam. In addition, fine carbide precipitates are fixed along the grain boundaries.

Naturally, this structure can be radically changed by means of appropriate heat treatment (to form the fine γ' phase). The average content of zirconium in the deposited metal is about 0.7 wt.%, whereas after heat treatment it is hardly fixed. The high content of aluminium and titanium persists both in the as-arc brazed state and after heat treatment.

The approach suggested by the authors of this study, i.e. providing minimal heat input in arc heating to prevent formation of liquid interlayers in the seam zone and ensure wetting of the base metal at temperatures much lower than its melting point, turned out to be a very efficient method to eliminate hot cracking. The important role in this case is played both by brazing parameters and by specific features of the filler metals of the new systems, such as Ni–Cr–Zr (high adhesion activity, increased ductility and lower hardness, compared with, for example, the Ni–Cr–B–Si system filler metals). It should be emphasised that the

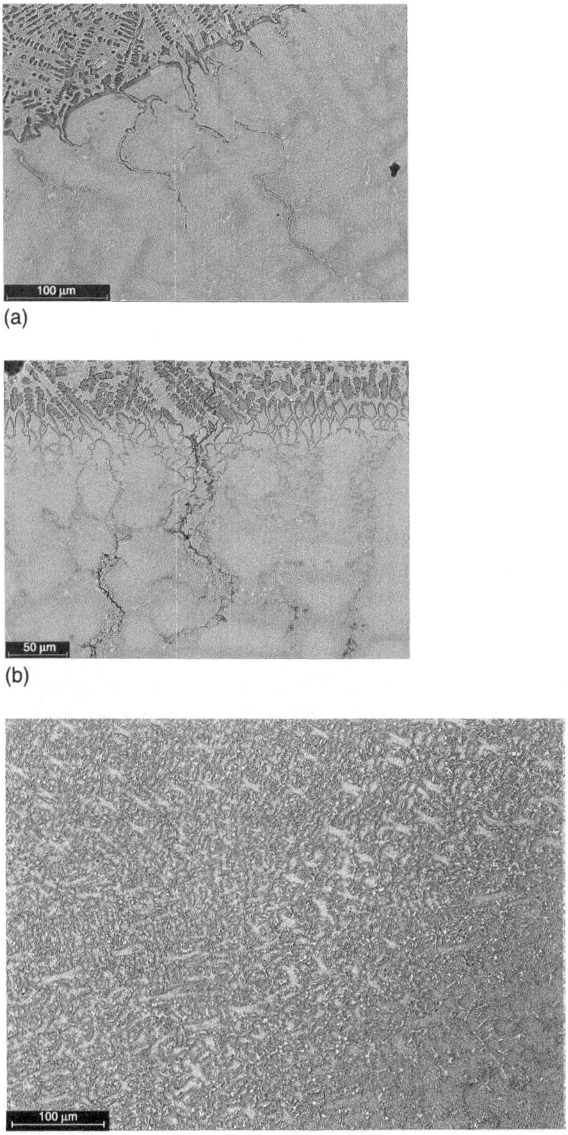

(c)

4.7 Structure of the joining boundary in brazing: using a cast filler metal of the Ni–Cr–Zr system with (a, ×350) and without (b, ×300) melting of the base metal, and (c, ×200) using a composite filler metal.

best results were obtained with the composite filler metals, rather than the cast ones, as the former make it possible to adjust the composition of the deposited metal. Moreover, the chemical composition of the deposited metal is hardly affected by the size of a defect.

Table 4.5 Chemical heterogeneity of the seams produced using arc heating and a composite filler metal

Low-melting point component to refractory component ratio	Structural components	Content of elements (wt.%) (Ni-base)							
		Co	Zr	Cr	Ti	Al	W	Mo	Nb
1:2	General	8.4	1.9	4.8	2.9	5.5	8.2	1.5	1.0
	Dendrite	9.9	0.2	5.9	2.3	4.9	9.6	0.8	1.0
	Metal in dendrite spacing	7.8	0.2	4.3	4.0	5.9	7.0	0.2	1.4
1:3	General	7.9	0.7	3.6	3.1	6.0	9.2	0.8	1.3

4.3 Brazing of titanium aluminides

Alloys based on intermetallics (IM) are a real alternative to heat-resistant metallic materials.[32–36] Titanium-based intermetallic alloys qualify as advanced materials due to their specific features, such as:

- the ability to retain high strength up to very high temperatures; moreover, strength properties of some intermetallics increase with an increase in temperature;
- the elasticity modulus of intermetallics decreases less intensively with an increase in temperature, compared with similar characteristics of alloys with disordered crystalline structure;
- the self-diffusion coefficient of intermetallics is several orders of magnitude lower than that of the disordered alloys at comparable temperatures, suggesting that a much lower creep rate may be attained in alloys with an ordered structure;
- low density caused by the presence of light elements (aluminium, boron), which is an important characteristic for aircraft and rocket engineering.[37]

The binary Ti–Al system,[30,31] the components of which differ greatly in the electron structure of the atoms and metal-chemical properties, forms a series of compounds, such as Ti_3Al (α_2 phase), TiAl (γ phase) and $TiAl_3$ (α phase). In view of their low specific weight, all of them are subject to close attention from researchers. In recent years, γ-TiAl base alloys have attracted considerable interest as high-temperature structural materials, which have a high heat resistance without using coatings They can be applied in aerospace engineering to manufacture lightweight compressors and turbines, and, most of all, to widen the field of application of non-cooled gas-turbine blades.

The titanium intermetallic TiAl retains its ordered structure up to the melting point (~1450°C).[30,31] Due to a high aluminium content, titanium aluminide TiAl has a low density (3.8 g/cm^3), which has the favourable effect of reducing the weight of structures made from this material.

Mechanical properties of alloys based on the Ti–Al system depend on the weight content of aluminium. The maximal characteristics of ductility are exhibited not by the single-phase γ-TiAl alloys, but by alloys of hypostoichiometric composition Ti-(46–48) at.% Al, which are the two-phase ($\alpha_2 + \gamma$) alloys, where the α_2 phase is intermetallic Ti_3Al. The alloys with a volume content of the α_2 phase ranging from 10 to 15% have the highest value of ductility.[38] The alloys with a fully lamellar coarse-grained structure (α_2 lamellae in a γ matrix) are characterised by maximal creep resistance at higher and low temperatures.

Traditional welding methods (heating with high heat input) or high deformation joining methods are unacceptable in the majority of cases for the production of permanent joints because of the probability of cracking or violation of geometric sizes of the fabricated structures. The method of choice for joining intermetallic alloys is brazing, which does not cause melting of the base metal. The brazing process allows avoidance of high residual stresses in the joints and cracking. Thus, it does not disturb the structural state of the joints and preserves the mechanical properties of the base material.

4.3.1 Filler metals

Much research is now being dedicated to brazing of intermetallic alloys. The range of available brazing filler metals is very wide, and just one joining method is used – diffusion brazing in vacuum. The use of copper as a brazing filler metal and a composite filler metal (a mixture of the base metal and copper powders) is suggested for brazing the alloy Ti–48Al–2Cr–2Nb.[39] Detailed consideration is given to the process of wetting and diffusion interaction of copper with the base metal. Also, it is shown that, after brazing at a temperature T = 1150°C and holding for 10 min, followed by heat treatment at 1350°C for 1 h, the resulting joints had a lamellar structure, similar to that of the base metal, but the strength of the joints was much lower compared with the base metal (350 MPa).

Foils of silver and aluminium, as well as the alloys Cu–Ni, Ti–Cu–Ni, Cu–Ni/Ti/Cu–Ni and Al/Ti/Al, have been used as filler metals for brazing alloy 45XD.[40] Based on the results obtained, the authors conclude that silver, aluminium and alloy Cu–Ni are not suitable for brazing this alloy. In the opinion of the authors, the rest of the filler metals can yield satisfactory results, providing that the filler metal has an optimal thickness and that the appropriate heat treatment is used after brazing. The most interesting result was obtained by the authors when using a combination of foils: 10 μm Al / 30 μm Ti / 10 μm Al. A fully lamellar structure was created by brazing at a temperature of 900°C with 1 h holding under a pressure of 20 MPa, followed by heat treatment at 1350°C for 1 h. Unfortunately, the strength value of the joints was not reported. At the same time, such a high temperature may cause undesirable changes in the base metal, leading to deterioration of its properties.

Shiue *et al.*[41] used the known filler metal BAg-8, which in fact is the Ag–Cu system eutectic alloy, for brazing γ-TiAl. They established that silver is inert with respect to the base metal, and copper actively interacts with TiAl to form an extensive reaction layer. Growth of two intermetallic phases, AlCuTi and $AlCu_2Ti$, was observed at the interface at a brazing temperature of 950°C, the AlCuTi phase growing more quickly with an increase in the brazing temperature and holding time. The strength of the joints produced at a brazing temperature of 950°C and holding time of 60 s was 343 MPa. Intensive growth of the AlCuTi phase took place with an extension of the holding time, which caused a substantial decrease in strength of the joints. So, the potential of this variant is questionable.

The use of BAlSi-4 as a brazing filler metal, where the silicon content was 11–13%, has been reported.[42] It is not clear what idea this experiment is based on. As might be expected, silicon-containing intermetallics (stable Al12Si3Ti5) were present in the seam, which determined the brittle fracture of the joints under low loads (maximum 86.2 MPa).

In one study[43] the filler metal was produced by collective mechanical milling of TiH_2–50% Ni in an argon atmosphere. The effect of this process on the microstructure and shear strength of the γ-TiAl joints brazed in vacuum furnaces was investigated. It was found that the grains of the powders were cleaned by the milling, and that the brazing temperature decreased. Sound joints were produced at a brazing temperature of 1140°C and holding for 15 min. A powder subjected to milling for 120 min was used. The highest values of shear strength, i.e. 256 MPa at room temperature and 207 MPa at 800°C, were obtained when the brazing temperature was increased to 1180°C. The increase of the brazing temperature to 1200°C led to considerable erosion of the base metal and deterioration of the strength of the joints.

An interesting experiment was carried out by Simoes *et al.*[44] They produced a filler metal NiAl in the form of thin layers by magnetron sputtering. Stimulated by temperature or pressure, these layers entered into an exothermic reaction, i.e. they acted as independent heat sources. The layers sputtered were 5, 14 and 30 nm thick. Brazing was performed in vacuum at a temperature of 900°C and holding for 30 and 60 min. Defect-free joints were produced using the 14 and 30 nm thick layers. Pores and cracks were detected in the seam at a layer thickness of 5 nm. The joints with the 14 nm thick NiAl interlayer, brazed at a temperature of 900°C, pressure of 5 MPa and holding for 60 min, showed the best shear strength (314 MPa).

Despite the wide variety of available filler metals for brazing of γ-TiAl, it should be noted that the most efficient method for improving homogeneity and strength of the joints is to use multi-component titanium-base filler metals. The following elements are used as depressants to decrease the melting point of titanium alloys: Ag, Cu, Ni, Pd, Mn, Be, Fe, Co, etc. It should be emphasised that intermetallic TiAl alloys are meant for operation at high temperatures. Hence, it is inexpedient to use filler metals based on Ag, Cu and Al because of the increased requirements for heat resistance. The application of beryllium is limited because of its high toxicity and negative effect on the ductility of titanium. For example, wide acceptance has been given to filler metals based on the Ti–Cu–Ni and Ti–Zr–Cu–Ni and other systems for brazing of titanium.[45–47] They are used in the form of rapidly quenched strips or powder mixtures (Table 4.6).

As concluded on the basis of analysis of constitutional diagrams,[30,31] it is reasonable to develop filler metals based on the Ti–Zr system, and use elements of group IV of the periodic table as depressants (V, C, Mn, Fe, Co, Ni). In particular, alloys of the Ti–Zr–Mn and Ti–Zr–Fe systems are promising for the development of filler metals, since in binary Ti–Mn, Ti–Fe, Zr–Mn and Zr–Fe systems there are relatively low-melting point eutectics with melting temperatures of 1180, 1085, 1135 and 928°C between solid solutions of β(Ti) and β(Zr), on the one hand, and titanium- and zirconium-rich intermetallics, on the other hand.[30,31]

The isostructurality of $TiMn_2$ and $ZrMn_2$ compounds, as well as the presence of a continuous series of solid solutions and fusion diagrams with a minimum in the Ti–Zr system, suggests the existence of a region of compositions with a lower solidus temperature than the temperatures of eutectics in the limiting binary systems in ternary alloys of the Ti–Zr–Mn system. A polythermal section connecting points of the titanium- and zirconium-rich eutectics in the limiting binary systems was plotted during investigations,[48] which confirmed the existence of a wide two-phase β(Ti,Zr)+(Ti,Zr)Mn_2 region in the Ti–Zr–Mn system.

Table 4.6 Filler metals for brazing titanium-base alloys

Filler metal	Composition (wt.%)	Brazing temperature (°C)	Melting temperature (°C)	
			Solidus	Liquidus
TiBraze375	Ti–37.5Zr–15Cu–10Ni	850–880	825	835
TiBraze240	Ti–24Zr–16Cu–16Ni–Mo	890–920	835	850
TiBraze260	Ti–26Zr–14Cu–14Ni–0.5Mo	880–920	840	860
TiBraze200	Ti–20Zr–20Cu–20Ni	870–900	848	856
TiBraze15-15	Ti–15Cu–15Ni	980–1050	902	950
TiBraze15-25	Ti–15Cu–25Ni	930–950	901	915
TiBraze70Ag	Ag–27Cu–(4–5)Ti	850–900	780	800
STEMET 1201	Ti–12Ni–12Zr–24Cu	900–1000	830	955
STEMET 1202	Ti–12Ni–12Zr–22Cu–1.5Be–0.8V	850–950	748	857
STEMET 1203	Ti–50Cu	1000–1050	950	990
STEMET 1204	Cu–28Ti	1000–1100	–	875
STEMET 1406	Zr–11Ti–14Ni–13Cu	900	770	833
STEMET 1409	Zr–11Ti–14Ni–12Cu–2Nb–1.5Be	750–859	685	767
VPr 16	Ti–(22–24)Cu–(8.5–9.5) Ni–(12–13.5)Zr	920–970	880	890
VPr 28	Ti–16.5Cu–15.5Ni–23Zr	850–870	830	840

Liquidus surfaces of the three-component systems Ti–Zr–Fe and Ti–Zr–Mn were plotted on the basis of experimental and literature data and using the simplex-lattice experimental design method.[49] Eutectic valleys comprising eutectics with a minimal melting point, which makes them suitable for use as basic filler metals in the development of filler metals for brazing intermetallic titanium alloys, were observed on the liquidus surfaces of the above systems. Promising compositions of filler metals were chosen on the basis of the investigations conducted.

The experiments were carried out using filler metals of the following systems: Ti–Zr–Fe, Ti–Zr–Mn, Ti–Zr–Cr, Ti–Hf–Fe and Ti–Zr–Cu–Ni (Table 4.7), as well as the intermetallic titanium alloys 47XD (Ti–45Al–2Nb–2Mn–0.8%TiB_2) and 48-2-2 (Ti–47Al–2Nb–2Cr) (Fig. 4.8(a), (b)). These intermetallic alloys contain an identical atomic percentage of aluminium, i.e. 48–49%, but they differ to some extent in the morphology of their structural components. The α_2 phase (Ti_3Al) in the form of light lamellae (Fig. 4.8(a)) is revealed in alloy 47XD along the boundaries of the main structural component – the ordered γ phase (TiAl). In addition, needle-like borides 1 μm wide and up to 30 μm long can be seen against the background of the lamellar structure. Their boron content amounts to 66.67–71.02%, which is proved by the results of X-ray spectrum microanalysis (Table 4.8, Fig. 4.8(a)). It should be noted that alloy 47XD was produced by the powder metallurgy method and is characterised by the presence of porosity, which makes the brazing process more complicated. Titanium aluminide alloy 48-2-2, produced by the electron beam remelting method, does not have a lamellar structure in the initial state. Dispersed particles of the α_2 phase (white colour), which contain an increased amount of chromium (6.7–9.3 at.%), can be seen in the alloy (Fig. 4.8(b)). Alloy 48-2-2 is a classic example of an intermetallic titanium alloy. It is characterised by increased ductility at room temperature ($\delta \sim 2.5\%$), compared with titanium aluminide TiAl ($\delta \sim 0.5\%$).[50]

Table 4.7 Chemical composition and melting ranges of brazing filler metals

Filler metal	Chemical composition (wt.%)								Melting temperature (°C)
	Ti	Zr	Fe	Mn	Cu	Ni	Cr	Hf	
1	45	30	–	–	–	–	25	–	1150/1180
2	45	–	25	–	–	–	–	30	1030/1070
3	45	30	25	–	–	–	–	–	940/970
4	45	30	–	25	–	–	–	–	1040/1090
5	52	12	–	–	24	12	–	–	890/910
6	45	24	–	–	16	15	–	–	820/850

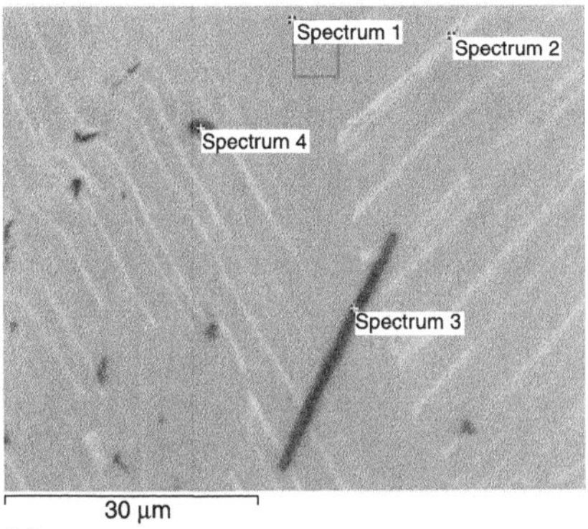

(a)

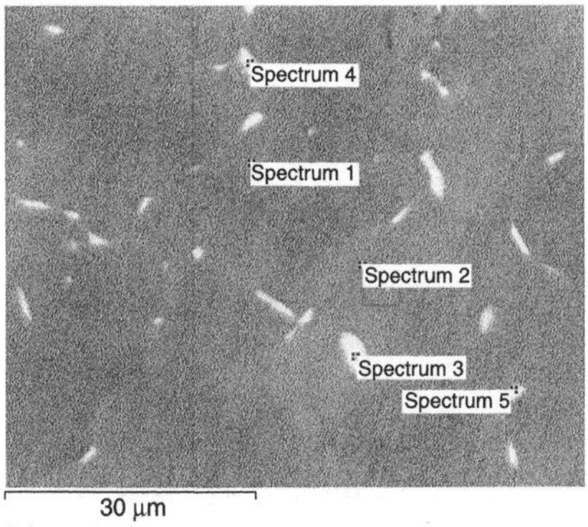

(b)

4.8 Microstructure of intermetallic alloys 47XD (a) and 48-2-2 (b) used as base materials.

Table 4.8 Chemical composition of structural components of base materials

Spectrum number	Chemical composition, atomic fraction of elements (%)					
	Ti	Al	Mn	Cr	Nb	B
47XD						
1	48.00	48.65	1.51	–	1.85	0.00
2	53.25	42.74	1.94	–	2.07	0.00
3	22.52	5.54	0.18	–	0.74	71.02
4	25.18	7.04	0.29	–	0.82	66.67
48-2-2						
1	48.89	47.70	–	1.8	1.61	–
2	49.36	47.54	–	1.57	1.52	–
3	54.49	34.33	–	9.31	1.87	–
4	52.87	38.47	–	6.7	1.96	–
5	51.22	43.36	–	3.72	1.7	–

The samples of overlap joints were brazed in a vacuum furnace (1.33×10^{-3} Pa) with radiation heating before the metallographic examinations. Filler metal in the cast state was placed near the gap, and capillary brazing was performed. In addition, the rapidly quenched thin strip (50 μm) was placed into the gap. The samples of butt joints were brazed to determine their mechanical properties. A chemical agent consisting of equal portions of hydrofluoric and nitric acids and glycerine was employed to reveal the microstructure.

4.3.2 Temperature and holding time

To select the temperature ranges for brazing an intermetallic titanium alloy containing about 45–47 at.% aluminium, it is necessary to take into account not only T_L of the filler metal, but also the chemical composition of the base metal, as well as the probable formation of intermetallic phases with filler metal components. To activate the diffusion processes and avoid formation of Fe and Mn aluminides, brazing was performed at 1250°C, which corresponds to the temperature of heat treatment of the given alloys. For comparison, some experiments were carried out at brazing temperatures of 1150, 1180 and 1200°C. The isothermal holding time during brazing was 5, 15 and 60 min. The maximal holding was used to ensure completion of the diffusion processes and homogeneous chemical composition of the molten metal of the brazed seam, fillet region and base metal.

Using a commercial (amorphous) filler metal of the Ti–Zr–Cu–Ni system (No. 5, Table 4.7) for brazing intermetallic titanium alloys at a temperature of 1250°C (for 5 min), despite the fit-up of samples for brazing without a gap, the seams formed were rather wide (~120–200 μm) and had a pronounced chemical heterogeneity. Eutectic in the form of a continuous strip 20–30 μm wide (Fig. 4.9(b)) solidified along the axis of the brazed seam and in the fillet regions. This eutectic acted as a centre of initiation and propagation of cracks (Fig. 4.9(a), (b)), causing deterioration in the mechanical properties of the brazed joints.[51]

Results of X-ray spectrum microanalysis of the chemical heterogeneity of the seams showed that the content of aluminium in the seams was 21–31 wt.% when using the commercial filler metal (No. 5, Table 4.7) and holding for 5 min. The titanium-base eutectic located on the seam axis contained 5.8% Zr, 12.9% Cu and 7.8% Ni. The maximal weight content of aluminium was detected in the faceted Ti–31Al–1.5Zr–2.6Cu–0.5Ni–3.8Nb–1.7Mn grains fringed with a light phase, the composition of which was not determined because of its small size (Fig. 4.9(c)).

In the case of brazing using filler metal No. 6 (T_b = 1250°C, τ_b = 5 min), the eutectic component in the fillet region consisted of two phases (wt.%): a light phase (Ti–23Al–18Zr–11Cu–16Ni–2Nb) and a dark phase (Ti–21Al–5.4Zr–1.8Cu–2.7Ni–2.3Nb). As seen in the central part of the microsection, two phases also solidified in the seam, in addition to the eutectic, but the concentration of the filler metal components in them was much lower (wt.%): light phase Ti–23Al–1.8Zr–5.9Cu–4.4Ni–6.3Nb and dark phase Ti–33.7Al–1.9Zr–0.8Cu–0.6Ni–3.7Nb–1.5Mn.

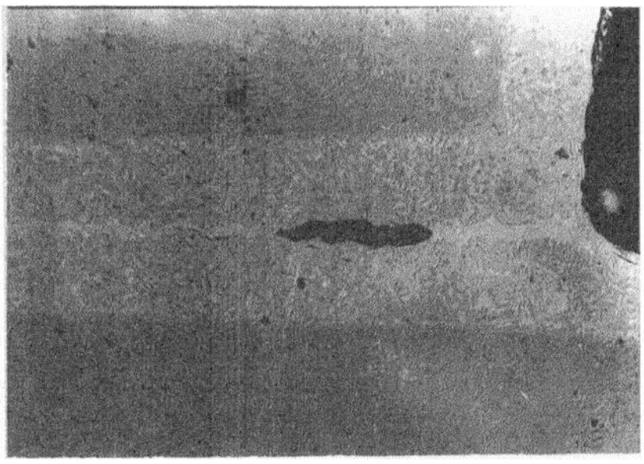

(a)

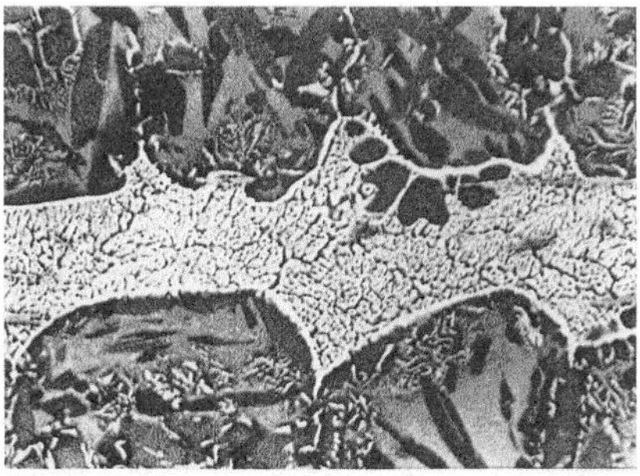

(b)

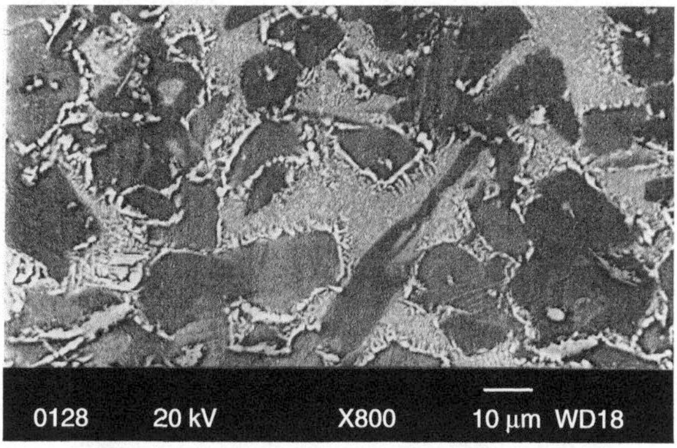

(c)

4.9 Microstructure of metal of the joint brazed using commercial filler metal Ti–12Zr–24Cu–12Ni (wt.%): (a) crack; (b) and (c) eutectic and structure of the central part of the seam.

Increasing the holding time to 15 min had hardly any effect on structure formation. Further increase of isothermal holding (to 60 min) led to a decrease in the concentration of the filler metal components both in the seam and in the fillet region, the quantity of the structural components remaining unchanged. Analysis of the metallographic examination results shows that interaction and formation of the structure of the brazed seams when using commercial filler metals Nos 5 and 6 were of an identical character in many respects, as might be expected. The seam metal was characterised by zone solidification and the presence of the eutectic located along the seam axis.

Brazing with experimental filler metal No. 2 (Table 4.7) (T_b = 1250°C, τ_b = 5 and 15 min) was also accompanied by zone solidification and formation of the eutectic in the form of a continuous strip along the seam axis. Increasing the holding time to 60 min stopped the formation of the eutectic in the brazed seam (Fig. 4.10(a)) and led to solidification of two phases with a concentration of aluminium corresponding to the α_2 and γ phases with a low content of other components of the filler metal and base material. The structure of the fillet region (Fig. 4.10(b), (c)) was inhomogeneous, consisting of two zones. The external zone comprised rather coarse light grains of hafnium-base phase Ti–49.8Hf–17Fe–13Al–0.9Nb–0.1Mn (wt.%), grains of grey titanium-base phase Ti-(17.2–25.3)Hf–6.2Fe–16.8Al–2.5Nb and a lamellar eutectic. The eutectic was the last to solidify, and was pressed out to the external zone of the fillet region (Fig. 4.10(c)). The second zone of the fillet region adjoining the base material (Fig. 4.10(c)) had a different morphological structure, which was

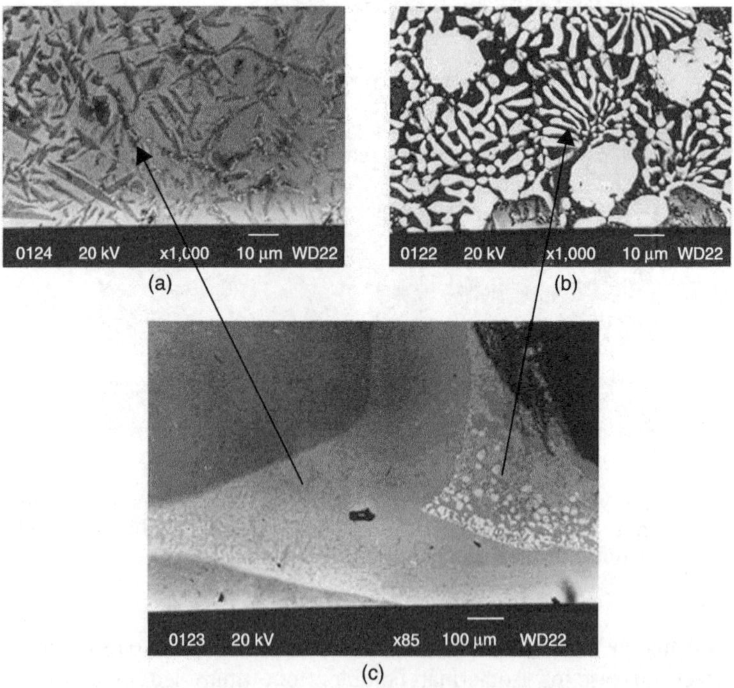

4.10 Zone solidification of fillet region in brazing with filler metal Ti–30Hf–25Fe: internal fillet region (a); external fillet region (b); general view (c).

closer to the structure of the brazed seam. It contained an increased amount of aluminium, i.e. 22.2 wt.%, the concentration of hafnium being decreased to 9.6 wt.%. A further increase in the concentration of aluminium (to 27.7 wt.%) and, hence, a decrease in the concentration of hafnium (to 3.9 wt.%) was observed in the seam with increasing distance from the fillet region. The structure of the seam consisted of two phases, rich in aluminium. The eutectic component was absent (Fig. 4.11).

Very similar results were obtained by brazing with the filler metal Ti–Zr–Cr. Phases corresponding to the γ-TiAl and Ti_3Al phases (Fig. 4.12(a), Table 4.9) solidified in the seam even at a short holding time (5 min) and brazing temperature of 1180°C. The content of aluminium in the base metal (in the zone adjoining the seam) and in dark grains of the seam metal (TiAl phase) was approximately identical (42.14–43.55 at.%). The fillet region consisted of two zones. As in brazing with a filler metal of the Ti–Hf–Fe system, the zirconium-rich eutectic solidified in the external part of the fillet (Fig. 4.12(b)).

The seams formed by brazing with a filler metal of the Ti–Zr–Fe system (No. 3, Table 4.7) (T_b = 1200°C, isothermal holding for 30 and 60 min) on intermetallic alloy 47XD had a width of about 100 μm and two-phase structure (Fig. 4.13(a), (b)). Grains of the γ-TiAl phase with an aluminium content of 45 at.% nucleated both on the substrate of the base metal (Fig. 4.6(a)), growing deep into the seam and preserving the orientation of the lamellae of the base metal, and in the central part of the seam. The Ti_3Al phase containing about 30 at.% aluminium solidified in the aluminium-depleted regions of the seam. The structure of the fillet

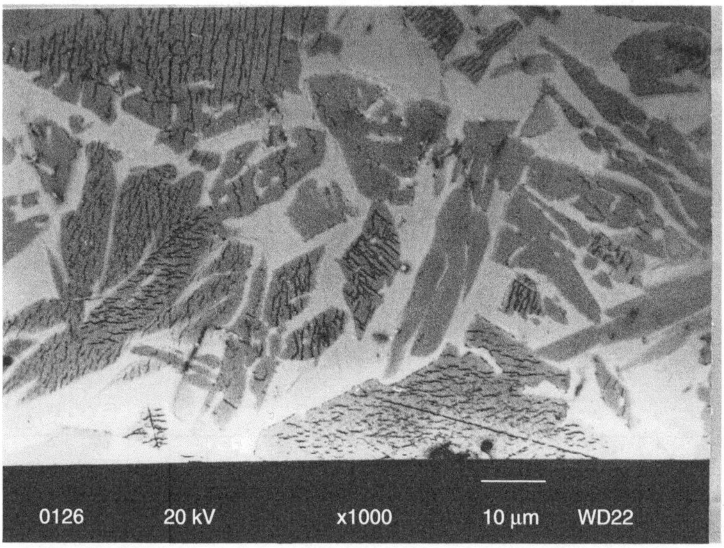

4.11 Microstructure of the seam brazed using filler metal Ti–30Hf–25Fe.

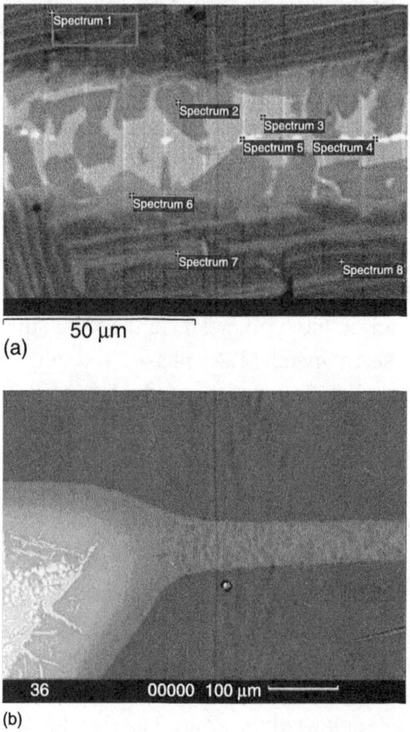

4.12 Microstructures of the brazed seam (a) and fillet region (b) resulting from using a filler metal of the Ti–Zr–Cr system.

Table 4.9 Content of chemical elements (at. %) in the joint brazed using a filler metal of the Ti–Zr–Cr system

Studied area (spectrum)	Al	Ti	Cr	Mn	Zr	Nb
1	42.14	55.21	–	0.71	–	1.94
2	43.55	49.14	1.43	0.54	3.60	1.75
3	32.20	56.83	4.45	1.70	2.83	2.00
4	32.64	52.07	2.51	0.63	10.97	1.17
5	33.40	51.68	2.50	0.55	10.82	1.05
6	37.55	55.71	1.69	0.80	2.21	2.04
7	37.96	58.72	–	1.12	–	2.19
8	43.88	53.09	–	0.94	–	2.09

regions was characterised by zone solidification. The external zone was primarily eutectic, which made it different from the internal zone and the structure of the seam, which is a characteristic feature of eutectic filler metals. Increasing the isothermal holding to 90 min (at T_b = 1200°C) also led to the formation of

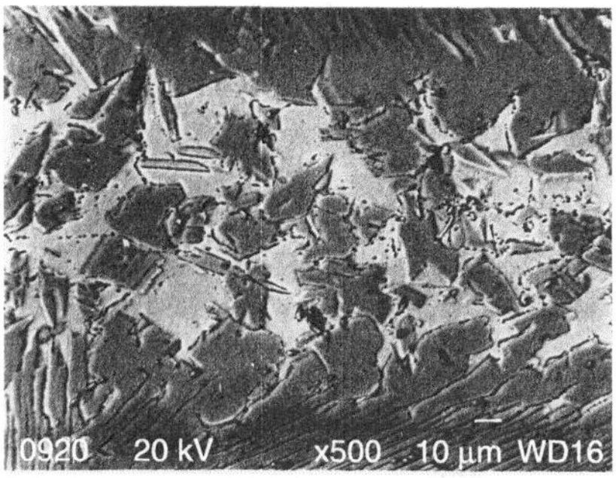

(a)

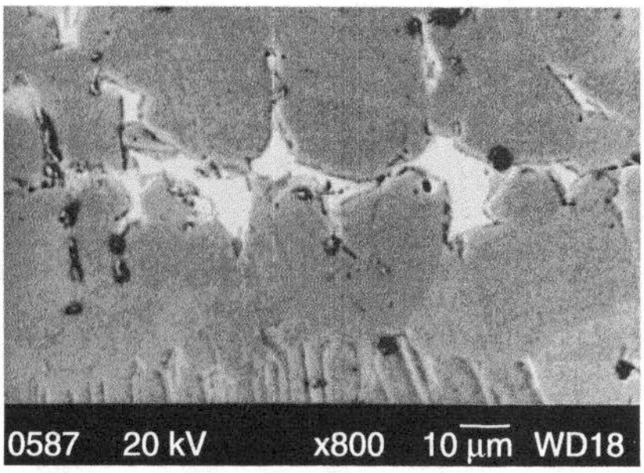

(b)

4.13 Formation of joints by brazing with a cast filler metal Ti–30Zr–25Fe (at.%) at T_b = 1200°C and τ_b = 30 min (a), and T_b = 1200°C and τ_b = 60 min (b).

two-phase seams. The primary phase consisted of Ti–43Al–2Nb–1.2Fe–1.43Zr grains. In addition, a large amount of the Ti–32Al–1.4Zr–3.4Fe phase was revealed. The seam width remained unchanged, i.e. about 100 μm. Attempts to avoid formation of eutectic in the fillet regions failed. Increasing the brazing temperature of titanium aluminide (47XD) to 1250°C with isothermal holding for 60 min allowed formation of the two-phase structure (γ-TiAl and Ti_3Al) both in the seams and in the fillet regions without the eutectic component (Fig. 4.14(a)–(d)). The content of aluminium in γ-TiAl was 43 at.%, and in the α_2 phase (Ti_3Al) it was 31 at.%.

Brazing of the intermetallic alloy 47XD using a filler metal of the Ti–Zr–Mn system (No. 4, Table 4.7) resulted in formation of regions where the seam had a lamellar structure close to that of the base metal. Some regions comprised coalesced grains of the base metal (Fig. 4.15). The chemical composition of the metal in these regions at the boundary of the joint was almost identical to that of the base metal.

The formation of a seam structure of similar character occurred in brazing (T_b = 1250°C, τ_b = 60 min) the cast intermetallic alloy 48-2-2 produced by electron beam remelting (Fig. 4.16 and 4.17).

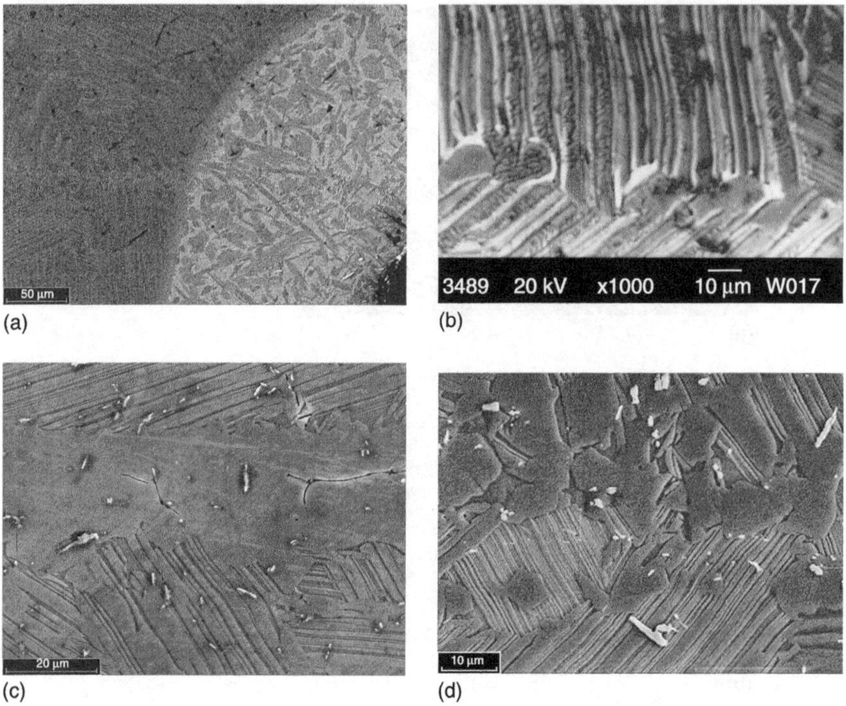

4.14 Microstructures of brazed joints on titanium aluminide 47XD produced using cast filler metal Ti–Zr–Fe at T_b = 1250°C and τ_b = 60 min: fillet region (a), seams (b–d).

The microstructure of the seam brazed with the Ti–Zr–Mn system filler metal consisted of two phases: primary TiAl grains containing 45.67 at.% aluminium, and a phase where the concentration of aluminium decreased to 31.9 at.%, corresponding to the α_2 phase (Ti_3Al). The range of homogeneity of this phase at room temperature is 22–35% aluminium. As shown by the results of X-ray spectrum microanalysis, both phases contained an insignificant atomic fraction of alloying elements of the base and filler metals. It should be noted that the chemical composition of the phases was consistent and did not depend on the location in the seam.

It is a known fact that heat treatment of cast alloy Ti–47Al–2Nb–2Cr should be conducted at T = 1250°C to give a lamellar structure. Due to optimal brazing parameters, after high-temperature heating the base metal (48-2-2) acquires a lamellar structure and consists of alternating lamellae of the α_2 and γ phases (Fig. 4.17(a)). Therefore, the technology for brazing cast intermetallic alloy Ti–47Al–2Nb–2Cr, developed on the basis of the investigations conducted, allows the thermal brazing cycle to be combined with the heat treatment operation. Also, it forms a lamellar structure and, hence, the joints have optimal mechanical,[52] thus there is no need for heat treatment of the base metal.[48] The amount of the TiAl

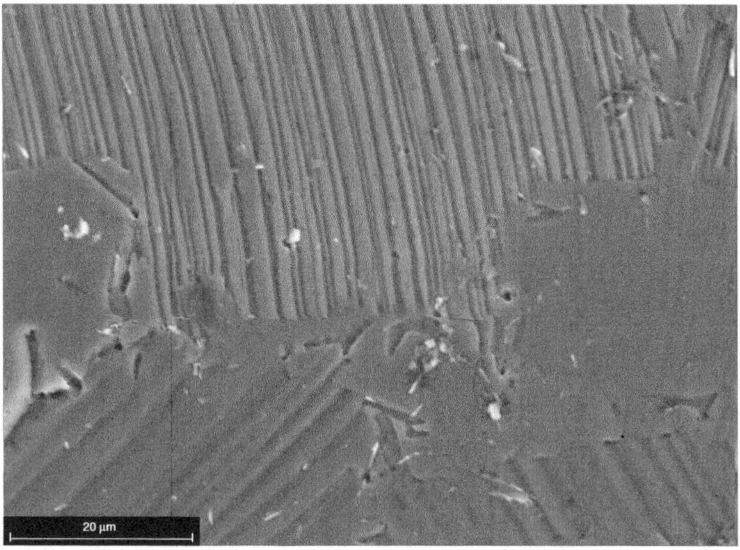

4.15 Microstructure of brazed joint on alloy 47XD produced using cast filler metal No. 4.

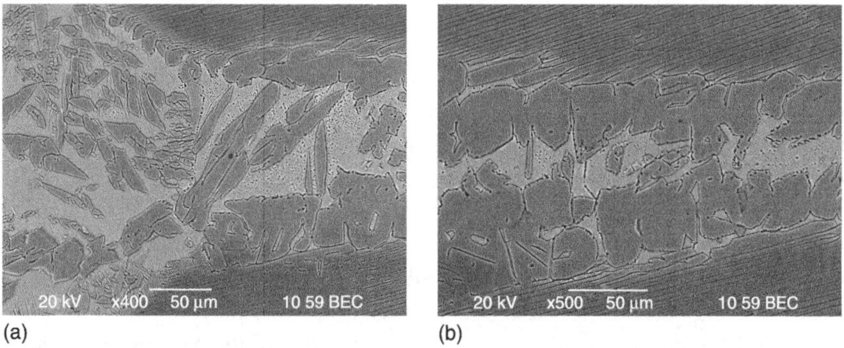

4.16 Microstructures of the fillet region (a) and seam (b) in the joint on alloy 48-2-2 brazed using a cast filler metal of the Ti–Zr–Mn system.

phase depends on the distance from the filler metal: the deeper the penetration of the filler metal into the gap due to capillary forces, the higher is the content of this phase in the seam (Fig. 4.17(b)).

The investigation results obtained by scanning electron microscopy and X-ray spectrum analysis show that the chemical composition and structure of the brazed seam differ substantially from those of the initial filler metal. The diffusion processes taking place at the solid base metal–liquid filler metal interface lead, in particular, to levelling of the concentration of aluminium in the base and seam metals. When using

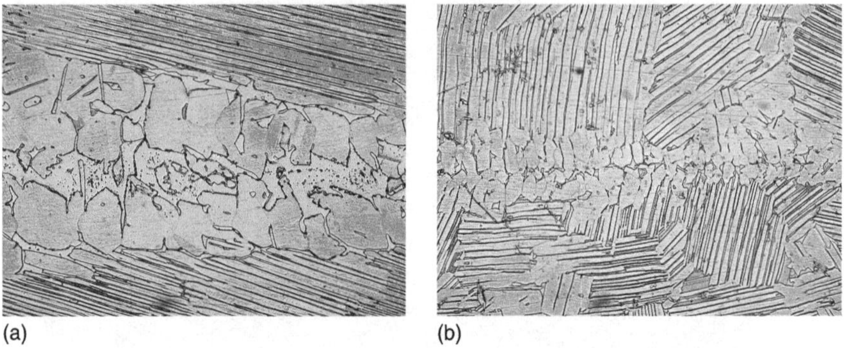

4.17 Brazed joints on alloys 48-2-2 (a) and 47XD (b) produced using a filler metal of the Ti–Zr–Mn system (optical microscope, ×500).

filler metals based on the Ti–Zr system, the content of zirconium in the seam metal decreases with an increase in the holding time at a brazing temperature, which is caused by mutual solubility in the Ti–Zr system.[31,32] The fillet regions contain a larger amount of liquid filler metal than the seams, the length of the filler to base metal contact line is smaller, and the diffusion processes occur less intensively. Therefore, the zirconium content is higher in the fillet regions than in the seam metal.

Butt joints were brazed and standard specimens (Fig. 4.18) were cut from these joints for mechanical tests at room and increased (700°C) temperatures. The test results are given in Table 4.10. Analysis of the results of tensile strength tests at room temperature shows that the joints on titanium aluminide 47XD produced using commercial filler metal (No. 5) Ti–12Zr–24Cu–12Ni are characterised by a minimal strength (Table 4.10). The low strength values are caused by structural peculiarities of the brazed seams and the presence of the eutectic component, which divides the seam into zones and promotes cracking. The best results were obtained by brazing with experimental filler metals Ti–30Zr–25Fe and Ti–30Zr–25Mn (wt.%), where the strength of the brazed joints was equal to that of the base metal, and the structure of the seams was close to that of the base metal. Results

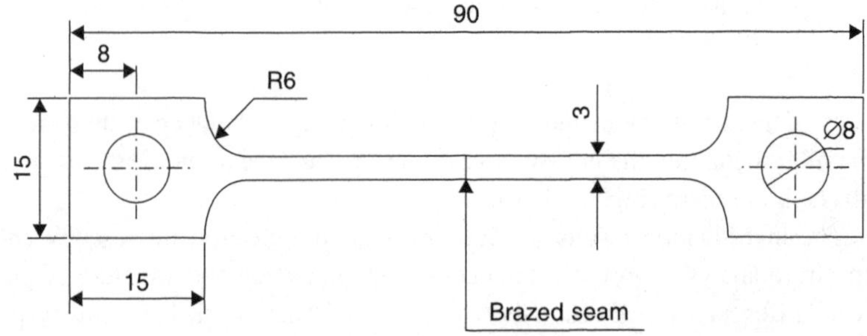

4.18 Drawing of a butt joint for mechanical and long-term strength tests. Dimensions are in mm.

Table 4.10 Properties of brazed joints

Alloy no.	Chemical composition (wt.%)	Strength (MPa) at:		Time to destruction of the samples at 700°C (h) under stress of:	
		20°C	700°C	140 MPa	200 MPa
1	Ti–30Zr–25Cr	490	315	216	–
2	Ti–30Hf–25Fe	592	280	480	–
3	Ti–30Zr–25Fe	651	284	483*	499*
4	Ti–30Zr–25Mn	693	316	500*	300*
5	Ti–12Zr–24Cu–12Ni	574	300	–	–
6	Ti–24Zr–16Cu–15Ni	468	–	–	–

Note: *Sample was not destroyed during the test.

of the evaluation of tensile strength at increased temperature (σ_t^{700}) were approximately at the same level, in a range of 280–316 MPa, for all filler metals (Table 4.10).

Long-term strength tests showing performance of the brazed joints under conditions as close as possible to those in service are an important indicator of heat resistance of the joints.[53] As determined by long-term tests of the brazed joints on γ-TiAl (47XD), the joints produced using filler metal Ti–Zr–Cr had the lowest heat resistance at a temperature of 700°C and stresses of 140 MPa. The specimens fractured at the seam after 216 h of testing. The use of filler metal Ti–Hf–Fe produced a longer operating time (480 h). It should be emphasised that the brazed joints (Table 4.5) produced using filler metals Ti–Zr–Mn and Ti–Zr–Fe did not fracture for 500 and 483 h, respectively, under a load specified by a customer (140 MPa). Increasing the load to 200 MPa did not lead to fracture either, and the specimens were removed from the tests after 300 and 499 h, respectively (Table 4.10).

Thus, the filler metals developed on the basis of the investigations conducted produced brazed joints on intermetallic alloy γ-TiAl with a structure and properties close to those of the base metal. The results obtained can serve as a basis for the manufacture of critical structures for different applications from new, advanced materials based on intermetallics using the above-considered filler metals.

4.4 Conclusion

1. Nickel-rich alloys of the Ni–Cr–Zr system were studied. Positions of the peritectic, univariant eutectics and ternary eutectic were determined. The liquidus surface of the alloys in the investigated range of concentrations was plotted, and promising ranges of alloys that can be used as filler metals for brazing nickel-base high alloys were identified.

2. Ductile intermetallic Ni_7Zr_2 was found to form, instead of brittle and hard intermetallic Ni_5Zr, in the structure of the alloys at a certain chromium content, leading to a decrease in hardness of the eutectic and making it possible to produce vacuum brazed seams close in hardness to the base metal.
3. Of special interest for brazing nickel-base high alloys is a combination of composite adhesion-active filler metals and arc heating, which creates conditions for regulating the seam composition and heat input.
4. Structures and melting ranges of alloys of the Ti–Zr–Fe and Ti–Zr–Mn systems were studied, liquidus surfaces for these systems were plotted, and ranges of concentrations for selection of the brazing filler metal bases were determined to develop filler metals for brazing titanium aluminide Ti–45Al–2Nb–2Mn+0.8 vol.% TiB_2.
5. Adhesion-active filler metals based on the Ti–Zr–Fe and Ti–Zr–Mn systems for brazing titanium aluminides form dense defect-free seams without the eutectic component. The filler metals make it possible to produce seams with a two-phase structure (TiAl and Ti_3Al) close to that of the base material. This structure had a short-term tensile strength of 651–693 and 284–316 MPa at room and higher temperatures (700°C), respectively, which is close to that of the base material. The long-term strength values at a temperature of 700°C and preset stresses (140, 200 MPa) were, respectively, 483 and 500 h without fracture.

4.5 Future trends

Of the many commercially used heat-resistant alloys, the superalloys used to manufacture components of hot sections of gas turbine engines, and turbine blades attract the most interest. At present, in the overwhelming majority of cases these are high nickel alloys.

One of the important challenges concerning these alloys is the development of consumables and technologies for repair of components, as well as manufacturing sectional parts from them. Superalloys are brazed using filler metals of the Ni–Cr–Si–B, Ni–Cr–Co–B, Ni–W–Co–B, Ni–Cr–Al–B and Ni–Cr–Al–Si–B systems (M.M. Shvarts, *Handbook*), in which boron and silicon act as the main depressants. As reported in many studies, it takes a significant amount of time to achieve acceptable properties for joints using such filler metals under conditions of diffusion brazing in vacuum, and it is economically inexpedient. Brazing with composite filler metals is much more efficient. In this case the filler metals are used as a low-temperature phase, and a powder of the metal brazed is used as a high-temperature phase. This process produces properties of the joints close to those of the base metal within an acceptable time. In coming years this process will find increasingly wide application. Nevertheless, problems with the grain boundary diffusion of boron are still a serious drawback of the filler metals under consideration, and the search for an alternative choice will continue.

Undoubtedly, new systems of filler metals, in which elements of groups IV and V of the periodic table (e.g. Ti, Zr, Hf, Nb, Ta) are used as depressants, will come into use. Filler metals with a high niobium content, for example VPr 24, are used commercially. It should be noted that this is a high alloy with a high content of tungsten, cobalt and a sum of aluminium with titanium, and a low content of boron (9.25–0.35%). Filler metals based on the Ni–Cr–Zr system have not yet found wide application, but they hold promise for both vacuum and arc heating, the latter being a particularly topical area.

Alloys based on intermetallics are a real alternative to metal alloys at working temperatures of approximately 700°C. Their primary purpose is to widen the application fields for non-cooled blades and provide high heat resistance without coatings. Moreover, they are less susceptible to changes of properties in operation at high temperatures and feature substantial advantages due to their lower specific weight. At present, alloys based on nickel and titanium aluminides are considered most promising. Of considerable interest are alloys based on titanium aluminide γ-TiAl, which has a low specific weight, high melting point and good combination of mechanical properties. It follows, on the basis of an analysis of the literature data, that the main joining method for these alloys is diffusion brazing. Hence, it can be concluded that filler metals based on silver, copper and aluminium are not recommended for joining intermetallic alloys. Even when a lamellar structure forms, as in the base metal, the mechanical properties of the joints are much inferior to those of the base metal.

The best structure and mechanical properties have been achieved up to now with filler metals based on titanium, and the Ti–Zr system in particular. These filler metals produce the lamellar structure of the weld metal and close strength characteristics at room temperature and 700°C. It should be emphasised that good results have been obtained in long-term strength tests.

A very interesting method for joining γ-TiAl is the use of foils: Al/Ti/Al or nanolayered coatings consisting of alternating titanium and aluminium layers. So far high strength values of the joints have not been achieved, but the area in general is promising. Another promising area, in our opinion, is the use of an exothermic reaction initiated in the sprayed layer of an Ni–Al system. The strength of the joints is approximately half that of the base metal.

It should be noted, in conclusion, that the technology for brazing intermetallic alloys is still in its early stages, and studies in this area will be expanded.

4.6 References

1. Paton, B.E., Stroganov, G.B., Kishkin, S.T. *et al. Heat resistance of casting nickel alloys and their protection from oxidation.* Kiev: Naukova Dumka, 1967. 256 pp.
2. Shalin, R.E., Svetlov, I.L., Kachanov, E.B. *et al. Single crystals of nickel superalloys.* Moscow: Mashinostroyeniye, 1997. 336 pp.
3. Casting superalloys. *S.T. Kishkin's Effect: Collected Scien.-Techn. Articles.* Ed. by E.N. Kablov. Moscow: Nauka, 2006. 272 pp.

4. Superalloys 2. *Heat-resistant materials for aerospace and industrial power units*. Ed. by C.T. Sims, N.S. Stoloff and W.K. Hagel. Transl. from English. In 2 books: Book 1. Ed. by R.E. Shalin. Moscow: Metallurgiya, 1995. 384 pp.
5. Superalloys 2. *Heat-resistant materials for aerospace and industrial power units*. Ed. by C.T. Sims, N.S. Stoloff and W.K. Hagel. Transl. from English. In 2 books: Book 2. Ed. by R.E. Shalin. Moscow: Metallurgiya, 1995. 384 pp.
6. Kablov, E.N. Cast Blades of Gas Turbine Engines. *Alloys, Technologies, Coatings* Moscow: MISiS, 2001. 632 pp.
7. Janke, B. High-Temperature Electron Beam Welding of the Nickel-Base Superalloy IN-738 LC. *Welding Journal* (1982) 11:343s–347s.
8. Xiaowei Wu, Chandel, R.S. and Hang Li. Evaluation of transient liquid phase bonding between nickel-based superalloys. *Journal of Materials Science* (2001) 36:1539–1546.
9. Arafin, M.A., Medrai, M., Turner, D.P. and Bocher, P. Transit liquid phase bonding of Inconel 718 and Inconel 625 with BNi-2: Modeling and experimental investigations. *Mater Sci and Eng: A* (2007) 447(1/2):125–133.
10. Ohsasa, K., Shinmura, T. and Narita, T. Numerical modeling of the transient liquid phase bonding process of Ni using Ni-B-Cr ternary filler metal. *J of Phase Equilibria* (1999) 20(3):199–206.
11. Chaturvedi, M.C., Ojo, O.A. and Richards, N.L. Diffusion brazing of cast Inconel 738 superalloy. *Advances in Technol: Materials and Materials Proc* (2004) 2(6):206–213.
12. Saha, R.K. and Khan, T.I. Effect of bonding variables on TLP bonding of oxide dispersion strengthened superalloy. *J Mater Sci* (2007) 42:9187–9193.
13. Liu, J.D., Jin, T., Zhao, N.R. *et al.* Effect of transient liquid phase (TLP) bonding on the ductility of a Ni-base single crystal superalloy in a stress rupture test. *Materials Characterization* (2008) 59:68–73.
14. Xiaowei Wu, Chandel, R.S., Hang Li *et al.* Induction brazing of Inconel 718 to Inconel X-750 using Ni–Cr–Si–B amorphous foil. *Journal of Materials Processing Technology* (2000) 104:34–43.
15. Mattheij, J.H.G. Role of brazing in repair of superalloy components – advantages and limitations. *Mater Sci Technol* (1996) 11(8):608–612.
16. Duvall, D.S., Owezarsks, W.A. and Paulonis, D.F. TLP bonding: a new method for jointing heat resistant alloys. *Welding J* (1974) 4:203–214.
17. Rabinkin, A. Brazing with (NiCoCr)-B-Si amorphous brazing filler metals: alloys, processing, joint structure, properties, application. *Sci and Technol of Welding and Joining* (2004) 9(3):181–199.
18. Khorunov, V.F. *Principles of brazing of thin-walled structures of high-alloy steels*. Kiev: Naukova Dumka, 2008. 240 pp.
19. Rabinkin, A., Wenski, E. and Ribaudo, A. Brazing Stainless Steel Using a New MBF-Series of Ni–Cr–B-Si Amorphous Brazing Foils. *Welding Journal* (1998) 2: 66s–75s.
20. Wu, X.W., Chandel, R.S., Seow, H.P. and Li, H. Wide gap brazing of stainless steel to nickel-based superalloy. *Journal of Materials Processing Technology* (2001) 113: 215–221.
21. Belyavin, A.F., Kurenkova, V.V., Malashenko, I.S. *et al.* Strength and microstructure of brazed joints on alloy JS6U produced by using boron- and boron-silicon-bearing filler metals. *Advances in Special Electrometallurgy* (2010) 2:40–51.
22. Kurenkova, V.V., Malashenko, I.S., Trokhimchenko, V.V. *et al.* Tensile strength of brazed joints on nickel alloy VJL12U at 20 and 950°C temperatures. *Advances in Special Electrometallurgy* (2006) 3: 30–40.

23. Malashenko, I.S., Kurenkova, V.V., Onoprienko, E.V. *et al.* Mechanical properties and structure of brazed joints on casting nickel alloy JS26VI. Part 1. *Advances in Special Electrometallurgy* (2007) 1:25–32.
24. Kurenkova, V.V., Onoprienko, E.V., Malashenko, I.S. *et al.* Mechanical properties and structure of brazed joints on casting nickel alloy JS26VI. Part 2. *Advances in Special Electrometallurgy* (2007) 2:23–34.
25. Kurenkova, V.V., Onoprienko, E.V., Malashenko, I.S. *et al.* Structure and strength properties of brazed joints on casting nickel alloy JS26NK. Part II. *Advances in Special Electrometallurgy* (2008) 1:26–35.
26. Henhoeffer, T., Huang, X., Yand, S., Au, P. and Nagy, D. Microstructure and high temperature tensile properties of wide gap brazed cobalt based superalloy X-40. *Materials Science and Technology* (2010) 26:4.
27. Khorunov, V.F., Ivanchenko, V.G. and Kvasnitskyy, V.V. *Investigation of Ni–Cr–Zr and Ni-Cr-Hf Alloys*. Hart- und Hochtemperaturloten und Diffusionsschweißen Vortrage und Posterbeitrage des 5. Intern. Kolloq. in Aachen: 59–61. Düsseldorf: DVS-Verl., 1998.
28. Khorunov,V.F., Maksymova, S.V. and Ivanchenko, V.G. Development of filler metals for brazing heat-resistant nickel – and titanium-base alloys. *The Paton Welding Journal* (2004) 9:26–31.
29. Khorunov, V.F., Maksimova, S.V. and Zvolinskii, I.V. Brazing alloys based on the Ni–Cr–Zr system for brazing creep-resisting nickel alloys *Welding International* (2010) 24(6):462–464.
30. Lyakishev, N.P. (Ed.) *Constitutional diagrams of binary metal systems: Handbook in 3 volumes*. Vol. 3, Book 1. Moscow: Mashinostroyeniye, 1999. pp. 672–675.
31. Massalski, T.B. *Binary Alloy Phase Diagrams*, 2nd edition. Metals Park, Ohio: ASM International, 1990.
32. Khorunov, V.F. Problems of brazing of metallic materials. In *Current Problems of Modern Materials Science*. In 2 volumes. Vol. 1. Kiev: Akademperiodika Publishing House, 2008. pp. 260–284.
33. Hua-Ping Hiong, Bo Cheng, Wei Mao *et al.* High-temperature brazing of alloy based on intermetallic Ti_3Al. *Transactions of National University of Shipbuilding (NUK)* (2009) 4 (427):43–49. Mykolaiv: NUK.
34. Ivanov, V.I. and Yasinsky, K.K. The efficiency of using superalloys based on intermetallic Ti_3Al and TiAl for operation at temperatures of 600–800°C in aerospace engineering. *Tekhnologiya Lyogkikh Splavov* (1996) 3:7–12.
35. Lukianychev, C.Yu., Shakhanova, G.V., Smirnova, T.R. and Goryunova, G.V. Structure and properties of semi-finished products of alloy Ti-48Al-2Nb-2Cr based on intermetallic TiAl, produced by the shaped casting method. *Tekhnologiya Lyogkikh Splavov* (1996) 3:16–19.
36. Kablov, E.N. VIAM – heart of the aviation materials science. *Vestnik Vozdushnogo Flota* (2003) 4:6.
37. Grinberg, B.A. and Ivanov, M.A. *Intermetallics: microstructure and deformation behaviour*. Ekaterinburg: UrO RAN, 2002. 360 pp.
38. Bannykh, O.A., Povarova, K.B., Braslavskaya, G.S. *et al.* Mechanical properties of cast alloys γ-TiAl. *Metallovedenie i Termicheskaya Obrabotka Metallov* (1996) 4:11–14.
39. Gale, W.F., Wen, X., Shen, Y., Xu, Y. and Fergus, J.W. Diffusion brazing of titanium aluminide. Wettability, microstructural development and mechanical properties. *International Brazing & Soldering Conference Proceedings*. 2–5 April 2000, Albuquerque, New Mexico, pp. 42–49.

40. Xu, Q., Chaturvedi, M.C., Richards, N.L. and Goel, N. Diffusion brazing of a TiAl alloy. *International Brazing & Soldering Conference Proceedings*. 2–5 April 2000, Albuquerque, New Mexico, pp. 57–64.
41. Shiue, R.K., Wu, S.K. and Chen, S.Y. Infrared brazing of TiAl intermetallic using BAg-8 braze alloy. *Acta Materialia* (2003) 51:1991–2004.
42. Shiue, R.K., Wu, S.K. and Chen, S.Y. Infrared brazing of TiAl using Al-based braze alloys. *Intermetallics* (2003) 11:661–671.
43. Peng He, Duo Liu, Erjing Shang and Ming Wang. Effect of mechanical milling on Ni–TiH_2 powder alloy filler metal for brazing TiAl intermetallic alloy: The microstructure and joint's properties. *Materials Characterization* (2009) 60:30–35.
44. Simoes, S., Viana, F., Vieira, M.F., Ventzke, V., Kocak, M. *et al*. Diffusion bonding of TiAl using Ni/Al multilayers. Springer Science+Business Media, LLC 2010, *Mater Sci* (2010) 45:4351–4357.
45. Kalin, B.A., Sevryukov, O.N., Fedotov, V.T., Plyushchev, A.N. and Yaikin, A.P. New amorphous filler metals for brazing of titanium and its alloys. *Svarochnoye Proizvodstvo* (2001) 3:37–39.
46. Schwartz, M.M. *Brazing*. Metals Park, Ohio 44073: ASM International, 1987. 455 pp.
47. Muller, H. and Breme, J. Brazing of Titanium with New Biocompatible Brazing Filler Alloys. *Titanium 99: Science and Technology* (1999). pp. 1758–1765.
48. Maksymova, S.V., Khorunov, V.F. and Ivanchenko, V.G. Brazing of intermetallic alloy of 48-2-2 grade by using filler metal of the Ti–Zr–Mn system. *Svarochnoye Proizvodstvo* (2011) 3:21–27.
49. Khorunov, V.F., Maksymova, S.V. and Zelinskaya, G.M. Investigation of structure and phase composition of alloys based on the Ti–Zr–Fe system. *Avtomaticheskaya Svarka* (2010) 9(689):14–19.
50. Ilyin, A.A., Kolachev, B.A. and Polkin, I.S. Titanium alloys. Composition, structure, properties. Ilyin, A.A. *Handbook*. Moscow: VILS – MATI, 2009. 520 pp.
51. Maksymova, S.V. Formation of brazed joints on titanium aluminide. *Avtomaticheskaya Svarka* (2009) 3:7–13.
52. Court, S.A., Lofvander, I.P.A., Loretto, M.H. *et al.* Plastic Deformation of Ti3Al. *Philos Magazine A* (1990) 61(1):109–139.
53. Khorunov, V.F., Maksymova, S.V. and Ivanchenko, V.G. Production of brazed joints on gamma titanium aluminide and investigation of their properties. *Adgezia Rasplavov i Paika Materialov* (2004) 37:100–108.

5

High-temperature brazing: filler metals and processing

A. RABINKIN, Metglas Inc., USA

DOI: 10.1533/9780857096500.2.121

Abstract: Huge technological progress in high-temperature brazing has made in the last few decades. This chapter reassesses the topic. It describes compositional and structural specifics of base metal (BM) classes at high temperatures. The classification of brazing filler metals (BFMs) is given, with some new compositions. In choosing optimal BFMs the metallurgical background is built on phase diagrams of BM and BFM and potential interactions. Novel methods of BFM placements, joint loading and heat treatment are covered. Rapidly solidified amorphous BFMs are described. The modern outlook on metallurgical paths of joint formation and an understanding of brazing mechanisms are provided. Finally, recent examples of advanced applications are given, particularly in the aerospace and automotive industries and in energy conservation and environment protection.

Key words: brazing, stainless steels, superalloys, brazing applications, brazing filler metals.

5.1 Introduction

A qualitative leap in brazing filler metal (BFM) technology occurred during and shortly after World War II as part of the considerable research into metallurgy. This progress was spurred on by the development of jet airplanes, atomic weaponry and energy systems, etc. The technological progress that was occurring required a dramatic increase in materials performance, particularly at high temperatures and in highly oxidizing/corrosive environments. It seems logical that in 1947, R. Peaslee discovered another application for Ni/Cr–B–Si hard-facing alloys as a new BFM class in his attempt to braze turbine disks on the recently developed jet airplanes (Peaslee, 2006). These hard-facing alloys had been previously used to enhance the gears in propeller-driven airplanes. The next 10–20 years saw the development of dozens of new and modified BFMs for practically all classes of steels and alloys. The major achievements in this respect were made mostly in the US, and also in Germany.

The most comprehensive overview of the status of high-temperature brazing technology was written by H. Pattee (1980). In the same year E. Lugscheider and K. Gundlfinger published a review of the mechanical properties of brazed joints produced from virtually all known iron- and nickel-based steels and alloys used as the base metals (BM) (Lugscheider and Gundlfinger, 1980). It is remarkable that very few, if any, new in principle and important BFM compositions and innovations in brazing technology development have been developed since then. On the positive side, it is worth noting that a much better scientific understanding of brazing mechanisms has developed. Brazing technology has also been employed in many new industrial applications.

The current chapter is not an attempt to give a complete and comprehensive review of what has been accomplished in the last few decades. Only a limited number of, subjectively chosen, joints manufactured for high-temperature applications will be presented. Titanium joining, considered in Shapiro and Rabinkin (2003), composites and ceramics are outside the scope of the chapter. The majority of examples and case histories will be drawn from work in which the current author has participated, and practically all of them are very strongly associated with practical industrial applications.

The chapter will start with a short review of the generic metallurgical specifics that should be considered for brazing operation when choosing both the most favorable BFM compositions and the optimal heat treatment conditions. Next, an outline of the existing classes of BFMs will be provided, with notes on the newest compositions that have been added to the existing ones. These newest compositions include two groups of alloys. The first group are Ni/Fe/Cr-based eutectic alloys in which phosphorus is the largest fraction of metalloid elements. In these new phosphorus-based alloys, boron and silicon play a minor role in creating general effects associated with metalloids, such as decreasing the melting temperature, improving BFM wettability, etc. Afterwards, the Ni-base alloys with either Ge or Zr/Hf/Cr additions, which are also eutectic, will be presented briefly. These alloys have been mostly developed to repair turbine blades and have to withstand high service temperatures. The major part of this chapter is devoted to a description of microstructures, properties and optimal brazing conditions of joints manufactured from steels and alloys. This description is given in a sequential way, starting from single-phased ferritic, then austenitic stainless steels, and concluding with multi-phased superalloys. After that, the modern outlook on metallurgical paths of joint formation will be presented. Finally, the recent expansion of brazing operations for advanced industrial applications will be described.

5.2 Features of base metal (BM) alloys used in high-temperature brazing

From the metallurgical point of view, modern high-temperature steel and alloy BMs can be divided very roughly into two groups. The first group consists of stainless steels having the body-centered cubic (b.c.c.) and face-centered cubic (f.c.c.) solid solution structure and the second group contains superalloys having a complex multi-phase structure. In spite of a great variety of compositions in both groups, the alloying elements, which added to iron, nickel and cobalt, yield materials with stabilized and/or preserved α b.c.c. and γ f.c.c structures, or structures possessing strength and resistance under high temperature/oxidation service conditions. Chromium, titanium and aluminum in particular are added for the latter reason. The superalloy BMs are mostly used to withstand ultimate temperature and stress conditions over a long service time. Therefore, it is imperative to choose brazing conditions and BFM compositions in such a way as to maintain the complex microstructures and high-performance properties of the superalloys in the aftermath of the brazing operation. Here one should preserve the following microstructures: namely, the γ'-prime strengthened matrix phase in the case of Ni-based superalloys, and the solid-solution strengthened phase in Co-based superalloys. It is particularly important to keep the amount of boron in filler metals to a minimum because boron may induce brittleness by causing potential chromium boride segregation in the base metal grain boundaries. Refractory metals used in superalloys, such as tungsten, molybdenum and others, dissolve readily in chromium borides and, therefore, do not prevent this brittleness effect.

5.3 Brazing filler metals (BFMs) for joining high-temperature base metals

All current BFMs can be split into two major classes according to their nature: namely, eutectic and solid solutions (Table 5.1). In 2010, after many years of discussion, the International Standards Organization published International Standards for Brazing Filler Metals. In these ISO/FDIS 17672 standards all filler metals are divided into four groups, each consisting of alloys having one major metal component. For example, classes Ni and Co according to these standards

Table 5.1 Commercial and developmental brazing filler metals (BFMs) for joining steels and metal alloys at T_{Br} >900°C

Alloy system	Alloy class	Major alloying elements	Forms	T_{Br} (°C)	Service temp. (°C)
Transition metal – metalloid alloy systems					
Ni/Fe/Co–(B)–(Si)–(C)–(P)	Eutectic	Cr, Mo, W, Ti, Al	Powder Paste Tape, RS foil	950–1200	≤1200
Ni/Pd–(Si)–(B)	Eutectic	Cr, Co, W, Mo	Powder Paste Tape, RS foil	900–1000	400–800
Transition metal – metal alloy systems					
Ni–(20–23)Ge	Eutectic	–	Powder, paste	~1200	>1200
Ni/Zr/Hf	Eutectic	Cr	Powder, paste	1200–1250	>1150
Noble metal systems					
Au/Pd/Ag	Solid Solution	Cu, Ni, Cr	Powder, foil, Wire	900–1300	≤1200

are alloys containing Ni and/or Co as the major components, and the same is true for the Au, Pd and Ag classes. These new BFM classifications include practically all AWS A5.8/A5.8m (2004), German, British and other national classifications, only under different designations. Thus, for example, BNi-2 AWS Classification is designated as Ni 620, etc.

5.3.1 BFM eutectic alloys

These alloys are produced from transition metals (TM) such as nickel, iron, cobalt and chromium, etc., in combination with metalloids, such as silicon, boron, phosphorus and carbon, which, most importantly, decrease the BFM melting temperature. These metalloids are also chemically active elements, which react strongly with oxygen and are thus reducing; they clean stainless steels and superalloys of their natural surface oxide films when they are not prohibitively thick. This is the reason why all these materials need no special treatment for removing the permanent natural oxide scale. These scales in most cases are thin, a few hundred angstroms, whereas a relatively thick liquid metal film of a few dozen microns, i.e. a few thousand angstroms, can dissolve oxides. In addition, they decrease the surface tension of BFMs due to their intrinsic low surface tension.

Table 5.2 presents a list of such alloys according to the ISO/FDIS 17672 Standards. The alloys are Ni and Co filler metals. In the conventional crystalline state all these materials are inherently brittle and cannot be produced in continuous forms such as foil, wire, etc. using traditional metallurgical processing. Therefore, they are available only as powders or their derivatives. The appearance of rapid solidification technology (see details below) eventually eliminated this disadvantage and led to completely ductile foils made from many of these alloys. These foils are well known in industry as Metglas Brazing Foil®, or MBF, developed and produced by Metglas Inc. Fortunately, these ductile foils are listed as one of the forms in which these compositions can be found, but without clarification of how they are produced or what atomic structure (crystalline or amorphous) they have.

The majority of Ni and Co filler metals are alloys having boron and silicon as metalloids. In spite of their long history of successful application, controversy still exists about the role of boron in the strength of the joints manufactured from these classes (see Section 5.3.5).

5.3.2 Solid solution BFMs

Solid solution alloys are made of silver, gold and palladium as the bases to which Pd, in the case of gold base, Ni, and Cu are added. These alloys have a very wide range of melting temperatures, are very malleable and can be produced in foil and wire forms (Tables 5.3 and 5.4). Outstanding resistance to oxidation and corrosion is their most remarkable characteristic. This group has rather limited application

Table 5.2 International standard ISO/FDIS 17672:2010(E) – Classes Ni and Co: nickel- and cobalt-bearing brazing filler metals

Code	Composition (% by mass)													Melting temperature (approximate) (°C)	
	Ni min./max.	Co max.	Cr min./max.	Si min./max.	B min./max.	Fe min./max.	C min./max.	P min./max.	W min./max.	Cu min./max.	Mn min./max.	Mo min./max.	Nb min./max.	Solidus	Liquidus
Ni–Cr–B alloys															
Ni 600	Rem.	0.10	13.0/15.0	4.0/5.0	2.75/3.50	4.0/5.0	0.60/0.90	–/0.02	–/–	–/–	–/–	–/–	–/–	980	1060
Ni 610	Rem.	0.10	13.0/15.0	4.0/5.0	2.75/3.50	4.0/5.0	–/0.06	–/0.02	–/–	–/–	–/–	–/–	–/–	980	1070
Ni 612	Rem.	0.10	13.5/16.5	–/–	3.25/4.0	–/1.5	–/0.06	–/0.02	–/–	–/–	–/–	–/–	–/–	1055	1055
Ni 620	Rem.	0.10	6.0/8.0	4.0/5.0	2.75/3.50	2.5/3.5	–/0.06	–/0.02	–/–	–/–	–/–	–/–	–/–	970	1000
Ni–Si–B alloys															
Ni 630	Rem.	0.10	–/–	4.0/5.0	2.75/3.50	–/0.5	–/0.06	–/0.02	–/–	–/–	–/–	–/–	–/–	980	1040
Ni 631	Rem.	0.10	–/–	3.0/4.0	1.50/2.20	–/1.5	–/0.06	–/0.02	–/–	–/–	–/–	–/–	–/–	980	1070
Ni–Cr–Si alloys															
Ni 650	Rem.	0.10	18.5/19.5	9.75/10.50	–/0.03	–/–	–/0.06	–/0.02	–/–	–/–	–/–	–/–	–/–	1080	1135
Ni 655	Rem.	0.10	21.0/23.0	6.0/7.0	–/0.01	–/–	–/0.16	3.5/4.5	–/–	–/–	–/–	–/–	–/–	960	1079
Ni 660	Rem.	0.10	18.5/19.5	7.0/7.5	1.0/1.5	–/0.5	–/0.10	–/0.02	–/–	–/–	–/–	–/–	–/–	1065	1150
Ni 661	Rem.	1.0	14.5/15.5	7.0/7.5	1.1/1.6	–/1.0	–/0.06	–/0.02	–/–	–/–	–/–	–/–	–/–	1030	1125
Ni–W–Cr alloys															
Ni 670	Rem.	0.10	10.0/13.0	3.0/4.0	2.0/3.0	2.5/4.5	0.40/0.55	–/0.02	15.0/17.0	–/–	–/–	–/–	–/–	970	1105
Ni 671	Rem.	0.10	9.0/11.75	3.35/4.25	2.2/3.1	2.5/4.0	0.30/0.50	–/0.02	11.5/12.75	–/–	–/–	–/–	–/–	970	1095
Ni–P alloys															
Ni 700	Rem.	0.10	–/–	–/–	–/–	–/–	–/0.06	10.0/12.0	–/–	–/–	–/–	–/–	–/–	875	875
Ni 710	Rem.	0.10	13.0/15.0	–/0.10	–/0.02	–/0.2	–/0.06	9.7/10.5	–/–	–/–	–/0.04	–/–	–/–	890	890
Ni 720	Rem.	0.10	24.0/26.0	–/0.10	–/0.02	–/0.2	–/0.06	9.0/11.0	–/–	–/–	–/–	–/–	–/–	880	950

(*Continued*)

Table 5.2 Continued

Code	Composition (% by mass)													Melting temperature (approximate) (°C)	
	Ni min./max.	Co max.	Cr min./max.	Si min./max.	B min./max.	Fe min./max.	C min./max.	P min./max.	W min./max.	Cu min./max.	Mn min./max.	Mo min./max.	Nb min./max.	Solidus	Liquidus
Ni–Mn–Si–Cu alloys															
Ni 800	Rem.	0.10	–/–	6.0/8.0	–/–	–/–	–/0.06	–/0.02	–/–	4.0/5.0	21.5/24.5	–/–	–/–	980	1010
Ni–Cr–B–Si–Cu–Mo–Nb alloys															
Ni 810	Rem.	0.10	7.0/9.0	3.8/4.8	2.75/3.50	–/0.4	–/0.06	–/0.02	–/–	2.0/3.0	–/–	1.5/2.5	1.5/2.5	970	1080
Co–Ni–Si–W alloys															
Co 900	16.0/18.0	Rem.	18.0/20.0	7.5/8.5	0.70/0.90	–/1.0	0.35/0.45	–/0.02	3.5/4.5	–/–	–/–	–/–	–/–	1120	1150

Note: Maximum impurity limits acceptable to all types are (% by mass): Al 0.05, Cd 0.010, Pb 0.025, S 0.02, Se 0.005, Ti 0.05, Zr 0.05; if elements other than those given in this table or this note are found to be present, the amount of these elements shall be determined; the total of other such elements shall not exceed 0.50%.

Table 5.3 International standard ISO/FDIS 17672:2010(E) – Class Pd: palladium-bearing brazing filler metals

Code	Composition (% by mass)						Melting temperature (approximate) (°C)	
	Ag min./max.	Cu min./max.	Pd min./max.	Mn min./max.	Ni min./max.	Co min./max.	Solidus	Liquidus
Pd 287	67.0/69.0	26.0/27.0	4.5/5.5	–/–	–/–	–/–	805	810
Pd 288	94.5/95.5	–/–	4.5/5.5	–/–	–/–	–/–	970	1010
Pd 387	57.0/59.0	31.0/32.0	9.5/10.5	–/–	–/–	–/–	825	850
Pd 388	67.0/68.0	22.0/23.0	9.5/10.5	–/–	–/–	–/–	830	860
Pd 481	64.5/65.5	19.5/20.5	14.5/15.5	–/–	–/–	–/–	850	900
Pd 483	–/–	81.5/82.5	17.5/18.5	–/–	–/–	–/–	1080	1090
Pd 484	51.5/52.5	27.5/28.5	19.5/20.5	–/–	–/–	–/–	875	900
Pd 485	74.5/75.5	–/–	19.5/20.5	4.5/5.5	–/–	–/–	1000	1120
Pd 496	–/–	–/–	20.5/21.5	30.5/31.5	47.0/49.0	–/–	1120	1120
Pd 587	53.0/55.0	20.5/21.5	24.5/25.5	–/–	–/–	–/–	900	950
Pd 597	73.0/75.0	–/–	32,/–33.5	2.5/3.5	–/–	–/–	1180	1200
Pd 647	–/–	–/–	59.5/60.5	–/–	39.5/40.5	–/–	1235	1235
Pd 657	–/–	–/–	64.0/66.0	–/–	–/0.06	34.0/36.0	1235	1252

Notes: For Pd 287, Pd 288, Pd 387, Pd 388, Pd 481, Pd 483, Pd 484, Pd 587 and Pd 657, maximum impurity limits applicable are (% by mass): Al 0.0010, P 0.008, Ti 0.002, Zr 0.002; total of all impurities = 0.15.

For Pd 485 and Pd 597, maximum impurity limits are (% by mass): Al 0.010, Ti 0.01, Zr 0.01; total of all impurities = 0.30.

Table 5.4 International standard ISO/FDIS 17672:2010(E) – Class Au: gold-bearing brazing filler metals

Code	Composition (% by mass)						Melting temperature (approximate) (°C)	
	Au min./max.	Cu min./max.	Ni min./max.	Pd min./max.	Ag min./max.	Others min./max.	Solidus	Liquidus
Au 295	29.5/30.5	69.5/70.5	–/–	–/–	–/–	–/–	995	1020
Au 300	29.5/30.5	–/–	35.5/36.5	33.5/34.5	–/–	–/–	1135	1165
Au 351	34.5/35.5	61.0/63.0	2.5/3.5	–/–	–/–	–/–	975	1030
Au 354	34.5/35.5	64.5/65.5	–/–	–/–	–/–	–/–	990	1010
Au 375	37.0/38.0	62.0/63.0	–/–	–/–	–/–	–/–	980	1000
Au 503	49.5/50.5	49.5/50.5	–/–	–/–	–/–	–/–	955	970
Au 507	49.5/50.5	–/–	24.5/25.5	24.0/26.0	–/–	Co –/0.06	1100	1120
Au 625	62.0/63.0	37.0/38.0	–/–	–/–	–/–	–/–	930	940
Au 700	69.5/70.5	–/–	21.5/22.5	7.5/8.5	–/–	–/–	1005	1045
Au 752	74.5/75.5	–/–	24.5/25.5	–/–	–/–	–/–	950	990
Au 755	74.5/75.5	11.5/13.5	–/–	–/–	12.0/13.0	–/–	880	895
Au 800	79.5/80.5	19.5/20.5	–/–	–/–	–/–	–/–	890	890
Au 801	79.5/80.5	18.5/19.5	–/–	–/–	–/–	Fe 0.5/1.5	905	910
Au 827	81.5/82.5	–/–	17.5/18.5	–/–	–/–	–/–	950	950
Au 927	91.0/93.0	–/–	–/–	7.0/9.0	–/–	–/–	1200	1240

Note: Maximum impurity limits applicable to all types are (% by mass): Al 0.0010, Cd 0.010, P 0.008, Pb 0.025, Ti 0.002, Zr 0.002; total of all impurities = 0.15.

due to relatively low joint strength at high temperatures and the high cost of the noble metal components.

5.3.3 New BFM development

Phosphorus-based BFMs with compositions modified from those previously known have appeared on the world market in the last 10 years. These alloys contain phosphorus as the major metalloid added to (Ni,Fe,Cr)-base (Sjodin *et al.*, 2005; Rassmus and Sjodin, 2006; Rangaswamy and Fortuna, 2008; Hartmann and Nuetzel, 2009).

As is well known, phosphorus decreases alloys' melting temperature more strongly than boron and silicon. For example, in Ni–B alloys the liquidus temperature, T_{liq}, drops by 21°C with the addition of 1 at.% B, whereas in Ni–P alloys it decreases by 31°C/at.% P. The partial or complete replacement of boron and silicon with phosphorus permits the amount of iron, chromium and molybdenum to be increased without increasing the melting temperature above 1150–1200°C. As a consequence, these alloys may dissolve relatively larger chromium concentrations, thus providing good corrosion resistance. They also have two economic advantages: (a) they have a lower cost because it is possible to replace part of the nickel with iron without increasing melting characteristics, and (b) they can be employed in low-cost belt furnaces with protective atmospheres instead of vacuum ones due to their low melting temperature. Some phosphorus-containing compositions can also be cast as amorphous ductile foil. These are produced by Metglas Inc. (Sexton and DeCristofaro, 1979) and Vacuumschmeltze (Hartmann and Nuetzel, 2009).

Recently, there has been growing interest in using Ni-base alloys with either Ge or Zr/Hf/Cr additions, which are also eutectic in nature (Zheng, 1990; Khorunov and Peshcherin, 1998; Khorunov *et al.*, 2001; Lugscheider *et al.*, 2001; Kvasnitski *et al.*, 2002; Dinkel *et al.*, 2008; Cretegny *et al.*, 2009). These alloys have been mostly developed to repair turbine blades and to withstand high service temperatures.

Ni-(20–23) Ge alloys were tested as new BFMs for diffusion brazing of single-crystal superalloys (Dinkel *et al.*, 2008). It was proved that germanium can be used as a new melting point depressant in these materials. It also assists in γ/γ'-phase microstructure formation throughout the whole superalloy braze. More impressively, the original single-crystal structure orientation is preserved, and it is the same in both parts and the joint. However, the brazing time needed for completing this process is substantially longer than that required when using boron-containing BFMs.

Ni–Zr/Hf alloys have a rather high melting temperature and their joints have high strength and good corrosion and oxidation resistance. Their eutectic structure consists of an Ni-based matrix phase and various intermetallic phases formed from Ni, Zr, Hf and Cr. These intermetallic phases have a lower hardness and are much more ductile in comparison with intermetallic phases formed from Ni–B–Si-based BFMs (Khorunov *et al*., 2001). The microhardness of phases varies slightly across the joint – another good feature of these alloys. So far, Ni–Zr/Hf alloys with up to 25 wt.% Hf have been used for repairing vanes and blades in which wide brazing gaps need to be filled (Cretegny *et al.*, 2009).

5.3.4 BFM forms and placement

Amorphous BFM: it is not an exaggeration to attribute the appearance of rapidly solidified amorphous ductile BFMs (Sexton and DeCristofaro, 1979) as the major step forward in the development of high-temperature brazing technology over the last 40 years (Rabinkin and Liebermann, 1993; Rabinkin, 2004). Indeed, most BFMs with metalloids are inherently brittle in the conventional crystalline form and cannot be produced in continuous shapes such as foil, wire, etc. Therefore, they were available only as powders or their derivatives. Starting from Sexton's and DeCristofaro's discovery of the rapid solidification (RS) potential in the manufacturing of flexible amorphous brazing foil, this new class of filler metals has found numerous applications during the last three decades. These amorphous filler metals are known as Metglas® Brazing Foil (MBF) alloys. Today, they are essential components of the AWS American Standard BFM classification. Their total annual market sales come close to a quarter of total sales of all Ni/B/Si-based alloys.

The most important advantage of RS amorphous and microcrystalline filler metal (FM) alloys is their flexibility and ductility. Because a ductile amorphous alloy brazing foil such as MBF may be used as a preplaced preform, there is no need for large brazing gaps, as required with pastes, to achieve a complete filling of the braze cross-section. MBF has a particular advantage over powder and polymer-bonded tape forms because of its superior flow characteristics (DeCristofaro and Bose, 1986).

Gas-atomized powder has a very large total surface area with consequently large amounts of surface oxides. These oxides prevent, to a certain degree, the fusion of individual powder particles into a uniform liquid pool. MBF flows more freely upon melting than any powder form. A smaller clearance associated with the amorphous foils also promotes improved retention of BM properties because of curtailed BM erosion through the use of a smaller volume of FM in the MBF form. It also results in better joint microstructures with a smaller amount of detrimental eutectic phases because of narrower brazing gaps. It is easy to automate assembly of parts and MBF preforms, and consequently it was introduced into industrial production without difficulty. From the economic point of view, only about one-half or even one-third of the filler metal weight is needed per joint square area when using MBF as a filler metal vs. that of a sprayed or pasted powder.

Powder placement via 'silk screening' and 'rolling' is a recent attempt by industry to respond to challenges posed by MBF amorphous foil, which may be used as preforms for covering large surfaces. New compositions were specifically developed for use in mass production of products such as heat exchangers (Sjodin *et al.*, 2005; Rassmus and Sjodin, 2006). In the silk screening process a powder mixed with a binder is delivered through multiple openings of the mask screen and is deposited on corrugated heat exchanger (HE) plate surfaces. The plates, with the powder attached to some local spots, are then placed on each other and brazed together. The multiple joints are formed in locations where there was contact between powder-covered spots. The rolling technique is also used to deposit powder mixed with binder on protruding local points that should be brazed. Both processes can deliver an increased amount of the powder into brazing gaps. This leads to an increase of the cross-section of each individual joint. As a result, the internal pressure that HEs can withstand can be increased. The rolling application also permits automation of powder deposition (Schmoor *et al.*, 2006; Schnee *et al.*, 2004; Koch and Schmoor, 1998). Both processes have been used on an industrial production scale but suffer from all the inherent deficiencies associated with powders.

5.3.5 The boron controversy

From the first day Ni–B–Si alloys were proposed for brazing, there were strong objections because at that time boron was considered as a harmful impurity, by analogy with its role in steels. Since then the views on boron have changed substantially. Still, some application specifications exist in which the use of boron is prohibited. Meanwhile, as mentioned in Section 5.3.1, boron has a very positive effect on BFM properties: the presence of boron decreases the liquid metal surface tension and, therefore, strongly increases wetting and flow. It increases the BM dissolution rate. More importantly, boron is uniquely responsible for the amorphability of Ni–B–Si alloys. Alloys should contain at least 1.4 wt.% B in addition to Si to preserve the liquid atomic structure upon rapid solidification.

In addition, boron rapidly diffuses in the solid BM, thus decreasing the amount of boron left in the joint which can form chromium borides in the joint eutectic. The rapid boron depletion assists joint isothermal solidification. Therefore, if a BFM contains silicon in addition to boron, the boron-depleted joints may have a ductile and strong (Ni,Cr,Si) solid solution matrix phase after a longer brazing time (Rabinkin, 2006). Such a structure may have similar strength to the virgin base metal. If Ni-based BFM contains no silicon, as with BNi-9 (Ni–15Cr–3.75B) in powder or foil (MBF-80) forms, then the ultimate joining goal may be achieved; namely, parts are joined with minimal or no traces of brazing. On the other hand, boron is practically insoluble intergranularly in highly alloyed BM steels with the f.c.c. γ-phase structure. It segregates at BM grain boundaries, thus making the BM prone to brittle failure. It also depletes BM grain boundary interface zones of Cr and concentrates Cr in Cr_xB_y, thus affecting BM corrosion and oxidation resistance. These negative effects can be corrected via optimization of heat treatment and employing compositions with only 1.3–1.5 wt.% of boron (Rabinkin, 2006).

Meanwhile, it is worth mentioning that 300B-series austenitic stainless steels with very high, >2 wt.%, B content on average are used safely for pipes in atomic reactors, where high neutron absorption of boron is needed (AST A 887–89). This is a good example of an application in which the huge presence of chromium borides creates no concerns if properly addressed.

5.4 High-temperature base metal brazing

Joint manufacturing, structure and brazement properties will be considered here in a sequence of increasing number of base metal elemental components and metallurgical structural complexity: namely, ferritic 400 AISI series Fe–Cr steels, austenitic 300 AISI series stainless steels, then multi-component superalloys, and, finally, single-crystal superalloys. The joint mechanical properties/structure relationships will be described, with a primary emphasis on the brazing cycle thermal conditions. Transient Liquid Phase (TLP) processing and loading of brazing assemblies will be presented at the end of this section together with joint loading during brazing as effective extensions to brazing conditions yielding joints with superior quality.

5.4.1 AISI 400-series Fe–Cr-based ferritic stainless steels

AISI 409 joints brazed with BNi-2, -3 (MBF-20, -30)

The mechanical strength and microstructure of joints made using BNi-2 and BNi-3 FM in foil form and as pastes for comparison were studied in Rabinkin and Murzyn (1987) and Rabinkin (1989).

Standard AWS C3.2M specimens and procedures were utilized for measuring joint mechanical properties. Brazing was performed at 1055°C for 10 min using five different furnace atmospheres: 100% hydrogen, 50:50 H_2:N_2, 25:75 H_2:N_2, 50:50 H_2:Ar, and 100% Ar. Brazability and wetting of all samples were best when the atmospheres had a low (about −40°C) freezing point (FP). The joint strength from these experiments exceeded the 409 steel yield limit, $\sigma_{0.2}$, namely, 263 MPa (38.2 ksi) for BNi-2 (MBF-20) and 221 MPa (32 ksi) for BNi-3 (MBF-30). The joints made with powder pastes with similar BNi-2 and BNi-3 compositions had lower values of shear strengths, 221 and 197 MPa, respectively. The MBF-20 and -30 joint qualities and strength were close to those of joints made under vacuum conditions. The most important practical consequence demonstrated by these results is the possibility of using BNi-series (MBF) alloys in foil form in highly productive and very economical (relative to vacuum furnaces) belt and semi-continuous batch furnaces.

The microstructure of all MBF brazements, both inside and adjacent to the joint interfaces, was unaffected by the atmospheric composition. The joint

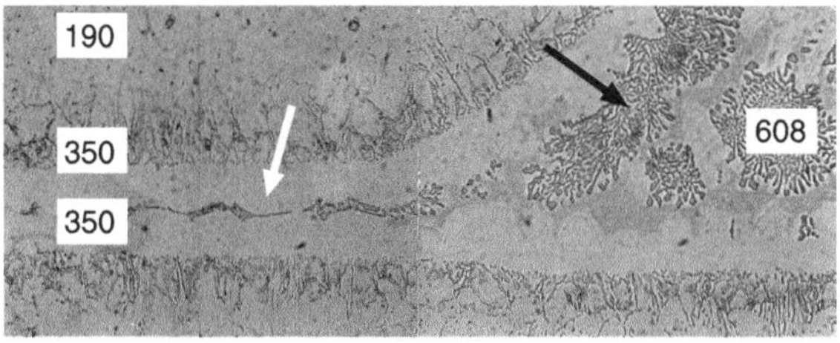

5.1 Micrograph of a 409/BNi-2 (MBF-20)/409 joint brazed at 1055°C for 10 min under 50:50 N_2/H_2 atmosphere having −40°C FP. Numbers indicate Vickers microhardness, kg/mm^2, of some joint areas, whereas the black arrow shows the fine eutectic mixture and the white arrow the solid solution matrix phase.

microstructure was comprised mostly of (Ni,Cr,Fe) solid solution phase (Fig. 5.1, white arrow) and a fine eutectic mixture of $(Ni,Cr)_{1-x}S_x$ and Cr_3B crystals (Fig. 5.1, black arrow). Boron strongly diffused into the base metal, forming a perlite-like structure, in spite of the short brazing time. Figure 5.1 also demonstrates the importance of keeping the fillet volume to a minimum, because a much larger volume of brittle eutectic forms there. Brazements with the same width as the BNi (MBF) foil preforms but made of powder pastes contained substantially more of the Cr_3B phase. This was probably the reason for their lower strength.

AISI 430 and 436 joints brazed with BNi-2 (MBF-20)

Brazed plates/sinusoidal fin specimens were extensively studied in Rabinkin (2000). Their shape design was similar to that of an industrial turbine heat exchanger. They were made of UNS436 stainless steel sheets of 100 μm and 50 μm thickness, respectively. BNi-2 (MBF-20) foil sheets, from which flat preforms were made, had a thickness of 25 μm, 37 μm, and 50 μm. These thicknesses were selected to find the optimal combination of base and filler metals. A load was placed on the top of each sample in the form of a metal or graphite block. The samples were placed in a vacuum furnace with the loading aligned in the direction of gravitational pull and were brazed at 1060°C for 30 min.

Metallographic observations showed that the joint thicknesses in the middle portion of all brazes were the same regardless of the thickness of the virgin amorphous foil, even when comparing samples manufactured using 25 μm and 50 μm foils, as Fig. 5.2(a) and (b) demonstrate. This effect was observed because the gaps between the parts were not constrained during brazing. Indeed, the excess liquid BNi-2 (MBF-20) alloy was partially expelled from the capillary gaps into the fillet areas upon melting. The liquid flow out of the gap continued until the surface tension forces at all gap surfaces became equal to the total load applied to the specimen. The excessive molten MBF-20 metal, particularly for the 50 μm thick FM foil, flowed out of the initial gaps, forming large fillets, and partially

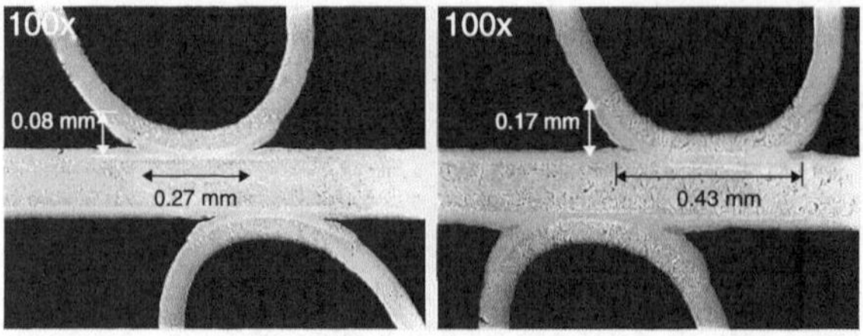

(a) (b)

5.2 Microstructure of two 436 stainless steel plate/fin samples brazed using 25 μm thick (a) and 50 μm thick (b) BNi-2 (MBF-20 foil). Note that, while inside the brazes the joint thickness is the same, the cross-section of the 50 μm joint is larger than that of the 25 μm joint. The former is wider and the fillet height is higher. Therefore, the 50 μm joint is substantially stronger.

climbed up the vertical walls of the fins. The initially thicker filler metal formed larger joint fillets with larger joint cross-sections, as depicted in Fig. 5.2(b). These larger fillets had advantageous shapes, without the narrow cavity-like crystallization shrinkage pattern seen in Fig. 5.2(a). As with the 409/BNi-2 (MBF-20) and BNi-3 (MBF-30) joints, the joint integrity was high. The total mechanical performance of units brazed with 50 μm thick foil was determined by the strength of the base metal fins, which was unaffected by brazing.

The microstructure and mechanical properties of 430 and 436 steel joints brazed using BNi (MBF-20) foils were also studied in Myazawa *et al.* (2000). They observed a chain of brittle chromium borides segregated in the central zone of the joint (Ni,Si)-solid solution matrix phase. An extensive network of elongated borides was also observed in the areas adjoining the braze. Some chromium carbides were also present, together with Cr_xB_y crystals. Joint failure observed in this work occurred in the brittle mode in the joint central area, which contained a substantial amount of the brittle eutectic phases because of the short brazing time.

It was shown that the joint shear strength is the same (about 200 MPa) regardless of the base metal grade if the brazing temperature is in the range 1050–1150°C and the brazing time is kept to about 10 min. The results are in some ways analogous to those obtained for 409 and 439 joints in Rabinkin (2000).

Fecralloy/BNi-5a (MBF-50)/Fecralloy joints

Fecralloy, with its Fe–20Cr–5Al–rare earth oxides composition, is a ferritic-class steel with extremely high oxidation resistance. It is used at service temperatures as high as 1200°C in highly corrosive environments. It has a single α-phase Fe–Cr-solid solution matrix structure containing completely dissolved aluminum at 5 wt.%, which is added to increase protection against corrosion. This α phase is similar to the b.c.c. α-Fe phase, and it has a crystal lattice, which is less densely packed than the γ phase (Villars *et al.*, 1995a). The diffusion rate of boron atoms in the α phase is higher than in austenitic stainless steels (Wang *et al.*, 1995). The presence of aluminum, with a much larger atomic radius than Fe, Ni and Cr, makes the Fecralloy lattice even less dense than pure iron. Therefore, the α phase can easily accommodate chromium boride crystals when they precipitate within the grains. This is the opposite of what happens in brazed stainless steels with a f.c.c. γ-phase structure, such as 300-series stainless steels, for example (see Section 5.4.2). It also makes the α phase receptive to fast intergranular boron diffusion. These specific circumstances seem to determine the formation of chromium borides as separate crystals, very often within BM grains.

Fecralloy joints with BNi-5a (MBF-50) were produced and studied in Leone *et al.* (2006). The sample used in this study was made of Fecralloy 50 μm thick corrugated/flat foils produced by Sandvik and BNi-5a (MBF-50) 25 μm thick foil, used as the BM and BFM, respectively. The general view of a cross-section is shown in Fig. 5.3(a). This sample was made with the materials and dimensions of the core of a standard metallic catalyst support used in automotive exhaust pipes. It was manufactured using the following braze cycle in a vacuum of about 1.33×10^{-5} Pa (1×10^{-5} torr):

(1) Heating to 950°C and holding for 15 min;
(2) Heating to 1175°C and holding for 15 min;
(3) Cooling in furnace under vacuum to 200°C with the power off;
(4) Final cooling to 25°C in a nitrogen atmosphere.

The major characteristic of this joint is a small difference in thickness of the BM and BFM. This led to the involvement of the total BM cross-section in the brazing process. The interaction with the BFM is particularly strong in the vicinity of and opposite to the fillets, where the liquid BFM volume is thicker than the BM itself. Therefore, a large part of the BM, up to 80%, was practically dissolved. The general appearance of the joint cross-section is shown in Fig. 5.3(a), while Fig. 5.3(b) gives a partial view of the joint area at high magnification. This structure is remarkably different from that of the 304LN/BNi-5b (MBF-51) joint. First of all, the joint contains the eutectic even in the narrowest joint segment. Chromium borides have segregated as large, separate, elongated, faceted crystals (Fig. 5.3(a)), and as small, more or less round-shaped particles. The latter formed a very regular 'chain' of crystals segregated within the braze in close proximity to the joint interface (Fig. 5.3(b,iii)). No 'decoration' of BM grain boundaries by borides is seen even in the interface zone. The dissolution of the BM is very intense and uneven, resulting in the formation of a 'wavy' interface (Fig. 5.3(b,ii and iv)). In some locations the liquid succeeded in penetrating, dissolving the BM and forming narrow channels all the way to the external surface (Fig. 5.3(b,i)). The metallurgical reaction involved the separation of chromium together with boron, accompanied by the formation of chromium boride intermetallic crystals. At the same time, part of the nickel, chromium, silicon and iron diffused from the BM, forming a solid solution matrix phase of the eutectic

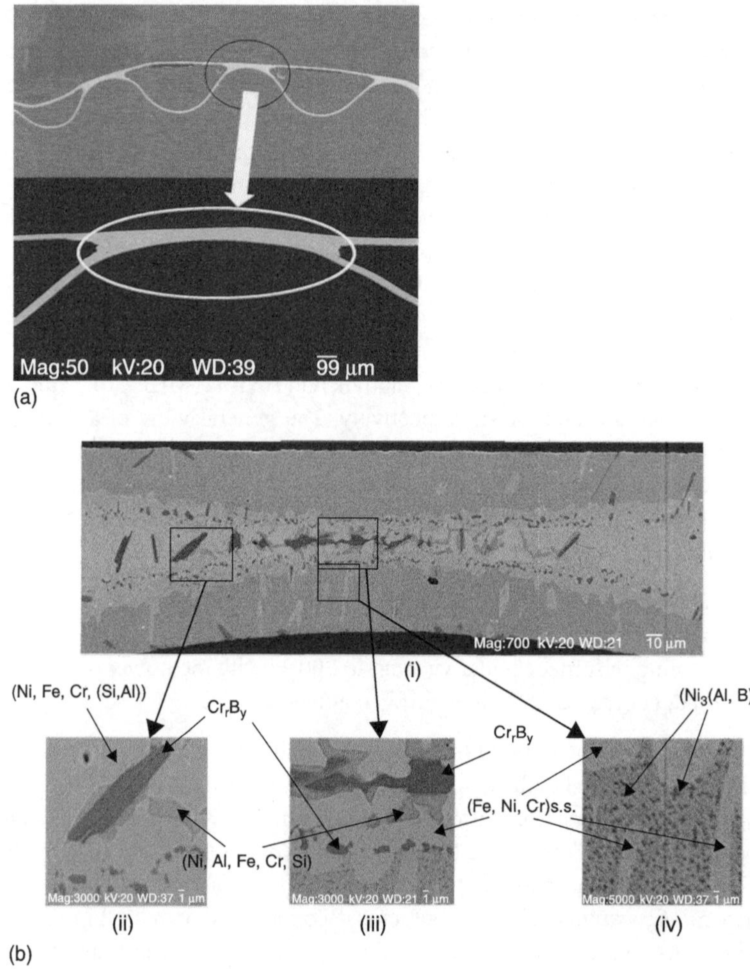

5.3 Fecralloy/BNi-5a (MBF-50) joint. (a) General view of the brazed parts; (b) Details of joint microstructure under different magnifications.

that encompasses the entire braze cross-section. Some of the nickel and most of the silicon crystallized in low-melting areas attached to large Cr_xB_y crystals, which were formed first as the prime eutectic component with the highest melting temperature (Fig. 5.3(b,iii)). No well-faceted intermetallic silicide crystals were observed.

The most remarkable microstructural feature of Fecralloy joints is the presence of small 'sub-micron-sized' $Ni_x Al_y$ cuboidal precipitates formed in all BM volumes in contact with BNi-5a (MBF-50) (Fig. 5.3(b,iv)). These particles consist only of nickel and aluminum. They formed arrays, rows that are more or less parallel to the profile of the joint interface. Their formation is a result of the diffusion of nickel atoms within the BFM accompanied by the simultaneous reaction of nickel with aluminum from the BM (Fe,Cr,Al) β-solid solution phase. This reaction will most probably yield an intermetallic compound having NiAl or Ni_3Al formula. Only these compounds can be formed at 1150°C, i.e. close to the brazing temperature, according to two Al–Ni–Fe and Al–Ni–Cr ternary paternal phase diagrams. Both components have a large enthalpy of formation, 118.5 and 153.2 J/mole, respectively (Kubaschewski and Alcock, 1978).

The size of the particles follows the diffusion profile of the nickel distribution: the larger particles are in close proximity to the interface and vice versa. In general, this new type of microstructure is similar to a typical superalloy structure of γ' precipitation-hardened alloys with one principal distinction: here the matrix has a b.c.c. crystal structure, whereas the vast majority of the superalloys have a f.c.c. Ni/Co-matrix phase in which γ'-Ni_3Al cuboidal precipitates segregate, providing long-term high-temperature material strength.

It is worth emphasizing the importance of Ni_xAl_y precipitation for Fecralloy BNi-5a/MBF-50 joint performance. In spite of the very extensive erosion of the BM, the newly formed brazements are, in fact, a strong precipitation-hardened material. Shear tests of brazed parts have shown their remarkable strength. This strength well exceeds the load that the parts encounter during many years of service in the extremely corrosive environment of engine exhaust gases. The joints have a very long service time in corrosive car exhaust gas environments, as was proved during industrial utilization of real brazed engine parts.

5.4.2 AISI 300-series austenitic stainless steels

AISI 304/BNi-5b (MBF-51) joints

Fin/plate joints manufactured from 304LN stainless steel and BNi-5b (MBF-51) at 1172°C for 10 minutes were studied by Leone *et al.* (2006). A detailed microstructural analysis was performed in this work. It was found that excellent wetting and low contact angle ($\theta \rightarrow 0°$) are characteristics of this BM/BFM interaction. All internal joint regions are free from porosity (Fig. 5.4). The entire structure can be divided morphologically into four segments: (I) the main central joint segment with a more or less even and minute thickness; (II) the BM areas adjoining the interface; (III) the remainder of the BM unaffected by brazing; and (IV) the fillets, with a much larger thickness that increases outwardly in both directions from the center of the joint. Measurements showed that the degree of BM erosion is not very large. Its magnitude varies from 5 to 10 µm in the central segment of the joint and up to 10–20 µm in the areas adjoining the fillets. The latter have a much larger liquid FM volume relative to the surface area of contacting BM than that in the main joint section. Therefore, the fillets have dissolved a larger amount of the BM without even being saturated. They also contain larger fractions of the initial amounts of boron and silicon.

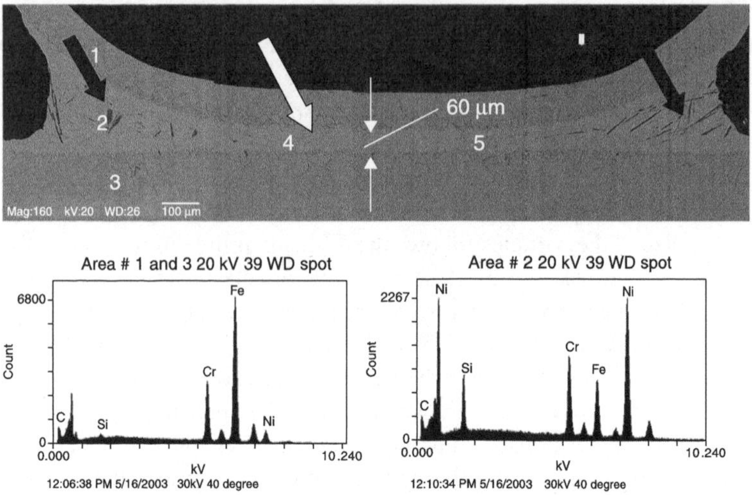

5.4 Total view of the microstructure and EDAX X-ray spectra for a 304LN/BNi-5b (MBF-51) U-shaped fin/plate joint. Spectra are taken from the spots marked by corresponding numbers.

The only phases observed in the longest internal joint regions, including the interface zone, are the (Ni,Cr,Si,Fe) solid solution matrix phase and $Cr_x B_y$ borides, as seen in Fig. 5.4. The latter are segregated mostly within the BM area adjoining the interface. The joint matrix phase has an even distribution of Ni, Si and Cr. The iron distribution decreases smoothly from the BM/joint interface to the middle of the joint. Interestingly, no separate nickel silicide particles are observed in either the central joint segment or the BM. The same phases were also observed in 316L SS/BNi-5b (MBF-51) (Rabinkin *et al.*, 1998).

It was found that the joint interface very sharply separates the areas formed by the BFM crystallization and the bulk of the BM (Fig. 5.5(b)). The microstructure of the BM parts is not too different across the joint circumference, even in proximity to the fillet. It consists of three consecutive morphological zones (Fig. 5.5(b)) consecutively located above the center of the joint. The first 5 to 8 μm-wide zone, adjoining the joint interface, contains fine, narrow chromium boride platelets formed within the original grains (Fig. 5.5(b)). This zone has the appearance of the perlite structure in carbon steels. The second 50 to 80 μm-wide zone contains chromium borides segregated as separate, narrow, island-like particles at the BM grain boundaries (Fig. 5.5(c), arrows). The amount of boron is substantial for segregation even at twin boundaries (Fig. 5.5(d), arrows) but no segregation in the bulk of the grains is observed. This pattern is similar to that in many other cases in which stainless steels were brazed with BNi-series alloys containing boron (Rabinkin *et al.*, 1998).

In the second zone, close to the interface, the borides formed complete shells covering most of the grain boundary surfaces (Fig. 5.5(c)). Because boride

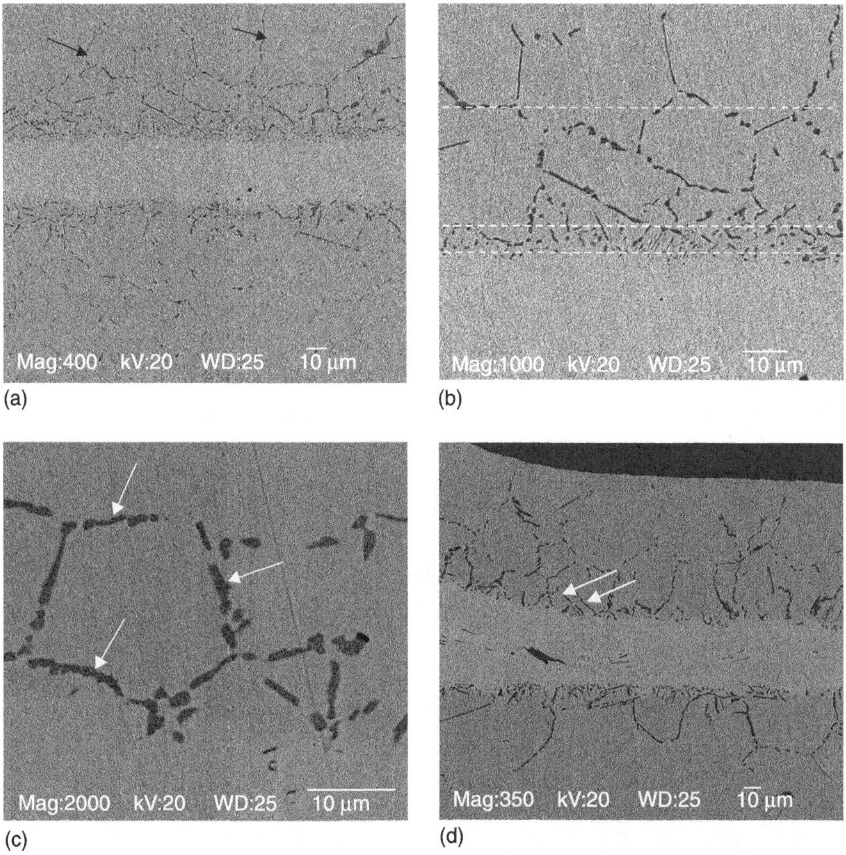

5.5 304LN/BNi-5b (MBF-51) fin/plate joint microstructure: (a) general appearance; (b, c) the interface area under high magnification. Two types of BM grains are clearly seen: small grains in the narrow zone immediately adjacent to the joint enclosed between two broken lines (b) and large grains in a wider zone decorated by continuous segregated chromium borides (arrows) (c). (d) Chromium boride segregations at twin boundaries (arrows).

segregation in 300-series stainless steels occurs only at the grain and the twin boundaries (Rabinkin *et al.*, 1998; Heikinheimo *et al.*, 2001), their width does not exceed 2–3 µm. Further from the interface, the borides are segregated only at some parts of the grain surface (Fig. 5.5(a), arrows). Starting at about 80 µm distance, they are not observed at all. Therefore, from the third (Fig 5.5(b), upper) zone outward, the BM is unaffected by interaction with the FM.

The low sensitivity of the BM microstructure to the joint thickness is due to the specifics of the diffusion processes during the rather short time of the brazing cycle. At the beginning of the metallurgical interaction between the liquid BFM and the solid BM, substantial gradients of elemental concentrations across the interface were created for boron, silicon, iron and nickel. The difference in the chromium concentrations in both interacting alloys is not very large. At the same time, the activity of boron in the BFM is lower due to the effect of the silicon. Therefore, boron has a high driving force to diffuse into the 304LN steel and form chromium borides at the grain boundaries. In addition, the rate of boron diffusion is two orders of magnitude higher relative to that of silicon, and significantly greater than nickel and iron (Karlson and Norden, 1988). The mutual diffusion mass-transfers of other elements are rather small. So, boron diffuses very fast and evenly across the interface into the solid stainless steel, lowering its concentration in the liquid FM and simultaneously forming chromium borides in 304LN.

The degree of boron depletion is inversely proportional to the volume (thickness) of the liquid in the gap. The 10 min brazing time was not sufficient to substantially change the initial elemental redistribution of Ni, Cr, Fe and Si in the 50 to 70 μm thick joint areas. Only boron diffused into the BM to a degree at which a single (Ni,Cr,Si,Fe)-solid solution phase, depleted in boron, was formed upon crystallization. It is evident that the joints are completely free of secondary segregated phases in areas of the gaps where the width did not exceed 60 to 80 μm (Fig. 5.4). The joints with a single-phase (Ni,Cr,Fe,Si)-solid solution microstructure possessed high strength and ductility, as shown in Rabinkin *et al.* (1998).

In regions where the joint width exceeded 70 to 80 μm, the chromium borides segregated upon cooling within the liquid phase. They have elongated platelet-like forms characteristic of solidified eutectic mixtures in which one of the components is an intermetallic phase (Fig. 5.4). The fillet areas of 200 to 250 μm width have a well-developed eutectic structure in which silicon and boron are both still present in large quantities (Fig. 5.4). This eutectic consists of the (Ni,Cr,Si,Fe)-solid solution matrix phase, segregated chromium borides and some regions enriched in silicon and nickel. They are intermixed with silicon-enriched regions. The total microstructure is similar to that of the core in cast ingots in which crystallization is accompanied by liquation. According to Karlson *et al.* (1988), boron segregation at grain boundaries does not cause substantial embrittlement. On the other hand, large plate- and rod-like boride crystals present in fillets, particularly those piercing the fillet surfaces, may play a substantially negative role as potential sources of crack nucleation (Rabinkin *et al.*, 1998).

AISI 316L/BNi-5b (MBF-51) joints

A comprehensive study of 316L plate/plate joint mechanical properties, corrosion and brazement microstructure via brazing cycle optimization was carried out using BNi-5b (MBF-51) 25 and 50 μm thick foil preforms and various brazing cycles (Rabinkin *et al.*, 1998). Brazing was carried out at a relatively high temperature, about 1175°C, because BNi-5b (MBF-51) contains low boron and high chromium concentrations.

Interaction of 316L austenitic steel with BNi-5b (MBF-51) yields brazements in which a (Ni,Fe,Cr,Si) solid solution phase constitutes a large part of the joint area microstructure. The joint also contains well-faceted chromium boride crystals. These crystals are absent if the amount of filler metal per joint area is small, i.e. for a thin foil, or, when the foil is thick, after long-term brazing or brazing combined with post-brazing annealing is applied (Fig. 5.6(a), (c)). This is due to the practically complete diffusion of boron into the base metal. On the other hand, these brazements are characterized by retention of almost all the silicon within the joint area as a component of the (Ni,Fe,Cr) solid solution phase (Fig. 5.6(c)). It is worth emphasizing one more time that boron has a very low solubility in the close-packed γ-phase lattice. When boron diffuses into an austenitic stainless steel BM during brazing, it segregates at the BM grain boundaries. It is also believed that 'unlike many other segregants it does not cause grain boundary embrittlement in steel' (Karlson and Norden, 1988). There is no intragranular precipitation of Cr_xB_y crystals at relatively low boron

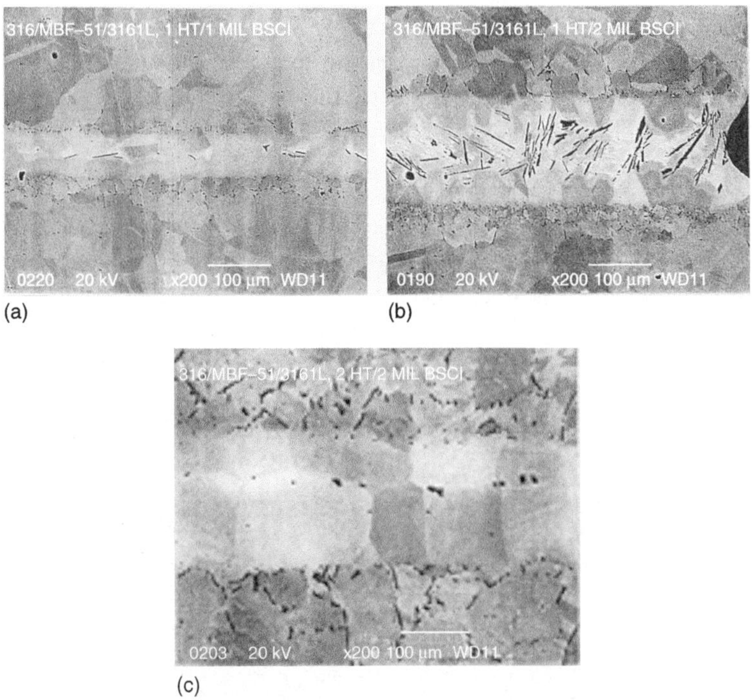

5.6 SEM micrographs of 316/BNi-5b (MBF-51)/316 joints brazed using 25 μm (a) and 50 μm (b) thick BNi-5b (MBF-51) ribbons and a short brazing cycle; 50 μm foil and a long brazing/annealing cycle were used in the joint shown in (c). Note that a strong ductile single-phase braze is formed when there is a small amount of boron in the joint and/or a long brazing time is used.

concentrations, which is clearly seen in Fig. 5.5(c). Therefore, chromium borides formed in the 316L base metal segregate at grain boundaries in the 30–50 μm deep zone and coagulate into particles with a rounded shape if the high-temperature exposure is sufficiently long. Again, a relatively low, 1.5%, boron concentration in BNi-5b (MBF-51) contributes to a favorable 316L/BNi-5b (MBF-51) joint microstructure and gives good joint ductility and strength. A very strong and ductile microstructure was obtained when brazing time extended well over 1 h. As a result, the joint itself was practically clean of any eutectic phases, as Fig. 5.7 illustrates. It is recommended that the plate/fin and plate/plate profiles should be designed in such a way that large joint areas, and particularly large fillet areas greater than 150 μm, are avoided. Moreover, it was strongly suggested that brazing gap widths should not exceed 50–70 μm. It was also shown that post-brazing annealing of 300 stainless steel/BNi-5b brazements might substantially improve the ductility and the maximum stress-to-failure values (Rabinkin *et al.*, 1998). Such secondary treatment is worth considering for these joints in the future.

The mechanical properties measured included the tensile strength, determined using a standard AWS C3.2M/C3.2:2001 procedure and specimens, and fatigue resistance. It was found that a joint tensile strength greater than 300 MPa (43.7 ksi) was achieved when a thin, 25 μm thick foil and a rather short 30 min brazing time were used. The thicker, 50 μm foil yielded a eutectic microstructure after a short, 30 min brazing time (Fig. 5.6(b)). A longer brazing time resulted in similar microstructure to the 25 μm foil case, and the same level of strength and joint ductility. It should be underlined that in both cases the joint strength was higher than the yield strength of the virgin 316L. Failure of the brazement occurred in the BM area adjoining the fillets, and not in the brazed area. Joint fracture started with the appearance of cracks in the fillet area, where an excessive amount of brittle eutectic phases was found. It is well known that large fillets usually have large chromium boride crystals, which can provide crack initiation sites. Such fillets should be avoided by optimizing the part geometry. These cracks proceeded

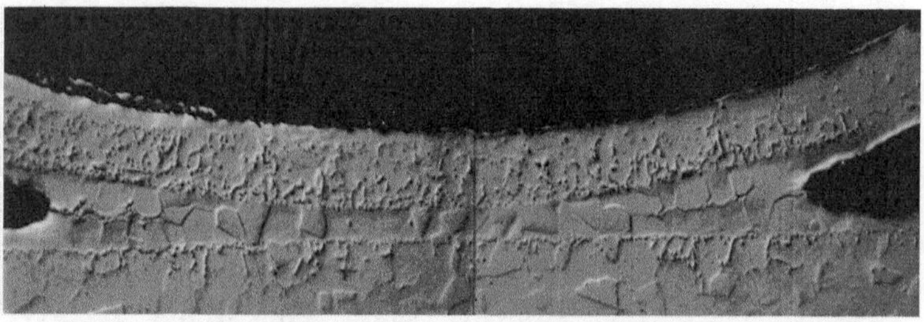

5.7 SEM micrograph of 316/BNi-5b (MBF-51)/316 joints brazed using 50 μm thick MBF-51 ribbons and a long, >60 min, brazing time. A strong ductile single-phase braze is formed.

through the base metal area adjoining the interface where chromium borides were formed as a result of boron diffusion into the base metal. In addition, separate data were obtained on the burst pressure of industrial heat exchangers manufactured using 316L plate/plate geometry and BNi-5b (MBF-51). Here again the pressure could be increased two to three times via proper brazing cycle choice. Today, in the best cases, heat exchangers made of 316L can be pressurized up to 12–14 MPa (120–140 atm) before failure.

5.4.3 Superalloy joints

These base metals are mostly used in critical locations in aerospace and land-based turbines and are required to withstand ultimate temperature and stress conditions over a long service time. Two Ni- and Co-based groups of superalloys are most commonly used in the aircraft and industrial gas turbine industries: Ni-base superalloys are used for both the blades and vanes of modern engines, whereas Co-base superalloys are preferred for vanes. The Ni-based superalloys are γ'-prime strengthened and possess remarkable mechanical properties at high temperatures up to typically 75% of their melting point in °C. Co-based superalloys are solid solution strengthened and have inferior mechanical properties when compared with Ni-based superalloys. To protect these parts from the high operating temperatures of the modern gas turbine, they are generally coated with an oxide-containing thermal barrier layer.

Therefore, it is imperative to choose brazing conditions and BFM compositions in such a way as to preserve the complex microstructures and high performance properties of the superalloys in the aftermath of the brazing operation. The following microstructures should be preserved: namely, a γ'-prime strengthened matrix phase in the Ni-based superalloys case, and a solid solution strengthened phase in Co-based superalloys. It is particularly important to keep the amount of boron in filler metals to a minimum because boron may induce brittleness by causing potential chromium boride segregation in the base metal grain boundaries. The properties of these joints, obtained under a variety of brazing conditions, were measured and their corresponding structures were studied comprehensively and described in detail in a series of papers (Rabinkin, 2000; Heikinheimo and Miglietti, 2000; Heikinheimo *et al.*, 2001; Nishimoto *et al.*, 1995, 1998). The resulting processing optimization conditions, which are described in Section 5.4.3, yield joints that have excellent mechanical properties, equal to or approaching those of the virgin superalloys. It should be noted that the composition of the majority of amorphous BFM chosen for testing in these papers contains a minimal amount (1.4 wt.%) of boron. As mentioned in Section 5.3.5, this amount is at the threshold of the alloy amorphability concentration limit that still allows production of these compositions in the foil form.

Superalloy Inconel 625/BNi-2 (MBF-20) and Inconel 625/BNi-5b (MBF-51) joints

Optimization of the joining procedure for Inconel 625 plate/fin heat exchangers used in aerospace applications was done by Hughes-Treitler Co. with the participation of this author (Rabinkin, 2000). Such plate/fin heat exchangers have a rigid frame, which provides a total unit strength which by far exceeds the service pressure. Therefore, the only developmental goal was to obtain sound plate/fin joints that were ductile and free from voids and cracks.

The Inconel 625 plate/fin heat exchanger joints were manufactured from 1 and 0.2 mm thick BM plates and 125 μm thick fins using 37 μm thick BNi-5b (MBF-51) foil. Powder-based BNi-2 and BNi-5 were also used as possible alternative BFMs. The brazing cycle applied to both kinds of filler metals consisted of slow heating and cooling stages and a long-term (up to 90 min) brazing stage at 1155°C. Such parameters were specifically chosen to stop the brazements from cracking due to high thermal stresses during cooling. The thermal stresses were created because the parts are big and their structural elements have different dimensions. The thermal cycle yielded brazes, which mostly had a single-phase joint microstructure similar to that of the AISI 300-series joints, and no cracks (Fig. 5.8). Only a small volume of eutectic phase mixture was present in the fillets (Fig. 5.8, white arrows) and, sometimes, in the fillet vicinity. Moreover, the observed chromium borides, segregated at the braze interfaces (Fig. 5.8, black arrows), had a favorable round morphology after such a long exposure. Joints made of the powder paste had a very pronounced porosity and poor integrity.

Inconel 738 and 939 joints brazed with BNi-5b (MBF-53)

IN738/MBF-53 joints have rather uniform and very favorable microstructures consisting mostly of the (NiCr)-based solid solution phase, in which small separate $(Cr,W)_xB_y$ borides are embedded (Fig. 5.9) (Heikinheimo and Miglietti, 2000;

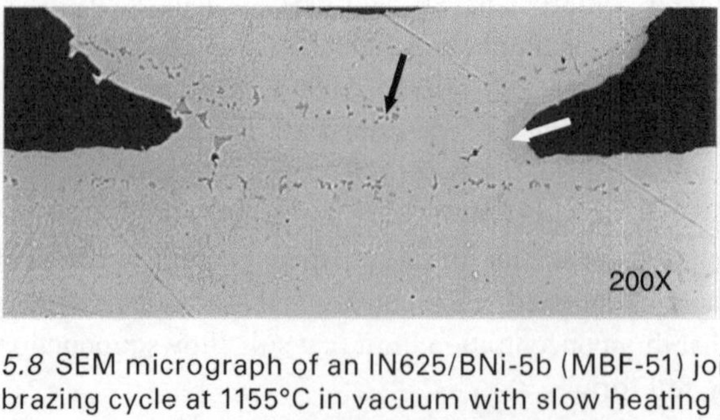

5.8 SEM micrograph of an IN625/BNi-5b (MBF-51) joint after a 90 min brazing cycle at 1155°C in vacuum with slow heating and cooling stages to avoid cracking induced by thermal stresses.

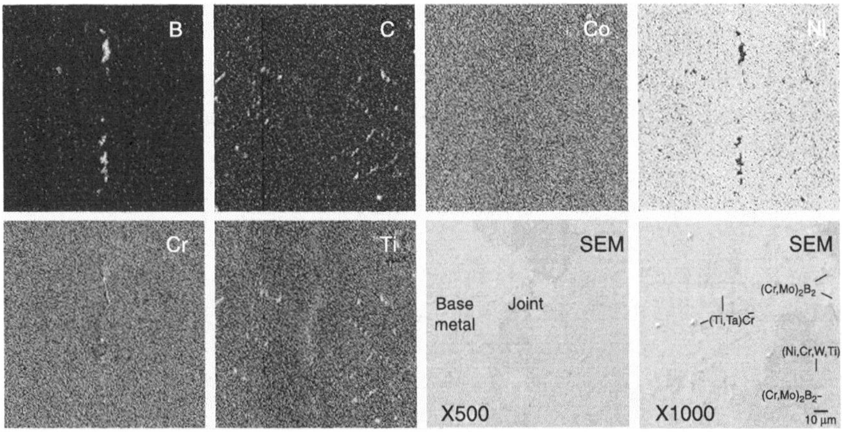

5.9 Microstructure and elemental distribution of IN738/BNi-5b (MBF-53)/IN738 joints brazed at 1170°C for 15 min and annealed at 1070°C for 6 h.

Heikinheimo *et al.*, 2001). This boride phase is composed of all metallic elements from both base and filler metals. Small, separated, and rounded borides, together with intrinsic $(Ti,Ta)_xC_y$ carbides, are segregated at the base metal grain boundaries in a rather narrow region. This region is adjacent to the interface.

IN939/MBF-53 joints were barely visible. Specimens displayed a complete recrystallization pattern through the entire joint area (Fig. 5.10). Isolated occasional binary Cr_xB_y borides with a narrow plate-like structure were observed

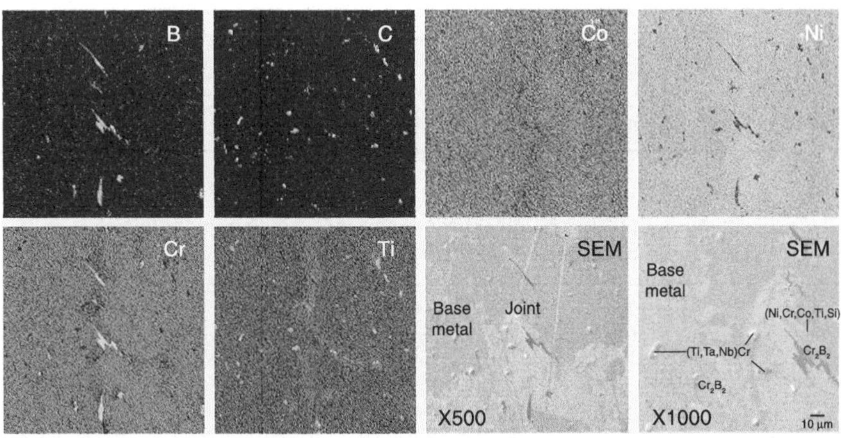

5.10 Microstructure and elemental distribution of IN939/BNi-5b (MBF-53)/IN939 joints brazed at 1170°C for 15 min and annealed at 1070°C for 6 h.

in the center of the joint area. Nickel, chromium and probably silicon were very evenly distributed across the brazement. Very small borides were segregated at grain boundaries in the 60–70 μm wide base metal zone adjacent to the joint. Small complex $(Ti,Ta,Nb)_xC_y$ carbides were present inside the base metal grains as well.

X40 and X45 joints

Joints manufactured from Co-based (X40 and X45) superalloys, together with Ni-based (IN738 and IN939) superalloys, which are commonly used in land-based power turbines, were studied by Heikinheimo and Miglietti (2000) and Heikinheimo *et al.* (2001). BNi-9, BNi-5b and Co-base amorphous alloys were used as the filler metals. Tables 5.5 and 5.6, taken from these two papers, show the base and filler metal compositions studied in these works. Details of the material combinations in the joints, the brazing cycle parameters and the mechanical testing (Table 5.7) of the samples can be found in Heikinheimo *et al.* (2001). Brazing was carried out under a high vacuum, typically 1×10^{-6} mbar, at 1170–1200°C and for various times.

X45/BNi-5b (MBF-53) joints have a narrow (~25 μm) molten zone. They are characterized by the presence of a broken eutectic central line formed mostly by (Ni,Cr,Mo) silicides and $W_x(SiC)_y$ silicocarbides, which are segregated inside the (NiCrWSi)-solid solution matrix phase. Most of the boron diffused into the base

Table 5.5 Composition (wt.%) of superalloys as base metals (BM)

Alloy	Composition (wt.%)														
	C	Co	Cr	Ni	Mo	Al	W	Ta	Nb	Ti	Fe	Si	Mn	Zr	B
X45	0.18	Bal.	26.1	10.6	0.38	0.01	6.7	–	–	–	0.93	0.95	0.37	–	–
X40	0.5	Bal.	25.5	10.5	–	–	7.5	–	–	–	–	0.75	0.75	–	–
IN738	0.17	8.5	16	Bal.	1.7	3.4	2.6	1.7	0.9	3.4	–	–	–	0.1	0.01
IN939	0.15	19.0	22.5	Bal.	–	1.9	2.0	1.4	–	3.7	–	<0.2	<0.2	0.1	0.01

Table 5.6 Compositions (wt.%), thicknesses and melting characteristics of MBF-series amorphous BFMs used in Heikinheimo *et al.* (2001)

Alloy	Thickness (μm)	Composition (wt.%)							T_{sol} (°C)	T_{liq} (°C)
		Ni	Co	Cr	Si	B	W	Mo		
MBF-100	37	–	Bal.	21	1.6	2.15	4.5	–	1136	1163
MBF-101	46	–	Bal.	20	4.5	1.70	4.5		1091	1143
BNi-5b (MBF-53)	25*	Bal.	–	15	6.5	1.35	–	5	1038	1127

*The foil was applied in two layers as a 50 μm thick preform.

Table 5.7 Tensile strength of X40 joints brazed with four different brazing amorphous foils

Filler metal	Tensile strength at *T*°C (MPa)			Yield strength at *T*°C (MPa)			Elongation at *T*°C (%)		
	25	540	650	25	540	650	25	540	650
MBF-100	717	531	497	499	257	246	6.5	14	11
MBF-101	711	522	474	487	242	236	5	10	7.5
MBF-103	653	508	448	465	223	201	4	8.5	6
MBF-104	655	501	446	460	222	203	4	8	6
X40 base metal	725	535	496	502	257	244	7	14	10

metal and formed rather large unfavorable $(Cr,W)_xB_y$ borides at grain boundaries in close proximity to the braze. Small borides are observed at BM grain boundaries some distance away from the joints together with the intrinsic $(Cr,W)_xC_y$ carbides, but the latter have segregated inside the base metal grains. The application of long-term annealing did not yield a decrease in the size of the borides.

The joint strength was measured in these works using both tensile and shear mode loading. Some results for the X40 joints, for example, are presented in Fig. 5.11. The strength of the X40 joints was 725 MPa at room temperature (RT)

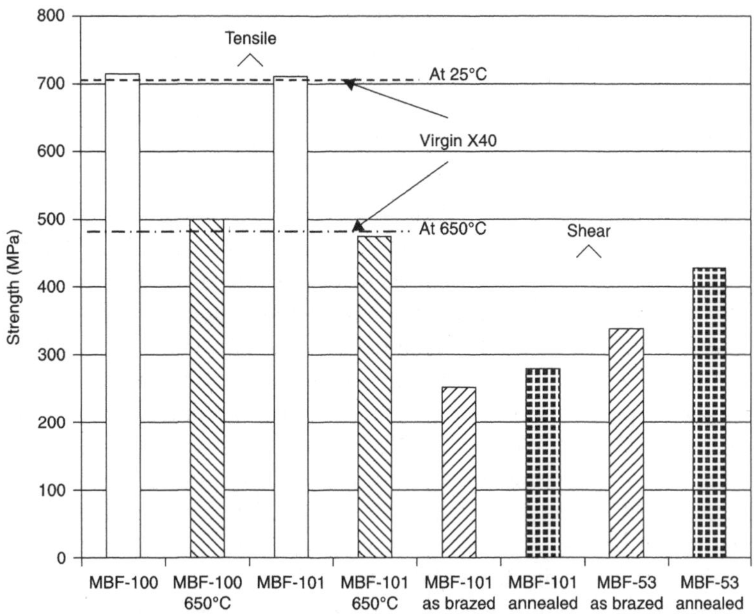

5.11 Tensile and shear strength of X40 joints brazed with BNi-5b (MBF-53) and Co-based amorphous MBF-100 series BFM at 25 and 650°C. Note that Co-based alloys are still pending for industrial applications.

and 496 MPa at 650°C. The elongation of the joints is 5–6.5% at RT (7% for an X40 joint) and 7.5–11% at 650°C (10% for an X40 joint). (More details may be found in these sources.)

5.4.4 Transient liquid phase (TLP) joining and joint loading during brazing

TLP processing

These methods are considered together because both have the same goal – creating joints with a composition as close as possible to that of the base metals. Moreover, substantial joint loading is now applied as a part of TLP when producing very large turbine parts, which will be used at high temperatures. To recap, TLP joining, in principle, is just a brazing process with an extended brazing plus annealing time. The result is dissolution of joint traces due to the solid state diffusion process equilibrating all elemental concentrations across the total brazement cross-section. It has attracted considerable attention in recent years because it is applied to the manufacturing of critical and very expensive parts of modern aerospace and stationary turbines (Gale and Butts, 2004; Saha and Khan, 2006; Arafin *et al.*, 2007).

Its progress is due to a much better modern understanding of the brazing mechanism (Cambell and Boettinger, 2000; Cambell and Kattner, 2002). Various TLP models involving real multi-phase alloy systems have been proposed in recent years (MacDonald and Eagar, 1998; Sinclair *et al.*, 2000). As a result, manufacturers are now provided with good instructions on how to carry out TLP and produce parts with excellent properties. At the same time, more sophisticated results have been published, describing real multi-phase Ni–Al–B (Cambell and Boettinger, 2000) and other complex alloys.

Out of many works on TLP, the work of K. Nishimoto's group, whose results were published in a series of papers, should be particularly noted (Nishimoto *et al.*, 1995, 1998). Both experimental and theoretical works showed a dramatic difference in the time needed for complete dissolution of the joint eutectic zone for different BFMs. Such behavior depends directly on the nature of the melting point depressants such as boron and germanium, for example, from the diffusion mechanism point of view. Thus, the time required for complete eutectic dissolution for the BNi-9 (MBF-80) vs. CMSX-2 superalloy at 1160°C is 2.8 h (Nishimoto *et al.*, 1998). On the other hand, the dissolution time for Ni-23Ge/Rene N5 at about 1160°C is at least an order of a magnitude longer (Dinkel *et al.*, 2008) because germanium diffuses as a substitutional species (Yeh and Chuang, 1997). Its diffusion rate is much lower than that of boron.

It is worth nothing that a thin preplaced BNi-9 (MBF-80) amorphous foil is used advantageously in a number of cases involving TLP. Indeed, the foil preform covers the gap cross-section 100%; it is thin and thus contains a small amount of boron per joint cross-section unit (Saha and Khan, 2006). Therefore, the filler metal can be completely dissolved in the BM within a reasonable heat-treatment time (Nishimoto *et al.*, 1998).

Single-crystal superalloy CMSX-2 and CMSX-4/BNi-9 (MBF-80) joints manufactured using TLP

These joints were extensively studied on specimens made of two separate single-crystal superalloy parts, which had a slight misorientation of their lattices (Nishimoto *et al.*, 1998). The parts had CMSX-2 (Ni–8.0Cr–4.6Co–0.6Mo–8.0W–1.0Ti–5.6Al–6.0Ta) and CMSX-4 (Ni–6.5Cr–9.0Co–0.6Mo–6.0W–1.0Ti–5.6Al–6.5Ta–3.0Re–0.1Hf) superalloy compositions. BNi-9 (MBF-80) 40 μm thick foil was used as a filler metal in brazing at 1100–1275°C for up to 5.5 h in vacuum under a slight load. The BFM and brazing conditions were essentially those which are used during TLP processing. Tensile tests of CMSX-2 joints bonded at ~1250°C in vacuum were carried out in the 650–900°C range. The tensile strength of all joints was equal to or greater than that of the base metal. The elongation and reduction of the joint area were observed to be either equal to or slightly lower than those of the virgin base metal. Most importantly, Nishimoto's group was able to produce single-crystal continuity of both base metal single-crystalline parts with virtually no traces of the braze and with a complete dissolution of BNi-9 (MBF-80) components in the CMSX-2 base metal parts. The effects of the BM and BFM compositions, and brazing temperature and time, were also elucidated. Basic parameters of TLP kinetics were also determined and the process was described using a rather simple model (for details see Nishimoto *et al.*, 1995, 1998).

Joint loading during brazing

Moderate loading, a few MPa, applied normal to the joint surface during brazing can strongly improve joint structure and mechanical properties. The appreciation of its positive effect on joint properties has been growing in recent years (Rabinkin and Pounds, 1988; Khorunov and Peshcherin, 1998; Kvasnitski *et al.*, 2002; Lugscheider *et al.*, 1989; Yeh and Chuang, 1997). The goal of loading is the same as TLP, namely, creating joints with a composition as close as possible to that of the base metals. Moreover, substantial joint loading is now applied with TLP in the production of very large turbine parts, which will operate at high temperatures.

Loading is widely used in mass production of polycrystalline diamond bits, injectors for huge stationary turbines with brazed cross-sections up to 1 m^2, and jet turbine blades (Khorunov and Peshcherin, 1998). Today hydraulic presses, as an intrinsic part of brazing furnaces, create and apply loads up to 1200 t(!) to large work pieces at about 1200°C. Another impressive example is the brazing of large gas pipes with wide brazing gaps (Lugscheider *et al.*, 1989). Loading works well with 100% metal preforms covering the whole gap cross-section. Therefore, BFMs in amorphous foil form are the best choice.

The loading effect can be explained using the 'ejection model' (Rabinkin and Pounds, 1988). The model assumes changes in the joint structure are due to mechanical rectification of the BFM composition from a fraction of the liquid metal formed at the beginning of melting. As follows from the general phase diagram constitution, this fraction is always enriched in metalloids. The applied load directly ejects this liquid fraction. Therefore, the amount of intermetallic phases in the crystallized joint decreases, whereas the fraction of solid solution matrix phase increases. The loaded joints have a very beneficial microstructure consisting mostly of the solid solution matrix phase. It was shown that moderate loading can increase joint strength by 100% or more, together with a particularly large increase in joint ductility (Rabinkin and Pounds, 1988; Yeh and Chuang, 1997). Very good and strong structures with a very small amount of eutectic were obtained, for example, when using Ni/Zr/Hf alloys under a 8–12 MPa load (Khorunov and Peshcherin, 1998; Kvasnitski *et al.*, 2002).

5.5 Metallurgical paths of joint formation

First, it is worthwhile giving a short description of the basic steps in the metallurgical path to joint formation. This description is rather reliable due to the wealth of experimental data accumulated over the years in the modern technical literature. During the brazing cycle, base metal dissolution starts as soon as the liquid phase of the BFM forms at the eutectic temperature/concentration levels. This dissolution completes in no longer than a few dozen seconds because the liquid phase becomes saturated with metallic components such as Co, Ni, Cr, W, etc. For example, in the case of the CMSX-2 superalloy brazed with the BNi-9 (MBF-80) alloy at 1250°C, this process is completed in 10 s (Nishimoto *et al.*, 1998; Wang *et al.*, 1995). The saturation process completes even more rapidly in BNi-series alloys, which have much lower boron concentrations than BNi-9 (MBF-80), in spite of the BNi-9 brazing temperature being lower than 1170°C. The next concentration equilibration stage, involving the entire brazement area, encompasses isothermal crystallization and extends well into the post-brazing annealing part of the brazing cycle when it is applied. At this stage the mass transfer proceeds with the elemental rates that are characteristic of diffusion processes in solid metals. The boron diffusion rate is unique among all the involved metallic elements and silicon as well; it is three orders of magnitude faster than that of chromium, silicon, etc. due to the relatively small atomic radius of boron (Kucera *et al.*, 1984; Wang *et al.*, 1995). On the other hand, boron has extremely low intergranular solubility in all highly alloyed materials because it has a large misfit with both interstitial and substitutional sites of metal crystal lattices. For example, in 316 stainless steel it is about 90 ppm at 1125°C and 30 ppm at 900°C (Karlson and Norden, 1988; Karlson *et al.*, 1988). As a result, boron diffusion proceeds effectively via grain and twin boundaries (Karlson *et al.*, 1988), where it decreases the interfacial free energy of loosely packed interfacial regions and reacts with iron, chromium and molybdenum (Karlson and Norden, 1988). The products of these reactions are multi-metal borides and carboborides, if the alloys which are being brazed contain carbon, such as $(Cr,Fe,Mo)_xB_y$ and $M_{23}(B,C)_6$ (Kucera *et al.*, 1984).

The strong driving force behind the creation of these borides is the high Gibbs free energy of formation: 295.9 and 242.7 kJ/mol for Cr_3B_4 and Cr_3B_5 respectively (Kubaschewski *et al.*, 1993). Chromium depletion of grain areas adjacent to grain boundaries accompanies this boride formation process and results in a well-known decrease in brazement corrosion resistance. The much slower rate of silicon diffusion results in delayed formation of silicides, if any, in the base metal interfacial areas. Sometimes, there is a complete retention of silicon as a substitutional component of the (Ni,Co,Cr)-base solid solution phase comprising the final joint (Rabinkin *et al.*, 1998). When long brazing and/or annealing are applied, boron diffuses deeper into the base metal and the number of borides in the interface area decreases. Simultaneously, boride phases acquire a rounded shape and coagulate, resulting in the improvement of mechanical properties. The joints are barely visible and are similar in appearance to TLP bonded joints when a sufficiently long annealing time is applied to IN939 joints (Fig. 5.10) and to 316 stainless steel joints (Fig. 5.6(c); Rabinkin *et al.*, 1998). The resulting mechanical properties, particularly ductility, increase substantially (Nishimoto *et al.*, 1995; Heikinheimo and Miglietti, 2000).

In the contemporary established analysis, the brazing processes are considered in terms of their inherent connection with paternal binary and multi-component phase diagrams of both the base metals and BFMs. In recent decades many thermodynamic assessments of multi-component diagrams have been made, for example in direct attempts to link these data with TLP parameters (Cambell and Boettinger, 2000; Cambell and Kattner, 2002). Due to the complexity of such assessments, the calculated results still differ from experimental ones (Cambell and Boettinger, 2000). Furthermore, no reliable phase diagrams showing alloy systems with more than four or five major components exist in the literature, whereas real metallurgical cases in modern superalloy brazing involve a minimum of five or six major components. However, even a simplistic analysis of existing binary and ternary phase diagrams can be useful in providing good guidance for the metallurgical paths by which the joint microstructures develop and in elucidating the reasons for the appearance of certain phases. Moreover, this is the most effective way to establish the conditions necessary for joint structure/properties optimization (Lugscheider and Sicking, 1998).

Thus, when considering, for example, the formation of transition metal borides in brazements, it is worth taking into account known paternal binary and ternary phase diagrams. Accordingly, Cr dissolves Ti, Mo and W in the solid solution state at a wide range of concentrations (Villars *et al.*, 1995c). The same is true for their borides. This readily explains why complex chromium borides $(Cr,Mo,W,Ti)_xB_y$ are formed in the joints of superalloys containing these elements when they are brazed with Ni/Co–B–Si (Heikinheimo *et al.*, 2001).

In the same vein, binary and ternary phase diagrams show that silicon dissolves significantly in the (Ni,Cr) γ phase, forming (Ni,Cr,Si) solid solution. This is in sharp contrast to boron, which has practically no solubility in Cr and Ni. When materials containing boron are brazed, they form a series of borides at BM grain boundaries. The fraction of boron in the Cr_xB_y compounds increases with BFM boron concentration.

One more impressive example of the intimate relationships between alloy phase diagrams and joint microstructure are Fecralloy joints brazed with BNi-5a (MBF-50) and described in Leone *et al.* (2006). These very unusual joints are strengthened by Ni_xAl_y precipitation. The formation of these particles is a result of the diffusion of nickel atoms from the BFM accompanied by the simultaneous reaction of nickel with aluminum from the BM (Fe, Cr, Al) β-solid solution phase. According to the Al–Ni–Fe and Al–Ni–Cr ternary paternal phase diagrams, this reaction should most probably yield an intermetallic compound having NiAl or Ni_3Al formula (Villars *et al.*, 1995b). Only these compounds can be formed at 1150°C, i.e. close to the brazing temperature. Indeed, these compounds were observed in the Fecralloy joints (Leone *et al.*, 2006).

Therefore, the following may be expected, and are indeed observed, in brazement structures arising when a Ni/Fe/Cr-based BM is brazed using a (Ni/Cr)–B–Si-based BFM:

- Formation of complex chromium/transition metal borides in both the affected BM and the joint solid solution matrix phase, but a lack of Ni borides.
- High sensitivity of the joint structure to the initial BFM silicon concentration: when the silicon concentration is low the (Ni,Si)-solid solution matrix phase provides high strength combined with ductility (Rabinkin *et al.*, 1998). This can occur because either there was a low initial silicon concentration in the BFM or/and the brazing process was carried out over a sufficiently long time. So, all of the silicon can dissolve in the braze solid solution matrix phase without the formation of silicides.

  A narrow brazing gap produces similar results because of the small combined concentration of B and Si in the braze volume, which becomes even lower after B and Si diffusion in the BM. The result is a limited 'B+Si' concentration in the joint for eutectic formation. Therefore, narrow gaps are strongly recommended for avoiding the formation of intermetallic phases with B and Si.
- If a filler metal contains no boron, then the amount of Si in it is usually enhanced to keep T_{Liq} low. The rate of Si diffusion in the BM is relatively low; silicides form in the joints even after an extended brazing time.
- A vast difference in the phase composition in the middle of joints (low B and Si amounts per joint cross-section unit) and fillets always exists. This occurs because the amount of metalloids in fillets per joint cross-section unit is too large to be dissolved completely in the BM over conventionally used brazing times. This causes the fillets to be the usual locations where stress-induced cracks originate, which is in the vicinity of large intermetallic phases.

- Optimization of brazing time and temperature not only provides improvement of the phase composition and the amount of intermetallic phases in the eutectic zone, but also improves the morphology of the chromium borides segregated at the BM grain boundaries. Such optimization yields a very substantial enhancement of a brazement's mechanical properties (Rabinkin *et al.*, 1998).

5.6 Industrial applications

At first, the prime applications of brazing for high-temperature/high-oxidation service conditions were confined to the aerospace industry, for joining aircraft structural parts, acoustical tail pipes (J-pipes), thrust reversers, turbine blades and seals (Assembly Eng., 1980; Irving, 1998; Yeaple, 1986; Rabinkin and Liebermann, 1993). It is also used in many other aerospace applications requiring high structural mechanical strength and demanding service conditions, such as the honeycomb parts of space vehicles, firewalls, torpedo hulls, missile fins and engine tail feathers. Ancillary engine components such as nacelles, heat exchangers, honeycomb seals, wear pads, seal shroud segments, diffusers, disks, corrugated interior hollow vanes, doublers, stiffeners and struts for frames were and still are widely produced with Ni-based alloys in powder and particularly foil forms. All these components have rigid dimensional tolerances and high strength/stiffness-to-weight ratio requirements, which are needed to handle the stresses inherent in aerospace applications. Ni-based amorphous (MBF) foils as thin as 18 μm (0.7 mil) used in these cases satisfy the low erosion requirements for joining structural aerospace honeycombs (Fig. 5.12). This gives consistently reliable and strong joints because of the accurately metered amount of brazing filler metal supplied in the joint area.

In recent years growing attention to energy conservation and pollution elimination has spurred the substantial growth of brazing in various industries outside aerospace applications. Brazing has been utilized in the production of plate heat exchangers for many industrial and utility functions, such as fuel cells, heat pumps, and catalytic converters for both gas and diesel internal combustion engines. Below are listed a few examples of the fastest-growing industrial areas, which have multi-digit percentage increases per year in the cost of both BFM sold as metallurgy products and, particularly, brazed finished manufactured goods.

An example is standard brazed plate/plate and plate/fin heat exchangers consisting of a large number of alternating corrugated/flat metal plates. These plates have multiple brazed, and thus sealed, contacts with each other. Inside these plates an elaborate system of channels is formed that comprises two separated liquid transport subsystems – one hot and one cool. Liquid and/or gas media flow through these subsystems, exchanging heat and thus saving energy (Fig. 5.13). Generally, these brazed heat exchangers are very compact, energy efficient, stronger and well suited to operating under high-temperature/high-pressure conditions. Because of the very large number of contact areas to be brazed, uniform pure metallic BFM amorphous foil is used, which provides much more uniform joints in comparison with powder filler metals.

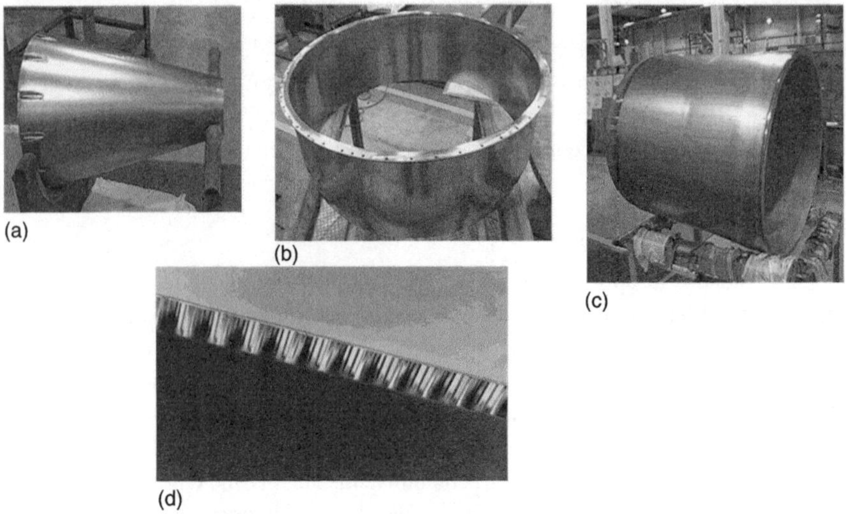

5.12 Boeing 727 and V2500 engine aircraft parts manufactured using Inconel honeycombs brazed using BNi-2 (MBF-20/AMS 4777) amorphous foil as BFM. (a), (b) V2500-CNA plug and jetpipe. (c) 727 nozzle assembly. (d) Typical panel.

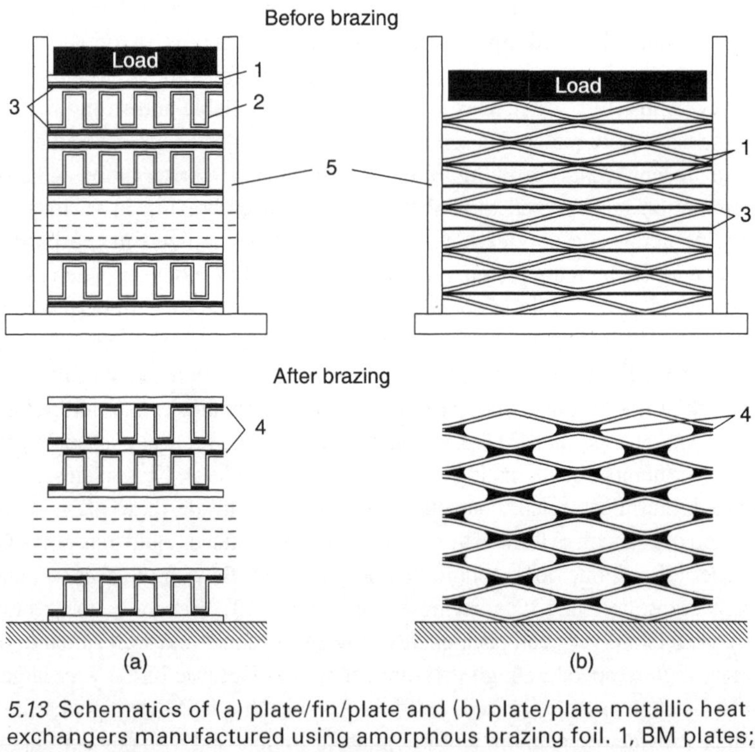

5.13 Schematics of (a) plate/fin/plate and (b) plate/plate metallic heat exchangers manufactured using amorphous brazing foil. 1, BM plates; 2, fins; 3, BFM foils; 4, joints; 5, guides.

Additional advantages of amorphous vs. powder BFM may be gained using robotic operation to assemble into a single package multiple plates and preforms. Amorphous brazing foils are produced in multi-tonnage quantities for these purposes. The foil materials enable the assembly of complex heat exchangers before brazing without the environmental contamination associated with powder organic binders. The size of the heat exchanger units brazed with amorphous MBF alloys has been growing rapidly, as Fig. 5.14(a) and (b) demonstrate. These developments, combined with improved brazing technology, created a rapid growth in usage of Ni-based amorphous BFM in new applications where there is a need for brazes with high corrosion resistance. New high-chromium-containing BFM foil also provides an opportunity to replace the much more expensively assembled plate/gasket heat exchangers with brazed heat exchangers, which possess a combination of high strength and superior corrosion resistance.

Stainless steel heat exchangers are currently produced in quantities of hundreds of thousands of pieces per year for the chemical, pharmaceutical, food processing and medical industries, fuel cells and refrigerators. In the latter case, a ban on the production of chlorofluorocarbons, a result of their depleting effect on the atmospheric ozone layer, considerably increased the need for non-corrosive stainless steel brazes containing no copper and capable of withstanding alternative heat exchanger media such as ammonia.

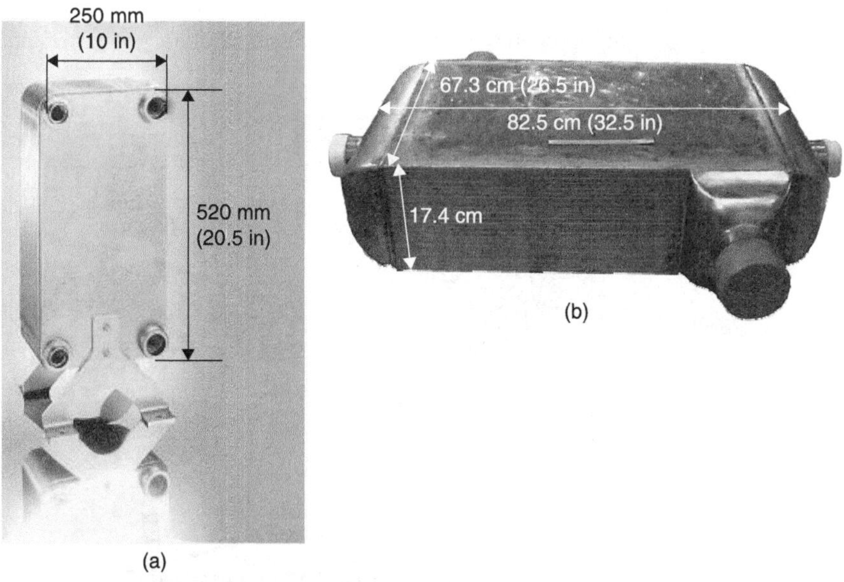

5.14 Two large heat exchangers brazed using side-by-side preplaced BNi-5b (MBF-51) 200 mm wide preforms. Each unit has up to 50 alternating plate/plate brazed compartments, as shown schematically in Fig. 5.13(b), in which a cold and a hot liquid medium flow in opposite directions.

The same intense efforts are now being applied to pollution elimination and increased engine efficiency for cars, motorbikes and working vehicles in construction, mining, the ship industry, etc. Examples of these three-dimensional brazed structures are shown in Fig. 5.15 (Rabinkin, 2009). Here, five brazed auxiliary units of a modern diesel engine are presented in a schematic way together with their photographs. These parts or units, such as Metallic Catalyst Substrates (MCS), Exhaust Gas Recirculator Coolers (EGRC), Oil Coolers (OC) and Diesel Particulate Filters (DPF), decrease the environmental impact of diesel and gas engines; they regulate the fuel/air gas mixture temperature and thus increase total efficiency.

A new fast-growing and very large brazing application is the production of so-called heat recuperators, which clean the exhaust gases of diesel engines and at the same time preheat the air coming into the diesel cylinders. Today diesel engines are fast displacing gas auto engines, particularly in Europe, where annual production of diesel-powered cars exceeds 50% of all cars. The number of diesel engine cars furnished with brazed EGRC units is well over 10 million. The same trend is expected soon in the USA as well. BNi-2, -3, -5a, -5b and -9 BFM in foil form containing various chromium concentrations have all been tested and used successfully in the production of various types of diesel cars and trucks.

BNi-5a (MBF-50) foil is also used in joining high-temperature-resistant Fecralloy honeycomb cores to the (Fe,Cr)-based steel cartridges of the catalytic converters (Leone *et al.*, 2006) in cars and motorcycles, and now in huge ship turbines with increasing frequency. Examples of such car catalytic converters are

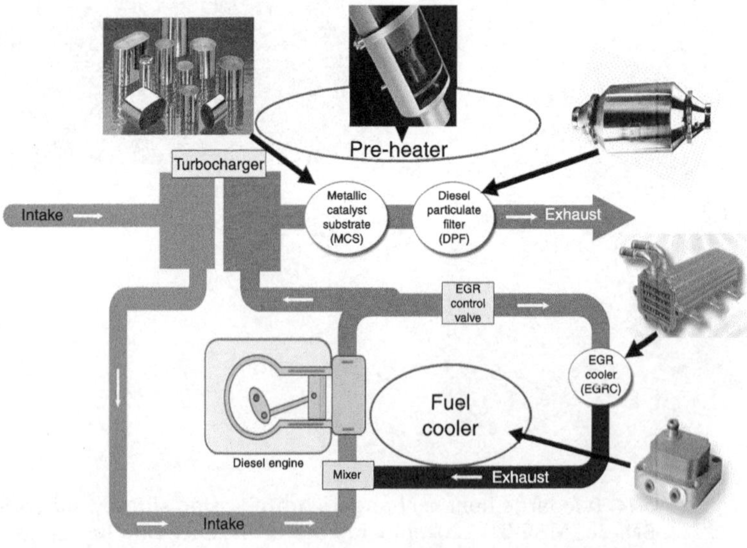

5.15 Topical applications of thin-gage brazed parts superimposed on a schematic of a modern diesel engine.

shown in top left of Fig. 5.15. In spite of the presence of a strong alumina scale on the surface of the Fe–Cr–Al core material, BNi-5a (MBF-50) wets and joins the core to the stainless steel cartridge due to the above-mentioned effect of oxide dissolution in liquid MBF-50.

A substantial reduction in the maintenance cost for crack and wear out defects in jet engines has been achieved via a brazing repair of enormously expensive air and stationary turbine blades using high alloyed BFM powders (Kennedy *et al.*, 1998).

Current efforts in high-temperature brazing development are directed toward finely tuning brazed products to specific application needs. In this respect, automotive applications – in which amorphous brazing foil is predominantly used, in hundreds of tonnes – have proven to be an area of particular growth, with an even more promising rate in the near future.

5.7 References

Arafin M., Medraj M., Turner D. and Bocher P. (2007), Transient liquid phase bonding of Inconel 718 and Inconel 625 with BNi-2: Modeling and experimental investigations, *Mat. Sci. Eng.*, A 447, pp. 125–133.

Assembly Eng. (1980), No. 3, *Air Space*, pp. 25–27.

AWS (2004), AWS A5.8/A5.8M:2004, *Specification for Filler Metals for Brazing and Braze Welding*.

AWS C3.2M/C3.2:2001, *Standard method for evaluating the strength of braze joints*.

AST A 887–89, *Borated stainless steel plate, sheet, and strip for nuclear application*.

Cambell C. and Boettinger W. (2000), TLP bonding in the Ni-Al-B system, *Met. Mat. Trans. A*, Vol. 31A, pp. 2835–2847.

Cambell C.E. and Kattner U. (2002), Assessment of the Cr-B system and extrapolation to the Ni-Al-Cr-B quaternary system, *Calphad*, 26, No. 3, pp. 477–490.

Cretegny L. *et al.* (2009), *Brazing alloys compositions and methods*. US Patent Appl. Publ. US 2009/0159645 A1.

DeCristofaro N. and Bose D. (1986), *Int. Conf. Rap. Sol. Mat.*, San Diego, ASM-AIME, pp. 415–424.

Dinkel M., Heinz P. and Pyczak F. (2008), New boron and silicon free single crystal-diffusion brazing alloys, TMS, *Superalloys*, pp. 211–220.

Gale W. and Butts D. (2004), Transient liquid phase bonding, *Sci. Tech. Weld. Join.*, Vol. 9, pp. 283–298.

Hartmann T. and Nuetzel D. (2009), New amorphous brazing foils for exhaust gas applications, *Proc. 4th International Brazing and Soldering Conference*, Orlando, AWS, pp. 110–118.

Heikinheimo L. and Miglietti W. (2000), Brazing of Co-based superalloy gas turbine vanes using novel amorphous filler metals, *Proc. IBSC 2000*, AWS-AMS, pp. 468–475.

Heikinheimo L., Miglietti W., Leone E., Kipsnis J. and Rabinkin A. (2001), Brazing of Co- and Ni-based superalloys using new amorphous brazing filler metals, LÖT 02001, pp. 30–34.

Irving B. (1998), Brazing and soldering: Facing new challenges, *Weld. J.*, N10, Appl., pp. 33–37.

Karlson L., Norden H. and Odelius H. (1988), Non-Equilibrium Grain Boundary Segregation Of Boron In Austenitic Stainless Steel. Large Scale Segregation Behavior-I, *Acta Metal.*, Vol. 36, No. 1, pp. 1–12.

Karlson L. and Norden H. (1988), Non-equilibrium grain boundary segregation of boron in austenitic stainless steel – I. Large Scale Segregation Behavior- IV, *Acta Metall.*, Vol. 36, No. 1, pp. 35–48.

Kennedy A., Cuthbert J. and Pappas C. (1998), Prolongation of service life of superalloy components of aero gas turbines by cost effective repair technology, LÖT 1998, *DVS* 192, pp. 1–4.

Khorunov V. and Peshcherin E. (1998), Brazing of nickel alloys, *DVS* 192, pp. 276–278.

Khorunov V., Maksimova S., Samokhin S. and Ivanchenko V. (2001), Brazing filler metals containing Zr and Hf as depressants, *Trans. JWRI*, Vol. 30, Special issue, pp. 419–424.

Koch J. and Schmoor H. (1998), BrazeSkin-application of nickel-based filler metals, *DVS* 192, pp. 231–233.

Kubaschewski O. and Alcock C. (1978), *Metallurgical Thermochemistry*, Univ. Toronto, 5th Ed., pp. 300–301.

Kubaschewski L., Alcock C. and Spencer P. (1993), *Materials Thermochemistry*, 6th Ed., Pergamon Press, Oxford, UK.

Kucera J., Buchal A., Rek A. and Stransky K. (1984), Redistribuce boru v okoli pajenych spoju typu Ni-B (Boron redistribution in joints brazed with Ni-B), *Kovove Materialy*, Vol. 22, No. 3, pp. 250–262.

Kvasnitski V., Timchenko V., Ivachenko V. and Khorunov V. (2002), Effect of depressant elements on properties of Ni-based filler metals and superalloys, *Adh. Met. and Braz.*, Vol. 35, pp. 129–139.

Leone E., Rabinkin A. and Sarna B. (2006), Microstructure of thin-gauge austenitic and ferritic stainless steels brazed using Metglas® amorphous foils, *Weld. World*, 50, N1, pp. 3–15.

Lugscheider E. and Gundlfinger K. (1980), Mechanical properties of brazed joints – Literature review 1950–1980, *Mat. Sci. Div.*, Working Document, Aachen University.

Lugscheider E. and Sicking R. (1998), Dissolution behavior of tungsten carbides in NiCrBSi-alloys, *DVS* 192, LÖT 1998, pp. 252–225.

Lugscheider E., Schittny T. and Harmoy E. (1989), Wide gap brazing of off shore oil field pipes, *DVS* 125, LÖT 1989, pp. 10–15.

Lugscheider E., Humm S. and Buschke I. (2001), Investigation of the chemical and mechanical-technological behavior of chromium bearing Ni-Hf-filler metals, *DVS* 212, LÖT 2001, pp. 367–370.

MacDonald W. and Eagar T. (1998), Transient liquid phase bonding, *Metall. Mater. Trans. A*, 29A, pp. 315–325.

Myazawa Y., Nanjyo T., Miyamoto Y., Ariga T. and Inoe J. (2000), Brazing of stainless steel by nickel based brazing filler metals, *Proc. IBSC 2000*, ASM-AWS, pp. 147–152.

Nishimoto K., Saida K., Kim D., Asai S. and Nakao Y. (1995), Transient liquid phase bonding of Ni-base single crystal superalloy, CMSX-2, *ISIJ, Int.*, 35, N10, pp. 1298–1306.

Nishimoto K., Saida K., Asai S., Furukaawa Y. and Nakao Y. (1998), Bonding phenomena and joint properties of transient liquid phase bonding of Ni-base single crystal superalloy, *Weld. World*, Vol. 41, No. 2, pp. 48–58.

Pattee H. (1980), High-temperature brazing, *Source Book on Brazing and Brazing Technology*, ASM, pp. 17–63.

Peaslee R. (2006), Interview with a legend, *IBSC2006*, AWS, San Antonio.

Rabinkin A. (1989), New Applications for Rapidly Solidified Brazing Foils, *Weld. J.*, No. 10, 68, pp. 39–46.

Rabinkin A. (2000), Optimization of brazing technology, structural integrity, and performance of multi-channeled, three dimensional metallic structures, *Proc. of the Int. Braz. Sold. Con.*, 2000, Albuquerque AWS/ASM, pp. 437–444.

Rabinkin A. (2004), Overview: Brazing with (NiCoCr)-B-Si amorphous alloys brazing filler metals: Processing, Joint Structure, Properties, Applications, *Sci. Tech. Weld. Join.*, Vol. 9, No. 3, pp. 181–199.

Rabinkin A. (2006), Amorphous brazing foil at age of maturity, *Proc. 3rd International Brazing and Soldering Conference*, AWS/ASM, San Antonio, pp. 148–154.

Rabinkin A. (2009), New brazing development in the automotive industry, *Proc. 4th IBSC*, Orlando, pp. 380–386.

Rabinkin A. and Liebermann H. (1993), *Rapidly Solidified Alloys: Processes, structures and properties, and applications*, Marcel-Dekker, pp. 691–726.

Rabinkin A. and Murzyn P. (1987), *Abstracts of 18th Int. AWS Braz. Conf.*, Chicago, pp. 265–266.

Rabinkin A. and Pounds S. (1988), Effects of load on brazing with METGLAS® MBF-2005 filler metal, *Weld. J*, Vol. 67, No. 5, pp. 33–43.

Rabinkin A., Wenski E. and Ribado A. (1998), Brazing stainless steel using a new MBF-series of Ni-Cr-B-Si amorphous foil, *Weld. J.*, Vol. 77, No. 2, pp. 66s–75s.

Rangaswamy S. and Fortuna D. (2008), Novel high chromium containing braze filler metals for heat exchanger application, *DVS* 243, pp. 12–16.

Rassmus J. and Sjodin P. (2006), Joining Aspects of Large Plate Heat Exchangers in Stainless Steel, *Proceedings of the 3rd International Brazing and Soldering Conference*, ASM and AWS, San Antonio, pp. 357–362.

Saha R. and Khan T. (2006), TLP Diffusion bonding of a ODS nickel alloy, *Azojomo*, 2 (July), pp. 1–15.

Sexton P. and DeCristofaro N. (1979), US Patent No. 4, 148,973.

Schmoor H., Schnee D. and Koch J. (2006), New selective application methods for brazing pastes, *IBSC2006 Proc.*, pp. 373–376.

Schnee D., Koch J., Schmoor H. and Weber W. (2004), Technique for brazing paste applications, *DVS* 231, pp. 21–23.

Shapiro A. and Rabinkin A. (2003), State-of-the-Art of Titanium-Based Brazing Filler Metals, *Weld. J.*, Oct. 2003, pp. 36–43.

Sinclair C., Purdy G. and Morral J. (2000), TPL Bonding in two-phase ternary systems, *Metal. Mater. Trans. A*, 31A, pp. 1187–1192.

Sjodin P., Wolfe C. and Wilhelmesson B. (2005), A novel type of all-stainless steel plate heat exchangers, *Proc. of 5th Int. Conf. on Enhanced, Compact, and Ultra-Compact Heat Exchangers; Sci. Eng. and Techn.*, Eds. R.K. Shah *et al.*, NJ, USA, pp. 215–220.

Villars P., Prince A. and Okamoto H. (1995a), *Handbook of Ternary Alloy Phase Diagrams*, ASM, 3, pp. 3105–3118.

Villars P., Prince A. and Okamoto H. (1995b), *Handbook of Ternary Alloy Phase Diagrams*, ASM, 3, pp. 3156 and 3538.

Villars P., Prince A. and Okimoto H. (1995c), *Handbook of Ternary Phase Diagrams*, ASM, 7, pp. 9066.

Wang W., Zhang S. and He X. (1995), Diffusion of boron in alloys, *Acta Metal. & Mat.*, 43, pp. 1693–1699.

Yeaple F. (1986), Design News, *Air*, pp. 25–26.

Yeh M. and Chuang T. (1997), Effects of applied pressure on the brazing superplastic INCONEL 718 superalloy, *Met. Mat. Trans.*, 28A, pp. 1367–1377.

Zheng Y. (1990), Microstructure and performance of Ni-Hf brazing filler metals, *Acta Met. Sinica*, B, No. 10, 3B-P, pp. 335–340.

6

Brazing of diamonds and cubic boron nitride

A. RABINKIN, Metglas Inc., USA, A. E. SHAPIRO, Titanium Brazing, Inc., USA and M. BORETIUS, Listemann AG, Liechtenstein

DOI: 10.1533/9780857096500.2.160

Abstract: A quiet revolution in industry has happened over the last 50 or so years due to the use of diamond and cubic boron nitride (CBN) in many applications. Joining of diamonds to various materials via brazing is very specific compared with conventional brazing due to the unique nature of diamond. This chapter describes the properties of diamond and CBN, and their wetting by and interaction with metals and alloys; factors that affect these interactions; and practical aspects of diamonds and CBN joining. Some properties of brazed joints of diamond and CBN with different metals, as well as cemented carbide inserts, are presented and discussed. Finally, application examples are provided.

Key words: diamond, cubic boron nitride, brazing, joining, brazing filler metals, joint properties.

6.1 Introduction

Diamond is among the very few substances that have been familiar to many billions of people for thousands of years. Its unique properties (Gauthier, 1998), first as the most valuable precious stone and then as an ultimately advantageous material exploited in various tools, have been attracting people for many hundreds of years. However, the explosive growth in the use of diamonds in many industries did not take place until the second half of the twentieth century. Diamond is one of the few critical factors necessary for the general progress of modern civilization, for many reasons, such as: a substantial increase in the efficiency and particularly accuracy when manufacturing machinery parts; expanding oil well drilling to places with hard drilling conditions; improving the cutting of semiconducting elements to manufacture computer chips; making possible the mechanical treatment of new materials, such as new plastics and graphite composites, in which conventional tools work badly; and so on.

The importance of machining with new tools should be particularly emphasized, because, despite the tremendous development in near-net-shape casting techniques, most technical components are still manufactured by machining. In addition to drilling, turning and milling operations in which tools with geometrically defined cutting edges are used, grinding with undefined cutting edges has gained increasing application. High-speed cutting, machining of hardened steels and the use of small amounts of coolants or even dry cutting have led to the development of an entirely new tooling concept requiring novel cutting or grinding materials. These materials, when in direct contact with the work pieces, must withstand extremely high mechanical stresses and maintain compositional stability under high thermal loads combined with a high resistance to chemical attacks.

In this respect, diamond has one fundamental deficiency: it cannot be used in iron-based steel and alloy machining operations due to catastrophic deterioration, occurring for reasons which will be outlined in detail below. Cubic Boron Nitride (CBN) is a super-hard material which, in contrast to diamond, can be used for machining iron-based alloys and steels and in many other important applications where diamond cannot be used. Moreover, its thermal stability and resistance to chemical attacks are superior to that of diamond.

Therefore, it seems logical to present diamond and CBN together in this chapter. CBN is similar to diamond in many respects, but it is not found in nature as diamond is. The history of boron nitride is much, much shorter than that of diamond. Its synthesis was a direct result of the high-pressure/high-temperature technology created for artificial diamond production, which was also applied successfully in this case. It was synthesized first in 1957 by Robert H. Wentorf (Wentorf, 1957), a member of the same group of General Electric (GE) scientists who developed diamond synthesis technology. It was introduced into the market under the trade name Borazon by GE. At that time it was more expensive than gold. Commercial production of CBN on a large scale began in the early 1960s (Greenville-Wells, 1957; General Electric, 2011).

The attachment of diamond or CBN to tools, either mechanically or via metallurgical joining methods, is the critical operation for its effective industrial utilization. This paper considers only metallurgical, mostly brazing, means of attachment. Substantial consideration of the unique physical and chemical properties of diamond is necessary to understand the specificity of its joining. A short review of the latter is given here, with the main emphasis on the interactions of diamond and CBN with a large number of elements, mostly metals, which are used as components or a base for the brazing filler metals for diamond attachment.

6.2 Physical properties of diamond and cubic boron nitride (CBN)

6.2.1 Diamond

Five forms of diamond are available as ultra-hard abrasive and cutting tool materials. They are as follows:

- Natural single-crystal diamonds (Prelas *et al.*, 1998).
- Synthesized single-crystal diamonds.
- Polycrystalline diamonds (PCDs) sintered under high-temperature/pressure conditions.
- Thick-film chemical-vapor-deposited (CVD) polycrystalline diamonds up to 1 mm.

- Thin-film chemical-vapor-deposited polycrystalline diamonds < 50 μm.

The physical and mechanical properties of these diamond forms are compared in Table 6.1. Thick-film CVD diamonds are manufactured in the form of wafers up to 1 mm thick and about 150 mm in diameter. These are cut by laser beam and brazed onto cemented carbide or a silicon nitride substrate. Thin-film CVD films are nucleated and deposited for cutting tools onto silicon nitride, silicon carbide, and WC-Co (with a low cobalt content, < 6 wt.%) cemented carbide substrates (Davis, 1995).

Diamonds are also classified by the presence of impurities or secondary phases. Most natural crystals belong to Type Ia, containing 0.1% nitrogen atoms in the form of small aggregates. Most synthetic diamonds are of Type Ib and contain dispersed nitrogen. Diamond crystals of Type IIa are substantially free of nitrogen and have enhanced optical and thermal properties. Very pure blue diamonds of Type IIb exhibit semiconducting properties due to a small content of boron.

Diamond have the highest hardness and highest thermal conductivity of any material at room temperature. The thermal conductivity of diamond crystals is as high as five times that of copper at room temperature. The unique combination of these two properties makes diamond the ultimate cutting and abrasive material. Diamond not only cuts the treated substance but also rapidly dissipates the heat from the cutting interface, thereby preventing thermal shock (Gauthier, 1998).

Brittleness is one of the few drawbacks of diamond crystals. They break on impact by cleavage on four (111) planes. However, contemporary sintering and CVD technologies allow the production of polycrystalline diamonds with a toughness sufficient for use in most industrial applications.

Since brazing involves heating, it is important that diamond is metastable under ambient pressure and temperature conditions. The graphitization of a single-crystal natural diamond, which is the process of returning it to a thermodynamically stable carbon structure, starts at 1600°C (2910°F) in a pure inert atmosphere and at 1500°C (2732°F) in vacuum 5×10^{-4} Pa (5×10^{-6} torr). Oxygen promotes the graphitization of diamond; thus, the temperature of graphitization decreases to 1200°C (2192°F) (in a vacuum of 1×10^{-4} torr (10^{-2} Pa)) and to only about 1000°C (1832°F) in air at atmospheric pressure (Naidich *et al.*, 1977).

Diamond is a dielectric material if it is free of boron and metal dopants.

6.2.2 Cubic boron nitride

Boron nitride is a chemical compound consisting of equal numbers of boron and nitrogen atoms. Its hexagonal crystal form (h-BN) has a crystal structure similar to graphite. The hexagonal boron nitride is the most stable and softest material among boron nitride polymorphs, and is, therefore, used as a solid lubricant and an additive to cosmetic products. Due to the very strong atomic bonds between atoms of boron and nitrogen, it has an outstanding chemical stability, which

Table 6.1 Physical properties of diamonds

Property	Units	Natural single crystal	Synthetic single crystal	Solid PCD*	Layered PCD†	CVD diamond films
Density	lb/in.^{3}	0.127	0.115–0.127	0.123	0.148	0.065–0.127
	g/cm^3	3.52	3.20–3.52	3.43	4.12	1.80–3.51
Transverse rupture	ksi	2380–4700	3130–4700	217	174	188
strength	GPa	16.4–32.4	21.6–32.4	1.5	1.2	1.3
Compressive strength	ksi	1260–2390	650–840	681	1102	2320
	GPa	8.7–16.5	4.5–5.8	4.7	7.6	16.0
Flexure strength	ksi	152	116–203	217	58	—
	MPa	1050	800–1400	1500	400	—
Young's modulus	ksi	150 724	115 940–134 060	134 058	112 464	77 680–171 014
	GPa	1040	800–925	925	776	536–1180
Poisson's ratio	—	0.07	0.20	0.086	0.07	—
Knoop hardness	ksi	8260–15 070	7826–12 170	7250	7250–10 870	12 320–14 500
	GPa	57–104	54–84	50	50–75	85–100
Fracture toughness	ksi × in$^{1/2}$	3.09	5.45–9.73	6.27	8.01	5.90
	MPa × m$^{1/2}$	3.40	6.0–10.7	6.90	8.80	6.50
Thermal conductivity	Btu/ft × h × °F	580–1160	580–1160	69	323	434–867
	W/m × K	1000–2000	1000–2000	120	560	750–1500
Coefficient of thermal	x10^{-6} F^{-1}	1.14–2.73	1.14–2.73	2.16	2.39	2.10
expansion	x10^{-6} K^{-1}	2.0–4.8	2.0–4.8	3.8	4.2	3.7

*Solid thermally stable polycrystalline diamond (PCD) used in mining tools.

results in its poor wettability by liquid metals. h-BN is also used as a stop-off material in brazing and sintering processes.

The essential physical properties of various polycrystalline cubic boron nitride (PCBN) grades in comparison with those of the other widely used cutting materials are shown in Table 6.2.

The crystal form of CBN has a crystal lattice analogous to that of diamond. Its hardness is inferior only to diamond, but still its thermal and chemical stability are superior to diamond. For example, h-BN does not decompose at temperatures up to 1000°C in air, 1400°C in vacuum and 2800°C in an inert atmosphere.

The thermal stability of CBN can be summarized as follows:

- In air or oxygen CBN does not undergo oxidation till ~1300 °C due to the formation of a thin protective layer of B_2O_3, which prevents further oxidation; also there is no conversion of CBN to the hexagonal form at 1400°C.
- In nitrogen some CBN conversion to h-BN occurs at 1525°C after holding for about 12 h.
- In vacuum of the order of 1×10^{-5} Pa, CBN conversion to h-BN starts at 1550–1600°C (Spriggs, 2002).
- Compared with conventional hard materials such as silicon carbide and aluminum oxide, CBN, in addition to being harder at room temperature, maintains its hardness over a wider temperature range.

Table 6.2 Properties of different cutting materials (Diamantschneidstoffe, 2011)

	Al_2O_3	Al_2O_3/TiC	Hard metal K10	PCD	PCBN Grade 1	PCBN Grade 2	PCBN Grade 3
Density (g/cm^3)	3.91	4.28	14.7	4.12	3.43	3.52	4.28
Knoop hardness (GPa)	16	17	17	50	32	30	28
Young's modulus (GPa)	380	395	590	780	690	650	610
Compression strength (GPa)	4.0	4.5	4.4	7.5	2.8	3.8	3.5
Fracture toughness (MPa m$^{1/2}$)	2.34	2.9	10.5	8.8	6.5	5.9	4.9
Bending strength (MPa)	650	600	1600	1260	660	700	280
Thermal expansion coeff. (10^{-6} K^{-1})	8.5	7.8	5.4	4.2	4.9	4.6	4.7
Thermal conductivity (W m^{-1} K^{-1})	8.4	9.0	100	550	110	90	55

*PCD, polycrystalline diamond ex WC with 10% Co.

6.3 Diamond's interaction with metals

6.3.1 Nature of the interactions

Both the diamond–metal interaction and the structure of the diamond/metal joint interface depend on the nature of the atomic bonding between the atoms of carbon and metal. This nature in turn is also determined by the crystallographic orientation of the diamond crystal lattice surface on which the carbon atoms are positioned when reacting with metals and metal compounds, forming mostly carbides as a result of this interaction. One of the important effects of crystal orientation is the different wettability of diamond planes with different crystal indexes. Detailed analyses of such interactions are given in Naidich *et al.* (1988) and Rabinkin (1981).

Alloy phase diagrams containing carbon as one of the elements under consideration are well constructed and reliably express what may be formed from these interactions, not only at the diamond/filler metal (FM) interface but in the metal volume outside the interface. They also help to predict what kind of carbon/metal compounds should be expected in the interface area. Practically all aspects of the interaction of any particular metal or alloy with diamond depend on whether or not it forms a stable carbide. Important transition metals have rather diverse free energies of carbide formation, as can be seen in Table 6.3. The second important factor is the solubility of carbon in the liquid brazing filler metal and in the base metal in the solid state as well. This solubility depends on the temperature and pressure applied.

From the practical, i.e. brazing, point of view, only elements from IVA to IIB of the periodic table are worth considering. They, in turn, can be separated into three different groups according to the type of diamond–element interaction. The first group consists of non-transition elements, such as Ag, Au, Cu, Zn, Al and others, which practically do not form any compounds with carbon. Most of the IB-group elements have a rather low energy of atomic bonding with diamond and, therefore, dissolve carbon in minuscule quantities in both the liquid and solid states. The

Table 6.3 Standard free energy of formation ΔF of some metal carbides per g-atom in kcal/g-atom

	Metal							
	Ni	Co	Fe	Cr	Nb	Ta	Ti	Zr
Highest carbide	Ni_3C	Co_2C	FeC	CrC	NbC	TaC	TiC	ZrC
ΔF at:								
1200 K	1.525	0.5	−0.09	−5.34	n/d	−18.45	−20.15	−20.35
1300 K	1.475	0.4	−1.62	−5.32		−18.4	−20.00	−20.15
Sign of $\partial(\Delta F)/\partial T$	<0	<0	>0	<0	n/d	~0	<0	<0

ΔF kcal/g-atom = ΔF kcal/mole/ (x+y), where x and y are taken from carbide formula Me_xC_y.

corresponding phase diagrams in these cases have no areas with any noticeable carbon solubility in the solid state of these metals. The second group of metals consists of transition metals forming carbides, such as V, Ta, Cr and others, whereas the third one contains transition metals from the VIIIA group, which includes not only Ni and Co but all metals from the Pt-group elements. The elements of the VIII group and part of the VII group dissolve diamond in its liquid phase but do not form carbides.

6.3.2 Diamond's interaction with elemental metals and alloys

One of the most important factors, unique to diamond, which should be underlined, is its inability to dissolve any substantial, from a practical point of view, amount of any element. On the other hand, diamond carbon atoms dissolve in and/or react with metals, forming a variety of metal carbides. In this respect, the diamond-metal interaction is literally one-directional: all joint formation, including morphology, phase composition and joint stresses arising from joining, occurs outside the diamond surface in the metal or alloy volume.

It has been shown (Rabinkin, 1981) that dissolution of both diamond and graphite proceeds at practically the same rate, being slightly faster in the case of diamond. It is very likely that the diffusion of carbon from diamond into the metal matrix at moderate temperatures occurs via direct diamond dissolution without the intermediate conversion-to-graphite stage. This dissolution starts at diamond crystal corners, edges, crystallization defects, dislocations, etc. For Ni, Co and Fe matrixes, the dissolution process occurs step by step along the planes of the diamond crystals. However, the selectivity of dissolution strongly decreases as the carbide-forming ability of the matrix metal increases. Thus, carbide 'cases' formed by Zr, Nb and Ta have equal thickness for all kinds of facets and edges (Fig. 6.1).

6.1 Carbide film formed on diamond sintered with Nb at 800°C, 5 h in vacuum. Film morphology depends neither on the metal matrix morphology nor on the crystal plane indexes. Niobium carbides bands spread around the crystal, which appears as if it is completely wetted.

Relative to the rate of diamond dissolution, all metals may be divided into two categories: carbide forming and non-carbide forming. The solubility in non-carbide forming metals depends on the limit of carbon solubility in the metal at a given temperature and pressure. Therefore, the higher the solubility, the higher the rate of carbon diffusion, and, therefore, the higher the rate of diamond dissolution. For carbide-forming elements, if the carbon solubility is low, then the rate of dissolution is nearly equal to the rate of carbide formation at the diamond–metal interface. Therefore, as was mentioned, the carbide case protects the diamond because the carbon diffusion rate through a carbide film is extremely low. When the solubility is high, the rate of dissolution is a compromise between the carbide build-up and the carbon diffusion from the carbide–metal interface into the Me(C) solid solution.

Since the solubility of diamond depends only on diffusion processes and capillarity effects occurring in the metal and carbide, it is worthwhile considering some parameters of diffusion of some metals and C in metals and carbides, presented in Table 6.4. These data clearly demonstrate an enormous difference in the rate of diffusion in carbides and metals by both metals and C. The coefficients of diffusion in carbides are of the order of 10^{-4} to 10^{-5} times smaller than those in metals. On the other hand, the stability of carbide films to dissolution in filler and base metals also depends to a certain extent upon the carbon solubility limit in both

Table 6.4 Parameters of carbon diffusion in various metals, and carbon and metal diffusion in carbides

Material	Temperature (°C)	Activation energy for diffusion, Q (kJ/mol)		Diffusion coefficient, D_0 (cm^2/sec)		Reference
		Metal*	C	Metal*	C	
Ni	700–1365	267–280 (Ni)	167	0.4–1.3 (Ni)	0.9–2.5	1
Fe						
α	900–1050	249 (Fe)	80.3	5.8 (Fe)	0.167–0.006	2
γ		280 (Fe)	135	1.3	0.1	
Nb			139 113		0.004–0.015	3
NbC		531 (Nb)	393	0.11 (Nb)	0.11	3
Ti						
α			182		5.06	1
β			203		1.08	
TiC		737 (Ti)	399	4×10^{-4} (Ti) (!)	7	3
ZrC		720 (Zr)	460	1×10^{-3} (Zr)	14	3

*Metals being referred to are given in parentheses.

References: 1, Smigelskas (1962); 2, Krishtal (1970); 3, Andrievsky and Umansky (1977).

materials. In all the various cases of diamond dissolution, still the diffusion processes between components of all virgin phases in base metals and newly formed phases, and their comparative rates, are the factors determining dissolution.

Dissolving carbon may precipitate in the metal matrix, forming small carbide particles. Usually this process is accompanied by strong dispersion hardening, which was observed in most of the metals and metal matrixes.

Brazing and annealing diamond/metal composites in nitrogen during sintering decreases the diamond dissolution rate in both cases, for it decreases carbon activity in γ-Fe, as was proven by direct measurements in the samples annealed in vacuum and in nitrogen atmosphere (Rabinkin, 1981).

6.3.3 Formation of carbides on a diamond surface

According to Naidich *et al.* (1988), transition elements of IV–VI groups and silicon and boron have very strong chemical bonds with carbon, forming carbide layers with a rather permanent composition. As is well known, the Me–C phase diagrams for b.c.c. Zr, Ti, Ta and Nb demonstrate the existence of the Me_xC_y carbides. Figure 6.2(a) shows a schematic diagram of carbide formation in the practically important Ti–C and Zr–C systems, while Fig. 6.2(b) presents the general outline of the carbon concentration profile at the diamond/metal interface during diamond–metal interaction at temperatures higher than 1000°C. Carbide formation in joints occurs via the reactive diffusion process.

The morphology of the carbide layer formed by Zr, Ta, Nb, 316L and Inconel on a diamond surface is the same as the morphology of the diamond surface. These

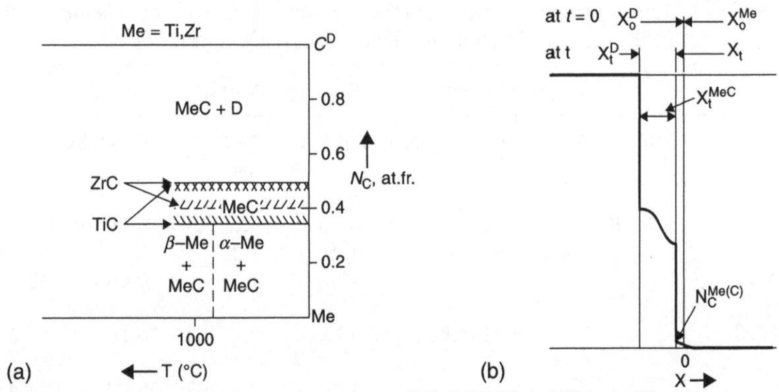

6.2 Formation of carbides according to Ti- and Zr–C partial phase diagrams (a) and general outline of the carbon concentration profile at the diamond–metal interface during diamond–metal interaction at temperatures higher than 1000°C. Here *t* is the time, x^D is the diamond dimension at times t=0 and t=t; x^{Me} is the metal in contact dimension at t=0 and t=t; $N_C^{Me(C)}$ is the carbon concentration; and C^D is the concentration of carbon in diamond, C^D = 1.

layers are very even and uniformly cover all diamond crystal facets (Fig. 6.3). In the case of titanium the attachment is not as good as that in Zr, Ta or Nb.

The strength of the interface between diamond and the formed carbide also depends strongly on how close are the epitaxial relationships of their crystal lattice parameters. These relationships depend on the misfit between atomic positions in both crystal lattices. The strength of the diamond–carbide bond will be a maximum when the misfit is a minimum. Moreover, due to differences in the atomic dimensions on different atomic planes, this strength is not constant on crystal planes with different indexes.

To understand the possible reasons for this, it is worth considering the structural patterns and lattice constants of diamond and the highest carbides (MeC) of Ti, Nb, Ta and Zr with the B1 structure. Figure 6.4 shows $(100)_{MeC}$ and $(110)_{MeC}$ projections of MeC (B1) carbide on which $(100)_D$ and $(110)_D$ diamond projections

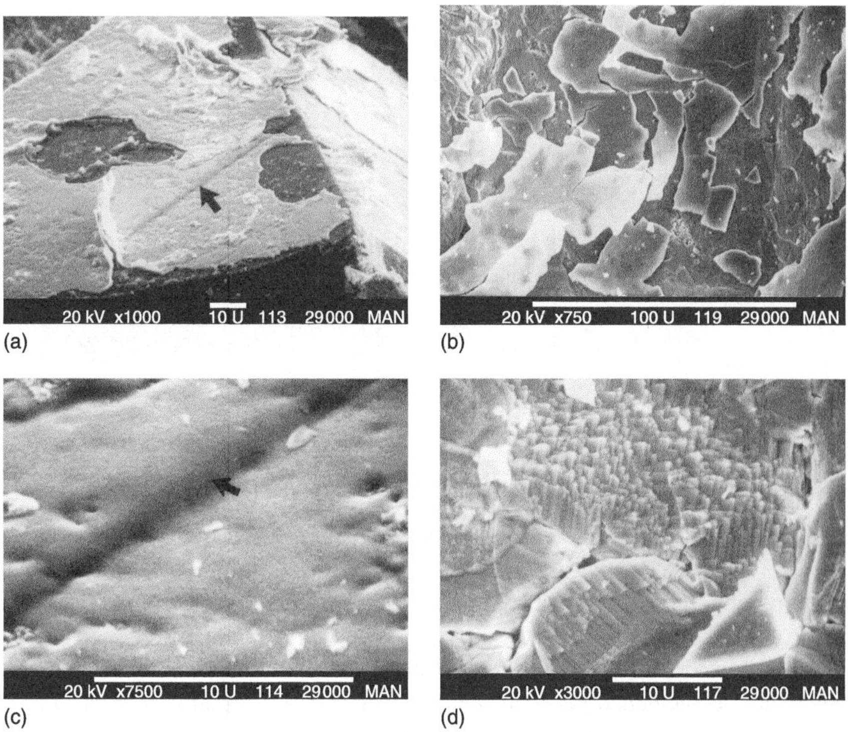

6.3 Epitaxial thin carbide film formed on a diamond surface during sintering of diamond in Zr powder at 1000°C, 5 h, in vacuum. The film is uniform and covers all crystal facets and even the original crystal growth steps (arrows on (a) and (c)). The longer annealing time results in overgrowth, which leads to loss of bonding with both the diamond (b) and the matrix (d). Fracture occurs on both D/ZrC and ZrC/Zr interfaces.

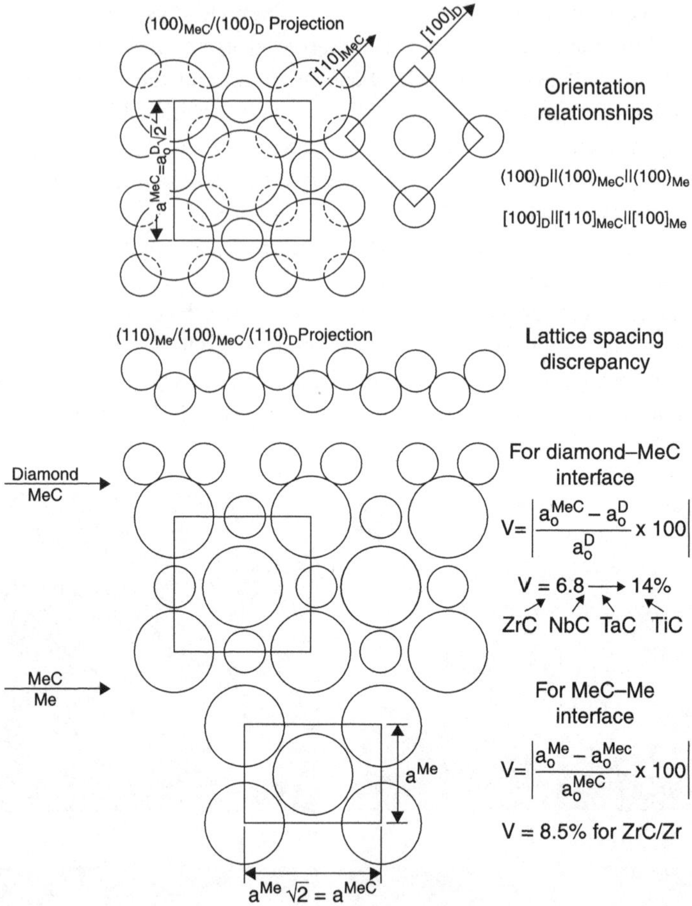

6.4 Schematic presentation of crystal lattices of D, some b.c.c. transitional metals, and their MeC carbides with B1 structure. Possible epitaxial relationships for $(100)_D$ and $(100)_{MeC}$ planes and lattice spacing discrepancy (Rabinkin, 1982).

are superimposed. The figure also shows $(110)_{Me}$ metal projection and lattice spacing discrepancies for four of the carbides and metals in question.

When considering these structural patterns and lattice constants, one finds a remarkable coincidence between the $(100)_D$ and $(111)_{MeC}$ planar nets of D and carbides when $(100)_D$ II $(110)_{MeC}$. Specifically, the closest proximity and the strongest atomic bonding calculated were found by Yamazaki *et al.* (2009) between D and V_4C_3 carbide lattices in the case of the interface with $(111)_D$ and $(111)[111]_{V4C3}$ indexes (Fig. 6.5).

The difference between the lattice spacings of diamond and carbides, *V*, is in the range favorable for epitaxy: that is, −7% for ZrC and up to 14% for TiC. Therefore, formation of the thin carbide epitaxial layer on a diamond surface

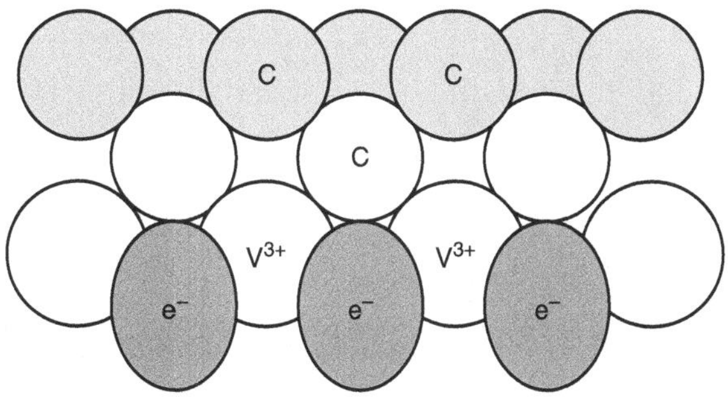

6.5 Schematic illustration of vanadium atoms binding to diamond in a NaCl type reaction product of V_4C_3 formed at 1080 K (Yamazaki *et al.*, 2009).

might be expected on the sites from which carbon atoms dissolve. On the other hand, in the case of graphite–metal, *V* is substantially higher and there is no convenient orientation relationship. These circumstances might be the reasons for the different contact behaviors of various forms of carbon. It is worthwhile mentioning that for Ti the difference, *V*, has the highest value among these four carbides. That may explain why titanium carbide has the worst attachment to diamond among all carbide-forming elements. Therefore, the use of a titanium coating before brazing for diamond grit may not be the best choice.

There are important details of the carbide formation process which are worth noting. First of all, the volume changes during formation of metal carbides by various metals and diamond are negative. Moreover, carbon and metal atoms have very different diffusivities. Therefore, the mass transfer is not capable of supporting volume continuity. This results in porosity within the carbide film and the interface region. This porosity was observed in composites made of Cr, Zr, etc. as shown in Fig. 6.6(c) (Rabinkin, 1982). Epitaxial formation of thin carbide films on the diamond surface might be expected during annealing in the solid state not only with Zr, Ta, Nb and Ti but with powders of alloys containing these elements. This formation is very important for sintering diamond/metal composite tools.

Nevertheless, in full agreement with the general theory, any crystal defects, such as edges or corners of crystals, dislocations, crystal growth steps, etc., are the places where the diamond–metal interaction starts. But the most interesting phenomenon found is the decrease of the selective sensitivity of the diamond–metal interaction with an increase of the metal affinity to carbon. Specifically, in diamond crystals/metal composites made via sintering with Ni, Co or Fe, the reaction starts and develops first at defects or on very uneven surfaces with numerous steps, pits, etc. For the cases of Zr, Nb and Ta, uniform and complete carbide films appear on all kinds of crystal faces, regardless of the presence of the

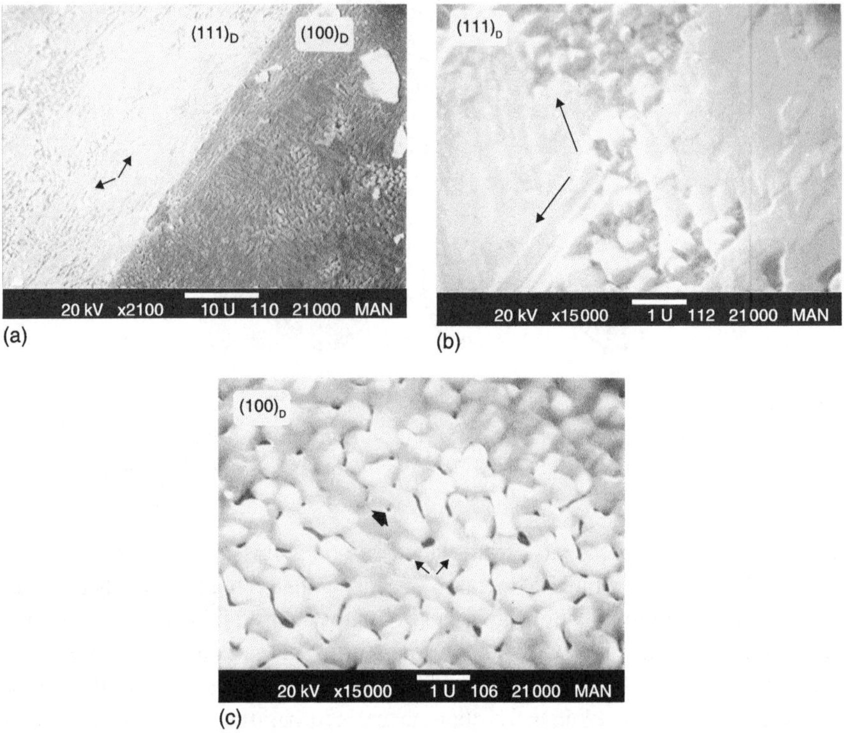

6.6 Wetting of, and carbide formation on, diamond in a diamond chromium composite after annealing in vacuum at 1000°C, 5h. The thin carbide film is attached well to the diamond crystal. Its morphology repeats the trigonal and square pattern characteristics of (111) and (100) crystal planes (see thin arrows on (a), (b) and (c)). Porosity formed on overgrown carbide is clearly seen in panel (c) (thick arrow). The film was poorly attached to the matrix socket.

same defects in diamond. Micrographs Nb (Fig. 6.1) or Zr (Fig. 6.3(a), (c)) composites with diamond show carbide films with uniform thickness in places where definite crystal growth steps exist. Therefore, dissolution proceeds uniformly, so that the majority of the crystals preserve their original shape during joining.

Carbide films formed by alloys are very different from those formed by their elemental constituents. Their formation starts at much lower temperatures.

6.3.4 Attachment of carbide films to diamond and to a metal matrix

Strong and regular atomic bonds formed on the diamond/carbide interface result in good attachment of the carbides to diamond when the carbide thickness is small (0.1–0.5 μm).

On the interface between the carbide films and the rest of the base metal matrix the difference, *V*, is higher than that for diamond–carbide. Consequently, strong stresses should arise at the interface, with subsequent crack formation. Indeed, during shear tests of metal–diamond composites sintered at low temperatures a fracture occurred along the metal/carbide interface, leaving a carbide film attached to the crystals (Fig. 6.1). With increasing temperature carbide layers grow further. This growth is accompanied by an increase in the stress level. Therefore, the film attachment deteriorates and fractures occur on both diamond/metal carbide (Fig. 6.3(a)) and metal carbide/metal (Fig. 6.3(d)) interfaces.

The appearance of fine and regular porosity in joint components after annealing and brazing in both vacuum and a nitrogen atmosphere might be an additional source of carbide film brittleness (Fig. 6.7). For elements which do not form

6.7 (a) Formation of carbonitride film on diamond sintered with 316L powder in nitrogen (1000°C, 5 h). As a result of the film overgrowth there is a poor attachment of the film to both the diamond and the metallic matrix ((a), (b)). Fine and regular porosity develops in the film (arrows in (c)) contributing to the joint brittleness.

carbide on diamond due to low affinity to carbon (Ni, Co, Fe), wetting practically does not occur. As a result, fractures occur easily along the diamond/metal interface, leaving the stones clean and transparent following a shear test.

6.4 Diamond graphitization during annealing and brazing

The diamond graphitization process starts at rather high temperatures in vacuum or in a medium that does not assist it. For pure natural grit this temperature is of the order of ~1500°C and even higher (Davies and Evans, 1972). Graphitization of the diamond/metal composites Ni+diamond, Co+diamond and Fe+diamond was observed in vacuum at 1000°C (Rabinkin, 1981). The retention of graphitized diamond is poor and, therefore, should be avoided during diamond joining. It was also shown that the use of nitrogen as an annealing atmosphere retards this process.

Since the activation energy of graphitization for different diamond crystal planes is very different, one would expect that if diffusion proceeds via graphitization, then the rate of diffusion would be strongly dependent on the crystallographic directions. The microstructures observed in sintered diamond/metal composites clearly display spherically shaped regions enriched by a carbon 'halo' surrounding single crystals with various crystallographic planes in contact with the matrix. Furthermore, strong carbide-forming elements yield uniform carbide films on different crystallographic planes, as can be seen in Fig. 6.6(a), (b). The same behavior was confirmed by direct measurements of the carbon concentration along different crystallographic directions (Rabinkin, 1981). These results suggest that the dissolution and diffusion rates on a macro level are affected weakly by the diamond crystal plane crystallographic indexes.

6.5 Wetting of diamond by metals and alloys

Naidich *et al.* (1988) found that wetting by and contact angles between diamond and any other elements are determined directly by the nature of their interaction, which correlates well with the work of adhesion of atomic species on a diamond surface. They also proved that wettability or work of adhesion, W, correlates well with the enthalpy of formation of the highest carbide existing in any particular metal–carbon system. Therefore, the following order of wettability may be postulated on the basis of thermodynamic data for the enthalpy of carbide formation for the important metals shown in Table 6.3: Ni, Co, Fe, Cr, Nb, Ta, Ti and Zr. Here Ni has the minimum work of adhesion whereas Zr should have the maximum value. It is important to underline that Ni and Co do not wet diamond under any of the temperature, pressure and environmental conditions applied in this study.

Non-transitional elements from Group IB of the periodic table interact with diamond in the liquid state only via physical interaction and thus have very large contact angles, θ_c. For example, Cu, Ag, Au, Ga, Sn and others and their alloys have θ_c~120–150°. In the case of the carbide-forming transition metals, wetting of diamond proceeds via an instant creation of carbide film, because, in fact, metal atoms have a tendency to build strong chemical bonds at the interface. Therefore, metallic layers that are in contact with diamond have a different nature and structure from that of the matrix. Thus, use of the word 'wetting' in the case of transition metals is rather conditional because of the practically instantaneous diamond/metal chemical bonding. It is worth mentioning that the closer the carbon atomic pattern on the diamond crystal to that of the forming carbide the better the wetting and the lower the contact angle, as confirmed by observation of a lower contact angle on a $(111)_D$ plane than that on $(100)_D$ (Naidich *et al.*, 1988).

The wetting of diamond by solid metals during annealing (sintering) of diamond/metal composites is important since metal–diamond bonds can exist only on the joint interface. This wetting occurs via metal vapor condensation, first in the form of small metal droplets, which grow into complete films over the sintering time (Rabinkin, 1982). It was also noticed that many alloys containing carbide-forming elements start to wet at much lower temperatures than corresponding elemental metals. An interesting phenomenon was observed by Rabinkin (1982) when considering diamond wetting by Cr-containing alloys and pure chromium. Thus, at low temperatures (about 900°C) chromium carbides appear on the socket surface like small drops of liquid with high surface tension (Fig. 6.8(b)). Wetting or carbide formation on diamond occurs only at 1000°C (Fig. 6.9).

When used as an annealing atmosphere, nitrogen contributes to the wetting by Fe (starting from 900°C); however, in the cases of Nb, Zr and Inconel it retards wetting and carbide formation substantially.

The 316L alloy also wets diamond like a liquid with high tension in a way similar to pure chromium (Fig. 6.8b).

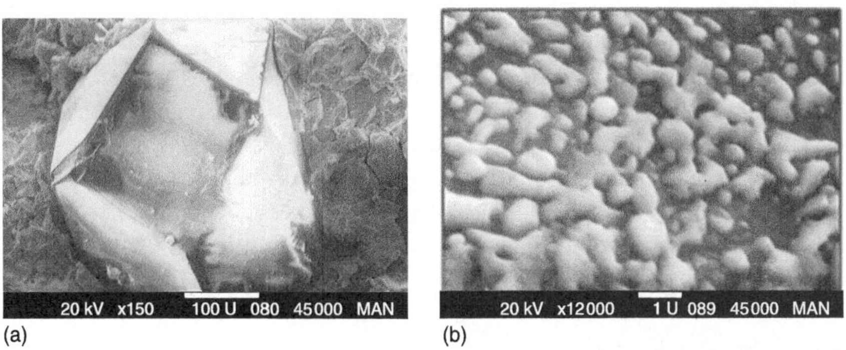

6.8 (a) Diamond–chromium interaction in vacuum (900°C, 5h). Carbide formation starts on a socket surface (b) and the diamond crystal is unwetted and clean. Carbide appears as frozen droplets of a liquid with high surface tension.

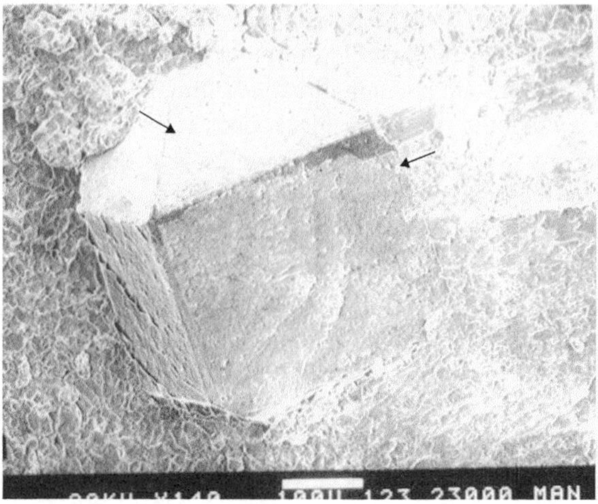

6.9 Diamond/316L composite sintered at 1000°C, 5 h. Good formation and attachment of the even chromium carbide film occurs on all crystallographic planes (arrows).

6.6 Wetting of cubic boron nitride (CBN)

Due to the chemical stability of CBN, in order to achieve some degree of wetting it is necessary either to coat preliminary CBN with materials containing nitride- or boride-forming elements or to add such materials as active components within the brazing filler metals (BFMs). The same researchers, Chattopadhyay and Hintermann (1993a), reported the wetting effectiveness of titanium additions to BFMs based on the silver–copper eutectic. Thus, when an AgCu alloy with 8% Ti was used for brazing at 900°C for 30 s under an argon atmosphere, perfect wetting of the CBN grits occurred. Naidich and Adamovskyi (2006) investigated wettability and work of adhesion of pure metals and alloys on CBN surfaces. Results of this work are shown in Table 6.5. Adding 10–15 wt.% titanium to a copper–tin alloy is very effective, resulting in a decrease of the wetting angle to about 30° (Fig. 6.10).

In contrast to diamond brazing, chromium is not active enough as an agent of wettability at brazing temperatures in the normally used range. Even at a fairly high brazing temperature of 1100°C, no reaction occurs between a Ni/Cr/P-alloy (BNi-7) and CBN. Switching to a BFM with an increased chromium content of 25% (Nicrobraz 51) did not show any improvement in wetting (Chattopadhyay and Hintermann, 1993a).

6.7 Brazing filler metals and techniques for diamond joining

Brazing filler metals for joining diamond are represented by two groups of alloys. The first group consists of alloys that do not form chemical bonds with carbon; the

Table 6.5 Contact angles, θ° and work of adhesion, W_A, of pure metals and alloys on CBN under vacuum of 2–3×10^{-3} Pa

Metal, alloy	CBN as substrate		
	T (°C)	θ°	W_A (mJ/m^2)
Cu	1100	137	360
Ag	1000	146	160
Sn	1100	137	115
Pb	700	124	190
Sn + 1.5 wt.% Ti	800	30	930
Cu + 10 wt.% Sn	950	136	290
Cu + 20 wt.% Sn	950	135	200
(Cu + 10 wt.% Sn) + 10 wt.% Ti	950	28	1800
(Cu + 20 wt.% Sn) + 15 wt.% Ti	950	21	1820

Sources: Naidich and Adamovskyi (2006).

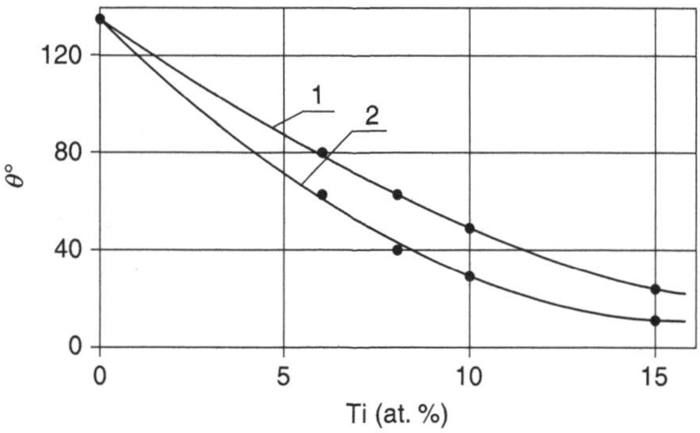

6.10 Wettability of CuSnTi alloys on CBN at 950°C under vacuum of 2–3×10^{-3} Pa. Alloy 1: Cu10 wt.%Sn. Alloy 2: Cu20 wt.%Sn. Naidich and Adamovskyi (2006).

second group consists of chemically active alloys containing carbide-forming elements such as titanium, zirconium, chromium, boron or silicon. Active alloys of the second group are commonly used now as BFMs to join diamonds with metals, ceramics or cermets. Active brazing alloys (ABAs) having a joining temperature below 1100°C are preferred to prevent graphitization of the diamond.

Commonly used BFMs are listed in Table 6.6, including ABAs suitable for joining diamonds (both grits and compacted polycrystalline inserts) to metals or cemented carbide substrates, which are usually used as diamond tool carriers. A number of diamond brazing techniques exist in industrial practice. The processes and equipment used for this purpose are almost the same as those used for brazing ceramics (especially boron nitride). Below, we describe processes and materials in order to demonstrate the specifics of the approach to brazing diamonds. In general, a diamond brazing technique includes preparation and application of grits and BFMs, metalizing the diamonds, choosing brazing temperatures which prevent graphitization and/or

Table 6.6 Commercial brazing filler metals for joining diamonds with metals, ceramics, and cemented carbides

Brazing filler metal	Composition (wt.%)	Solidus temperature (°C)	Liquidus temperature (°C)	Brazing temperature (°C)
BNi-2 (Amdry 770)	Ni–(6–8)Cr–(4–5)Si–3B–3Fe	971	999	1010–1177
BNi-3 (Amdry 780)	Ni–(4–5)Si–(2.8–3.5)B	982	1038	1010–1177
BNi-7	Ni–(13–15)Cr–10P	888	888	927–1093
BNi-10	Ni–(10–13)Cr–3.5Si–2.5B	970	1105	1149–1200
Nicrobraz LM	Ni–7Cr–3Fe–4.5Si–3B	970	1000	1010–1170
Metglas MBF-1002*	Ni–32.2Pd–8.6Cr–2.7B–0.9Fe	900	1010	1020–1060
Metglas MBF-1011*	Pd–40Ni–5Co–4.5Mo–5Si	847	895	900–950
Nicro-B™	Ni–15.2Cr–4B	1048	1091	1100–1140
Cusil®ABA	Ag–35.2Cu–1.75Ti	780	815	830–850
Incusil®ABA	Ag–27Cu–12.5In–1.25Ti	605	715	740–750
Ticusil®	Ag–26.7Cu–4.5Ti	780	900	920–960
BrazeTec® CB10	Ag–25.2Cu–10Ti	780	805	850–950
TiBraze®375	Ti–37.5Zr–15Cu–10Ni	825	835	850–900
TiBraze®200	Ti–20Zr–20Cu–20Ni	848	856	870–900
TiBraze®1200	Ni–27Ti–10Al	1140	1180	1180–1220
CuSnTi (non-commercial alloy)	Cu–(20–25)Sn–(15–20)Ti	860	890	890–950
TiBraze®Al-600†	Al–12Si–0.8Fe–0.3Cu	577	580	650–720
TiBraze®Al-665†	Al–2.5Mg–0.2Si–0.4Fe–0.2Cr	610	645	700–720
BAg-34‡	Ag–(31–33)Cu–(26–30)Zn–2Sn	649	721	721–843

*Brazing of cemented carbide substrates and cemented carbide with diamonds sintered in carbide matrix.

†Brazing of diamond coated by Ti or Mn film.

‡Brazing in air with flux.

formation of brittle intermetallics, and so forth. Every time the brazing technique is considered, the particular diamond material and design of the article must be analyzed first in order to select a suitable BFM and adjust the brazing process.

If the clearances between diamond grits are close to or larger than the average size of the diamond crystals, the application of hard BFMs such as Ni–Cr–Si–B or Ti–Zr–Cu–Ni provides better wear resistance and durability for the brazed abrasive tool than that of Ag–Cu–Ti or Cu–Sn–Ti-based BFMs. On the other hand, ductile Ag–Cu–Ti or Cu–Sn–Ti-based BFMs better reduce thermal stresses between the diamond and the substrate, especially when brazing relatively large diamond wafers or inserts to metals (Shapiro, 2007).

Joint strength essentially depends not only on the joint design and the selected filler metal but also on the brazing temperature and the holding time, both affecting reaction of the BFM with the diamond and the base material. Therefore, adjustment of the thermal cycle in vacuum brazing is also important for achieving better joint strength.

For example, a diamond wafer 0.5 mm thick and 50 mm in diameter was brazed to a cemented carbide WC-6Co substrate in vacuum using brazing filler metal Ag–Cu–Ti. The best joint strength of 120 MPa to 130 MPa was achieved for a brazing temperature range of 880–920°C and holding time 20 min, while the joint strength decreased by 25% at brazing temperatures above 920°C due to the formation of a thicker and brittle carbide layer at the interface. Brazing with a holding time of 3 min resulted in a drop of the strength to 50 MPa due to insufficient wetting of the diamond wafer (Dos Santos and Casanova, 2004).

Of experimental active filler metals, alloys of the Cu–Sn–Ti system are among the most promising. These alloys usually comprise 10–15 wt.% of titanium as an active component, which is a significantly higher content compared with that in silver-based ABAs. Brazing alloys of the Cu–Sn–Ti system possess high activity in the liquid state and better ductility in the solid state than that of the Ag–Cu–Ti alloy system.

As an illustration of the industrial potential of the Cu–10Sn–15Ti alloy, there is an example of its usage in fabricating a cutting tool equipped with a single layer of diamond grit consisting of 150–180 μm-sized particles. This tool was made of AISI 1045 steel substrate and brazed at 880°C in vacuum or by CO_2-laser beam for heating the tool surface (Huang *et al.*, 2004). The following procedure was used. A mixture of 10 vol.% of diamond crystals with 90 vol.% of the Cu–10Sn–15Ti powder was deposited on the 1.5 mm thick steel substrate before brazing and heated for 10 s by a 450 W laser beam. The beam had an effective diameter of 7 mm. Argon with a flow rate of 100 cm^3/s was used as a shielding gas to prevent oxidation of both diamond grit and BFM. A continuous TiC layer was formed on the diamond grit after a 5 min hold in vacuum at the brazing temperature, while laser heating resulted only in the appearance of discrete TiC islands.

The composition of the Cu–Sn–Ti alloys, which contain from 5 wt.% to 20 wt.% of titanium, essentially affects both the brazing temperature to be applied and the strength of the diamond brazed joints (Klotz *et al.*, 2008). Thus, the liquidus of Cu–14.4Sn–10.2Ti–1.5Zr (wt.%) BFM is 925°C, whereas it is 1110°C for the Cu–25Sn–15Ti alloy. Due to the high concentration of titanium, the molten filler metal always reacts with diamond, and the reaction product formed on the diamond surface is a solid layer of titanium carbide (TiC) having different thicknesses: from 50–100 nm for the alloy Cu–15Sn–20Ti to 100–250 nm for the alloy Cu–25Sn–15Ti and 150–300 nm for the alloy Cu–14.4Sn–10.2Ti–1.5Zr (in wt.%). This carbide layer in turn has a complex microstructure consisting of two types of TiC crystals: the first type is formed in the immediate proximity to the diamond and has 'cuboidal' morphology, whereas the other type is columnar. A maximum thickness of this cuboidal+columnar titanium carbide layer of about 600 nm was observed after brazing at 930°C for 100 min.

Very promising results were recently obtained by substituting titanium with vanadium as an active component in silver–copper-based filler metals. The eutectic Ag–27.8Cu braze alloy containing less than 1 wt.% of vanadium had a stable shear strength for diamond joints of over 200 MPa (Yamazaki and Suzumura, 2006). This effect is caused by the fact that silver and vanadium carbide have only a slight difference in crystal lattice mismatch. Microstructural study of the interface between the diamond and the joint metal revealed that silver crystal grains grew from vanadium carbide islands formed on the diamond surface. X-ray diffraction indicated that different types of vanadium carbides, V_2C, V_4C_3 and V_8C_7, were formed on the diamond, and the V_4C_3 reaction product especially provides good adhesion of filler metal to diamond due to preferable solidification of silver on islands of this carbide.

As mentioned above, one of the important factors detrimental to integrity of the join is the formation of graphite on the diamond surface. The graphitization process was sometimes observed when Ni–12Cr–4Fe–3Si–2.5B BFM was used for joining the same grit in vacuum at 970°C. A graphite layer on the diamond surface is the weakest phase in the brazed structure. At the same time, a chromium carbide layer may also appear at the interface between the graphite film and metal matrix, whereby conversion into graphite proceeds faster than growth of chromium carbide phases (Huang *et al.*, 2004).

In order to prevent graphitization, titanium, chromium or manganese films can be deposited onto the diamond surface using CVD or physical vapor deposition (PVD) techniques. Manganese or titanium coating films 1–3 μm thick decrease the contact angle of molten Al-based alloys such as TiBrazeAl-600 to about 25–40°, vs. 75–90° for the uncoated diamond (Table 6.7). Low-temperature brazing at 700–750°C using aluminum filler metals made it feasible to produce diamond-brazed drills with twice as long a service life due to the prevention of stresses in joints and diamond graphitization. Other low-temperature BFMs such as Al–2.5Mg–0.3Cr or Al–4.2Cu–1.5Mg–0.5Si are also suitable for joining titanium-coated diamond grit to nickel alloys or titanium alloys (Shapiro and Flom, 2010). Single-crystal diamond grits and titanium-coated diamond grits of −40/+50 mesh can be brazed to the

Table 6.7 Contact angles of wetting diamonds with some pure metals and different 8FM in vacuum

Liquid metal	Temperature (°C)	Contact angle, θ
Tin	800	130
Sn–2.9%Ti	800	40
Sn–2.9%Ti	1100	7
Silver	1100	120
Ag–28%Cu	850	120
Ag–27%Cu–1%Ti	850	15
Ag–26%Cu–2%Ti	850	12
Ag–34.5%Cu–1.5Ti	810	45
Ag–34.5%Cu–1.5Ti	850	18
Copper	1150	140
Cu–20%Sn–2%Ti	1150	35
Cu–20%Sn–7%Ti	1150	5
Cu–1%Cr	1150	30
Cu–4%Cr	1150	5
Cu–40%Mn	1100	30
Ag–30%Mn	1100	20

Sources: Naidich *et al.* (1988); Palavra *et al.* (2001).

periphery of a circular steel saw or cemented carbide WC-Co cutting tool inserts using torch brazing in air and BFM wire similar to AWS BAg-22, with the fluxes containing a combination of either (a) 42 wt.% KF, 23 wt.% KBF_4, and 35 wt.% B_2O_3 or (b) 25 wt.% KBF_4, 5 wt.% K_2CO_3 and 70 wt.% borax (Wang *et al.*, 2009).

A chromium–tungsten–copper multilayer film deposited onto the polycrystalline diamond wafer by magnetron sputtering can effectively diminish residual thermal stresses in the metal–diamond brazed joint. The 0.12 μm-thick top layer of copper aims to improve wetting by the brazing filler metal Cusil-ABA® (Ag–35.2Cu–1.75Ti), which is mounted between the metalized surface of the diamond and a substrate comprising nickel or Fe–42Ni alloy before brazing. Chromium film 0.02 μm and tungsten film 0.03 μm thick, which have a coefficient of thermal expansion between diamond and nickel, are used as stress compensators (Petkie, 2003). In this work the assembly was placed in a furnace with an atmosphere containing either an argon and nitrogen mixture, or nitrogen with 4% hydrogen, to braze at 825°C for 5 to 10 min. The adhesion between the components was tested using the peel test and found to be very strong. Moreover, the joint strength was improved when 50 μm copper foil was used as the BFM for joining the same coated diamond wafer at 1100°C. No delamination was observed after four thermal cycles carried out between room temperature and 800°C.

In another case it was found that the microstructure of a diamond–steel brazed joint manufactured with titanium-containing active filler metals (e.g. Cusil-ABA®) comprises five different phase layers: (a) two interlayers L1 and L2 directly attached to the steel, (b) two solid solution (Cu-rich and Ag-rich) layers in the middle of joint 1, and (c) a thin TiC intermetallic reaction layer 50–100 nm thick at the diamond interface (Buhl *et al.*, 2010). The interlayer L1 has composition $(Fe,Cr,Ni)_2Ti$ whereas the interlayer L2 is a (Fe,Ni,Cu)Ti phase. The thickness of the brazed joint in this case does indeed decrease with increased brazing temperature and extended dwell time: the final thickness of the joint metal is 36 μm at brazing parameters 850°C for 10 min, and 15 μm after brazing at 910°C for 30 min. This can be explained by the decreasing viscosity of the liquid filler metal with an increase of the process temperature. This means that brittle interlayers make up a higher percentage of the total thickness of the joint metal in the latter case, affecting the shear strength. A ductile fracture of the joint is favored when lower brazing temperatures are applied, whereas a fracture will occur in the brittle interlayers at higher brazing temperatures. Residual stresses are almost independent of the brazing parameters, but they strongly affect the microstructure and therefore the shear strengths of the joints. For this reason, the best joint properties, i.e. low residual stresses and high shear strength up to 390 MPa, were obtained by brazing at the lower temperature 830°C for 10 min.

It is worth underlining that whereas carbide formation creates a strong chemical bond between the diamond and the joint metal, on the other hand this reaction may degrade the diamond surface, and the diamond impact strength may be reduced after brazing. In order to diminish the negative effect of the carbide-forming reaction, it is recommended depositing a thin metal layer on the diamond surface to be brazed. For example, deposition of a 10 μm-thick copper coating film provides good wetting with BFMs without degradation of the diamond surface. The adherence of diamond grains surrounded by the liquid filler metal is satisfactory without compromising the impact strength (Sung and Sung, 2009). The positive effect of a Ti coating on diamond before brazing with Ni–Cr filler metals has also been experimentally corroborated. Fracture of Ti-coated diamond abrasive grit during grinding occurs without breaking the diamond grain (Shao *et al.*, 2008).

A copper film can also be deposited onto diamond grit using the chemical metalization technique, for example using a water solution of potassium–sodium tartrate 80 g/l, copper sulfate 20 g/l and formaldehyde 10 g/l at pH = 12.8 (Yakubovsky *et al.*, 2010). The diamond powder is preliminarily treated in an activated 20% solution of palladium chloride or in a mixture of palladium and tin chlorides at 50–80°C. Stabilizers, such as potassium hexacyanoferrate $K_3[Fe(CN)_6]$ 1 g/l and sodium thiosulfate $Na_2S_2O_3$ 0.01 g/l, cation-active surfactants 0.01 g/l, adhesion-enhancing agents such as nickel chloride $NiCl_2$ 3 g/l, and accelerators such as sodium carbonate Na_2CO_3 3 g/l are also needed to deposit a copper film on the diamond surface after 6–20 min at room temperature and continuous slow stirring.

6.8 Mechanical testing of diamond joints

The strength of diamond-to-diamond and diamond-to-metal brazed joints cannot be measured directly using standard methods for testing mechanical properties due to small size of the brazed diamond particles or inserts. Therefore, indirect methods are used for the mechanical characterization and quality assessment of brazed joints. For example, the three-point bending test of a steel plate coated with brazed diamond grains gives statistically proven characteristics of the mechanical behavior of brazed joints made in vacuum by different filler metals (BNi-2, BNi-3 and BNi-6) at different temperatures (Tillmann *et al.*, 2008). Specimens for the bending test were manufactured from chromium–nickel steel plates with brazed-on 300–400 μm diamond grains situated in the region expected to have the biggest deformation during the test. The number of diamond grains still remaining in the brazed metal coat after the test is a measure of the joint strength, which depends on brazing temperature, holding time and bending angle. Diamonds with cracks were rated as intact. Brazed joints made with BNi-2 exhibited good results at brazing temperatures in the range of 1005–1080°C and holding time 10–15 min: almost 100% of the diamond grains were intact, while lower brazing temperatures were associated with only 50% intact diamonds. The BNi-6 filler metal demonstrated the opposite behavior: increasing the brazing temperature over 1000°C results in worsening quality, to 55% intact diamonds at 1100°C. BNi-6 containing 3–5% of chromium exhibits good wetting and high quality when brazing diamond, with 96% of the diamond grains being intact after being subjected to joining at 950–1000°C for only 5 min of holding time.

Interfacial bonding strength between various brazing filler metals and CVD-formed diamond was tested by Hsien and Lin (2009). Nine BNi-series alloys, including the very frequently used BNi-2, -7 and Cu- and Ag-based alloys, were used. The best bending strength, about 280–400 MPa, was achieved for joints prepared using Cu–10Sn–15Ti and particularly silver-ABA FM. The brazed interface was stronger than the bulk of the joints. The popular BNi-series alloys containing B and P as melting point depressants caused the catalytic degradation of diamond into sp^2-bonded carbon because of the presence of Ni in the filter metal base. As a result, the joints had a much lower strength of 80–120 MPa.

The results of the bending test provide only limited information about the mechanical behavior of diamond grinding tools in real application conditions. In order to check the grinding ability of a diamond–steel or diamond-cemented carbide tool, specimens can be subjected to the pin-on-disk test (Tillmann *et al.*, 2008). The specimen, fixed in a grip and loaded in the vertical direction, is brought into contact with rotating steel or an abrasive counter-body. The test is interrupted after a certain running time or path intervals, and the loss of counter-body mass is measured, as well as the number of diamond grains removed or destroyed. This test provides information about the removal rate, in other words about the wear resistance of diamond–steel brazed joints.

Both tests (three-point bending and pin-on-disk) allow confident evaluation of the quality of brazed joints of diamond or boron nitride grits from the functional point of view. They showed that the specimens brazed with nickel-based filler metals exhibit a substantially higher removal rate of diamond grains and higher masses of the counter-body than in specimens brazed with silver-based filler metals. The amount of material removed from ceramics is greater with grinding tools brazed with BNi-6 in comparison with tools brazed with the BNi-2 alloy. However, the BNi-2 filler metal has better abrasion ability due to higher microhardness than that of BNi-6.

6.9 Brazing of cubic boron nitride (CBN)

Most grinding tools are manufactured by attaching CBN grits to a tool body normally made of steel or a hard metal. A hard metal is the better choice because it exhibits higher stiffness and less distortion, which is of benefit when using a high-temperature joining processes like brazing. The expansion behavior of hard metal fits perfectly with that of CBN, which results in little residual stress within the brazed components.

Two methods can be used to braze CBN grits. In the first method, the grits are coated by metal(s) such as nickel or titanium and afterwards are brazed with a conventional BFM. These coated grits are commercially available. Nickel as a coating material is the preference for electroplating joining, whereas titanium is preferred for brazing. Brazing titanium-coated grits results in higher bonding strength between grit and BFM due to the chemical reaction of titanium with the CBN interface at elevated brazing temperatures. Chattopadhyay and Hintermann (1993b) proved that a TiC coating, processed by CVD, is also sufficient to promote wetting of standard nickel-based BFMs.

The second joining method is so-called 'active brazing', which means using 'active' brazing alloys containing metals that react with BN, such as titanium and zirconium. Today all BFMs used in industry for brazing CBN to a tool body contain these elements. Active brazing foil compositions are chosen from AgCuTi-based systems, whereas pastes consist of AgCuTi or CuSnTi powders and an appropriate binder. During production of these BFMs, titanium can be added to their master alloys as a pure element or as an additional ingredient in the form of a titanium hydride. CuSnTi BFMs are much cheaper than AgCuTi alloys and possess a higher wear resistance, important when manufacturing grinding tools with CBN grits.

An important factor in CBN tool reliability is joint wear resistance, which depends on the nature and amount of binder within an applied braze alloy paste. In this respect Elsener *et al.* (2005) showed the beneficial effect of a cellulose nitrate binder addition. Indeed, binder residues and graphite react with titanium-containing phases of a CuSnTi BFM, forming nanosized titanium carbide particles and thus enhancing erosion resistance.

Shiue *et al.* (1997) tried to improve the wear resistance of joints obtained using AgCuTi and CuSnTi BFMs by adding powders of hard materials such as TiC, SiC and WC as well as pure metals like Mo and W. They reported that the abrasion resistance of BFMs can be greatly improved by introducing these hard and blocky particles. Wear tests showed that TiC addition is superior to all other powder materials studied in this work.

Another major group of CBN applications is in the production of cutting tools, which are used for machining pearlitic cast irons, Cr and Ni-containing alloys, hardened steel, iron-based powder metal parts, hard facing alloys and superalloys (Diamond Innovations, 2004). In all these cases polycrystalline CBN compacts with a defined cutting edge are exploited successfully. Such compacts are available either as tool parts having a sandwich-like structure containing a hard metal substrate supporting PCBN or as solid piece, compacts made of 100% CBN. In order to use such compacts only in areas where their advantageous properties are required and to decrease the tool cost, PCBN tips are cut out of the compacts by wire electrical discharge machining (EDM) or laser and brazed to the metallic tool bodies. Figure 6.11 shows an indexed cutting tool in which CBN tips (arrows) have neen brazed to a hard metal body (Boretius and Klose, 1999; Boretius, 2001).

The wettability of PCBN is more or less comparable to that of CBN grits. That means that an active metal coating has to be applied before brazing with conventional BFMs, or an active BFM has to be used. Again, AgCuTi and CuSnTi alloys are used for the active brazing process. Brazing PCBN tips to a hard metal body using CuSnTi results in higher bonding strengths compared with AgCuTi BFM.

From the brazing point of view, indexed parts in a sandwich-like form with a hard metal substrate base can be considered as conventional hard metals to which the usual brazing technology can be applied. In these cases silver-based BFMs are frequently used with manual torch and induction heating sources. Vacuum brazing is also used. Although the mismatch in the coefficient of thermal expansion

6.11 Vacuum-brazed indexable insert with two solid PCBN tips.

between PCBN and hard metal is rather low, brazing temperatures should be kept as low as possible to avoid delaminating of the PCBN layer from the hard metal back-up.

Figure 6.12(a) shows a vacuum-brazed indexable insert consisting of a hard metal body and a solid PCBN tip together with micrographs of the joint between them. The reaction layer formed in the joint (Fig. 6.12(b), (c)) has a pattern similar to that observed and well known in ceramic–metal joints brazed with active metal additions (Boretius *et al.*, 2006).

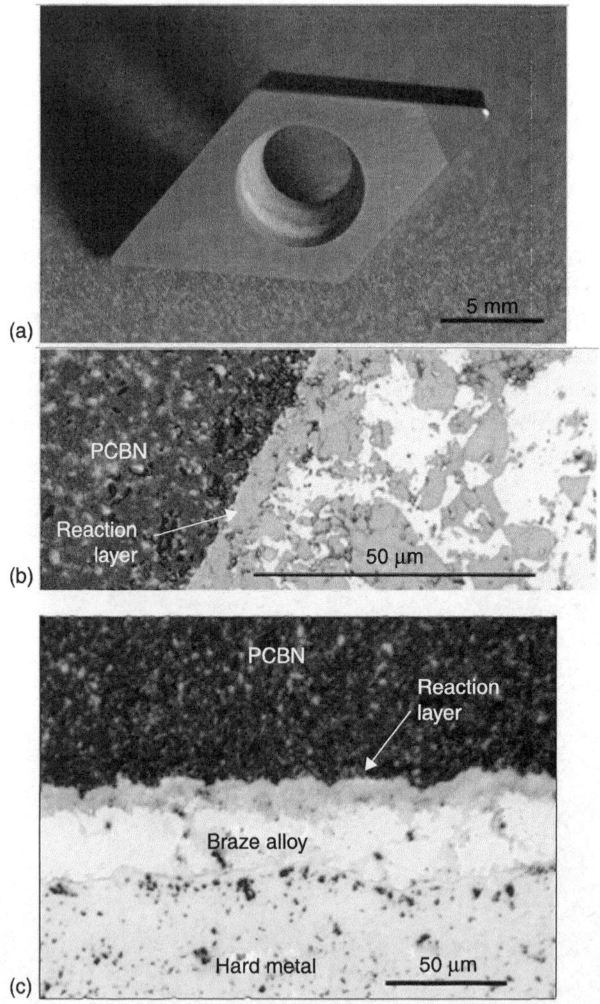

6.12 (a) Vacuum-brazed CBN indexable insert; (b), (c) micrographs of vacuum-brazed CBN-hard metal joint. Braze alloy: CuSnTi; brazing temperature: approximately 950°C.

6.10 Brazed cubic boron nitride (CBN) products

Abrasive CBN grits are used for grinding applications, whereas PCBN – a compound consisting of CBN particles, 1 to 10 μm in size, sintered together with a ceramic matrix – is used for cutting applications. PCBN can also be produced in a sandwich-like form with a hard metal substrate base or as a solid piece. PCBN is used generally for rough and finishing machining of certain steels, gray cast iron and hard cast irons and/or for finish turning of sintered irons and powder metallurgy components.

6.10.1 Indexable inserts

Active brazing is the most promising way to join uncoated, solid CBN tips to a tool body. Due to the chemical reaction between the BFM and the CBN surface a high joint strength can be achieved. Using solid CBN gives more flexibility in the selection of BFMs. Even high melting alloys like CuSnTi can be used without negative impact on CBN properties while guaranteeing high service temperatures. Furthermore, CuSnTi alloys are more wear resistant compared with AgCuTi alloys. The properties of CuSnTi alloys are particularly beneficial for machining applications where high temperatures and mechanical loads occur, like dry, high-speed and hard machining. By brazing solid CBN tips to the tool body the number of cutting edges can be increased easily. Figure 6.13 shows indexable inserts with horizontal (a–c) and vertical (d–h) joint configurations (Becker, 2011).

6.10.2 Grinding tools with designed CBN or diamond grit distribution

Tools with a single layer of abrasive grits are produced mainly by electroplating. Grits are embedded within a nickel-alloy matrix and fixed mechanically only. Grit

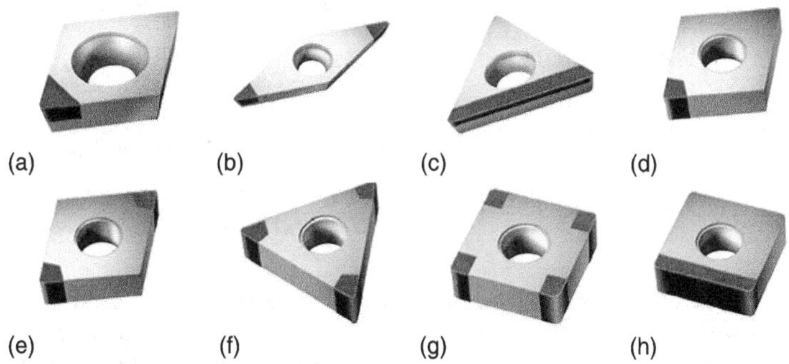

6.13 (a)–(h) Indexable inserts with horizontal (a)–(c) and vertical (d)–(h) joint configurations (Becker, 2011).

protrusion is less than 40–50% of the average grit diameter. During a grinding operation, matrix wear can result in early pull-out of grit and short service times. Vacuum-brazed grinding tools with diamond or CBN grits are promising alternatives. Different brazing alloys and modification of the base alloys, together with adapted brazing parameters, allow chemical interaction and bonding strength to be controlled.

For high removal rates and long service times, two design objectives have to be achieved:

- grit protrusion to ensure sufficient cutting depth;
- grit clearance to transport chipping material out of the working area, to avoid blocking.

Recently a new technique has been developed to place grits with a controlled distribution (Burghard *et al.*, 2002). According to this technique abrasive tools with designed grit protrusion and clearance can be fabricated using vacuum brazing (Fig. 6.14).

Practically the whole production process can be divided into three successive steps:

1. Cleaning and preparation of tool body.
2. Setting of glue spots and sprinkling of grits.
3. BFM application and brazing.

Step 2 is the most innovative one. Using a special dispensing system glue drops are 'fired' to the tool surface (Fig. 6.15). Afterwards grits are sprinkled over the surface and stick where the glue spots are located. Finally a slurry of BFM is

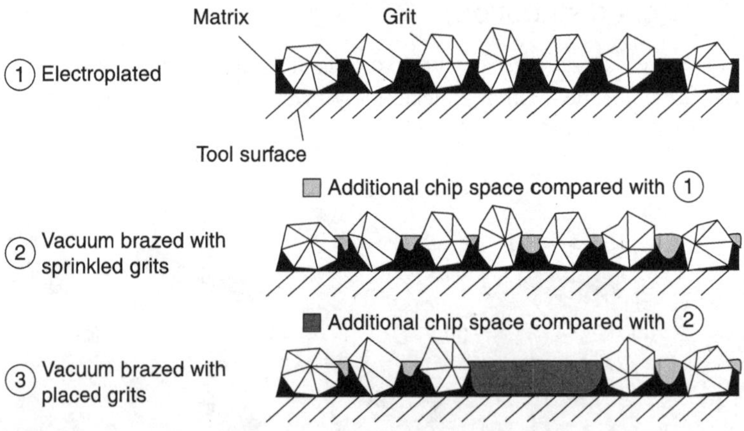

6.14 Grit clearance for different production methods for use in abrasive tools. 1. Cleaning and preparation of the surface of tool body to be joined with diamond grit. 2. Setting of glue spots on the tool surface and sprinkling of diamond grit on top of glue-coated tool surface. 3. Application of brazing filler metal and brazing.

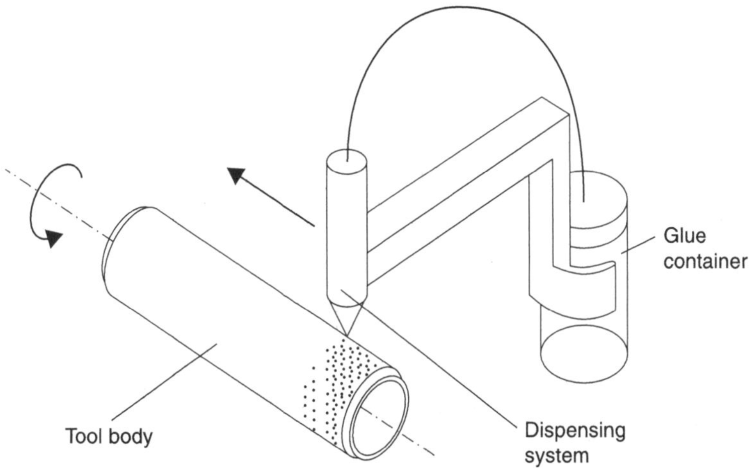

6.15 Schematic view of new grit placement technique (Burghard *et al.*, 2005).

brushed or sprayed over the grits. The amount of braze alloy applied gives the desired grit embedment after brazing.

Figure 6.16(a) shows a surface section of a honing tool $\phi_a 11\times50$ mm after placement of CBN grits B107, and Fig. 6.16(b) shows the same tool after vacuum brazing. Figure 6.16(c) shows the ready-to-use honing tool $\phi12\times50$ mm with CBN B126 (Burghard *et al.*, 2005).

To evaluate the performance of the new CBN honing tools compared with electroplated tools, they were tested in a real production process. Using vacuum-brazed tools cuts down total operation time by 33% compared with electroplated tools. More impressive is an improvement in the tool lifetime (Fig. 6.17). For example, electroplated tools are used to machine about 2000 parts, whereas vacuum-brazed tools can be used for machining up to 20 000 identical parts.

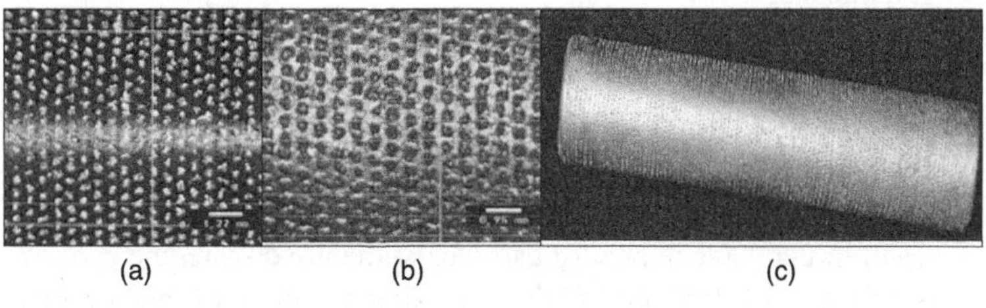

6.16 (a) Surface section of a honing tool $\phi_a 11\times50$ mm after placement of CBN grits B107; (b) the same tool after vacuum brazing; (c) ready-to-use honing tool $\phi12\times50$ mm with CBN B126 (Burghard *et al.*, 2005).

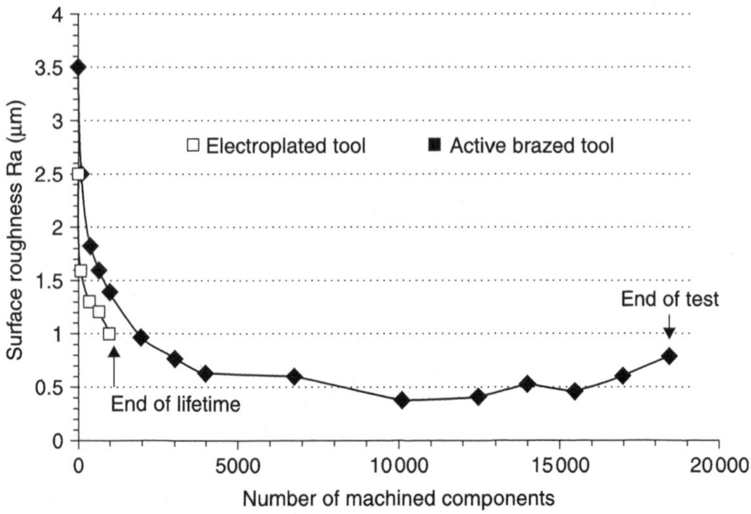

6.17 Number of machined components produced using electroplated and vacuum-brazed honing tools (Boretius *et al.*, 2006).

6.11 Conclusion

1. Diamond strongly reacts with and dissolves in many metals, particularly in all transition metals, which form carbides as a result of chemical bonding with carbon atoms. Non-transition metals such as copper, gold, silver and others do not bond with carbon atoms and as a result wet diamond poorly.

   Importantly, no atoms dissolve in diamond, due to its unique crystal structure. Therefore, all joint formation occurs outside the diamond crystals, and it is regulated by processes taking place at the diamond/metal interface.
2. Diffusion of carbon from diamond into the metal matrix at brazing temperatures occurs via direct diamond dissolution without the intermediate conversion-to-graphite stage.
3. As soon as a thin (a fraction of a micron) carbide film is formed on the diamond surface, the rate of further diamond dissolution drastically decreases due to the very low rate of diffusion of carbon atoms through the carbide lattice compared with the rate of carbon diffusion in all brazing and base metals.
4. Strong and regular atomic bonds formed on the diamond/carbide interface result in good attachment of carbides to diamond when the carbide thickness is small, ~0.1–0.5 μm. With increasing temperature and/or reaction time carbide layers grow farther. This growth is accompanied by an increase in the stress level across the joint. Therefore, the film attachment deteriorates and fractures occur on both diamond/metal and carbide/metal interfaces.

5. The development of brazing process parameters and selection of BFMs strongly depend on crystal characteristics and physical properties of diamonds, as well as the choice of the carbide-forming element as the critical component of the BFM composition. The best and strongest diamond–metal bonding occurs with zirconium and vanadium.
6. Active BFMs containing such strong carbide-forming elements as Cr, Nb, Ta, V, Ti and Zr provide wetting of diamonds and CBN during brazing in vacuum. These elements are equally effective in silver-, copper- and nickel-based alloys.
7. High melting, silver-free alloys like Cu–20Sn–10Ti can be used without a negative impact on CBN properties while guaranteeing high service temperatures. Moreover, the Cu–Sn–Ti alloys are more wear resistant compared with Ag–Cu–Ti alloys.

6.12 References

Andrievsky P, Umansky Ja (1977), *Phazy Vnedrenia (Interstitial Phases)*, Nauka, Moscow, p. 191.

Boretius M (2001), 'Vacuum brazed diamond and CBN-tools', *Proceedings of 6th Int. Conference Brazing, High Temperature Brazing and Diffusion Welding*, Aachen, Germany, pp. 73–75.

Boretius M, Klose H (1999), 'Vacuum brazed diamond and CBN tools', *Proceedings of EPMA Int. Workshop on Diamond Tool Production*, Turin, Italy, pp. 143–149.

Boretius M, Sonderegger M, Kuntzmann B (2006), 'Vacuum brazing of hard materials including cBN and diamond', *Proceedings of European Conference on Hard Materials and Diamond Tooling*, Ghent, Belgium.

Becker Diamantwerkzeuge GmbH (2011), *Main product catalogue*, Puchheim, Germany.

Buhl S, Leinenbach C, Spolenak R, Wegener K (2010), 'Influence of brazing parameters on microstructure, residual stresses, and shear strength of diamond-metal joints', *Journal of Materials Science*, Vol. 45, 4358–4368.

Burghard G, Zigerlig B, Boretius M (2002), 'Spanen mit definiert angeordneten Diamant- oder cBN Körnern', *Industrie Diamanten Rundschau (IDR)* 36, No. 2, 116–120.

Burghard G, Boretius M, Zigerlig B (Listemann AG) (2005), *Method of making an abrasive tool*, European Patent EP 1 208 945 8.

Chattopadhyay A K, Hintermann H E (1993a), 'On brazing of cubic boron nitride abrasive crystals to steel substrate with alloys containing Cr or Ti', *Journal of Materials Science*, Vol. 28, No. 21, 5887–5893.

Chattopadhyay A K, Hintermann H E (1993b), 'On improved bonding of TiC-coated CBN grits in nickel-based matrix', *CIRP Annals – Manufacturing Technology*, Vol. 42, Issue 1, 413–416

Davies G, Evans T (1972), *Proc. Roy. Soc.*, A328, 413–427.

Davis R J, ed. (1995), 'Ultrahard Materials', in *Tool Materials, ASM Specialty Handbook*, Materials Park, Ohio: ASM International, pp. 85–100.

Diamantschneidstoffe (2011), *Physikalische Eigenschaften verschiedener Schneidstoffe*. Available from: www.diamantschneidstoffe.de (Accessed July 2011).

Diamond Innovations (2004), Product information: *BZN* Compacts – Tool Blanks and Inserts – Machining of Ferrous Materials*, Worthington, USA.

Dos Santos S I, Casanova C A (2004), 'Evaluation of Adhesion Strength of Diamond Films Brazed on K-10 Type Hard Metal', *Materials Research*, Vol. 7(2), 293–297.

Elsener H R, Klotz U E, Khalid F A, Piazza D, Kiser M (2005), 'The role of binder content on microstructure and properties of a Cu-based active filler metal for diamond and cBN', *Advanced Engineering Materials*, Vol. 7, No. 5, 375–379.

Gauthier M (1998), *Engineered Materials Handbook, ASM, Testing of Ceramics*, pp. 967–968.

General Electric (2011), *Dedicated to Science & Research*. Available from: www.ge.com/innovation/timeline/eras/science_and_research.html (Accessed July 2011).

Greenville-Wells J (1957), 'Harder than diamond?', *The New Scientist*, 16–18.

Hsien Y-C, Lin S-T (2009), 'Interfacial bonding strength between brazing alloys and CVD diamond', *J Mat Eng Per*, Vol. 18, No. 3, 312–318.

Huang S-F, Tsai H-L, Lin S-T (2004), 'Effects of Brazing Route and Brazing Alloy on the Interfacial Structure between Diamond and Bonding Matrix', *Materials Chemistry and Physics*, Vol. 84 (2–3), 251–258.

Klotz U E, Liu C, Khalid F A, Elsner H R (2008), 'Influence of brazing parameters and alloy compositions on interface morphology of brazed diamonds', *Materials Science and Engineering*, A495, 275–270.

Krishtal M (1970), *Diffusion Processes in Iron Alloys*, 2, P.S.T. Jerusalem, Israel.

Naidich Y V, Kolesnichenko G A, Lavrinenko I A and Motsak Y F (1977), *Brazing and metallization of ultrahard tool materials*, Kiev, Naukova Dumka.

Naidich Y V, Umanski B P, Lavrinenko I A (1988), 'Strength of diamond/metal contact and diamond brazing' (Russian), Kiev, Naukova Dumka, pp. 1–136.

Naidich Y V, Adamovskyi A A (2006), 'Brazing of cubic boron nitride base superhard materials', *Proceeding of 3rd Int. Brazing and Soldering Conference*, 24–26 April, San Antonio/USA, pp. 133–135.

Palavra A, Fernandes A J S, Serra C, Costa F M, Rocha L A, *et al.* (2001), 'Wettability studies of reactive brazing alloys on CVD diamond plates', *Diamond and Related Materials*, Vol. 10, 775–780.

Petkie R (2003), *Brazeable Metallization for Diamond Components*, US Patent 6,531,226.

Prelas M, Popovichi G, Bigelow L K (1998), *Handbook of Industrial Diamonds and Diamond Films*, New York, Marcel Dekker, pp. 1135–1144.

Rabinkin A (1981), 'Diamond Interaction with Various Metals and Alloys under Different Environmental Conditions', *8th AIRAPT Conference on High Pressure in Research and Industry*, Uppsala, 1, pp. 361–364.

Rabinkin A (1982), 'The stability of diamond in contact with various metals under different temperatures, pressures, and environmental conditions', *ManLabs Report*, 12-11281, 1–135.

Shao H-M, Wang J-B, Sun H (2008), 'Experimental research on vacuum brazing of Ti-coated diamond', *Superhard Material Engineering*, Vol. 20, No. 8, 7–10.

Shapiro A E (2007), 'Diamonds', Chapter 36 of the *AWS Brazing Handbook*, 5th edition, American Welding Society, Miami, FL, pp. 623–636.

Shapiro A E, Flom Y A (2010), 'Characterization of low-melting brazing foils of Al-Mg, Al-Ag-Cu, and Al-Cu-Si alloys designed for joining titanium', *DVS-Berichte*, Vol. 263, 13–16.

Shiue R K, Buljan S T, Eagar T W (1997), 'Abrasion resistant active braze alloys for metal single layer technology', *Science and Technology of Welding and Joining*, Vol. 2, No. 2, 71–78.

Smigelskas C (1962), *Metal Reference Book*, Butterworths, v. 2.

Spriggs G E (2002), ed. by Beiss P, Ruthardt R, Warlimont H, '13.5 Properties of diamond and cubic boron nitride', *Landolt-Börnstein Group VIII Advanced Materials and*

Technologies Numerical Data and Functional Relationships in Science and Technology, Volume 2A2: Powder Metallurgy Data. Refractory, Hard and Intermetallic Materials, Berlin: Springer, pp. 118–139, ISBN 978-3-540-42961-6.

Sung J C, Sung M (2009), 'The brazing of diamonds', *Int. Journal of Refractory Metals & Hard Materials*, Vol. 27, 382–393.

Tillmann W, Osmanda A M, Yurchenko S, Boretius M H (2008), 'Properties of nickel-based joints between diamond and steel for diamond grinding tools', *Welding and Cutting*, Vol. 7, No. 4, 228–235.

Wang L, Li M, Jia Q, Zhang H, Dong H (2009), 'Study on brazing diamond compact and cemented carbide', *Zhongguo Jixie Gongcheng*, Vol. 20, No. 3, 365–369.

Wentorf R (1957), 'Cubic Form of Boron Nitride', *J. Chem. Phys.*, Vol. 26, 956.

Yakubovsky S V, Kulbitsky L V, Sudnik L V, Gurbo N M, Budeyko N L, *et al.* (2010), 'Effect of stabilizing agents on formation of copper-titanium dioxide coating onto the surface of diamond powder' (in Russian), *Trans. of National Academy of Sciences of Belorussia, Chemical Sciences*, No. 3, 107–111.

Yamazaki T, Suzumura A (2006), 'Reaction products at brazed interface between Ag-Cu-V filler metal and diamond (111)', *J. Materials Science*, Vol. 41, 6409–6416.

Yamazaki T, Uzumura A, Ikeshogji T-T, Ishiguro T (2009), 'Role of valence electrons of vanadium carbide in metallization of diamond surface', in *Proc. Int. Braz. Sold Conf.*, Orlando, FL, pp. 291–295.

7
Brazing of oxide, carbide, nitride and composite ceramics

H. PENG, Harbin Institute of Technology, China

DOI: 10.1533/9780857096500.2.194

Abstract: Ceramics are inorganic non-metallic materials made from natural or synthetic compounds using shaping and high-temperature sintering processes. At present, the poor plasticity and low impact resistance of engineering ceramics have limited their further application. Joining ceramics or composites to metal is challenging because of their chemical dissimilarity and differences in coefficient of thermal expansion. The complexity and high cost of procedures that can overcome these problems limit their application. In this chapter, the difficulties of brazing ceramics and solutions are presented. The brazing of oxide ceramics, nitride ceramics, carbide ceramics and C/C composites is discussed in more detail.

Key words: brazing, oxide, carbide, nitride, composite ceramics.

7.1 Introduction

Ceramics are inorganic non-metallic materials made using shaping and high-temperature sintering processes from both natural and synthetic compounds. The family includes oxide ceramics, carbide ceramics, nitride ceramics and ceramic-based composites. At present, engineering ceramics, which have excellent thermostability, corrosion resistance and abrasion resistance, have been developed as high-performance structural materials. However, their poor plasticity and low impact resistance have limited further applications. Joining ceramics with tougher materials like metals to form high-performance structural components is one way to broaden the applications of ceramics. However, the strong ion/covalent bonding in ceramics results in low heat conductivity and weak heat shock resistance. This means that flaws are easily formed in ceramics when they are heated. Furthermore, most ceramics are insulators or have very low electrical conductivities. Conventional welding methods are thus unsuitable for joining ceramics. The brazing process is regarded as one of the most effective approaches for joining ceramics.

7.2 Difficulties of brazing with ceramics and solutions

As ceramics have stable ionic and covalent bonding as well as other special physical and chemical properties, they are incompatible with brazing fillers during the brazing process. On the one hand, the molten brazing filler cannot infiltrate the ceramic; it is difficult for interatomic metallurgy bonding between ceramics and the molten metal filler to occur. On the other hand, the coefficient of thermal expansion (CTE) mismatch between ceramics and brazing fillers leads to a high stress gradient in the brazed joint, which can induce stress concentrations or cause residual stress in the joint and consequently weaken the mechanical properties of the joint.

The good wettability of brazing fillers when applied to ceramics provides a potential solution. The mechanism for ceramic brazing is infiltration, which can be divided into non-reactive wetting and reactive wetting. The primary driving forces of non-reactive wetting are van der Waals and dispersion forces. In this case, the molten metal can quickly spread out over the ceramic surface. The wettability of the molten brazing filler on the ceramic surface is related to ceramic grain orientation, and the addition of alloying elements in the filler. In reaction wetting, wettability depends on wetting time, the addition of alloy elements into the brazing filler and the reaction temperature. Generally, wettability in reaction wetting will be improved with a higher reaction temperature, longer brazing time and the addition of alloying elements. The work of adhesion will increase, whereas the contact angle will decrease, when surface wettability is allowed to increase.

There are two processes that can be used to improve wettability on a ceramic surface:

- indirect brazing
- reactive brazing.

The indirect brazing process includes two steps: ceramic surface metallization and ceramic brazing. Ceramics metallized with a thin metal film can be brazed with a standard brazing process. During the reactive brazing process, the brazing filler is modified with some active alloy elements. A chemical reaction will happen between the ceramic and the active elements and then a stable reaction gradient layer will form. The metal base will be brazed to the ceramic by the stable reaction gradient layer.

There are two ways to solve the problem of the physical property mismatch between the ceramic and brazing filler or the ceramic and metal substrate. One way is to insert a buffer layer between the ceramic and brazing filler. Some examples of buffer layers include:

- soft buffer layer
- rigid buffer layer
- soft/rigid double-deck buffer layer.

The buffer layer helps to decrease residual thermal stress in the joint during cooling. The residual stress is absorbed by elastic deformation or plastic deformation of the buffer layer. However, a soft buffer layer, which has a high CTE, can cause residual stress in the joint because of its CTE mismatch with a ceramic surface. A rigid buffer layer normally has a high elastic modulus and a low CTE. Ceramics have a similar elastic modulus and CTE, meaning that any residual stress in the joint will be relatively low. A soft/rigid double buffer layer works in two ways. The soft buffer layer joins with the ceramic surface and the rigid buffer layer is used to decrease the residual stress in the joint. The use of a soft/rigid double buffer layer can therefore significantly improve the mechanical properties of a joint.

Another way to compensate for the physical property mismatch between the ceramic surface and the brazing filler is to add high Young's modulus and low CTE ceramic particles, hard metal particles and fibers into the brazing filler metal to form a composite brazing filler. This allows joints with low residual stress to be obtained as the ceramic and the composite brazing filler have similar physical properties (high elastic modulus and low CTE). Furthermore, the additives in the composite filler can strengthen and toughen the joint and absorb thermal stress, which can lower the residual stress in the joint and improve joint quality.

7.3 Brazing of oxide ceramics

Oxide ceramics, which include Al_2O_3, SiO_2, ZrO_2 and BeO, are widely used in industry. Oxide ceramics are typically joined with metals that have high heat resistance, high corrosion resistance and high electrical conductivity, such as Al, Cu, Ni, Au, Ti and their alloys, and stainless steel.

Figure 7.1 shows the brazing process for 5A05 aluminum alloy, the composition of which is 0.5 wt.% Si, 0.5 wt.% Fe, 0.1 wt.% Cu, 0.3–0.6 wt.% Mn, 4.8–5.5 wt.% Mg, 0.2 wt.% Zn and 92.6–93.6 wt.% Al, and Al_2O_3. The active metal powder is applied to the Al_2O_3 surface and then treated in a vacuum furnace. This forms active metallized layers, which include the reaction layer and the metallization layer, on the Al_2O_3 surface. Finally, the metallization layer is used as the interlayer to braze Al_2O_3 with 5A05 by diffusion bonding.[1] The detailed processes are discussed below.

TiH_2+AgCu+B mixed alloy powder is pasted onto the Al_2O_3 surface (Fig. 7.1(a)). As the mixed powder contains some Ti atoms (which are the active element), it can react and join with the Al_2O_3 surface (Fig. 7.1(b)). On the ceramic side of the joint, a continuous compact reaction layer of Al_2O_3/Ti_3Cu_3O/Ag(s.s.)+Cu(s.s.)+TiB+Ti–Cu compounds is formed (Fig. 7.1(c)). This reaction layer will be the interlayer for the diffusion brazing of the Al_2O_3 and the 5A05 alloy (Fig. 7.1(d)). The resulting joint is shown in Fig. 7.1(e). Figures 7.2(a) and (b) show the reactant's distribution in the joint, which is $Al_2O_3/Ti_3Cu_3O/Al_3Ti/\alpha$-Al+$\theta$-$Al_2Cu$+$\xi$-$Ag_2Al$+TiB/5A05 alloy. The composition of the brazing filler, the active metallization temperature and time can greatly affect the formation of the interfacial microstructures, while the interfacial microstructures will affect the joint properties in return. If the Ti_3Cu_3O layer is too thin or is discontinuous, it is difficult to create a compact joining interface. Conversely, if the Ti_3Cu_3O layer is too thick, there will be more defects in the layer, which will cause the joint

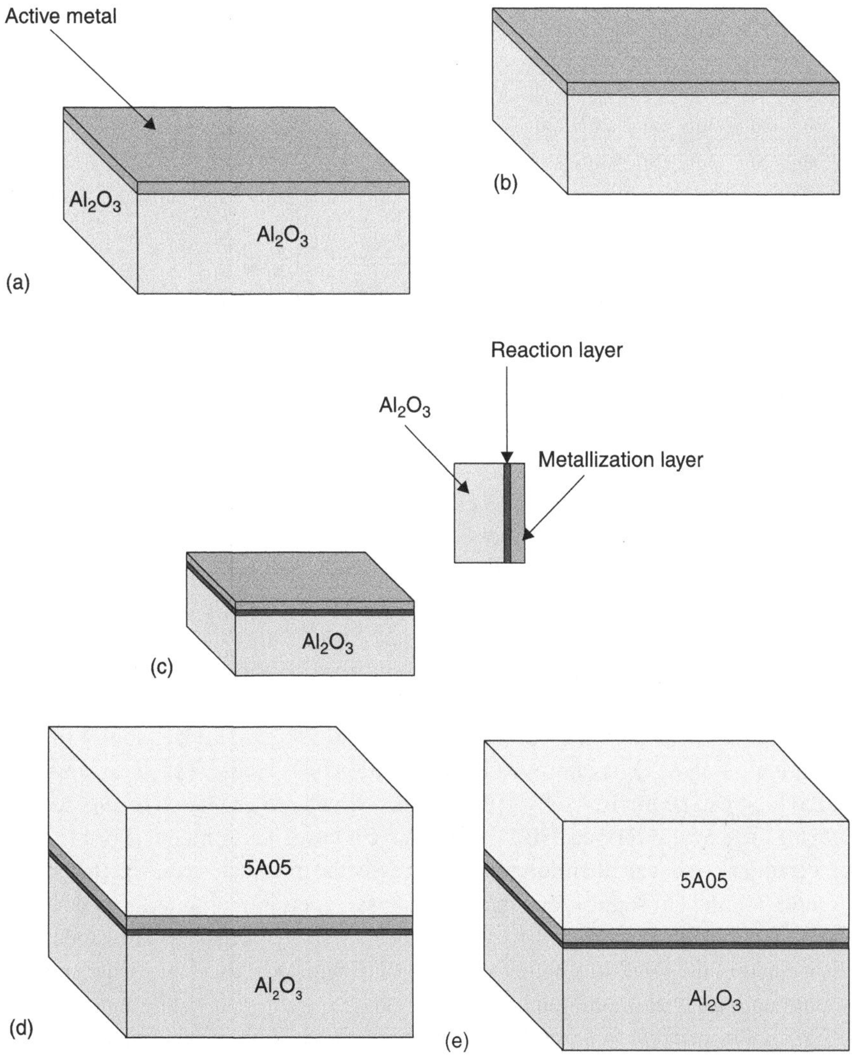

7.1 Schematic diagram of active metallization of Al_2O_3 and diffusion brazing with 5A05 alloy.[1] (a) Placing active metal on Al_2O_3 surface; (b) active metallization of Al_2O_3 surface in vacuum furnace; (c) active metallized alumina ceramic; (d) diffusion brazing of active metallized Al_2O_3 and 5A05 alloy in vacuum furnace; (e) Al_2O_3/5A05 alloy joint.

properties to deteriorate. The *in situ* formed TiB in the active metallization layer can regulate the residual stress in the joint and improve the joint gas tightness and the joint mechanical strength, as it has a low CTE and high elastic modulus. The dissolved quantity of the 5A05 alloy during the diffusion brazing process should

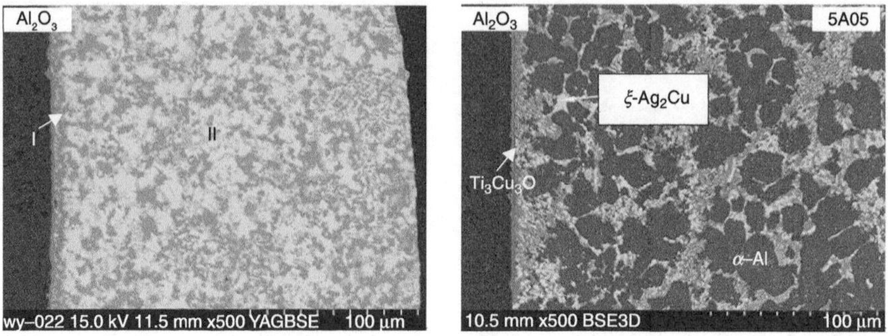

(a) (b)

7.2 Interface microstructures of an Al_2O_3/5A05 alloy joint.[1] (a) Metallization layer on Al_2O_3 side; (b) Al_2O_3/5A05 alloy joint.

be well controlled. When the dissolved quantity is too small, the oxidation film on the Al alloy surface cannot be removed effectively, hindering the brazing process, while, if the dissolved quantity is too large, the joint will be over-corroded.

Al_2O_3/Al_2O_3 and Al_2O_3/Ti–6Al–4V alloy[2,3] can be directly brazed using an Ag–Cu–Ti+B composite brazing filler, as shown in Figs 7.3 and 7.4. Figure 7.3 shows the microstructure of the Al_2O_3/Al_2O_3 joint and Fig. 7.4 shows the microstructure of the Al_2O_3/Ti–6Al–4V joint. In the two figures, Figs 7.3(b) and 7.4(b) are the magnification of area A shown in Figs 7.3(a) and 7.4(a). The wettability of the brazing filler on Al_2O_3 is improved by the addition of Ti atoms. The shear strength of the joint can be improved by TiB whiskers, which are generated *in situ* during brazing. The newly formed TiB whiskers have a low CTE (which is close to that of ceramics) and can therefore help decrease the residual stress in the joint. Figures 7.3 and 7.4 suggest that these joints have good joining interfaces with no flaws. As indicated in Figs 7.3(b) and 7.4(b), the TiB whiskers are randomly distributed in the joint. It is believed that the TiB whiskers also have a dispersion-strengthening effect on the joints.

SiO_2 glass-ceramic is a porous material, which has high heat resistance, good shock resistance, adjustable thermal expansion performance, high corrosion resistance, high heat stability and superior high-temperature viscosity. In the astronautics industry, it is joined with a Ti–6Al–4V alloy and used for the production of aircraft engine bays. An AgCu/Ni compound has been proved to be a suitable braze for SiO_2 ceramics and Ti–6Al–4V alloys.[4] Figure 7.5 shows the typical microstructures of a Ti–6Al–4V alloy/SiO_2 joint. Figure 7.5(a) shows the microstructure of the SiO_2/AgCu/Ni//Ti–6Al–4V alloy joint. Figure 7.5(b) shows the interface between the braze and SiO_2 and the electron diffraction pattern for SiO_2. The joint comprises:

- Ti–6Al–4V alloy
- a needle-like α-Ti/Ti(s.s.)+ Ti_2(Cu,Ni)+Ti_2(Ni,Cu) hypereutectoid structure

- a Ti(s.s.) + $Ti_2(Cu,Ni)$ + $Ti_2(Ni,Cu)$ hypereutectic structure
- $Ti_2(Ni,Cu)$ + $Ti_2(Cu,Ni)$ compound
- a Ti(s.s.) + $Ti_2(Cu,Ni)$ + $Ti_2(Ni,Cu)$ hypereutectic structure
- $Ti_4O_7 + TiSi_2$
- SiO_2

The joint shear strength is up to 110 MPa when the brazing temperature is 970°C and the holding time is 10 min. Fractures will occur in the SiO_2 substrate.

The Ti in a Ti–6Al–4V alloy plays a very important role in the brazing process. Ti can react with the interlayer metal to form a liquid eutectic, which can easily wet the SiO_2 surface. The liquid eutectic can quickly spread out over the ceramic surface and react with the ceramic to form a reaction layer. This allows a reliable metallurgical bond to be formed between the ceramic and the brazing filler. As the

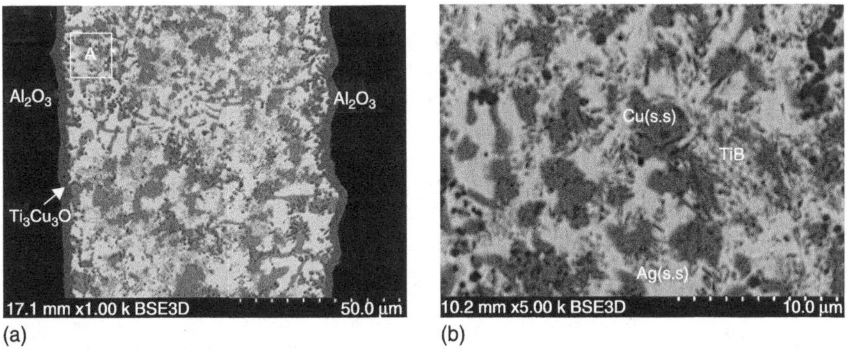

7.3 Microstructure of Al_2O_3/Al_2O_3 joint.[2] (a) Al_2O_3/Al_2O_3 joint; (b) magnification of area A.

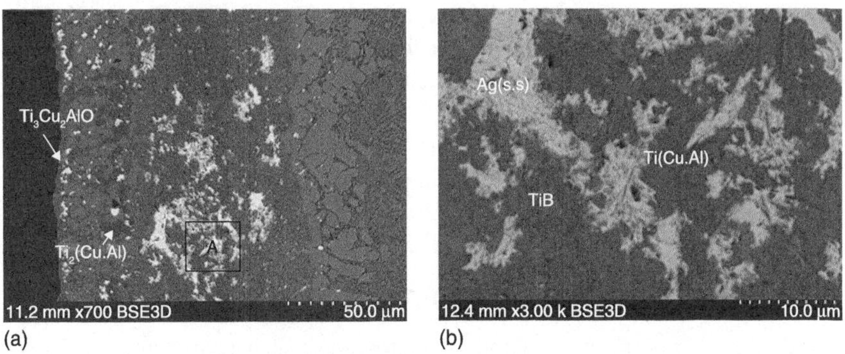

7.4 Microstructure of Al_2O_3/Ti–6Al–4V joint.[3] (a) Al_2O_3/Ti–6Al–4V joint; (b) magnification of area A.

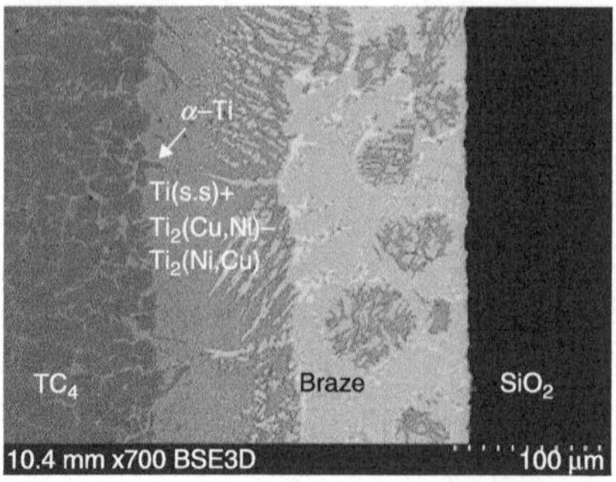

7.5 Microstructures of SiO_2/AgCu/Ni//Ti–6Al–4V alloy joint.[4] (a) SiO_2/AgCu/Ni//Ti–6Al–4V alloy joint; (b) interface of braze and SiO_2 and electron diffraction pattern of SiO_2.

diffusion coefficient of Ni to Ti is larger than that of Ti to Ni, the liquid eutectic is formed by the Ti–6Al–4V alloy. The thickness of a Ti–6Al–4V molten alloy can be determined by the concentration of Ti in the liquid brazing filler, as shown in Eq. 7.1.[4]

$$\begin{aligned} &0.9SX\rho_{\text{Ti-6Al-4V}} = 48C_S V[1-\exp(-KSt/V)+48C_A V] \\ &K(T) = 6.936\times10^{-7}T - 0.0006299 \\ &C_A(T) = 1.988\times10^{-4}T - 0.1377083 \end{aligned} \qquad [7.1]$$

where: S is the contact area of Ti–6Al–4V alloy/brazing alloy; X is the thickness of Ti–6Al–4V molten alloy; $\rho_{\text{Ti-6Al-4V}}$ is the density of Ti–6Al–4V alloy; C_s is the saturated solubility of Ti in liquid brazing alloy; V is the volume of liquid brazing filler; K is the dissolution rate constant; t is the holding time; and T is the brazing temperature.

ZrO_2 ceramic components, or the components made from ZrO_2 and other metals, including Au, Ni, Pt and Ti, have found wide application in the electronics industry. Ag-base or Cu-base brazes, which contain some active elements, such as Ag–Cu–Ti, Cu–Sn–Ti, Cu–Ga–Ti, Cu–Sn–Pb–Ti, are widely used in vacuum brazing. ZrO_2/malleable cast-iron joints brazed with Cu–Ga–Ti at 1150°C for 10 min and Cu–Sn–Pb–Ti at 950°C for min have a shear strength of 277 ± 37 MPa and 156 ± 25 MPa, respectively.

7.4 Brazing of nitride ceramics

Nitride ceramics perform excellently with properties such as high intensity, super hardness, high thermal shock resistance, high wear resistance, high corrosion resistance and they are self-lubricating. Among the most widely used ceramics, such as Si_3N_4, BN and AlN, Si_3N_4 is an advanced engineering ceramic and is the best at high temperatures. As Si_3N_4 ceramics are so widely used in industry, research activities have focused on the joining of Si_3N_4 ceramics to Fe, Ni, Co, Cu, W, Mo, Ta, Nb, Cr, V, Ti and its alloys, stainless steel and TiAl alloy. It is well known that Si_3N_4 ceramics have a typical covalent bonding structure and that the bonding of Si–N is very firm. Hence, the wettability of a brazing material onto a Si_3N_4 ceramic surface is very poor. In order to produce reliable Si_3N_4 ceramic joints, the difficulty of wetting needs to be solved. At present, much research has focused on the wettability of Ti-containing brazing fillers onto a Si_3N_4 ceramic surface.

Nomura *et al.*[5] studied the microstructures produced by wetting a three-phase line in the Ag–Cu–Ti/Si_3N_4 system. Figure 7.6 shows a TEM image of interface microstructures. Luz and Ribeiro[6] investigated the wetting behavior of Ti–Cu alloys with different compositions on Si_3N_4 ceramic surfaces using the sessile drop method. They suggest that the amount of Ti present in the alloy is the key influence on the wettability of the solder alloy on the Si_3N_4 surface. The chemical reactions at the interface, which affect wettability are:

$$Si_3N_4 + 4Ti \leftrightarrow 4TiN + 3Si \qquad [7.2]$$

$$\Delta G = -139 \text{ kJ/mol } (1100\,\text{K})$$

$$Si_3N_4 + 5Ti \leftrightarrow Ti_5Si_3 + 2N_2 \qquad [7.3]$$

$$\Delta G = -42 \text{ kJ/mol } (1100\,\text{K})$$

It is noted that a Si_3N_4 ceramic has a relatively low coefficient of linear expansion of $2\text{–}3\times10^{-6}/°C$ (compared with those of metals). This means that certain processes should be applied to alleviate stress in the joint. One effective

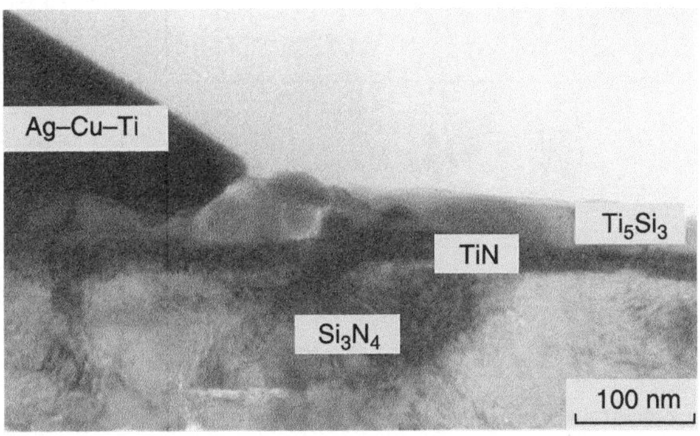

7.6 TEM image of an Ag–Cu–Ti/Si_3N_4 joint.[6]

way to alleviate stress in the joint is to use soft metals, such as Cu, Al and Ag, or a metallic multi-layer, as an interlayer to help compensate for the CTE mismatch. Another way is to add low CTE materials into the selected brazing materials to create composite brazing fillers. These modified composite brazing fillers will help decrease stress in the joint caused by the CTE mismatch.

Figure 7.7 shows the microstructures (a) and the line scanning results (b) of a Si_3N_4/Cu interface.[7] The interface is well joined and reveals that the cushioning effect of a Cu foil is significant. The whole Si_3N_4/Cu interface is composed of two kinds of compounds: TiN and Ti_5Si_3. A four-point bend test carried out on the joint

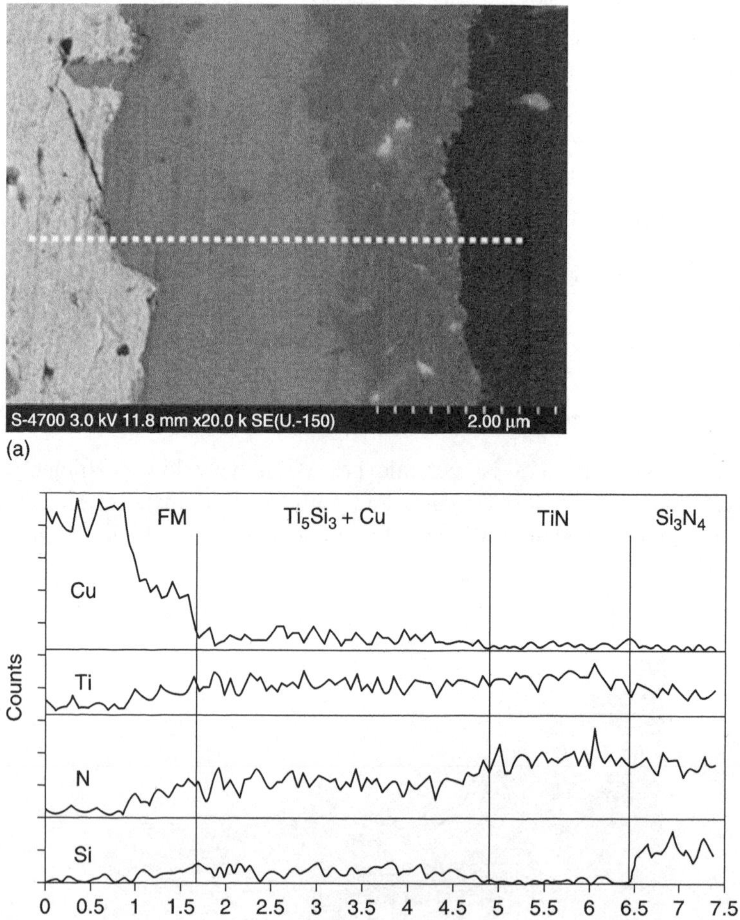

7.7 (a) Back-scattered micrograph of the interface between Si_3N_4 and the Cu interlayer brazed at 1025°C for 1 min; (b) low voltage line scanning of the reaction layer observed in (a).[7]

revealed that the maximum mechanical strength of the joint approaches 160 MPa. Other pure metal foils, such as Ti foil,[3,4] Nb foil[8] and Ni foil,[9] can also be used as the interlayer for the brazing of Si_3N_4 ceramics.

Ag–Cu–Ti+Mo composite fillers have been used to braze Si_3N_4 and 42CrMo steel at 900°C for 10 min. Mo, which has a low CTE of 5.1×10^{-6}/°C, can decrease the CTE of the composite fillers, which is beneficial for the brazing of Si_3N_4 and 42CrMo steel. It has been found that different Mo contents have different effects on joint microstructures and joint bending strengths, as shown in Fig. 7.8 ((a) 0 vol.%, (b) 5 vol.%, (c) 10 vol.%, (d) 15 vol.%) and Fig. 7.9, respectively. The continuous reaction layer near the Si_3N_4 side of the joint is composed of TiN and Ti_5Si_3. The thickness of the reaction layer decreases with an increased Mo content.

When the Mo content is about 10 vol.%, the joint bending strength reaches a maximum value of 587.3 MPa, which is more than four times higher than that of a joint brazed with a filler without the addition of Mo. Ag–Cu–Ti+SiC_p,[8,9] Cu–Pd–Ti,[10] Au–Ni–V,[11] Au–Ni–Pd–V[12] and Cu–Zn–Ti[13] can also be used to braze Si_3N_4.

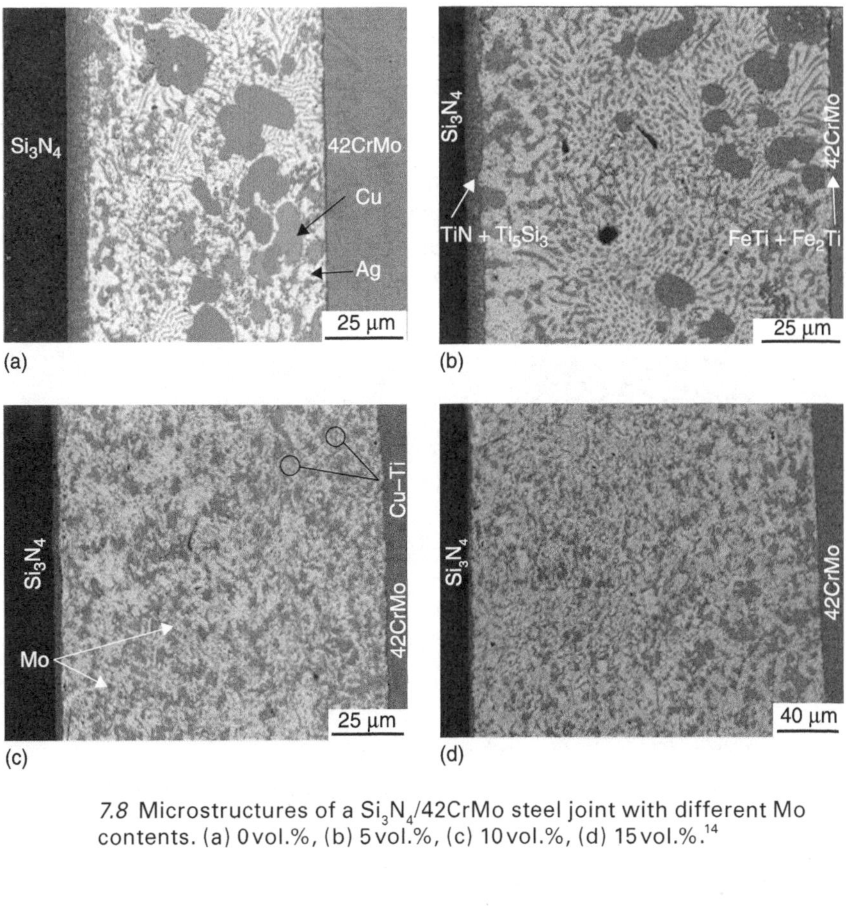

7.8 Microstructures of a Si_3N_4/42CrMo steel joint with different Mo contents. (a) 0 vol.%, (b) 5 vol.%, (c) 10 vol.%, (d) 15 vol.%.[14]

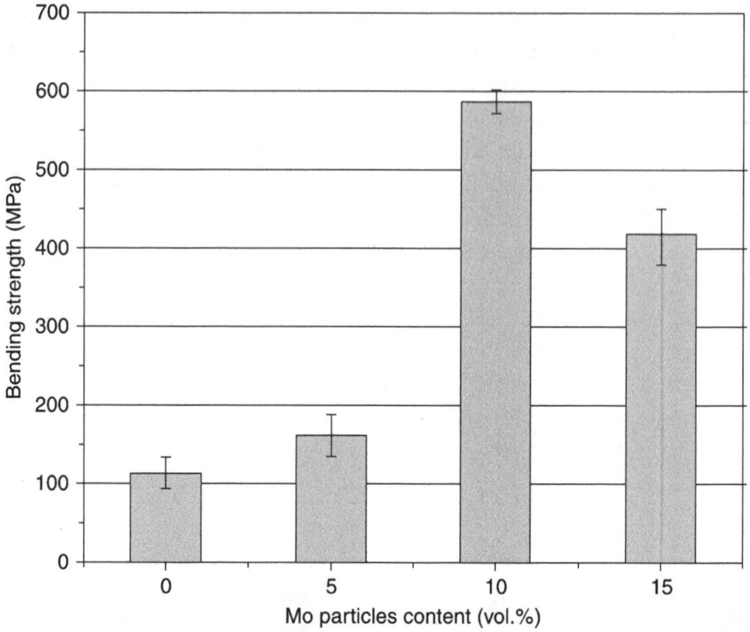

7.9 Effect of Mo content on joint bending strengths.[14]

AlN is a non-toxic ceramic, which has a similar thermal conductivity to BeO and hence has been regarded as an alternative to BeO. It has been used in microelectronic and microwave vacuum tubes. Active brazing fillers can be directly used for the brazing of AlN (without metallization). Figure 7.10 shows

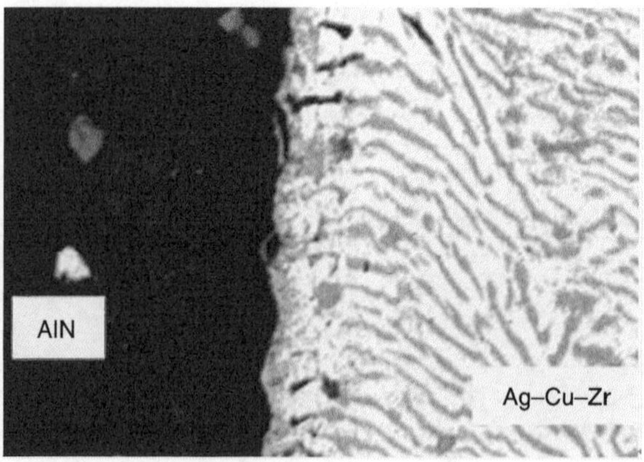

7.10 Microstructure of AlN/Ag–Cu–2%Zr interface (T = 1100°C, t = 60 min).[15]

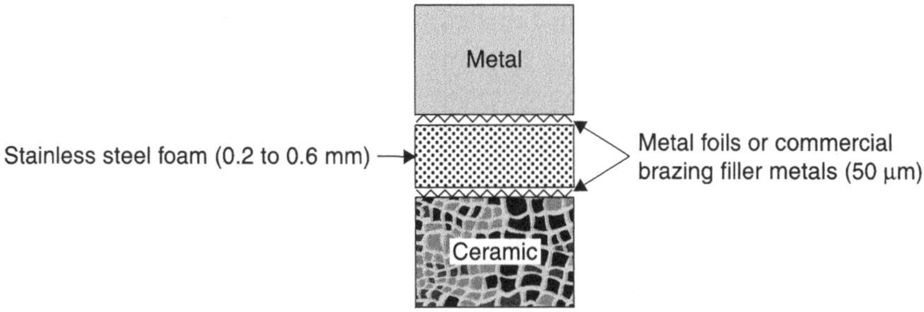

7.11 Joint configuration using stainless steel foam as flexible interlayer (not to scale).[16]

the microstructure of an AlN/Cu joint brazed with AgCuZr at 1100°C for 60 min. Furthermore, AlN can be joined with FeNi alloy and Pt by brazing.

Metallization is widely used for ceramic joining at present. However, the conventional metallization process is difficult to perform on an AlN surface. To improve the wettability of brazing materials on an AlN surface, active elements, such as Ti, Zr and Hf, are added to the brazing filler. Up to now, the most commonly used metallization methods include the Ti–Ag–Cu method, the active Mo–Mn method and electroless plating. Norton *et al.*[15] have used TiH_2 and ZrH_2 to metallize AlN surfaces and then brazed the metallized AlN to metallized AlN with an Ag–22Cu–22Zn brazing filler. It was observed that Ti_2N, Ti_3Al and Zr_2N, and Zr_3Al form on the AlN side of the joint.

With the above mentioned processes, residual stress in the joint due to the physical properties' mismatch between the metallization layer and the AlN surface cannot be avoided. Shirzadi[16] has used foam metal as the interlayer when brazing AlN with metals, as shown in Fig. 7.11. The foam metal can effectively absorb the residual thermal stress in the joint.

7.5 Brazing of carbide ceramics

7.5.1 Brazing of SiC ceramics

SiC ceramics are covalent compounds with a diamond lattice structure. Their crystal systems include a cubic system, hexagonal system and rhombohedral system. α-SiC and β-SiC are the most common crystal systems. These crystal structures mean that SiC ceramics have a high melting point, high hardness and excellent chemical inertness and are therefore widely used in the petroleum, chemical, automotive, aviation, aerospace, machinery, and microelectronics industries. Cu–Ti, Ni–Ti and Ag–Cu–Ti filler alloys have been used to braze SiC ceramics. Furthermore, new filler alloys and joining methods have been developed for SiC joining.

A composite joining technique is used to join SiC ceramics and Kovar. The technique uses Ni–56Si filler alloy and Mo for the joint interlayers.[17] First, a Mo foil is joined to Kovar on one side by means of transient liquid phase bonding (TLP) and then brazed with SiC on the other side using a Ni–Si coating as a filler alloy. As Mo has a similar CTE (5×10^{-6} °C^{-1} at 0–100°C) to SiC (3×10^{-6} °C^{-1} at 0–100°C),[17] it can effectively relieve the residual stress caused by the CTE mismatch during the brazing process. In the wetting process, the liquid Ni–Si alloy can reach equilibrium within 400 s on the surface of the SiC with a contact angle of 23°.[17] When pure Ni contacts the SiC surface at 1300°C, some Si will be dissolved into the liquid Ni and C will be precipitated. When a balance in the chemical activity between the Si and the SiC is obtained, the Si dissolution will stop. According to the ternary phase diagram of Ni–Si–C, which is shown in Fig. 7.12, Si reaches saturation in Ni and will not react further with the SiC surface when X_{Si}=0.4. At more than 40 at.% Si addition, the reaction between the SiC and the Ni can be prohibited and the reactant of the brittle C decreases. The addition of a layer of Mo can improve the stability of the C-containing liquid phase. Mo atoms diffuse into the liquid phase and get enriched on the SiC surface and then reduce the solubility of the C by decreasing the chemical activity of the Si. The microstructures of a Mo/Kovar joint brazed by the TLP method show that a thin liquid layer forms in the joint interface (Fig. 7.13(a)) with a composition of 29 at.% Fe, 9 at.% Co, 14 at.% Ni and 48 at.% Mo. It is suggested in Fig. 7.13(b) that

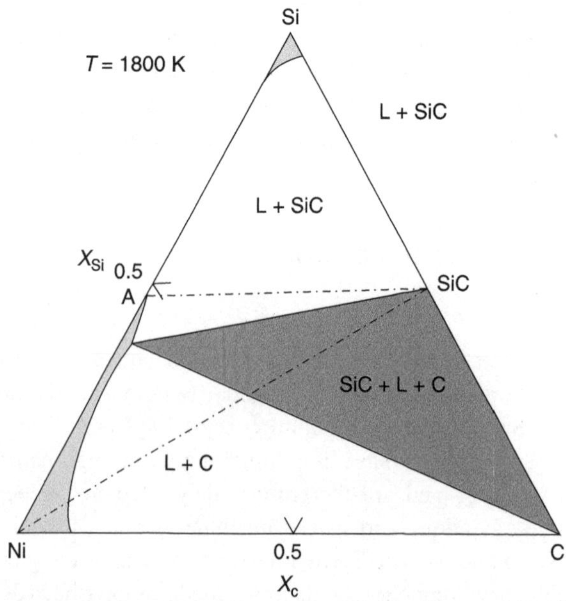

7.12 Ni–Si–C phase diagram at 1800K.[17]

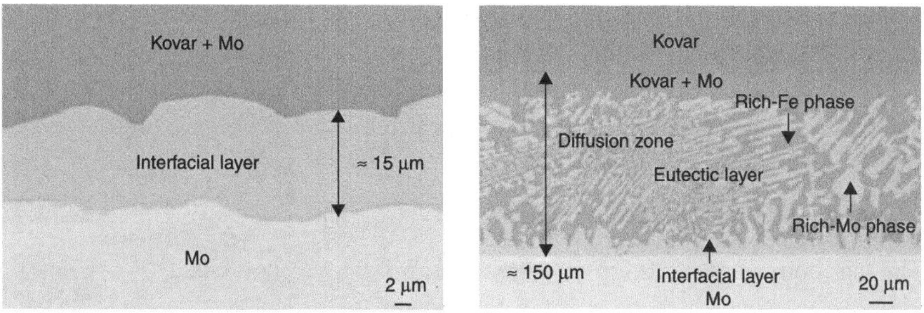

(a) (b)

7.13 SEM micrographs of two cross-sectioned Mo/Kovar joints prepared at 1350°C for different holding times: (a) 10 min and (b) 20 min.[17]

a wide diffusion zone forms between the Mo layer and the Kovar alloy and that Mo atoms have diffused into the Kovar alloy with a long diffusion path. However, no elements from the Kovar alloy have diffused into the Mo layer. With a longer holding time, a wide range of eutectic structures form in the interface, as shown in Fig. 7.13(b).

The ceramic precursor polymer (methylpolysilsesquioxane-MK) can be added to the solder for the brazing of SiC ceramics.[18] The precursor is mixed with Al powder by dry ball milling for 1 h. The mixture is further mixed with isopropanol to form the pasty solder. Before the brazing process begins, the pasty solder is uniformly pasted onto the SiC surface. During brazing, Si atoms react with the organic material. The product of the reaction, vitrescent Si–O–C, inhibits the reaction between the Al and the SiC, meaning that a joint made using a MK brazing alloy has a thin reaction layer at the interface. Furthermore, the newly formed vitrescent Si–O–C will restrain the fluidity of the molten Al and may cause micropores to appear in the joint. If excessive MK is added too many micropores will occur, which results in a decrease in the strength of the joint.

In addition to the diffusion of the Si into the solder, the Al atoms also diffuse into the SiC to replace the Si atoms. Brazed with a 20 vol.% MK brazing alloy, the highest bending strength (206 MPa) can be obtained at a temperature of 1200°C for 30 min. With a longer brazing time, the possibility of forming brittle Al_4C_3 will increase.

Many high melting point metal alloys, such as the alloys of Ni-base, Fe-base and Pt-base, will strongly react with SiC ceramics to form a series of bulk carbides. Hence, a family of high melting point brazing alloys based on transition metal silicides, such as Co, Pd or Pt, have been developed. The presence of Si in these alloys can effectively inhibit the reaction between the alloy and the SiC ceramic and consequently improve the wettability of the brazing alloys and the high temperature oxidation resistance of the brazing seam. However, due to their high melting points (higher than 1300°C), these silicides are not suitable for the brazing process. Alloys with a low melting point, such as $MnSi_2$–Si (1150°C) and $PrSi_2$–Si (1212°C) eutectics, are used instead as this allows the brazing temperature to be decreased.[19]

The formation of SiC can be expressed as follows:[19]

$$Si(l) + C(s) \rightarrow SiC(s) \qquad [7.4]$$

The thermodynamic equilibrium of reaction 7.4 can be expressed as:

$$\Delta_f G^0_{SiC} = \overline{\Delta G}^{XS}_{Si} + RT \ln x^l_{Si} \qquad [7.5]$$

The x^l_{Si} refers to the mole fraction of silicon in a $PrSi_2$–Si eutectic. *R* and *T* are the gas constant and absolute temperature respectively. The Gibbs energy of reaction 7.4 at 1250°C is −56.389 J/mol. The partial excess Gibbs energy of Si in a Pr–Si alloy can be estimated by a regular solution approximation, which can be expressed as a function of the exchange energy of the solution (λ):

$$\overline{\Delta G}^{XS}_{Si} = \lambda(1 - x_{Si})^2 \qquad [7.6]$$

The λ can be calculated by means of the three-phase thermodynamic equilibrium equation among liquid Pr–83 at.%Si alloy, Si and $PrSi_2$ at 1212°C:

$$2Si(s) + Pr(l) \rightarrow PrSi_2(s) \qquad [7.7]$$

The equilibrium reaction condition is:

$$\lambda = \frac{\Delta G^0_4 - RT \ln x_{Pr}}{(1 - x_{Pr})^2} \qquad [7.8]$$

In Eq. 7.8, x_{Pr}=0.17 and ΔG^0_4 is the standard Gibbs energy in reaction 7.7, which can be estimated from the standard formation enthalpy of $PrSi_2$ (= −184.5 kJ/mol). The entropy can be reasonably ignored as its contribution to the Gibbs free energy is less than 10%. With this method, the estimated λ is −225 kJ/mol, while the λ calculated from the formation enthalpies and melting points of $PrSi_2$ and PrSi is −222 kJ/mol and −260 kJ/mol, respectively. These three λ calculated by different methods are consistent. With the λ of −222 kJ/mol and −260 kJ/mol, the calculated x_{Si} is between 0.53 and 0.57, which is much lower than the Si content in Pr–83 at.%Si. This indicates that SiC does not react with Pr–83 at.%Si during brazing. The wettability of Pr–83 at.%Si on a SiC surface at 1250°C is shown in Fig. 7.14. It was observed that the bonding between a PS brazing alloy and the SiC surface is firm. The PS alloy mainly consists of Si crystals and a Si–$PrSi_2$ eutectic. No reaction layer was found at the interface, which had been predicted by the above thermodynamics analysis. It was also found that the droplet spreading process on a SiC surface takes only 200 s, which is thousands of times faster than for other liquid alloys (assuming the liquid alloys do not react with the matrix). Generally, there is an oxide layer with a thickness of several nanometers on the surface of the SiC ceramic. The spreading

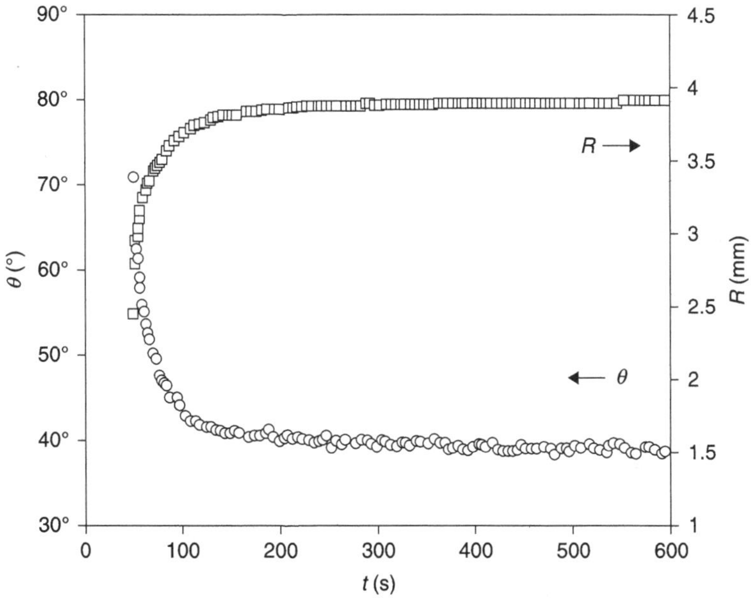

7.14 Contact angle θ and drop base radius R as functions of time for Pr–83at.% Si alloy on a SiC substrate at 1250°C in high vacuum (1×10^{-4} Pa); t=0 is taken to be the beginning of melting.[19]

process is controlled by the reduction reaction of the oxide layer. Gaseous SiO is generated by the reaction between the Si and the SiO_2, as expressed by:[19]

$$(Si) + \langle SiO_2 \rangle \rightarrow 2SiO \qquad [7.9]$$

Under a high vacuum condition, the gaseous SiO evaporates quickly and a rapid spreading process is obtained.

7.5.2 Brazing of C/SiC composites

Recently, C/SiC composites, combining the advantages of carbon fibers and SiC ceramics, have exhibited a series of unique properties, such as low density, high mechanical strength, high toughness, high thermostability, high ablative resistance, high anti-scouring performance and high hardness. C/SiC composite ceramics have been used in aerospace, weapons and other industries. However, there are still some joining problems, which are outlined below:

- C/SiC composites are various and anisotropic due to the different weaving methods of carbon fibers and matrix formation technologies. Hence it is difficult to determine the basic properties of the composite ceramics.
- The elastic modulus and the CTE of C/SiC composites are higher than those of metals, which can cause residual stress in the joint after brazing. The

internal stress caused by the CTE mismatch between the carbon fiber and the SiC matrix in C/SiC composites also cannot be neglected.

- The chemical structures of C/SiC composites are different from those of metals. Thus the chemical compatibility between C/SiC composites and metals is poor.

The main joining approaches for C/SiC composites are brazing and diffusion bonding. TiZrNiCu foil and Ni foil have been used as filler alloys to join C/SiC composites with Ti–6Al–4V alloy.[20] In Fig. 7.15(a) it can be seen that a joint interface brazed at 900°C for 10 min consists of three reaction layers. Layer I, which is close to the C/SiC side of the joint, is a discontinuous gray layer. Layer II, in the center of the joint, contains the gray phases doped by some black phases. Layer III is composed of black strip phases and some white block phases. The microstructures in the joint are: C/SiC composites/Ti_5Si_3+TiC+ZrC/Ti_2Cu+TiCu+CuZr/ [Ti(s.s.)+(Ti_2Cu+TiCu+CuZr)]/Ti-6Al-4V alloy. At a higher brazing temperature of 960°C, the reactions between different elements will be more severe and the obtained joint microstructures will be more uniform, as shown in Fig. 7.15(b). The TiZrNiCu layer has been replaced by Ti(s.s.), which appears as black strip-like sticks. Compounds of Ti_2Cu, TiCu and CuZr have penetrated into the Ti–6Al–4V alloy along the α-Ti grain boundary. Moreover, a β-Ti and Ti_2Cu/α-Ti eutectic structure forms when the temperature is above 790°C. With a higher brazing temperature, the growth rate of the β-Ti structure will be much higher. The microstructures in the joint brazed at 960°C for 10 min are C/SiC composites/TiC/[(TiCu+Ti_2Cu+CuZr)+Ti(s.s.)]/Ti(s.s.)/Ti–6Al–4V alloy.

A C/SiC composite and Ti–6Al–4V alloy can also be joined with an Ag–Cu–Ti filler alloy.[21] The highest joint strength is obtained at 900°C for 5 min. Multi-layer intermetallic compounds are formed in the joint, as shown in Fig. 7.16. Ti atoms in the brazing alloy can diffuse into the SiC to form Ti_3SiC_2+TiC/Ti_2Cu+Ti_5Si_3 (Fig. 7.16(b)). Ti_3Cu_4/TiCu/Ti_2Cu/Ti_2Cu+Ti structures form at the Ti–6Al–4V

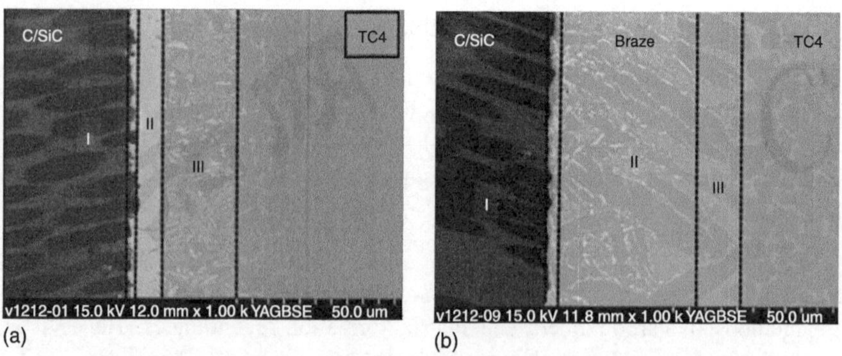

7.15 Microstructures of C/SiC composites/TiZrNiCu/TC4 alloy brazed at different temperatures: (a) 900°C and (b) 960°C.[20]

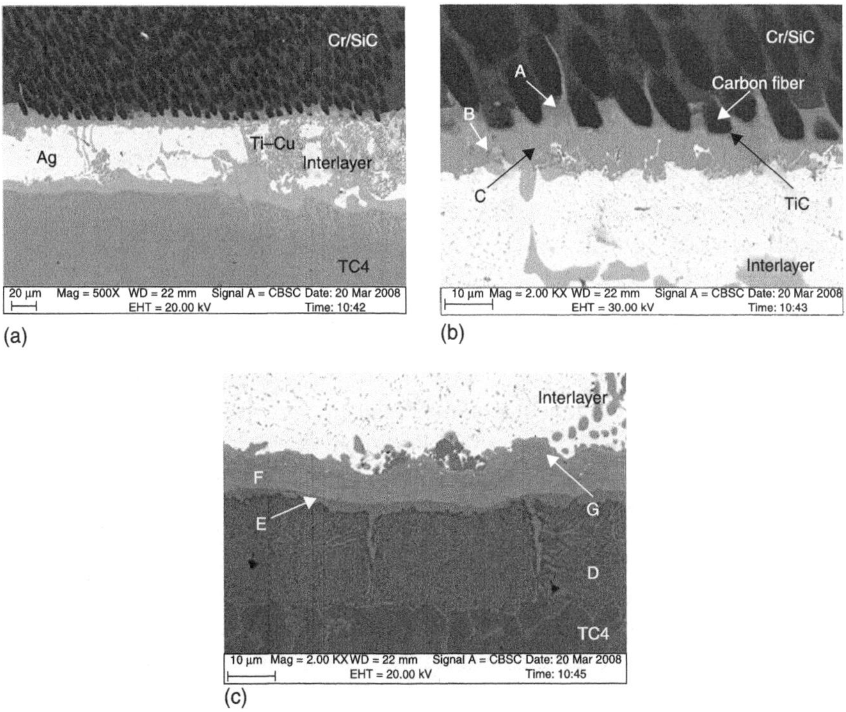

7.16 SEM images of C/SiC composite/TC4 joint using Ag–Cu–Ti filler alloy. (a) Cross-section of joint. (b) Interface of C/SiC composite. (c) Interface of TC4 alloy. The labels A–G represent corresponding phases: A: Ti_3SiC_2; B: Ti_5Si_3; C: Ti_2Cu; D: Ti_2Cu+Ti; E: Ti_2Cu; F: TiCu; G: Ti_3Cu_4.[21]

alloy side of the joint (Fig. 7.16(c)). The interfacial evolution of a C/SiC composite/Ag–Cu–Ti/Ti–6Al–4V alloy joint can be illustrated by the equations below.

Ti atoms react with SiC to produce TiC and Si:[21]

$$SiC + Ti \rightarrow TiC + Si \qquad [7.10]$$

TiC compounds form on the SiC substrate surface and grow inwards toward the liquid phase. Then Ti and Si react with TiC to produce Ti_3SiC_2:[21]

$$Ti + Si + 2TiC \rightarrow Ti_3SiC_2 \qquad [7.11]$$

These Ti_3SiC_2 compounds cover the SiC substrate surface and form a reaction layer. Furthermore, except for the reaction with Ti and TiC, the Si atoms produced by Eq. 7.10 further react with the Ti under the driving force of the concentration gradient, as expressed by:[21]

$$3Si + 5Ti \rightarrow Ti_5Si_3 \qquad [7.12]$$

On the Ti–6Al–4V alloy side of the joint, Ti elements dissolved from the Ti–6Al–4V alloy continuously diffuse into the liquid alloy and react with Cu. As the Cu atoms gradually deplete from the Ag–Cu eutectic, Ag-rich phases form in the solder alloy. XRD analysis indicates that Ti_3Cu_4, TiCu, Ti_2Cu and Ti+Ti_2Cu form in the interface between the Ti–6Al–4V alloy and the interlayer.

7.6 Brazing of carbon–carbon (C/C) composites

C/C composites are composite materials composed of carbon fibers and matrix phases (such as coke, sintered carbon and graphite) and have the characteristics of low density, high mechanical strength, high thermostability, high electrical conductivity, high thermal conductivity, low CTE, excellent fracture toughness and high friction/wear resistance. C/C composites are the primary thermotolerant structural materials for advanced aircraft. However, it is expensive and time consuming to prepare C/C composites. Using joining techniques to join C/C composites with metal components to create a complete composite structure is the preferred way to overcome the above shortcomings. There have, however, been some problems with the joining of C/C composites:

- The CTE of a C/C composite is lower than that of most metals. Therefore considerable residual stress will be generated in the joint during the cooling process. Furthermore, cracks and even fractures may occur in the joint.
- Most filler alloys are difficult to spread onto the C/C composite surface during brazing.
- A large amount of gas will be released during the brazing process, which leads to a large number of pores in the joint.
- C/C composites contain a number of pores, which will consume some molten filler alloy and lead to weak joints.
- C/C composites will oxidize when the environmental temperature is above 673K. Hence, the joining process must be performed under vacuum or with the protection of inert gases.
- Some fiber and matrix debris will be left on the surface of a C/C composite during the contour machining process, which is detrimental to joints.

C/C composites and Ti–6Al–4V alloys can be successfully joined using a TiZrNiCu filler alloy with Cu and Mo layers as buffer layers.[22] The TiZrNiCu filler alloy is inserted between each pair of layers. The assembly pattern is shown in Fig. 7.17. When the TiZrNiCu alloy is used directly to join a C/C composite and a Ti–6Al–4V alloy (as shown in Fig. 7.17(a)), the interfacial microstructures in the joint will be C/C composite/TiC/$(Ti,Zr)_2(Cu,Ni)$/Ti(s.s.)+$(Ti,Zr)_2(Cu,Ni)$/ Ti–6Al–4V alloy. The shear strength of the joint will be less than 5 MPa. Shear fractures mainly occur in the zone of the C/C composite and TiC. When a Mo monolayer is used as the buffer layer (as shown in Fig. 7.17(b)), the interfacial microstructures will be Mo/$(Ti,Zr)_2(Cu,Ni)$+Ti compound/C/C composite. Although the presence of a Mo monolayer can release a certain amount of residual stress, the shear strength is still very low (6 MPa). When a 0.3 mm Cu / 0.1 mm Mo

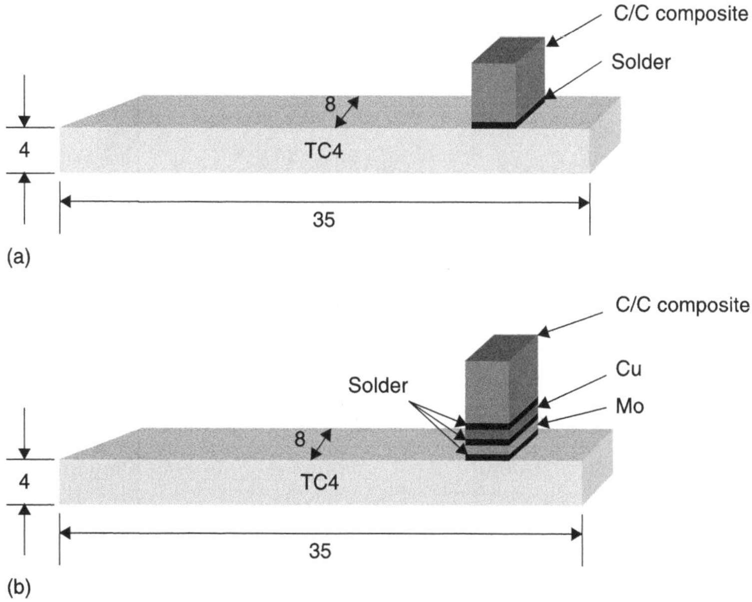

7.17 Assembly pattern of C/C composite/TC4 alloy joint before brazing. (a) C/C composite/Ti–6Al–4V alloy joint with TiZrNiCu filler alloy; (b) C/C composite/Ti–6Al–4V alloy joint with Cu, Mo buffer layer and TiZrNiCu filler alloy. Units are in millimeters (mm).[22]

multi-layer is used as the buffer layer, the joint shear strength will increase with the brazing temperature or holding time. The highest shear strength (21 MPa) is obtained at 1173 K for 300 s. The interfacial microstructures are $Cu/Cu_{51}Zr_{14}/Cu(s.s.) + Ti(Cu,Ni)_2/TiC/C/C$ composite. Fracture analysis results suggest that when the carbon fiber axis in the C/C composite is parallel to the joint interface, fractures usually occur in the C/C composite. When the carbon fiber axis is perpendicular to the joint interface, fractures occur in the TiC layer between the C/C composite and the filler alloy.

SiC particles can be used to reinforce an Ag–26.7Cu–4.6Ti (wt.%) brazing alloy to join a C/C composite and a Ti–6Al–4V alloy.[23] SiC particles with an average diameter of 4.6 μm were uniformly mixed with the brazing alloy powder to create the reinforced brazing alloy. Figures 7.18 and 7.19 show that the SiC particles react with the filler alloy. The SiC particles are surrounded by new black phases comprising Ti, C and Si. The black phase is assumed to be a Ti–Si–C compound. With more SiC particles, more Ti atoms in the filler alloy will be consumed to form the black phase, which results in a thinner TiC/TiCu reaction layer on the surface of the C/C composite. The highest joint strength was obtained when the SiC addition was 15 vol.%. Excessive addition of SiC particles can cause the formation of many pores in the joint.

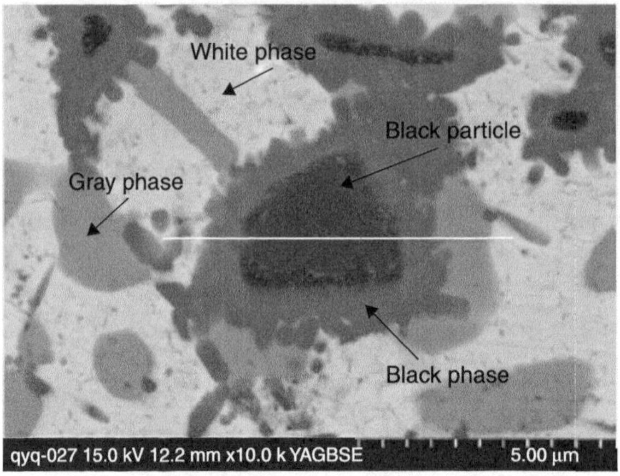

7.18 Microstructure of C/C composite/Ti–6Al–4V alloy brazed joint with Ag–Cu–Ti–15vol.% SiC particles.[23]

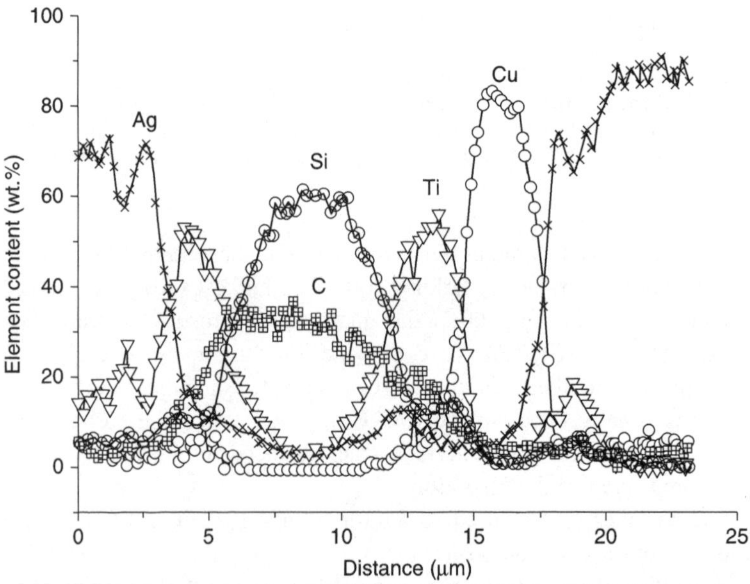

7.19 Major element content distribution of the brazing alloy with SiC particles.[23]

Joining a C/C composite with a carbon fiber reinforced C–SiC dual matrix composite has been realized through a reaction joining process using boron-modified phenolic resin with B_4C and SiO_2 powder additives.[24] First, B_4C powder and boron-modified phenolic resin with SiO_2 powder are uniformly mixed to form a high-temperature-resistant adhesive. The adhesive is then evenly pasted onto the

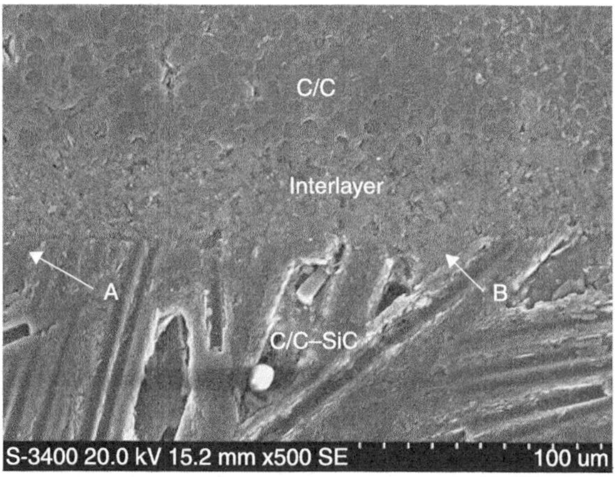

7.20 SEM image of C/C to C/C–SiC joint prepared under optimized conditions (holding for 2 h at 200°C under a pressure of 20 kPa). The regions indicated by arrows demonstrate the penetration of adhesive into the open pores of the substrate.[24]

C/C composite surface. Next, the bonding process is carried out in an oven and the assembled specimens are incubated at 200°C for 2 h at a pressure of 20 kPa. The joint shear strength was 12.4 MPa, which is 82.6% of the C/C matrix strength. The interfacial microstructures of the joint are shown in Fig. 7.20. It was found that the boron modified phenolic resin continuously distributes around the B_4C particles. The adhesive also penetrates the exposed holes in the matrix, which is beneficial for the strength of the joint. The adhesive is mainly composed of B_4C, SiO_2 and boron modified phenolic resin. The addition of B_4C can reduce volume shrinkage at the joining interface and can also help to reinforce the joints.

The cured joints can be enhanced by heat treatment. This involves treating the specimens at 1000°C, 1100°C, 1200°C, 1300°C or 1400°C in vacuum for 30 min. The joints are then cooled to room temperature within the furnace. Figure 7.21 shows the mechanical strengths of the joints after heat treatment at different temperatures. The highest joint mechanical strength was obtained with a heat treatment of 1200°C.

Some B_4C may be oxidized during heat treatment. As shown in Fig. 7.22 and 7.23, the low binding energy of 188.0 eV corresponds to the B–C bond in B_4C and the binding energy of 190 eV corresponds to the B–C bond in BC_2O. The highest binding energy of 193.0 eV indicates the presence of oxygen. The high binding energy corresponds to the B–C bond in the amorphous B_2O_3. B_4C has been oxidized at a high temperature to create B_2O_3 and then the B_2O_3 has melted into the glassy phase. As B exists in this glass phase of BO_3, the B_{1s} binding energy is increased. It can also be seen from Fig. 7.22 that the B_{1s} peak intensity in the B–C bond is reduced with the increase of temperature (1000°C–1200°C), which may

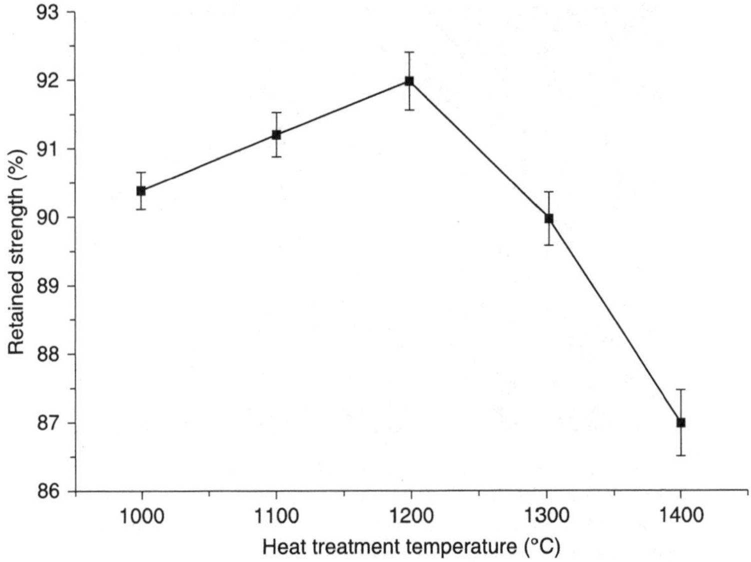

7.21 Effect of heat treatment temperature on the retained strength of joints with a holding time of 30 min in vacuum.[24]

be due to the decomposition and carbonation of boron-modified phenolic resin at higher temperatures. The decomposed oxygen-containing compounds react with the B_4C, which leads to the decrease of the B–C bond strength. Furthermore, the amount of B_2O_3 will increase as the temperature increases. When the temperature reaches 1200°C, the organic substances are completely carbonized and become amorphous carbon, which has higher strength. In addition, the SiO_2 will react with the B_2O_3 at 1200°C to form a borosilicate glass. The borosilicate glass can prevent volatilization of the B_2O_3 and help prevent the B_2O_3 and SiO_2 from reacting with the amorphous carbon, which can increase the strength of the joint. When the temperature reaches 1400°C, the strength of the B–C bond increases while the strength of the B_2O_3 bond decreases. This indicates that the reaction below occurs between the C and the B_2O_3 at this temperature:[24]

$$2B_2O_3(l) + 7C(s) = B_4C(s) + 6CO(g) \quad [7.13]$$

This reaction will consume the amorphous carbon and damage the integrity of the amorphous carbon structure. Hence, the joint strength is decreased at 1400°C.

7.7 Conclusion

Joining ceramics or composites to metal is challenging due to their chemical dissimilarity and the mismatch in their CTE. In this chapter the brazing of oxide, carbide, nitride and composite ceramics was discussed. In typical vacuum brazing,

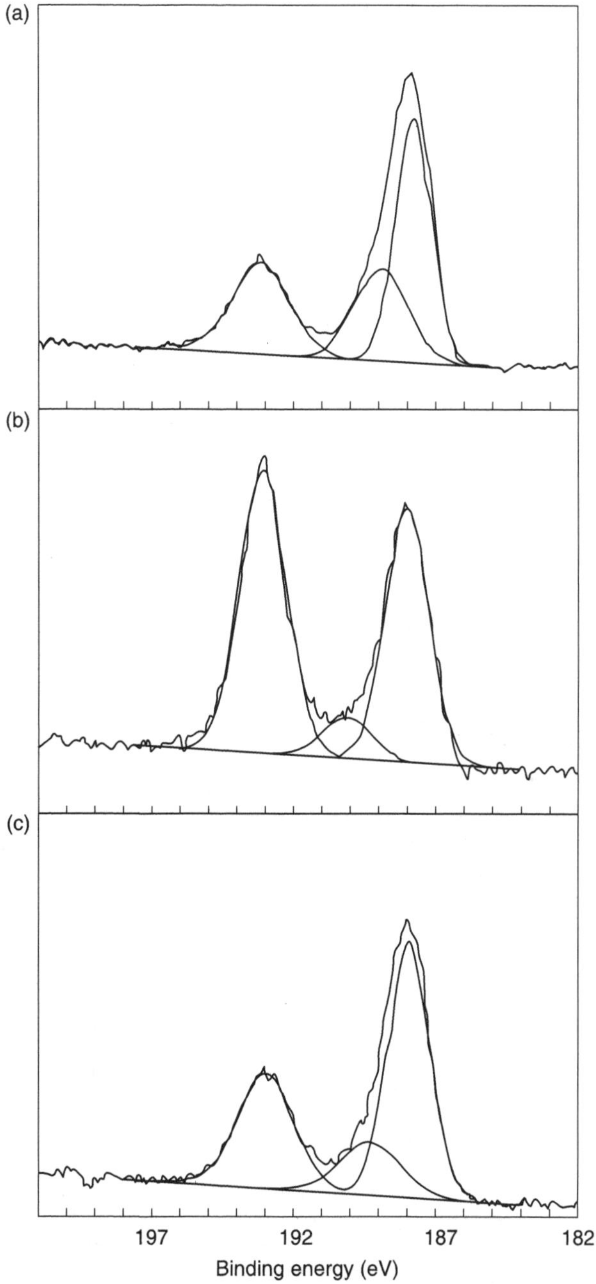

7.22 The XPS B_{1s} spectra of the pyrolysates of the adhesive pyrolyzed at various temperatures: (a) 1000°C, (b) 1200°C and (c) 1400°C.[24]

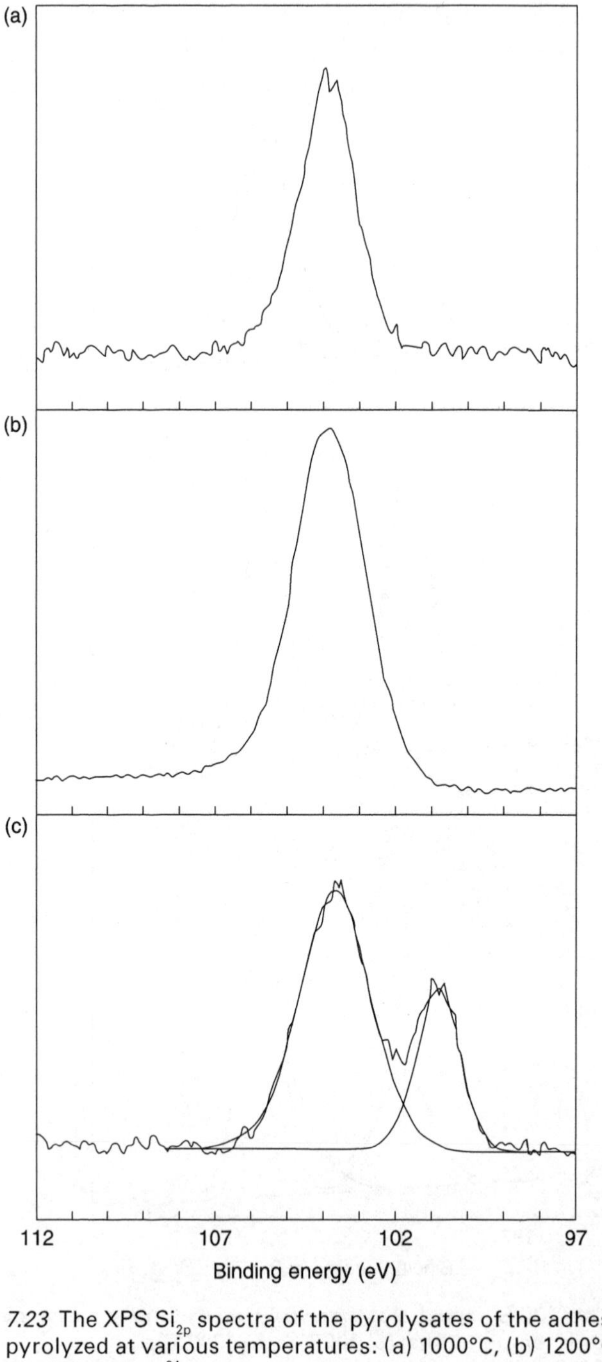

7.23 The XPS Si_{2p} spectra of the pyrolysates of the adhesive pyrolyzed at various temperatures: (a) 1000°C, (b) 1200°C and (c) 1400°C.[24]

filler alloys with reinforcements or buffer layers are employed to relieve the stress induced in the joint by the CTE mismatch. The volume and distribution of reinforcement and the interaction between the filler alloy and the buffer layer greatly influence the properties of the joint. Other brazing methods, such as ceramic precursor polymer joining or organic reaction joining, can realize firm joints by forming a covalent compound, which exhibits similar chemical properties to the ceramics or composites. However, the high expense and complex procedure of these methods limit their application. It is hoped that this chapter will improve the progress made in joining techniques and expedite an even greater uptake in the application of ceramics and composites.

7.8 References

1. Yin, W. 2009. *Research on mechanism of alumina ceramic active metallization and diffusion brazing with 5A05 alloy*. Harbin Institute of Technology. Dissertation for the Doctoral Degree.
2. Yang, M. X., Lin, T. S. and He, P. 2011. Microstructure evolution of Al_2O_3/Al_2O_3 joint brazed with Ag-Cu-Ti+B+TiH_2 composite filler. *Ceramics International*, 38(1), 289–294.
3. Yang, M. X., Lin, T. S., He, P. and Huang, Y. D. 2011. *In situ* synthesis of TiB whisker reinforcements in the joints of Al_2O_3/TC4 during brazing. *Materials Science and Engineering: A*, 528, 3520–3525.
4. Duo, L. 2009. *Technology and mechanism of brazing SiO_2 ceramic to TC4 alloy using AgCu/Ni composite interlayer*. Harbin Institute of Technology. Dissertation for the Doctoral Degree.
5. Nomura, M., Iwamoto, C. and Tanaka, S. I. 1999. Nanostructure of wetting triple line in a Ag-Cu-Ti/Si_3N_4 reactive system. *Acta Materialia*, 47, 407–413.
6. Luz, A. P. and Ribeiro, S. 2008. Wetting behavior of silicon nitride ceramics by Ti-Cu alloys. *Ceramics International*, 34, 305–309.
7. Brochu, M., Pugh, M. D. and Drew, R. A. L. 2004. Brazing silicon nitride to an iron-based intermetallic using a copper interlayer. *Ceramics International*, 30, 901–910.
8. Lemus-Ruiz, J., León-Patiño, C. A. and Aguilar-Reyes, E. A. 2006. Interface behavior during the self-joining of Si_3N_4 using a Nb-foil interlayer. *Scripta Materialia*, 54, 1339–1343.
9. Osendi, M. I., De Pablos, A. and Miranzo, P. 2001. Microstructure and mechanical strength of Si_3N_4/Ni solid state bonded interfaces. *Materials Science and Engineering: A*, 308, 53–59.
10. Liu, C. F., Zhang, J., Zhou, Y., Meng, Q. C. and Naka, M. 2008. Effect of Ti content on microstructure and strength of Si_3N_4/Si_3N_4 joints brazed with Cu-Pd-Ti filler metals. *Materials Science and Engineering: A*, 491, 483–487.
11. Zhang, J. and Sun, Y. 2010. Microstructural and mechanical characterization of the Si_3N_4/Si_3N_4 joint brazed using Au-Ni-V filler alloys. *Journal of the European Ceramic Society*, 30, 751–757.
12. Sun, Y., Zhang, J., Geng, Y. P., Ikeuchi, K. and Shibayanagi, T. 2011. Microstructure and mechanical properties of an Si_3N_4/Si_3N_4 joint brazed with Au-Ni-Pd-V filler alloy. *Scripta Materialia*, 64, 414–417.

13. Zhang, J., Zhang, X. M., Zhou, Y., Naka, M. and Svetlana, A. 2008. Interfacial microstructure of Si_3N_4/Si_3N_4 brazing joint with Cu-Zn-Ti filler alloy. *Materials Science and Engineering: A*, 495, 271–275.
14. He, Y. M., Zhang, J., Sun, Y. and Liu, C.F. 2010. Microstructure and mechanical properties of the Si3N4/42CrMo steel joints brazed with Ag-Cu-Ti + Mo composite filler. *Journal of the European Ceramic Society*, 30(15), 3245–3251.
15. Norton, M. G., Kajda, J. M. and Steele, B. C. H. 1990. Brazing of aluminum nitride substrates. *Journal of Materials Research*, 5, 2172–2176.
16. Shirzadi, A. A., Zhu, Y. and Bhadeshia, H. K. D. H. 2008. Joining ceramics to metals using metallic foam. *Materials Science and Engineering: A*, 496, 501–506.
17. Liu, G. W., Valenza, F., Muolo, M. L. and Passerone, A. 2010. SiC/SiC and SiC/Kovar joining by Ni-Si and Mo interlayers. *Journal of Materials Science*, 45, 4299–4307.
18. Lee, D. H., Jang, H. W., Kim, D. J., Chun, D. I. and Kim, Y. S. 2009. Joining of RBSiC using a preceramic polymer with Al. *Journal of Ceramic Processing Research*, 10, 263–265.
19. Koltsov, A., Hodaj, F. and Eustathopoulos, N. 2008. Brazing of AlN to SiC by a Pr silicide: Physicochemical aspects. *Materials Science and Engineering: A*, 495, 259–264.
20. Dong, Z. H. 2008. *Brazing of C/SiC composite to TC4 alloys*. Harbin Institute of Technology. Dissertation for the Doctoral Degree.
21. Xiong, J. H., Huang, J. H., Zhang, H. and Zhao, X. K. 2010. Brazing of carbon fiber reinforced SiC composite and TC4 using Ag-Cu-Ti active brazing alloy. *Materials Science and Engineering: A – Structural Materials Properties Microstructure and Processing*, 527, 1096–1101.
22. Qin, Y. Q. 2007. *Study on microstructure and mechanical properties of C/C composite TC4 brazed joint*. Harbin Institute of Technology. Dissertation for the Doctoral Degree.
23. Qin Y. Q. and Yu, Z. S. 2010. Joining of C/C composite to TC4 using SiC particle-reinforced brazing alloy. *Materials Characterization*, 61, 635–639.
24. Li, S. J., Chen, X. F. and Chen, Z. J. 2010. The effect of high temperature heat-treatment on the strength of C/C to C/C-SiC joints. *Carbon*, 48, 3042–3049.

8
Brazing of nickel, ferrite and titanium–aluminum intermetallics

H. PENG, Harbin Institute of Technology, China

DOI: 10.1533/9780857096500.2.221

Abstract: Intermetallics have many metallic characteristics, including luster, and electrical and thermal conductivity. Ordered intermetallic structural materials have recently been the focus of many investigations. The physical and brazing properties of Ni–Al system intermetallics, Fe–Al intermetallics and Ti–Al intermetallics are presented in this chapter, including brazing methods, joint microstructure and corresponding mechanical strengths. The suitability of a variety of brazing filler metals for Ti–Al intermetallics is also discussed, and an extensive comparison is made between the brazing of traditional materials and those including more than three intermetallics.

Key words: physical properties, brazing properties, nickel, ferrite, TiAl intermetallics.

8.1 Introduction

Intermetallics are compounds synthesized either from two metal elements or from one metal and one metalloid with integer ratios (chemometry). Intermetallics have many metallic characteristics, such as luster, and electrical and thermal conductivity. Intermetallics maintain structural stability even when combined with other chemicals, and appear as ordered solid solutions in phase diagrams.

The chemical formulas of intermetallics formed with the constituent elements A and B will usually be AB, A_2B and A_3B. There are also some special structures, such as A_5B_3 or A_7B_6. The existing A_5B_3 or A_7B_6 intermetallics include Mo_5Si_3, Ti_5Si_3, Nb_6Fe_7 and W_6Co_7, among others.[1] Intermetallics can also be formed from three or more constituent elements, so the intermetallic family is very large.

Ordered intermetallic structural materials have recently been the focus of many investigations. Remarkable progress has been made using A_3B and AB intermetallics from the Ti–Al, Ni–Al and Fe–Al systems. Some physical properties of several important intermetallics are listed in Table 8.1.

Table 8.1 Physical properties of several important intermetallics

Intermetallic		Structure	Young's modulus (GPa)	Melting point (°C)	Orderly critical temperature (°C)	Density ($g.cm^{-3}$)
Ti–Al	Ti_3Al	DO19	110–145	1600	1100	4.20
	TiAl	L10	176	1460	1460	3.90
Ni–Al	Ni_3Al	L12	178	1390	1390	7.50
	NiAl	B2	293	1638	1638	5.86
Fe–Al	Fe_3Al	DO3	140	1540	540	6.72
	FeAl	B2	259	1250–1400	1250–1400	5.56

8.2 Physical properties and brazing properties of Ni–Al system intermetallics

8.2.1 Physical properties of Ni–Al system intermetallics

There are many kinds of intermetallics in the Ni–Al system, including Al_3Ni, Al_3Ni_2, Ni_5Al_3, Ni_3Al and NiAl. Attention has recently been focused on Ni_3Al and NiAl.

Ni_3Al has an ordered FCC L12 superlattice structure. Polycrystalline Ni_3Al is very brittle, which has previously limited its applications. However, recent studies have found that the addition of trace elements, such as Mn, Zr, Cr, Pd, B, etc., can significantly improve the plasticity of Ni_3Al. A microalloying process can be used to obtain better plasticity at room temperature and higher mechanical strength at high temperatures for Ni_3Al. Modified Ni_3Al has been successfully used in both civil and military applications.

NiAl has an ordered BCC B2 superlattice structure, which makes it a Hume-Rothery β phase electron compound. The structure of NiAl is stable due to the ratio of its valence electron to atom number, which is 3/2. The ratio of Ni/Al atoms in single-phase NiAl has a wide range; the proportion of Ni atoms can be between 45 and 60%. The melting point of NiAl is 1638°C. NiAl normally has a high melting point and good inoxidizability. However, the crystal structures of NiAl alloys lack sufficient independent slip systems at room temperature, and hence NiAl alloys exhibit low plasticity and are very brittle. Moreover, NiAl alloys are easily oxidized to produce continuous Al_2O_3 thin films on their surfaces, which cause their weldability to deteriorate. In most cases, therefore, separate NiAl components are joined with the major structures made from Ni-based alloys via the brazing process.

8.2.2 Brazing of Ni–Al intermetallics

Co-based alloys have excellent wettability with Ni_3Al intermetallics. A compact consecutive brazed joint of Ni_3Al intermetallics can therefore be obtained using

Co-based alloys, as shown in Fig. 8.1.[3] In Fig. 8.1(a), Phase 1 and Phase 2 are two different compounds, which consist of different carbides, borides and silicides. Phase 3 is considered to be a kind of boride: M_3B_2. The element M could be Cr, W, Mo, Ni or Co. In Fig. 8.1(b), Phase 1 is the carbide M_6C, which is rich in W and Mo. Phase 2 is M_3B_2, which is rich in Cr. Phase 3 is a Ni–Co based solid solution. In Fig. 8.1(c), the white Phase 1 is W_3B_2. Phase 2 in the seam is also a kind of Ni–Co based solid solution.

To lower the melting point of a brazed material, the element Si or B can be added to the filler metal. However, many kinds of brittle silicon boron compounds will be formed in the brazed joints, making them less reliable. Si has more influence than B on brazed joint formation, because Si atoms have a bigger atomic radius and consequently a lower diffusion velocity in the liquid brazing system. The diffusion velocity of Si or B atoms can be improved with a relatively high brazing temperature. Furthermore, brazed joints with uniform textures and stable

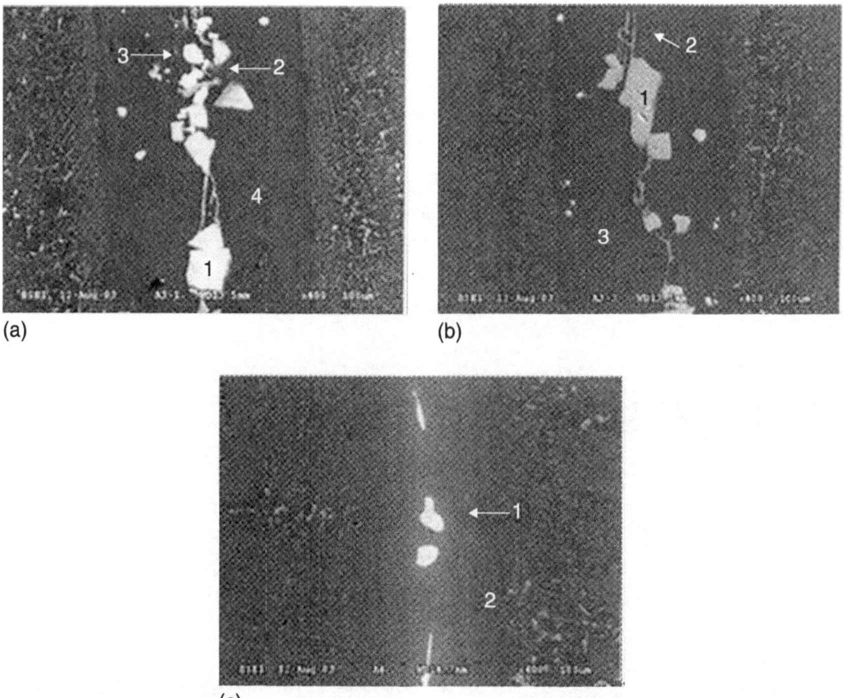

8.1 Microstructures of Ni_3Al (Ic6A alloy) braze joint with Co-based alloys. (a) Co45CrNiWBSi, 1180°C for 4h; (b) N300E, 1180°C for 4h; (c) Co45NiCrWB 1220°C for 4h.

mechanical properties can be obtained if the contents of those melting point depressant elements are relatively low.

The corresponding creep rupture lives of Ni_3Al based alloy brazed joints with different Co-based alloys are shown in Table 8.2. The creep tests were performed at 900°C under 160 MPa.

Pure Cu foil can be used as interlayer in brazing NiAl.[4] The brazing process is controlled by the diffusion of Cu into the substrate and the dissolution of Al into the melted filler metal. An isothermal solidification process can be used to obtain good joints. The microstructure of NiAl/Cu/Ni joints with different holding times is shown in Fig. 8.2. In both images, the different phases from (a) to (e) are the NiAl substrate, γ' precipitates, γ/γ', β-NiAl and the Ni substrate, respectively.

In the NiAl/Cu/Ni system, the microstructural features are primarily determined by the ratio of Al to Cu. Because Cu diffusion into the substrates is slow and Al transfer across the joint is rapid, a repeated sequence of microstructural features forms across the joint. Depletion of Al from the NiAl substrate resulted in the formation of regions of $L1_0$ type martensite. Conversely, the transfer of Al from the NiAl substrate to the Ni substrate led to extensive γ' precipitation. Further Al enrichment caused the formation of β regions near the Ni substrate-bond-line

Table 8.2 Creep rupture life of Ni_3Al based alloy brazed joints

Brazing filler metal	Brazing condition	Creep rupture life under 900°C, 160 MPa/h
Co45CrNiWBSi	1180°C for 4 h	7–23
N300E (Co–Cr–Ni–W–B)	1180°C for 4 h	61–73
Co45CrNiWB	1220°C for 4 h	110–142

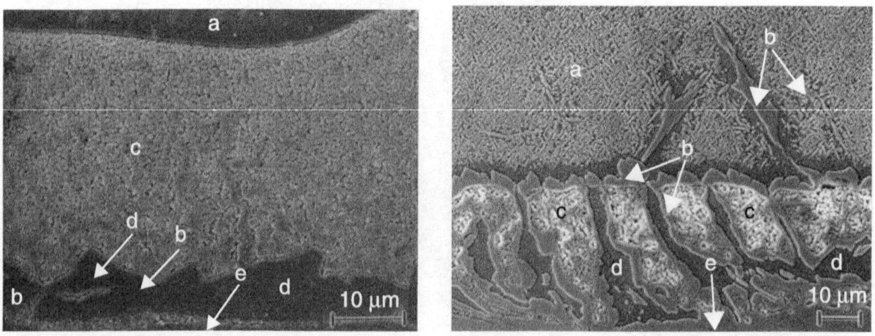

(a) (b)

8.2 Microstructure of NiAl/Cu/Ni joints brazed at 1150°C for different holding times.[4]

interface. The isothermal solidification time can be reduced dramatically when pure Cu foil is replaced by a composite interlayer consisting of Cu and NiAl.[4]

Ni-based alloys are also used as filler metals for brazing NiAl or Ni–Al alloys. With no holding time, or with a short holding period, many eutectic structures form in the brazed joints. This is due to limited interdiffusion between the liquidoid and the solidoid during the brazing process. With a longer holding time, NiAl will start to dissolve from the base material into the liquid brazing filler metal. The content of Al element in the liquid brazing filler metal then increases gradually. With the addition of Al, the eutectic structures dissolve over time. However, because many Al atoms have diffused into the liquid brazing filler metal, an Al depletion region forms at the NiAl base, close to the filler metal.

8.3 Physical properties and brazing properties of Fe–Al intermetallics

8.3.1 Physical properties of Fe–Al intermetallics

Fe–Al alloys have high elastic modulus, high melting point and high specific strength. The Fe–Al alloys with low Al content exhibit severe environmental embrittlement. The Fe–Al alloys with high Al content present extremely low plasticity and high brittleness due to weak bonding between the grain boundaries.

Among the Fe–Al alloys, Fe_3Al is the most promising engineering material. Fe_3Al has an ordered monocrystal DO3 superlattice structure, with a lattice constant of 0.578 nm. It also has a high elasticity modulus, high melting point and low density. The chemical constituents and mechanical properties of typical Fe_3Al intermetallics are shown in Table 8.3.

8.3.2 Brazing of Fe–Al intermetallics

There are several ways to braze Fe_3Al-based alloys, including torch brazing, vacuum brazing and infrared brazing. It is easy to obtain high-quality brazed joints using the vacuum brazing and infrared brazing processes. However, during the torch brazing process, microcracks, pores and slag inclusions are often formed

Table 8.3 Chemical constitution and mechanical properties of typical Fe_3Al intermetallics

Alloy	Tensile strength (MPa)	Yield strength (MPa)	Elongation (%)	Hardness HRC	Elastic modulus (GPa)
Fe–Al27	483	831	2.8	–	–
Fe–Al28–Cr1	455	75	2	29	140

in the brazed joints. The high heating and high cooling rates during the torch brazing process are responsible for these defects.

Vacuum brazing of Fe_3Al alloy

A number of materials are suitable for Fe_3Al alloy vacuum brazing, including pure aluminum foil, Cu-based alloys, Ag-based alloys and Ni-based alloys.[5] When pure aluminum foil is selected as the interlayer for Fe_3Al alloy vacuum brazing, the diffusion reaction area occurs in the joint, and the hard, brittle Fe_2Al_5 phase will also form in this area. Diffusion treatment at high temperatures for a long time improves the mechanical properties of such joints. After undergoing thermal treatment, the compositions in the joint will be more uniform and the quantity of the crispy hard IMCs will be reduced or eliminated. When Cu-based alloys are used in vacuum brazing, intercrystalline diffusion of Cu occurs at the Fe_3Al interfacial region. BNi-2 is one of the best materials for Fe_3Al brazing. Joints brazed with BNi-2 have relatively high hardness, high strength and good mechanical performance at high temperatures. Fe_3Al/18-8 stainless steel joints brazed with BNi-2 through vacuum brazing are shown in Fig. 8.3.

When Ni-based alloys are used to braze Fe_3Al intermetallics, a strong diffusion reaction occurs between the braze and the base metal. Strong metallurgical bonds form between the steel and the Ni-based alloys, especially when 18–8 stainless steel is used. Further diffusing heat treatment reduces the crispy hard phases in the joints and results in a more uniform texture.

Infrared brazing of Fe_3Al alloy

Materials including Al foil and Cu-based, Ag-based and Ni-based alloys can be used for infrared brazing of Fe_3Al intermetallics. When Al foil is used, Fe_3Al

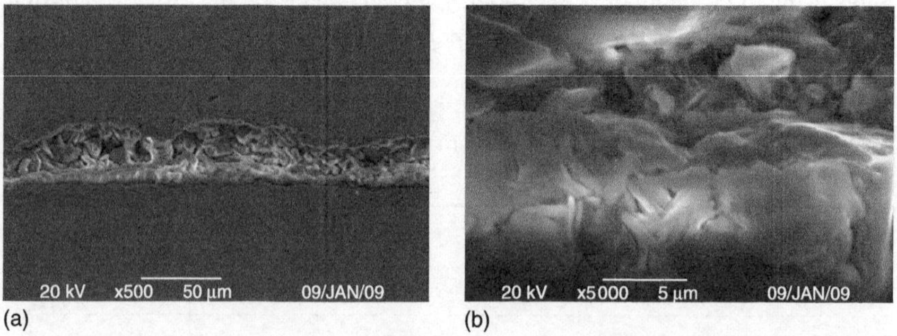

8.3 Fe_3Al/18-8 joint with BNi-2 filler metal after vacuum brazing. (a) Brazing line; (b) fractionated gain of brazing line.

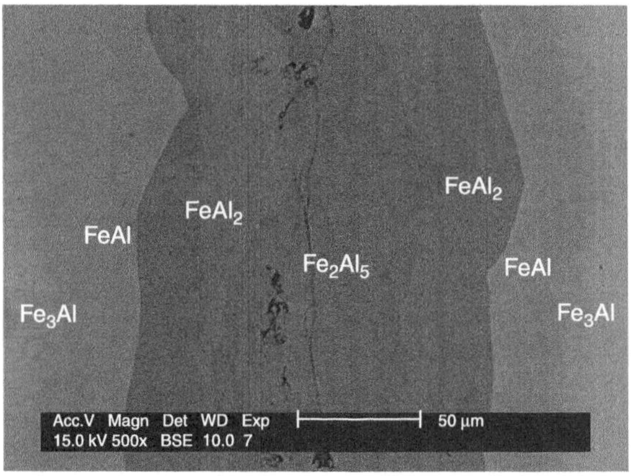

8.4 SEM image of $Fe_3Al/Al/Fe_3Al$ joint infrared brazed at 1050°C for 180 s.

dissolves into the molten braze quickly. Meanwhile, Al from the Al foil dissolves into the base metal. Furthermore, with different brazing temperatures and times, the eutectic structure of FeAl–$FeAl_2$ and a series of compounds such as $FeAl_3$, Fe_2Al_5, $FeAl_2$ and FeAl form in the joints, as shown in Fig. 8.4.[6] Raising the brazing temperature or prolonging the holding time aggravates interdiffusion and causes the brazing interface to become fuzzy. The textures in the brazing seam gradually evolve from the Fe_2Al_5 phase to FeAl phase and the eutectic structure of FeAl–$FeAl_2$. Finally, the texture in the brazing seam becomes FeAl phase only, and the interface between the brazing seam and the base metal vanishes.

When Cu-based alloys are used for infrared brazing of Fe_3Al intermetallics, b10 martensite and a copper-rich phase form in the brazing seam. With higher brazing temperatures or longer holding times, the Fe_3Al intermetallics gradually dissolve into the brazing seam. The Al content in the brazing seam then increases dramatically, which leads to more b10 martensite in the joint, as shown in Fig. 8.5.

If Ag foil is used to braze Fe_3Al intermetallics, the brazing seam will include Al-containing Ag-rich phases. In this case, the effect of brazing time on the shear strength of the brazed joints is negligible. When BAg72Cu28 is used to braze Fe_3Al intermetallics, the main structure in the brazing seam will be Ag-Cu eutectic, as shown in Fig. 8.6.[8] Because no brittle compounds form, the joints will be highly ductile.

When BNi-2 materials are used to braze Fe_3Al intermetallics, a series of transition phases including $(Ni,Fe)_3Al$, $(Ni,Fe)_3(Si,Al)$, $(Fe,Ni,Cr)_3B$ and BCr, etc., can be observed in the brazing seam.[8] With a longer homogenizing treatment at 1000°C, the previously formed $(Ni,Fe)_3Al$ and $(Ni,Fe)_3(Si,Al)$ transform into $(Ni,Fe)_2Al$ and $(Ni,Fe)_2(Si,Al)$. Meanwhile, the quantity of $(Fe,Ni,Cr)_3B$ decreases

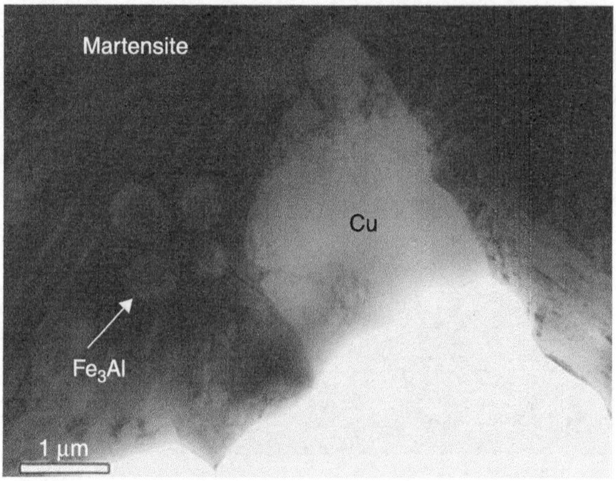

8.5 TEM image of the infrared brazed Fe_3Al joint with Cu foil at 1100°C for 300 s.

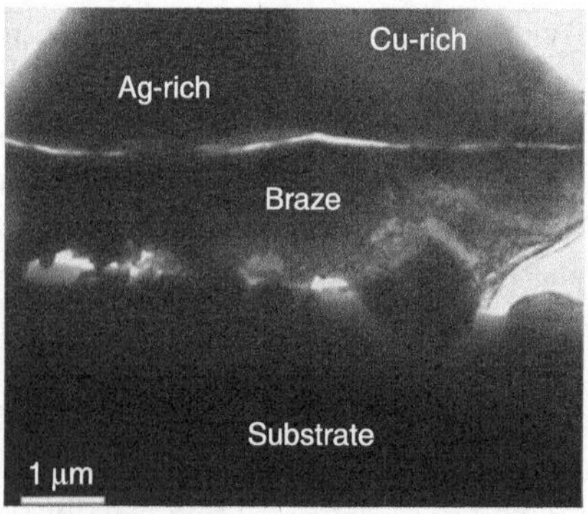

8.6 TEM image of Fe_3Al phase in a joint infrared brazed at 1123 K for 300 s using BAg-8 braze alloy.

dramatically. With sufficient homogenizing treatment, a joint composed of a single phase can be obtained, as shown in Fig. 8.7. At first, the base metal (marked 1 in Fig. 8.7) is primarily a Fe_3Al substrate solid solution with small amounts of Ni, due to the high brazing temperature. A continuous layer close to the interface of Fe_3Al/BNi-2 is shown in Fig. 8.7, marked 2 in the SEM BEI photograph. There are three different phases in the central part of the joint. The small black spots (marked 5 in the figure) are chromium boride, BCr. Similarly, the gray matrix in the middle part of the joint (marked 3 in the figure) can be identified as $(Fe,Ni)_3(Si,Al)$ according to the stoichiometric ratio of Fe, Ni, Si and Al. Finally, the dark gray streak-like phase (marked 4 in Fig. 8.7) contains B, Cr, Fe and Ni.[9]

(a) (b)

8.7 (a) SEM image of Fe_3Al/BNi-2/Fe_3Al joint brazed at 1150°C for 240 s and (b) subsequently followed by 1000°C homogenizing treatment for 72 h.

Table 8.4 Infrared brazing processes for Fe_3Al alloy with different filler metals and the corresponding strengths

Brazing filler metal	Brazing process	Shear strength
Cu foil (50 μm)	1100°C for 300 s	291 MPa
BAg-8	800°C for 600 s	181 MPa
Al foil (100 μm)	700–1200°C for 15–240 s	–
BNi-2	1150°C for 2–240 s	–

The infrared brazing processes for Fe_3Al and the corresponding joint strengths are listed in Table 8.4.

Compared with traditional furnace brazing, the infrared brazing process has a very short thermal cycle. The interdiffusion between the braze materials and the base metal is therefore very limited. The subsequent homogenizing treatment can improve interdiffusion between the base metal and elements in the solidified brazing seam. When B is involved in the brazing process, the Kirkendall effect occurs in the brazing joints because B has a relatively high diffusion velocity, and this leads to asymmetrical diffusion or substance migration in the joints.

8.4 Physical properties and brazing properties of Ti–Al intermetallics

8.4.1 Characteristics of Ti–Al intermetallics

Ti–Al intermetallics have several excellent characteristics: low density, high elastic modulus, good oxidation resistance and high-temperature strength. They are therefore very suitable for use as high-temperature structural materials in the aviation and aerospace fields.

Ti–Al intermetallics mainly include γ-TiAl, α_2-Ti_3Al and θ-$TiAl_3$. Their room-temperature ductilities can be improved by alloying and microstructure controlling processes. For example, the plasticity of Ti_3Al is low at room temperature, which makes it difficult to deform. With the addition of β stability elements, such as Nb, V and Mo, etc., the martensite transformation point (Ms) of Ti_3Al is decreased and its α_2 phases are refined, and the slip length is also reduced. Furthermore, α_2+β phases are generated, which have better plasticity and mechanical strength.

The mechanical properties of Ti–Al intermetallics are strongly affected by their microstructures. If alloying elements are used excessively, the linear expansion coefficient of Ti–Al intermetallics will be relatively low. High internal stress is easily produced in the joint if Ti–Al intermetallics are used in brazing dissimilar materials.

In the fusion welding process, joint composition is complicated, and brittle intermetallics and hot cracking often appear in the joint. Hence, the commonly used welding methods for joining Ti–Al intermetallics to dissimilar materials are brazing and diffusion bonding.

8.4.2 Brazing of Ti–Al intermetallics

Joining techniques currently used for Ti–Al intermetallics include fusion welding (arc welding, electron beam welding and laser welding), solid-state welding (friction welding, diffusion bonding, self-propagating high-temperature synthesis reaction welding (SHS), etc.), and brazing (vacuum brazing, induction brazing and infrared brazing, etc.). There are two outstanding problems with fusion welding between Ti–Al intermetallics. One is the hot cracking that tends to occur in the welding zone, while the other is that microstructures in the heat affected zone are different from those of the matrix, which reduces the joint mechanical strength. The fusion weldability of Ti–Al intermetallics is therefore poor, and the joint strengths are inconsistent. If a solid-state welding process is employed, welding defects related to melting and solidification processes can be avoided. With friction welding processes, the matrix can easily crack under the forging force. In brazing, the filler metal can reduce the stress in the joint. The amount of brittle intermetallics and their distribution in the joints can also be effectively controlled by optimizing the brazing parameters. In summary, the brazing process is a suitable method for joining Ti–Al intermetallics.

8.5 Brazing between Ti–Al intermetallics

Commonly used filler metals and parameters for the brazing of Ti–Al intermetallics, and the resulting joint mechanical strengths, are shown in Table 8.5.[10–27] It can

Table 8.5 Commonly used brazing filler metals and brazing parameters for the brazing of Ti–Al intermetallics and the mechanical strengths of the joints obtained

Brazing filler metal (wt.%)		Ti–Al intermetallics (at.%)	Brazing parameters		Joint mechanical strength (MPa)	References
			Brazing temperature (°C)	Holding time (s)		
Ti-based	Ti–15Cu–15Ni	$Ti_{50}Al_{50}$	1100–1200	30–60, 18–60	—	10,11
		Ti–47Al–2Cr–2Nb	980, 1000	600	—	12,13
		Ti–48Al–2Cr–2Nb	1100–1200	30–60	Shear 322	14
		Ti–48Al–2Cr–2Nb	950	480–2400	Tensile 295	15
		Ti–48Al–2Cr–2Nb	1040, 1000	600, 1800	Shear 220	16
	Ti–33Ni	Ti–47Al–2Cr–2Nb	1100–1200	600	—	17
	Ti–50Ni	Ti–43Al–9V–0.3Y	1140–1200	900	Shear 256, 207 (600°C)	18
	Ti–Zr35–Ni15–Cu15	TiAl	950–1100	600–3600	248	19
Ag-based	BAg–8	$Ti_{50}Al_{50}$	900–1150	15–180	Shear 343	20
	Pure Ag	$Ti_{50}Al_{50}$	1000–1100	15–180	Shear 385	21
	Ag–Cu eutectic	Ti–48Al–2Cr–2Nb	850–1000	300–3600	Tensile 225	22
	Ag34Cu16Zn	Ti–48Al–2Cr–2Nb	850–900	300–3600	Tensile 210	22
Al-based	63Ag–35.2Cu–1.8Ti	Ti–47Al–2Cr–2Nb	750	600	—	23
	Al foil	$Ti_{50}Al_{50}$	800–900	15–300	63.9	24
		Ti–48Al	900		Tensile 220	25
	BAlSi–4	$Ti_{50}Al_{50}$	800, 900	120–300	86.2	26
Others	Zr65Al7.5Cu27	Ti–48Al–2Cr–2Nb	950	1200	—	27

be seen that most of the suitable brazing filler metals are Ti-based, Ag-based or Al-based alloys.

8.5.1 Brazing of Ti–Al intermetallics using an Ag-based brazing filler metal

Ti–Al intermetallics have been successfully infrared brazed using pure Ag as the filler material, with brazing parameters of 1000–1100°C for 30–180 s. Three phases form in the reaction layer: Ti (Al, Ag), Ti_3 (Al, Ag) and Ag-rich phases. During the brazing process, part of the TiAl base metal dissolves into the pure Ag filler metal. Since the solubility of Al in Ag is higher than that of Ti in Ag, more Al atoms are consumed. Ti-rich areas occur at the interface, which results in the formation of Ti_3 (Al, Ag) phases.

Joints brazed with a pure Ag filler have high shear strengths. When the brazing temperature is 1050°C, all joints break at the TiAl base metal during shear testing. With a brazing temperature and holding time of 1100°C and 60 s, the highest joint shear strength of 385 MPa is achieved. The shear fracture is located at the brazing seam.

Ag–28Cu eutectic also can be used to infrared braze Ti–Al intermetallics. Figure 8.8[14] shows the microstructures of a TiAl/Ag–28Cu/TiAl joint brazed at 950–1150°C for 15–180 s. Here, Cu atoms have replaced Ag atoms and reacted with the TiAl base metal to form a continuous reaction layer. When the brazing temperature is 950°C, AlCuTi and $AlCu_2Ti$ form in the brazing seam. With higher brazing temperatures and longer holding times, the growth rate of AlCuTi is obviously higher than that of $AlCu_2Ti$. However, excessive AlCuTi will lead to deterioration of the joint strength. With a brazing temperature of 950°C and a holding time of 60 s, joint shear strengths of up to 343 MPa can be achieved.

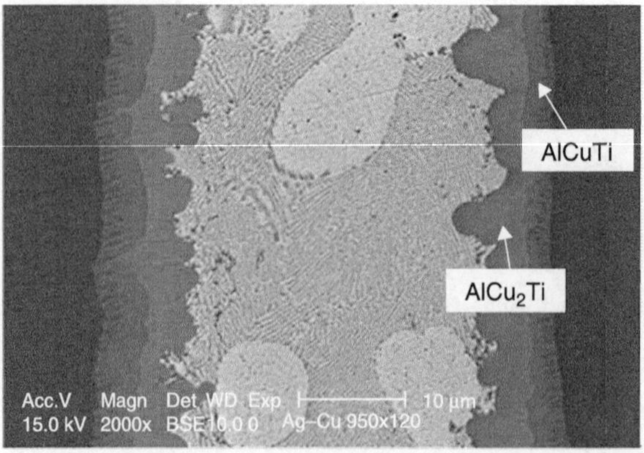

8.8 Brazing interface of TiAl/Ag–28Cu/TiAl joint (950°C for 120 s).

8.5.2 Brazing of Ti–Al intermetallics using an Al-based brazing filler metal

Pure Al foil and BAlSi-4 are commonly used as the brazing filler metals for Ti–Al intermetallics.[24–26] When pure Al foil is used, a stable, brittle phase of $TiAl_3$ forms in the brazing seam, which causes inherent brittleness in the joints. If BAlSi-4 is used, $Al_{12}Si_3Ti_5$ forms in the brazing seam. The maximum shear strength of such joints is 86.2 MPa.

8.5.3 Brazing of Ti–Al intermetallics using a Ti-based brazing filler metal

The shear strengths of Ti–Al intermetallic joints brazed with an Al-based filler are normally lower than those of joints brazed with an Ag-based filler. However, the working temperature of joints brazed with an Ag-based filler is usually below 400°C. Thus, joints brazed with a Ti-based filler, which have higher working temperatures, are more popular. Commonly used Ti-based fillers for the brazing of Ti–Al intermetallics include Ti–Cu–Ni, Ti–Zr–Ni–Cu and Ti–Ni.

Ti–15Cu–15Ni filler metal can be used to infrared braze Ti–48Al–2Cr–2Nb and Ti50Al50 casting alloy at temperatures of 1100–1200°C and holding times of 30–60 s.[10,11] It has been reported that the dissolution and diffusion of Al and Ti from TiAl into the brazing seam are the main controlling factors for the formation of interfacial microstructures. The existence of Nb and Cr in the TiAl base metal also has some impact on the evolution of a 2-phase-columnar zone and an $\alpha 2$-continuously-reacting zone in the reaction layer.

During isothermal solidification and solid-state diffusion, a multi-layer interfacial microstructure forms in the joint. The interfacial microstructures and formation mechanisms vary with the bonding temperature, holding time and type of base metal.

With a brazing temperature of 1150°C and a holding time of 30 s, the joint interfacial microstructures contain γ-TiAl/α-Ti/α+β/β-Ti/residual brazing filler. At the same brazing temperature and with a holding time of 60 s, the interfacial microstructures are γ-TiAl/α-Ti/$\alpha 2$–Ti3A1/β-Ti/residual brazing filler.

The evolution of interfacial microstructures, which vary with holding time, can be described by the following five stages: (a) β-Ti layer formation, (b) formation of columnar α+β phase zone, (c) $\alpha 2$-nucleation, (d) rich α phase formation, (e) $\alpha 2$ phase formation, as shown in Fig. 8.9.[10,11]

Figure 8.10[18] shows the interfacial microstructures of Ti–Al intermetallic joints brazed with a TiH_2-50Ni filler, under a range of brazing temperatures with a holding time of 900 s. The TiH_2-50Ni brazing filler was prepared via the mechanical ball milling process for 28 800 s. The joint comprised a diffusion layer and a brazing seam zone. The diffusion layer was formed by the crossing-lamellar layer of Ti_3Al + TiAl, and the brazing seam zone was formed by Al_3Ti, Ni_4Ti_3 and Ti_3Al.

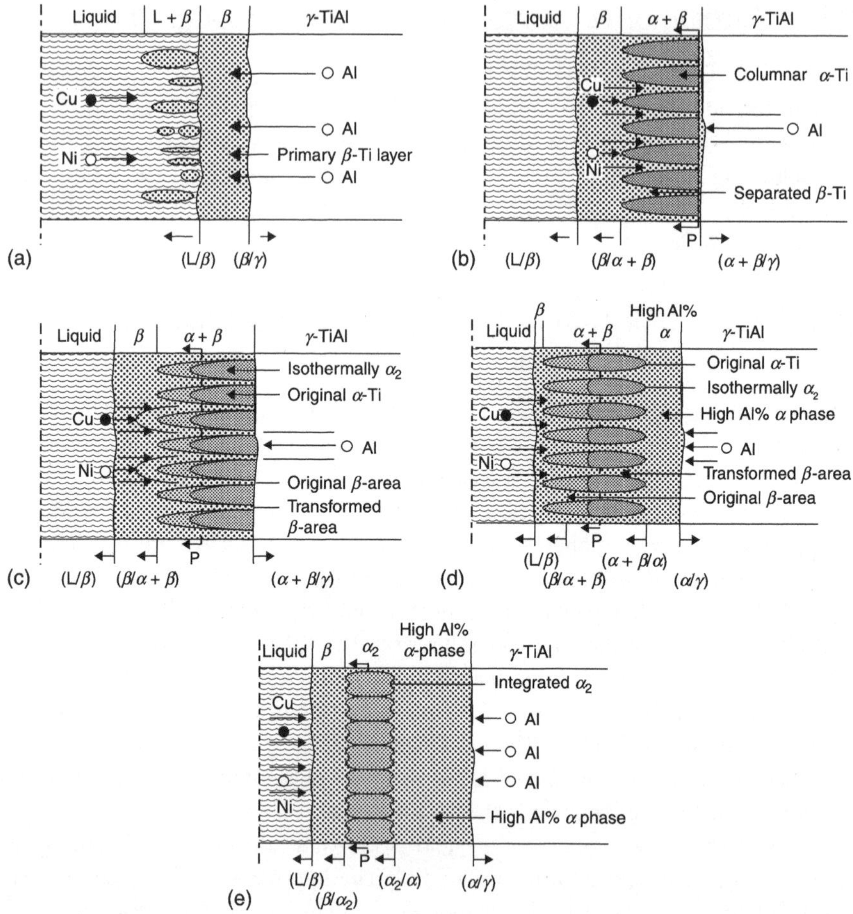

8.9 Five stages during joint interface formation. (a) β-Ti layer formation; (b) columnar $\boldsymbol{\alpha}$+$\boldsymbol{\beta}$ two-phase formation; (c) α2-phase nucleation; (d) rich Al $\boldsymbol{\alpha}$ phase formation; (e) $\boldsymbol{\alpha}$2 phase formation.

With higher brazing temperatures or longer holding times, the Ti_3Al in the diffusion layer gradually changes into TiAl, and the microstructures in the diffusion layer become finer and more homogeneous.

The interfacial microstructures in joints brazed with TiH_2-30Ni are similar to those in joints brazed with TiH_2-50Ni. It has also been found that the milling time of the brazing filler metal has a significant effect on the melting point of the brazing filler. With a holding time of 900 s, an unmilled brazing filler metal can melt completely at 1180°C, while a brazing filler metal milled for 7200 s melts at 1140°C, and a brazing filler metal milled for 28 800 s melts at 1150°C. The surface energy difference, caused by the different powder

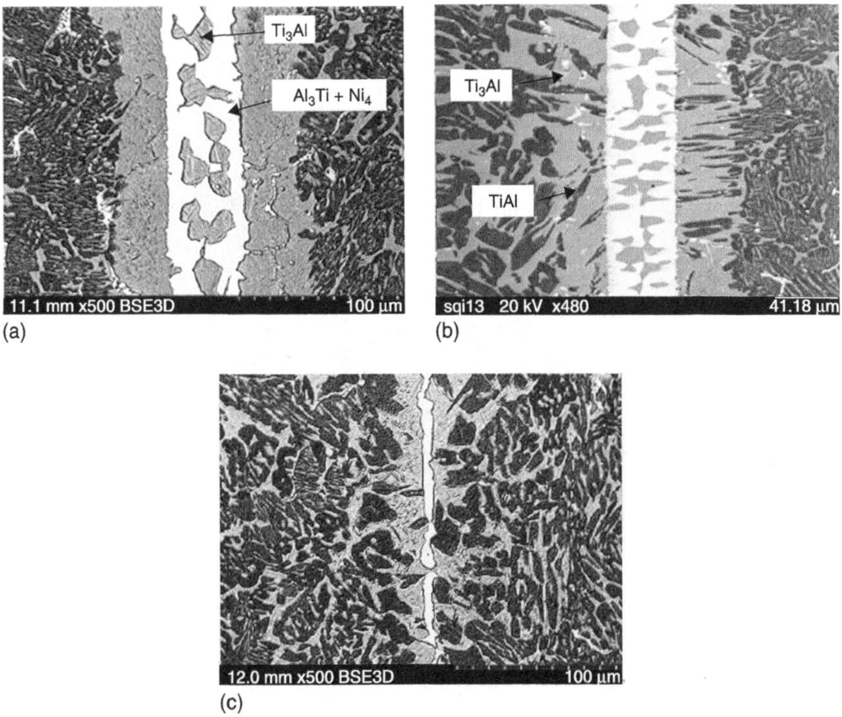

8.10 Influence of brazing temperature on the brazing seam microstructures in a joint brazed with TiH_2-50 wt.%Ni (milling 480 min, holding time 5 min). (a) 1150°C; (b) 1180°C; (c) 1200°C.

particle sizes, the different lattice distortion energy and phase changes during the milling process, is believed to be the main reason for the variations in melting point.

Heat treatment can be used to modify the microstructures in the joints. When joints brazed at 1200°C for 900 s are subjected to a heat treatment at 1200°C for 3600 s, the microstructures in the brazing seam become finer and more similar to those in the diffusion layer, as shown in Fig. 8.11.[18]

The milling time of the brazing filler metals has a significant influence on joint shear strength. At a brazing temperature of 1150°C, the maximum shear strength of a joint brazed by unmilled TiH_2-50Ni filler is 186 MPa (t = 1800 s). The maximum joint shear strength after 7200 s of milling is 244 MPa (t = 300 s), and the maximum shear strength after 28 800 s of milling is 276 MPa (t = 600 s). The high-temperature shear strength (800°C) of a joint in which the brazing filler has undergone 28 800 s of milling is up to 211 MPa, and then fractures mainly occur in the $Al_3Ti + Ni_4Ti_3$ phases in the brazing seam.

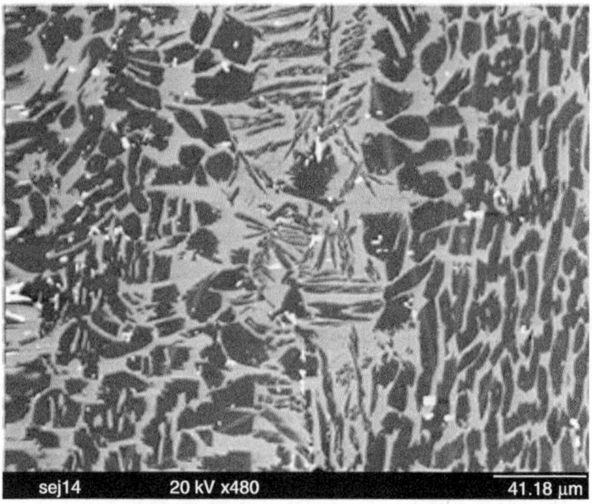

8.11 Microstructures in a brazed joint after the heat treatment 1200°C for 60 min (T = 1200°C, t = 15 min).

8.5.4 Brazing of Ti–Al intermetallics with steels

Turbine blades made from Ti–Al intermetallics can be joined to steel bearing bars to create a turbo engine rotor, which improves the engine toggle speed, reduces emissions and greatly increases thermal efficiency. Joining Ti–Al intermetallics and steel is therefore a hot topic. Table 8.6 lists commonly used brazing materials, brazing parameters and joint mechanical strengths. Ag-based brazing materials are commonly used and their joints normally have higher mechanical strengths.[28–36]

Figure 8.12(a) and (b)[28] show the interfacial microstructures of Ti–Al intermetallics/AISI4340 steel joints that have been induction brazed with 63Ag–35.2Cu–1.8Ti (wt.%) and 70Ti–15Cu–15Ni (wt.%), respectively. It was found that the reaction between Ti–Cu–Ni and steel is more violent than that of Ag–Cu–Ti and steel. When Ti–Cu–Ni is used as a filler, a TiC reaction layer forms at the brazing interface, which leads to a reduction in the mechanical strength of the joint. Joints brazed with the Ag–Cu–Ti filler have shear strengths of 320 MPa at room temperature, and 310 MPa at 500°C, which are higher than those of joints brazed with the Ti–Cu–Ni filler.

When Ag–Cu is used as the brazing filler for TiAl intermetallics, hard and brittle Al–Cu–Ti intermetallics[22] will be generated at the interface. When the intermetallics layer reaches a certain thickness, it becomes prone to cracking under the effect of an external force, which affects the joint strength. If the reaction layer is kept to a suitable thickness, the dispersed intermetallic particles or blocks actually improve joint strength.

TiAl intermetallics and 42CrMo steel can be brazed with Ag–Cu35.2–Ti1.8 (wt.%).[36] A thin Ti foil is sandwiched between two Ag–Cu foils using the

Table 8.6 Commonly used brazing filler metals, brazing parameters and joint strength of the brazing of Ti–Al intermetallics with steels

Brazing filler metal (wt.%)		Ti–Al intermetallics (at.%)	Steel	Brazing parameters		Mechanical strength (MPa)	References
				Brazing temperature (°C)	Holding time (s)		
Ti–based	Ti–15Cu–15Ni	Ti33.5Al1.0Nb0.5Cr0.5Si	AISI4340	1075	30	Tensile 260, (210 (500°C))	20
Ag–based	B–Ag72Cu	Ti–46.5Al–5Nb	42CrMo	870	1200		21
	Ag–Cu–Ti	Ti–47Al–2Cr–2Nb	40Cr	900	600	Tensile 426	22,23
	Ag–Cu plated with Ti	TiAl	AISI4140	800	60	Tensile 294	24
	Ag–Cu–Ti	Ti–46.5Al–2.5V–1Cr	40Cr	870–910	300	Tensile 267	25
	Ag–Cu35.2–Ti1.8	Ti–47.5Al–2.5V–1Cr Ti–48Al–2Cr–2Nb	35CrMo	850–970	60–600	320	26
	CUSIL–ABA (63Ag–35.2Cu–1.8Ti)	Ti–47Al–2Cr–2Nb	AISI4340	845	30	Tensile 320, (310 (500°C))	20
	Ag–Cu–Ni–Li	TiAl	35CrMo	850–970	60–600	Tensile 324	27
	Ag–Cu–Ti	TiAl	42CrMo	850–1000	2–25	Tensile 347 Shear 229 (186 (400°C))	28

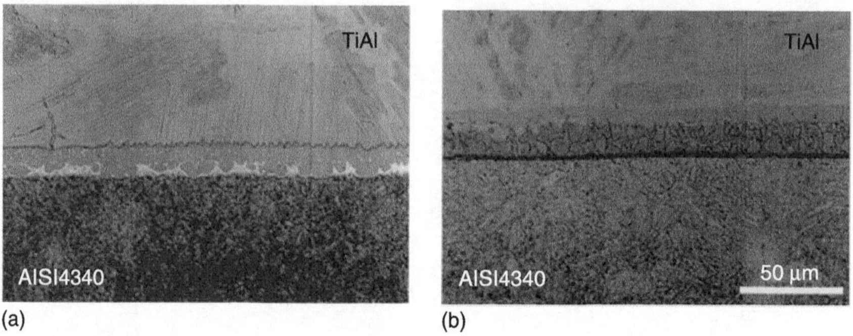

8.12 Joints brazed with different filler metals. (a) Ag–Cu–Ti; (b) Ti–Cu–Ni.

composite rolling method. The melting point of this brazing filler is 790–830°C. During a brazing process similar to the one shown in Fig. 8.13, the newly formed $AlCu_2Ti$ intermetallics that are close to the TiAl side are broken into pieces. The broken pieces are then pushed by the liquid phase into the brazing seam, where they act as dispersed particles.

A holding time of 300 s at a brazing temperature of 800°C is applied during the process. From the moment the brazing filler starts to melt to the end of the holding time, it reacts with the base material TiAl to generate $AlCu_2Ti$. The interfacial structures at this stage are shown in Fig. 8.14(a). Between 800°C and 900°C, TiAl reacts with Ag from the filler metal to generate a liquid phase at the boundary of the TiAl and $AlCu_2Ti$ intermetallics, as shown in Fig. 8.14(b). As the heating temperature increases, the $AlCu_2Ti$ layer breaks up and is pushed into the brazing seam, as shown in Fig. 8.14(c). Continuous heating leads to homogenization, generating dispersed intermetallics particles in the brazing seam, as shown in

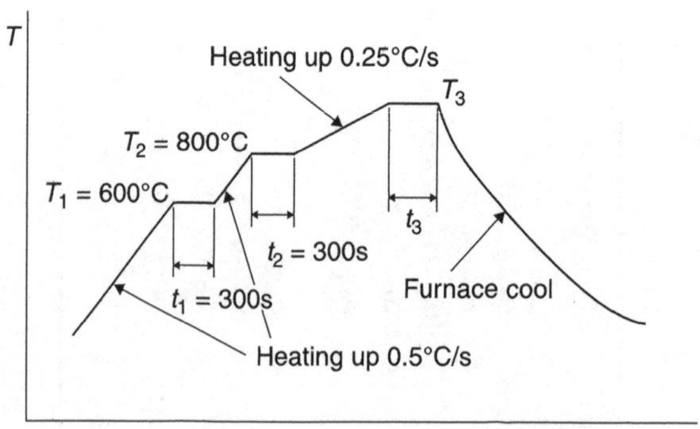

8.13 Brazing process for brazing TiAl 40 42CrMo steel with Ag–Cu35.2–Ti1.8.

Fig. 8.14(d). A reaction layer of Al–Cu–Ti-based intermetallics is then generated at the side of the TiAl base metal.

An interfacial microstructure typical of brazed joints, achieved by controlling the interface reactions, is shown in Fig. 8.15. It can be seen from the figure that the reaction zone has three distinct regions, which are marked as A, B and C. Zone A is composed of at least two kinds of continuous reaction layers, while Zone B is composed of three phases. Some eutectic phases and irregular gray blocks and strips are dispersed in the white matrix. Zone C is a continuous thin reaction layer adjacent to the steel.

Figure 8.16 shows that the reaction layer A mainly consists of the elements Ti, Al and Cu, while layer C consists mostly of C and Ti. In layer B, the black mass and irregular strip reaction phase mainly comprise the elements Ti, Al and Cu. The white matrix phase is the Ag-rich phase. Combining these results with the XRD results shown in Fig. 8.17, the joint interfacial microstructures are:

(1) the reaction layer near TiAl includes Ti_3Al + AlCuTi and an $AlCu_2Ti$ reaction layer;

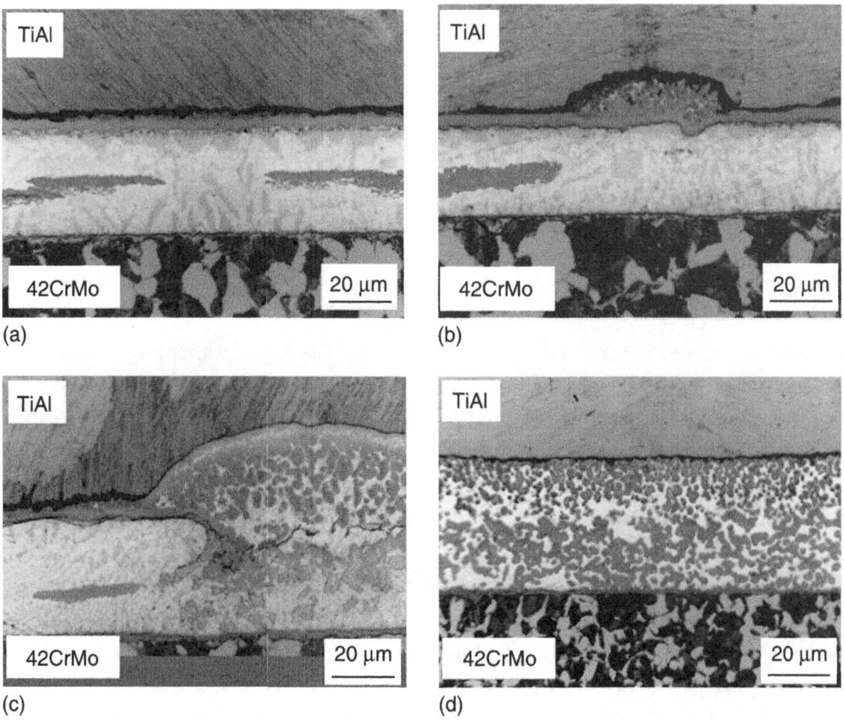

8.14 Joint interfaces at different stages: (a) 800°C for 300 s; (b) 820°C for 100 s; (c) 830°C for 100 s; (d) 860°C for 100 s.

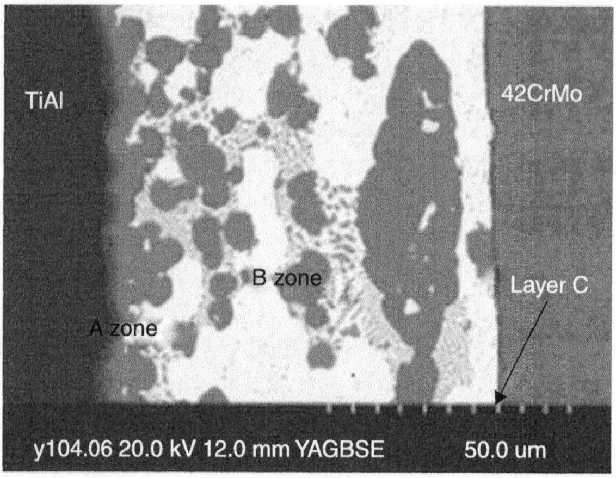

8.15 Typical joint interface (brazing temperature 900°C, holding time 300 s).

(2) the middle zone of the brazing seam consists of Ag [s.s.] and an Ag–Cu eutectic phase, and irregular $AlCu_2Ti$;
(3) the TiC reaction layer is near the 42CrMo steel side.

Mechanical property tests have shown that the room-temperature tensile strength, high-temperature tensile strength (400°C), room-temperature bending strength, high-temperature bending strength (400°C), room-temperature shear strength and high-temperature shear strength (400°C) are 347 MPa, 270 MPa, 501 MPa, 351 MPa, 229 MPa and 186 MPa, respectively.[36]

8.5.5 Brazing of Ti–Al intermetallics with ceramics

Ceramics have many excellent characteristics, such as high-temperature resistance, high corrosion resistance and good wear resistance, and so are widely used as high-temperature structural materials. Brazing could potentially be used to join Ti–Al intermetallics with ceramics to obtain a composite structural component with the properties mentioned above. Future research should focus on joining Ti–Al intermetallics and ceramics, whereas most studies currently concentrate on the brazing of TiAl and SiC.[37]

Ag–27Cu–4.5Ti (wt.%) is widely used as a filler metal for brazing TiAl with SiC. Four different diffusion reaction layers are generated in the joint. In the layer close to the TiAl side, Cu atoms in the Ag–Cu–Ti filler continually diffuse into the TiAl base metal to form the TiAlCu alloy. This reaction layer is formed by liquid phase solidification. The composition and thickness of this layer vary with the holding time. When the holding time was increased from 600 s to 2400 s, Ti content decreased and Cu content increased. The layer thickness increased from

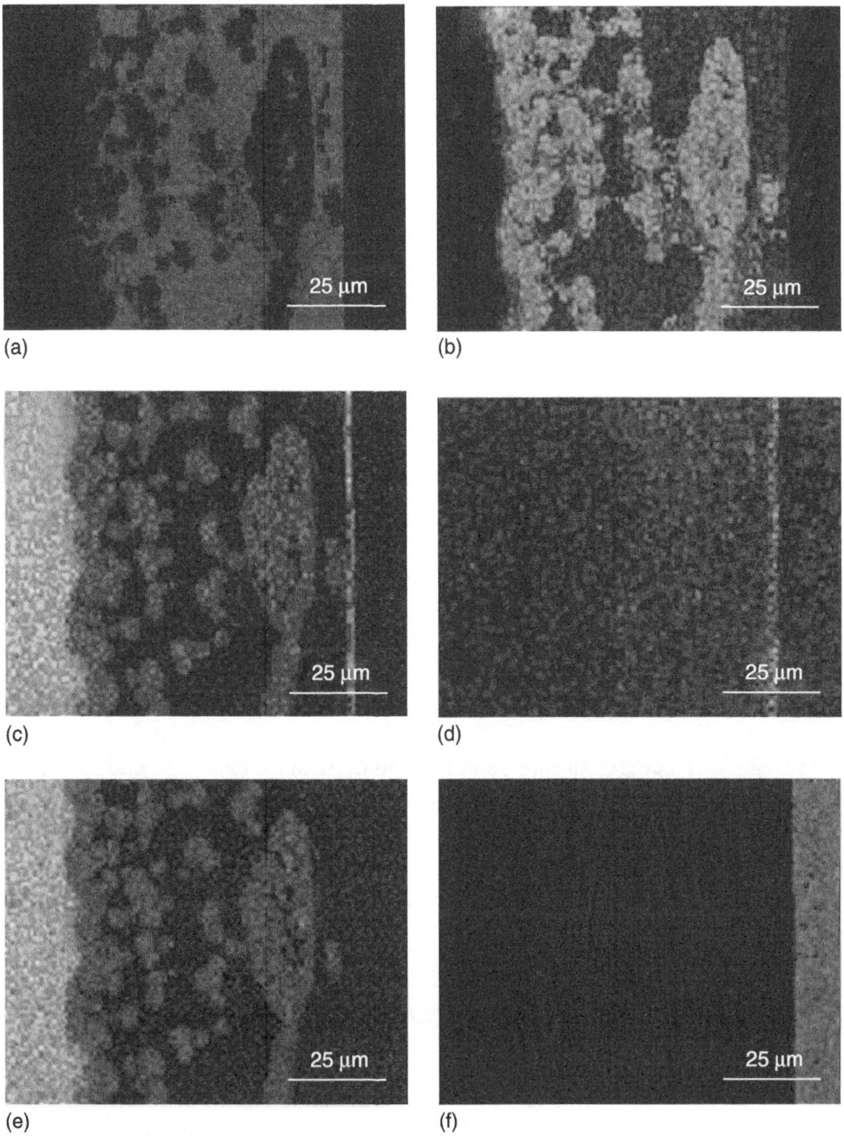

8.16 Surface scan of the main elements in the joint interfaces. (a) Ag; (b) Cu; (c) Ti; (d) C; (e) Al; (f) Fe.

8 μm to 15 μm. The reaction layer next to the SiC side is a compound layer, which includes the elements Ti, Cu and Si. Its composition and thickness also vary with the holding time. When the holding time increased from 600 s to 2400 s, Ti content increased from 32 at.% to 51 at.% while Si content decreased from 29 at.% to 13 at.%. The thickness of the layer increased from 0.4 μm to 2.9 μm.

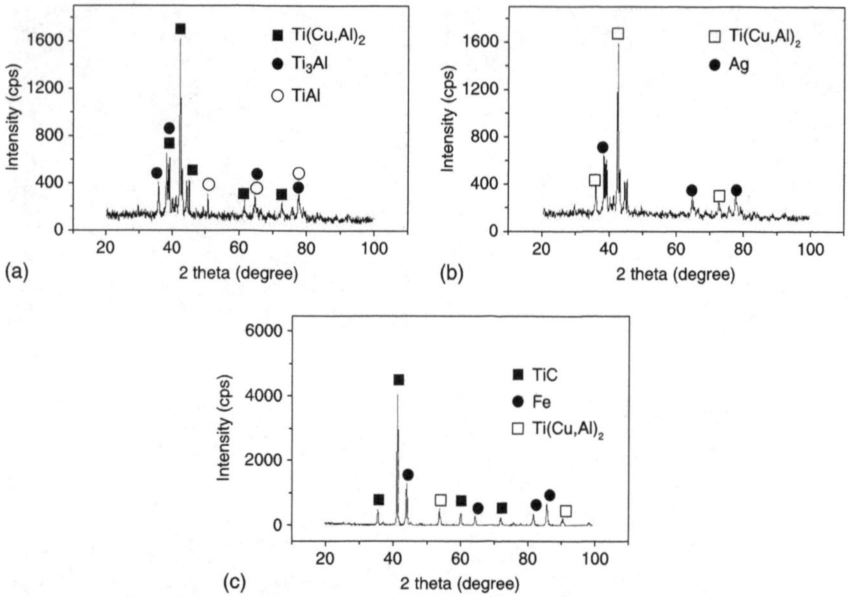

8.17 Interface layer X-ray diffraction. (a) Layer A; (b) layer B; (c) layer C.

Segregation occurs during solidification, when the Ag- and Cu-rich microstructure is formed. Two reaction layers can therefore be found in the middle of the brazing seam. Here, the holding time has no significant influence on the compositions. An increased holding time leads to a small decrease in joint shear strength. The maximum shear strength of 173 MPa was obtained at a holding time of 600 s.

8.5.6 Brazing of Ti–Al intermetallics and C/SiC composites

C/SiC composites have low densities, high-temperature resistances, high strengths and high-thermal shock resistances, and are not easily eroded or fractured. They therefore have a wide range of applications in aerospace, weapons manufacture and other industries.

Complex components that meet the requirements of high-temperature performance can be made from Ti–Al intermetallics and C/SiC composites. It is therefore worth investigating how to obtain reliable joint between TiAl and C/SiC composites.

Figure 8.18[38] shows a TiAl/C/SiC composite joint brazed with Ag–72Cu filler metal at 900°C for 600 s. The phases closest to the TiAl side are β-Ti+AlCuTi. The phases in the middle of the brazing seam are gray $AlCu_2Ti$, white Ag(s.s.), and a few

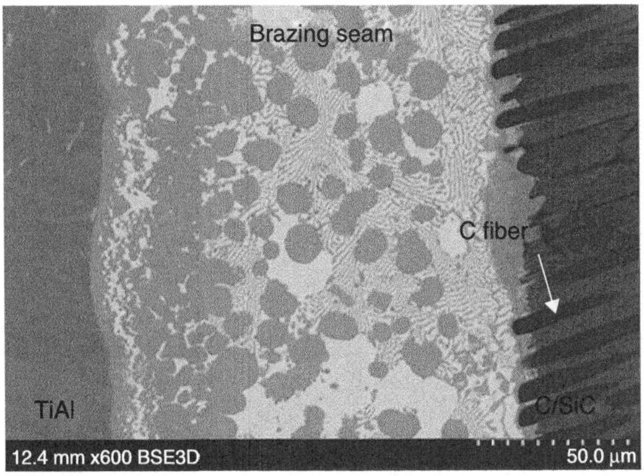

8.18 TiAl to C/SiC joint brazed with Ag–72Cu.

Cu-rich phases. The phases close to the C/SiC side are $TiSi_3$ and TiC. The dissolved quantity of TiAl is the governing factor in the evolution of interface microstructures. As the amount of dissolved TiAl increases, more Al–Cu–Ti compound is formed in the brazing seam. Meanwhile, the amount of Ag-base solid solution decreases, and the reaction layer next to the TiC composite side becomes thicker.

The TiC reaction layer formed by Ti, dissolved from TiAl, and the C/SiC composite, is the weakest region in the whole joint. Thermal stress in the joint can be relieved by controlling the thickness of the TiC reaction layer. The joint shear strength was increased up to 85 MPa at 900°C for 600 s.

TiAl and C/SiC composite joints brazed with an Ag–Cu filler alloy perform well at room temperature but poorly at high temperatures. TiNi can be used as a filler metal in order to obtain a joint with good high-temperature performance. However, the CTE mismatch between TiAl and the C/SiC composite is large, which leads to high thermal residual stress in the joint. To solve this problem, B as an element can be added to the filler metal. During the brazing process, TiB whiskers are generated in the brazing seam. The formation of TiB whiskers can help to achieve a CTE transition gradient between the base metals and the brazing seam, which can alleviate the residual stress in the joint.

The interfacial microstructure of a joint brazed with Ti–Ni–B filler metal is shown in Fig. 8.19(a). The interface can be divided into three layers. The long, stripe-like TiB whiskers are dispersed in layer II, which can be seen in Fig. 8.19(b). The microstructures in the joint from the TiAl side to the C/SiC side are: $\beta/\beta+\tau3/\beta+\tau3+TiB/\tau3/TiC$. A joint brazed at 1180°C for 10 min achieved the highest room-temperature and high-temperature (600°C) shear strengths: 99 MPa and 63 MPa, respectively.[38]

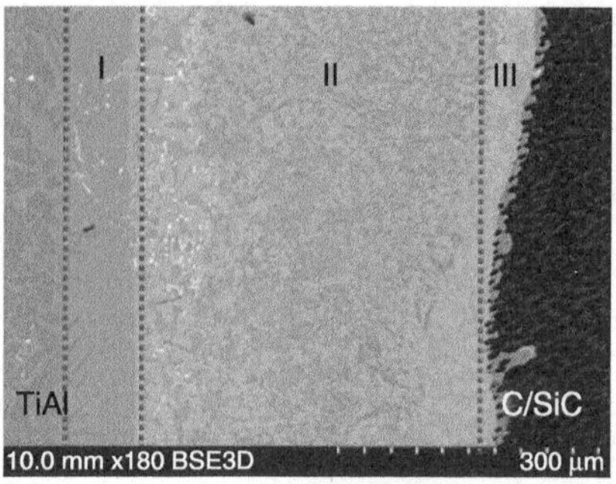

(a)

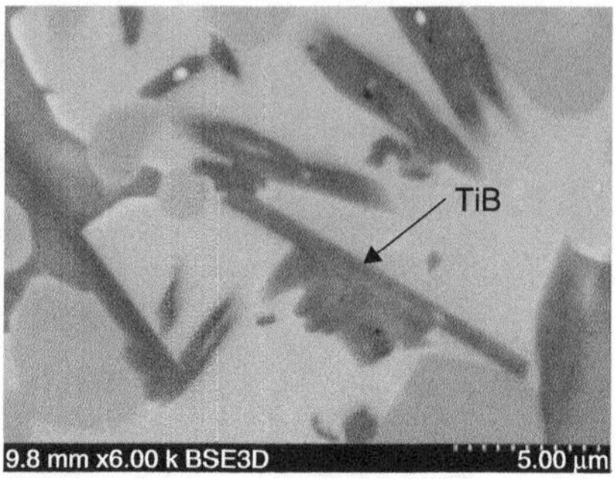

(b)

8.19 TiAl and C/SiC joint brazed with Ti–Ni–B filler metal. (a) Interface zone; (b) magnified brazing seam zone.

8.5.7 Brazing of Ti–Al intermetallics and C/C composites

C/C composites have many excellent characteristics, including low density, good friction resistance, high thermal conductivity, high thermal shock resistance and good fatigue performance. They also perform well at temperatures of up to 2000°C, and are therefore considered to be ideal high-temperature structural materials.

Making a rocket engine nozzle from CC and TiAl would effectively reduce the component weight while achieving the necessary high-temperature performance. It is therefore necessary to thoroughly investigate the joining of these two materials.

Figure 8.20 shows the microstructure of a joint formed by TiAl intermetallics and C/C composites using Ag–26.7Cu–4.6Ti (wt.%).[39] Phases of Ti_3Al and AlCuTi have formed between the TiAl and the brazing seam, as has the gray phase of $AlCu_2Ti$. Because of the violent reaction between Cu and TiAl, many Cu atoms have been consumed and many Ag atoms have been pushed to the C/C side, and consequently white Ag(s.s.) has formed in the brazing seam. The irregular TiC reaction layer has formed close to the C/C side.

A non-straight joining interface at the C/C composite surface can solve the problem of high residual stress in the brazing seam, caused by the mismatched CTE between the C/C composite and TiAl. Figure 8.21 shows joints with straight and non-straight brazing interfaces. After brazing, the penetration depth from the brazing seam into the C/C composite ranges from 300 µm to 600 µm. Figure 8.22 shows the contrast curves of the joint shear strengths. It has been proved that a non-straight interface can significantly improve shear strength by increasing the joining area and relieving the residual stress.

When Ag–Cu–Ti is used as a filler metal, the joint between the C/C composite and TiAl is mechanically strong at room temperature, but has a low mechanical strength at high temperatures. When a Ti–Cu–Ni filler metal is used, the high-temperature performance improves. Figure 8.23 shows a joint brazed with a

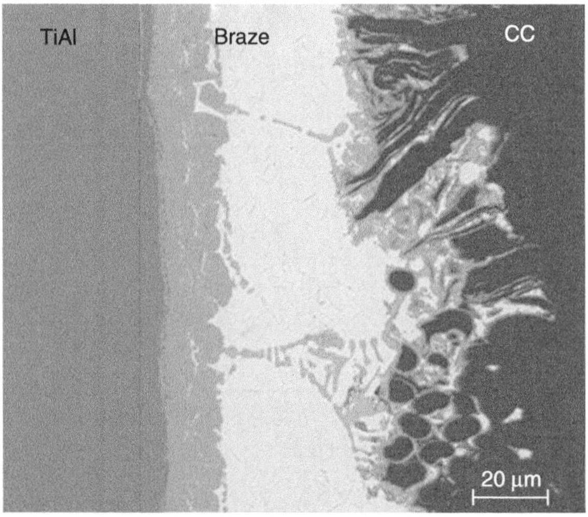

8.20 Interfacial microstructures of a C/C/Ag–Cu–Ti/TiAl joint (T = 920°C, t = 600 s).

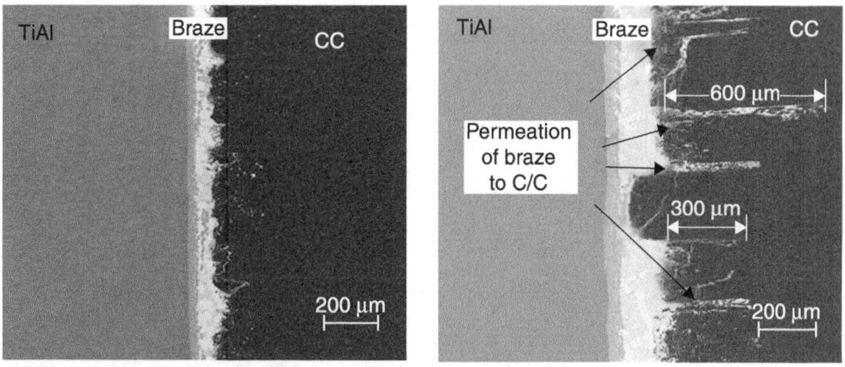

8.21 Joints formed by C/C composites with different interfaces. (a) Straight interface; (b) non-straight interface.

Ti–Cu–Ni filler metal at 980°C with a holding time of 600 s. No cracks or voids can be seen in the joint.[39]

In Area I, the main phases are Al(Cu,Ni)Ti+Ti_3Al. In Area II, the main phases are Ti(Cu,Ni)+Al$(Cu,Ni)_2$Ti. In Area III, the main phases are TiC+Al_2(Cu,Ni)Ti_3C. With a brazing temperature of 980°C and a holding time of 10 min, the average shear strength was 18 MPa at room temperature, but 22 MPa at 600°C. The high-temperature mechanical performance of the joint has been significantly improved through the use of Ti–Cu–Ni as a filler metal.

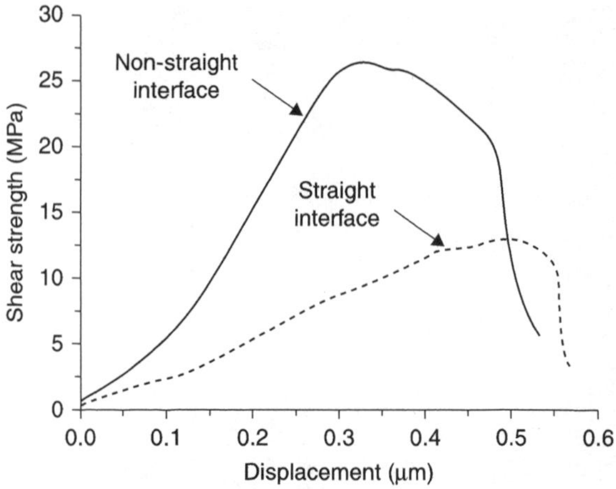

8.22 Shear strength curves of a straight interface and non-straight interface.

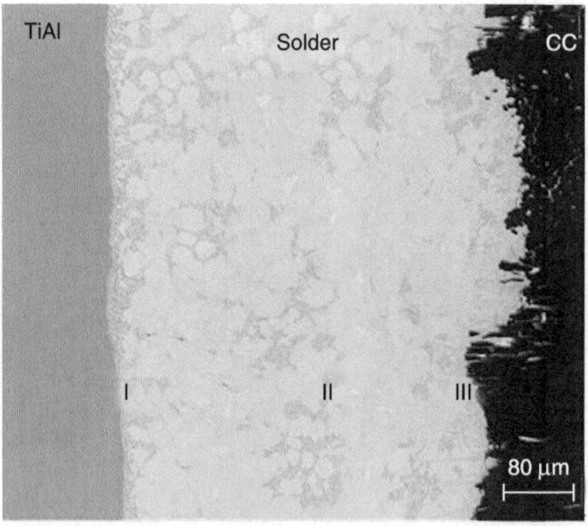

8.23 CC/Ti–Ni–Cu/TiAl joint ($T = 980°C$, $t = 600\,s$).

8.6 Conclusion

The brazing of nickel aluminides, iron aluminides and Ti–Al intermetallics has been described in this chapter. The brazing methods, brazing filler metals, microstructures of the joints and corresponding mechanical strengths were also introduced. An extensive comparison was made between the brazing of traditional materials and those including more than three intermetallics.

8.7 References

1. Zang Y. G., Han Y. F., Chen G. L., Guo J. T. and Wan X. J. 2001. *Structural Intermetallics*, National Defence Industry Press. 2–17.
2. Li Y. J., Wang J. and Liu P. 2003. *Welding and Application of Different Difficult Welding Metal*, Chemical Industry Press. 12, 212–232.
3. Mao W., Li X. H. and Ye L. 2006. Vacuum Brazing of a Directionally Solidified Ni3 Al-Base Superalloy IC6A. *Journal of Aeronautical Materials*, 3(26), 103–105.
4. Abdo Z. A. M., Guan Y. and Gale W. F. 1999. *Joining of NiAl to Nickel-based Alloys by Transient Liquid Phase Bonding*, Materials Research Society. 941–947.
5. Zhang T. 2009. *The Experimental Research on Brazing Process of Fe3Al and High Temperature Heat-resistant Alloy*. Master's Thesis. Hefei University of Technology. 3, 34–36.
6. Shiue R. K., Wu S. K. and Lee Y. L. 2005. Transient microstructural evolution of infrared brazed Fe_3Al intermetallics using aluminum foil. *Intermetallics*, 13, 818–826.
7. Li Y., Shiue R. K., Wu S. K. and Wu L. M. 2010. Infrared brazing Fe3Al intermetallics using the Cu filler metal. *Intermetallics*, 18, 422–428.
8. Shiue R. K., Li Y., Wu S. K. and Wu L. M. 2010. Infrared Brazing Fe3Al Using Ag-Based Filler Metals. *Metallurgical and Materials Transactions A*, 41, 2836–2843.
9. Lee Y. L., Shiue R. K. and Wu S. K. 2003. The microstructural evolution of infrared brazed Fe3Al by BNi-2 braze alloy. *Intermetallics*, 11, 187–195.
10. Lee S. J., Wu S. K. and Lin R. Y. 1998. Infrared Joining of TiAl intermetallics using Ti-15Cu-15Ni Foil-I. The Microstructure Morphologies of Joint Interfaces. *Acta Materialia*, 46(4), 1283–1295.
11. Lee S. J., Wu S. K. and Lin R. Y. 1998. Infrared joining of TiAl intermetallics using Ti-15Cu-15Ni Foil-II. The Microstructural Evolution at High Temperature. *Acta Materialia*, 46(4), 1297–1305.
12. Guedes A., Pinto A. M. P., Vieira M. F. and Viana F. 2003. The influence of the processing temperature on the microstructure of γ-TiAl joints brazed with a Ti-15Cu-15Ni alloy. *Materials Science Forum*, 426–432(5), 4159–4164.
13. Guedes A., Pinto A. M. P., Vieira M. F. and Viana F. 2003. Joining Ti-47Al-2Cr-2Nb with a Ti/(Cu,Ni)/Ti clad-laminated braze alloy. *Journal of Materials Science*, 38(11), 2409–2414.
14. Lee S. J. and Wu S. K. 1999. Infrared joining strength and interfacial microstructures of Ti–48Al–2Nb–2Cr intermetallics using Ti–15Cu–15Ni foil. *Intermetallics*, 7(1), 11–21.
15. Ming H., Hui K. and Ping Q. 2004. Ti-15Cu-15Ni vacuum brazing TiAl alloys. *Aeronautical Manufacturing Technology*, 4, 22–24.
16. Wang Y. F., Wang C. S., Gao Q., Dong C., Huo S. B. and Wang J. J. 2004. Brazing of a TiAl-based alloy with Zr56Al7.5Cu27.5 amorphous ribbon. *Transactions of the China welding institution*, 25(2), 111–114.
17. Guedes, Pinto A. M. P., Viera M. F. and Viana F. 2004. Joining Ti-47Al-2Cr-2Nb with a Ti-Ni braze alloy. *Materials Science Forum*, 455–456(6), 880–884.
18. He P., Liu D., Shang E. *et al.* 2009. Effect of mechanical milling on Ni-TiH_2 powder alloy filler metal for brazing TiAl intermetallic alloy: The microstructure and joint's properties. *Materials Characterization*, 60, 30–35.
19. Liu D. 2006. *Study on connecting of directional-organization TiAl alloys*. Harbin Institute of Technology. Dissertation for the Master Degree.

20. Shiue R. K., Wu S. K. and Chen S. Y. 2003. Infrared brazing of TiAl intermetallic using BAg-8 braze alloy. *Acta Materialia*, 51(7), 1991–2004.
21. Shiue R. K., Wu S. K. and Chen S. Y. 2004. Infrared brazing of TiAl intermetallic using pure silver. *Intermetallics*, 12(7–9), 929–936.
22. Yan P. 2002. *Study on vacuum brazing of TiAl intermetallics*. Harbin Institute of Technology. Dissertation for the Master Degree.
23. Guedes A., Pinto A. M. P., Viera M. F., Ramos A. S. and Vieira M. T. 2002. Microstructural characterisation of γ-TiAi joints. *Key Engineering Materials*, 230–232(9–11), 27–30.
24. Shiue R. K., Wu S. K. and Chen S. Y. 2003. Infrared brazing of TiAl using Al-based braze alloys [J]. *Intermetallics*, 11(7), 661–671.
25. Uenishi K., Qumi H. and Kobayashi K. F. 1995. Joining of intermetallic compound TiAl by using Al filler metal. *Zeitschrift fuer Metallkunde*, 86(4), 270–274.
26. Shiue R. K., Wu S. K. and Chen S. Y. 2003. Infrared brazing of TiAl using Al-based braze alloys. *Intermetallics*, 11, 661–671.
27. Wang Y. F., Wang C. S, Gao Q. *et al.* 2004. Amorphous brazing of TiAl alloy. *Welding Technology*, 25(2), 111–114.
28. Noda T., Shimizu T., Okabe M. and Iikubo T. 1997. Joining of TiAl and steels by induction brazing. *Materials Science and Engineering: A*, 239–240(1–2), 613–618.
29. Gao Q., Guo J. T., Liu W. and Zhang J. S. 2003. Study on interface and formation mechanism of diffusion brazing TiAl alloys to 42CrMo. *Journal of Aeronautical Materials*, 23(S1), 52–54.
30. Zhang K., Wu L. H., Lou S. N. and Ruan H. 2002. Diffusion brazing TiAl to 40Cr. *Welding*, 4, 35–38.
31. Xue X. H., Wu L. H. and Mao J. F. 2003. Vacuum brazing of TiAl alloy to 40Cr steel. *Journal of Aeronautical Materials*, 23(S1), 136–138.
32. Ja-Myeong Koo, Won-Bae Lee and Myoung-Gyun Kim. 2005. Induction brazing of γ-TiAl to alloy steel AISI 4140 using filler metal of eutectic Ag-Cu alloy coated with Ti film. *Materials Transactions*, 46(2), 303–308.
33. Li Y. L., He P., Feng J. C. *et al.* 2005. Analysis on interface and mechanical performance of induction brazing TiAl/40Cr joint. *Mechanical Engineering Journal*, 41(10), 93–96.
34. He P., Feng J. C. and Xu W. 2006. Mechanical property of induction brazing TiAl-based intermetallics to steel 35CrMo using AgCuTi filler metal. *Materials Science and Engineering A*, (418), 45–52.
35. He P., Feng J. C. and Xu W. 2005. Mechanical property and fracture characteristic of induction brazed joints of TiAl-based intermetallics to steel 35CrMo with Ag–Cu–Ni–Li filler. *Materials Science and Engineering A*, (412), 214–221.
36. Li Y. L. 2007. *Study on brazing mechanism and technology of TiAl/AgCuTi/42CrMo*. Harbin Institute of Technology. Dissertation for the Doctoral Degree.
37. Liu H. J., Li Z. R., Feng J. C. *et al.* 1993. Vacuum brazing of SiC ceramics and TiAl alloys. *Welding*, 3, 7–10.
38. Yang Z. W. 2010. *Study on mechanism of in situ reaction assisted brazing of C/SiC composites to TiAl alloy*. Harbin Institute of Technology. Dissertation for the Master Degree.
39. Wang H. Q. 2010. *Study on brazing and self-propagating reaction joining of C/C Composites and TiAl alloy*. Harbin Institute of Technology. Dissertation for the Master Degree.

9
Brazing of aluminium and aluminium to steel

V. F. KHORUNOV and O. M. SABADASH,
E. O. Paton Electric Welding Institute, Ukraine

DOI: 10.1533/9780857096500.2.249

Abstract: The chapter focuses on the problems of reactive-flux brazing and soldering of aluminium and aluminium to steel. High-temperature fluxes of the K,Al,Si/F salt system improve wetting and capillary properties of filler metals during brazing. The possibility of brazing aluminium using a reactive flux without a filler metal is established. Brazed joints on aluminium have strength equal to that of the base metal, and joints between aluminium and steel preserve their performance after thermal cycling tests. Reactive fluxes can be used for furnace, induction and arc brazing. Low-temperature fluxes based on polyatomic alcohols, which contain synthesised complex tetrafluoroborates of metals with nitrogen-bearing bases, improve conditions for formation of brazed joints on aluminium.

Key words: aluminium brazing and soldering, reactive flux, K,Al,Si/F system salt melts, solderless brazing, aluminium to stainless steel joints, low-temperature fluxes, synthesised complex tetrafluoroborates ($Me \cdot 2L(BF_4)_2$).

9.1 Introduction

The chapter has three sections. The first section is dedicated to the problem of high-temperature brazing of aluminium alloys using reactive fluoride fluxes. The ternary salt system K,Al,Si/F was investigated in the development of reactive fluxes. It was found that stable salt mixtures are limited to a triangle of the $KF–K_3AlF_6–K_3SiF_7$ compositions. The salt mixtures are unstable in other regions and release SiF_4 into the gas phase. The temperature dependence of the free energy was calculated. It is shown that K_2SiF_6 can perform two functions: it can destroy oxide Al_2O_3 and reduce silicon in interaction with aluminium. It was experimentally proved that aluminium can form an alloy with silicon, and that this alloy can function as a filler metal. The experience of brazing aluminium structures without a filler metal and with flux FAF540 of the $KF–AlF_3–K_2SiF_6–KZnF_3$ salt system is described.

The second section considers peculiarities of induction and arc brazing of aluminium to stainless and galvanised steels by using reactive flux FAF540, as well as temperature–time conditions of formation of a transition intermetallic layer. Results of thermal cycling tests of the joints between alloy 1050 or 3003 and steel 12Kh18N10T are presented, proving the serviceability of these joints at increased and ultra-low temperatures. It is shown that aluminium can be joined to galvanised steel by arc brazing through the melted filler, the thickness of the transition intermetallic layer formed in this case being no more than 3 μm.

The third section considers a new class of synthesised complex tetrafluoroborates ($L \cdot HBF_4$ and $Me2L(BF_4)_2$, where L is the ligand, i.e. the nitrogen-bearing base), which exhibit a differing thermal stability in a composition of low-temperature fluxes based on high boiling alcohols (e.g. glycerol, triethanolamine). A complex-forming metal coating, which substantially improves wetting of aluminium with the filler metal, is formed during brazing. The temperature of melting and decomposition of the synthesised complex tetrafluoroborates can be changed by changing the ligand or metal. The experience gained in brazing aluminium using low-temperature fluxes under conditions of resistance and flame heating is described.

9.2 Brazing aluminium and its alloys using reactive fluxes

High-temperature fluxes of different salt systems can be divided into two main groups: fluxes on a chloride–fluoride base, which are corrosive and require mandatory washing, and fluxes on a fluoride base, which are non-corrosive. It should be noted that modern methods used for brazing aluminium and its alloys are based on the elimination of corrosive fluxes and, in a number of cases, of vacuum technologies, and the complete replacement of these processes by the high-productivity methods of brazing in a protective atmosphere (nitrogen, argon, helium) using non-corrosive fluoride fluxes.

In the heterogeneous flux (ion melt), brazing filler metal (metal melt) and aluminium system, non-corrosive fluoride fluxes of salt systems consisting of elements of groups IA, IIA and IIIA (AlF_3) of the periodic table can be divided into non-reactive fluxes, the melts of which destroy oxide Al_2O_3 (brazing with these should be performed using a brazing filler metal), and reactive fluxes (interaction with these leads to destruction of the oxide and an aluminothermic reaction, which forms elements reduced from the flux, as well as to contact-reactive melting to form a low-melting point alloy with aluminium, which can fill the gap between the pieces joined, thus acting as a filler metal).

Among a large group of non-corrosive fluxes that do not interact with aluminium, one worthy of note is NOCOLOK™ of the eutectic KF–AlF_3 salt system. It consists of a mixture of potassium fluoroaluminates, which, starting from the moment of melting, actively destroy the oxide film during brazing in a temperature range of 585–610°C. The melting point of the eutectic is 560±2°C, and its composition is 45 mole% AlF_3.[1] For example, brazing car radiators in continuous furnaces in a pure nitrogen atmosphere using the flux NOCOLOK is characterised by higher (1.3 times) productivity and low cost compared with brazing in vacuum.[2] Flux NOCOLOK-SIL, consisting of a mixture of NOCOLOK powder and fine silicon particles, is reactive, and can be used without a brazing filler metal.[3]

Fluxes consisting of simple or binary fluorides, e.g. silicon, zinc or germanium fluorides, which improve wetting and the capillary properties of brazing filler

Table 9.1 Monovariant reactions of binary Al-Me (Me=Si,Ge,Zn)[4] or multi-component alloys Al-Si-Ge[5]

System	Point	Temp. (°C)	Reaction	Elements in liquid (wt.%)		
				1	2	3
Al–Si	E	577	L→Al+Si	77.5	12.5	–
Al–Ge	E	424	L→Al+Ge	46	54	–
Al–Zn	E	382	L→Al+Zn	5.5	94.5	–
Al–Si–Ge	Monovariant valley	578→424	L⇔(Al)+(SiGe)	–	12.7	53

metals at the interface between the solid metal (aluminium) and ion melt (flux), hold more promise in the development of reactive flux brazing. All these elements, which act as depressants, combine with aluminium to form binary or multi-component alloys with a melting point below 580°C (see Table 9.1) and good corrosion properties.

The patent literature gives methods for brazing aluminium at a temperature of 600–620°C without a filler metal using fluxes of the KF–AlF_3 salt system, consisting of potassium fluoroaluminates $KAlF_4$ and K_2AlF_5 with additions of 6–50 wt.% potassium fluorosilicate[6] or 5–50 wt.% potassium and caesium fluorozincanates.[7] However, the phase equilibria and physical–chemical properties of melts of the fluoride salt systems with these elements have been little studied as yet.

Andreiko *et al.*[8] investigated the chemical interaction and fusibility of mixtures in the part of the K,Al,Si/F ternary system limited by the KF–K_2SiF_6–AlF_3 triangle. The investigations used thermal, X-ray phase and chemical analyses (Fig. 9.1).

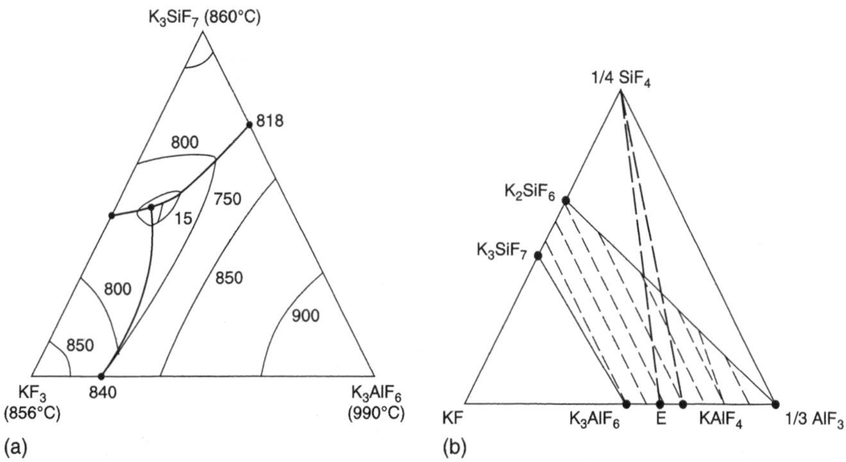

9.1 Diagram of fusibility of the KF–K_3AlF_6–K_3SiF_7 system (a) and general constitutional diagram of KF–AlF_3–SiF_4 (b).

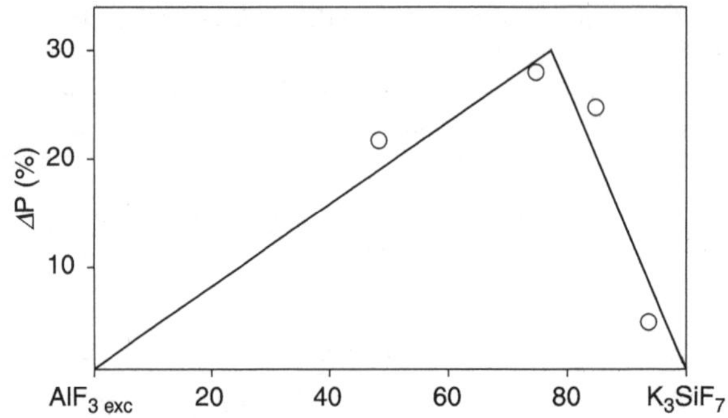

9.2 Weight loss (P) calculated from reaction 9.3 (straight line) and actual (points) for the mixture (0.44 AlF_3 + 0.56 K_3AlF_6) – K_3SiF_7 at 900°C.

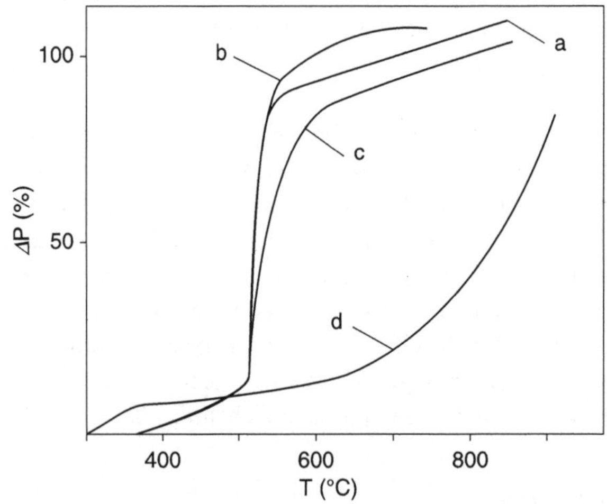

9.3 Degree of completion of reaction 9.1 in mixtures of Fig. 9.2 for content of K_3SiF_7 (wt.%). Line a: 3; b: 54.5; c: 73.1; d: 88. Heating rate 12°C/min.

As shown by the experiments conducted to plot the diagrams, the interaction is accompanied by a substantial loss in weight during heating with an excess amount of aluminium fluoride relative to the stoichiometry of cryolite. The interaction begins at about 550°C, reaches the maximal rate at 580–590°C, and is accompanied by considerable absorption of heat.

The loss in weight for compositions belonging to mixtures within the K_3SiF_7–K_3AlF_6–AlF_3 concentration triangle was investigated. The data obtained are presented in Fig. 9.2 and Fig. 9.3 as the dependence of the weight loss on the excess AlF_3 to K_3SiF_7 ratio. This diagram shows that the interaction occurs up to complete utilisation of the excess aluminium fluoride to form end products SiF_4 and K_3AlF_6 by the following reaction:

$$K_3SiF_7 + AlF_3 = K_3AlF_6 + SiF_4 \quad [9.1]$$

Therefore, all mixtures containing excess aluminium fluoride relative to the stoichiometry of K_3AlF_6 are unstable in the $KF–AlF_3–K_3AlF_6$ ternary system, and during melting they transform with time into two-component compositions of the $K_3AlF_6–AlF_3$ (for an excess amount of AlF_3) or $K_3AlF_6–K_3SiF_7$ systems. In other words, the stable mixtures in the system considered are located within the $KF–K_3SiF_7–K_3AlF_6$ triangle.

The fusibility diagram of the $KF–K_3SiF_7–K_3AlF_6$ system was investigated. Two limiting binary systems ($KF–K_2SiF_6$ and $KF–AlF_3$) are known. The third binary system, i.e. $K_3SiF_7–K_3AlF_6$, was investigated using the differential thermal analysis of cooling curves recorded by a derivatograph. The data obtained[8] indicate that the system is a simple eutectic one. The eutectic contains 19 mole% K_3AlF_6 (18 wt.%) and melts at 817°C. A polymorphic transformation in the solid phase (K_3SiF_7) takes place at 760°C.

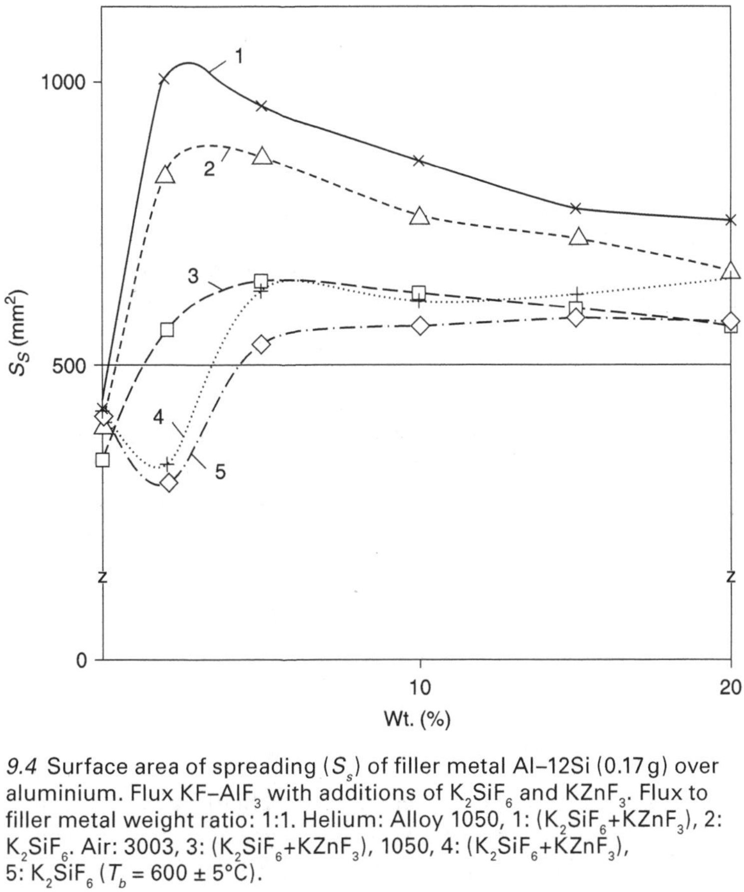

9.4 Surface area of spreading (S_s) of filler metal Al–12Si (0.17 g) over aluminium. Flux $KF–AlF_3$ with additions of K_2SiF_6 and $KZnF_3$. Flux to filler metal weight ratio: 1:1. Helium: Alloy 1050, 1: (K_2SiF_6+$KZnF_3$), 2: K_2SiF_6. Air: 3003, 3: (K_2SiF_6+$KZnF_3$), 1050, 4: (K_2SiF_6+$KZnF_3$), 5: K_2SiF_6 (T_b = 600 ± 5°C).

Table 9.2 The temperature dependence of the free energy (ΔG) of interaction in the Al_2O_3–K_2SiF_6 and Al–K_2SiF_6 systems

		$\Delta G_{900\,°C}$ (kJ·mol^{-1})
1	$\Delta G(T) = \Delta H_{298,15} - T\Delta S_{298,15} + \Delta C_{298,15}\left[(T-298) - T\ln\frac{T}{298}\right]$	
2	$2Al_2O_3 + 3K_2SiF_6 = 2\,K_3AlF_6 + 2AlF_3 + 3\,SiO_2$ $\Delta G_1(T) = -437 - 0.0404T - 0.0318\,[(T-298) - T\ln(T/298)]$	−1237
3	$4Al + 3K_2SiF_6 = 2K_3AlF_6 + 2AlF_3 + 3Si$ $\Delta G_2(T) = -1207 - 0.0418T - 0.0449\,[(T-298) - T\ln(T/298)]$	−464

*Thermodynamic data for solid aluminium-containing and silicon-containing compounds according to JANAF.[9]

The temperature dependence of the free energy (G) of interaction in the Al_2O_3–K,Al,Si/F and Al–K,Al,Si/F systems allows evaluation of the activity of the fluxes depending on the brazing temperature to a sufficiently good approximation. For example, if we assume that the difference in heat capacities does not depend on the temperature: $\Delta C = \Delta C_{298,15}$ = const, then the thermodynamics of KF–AlF_3–K_2SiF_6 can be calculated from Eq. 9.1 (Table 9.2).

As shown by the equations in Table 9.2, potassium hexafluorosilicate performs two functions in reactions 9.2 and 9.3: it destroys Al_2O_3 (reaction 9.2) and, probably, causes the aluminothermic reduction of silicon in a composition of a binary fluoride through its interaction with aluminium (reaction 9.3). The equilibrium of both processes is almost fully shifted toward the interaction products.

The spreading of aluminium–silicon brazing filler metals over chemically treated aluminium alloys 1050 (99.5% Al) and 3003 (Al–1.6 Mn) was investigated while heating to 615°C in inert and active gas atmospheres (Fig. 9.4).

It was found that the surface area of spreading of the filler metal over aluminium 1050 greatly depends on the content of potassium fluorosilicate and fluorozincanate in the salt melt, as well as on the activity of the gas atmosphere under which brazing is performed. In a temperature range of 580–630°C, the salt melts of the KF–AlF_3–K_2SiF_6 and the salt melt of the system KF–AlF_3–$KZnF_3$ in contact with aluminium reduce silicon and zinc from complex fluorides by reactions 9.3 and 9.4, which are then deposited on the surface in the form of a thin layer of low-melting point alloy of the Al–Si–Zn metal system (Fig. 9.5).

$$3K_2SiF_6 + 4Al \rightarrow 3Si + 3KAlF_4 + K_3AlF_6 \quad [9.3]$$

$$6KZnF_3 + 4Al \rightarrow 6Zn + 3KAlF_4 + K_3AlF_6 \quad [9.4]$$

Photographs of the spreading filler metal Al–12Si (0.2 g in weight) revealed that the mean speed of movement of the perimeter of a filler metal drop over this layer is 1.5–2.2 times higher for a flux of the KF–AlF_3–K_2SiF_6 salt system with K_2SiF_6

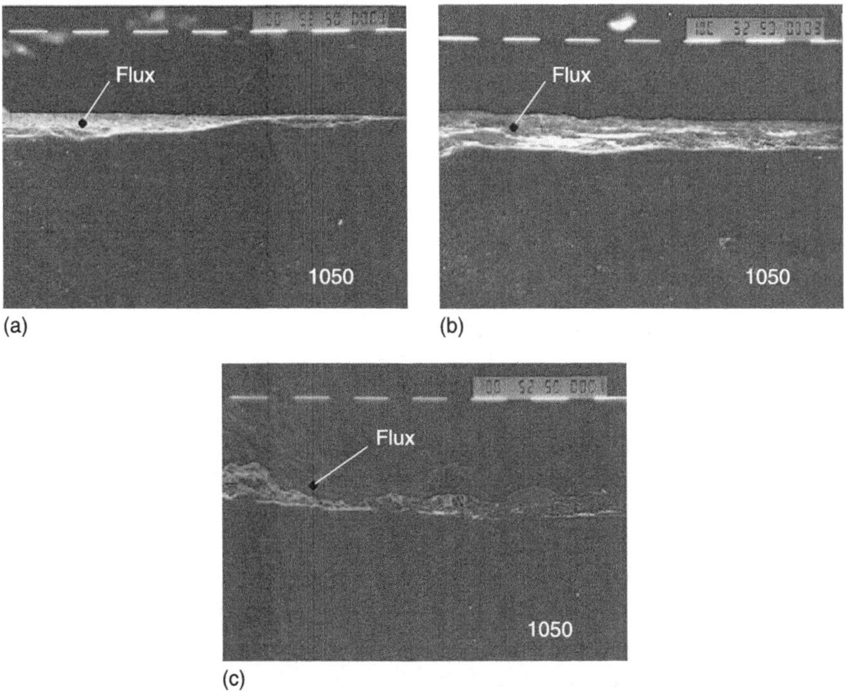

9.5 Microstructure of an aluminium (1050) sample after spreading of the reactive flux of the $KF–AlF_3–10K_2SiF_6$ system: (a) final, (b) middle and (c) initial regions of the dispersed flux.

additions, compared with a flux of a eutectic composition of the $KF–AlF_3$ salt system. However, attempts to reliably determine the dependence of the speed at the initial stage of spreading failed because of the complex character and the short time for the filler metal to spread under a layer of the salt melt. The contact angle for the filler metal measured from microsections after solidification was 1–4° for aluminium of the 1050 grade, and 2–6° for alloy 3003.

The formation of joints due to the filler metal filling the capillary gaps between the aluminium plates in a temperature range of 570–630°C can be divided into four stages: (1) melting, spreading over aluminium and rapid filling of the brazing gaps by fluid salt melts (at a temperature above 575°C), (2) active destruction of the surface oxide film as a result of chemical interaction with the salt melt (above 585°C), (3) reduction of silicon at the interface due to an aluminothermic reaction and formation of the liquid melt with aluminium (Al–Si) as a result of contact melting (above 585°C), and (4) distribution of the filler metal over the surface and in the capillary gaps (above 595°C). The reduction of zinc occurs in a similar way through reaction 9.4. The optimal salt compositions are usually chosen on the basis of providing the specified range of melting of fluxes and conditions leading to removal of SiF_4 from the brazing zone.

The methods used to prepare the surfaces for brazing being similar, the quality of the brazed joints is highly affected by the brazing atmosphere (active–air, non-active–helium, nitrogen or argon), temperature and time. Experiments show that fluxes with potassium fluorosilicate and fluorozincanate additions are active for longer in a shielding gas atmosphere than in air. For example, flux KF–AlF_3–10 K_2SiF_6 in air is active for about 60 s, which is approximately 1.5 times shorter than in helium. It was found that during the brazing process the moisture from the air atmosphere actively destroys the salt melts of the KF–AlF_3 fluoride system by the following reaction:[10]

$$2AlF_3 + 3H_2O = 6HF + Al_2O_3 \qquad [9.5]$$

The negative effect of moisture can be decreased by using more flux and by reducing the brazing time.

An analysis of the microstructure of brazed joints on aluminium, produced under non-equilibrium solidification conditions, revealed the following peculiarities of the eutectic structure (Fig. 9.6).

Joints brazed with a filler metal (Fig. 9.6(a)) are characterised by the formation of a large amount of the eutectic phase (seam), which is located between columnar dendrites of the aluminium-base α-solid solution. The seam boundaries on the side of the aluminium alloy 1050 are relatively smooth. The silicon content (see Table 9.3) of the solid solution based on aluminium and the eutectic phase is close to the equilibrium one.[11] A relatively small amount of the lamellar-type eutectic phase is formed in joints brazed without a filler metal (Fig. 9.5(b)). Eutectic colonies are located between grains of the aluminium-base α-solid solution, and the seam boundaries have a clearly defined relief. Substantial changes in the concentration of silicon from 0.4 to 13 wt.% were revealed in microvolumes of the seam (Table 9.3), depending on the amount of the active reagent (K_2SiF_6) and the occurrence of silicon reduction. Zinc reduced from the corresponding fluoride

(a)

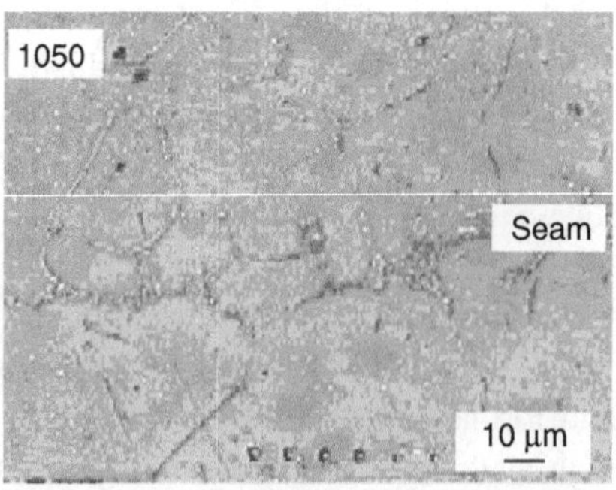

(b)

9.6 Microstructure of brazed joints on aluminium 1050 produced: (a) with filler metal Al–12Si and a flux of the KF–AlF_3–10K_2SiF_6 system; (b) without a filler metal (T = 600 ± 3°C, τ = 30–40 s).

Table 9.3 Analysis of elements in microvolumes of aluminium joints (wt.%)

	Al	Si	K	F
	99.48–86.45	0.4–13.5	0–0.01	0, traces

Aluminium 1050 flux 600°C, 60 s.

during brazing enriches mostly the aluminium-base α-solid solution, the concentration measured in microvolumes was 0.6–0.9 wt.%.

The reactivity of fluxes of the $KF–AlF_3–K_2SiF_6$ system is so high that the fluid salt melts can fill up the 'zero' and irregular gaps to form the joining metal alloy, i.e. with the filler metal, at all points of the contact surfaces.

Flux FAF540 was developed on the basis of the investigation results. This flux has a higher reactivity compared with the flux NOCOLOK, and improves spreading of filler metals over aluminium in air and helium atmospheres, the spreading area being 1.5 and 1.7 times bigger, respectively. This flux can be used for brazing aluminium and its alloys (series 1XXX, 3XXX, 6060, 6063, 6005) both with and without a filler metal to join structures with wall thicknesses differing by more than three orders of magnitude. Butt and overlap joints on aluminium 1050 have a strength equal to that of the base metal, and the strength of the joints on alloy 3003 is 0.85–0.9τ_{shear} of the strength of the alloy in standard shear tests.[12] The flux and its residue after brazing do not cause corrosion of the materials brazed.

For example, flux FAF540 was used in high-temperature brazing in through-type furnaces in a pure nitrogen atmosphere without a filler metal to join a thermoelectric module radiator made from aluminium 1050A (Fig. 9.7(a) and 9.8), and with a filler metal in the form of foil 0.3 mm thick to join an SHF (superhigh frequency) communication slit antenna made from alloy 3003 (Al–1.6 Mn) (Fig. 9.7(b)).

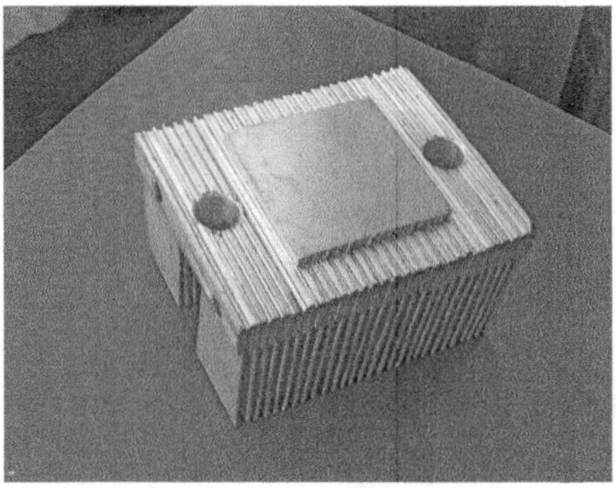

(a)

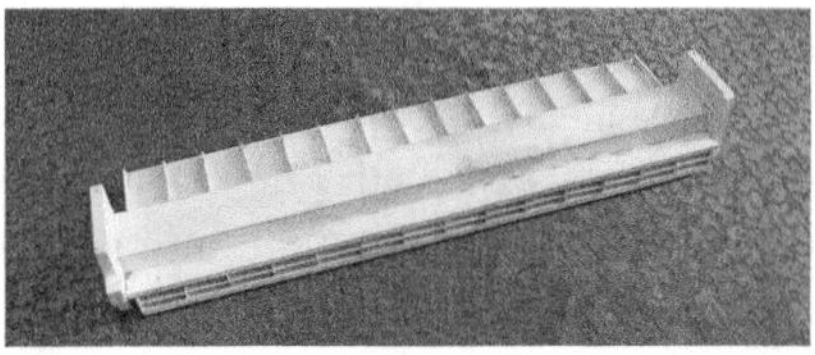

(b)

9.7 Aluminium parts brazed using flux FAF540: (a) plate-type radiator, set of 1.0 mm thick plates of alloy 1050A (without a filler metal); (b) elements of a slit antenna of alloy 3003 with thicknesses of the elements joined ranging from 0.3 to 4 mm (powder filler metal).

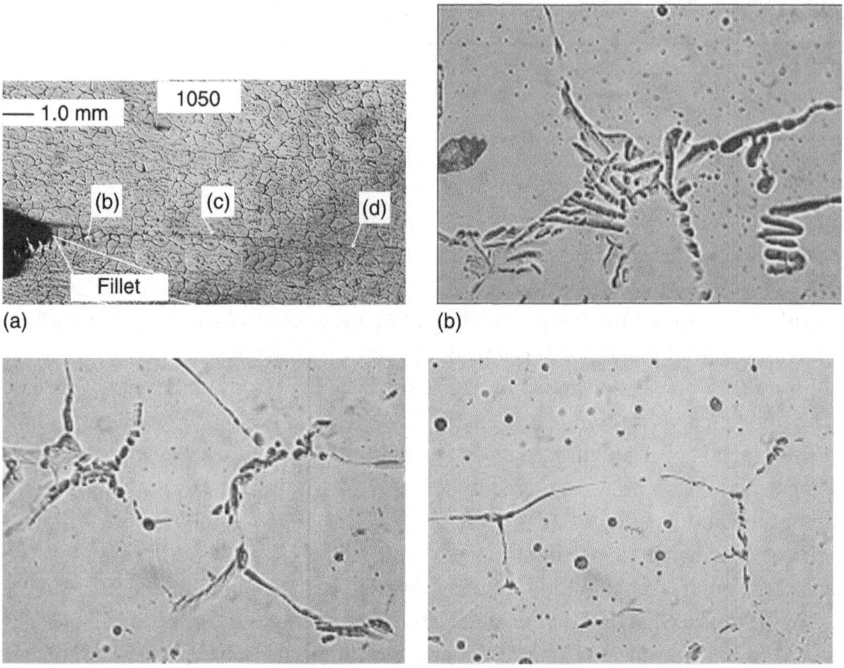

(a) (b) (c) (d)

9.8 Macro- and microstructures of regions of the 'monolithic' (d) joint on aluminium 1050A for a thermoelectric module radiator (Fig. 9.7(a)) brazed using reactive flux FAF540.

The reactive flux is a good base for developing new brazing technologies. In particular, it can widen the temperature (+300…–196°C) range for the application of aluminium and its alloys to stainless steel joints.

9.3 Brazing of aluminium to stainless steel

One of the most important problems that can be addressed by using materials with different properties is a decrease in the weight of structures with a simultaneous increase in their technical characteristics and reliability in operation under different types of load. The most common joints made from dissimilar materials are joints between aluminium and galvanised or stainless steel. The examples include thin-sheet car structures, tubular transition pieces in cryogenic units and bimetal ware, which operate under alternating loads in different temperature ranges. The required quality of the joints can be achieved with relatively low external effects exerted on the materials joined: a temperature below the solidus temperature of the aluminium alloy and an insignificant compressive force for producing a capillary gap between workpieces. Wetting of steel and filling the capillary gap with a liquid filler metal result in a reactive intermetallic layer, which has a significant effect on the properties of the brazed joint. The chemical composition and thickness of this brittle transition layer depend on many factors, such as the composition of the brazing consumables (filler metals and reactive fluxes), temperature–time conditions for the brazed joint, etc. Different types of joint between aluminium and stainless or galvanised steel can be produced by high-temperature brazing using filler metals of the Al–Si system, which feature high strength and good corrosion properties.

Intermetallic phases of different compositions, such as $FeAl_3$, $FeAlSi_5$ and Fe_2AlSi_8, can be found in the Al–Fe–Si metal system[4] over a temperature range of 576–630°C in aluminium-rich alloys in a state of equilibrium with aluminium solid solution.

In brazing under non-equilibrium conditions, a brittle intermetallic layer, which hampers or arrests spreading, is formed on the steel as a result of interaction with the aluminium–silicon melt.[13] In practice, one of the methods to decelerate the growth of the brittle transition layer at the interface is to introduce into the liquid or solid phase additions that retard the mutual diffusion of components due to the formation of a solid solution or ternary intermetallic phases. Any change in properties of the melt or solid phase, e.g. by adding silicon to aluminium, or chrome, nickel, silicon and manganese to steel, as well as deposition of an aluminium or zinc coating can substantially change the character of the interaction at the interface between the liquid and solid phases. This involves a change of the interface surface energy, thus allowing the regulation of the spreading rate and reactive diffusion processes. Also, it is necessary to limit the time for the reactive interaction of the liquid filler metal with the base materials within the temperature range close to the solidus temperature of the aluminium alloy because of rapid growth in the thickness of the brittle transition layer.

Capillary properties of the Al–Si filler metals and results of brazing of aluminium to stainless steel can be improved by using non-corrosive fluxes based on the KF–AlF_3 eutectic. For example, the use of flux NOCOLOK in the technological process of induction-pressure brazing of bimetalware makes it possible to produce steel–aluminium joints with a large joining surface area during a relatively short time (up to 2 min). According to the study by Roulin *et al.*,[14] increasing the holding time at a brazing temperature of 600°C leads to growth of the brittle transition layer and, as a result, to a considerable decrease of strength of a brazed joint between aluminium 1100 and stainless steel 304.

Modification of the properties of the metal melt, e.g. by adding potassium fluorozincanate, which is reduced by aluminium during brazing to reactive fluxes of the K,Al,Si/F salt system, can provide the optimal capillary properties of the Al–Si filler metals. $KZnF_3$ and K_2SiF_6 will react with Al and Al/Si and generate Zn and Si, resulting in joints with higher Zn and Si levels.

For instance,[15] when using a compact Al–12Si filler metal with salt mixtures of the KF–AlF_3–(3–30)K_2SiF_6 system, a thin refractory layer of complex composition

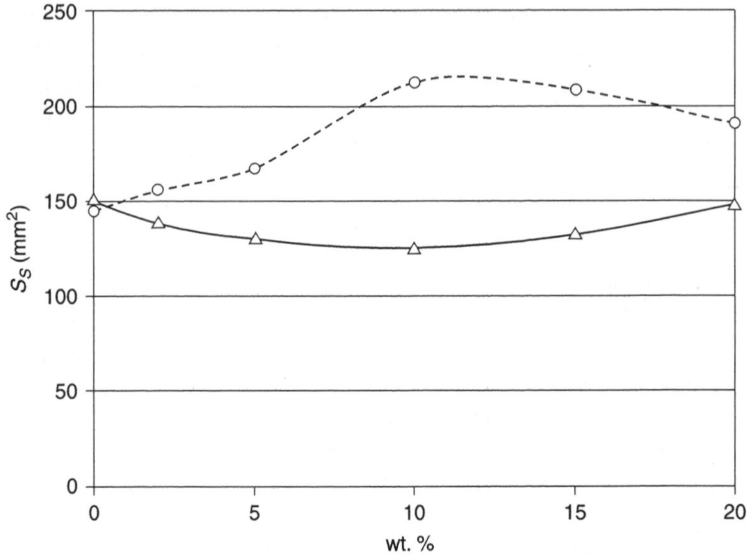

9.9 Area of spreading (S_s)of filler metal Al–12Si over stainless steel 12Kh18N10T depending on the content (wt.%) of K_2SiF_6 (△) and $KZnF_3$ (○) in 45KF–55AlF$_3$ mixture.

of the Fe–Al–Si system is formed at the interface with steel 12Kh18N10T, which prevents spreading of the filler metal over the steel surface (Fig. 9.9).

The use of flux FAF540 of the KF–AlF_3–K_2SiF_6–$KZnF_3$ salt system improves spreading of filler metal Al–12Si over steel within a temperature range of 600–620°C due to zinc reduced by aluminium from $KZnF_3$. An increase in the concentration of zinc at the aluminium–flux interface increases fluidity of the liquid reactive layer as a result of exothermic reaction 9.4 (see Section 9.2), thus improving the capillary properties of the aluminium filler metal. When using powdered flux FAF540 and powdered silumin AKD12 (Al–12Si), the ratio of their weight contents in a brazing mixture should range from 40/60 to 60/40.

Joints made from aluminium 1050 brazed to corrosion-resistant steel with a powdered filler metal (Fig. 9.10) are heterogeneous in structure and composition. The 8–10 μm thick transition layer on the steel side consists of the $FeAlSi_5$–(4.4Cr+0.4Ni) intermetallic phase. Adjoining this layer is a layer of an aluminium-base α-solid solution, 20–26 μm thick, with interlayers of skeletal eutectic Al–Si (Fig. 9.10, Table 9.4). Noteworthy are structural peculiarities of the seams brazed

Table 9.4 Quantitative analysis of transition layer in microvolumes (wt.%)

	Al	Si	Ti	Cr	Mn	Fe	Ni	Total
Transition layer	52.8	0.51	0.56	8.44	0.14	33.89	3.66	100

12Kh18N10T/1050, 610°C, 40c.

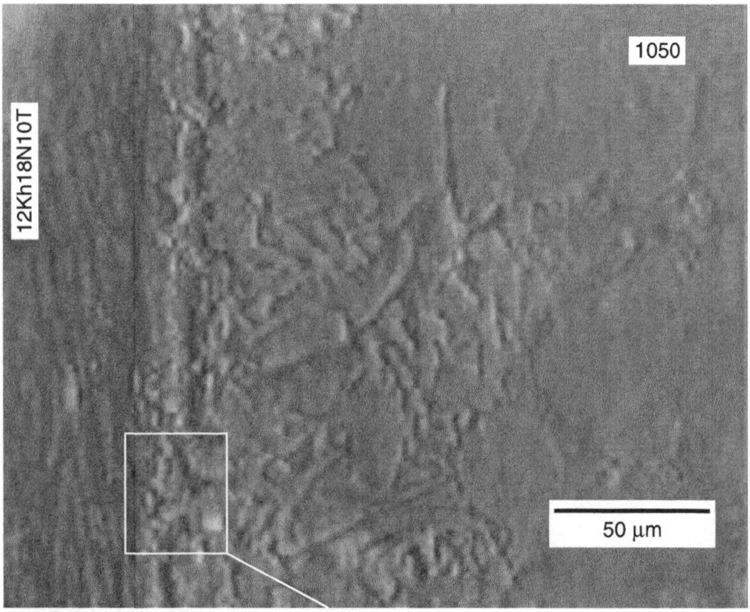

9.10 Microstructure of a brazed joint between aluminium 1050 and stainless steel 12Kh18N10T (filler metal: mixture of flux FAF540 and powder AKD12 (Al–12Si) in a ratio of 50/50, wt.%; brazing temperature T_b = 605 ± 5°C).

with either filler metal or flux alone: in brazing using a filler metal the 8–12 μm thick intermetallic layer is formed in the transition zone, while when using the flux alone this layer is 3–8 μm thick (Fig. 9.11).

The brazed joints between aluminium alloy 3003 and steel 12Kh18N10T have a heterogeneous intermetallic layer in the transition zone (Fig. 9.12, Table 9.5). No structural changes were seen in the steel.

As established from X-ray spectrum microanalysis (Table 9.5) of the brazed joints between stainless steel and aluminium alloys, the transition zone contains chromium, nickel and titanium, which transferred from the steel, and silicon, which transferred from the flux. Zinc is contained in a composition of aluminium-base α-solid solution.

Brazing of standard specimens for mechanical tests was performed using a powdered flux (flux/filler metal: 50/50, wt.%) at a temperature of 600–610°C (Fig. 9.13). It was established that the strength of the brazed joints after thermal cycling did not exceed a limiting value of ±10% of the initial one. Variations in strength of the brazed joints can be explained by an inconsiderable change in width (increase of about 3–5%) of the transition intermetallic layer as a result of a long temperature effect during thermal cycling.[16]

These results were used to develop a computer-controlled process for induction flux brazing of tubular steel–aluminium transition pieces (TSATP) in an argon

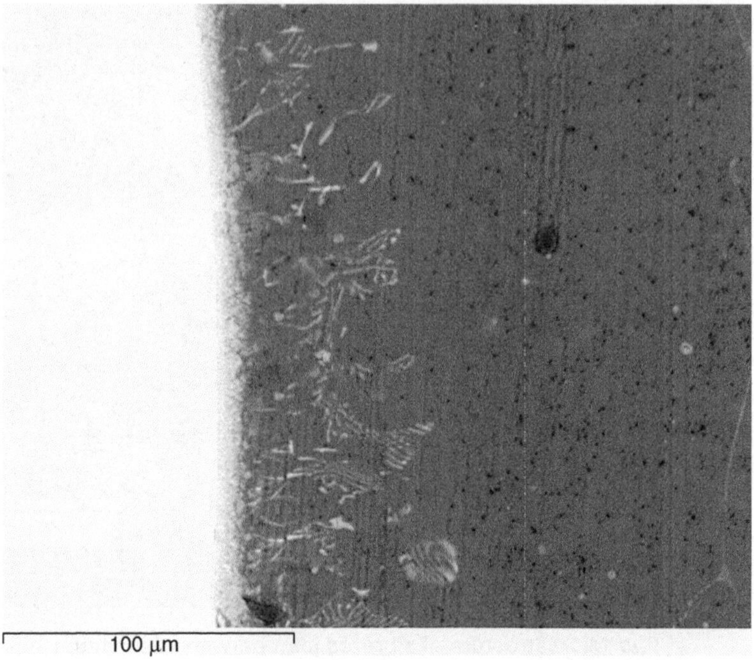

9.11 Microstructure of a brazed joint between aluminium 1050 (black) and steel 12Kh18N10T (without filler metal, flux FAF540, $T_b = 605 \pm 5°C$).

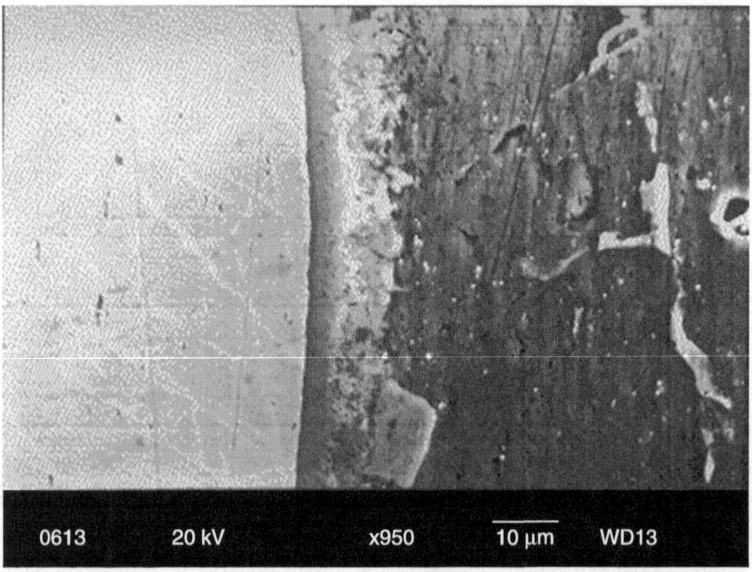

9.12 Microstructures of a joint between alloy 3003s (black) and steel 12Kh18N10T brazed using filler metal Al–12Si and a reactive flux of the K,Al,Si/F system with 15 wt.% additions of K_2SiF_6+$KZnF_3$ (etching: Keller reagent).

Table 9.5 Quantitative analysis of transition layer in microvolumes (wt.%)

	Al	Si	Ti	Cr	Mn	Fe	Ni
Transition layer	77.4	5.72	0.5	1.73	7.25	7.2	0.2

12Kh18N10T/3003, 610°C, 45c.

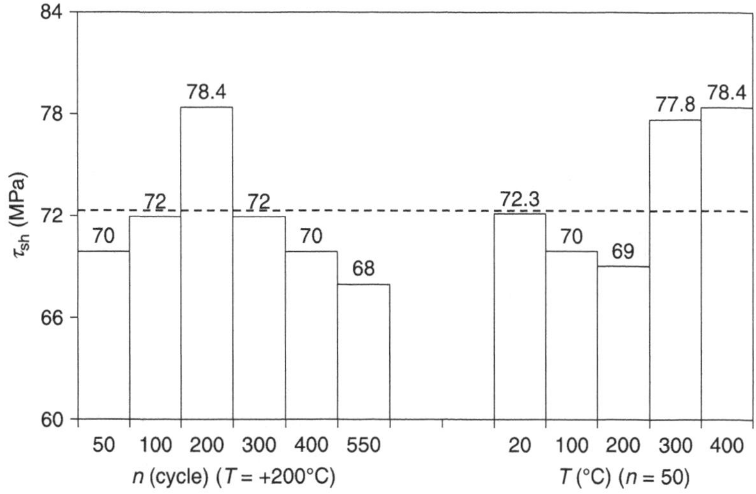

9.13 Variation in shear strength of brazed overlap joints between alloy 1050 and steel 12Kh18N10T after thermal cycling: heating–cooling in water (filler metal: mixture of flux FAF540/powder AKD12, 50/50, wt.%; T_b = 600–610°C).

atmosphere. The brazed joints on TSATPs with dissimilar 3003+12Kh18N10T materials are characterised by strength and tightness under an air pressure of 40 atm after 40 cycles (cooling in liquid nitrogen at −196°C and heating in warm water at +35°C).

9.4 Arc flux brazing of aluminium to galvanised steels

Upgrading the arc brazing process and widening its field of application rely on the development of brazed steel–aluminium thin-sheet structures with a lighter weight, such as high-strength extra low-carbon (interstitial-free (IF)) steels,[17] most often with aluminium and zinc coatings, and aluminium alloys. An example is motor car construction, where decrease in weight makes it possible to achieve substantial fuel savings, extend the service life of cars and decrease the negative effect on the environment. A problem of current importance is the improvement of the quality of bimetal joints on thin-walled structural members by using feedback[18] and low-heat input welding units. Both metals can be joined by braze welding, where the aluminium is partially melted and the steel is wetted with a filler metal.[19] In this case, to avoid the formation of solidification cracks, the content of brittle phases in the weld pool should not exceed 10 wt.%.[20] The other process for this application, i.e. arc brazing, is becoming accepted. It is performed at a temperature below the solidus temperature of the aluminium alloy, and the level of thermal deformation of the materials being brazed is much lower in this process than in welding. Different types of joints (T-, overlap, butt joints) between thin sheets of steel and aluminium are most often brazed using aluminium filler metals of the Al–Si system, and non-corrosive fluoride fluxes of the KF–AlF_3 system are applied to activate wetting and improve the capillary properties of the aluminium filler metals. Arc brazing using activating fluxes can be performed with a direct current with welding units designed for constricted arc welding or TIG processes in manual or automatic filler wire feed modes.

The kinetics of the spreading of the filler metal in arc heating depends on the complex chemical interaction of the salt and metal melts at the interface with the base metal.

The reactivity of flux FAF540 of the basic salt system in arc heating was studied by investigating the spreading of aluminium filler metal Al–12Si over aluminium alloy 1050 and galvanised steel 08Yu in a high-purity argon atmosphere (Fig. 9.14). The current and time of holding at the arc brazing temperature were determined

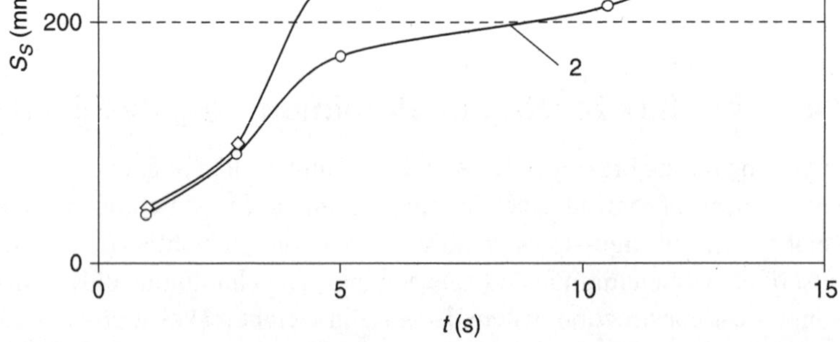

9.14 Surface area of spreading (S_s) of filler metal Al–12Si (0.17 g) over steel 08Yu (1) and Z-alloy 1050 (2) argon atmosphere. Arc heating of specimens using 'KEMPPI' unit Mastering MSL™ 2300 ACDC. Brazing parameters: I = 25 A, U = 12 V. Content of flux FAF540 in coating: 5 g/m^2.

experimentally in order to prevent local melting of the aluminium coating or destruction of the zinc coating on the steel specimens.

The spreading of the filler metal over the low-carbon steel improved due to the active effect of the flux and the additional molten ($T_{melt} \geq 420°C$) zinc coating formed on the specimen surface heated to a temperature of 600–630°C. For instance, in a stable mode of arc heating (direct current 25 A, voltage 12 V) the maximal area of spreading of the filler metal over galvanised steel during 5 s is 540 mm^2. A close spreading area (560 mm^2) on aluminium can be achieved only during a much longer (approximately three times) time of 14 s. Reactive flux FAF540 performs the following main functions in arc brazing: (1) it actively destroys oxides and cleans them from the base and filler metals; (2) it promotes wetting of the aluminium and galvanised steel with the filler metal; and (3) it improves spreading of the filler metal over the aluminium due to the formation of a low-melting point aluminium–silicon alloy on its surface.

Flux arc brazing of thin-sheet aluminium to galvanised steel was performed on overlap specimens measuring 120 × 40 × 1 mm. 1.6 mm diameter filler wire AK12 (Al–12Si) and non-corrosive flux FAF540 were used as filler metals.

The investigation of arc brazing with a direct current of straight polarity showed that the reactive flux creates good quality joints between aluminium and galvanised steel at the optimal parameters of automatic brazing.[21] Arc heating through the melted filler metal (the arc is short circuited only to a filler wire) decreases the thermal effect of the arc on the materials brazed and preserves continuity of the zinc coating on the steel. In this case the brazed seams have, as a rule, a segmental shape, smooth surface and smooth transition to aluminium and steel (Fig. 9.15, Table 9.6). A concave meniscus of filler metal with a radius of less than 0.1 mm is formed on the reverse side of the joint. The smooth surface of the seam and smooth fillets reduce the stress concentration, which is particularly important in the manufacture of rigid structural members of these dissimilar metals.

Analysis of the microstructure of brazed joints between aluminium 1050 and galvanised steel 08Yu showed (Fig. 9.16) that a homogeneous transition layer (TL)

Table 9.6 Parameters (approximate) for arc brazing thin sheets of aluminium 1050 to galvanised steel 08Yu

Joints	Rod AK12 (Al–12Si)	I_b (A)	U_a (V)	V_b (mm/s)	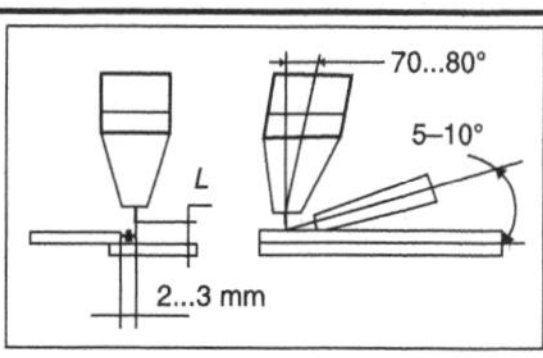
Overlap 1–3 (mm)	d_f = 1.6 mm	50–52	9.0–9.5	12	Brazing scheme

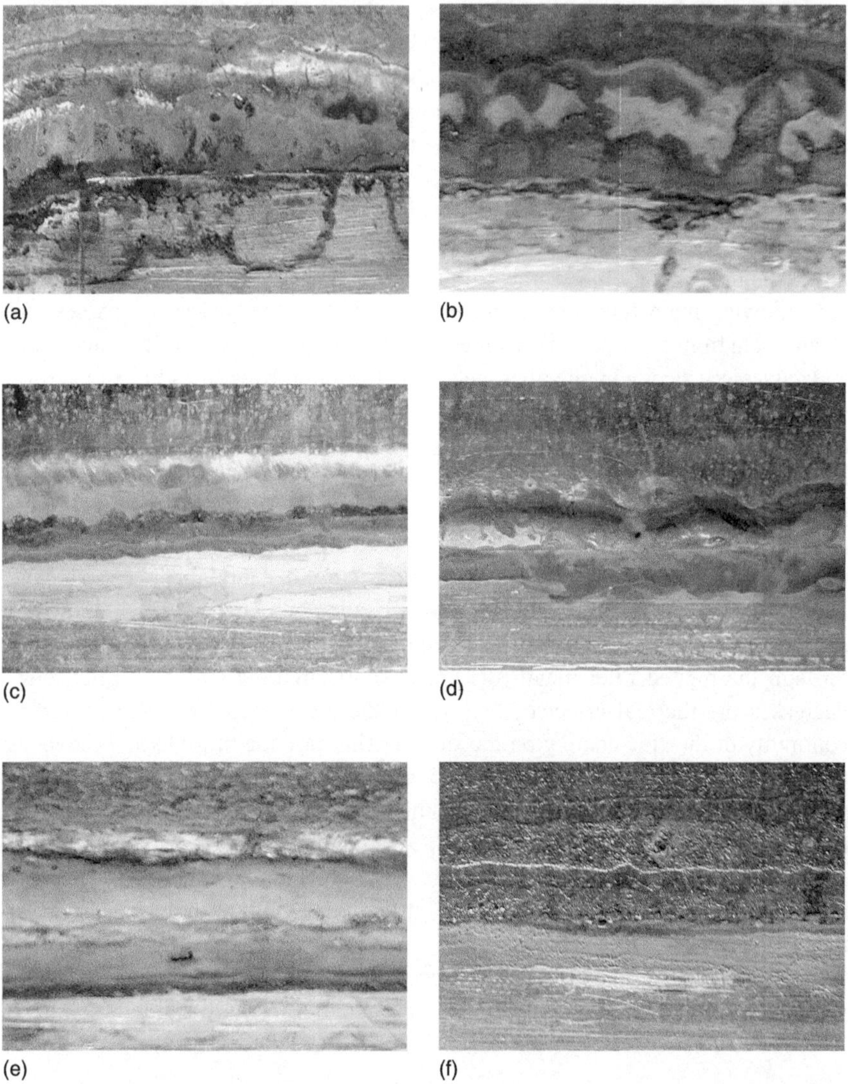

9.15 Automatic DC brazing (V_b = 12 mm/s) of overlap joints between aluminium 1050 and steel 08Yu, 1.0 mm thick. (a), (b) Excessive amount of flux (30 g/m^2). (c), (d) Lack of fusion and laps of filler metal on aluminium and steel. (e), (f) Sound formation of brazed seam, flux consumption: 10 g/m^2.

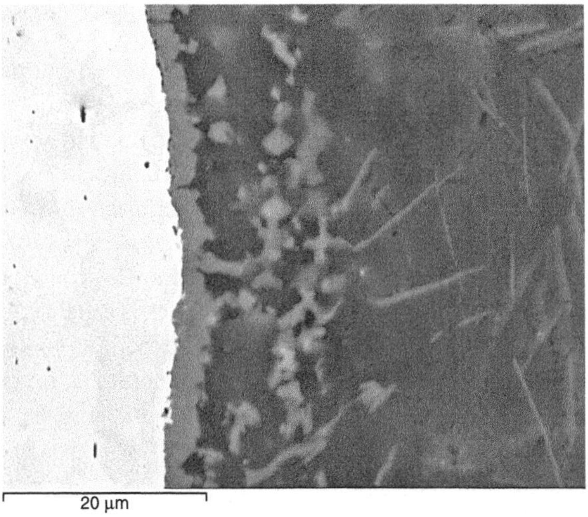

9.16 Microstructure of brazed joint between alloy 1050 and galvanised steel.

of the Al–Fe–Si system, less than 2 μm thick, forms at the interface with the steel. The aluminium-base α-solid solution is located in the central part of the seam. Lamellar dispersed binary eutectic Al–Si containing small amounts of zinc is found in interlayers between the solid solution grains. The interface with aluminium is wavy, having traces of deep (down to 300 μm) intergranular penetration of the eutectic phase into the aluminium. A low-melting point alloy of the Al–Zn–Si system with a zinc content of up to 40 wt.% forms in fillets (Table 9.7). The contact angle for the aluminium is 8–10° and for the galvanised steel 28–33°.

As already noted, the zinc coating, which forms in arc brazing of steel–aluminium joints, functions as a diffusion barrier. It hampers interaction of the steel with the aluminium and the growth of brittle interlayers of intermetallics of the Al–Fe–Si system at the interface between the solid aluminium and liquid metal melt.

The aluminium and steel–aluminium brazed joints fracture at room temperature in the aluminium or its alloys. The shear strength of overlap joints between aluminium alloys and steel 08Yu is not less than 0.95 τ_{1050} and 0.75–0.8 τ_{AD31} of their initial strength.

Table 9.7 Quantitative analysis of elements in microvolumes (wt.%) of a brazed joint between aluminium 1050 and galvanised steel 08Yu

	Al	Si	Mn	Fe	Zn	Sum
Transition layer (~2 μm)	63.62	4.56	–	31.23	0.59	100

TIG process, *I*=24 A, *V*=1.5 mm/s.

9.5 Soldering of aluminium

Low-melting point filler metals (alloys based on tin, cadmium, lead and zinc) do not wet aluminium because of the presence of a strong and dense oxide on its surface. There is a group of flux systems that destroy the aluminium oxide and enable wetting of the aluminium surface by the liquid filler metal at temperatures below 450°C.

For example, reactive low-temperature fluxes are often used to destroy oxide Al_2O_3. Metals reduced by aluminium from the fluxes activate the surface of the base metal and dope the filler metal, while chemical compounds dissociating at the soldering temperature can form a gas atmosphere, which is capable of reducing the oxide and protecting the seam from oxidation.[22–24] The required components of inorganic reactive fluxes are salt mixtures of ammonium, tin and zinc chlorides[25–27,29–31] and bismuth bromide.[28] Chemical interactions between salt melts of halides and aluminium at temperatures above 300°C in gaps in the Al_2O_3 film takes place by the following reactions:

$$2Al_S + 3MeCl_{2L} = 2AlCl_{3G}\uparrow + 3Me_L \qquad [9.6]$$

$$Al_S + MeCl_{3L} = AlCl_{3\,G}\uparrow + Me_L \qquad [9.7]$$

$$Al_S + MeBr_{3L} = AlBr_{3\,G}\uparrow + Me_L \qquad [9.8]$$

where Me = Zn, Sn in reaction 9.6 and Me = Bi in reactions 9.7 and 9.8.

Liquid metals penetrate under the oxide film and cause it to disperse. As a result, they improve wetting and the filling of the capillary gap with filler metal. Almost all reactions occur with an intens release of heat and gaseous aluminium halides. The fluxes and their residue are corrosive. Hence, extra costs are required to remove them.

The second group of fluxes includes reactive fluxes based on organic liquids containing active materials in the form of tin halides[32,33] and simple tetrafluoroborates (TFBs) of ammonium, tin, zinc, cadmium, lead or bismuth.[34–36] An important advantage of this group of fluxes is that they are non-corrosive.

The efficiency of the TFB fluxes based on organic liquids (amino- or polyhydric alcohols, organic acids) in soldering aluminium is shown by the ability of gaseous boron trifluoride (BF_3) to activate the removal of aluminium oxide film.[24] The aluminium oxide undergoes the following reaction at a temperature of 450°C:[37]

$$2BF_3 + Al_2O_3 = 2AlF_3 + B_2O_3 \qquad [9.9]$$

Boron oxide fully evaporates in the form of $(BOF)_3$ in an excess of boron fluoride.[37*] The spontaneous reduction of metals and their deposition on the aluminium surface takes place at a temperature of 240–300°C, much higher than

* Borates – including B_2O_3 – are rated as reprotoxic in the EU. Processes releasing borates will therefore have health, safety and environmental restrictions placed on their use. Fluoroborates, however, are not reprotoxic. (Anonymous reviewer, personal communication.)

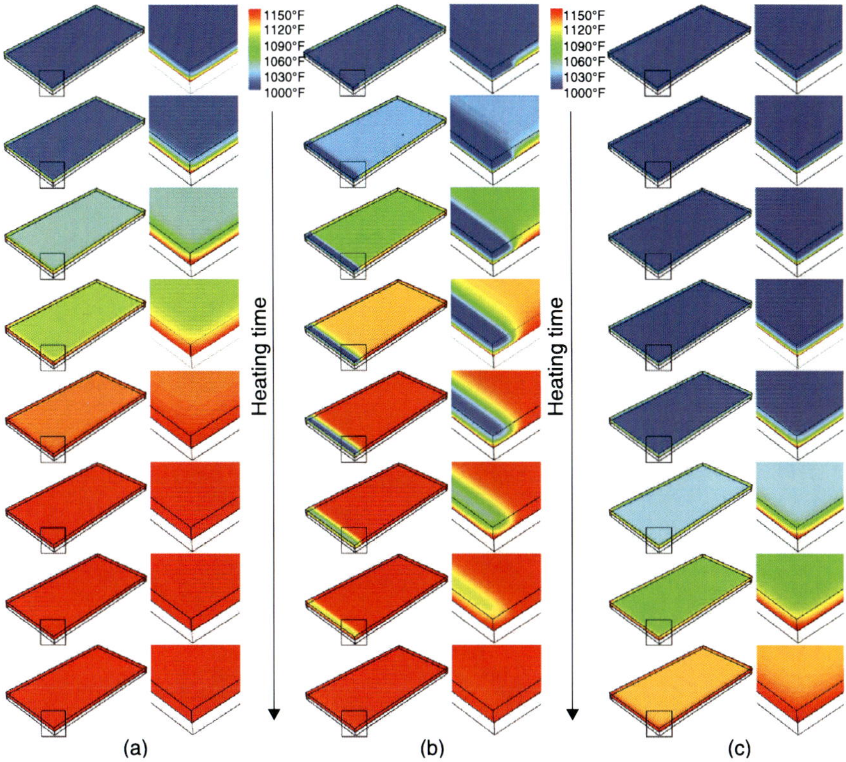

Plate I (Chapter 3) Simplified models of temperature distributions during the last phase of the ramp-up heating (CAB of an aluminum heat exchanger): (a) uniform (averaged lumped thermal properties of the core and header domains), (b) distinct properties for header vs. core domains, (c) uniform lumped properties of the entire unit (based on core properties). Note a dramatic loss of local temperature gradients in (c) vs. (a) and (b).

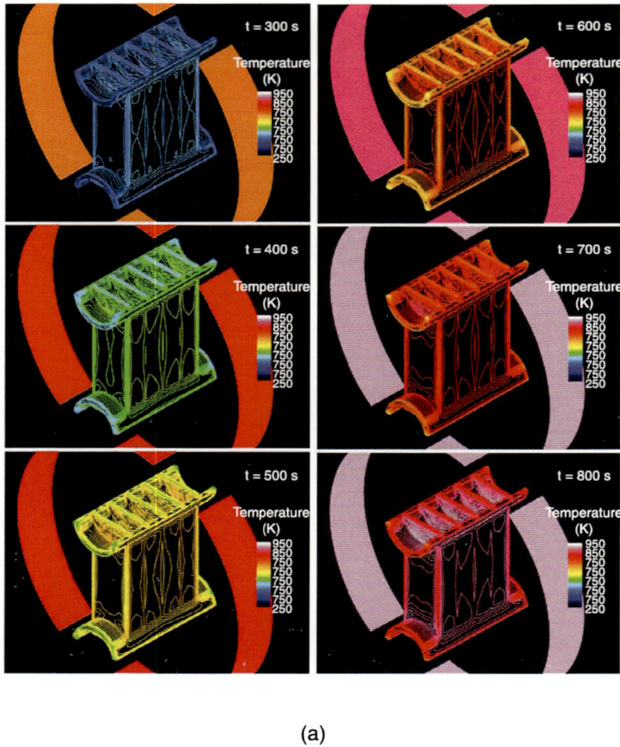

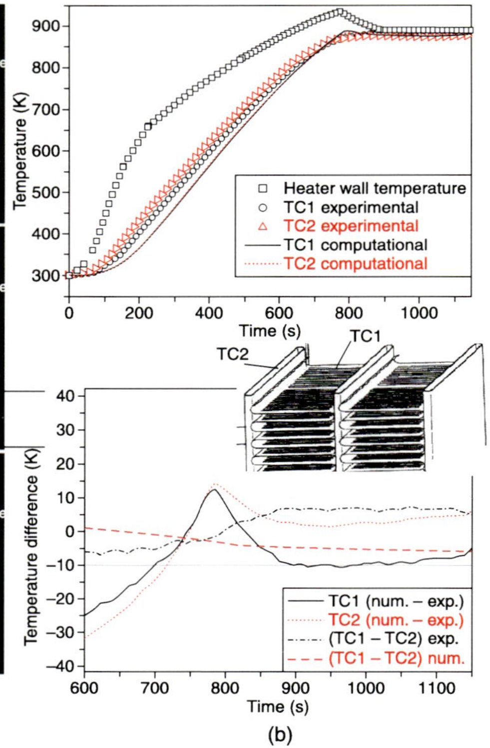

Plate II (Chapter 3) FEM model of a brazed fin-tube-header assembly made of aluminum during CAB brazing. (a) Temperature contours separated for 5 K. Assembly peak temperature 878 K and heating zone peak temperature 936 K. 66 460 blocks in the computational model (CPU time: 72 h for 1150 s of ramp-up on a HP N-4000 cluster. Node: 440 MHz, 8 GB RAM 8 PA-RISC 8500 at 440 MHz) (Sekulic *et al.*, 2001). (b) Comparison between modeled temperatures/differences and the experimental data.

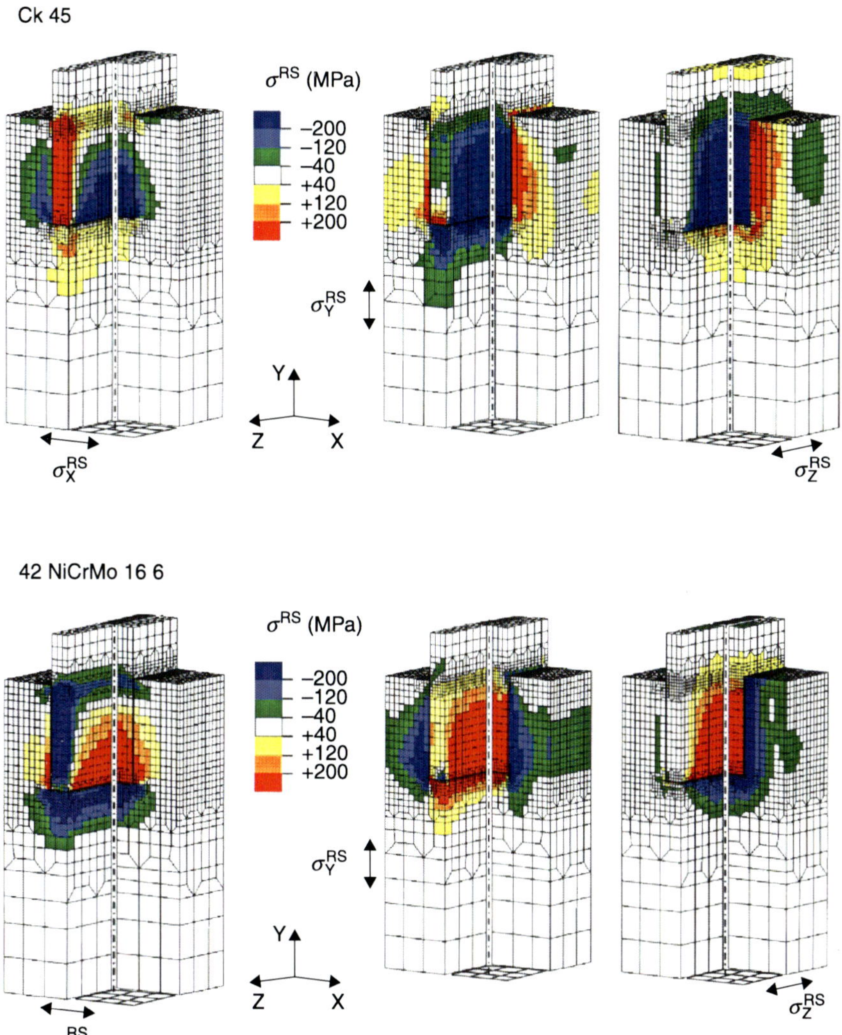

Plate III (Chapter 3) Cemented carbide inset brazed to carbon or tool steels (Pintschovius *et al.*, 1999) with the residual stress fields shown in three coordinate directions. The WC material features the smallest thermal expansion. (Courtesy Dr Pintschovius.)

(a) (b) (c)

Plate IV (Chapter 10) *In situ* filming of liquid filler metal blocking micro-channels: (a) molten filler has formed a joint between the header and extruded tube; (b) micro-channels are now being blocked by the molten filler; (c) blocked micro-channels after brazing (courtesy of Creative Thermal Solutions, Inc.).

100–160°C, the temperature at which thermal decomposition begins for the majority of simple TFBs of metals and amines.[38,39] For instance, in the thermal decomposition of zinc tetrafluoroborate hydrate melting occurs at 87–107°C, and destruction at 152–195°C:[38]

$$Zn(BF_4)_2 \cdot 6H_2O_S \rightarrow Zn(BF_4)_2 \cdot 6H_2O_L \rightarrow ZnF_2s + 2BF_{3G}\uparrow + 6H_2O_G\uparrow \quad [9.10]$$

The liquid filler metal is distributed during heating only over the surface covered by the deposited metal and fills the capillary gap poorly, prevented by the gaseous products of decomposition of simple TFBs of metals. Therefore, overlap joints on aluminium are soldered in the following sequence: first the surface is fluxed and plated with filler metal, and then soldering is performed over the plated surface.

The drawbacks of the fluxes containing simple metal TFBs can be eliminated by using newly synthesised non-corrosive tetrafluoroborates of nitrogen-bearing bases and metal tetrafluoroborates with organic nitrogen-bearing bases, which have different thermal stability.[40–44]

Gaseous products evolve during soldering by thermolysis of the above compounds. The chemical composition of these products determines the technological applicability of tetrafluoroborates as filler metals with different melting points, their fluxing ability in particular, as well as the required safety measures.

Complex compounds of metal TFBs were produced by preparatory synthesis. They were white powders that hydrolyse poorly in air and have limited solubility in high boiling solvents containing a hydroxyl group, such as glycerol, or an amine group, such as triethanolamine. According to elemental, atomic adsorption and titrimetric analyses, the composition of complex $HBF_4 \cdot L$ with a HBF_4 to organic ligand mole ratio of 1:1 was determined for TFBs with nitrogen-bearing bases of benztriazole, benzimidazole, morpholine, benzylamine, tert-butylamine and piperidine, while the composition of complex $Me_2L_2(BF_4)_2$ with a metal to organic ligand mole ratio of 1:2 was determined for compounds of Pb(II) with benztriazole and Zn(II) and Cd(II) with morpholine.[40–42] The thermal dissociation of tetrafluoroborates of nitrogen-bearing organic bases was studied using thermogravimetric analysis. A mechanism was suggested for the thermolysis of the compounds.[44] It was shown that the electron-donor ability of the organic base determines the sequence and character of the thermal transformations, as well as the thermal stability of the compounds studied.

Synthesised complex compounds are characterised by the following complex thermochemical processes occurring during heating within a temperature range of 100–350°C: melting, sublimation of hydrogen fluoride and boron trifluoride, decomposition of the organic ligand and oxidation of the reaction products.[43,44]

Comparison of the above schemes for the thermochemical transformation of simple tetrafluoroborates of metals

$$Me(BF_4)_{2S} \rightarrow MeF_{2S} + 2BF_{3G} \quad [9.11]$$

and tetrafluoroborates of amines

$$L_* HBF_{4\,S} \rightarrow L_* HBF_{4\,L} \rightarrow L_G + HF_G + BF_{3\,G} \qquad [9.12]$$

allows the scheme for the thermolysis of the complexes investigated to be presented as follows:

1. For complexes Cd, Zn with morpholine and benztriazole, and Pb with piperidine:

$$Me\,L_*\,(BF_4)_{2\,S} \rightarrow Me\,L_*\,(BF_4)_{2\,L} \rightarrow Me\,L_*\,F_{2\,L} + 2BF_{3\,G} \rightarrow MeF_{2\,S} + L; \qquad [9.13]$$

2. For complexes of Pb with morpholine, and Cd, Zn with piperidine:

$$Me\,L_*\,(BF_4)_{2\,S} \rightarrow Me\,L_*\,(BF_4)_{2\,L} \rightarrow Me\,F_{2\,S} + L_G + 2BF_{3\,G} \qquad [9.14]$$

Tables 9.8 and 9.9 give the temperatures for the destruction of activators of low-temperature fluxes: simple TFBs of metals, complex TFBs with nitrogen-bearing bases, and TFBs of Zn(II), Cd(II) and Pb(II) with nitrogen-bearing bases.

Table 9.8 Temperature of beginning of decomposition ($T_{b.d.}$) of complex TFBs with different bases[43]

Base	Benztriazole	Benzimidazole	Morpholine	Benzyl amine	t-Butylamine	Piperidine
p*Kα**	1.61	5.50	8.33	9.33	10.78	11.12
Ligand boiling point (°C)[42]	201–204	172	128.9	184.5	44.5	106
HBF_4 L $T_{b.d.}$ (°C)	130	145	290	120	220	200

*p*Kα* – Module of indicator of the degree of the constant of acid ionisation.

Table 9.9 Temperature of beginning of decomposition (°C) of simple and complex TFBs of metals with different bases[38,40,43,44]

TFBs of metals, ligand	Boiling temperature of ligand (°C)	Complex-forming metal		
		Zn(II)	Cd(II)	Pb(II)
$MeBF_4$ (ref. 24)	–	152	160	100
Benztriazole	201–204	215	140	220
Morpholine	128–130	215	280	300

The tetrafluoroborate-containing compounds have different temperature ranges for thermal decomposition: simple TFBs of metals: 100–160°C, TFBs with nitrogen-bearing bases: 120–290°C, and complex TFBs of metals: 140–300°C.

The fluxes investigated are liquid (at room temperature) immersion media (solvents: mono- and polyatomic alcohols, activators: heat-resistant complex TFBs of metals with nitrogen-bearing bases).

Aluminium of the 1050 grade (99.5 % Al), according to DSTU ISO 209-1:2002, was soldered in a temperature range of 220–320°C using filler metals (Table 9.10) and fluxes (Table 9.11). The activity of the fluxes was evaluated from the results of five tests for specified soldering temperatures and compositions of the filler metals and fluxes using the spreading factor for the filler metal and the ability to form soldered joints.

According to standard JIS Z 3197-1986 (Japan), the spreading factor (F_s) is determined from: $F_s = (D - H)/D \cdot 100\%$, where D is the calculated diameter of a sphere equal in volume to a sample of the filler metal tested (mm); and H is the height of the solidified drop measured after spreading of the filler metal (mm).

The solidified drop of the filler metal on the aluminium surface after removal of the residue of the flux, as well as the soldered T- and overlap joints, were examined by optical and electron microscopy.

In soldering aluminium in a temperature range of 230–320°C, the values of the spreading factor of the filler metals and the properties of the soldered joints

Table 9.10 Low-temperature filler metals for soldering aluminium

Filler metal	Composition (wt.%)			T_{melt} (°C)
	Sn	Pb	Zn	
SnPbZn	71	24	5	177
POS61	61	39	–	183
P200A	90	–	10	199–200

Table 9.11 Composition of low-temperature TFB fluxes

No.	Flux	Solvent	Activator
1–6		Triethanolamine	$LHBF_4$, (L)-benztriazole, benzimidazole, morpholine, benzylamine, t-butylamine, piperidine
7	F59A	Triethanolamine	Cadmium and zinc TFBs
8	FTFA220	Triethanolamine	Complex of lead with benztriazole
9	FTFA280	Glycerine	Complexes of cadmium and zinc with morpholine

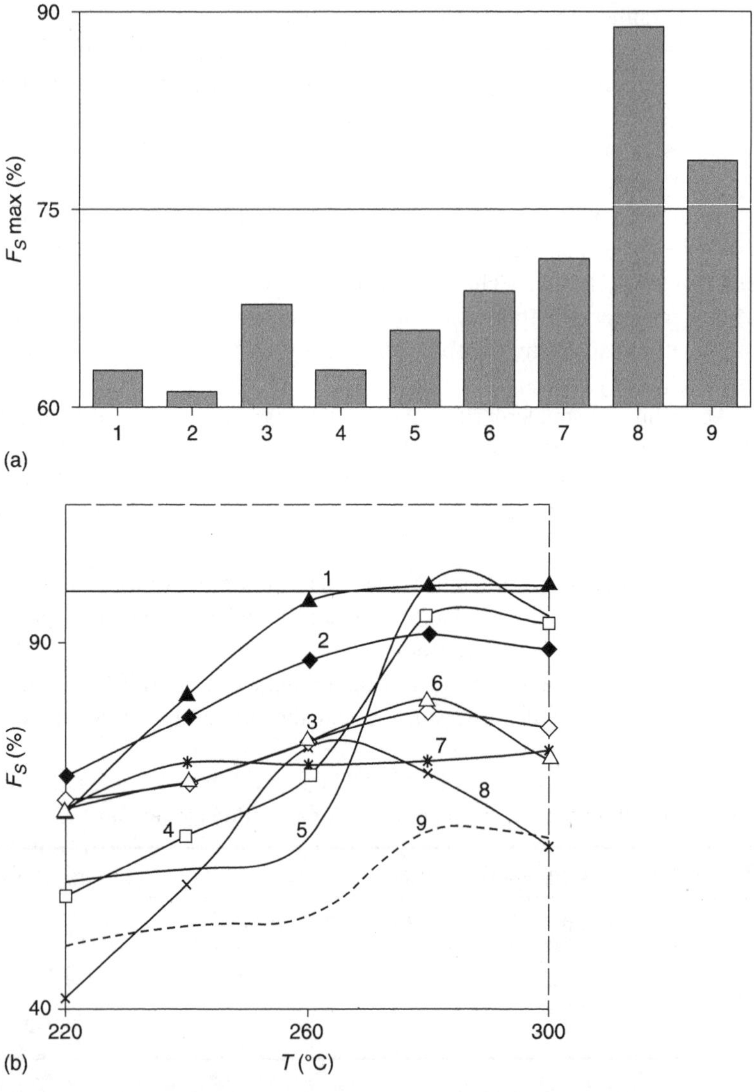

9.17 Spreading factor (F_s) of tin solders over aluminium 1350. (a) Sn10Zn, activator (g-mole/l) – from 0.063 to 2.5 in fluxes (1–9, Table 9.11); (b) 1–4 – Sn10Zn, Sn24Pb5Zn, Sn39Pb,[45] Sn25Cd20Zn,[46] FTFA200; 5 – Sn24Pb5Zn, FTFA280; 6–9 – Sn10Zn, Sn24Pb5Zn, Sn39Pb,[45] Sn25Cd20Zn,[46] F59A, activator (0.3 g-mole/l).

substantially differ for different compositions of the TFB fluxes (Fig. 9.17). As shown by the investigations, the spreading of filler metals of different compositions is greatly affected by heat resistance and the composition of active chemical compounds in a flux. For example, the spreading factor (%) for filler metal P200A is 71.2 for flux (7); for simple TFBs of metals it ranges from 61.2 to 69; for fluxes (1–6) – complex TFBs with nitrogen-bearing bases of benztriazole, benzimidazole, morpholine, benzylamine, tert-butylamine and piperidine – it is 89.1; for flux (8) – complex TFBs of Pb(II) with the benztriazole bases – 78.8; and for flux (9) – complex TFBs of Zn(II) and Cd(II) with the morpholine bases – 86.9. Metal coatings formed during the thermochemical interaction of the flux and aluminium considerably improves the wetting and spreading of the filler metals, and promotes the formation of sound joints.

In addition, it was found in the course of the experiments that filler metals wet the upper plate of a T-sample in a region of less than 10% of its length, but do not fill up the horizontal gap 0.1–0.5 mm wide because of the low heat resistance of the activating fluxes (1–6) and the formation of gaseous products when they decompose at different temperatures by reaction 9.11.

It has been established[40,42] that no HF gas is formed even on complete decomposition of complex TFBs of Pb(II), Zn(II) and Cd(II) with the benztriazole and morpholine bases. At the same time, the temperature ranges for the active interaction of fluxes FTFA200 and FTFA280 with aluminium are higher than those of fluxes 1–7 and lie within 240–300°C, whereas boron trifluoride can be partially absorbed by a solvent. Fluxes 8 and 9, which contain heat-resistant complex TFBs of metals with the benztriazole and morpholine bases, substantially improve conditions for wetting of aluminium and filling the capillary gaps with filler metal.

Optical and electron micrographs of the surfaces of flat samples after soldering showed that chemical interaction of the flux with aluminium during heating results in a coating of metal (alloy) deposited from the corresponding TFBs of Pb(II), Zn(II) and Cd(II) with benztriazole and morpholine bases on the aluminium surface.[47] The coating of a complex-forming metal forms at the interface between the filler metal and aluminium, i.e. on the surface of the base metal around a drop (Fig. 9.18). The maximal width of individual regions of the coating on the aluminium amounts to several millimetres. Characteristic regions of the coating differ in elemental composition and contain lead, cadmium and zinc reduced by aluminium from the corresponding complex TFBs.

X-ray spectrum microanalysis of soldered joints revealed the complex-forming elements reduced from the TFB fluxes in contact with aluminium at the filler metal–aluminium interface (Table 9.12).

X-ray spectrum microanalysis revealed considerable concentrations of lead reduced by aluminium from the flux at the filler metal–aluminium interfaces (2.41 wt.% on the surface and 15.8 wt.% in a joint) (Fig. 9.19, Table 9.13). The possibility of changing over wide ranges the melting and decomposition temperatures of synthesised complex tetrafluoroborates, which have better fluxing properties due to replacement of the organic ligand or metal, allows the optimal technological compatibility to be achieved in different filler metal–flux–base

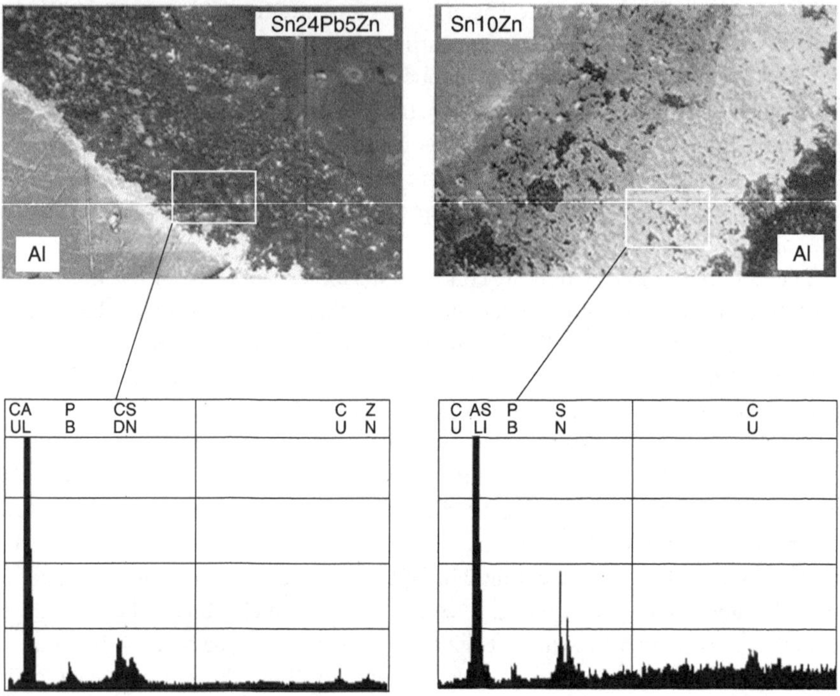

(a) (b)

9.18 Analysis of a region within a solidified solder droplet on the surface of aluminium, and spectra of elements of the coating: (a) solder Sn24Pb5Zn and flux with complex TFB Cd(II) with morpholine base; (b) solder Sn10Zn and flux with complex TFB Pb(II) with benztriazole base.

Table 9.12 Quantitative analysis of the content of elements in microvolumes of the solder–aluminium transition zone at T_s = 280°C (wt.%)

Filler metal	Flux	Sn	Zn	Pb	Cd	Al
P200A Sn10Zn	FTFA200	94.28	4.48	0.73	–	0.54
P200A Sn10Zn	FTFA280	94.43	4.91	–	0.12	0.54
Sn24Pb5Zn	Complex TFB of Cd(II) with Mr	72.81	2.93	23.66	0.1	0.50

Table 9.13 Content of elements in microvolumes of a soldered joint

Point no.	Chemical compositions in microvolumes (wt.%)				
	Sn	Cd	Zn	Al	Pb
1	39.2	43.3	1.3	0.3	15.8
2	77.0	19.0	3.9	0.1	0.0
3	0.06	0.74	99.2	0.0	0.0

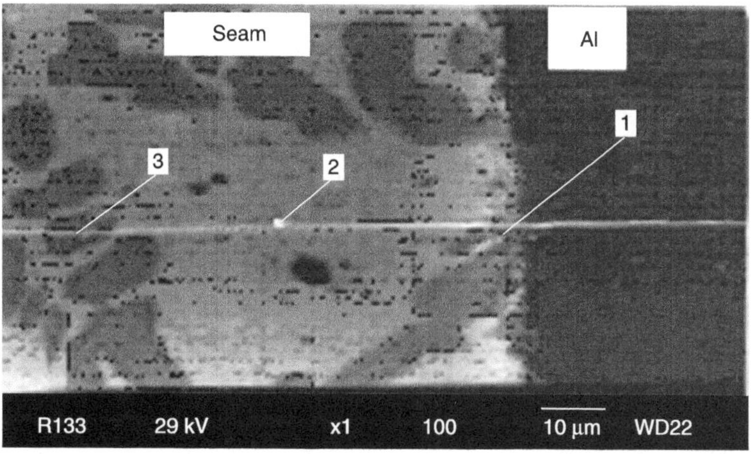

9.19 Microstructure of soldered joint on aluminium 1050A (filler metal Sn–20Cd–25Zn, flux FTFA200, T_s = 280°C).

metal systems. The shear strength of joints on aluminium 1050 soldered with filler metal P200A at room temperature was 0.4–0.5 τ_{shear} of the base metal.

The results obtained led to the development of fluxes FTFA200 and FTFA280, and the technology for soldering aluminium foil and radio elements of signalling systems through heating with a soldering iron, and soldering pressure seals of electron module casings, as well as the technology for repair soldering of thin-walled radiator tubes of aluminium alloys through heating with hot air or a torch flame. The soldered joints of the radio elements had the required electrical conductivity. The aluminium casings successfully passed tightness tests under a pressure of 10 atm. The fluxes and their residue are non-corrosive to aluminium (pH is 7 according to the PND50-975-84 scale, Ukrainian standard) and can be readily removed by water.

9.6 Conclusion and future trends

The creation of conditions for wetting and the formation of sound joints between parts is a very important scientific and technical problem for brazing aluminium covered with a strong and refractory oxide film.

As seen from the materials presented in this chapter, one of the main current trends in the high-temperature brazing of aluminium is the elimination of chloride fluxes, which require thorough washing of workpieces to remove the residue of corrosive fluxes, and their replacement by fluoride fluxes, the residue of which do not cause corrosion fracture. Flux NOCOLOK, which is successfully used in the manufacture of high-volume products in many countries throughout the world, is a good example of this trend. Fluoride fluxes, which have a number of advantages, will continue to be developed. In particular, in some cases they can be used without a filler metal due to the reduction of silicon from the flux, while silicon forms a filler metal from the Al–Si system as a result of its interaction with aluminium. Uses of fluoride fluxes of both of the above types will grow in the future.

The number of uses of flux-free vacuum aluminium brazing for high-volume products will not be greatly increased. This is due to the complexity of the equipment and technological processes, and, most importantly, by low productivity. Despite the huge efforts made by a number of countries, they have had to draw back from the use of through-type furnaces for brazing aluminium, which could have provided high productivity.[48] Hence, this process will be used only for the manufacture of single and small-lot products.

Flux-free brazing without a vacuum requires a high-purity inert atmosphere. So far it has been used to manufacture unique products, such as large heat exchangers.

The 300–500°C temperature range is of interest, as it is applicable for soldering practically all wrought, heat-hardened and casting aluminium alloys. A new class of non-corrosive fluxes containing tetrafluoroborates of metals with nitrogen-bearing bases was developed for soldering aluminium in a temperature range close to 300°C. The breaking of acceptor–donor bonds of such compounds and the evolution of a gaseous reducing agent (BF_3) within the required temperature range, as well as the formation of a low-melting point interlayer of metals, which facilitates production of sound joints, take place during soldering. In our opinion, this area will be actively developed in the future. In particular, these fluxes are advantageously used in the soldering of aluminium to different grades of steels using high-frequency and arc heating.

Arc soldering of aluminium and aluminium, e.g. to galvanised steel, using arc heating and zinc-base filler metals, is a serious competitor to fluxes for soldering aluminium. This process requires no fluxes and is carried out using local heating, i.e. it is very cost-effective, and creates satisfactory joints. However, it should be noted that specialised advanced equipment is required to implement this process.

It should be noted in conclusion that a number of fluxes, filler metals and processes developed in recent years successfully solve several different problems in the brazing and soldering of aluminium to steel.

9.7 References

1. B. Phillips, C.M. Warshaw, and I. Mockrin. Equilibria in $KAlF_4$-Containing Systems. *Journal of The American Ceramic Society*, Vol. 49, No. 12 (1966), pp. 631–634.
2. Solvay Fluor und Derivate. THE NOCOLOK® Flux Brazing Process (1997). 20 pp.
3. R.S. Timsit, B.J. Janeway. A Novel Brazing Technique for Aluminum. *The Welding Journal*, Vol. 73, No. 6 (1994), pp. 119s–128s.
4. L.F. Mondolfo. *Structure and Properties of Aluminium Alloys*, Translated from English. M.: Metallurgiya, 1979. 640 pp.

5. H. Song, A. Hellawell. Solidification in the system Al-Ge-Si: The phase diagram, coring patterns, eutectic growth, and modification. *Metallurgical and Materials Transaction A*, Vol. 21, No. 2 (1990), pp. 733–740.
6. US Pat. 6019856, 1 February 2000. *Solderless aluminum brazing*. T. Born (Holle, DE), H.-J. Belt (Burgwedel, DE), and Solvay Fluor und Derivate GmbH (Hannover, DE).
7. US Pat. 6432221, 13 August 2002. *Fluxing agents*. U. Seseke-Koyro (Vellmar, DE), J. Frehse (Hannover, DE), A. Becker (Lachendorf, DE), and Solvay Fluor und Derivate GmbH (Hannover, DE).
8. A.A. Andreiko, E.V. Panov, B.V. Yakovlev, V.F. Khorunov, and O.M. Sabadash. Fusibility and chemical interaction in the K, Al, Si/F salt system. *Ukr. Chem. Journal*, Vol. 63, No. 9 (1997), pp. 121–124.
9. NIST-JANAF Thermochemical Tables. 3rd ed. M.W. Chase and others. Thermal Group. Dow Chemical USA, Midland, Michigan. *J. Phys. Chem. Ref. Data*, Vol. 14, Suppl. 1 (1985), 1856 pp.
10. T. Takemoto, A. Matsunawa, and A. Kitagawa. Decomposition of non-corrosive aluminium brazing flux during heating. *J. Mater. Sci. Lett.*, Vol. 15, No. 4 (1996), pp. 301–303.
11. V.F. Khorunov, O.M. Sabadash. Reactive fluoride flux for brazing of aluminium and dissimilar joints. *Adhesion of Melts and Brazing of Materials*, Ed. 39, 2006. pp. 68–75.
12. V.F. Khorunov, O.M. Sabadash. Peculiarities of interaction of fluxes of the KF-AlF_3-K_2SiF_6 system with aluminium during brazing. *Proc. of Conf. 'Brazing. Modern Technologies, Materials, Structures, and Experience of Operation of Brazed Structures'*. Society 'Znaniye', 23–24 April 2003. M.: TsRDZ. pp. 126–129.
13. V.N. Eremenko, N.D. Lesnik, T.S. Pestun, and V.R. Ryabov. Kinetics of spreading of aluminium-silicon melts over iron. In: *Wetting and Surface Properties of Melts and Solids*. Kiev: Naukova Dumka Naukova Dumka, 1972. pp. 39–41.
14. M. Roulin, J.W. Luster, G. Karadeniz, and A. Mortensen. Strength and Structure of Furnace-Brazed Joints between Aluminium and Stainless Steel. *The Welding Journal*, May 1999, pp. 1s–5s.
15. V.F. Khorunov, O.M. Sabadash. High-temperature flux brazing of aluminium to steel 12Kh18N10T. *Proc. of Scient.-Pract. Seminar 'Brazing in Instrument Making and Machine Building. Technology and Materials'*, 9–10 December 2003, St Petersburg, 'IVA'. pp. 13–17.
16. O.M. Sabadash, V.F. Khorunov. Technology for flux brazing of aluminium to corrosion-resistant steel. *Proc. of Seminar 'Brazing-2004'*. M.: TsRDZ. 2004. pp. 53–56.
17. M. Kuroda, A. Uenishi, H. Yoshida, and A. Igarashi. Ductility of interstitial-free steel under high strain rate tension: Experiments and macroscopic modeling with a physically-based consideration. *International Journal of Solids and Structures*, Vol. 43 (2006), pp. 4465–4483.
18. T. Siewert, I. Samardzic, and S. Klaric. Application of an On-Line Weld Monitoring System. *1st International Conference on Advanced Technologies for Developing Countries*, 12–14 September 2002. Slavonski Brod, Croatia. pp. 1–6.
19. J. Brukner. Arc welding of steel to aluminium. *Avtomaticheskaya Svarka*, No. 11 (2003), pp. 185–187.
20. V.R. Ryabov, D.M. Rabkin. *Welding of dissimilar metals and alloys*. M.: Mashinostroyeniye, 1984. 239 pp.

21. O.M. Sabadash, V.F. Khorunov. Flux brazing of aluminium to steel by using arc heating. *Proc. of Seminar 'Brazing-2010'*. M.: TsRDZ. 10102010. pp. 73–77.
22. A.M. Nikitinsky. *Soldering and brazing of aluminium and its alloys*. M.: Mashinostroyeniye, 1983. 192 pp., il il.
23. N.F. Lashko, S.V. Lashko. *Brazing of metals*. 2nd ed. M.: Mashinostroyeniye, 1967. 367 pp., il il.
24. B.A. Maksimikhin. *Brazing of metals in instrument making*. Ed. by P.I. Petrov. L.: TsBTI, 1959. 116 pp.
25. US Pat. 3988175, In.Cl.[2] B23K 35/34 *Soldering flux and method*. D.L. Rutledge, G.T. Ozaki (USA); Bethlehem Steel Co (USA). No. 499959; Filed 23 August 1974; Publ. 26 October 1976; Nat.Cl. 148–26. 6 pp.
26. US Pat. 4496612, In.Cl.[3] B05D 3/04, B23K 35/34 *Aqueous flux for hot dip metallizing process*. J.E. McNutt, R.J. Scott (USA); E.I. Pont de Nemous and Co. (USA). No. 565773; Filed 27 Desember 1983; Publ. 29 January 1985. Nat.Cl. 427–310. 4 pp.
27. US Pat. 4802932, In.Cl.[4] B23K 35/34 *Fluoride-free flux compositions for hot galvanization in aluminum-modified zinc bath*. J. Billiet (Belgium); Dionne & Cantor (USA). No. 27293; Filed 4 November 1987; Publ. 7 February 1989; Nat.Cl. 148–23. 3 pp.
28. US Pat. 3074158, In.Cl.[2] B23K 35/34 *Flux composition and method of using same to solder aluminum*. W.D. Finnagan (USA); Kaiser Aluminum & Chemical Co. (USA). No. 494573; Filed 15 March 1955; Publ. 22 January 1963; Nat.Cl. 29–495. 4 pp.
29. US Pat. 3330028, In.Cl.[3] B05D 3/04, B23K 35/34 *Soldering flux and method of soldering with same*. C.H. Elbreder (USA), Frontenac Mo. (USA), D. Berryman, G. Garner (USA); both of Lemay. Mo. No. 317030; Filed 17 October 1963; Publ. 11 July 1967; Nat.Cl. 29–495. 2 pp.
30. US Pat. 2801943, In.Cl.[2] B23K 35/34 *Composition of matter for soldering aluminium*. M.L. Freedman (USA); Horizons Inc. (USA). No. 501730; Filed 15 April 1955; Publ. 6 August 1957; Nat.Cl. 148–23. 2 pp.
31. V.E. Khryapin. *Solderer's Handbook*. 5th ed. Reviewed and supplemented. M.: Mashinostroyeniye, 1981. 348 pp., il il.
32. US Pat. 2788303, In.Cl.[2] B23K 35/34 *Soldering flux*. R.L. Ballard, D.C. Burch (USA); Essex Wire Co. (USA). No. 441154; Filed 2 February 1954; Publ. 9 April 1957; Nat.Cl. 148–23. 1 p.
33. S.V. Lashko, N.F. Lashko. *Brazing of metals*. 4th ed. Reviewed and supplemented. M.: Mashinostroyeniye, 1988. 376 pp., il il.
34. *Brazing manual*. 3rd ed. Ed. by I.E. Petrunin. Reviewed and supplemented. M.: Mashinostroyeniye, 2003. 480 pp., il il.
35. US Pat. 22286298, In.Cl.[2] B23K 35/34 *Aluminum soldering flux*. M.E. Miller (USA); Aluminum Company of America (USA). No. 410041; Filed 8 September 1941; Publ. 16 June 1942; Nat.Cl. 148–23. 1 p.
36. US Pat. 3655461, In.Cl.[2] B23K 35/34; B23K35/36 *Flux for aluminum soldering*. D.L. Rutledge, G.T. Ozaki (USA); Sanyo Electric Works Ltd. (USA). No. 05/069131; Filed 2 September 1970; Publ. 26 October 1976; Nat.Cl. 148–23. 4 pp.
37. I.G. Ryss. *Chemistry of fluorine and its inorganic compounds*. M.: Gostekhnauchizdat, 1956. 718 pp.
38. T.V. Ostrovskaya, S.A. Azmirova. Chemical transformations of magnesium, calcium, strontium, zinc and cadmium tetrafluoroborates in heating. *Journal of Inorganic Chemistry*, Vol. 4 (1970), No. 3, pp. 657–660.

39. *Manual of Chemistry*. 3rd ed. Reviewed and supplemented, in 5 volumes. Main properties of inorganic and organic compounds. Vol. 2. Leningrad: Khimiya, 1971. 1168 pp.
40. A.N. Chebotarev, M.V. Shestakova, and T.M. Shcherbakova. Complex tetrafluoroborates of copper (II) and zinc (II) with nitrogen-bearing organic bases. *Journal of Inorganic Chemistry*, Vol. 38 (1993), No.2, pp. 273–275.
41. A.N. Chebotarev, M.V. Shestakova, V.F. Khorunov, and O.M. Sabadash. Complex tetrafluoroborates of lead (II) with nitrogen-bearing organic bases. *Ukr. Khim. Zhurnal*, Vol. 66 (2000), No. 6, pp. 88–92.
42. A.N. Chebotarev, M.V. Shestakova, V.F. Khorunov, and O.M. Sabadash. Synthesis and physical-chemical properties of complex tetrafluoroborates of zinc (II) and cadmium (II) with nitrogen-bearing organic bases. *Ukr. Khim. Zhurnal*, Vol. 66 (2000), No. 8, pp. 81–86.
43. V.F. Khorunov, V.G. Samoilenko, A.N. Chebotarev, I.P. Bezhaeva, and A.P. Slonimsky. Tetrafluoroborate fluxes for soldering. *Avtomat. Svarka*. (1990), No. 6, pp. 64–66.
44. A.N. Chebotarev, M.V. Shestakova, V.F. Khorunov, and O.M. Sabadash. Peculiarities of thermal transformations of tetrafluoroborate salts of nitrogen-bearing organic bases – promising components for soldering fluxes. *Proc. of the 1st Intern Scient.-Pract. Conf. 'Protection of Environment, Health and Safety in Welding Production'*, 11–13 September 2002, Odessa. Odessa: Astroprint, 2002. pp. 657–664.
45. V.F. Khorunov, O.M. Sabadash, and A.N. Chebotarev. Soldering of aluminium using low-temperature fluxes containing tetrafluoroborate activators. *Proceedings of the 3rd Int. Seminar in Precision and Electronic Technology 'INSEL-99'*, Warsaw, 1999. pp. 185–189.
46. O.M. Sabadash, V.F. Khorunov. Materials and technology for flux soldering of aluminium and aluminium to stainless steel. *Avtomat. Svarka*. (2005), No. 8, pp. 69–74.
47. O.M. Sabadash. Soldering aluminium by using tin-base solders and reactive flux. *Adhesion of Melts and Brazing of Materials*. 40th Ed. 2007. pp. 82–90.
48. E. TolesGeorge. Aluminum brazing for high production rates. *Can. Mach. and Metalwork*, 83 (1972), No. 10, pp. 72–73, 106.

10

Controlled atmosphere brazing of aluminum

H. ZHAO, Creative Thermal Solutions, Inc., USA
and R. WOODS, Pleasanton, California, USA

DOI: 10.1533/9780857096500.2.280

Abstract: Compact aluminum heat exchangers are widely used in the automotive industry. The market in the heating, ventilation, air conditioning and refrigeration (HVAC&R) industry is expected to grow. Controlled atmosphere brazing (CAB) is the state-of-the-art mass production technology. This chapter presents the manufacturing process for compact aluminum heat exchangers (e.g. micro-channel heat exchangers) using CAB. Suitable aluminum alloys are introduced and the principles of brazing sheet metallurgy, filler metal selection and application are described. Fluoride fluxes and recent developments in flux modification and application methods are discussed, and wetting and flow behavior of molten aluminum filler metals during brazing are explored. Some phenomena due to filler/base metal interactions are included. Recent developments of CAB heat exchanger materials with enhanced corrosion resistance are reviewed, and corrosion testing methods are discussed.

Key words: aluminum, controlled atmosphere brazing, fluxes, heat exchangers, brazing alloys, wetting, corrosion.

10.1 Introduction

Over the past 30 years the traditional copper/brass heat exchangers in automobiles and trucks have been almost completely replaced by brazed aluminum units. Weight savings and cost advantages were the motivators for this change, which was enabled by advances in aluminum vacuum-brazing technology. With the advent of aluminum controlled atmosphere brazing (CAB) technology, further savings in production costs were possible, and this has now become the state-of-the-art manufacturing process for the high-volume joining of most automobile heat exchanger assemblies. Current predictions are that CAB will play an increasing role in the growing market for brazed aluminum condensers in the heating, ventilation, air conditioning and refrigeration (HVAC&R) sector, where aluminum is expected to once again replace copper.

The CAB aluminum brazing process is characterized by: (1) a tightly controlled low-moisture, low-oxygen working atmosphere of nitrogen gas and (2) usage of fluoride-based fluxes for surface oxide removal. Historically, radiator and heater core tubes were, and still are, made very cost-effectively from roll-formed sheet clad with a thin layer of filler metal to make the joints to the bare fins. Recent conversion of condenser production to CAB and the increasing production of charge-air coolers, both using bare extruded micro-channel tubes, has necessitated the use of filler metal clad onto the fin stock. Despite the success of the aluminum automobile heat exchanger in service, the drive is still towards lower cost, lighter units with higher strength and excellent corrosion resistance. Research and advances in these areas are crucial to the continued development and predominance of the brazed aluminum heat exchanger.

10.2 Applications of controlled atmosphere brazing (CAB) of aluminum

As mentioned above, over the past two decades, CAB aluminum brazing technology has largely replaced vacuum brazing and is now the commonest method used in the manufacturing of automotive heat exchangers. The newer CAB micro-channel heat exchangers combine superior heat transfer performance with high compactness and significant weight reduction based on the development of multiport extruded tube technology and novel designs of the air side heat transfer surface structures, such as the multilouver fins. These attractive characteristics are expected to promote the market penetration of micro-channel heat exchangers into the HVAC&R industry, replacing the traditional mechanically assembled round copper tube/aluminum fin condensers in stationary systems. Figure 10.1 shows an aluminum micro-channel condenser and a copper tube condenser of an identical size for a residential air conditioning system. Experimental studies show that the system with the micro-channel condenser achieves better thermal performance and lower refrigerant charge than the one

(a) (b)

10.1 Examples of condensers for an air conditioning system: (a) a CAB aluminum micro-channel heat exchanger; (b) a copper round tube/aluminum fin heat exchanger (courtesy of the Air Conditioning and Refrigeration Center, University of Illinois).

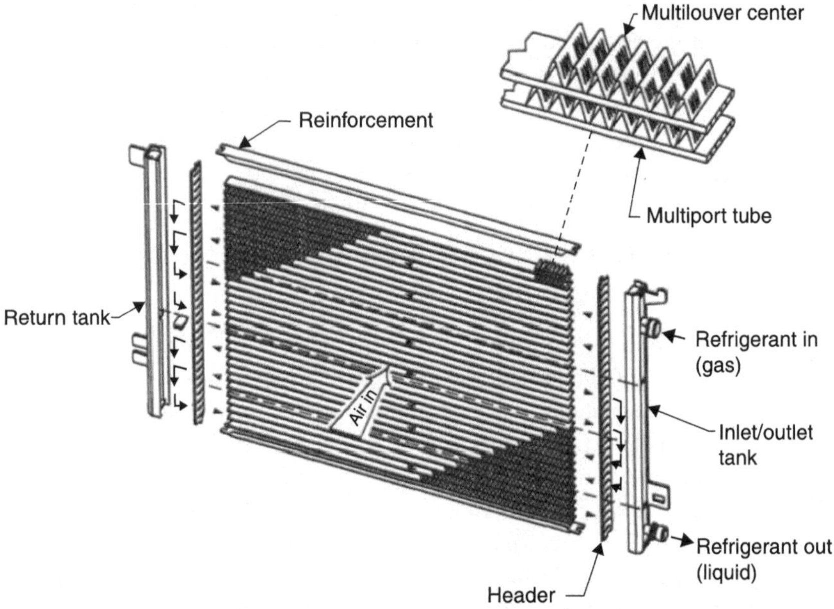

10.2 Micro-channel heat exchanger component (Shah and Sekulic, 2003) (courtesy of Delphi Thermal Systems, Lockport, NY).

with the copper tube condenser (Park and Hrnjak, 2002). Advances in CAB furnace design have also made possible the mass production of large micro-channel heat exchangers for HVAC systems. An example of a micro-channel heat exchanger component is illustrated in Fig. 10.2 (Shah and Sekulic, 2003).

Continuous aluminum alloy developments have significantly improved the extrudability, brazability, corrosion resistance and mechanical strength of materials at both room and elevated temperatures (Nordlien *et al.*, 2011). These property improvements have allowed the progressive thinning of heat exchanger tubes and reductions in the micro-channel tube diameters. Innovations such as these not only provide the opportunity for refrigerant charge reduction, but also make extruded tubes suitable for applications in heat exchangers working under high-pressure conditions such as condensers and evaporators in cooling systems with CO_2 as the refrigerant (Hrnjak, 2011). Further progress is needed in (a) improving the high-temperature strength of alloys brazed for applications such as charge air coolers, (b) the fatigue performance of the thinner tubes and (c) sag resistance (rigidity) during the brazing process. With the emphasis being placed upon weight reduction and the progressive down-gauging that is taking place, deformation of the very thin gauge components during CAB is becoming of concern. Proper control of the brazing conditions for joining such thin-wall structures is also more critical than in traditional heat exchanger fabrication.

The assembly of heat exchanger units for brazing is usually done automatically, though there remains some manual assembly of more complex components and in factories specializing in small volume runs (e.g. after market production). Once assembled, the aluminum heat exchangers are introduced into the hot zone of a brazing furnace following several pre-braze steps, e.g. cleaning, degreasing, fluxing, and finally drying off (Swidersky, 2001). Proper control of each step is very important for the quality of the final products. Pre-braze assembly and fixturing of heat exchangers should provide good contact between the clad and bare components to enable good fillet formation. However, pressures applied to the parts should not be too high, in order to avoid fin collapse. Solid contaminants and fabrication lubricants on the parts should be thoroughly removed in order to obtain good surface conditions for uniform coating of the flux. This is necessary to ensure good joint integrity and an acceptable post-braze cosmetic appearance (Swidersky, 2001). Traditionally, aqueous and nowadays predominantly thermal degreasing are the most common cleaning methods used in production. A uniform and optimal flux loading is essential to ensure good wetting of the molten filler metal on the base metal. Although flux loadings down to 2 g/m^2 are used in some cases, more common practices utilize a flux load nearer to 5 g/m^2, and occasionally up to 10 g/m^2. These flux coatings produce a surface appearance similar to that of a very dusty car (Solvay Fluor document, 2012a). In cases where the higher flux loadings are needed, they compensate for other non-optimized conditions. Examples are joint designs with complex geometries or when using a difficult-to-braze alloy (e.g. Mg-bearing alloys) as a heat exchanger component material.

The basic types of CAB furnaces include (1) batch furnaces, which provide the flexibility needed for different brazing products with various dimensions and

10.3 Continuous CAB furnace configurations (pre-braze processing not included) (courtesy of AFC-Holcroft).

geometries; and (2) continuous furnaces, which are more suitable for the high-volume production of heat exchangers with similar or fixed designs and dimensions. Figure 10.3 shows a continuous CAB furnace. The heat exchanger assemblies are delivered into the CAB furnace by a conveyor and exposed to radiation/convective heating. The proper temperature distribution within the furnace is achieved by using multi-zone (five to six zones) heating; the temperature of each zone can be adjusted by automatic or manual control. An inert gas atmosphere is achieved by a continuous flow of nitrogen gas through the furnace to maintain an oxygen level lower than 100 ppm and a dew point below −40°C (Garcia *et al.*, 2010). The peak brazing temperature should be within the melting range of the filler material and it must be well below the core melting point and sufficiently above the melting point of the flux to ensure complete oxide removal. The optimal brazing temperature for aluminum heat exchangers is in the range of 595°C to 600°C. Brazed joint quality is directly related to the heating profile of the brazing process and temperature uniformity throughout the heat exchanger core and manifolds with very different thermal masses. The American Welding Society (AWS) specification for aluminum brazing (AWS C3.7M/C3.7, 2011) requires that, during furnace brazing, 'After thermal equilibrium has been reached, the temperature measure by any test thermocouple shall not vary from the set temperature by more than ± 3°C.'

10.3 Materials involved in controlled atmosphere brazing (CAB) of aluminum

Light weight and high thermal conductivity are two of the properties that make Al alloys ideal materials for heat transfer devices. Additionally, aluminum has the advantages of good formability, fairly good corrosion resistance and being relatively cheap.

In the commonly used Aluminum Association four digit designation system for wrought Al alloys, the first digit of the four represents the major alloying element (see Table 10.1). Wrought alloys can be divided into two categories: non-heat treatable and heat treatable (Hatch, 1984). For heat treatable alloys, the alloy strength can be increased by processing through the proper sequence of thermal treatments. These aluminum alloys usually contain one or more of the elements Cu, Mg, Si or Zn, which have a significant solubility difference in an Al matrix at high and low temperatures (Hatch, 1984). The precipitation at lower temperatures of such intermetallic compounds as $CuAl_2$, Mg_2Si or $MgZn_2$ makes the alloys responsive to heat treatment. More details of brazing and heat treatment will be discussed in Section 10.5.4. For non-heat treatable alloys, mechanical strength must be increased by solid solution hardening and strain hardening from the annealed temper (Hatch, 1984). Table 10.1 lists commonly used wrought Al alloys, including examples of their application (Kaufman, 2000).

Table 10.1 Aluminum alloy designations and applications (Kaufman, 2000)

AA designation	Alloy composition	Main characteristics	Applications, examples
Non-heat treatable alloys			
1xxx	Pure Al	High corrosion resistance, formability, electrical conductivity	Packaging, chemical equipment, electrical applications
3xxx	Al–Mn	High formability, corrosion resistance, medium strength	Heat exchangers, cooking utensils, chemical equipment
4xxx	Al–Si	High fluidity, medium strength	Casting, metal joining
5xxx	Al–Mg	Excellent corrosion resistance, toughness, moderate strength	Building and construction, automotive, cryogenic, marine
Heat treatable alloys			
2xxx	Al–Cu	High strength at elevated temperatures	Aircraft and truck bodies
6xxx	Al–Mg–Si	High corrosion resistance, excellent extrudability, moderate strength	Architecture, structural members, marine, truck frames, pipelines
7xxx	Al–Zn–Mg	Very high strength, some alloys have especially high toughness	Aircraft industry

Note: AA, Aluminum Association.

10.3.1 Core alloys

Table 10.2 lists some common aluminum core alloys used in the CAB of Al heat exchangers, including both heat treatable and non-heat treatable alloys. The high purity aluminum (AA1xxx) and AA3003 (Mn-containing) alloys are non-heat treatable. Mechanical properties of these alloys can be improved by cold working (Kaufman, 2000; Schwartz, 2003). AA3003 (containing slightly over 1% of Mn by weight) is among the most widely used alloys in CAB aluminum heat exchangers. Mn increases the strength of the alloy, but is often combined with iron as an intermetallic compound, which may lead to some decrease in the material ductility (Hatch, 1984). The addition of Mn has little effect on the melting point of an Al alloy. Addition of a small quantity of Mg can further increase the alloy strength, such as in AA3005 (Magnusson *et al.*, 1997; Miller *et al.*, 2000). The trade-off is in the increased difficulty of brazing the Mg-bearing alloys due to the reduction in efficiency of the fluoride flux and the formation of a more complicated oxide (Bolingbroke *et al.*,

Table 10.2 Composition of commonly brazed core alloys for heat exchangers (Zhang and Zhuang, 2008)

Alloy	Alloying element (wt%) (nominal)					Melting temperature (°C)	
	Si	Cu	Mn	Mg	Al min%	Solidus	Liquidus
AA1100		0.12			Bal.	643	657
AA1145	0.3				Bal.	646	657
AA3003		0.12	1.2		Bal.	643	655
AA3005	0.6	0.3	1.2	0.4	Bal.	629*	654*
AA6063	0.4			0.7	Bal.	615	655
AA6951	0.3	0.3		0.4	Bal.	616	654

*Data from Aleris document (2010).

1997). This more tenacious oxide prevents good wetting of the alloy surface by the filler metals.

Heat treatable Al alloys used for heat exchangers will often contain both Mg and Si, e.g. AA6000 series alloys. These alloys have a lower solidus temperature (see Table 10.2) than pure aluminum (Brazing Handbook, 2007; Schwartz, 2003). However, alloys suitable for furnace CAB contain very limited amounts of Mg; therefore, overheating of the Mg-bearing base metal may not be a concern. In general, more flux is required for CAB and torch brazing of aluminum alloys containing Mg. Alloys that contain more than 0.7% Mg are usually not recommended for CAB with regular fluoride fluxes (Johansson *et al.*, 2003). When a much faster heating rate than in the CAB process is used in the brazing operation, such as in flame brazing, higher Mg contents (0.5%–0.7%) in the core alloy may be acceptable. The mechanism for the detrimental effect of Mg on the flux action will be discussed in Section 10.4.2.

10.3.2 Filler metals

The Al–Si series alloys are commonly used as filler metals for aluminum brazing. They are classified as the BAlSi series of filler alloys by the AWS (Brazing Handbook, 2007). The following characteristics of the Al–Si series of alloys make them suitable as filler metals for brazing aluminum alloys (Zhang and Zhuang, 2008): (1) the basic composition of the filler metal is similar to the core or base metal, which ensures good wettability, material compatibility and good metallurgical bonding; (2) the solidus temperature of the filler is at least 40°C lower than that of the base metal to avoid melting of the base metal during the heating process; (3) the difference between the liquidus and solidus temperatures is relatively small, which ensures better control of the brazing conditions and good flowability of the molten filler; (4) the corrosion potentials of the filler alloys are close to the base metal, resulting in good resistance of the brazed joint to galvanic corrosion; (5) the filler metals also have good formability and can be supplied in various forms such as wire, foil or powder depending on the process requirements. Table 10.3 lists the compositions and melting temperatures of commonly used filler metals for CAB aluminum brazing.

Table 10.3 Composition of common filler metals (Brazing Handbook, 2007)

		Alloying element (wt%)						Melting temperature (°C)	
Alloy	AWS classification	Si	Cu	Mn	Mg	Zn	Al min%	Solidus	Liquidus
AA4343	BAlSi-2	6.8–8.2	0.25	0.1		0.2	Bal.	577	617
AA4145	BAlSi-3	9.3–10.7	3.3–4.7	0.15	0.15	0.2	Bal.	521	585
AA4047	BAlSi-4	11.0–13.0	0.3		0.1	0.2	Bal.	577	582
AA4045	BAlSi-5	9.0–11.0	0.3	0.05	0.05	0.2	Bal.	577	599

Figure 10.4 illustrates the aluminum end of the binary Al–Si phase diagram (Massalski, 1986). In the dual-phase liquid plus solid region between the solidus and liquidus temperatures, the liquid fraction of molten filler metal at a certain brazing temperature (T_{peak}) and with a known Si content (C_0) can be calculated. The equation (10.1) for calculating the liquid fraction of a partially molten filler alloy is the so-called 'lever rule'. This is based on the mass balance of the Si amounts before and after melting. C_0 is the alloy initial Si concentration; C_S and

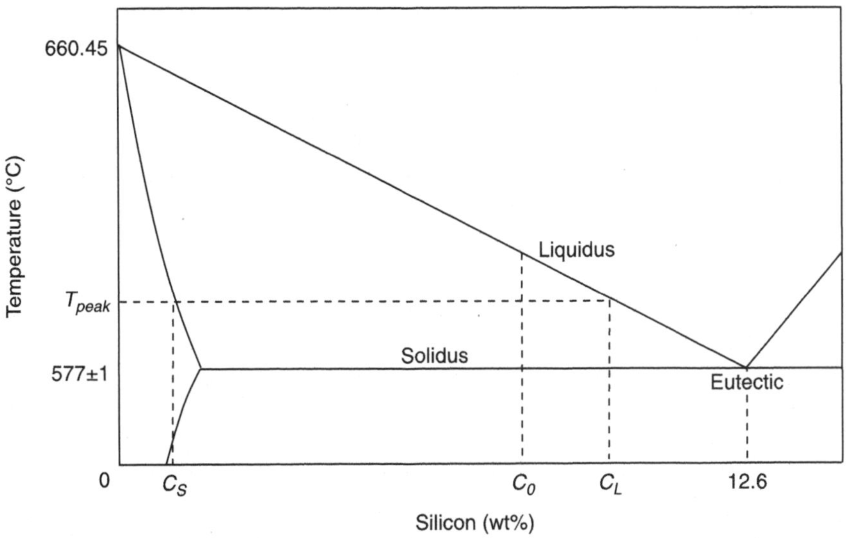

10.4 Phase diagram for Al–Si brazing alloys (Massalski, 1986).

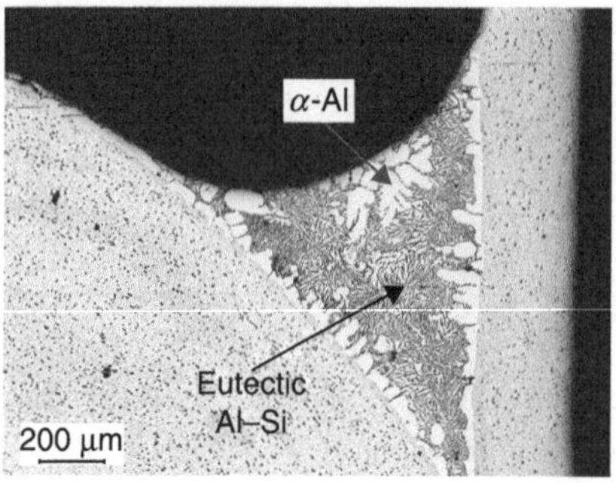

(a)

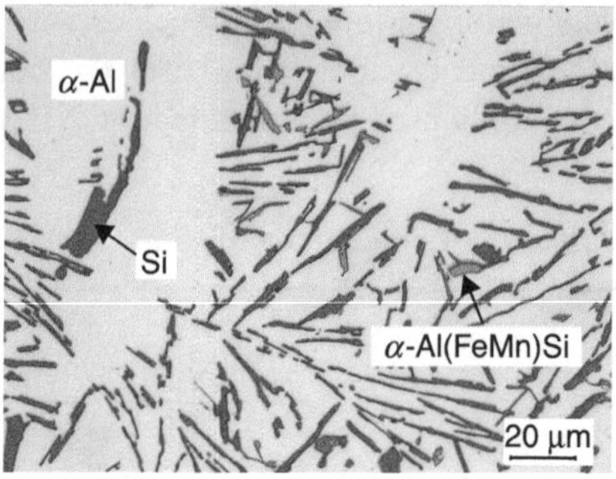

(b)

10.5 Microstructure of resolidified filler metal after brazing: (a) macro view of a brazed joint; (b) detailed microstructure at higher magnification (courtesy of Sapa Technology).

C_L are Si concentrations of the solid and liquid portions of the partially molten alloy at the brazing temperature T_{peak}, respectively.

$$f_L = (C_0 - C_S)/(C_L - C_S) \quad [10.1]$$

Figure 10.5 illustrates the macro- and microstructures of resolidified Al–Si filler metal after brazing. In this example, the eutectic Al–Si mixture is the major component of the brazed joint, indicating that the filler metal was close to the eutectic Al–Si alloy composition. Figure 10.5(b) shows a detailed view of the eutectic structure at a higher magnification. The white phase is the Al solid solution matrix. Small quantities of the intermetallic phase (α-Al(FeMn)Si) can be seen as light gray particles. The Fe- and Mn-bearing intermetallics (from substrate dissolution) are formed during filler metal solidification. Some studies have provided detailed analyses of the structure and composition of these intermetallics (Dehmas *et al.*, 2006; Lacaze *et al.*, 2005). The dark colored Si particles are distributed in the matrix in the form of thin flakes (Gray *et al.*, 2006). Figure 10.6 illustrates a '3-D' view of the eutectic area in a brazed joint between fin and tube.

10.3.3 Brazing sheet

The development of a brazing sheet in the form of a rolled composite, which is a sandwich of two or more layers, as illustrated in Fig. 10.7, provides a convenient and effective method of supplying filler metal to joints hidden deep within the heat exchanger assemblies (Brazing Handbook, 2007). The Al–Si filler metal (clad) is roll bonded to the core, first by hot rolling followed by cold rolling to gauge, usually with some in-process annealing. An intermediate cladding layer, usually of AA3003 or a similar alloy, is sometimes incorporated in roll-clad sheet to improve brazability and corrosion resistance.

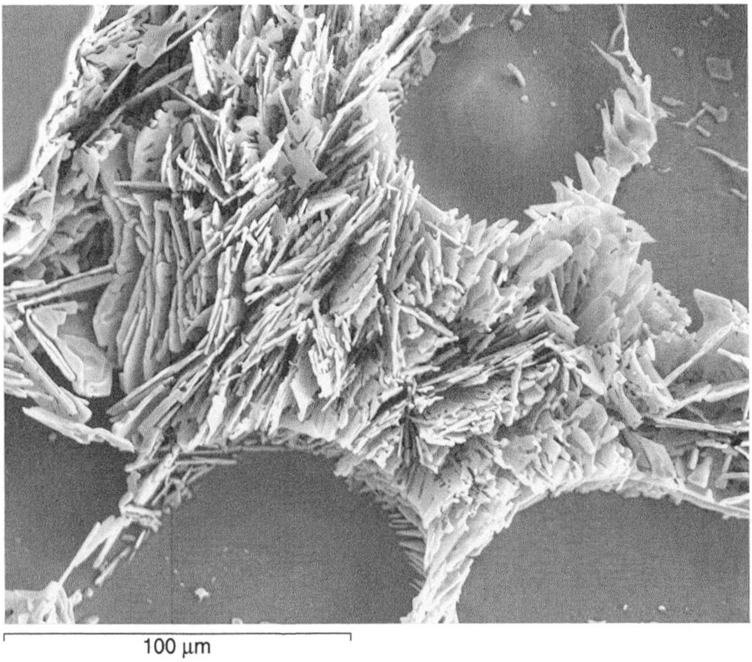

10.6 Scanning electron microscopy (SEM) image of the eutectic area of a brazed joint (Gray *et al.*, 2006) (courtesy of Innoval Technology).

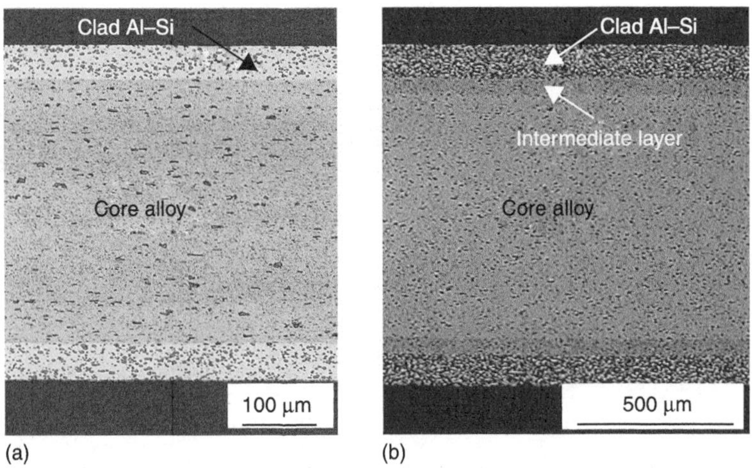

10.7 Etched aluminum brazing sheet (a) double-sided clad; (b) brazing sheet with an intermediate layer between the clad and core (courtesy of Sapa Technology).

The most extensive use of brazing sheet in heat exchanger manufacturing is for fin stock to be brazed with extruded tubes. Other applications include seam welded tubes, folded sheet tubes, and separator and cup plates for evaporators. Commonly used clad alloys include the hypoeutectic AA4045 and AA4343 alloys. When the application requires good flowability of the molten clad, AA4047 filler, with a composition close to the eutectic Al–Si binary alloy, can be used as the clad layer because of the narrow melting range of this alloy.

During brazing, there is solid-state diffusion of Si from the filler alloy into the adjacent core regions due to the high diffusion coefficient at elevated temperatures (Gao *et al.*, 2002). Figure 10.8 includes optical microscope images that illustrate the state of the interface between the core and cladding before and after melting begins, i.e. the samples were heated to 570°C and 620°C, respectively, with a dwell time of 5 minutes, then quenched to room temperature. Si depletion in the clad leads to an increase in the liquidus temperature of the cladding and an increased amount of unmelted filler metal remaining on the surface after brazing (Gao *et al.*, 2002; Terrill, 1966). This clad residue is predominantly the α-phase (Al–Si) solid solution, while silicon-rich liquid flows away to the joint location during brazing (Gao *et al.*, 2002). Recently, DSC (differential scanning calorimetry) analysis has also been used to monitor the effect of Si diffusion from the clad into the core (Turriff *et al.*, 2010).

The mass of clad flowing to form the joint can be influenced by many factors, such as the initial clad thickness, Si concentration, heating conditions, etc. For example, a slow heating rate, especially through a high temperature range, can prolong the Si diffusion process and lead to more depletion of Si from the clad layer (Gao *et al.*, 2002). The amount of residue increases when brazing sheet with a thicker clad is used for brazing (Zhao and Sekulic, 2006). Experiments with aluminum brazing sheet in a vacuum environment showed that increasing the Si

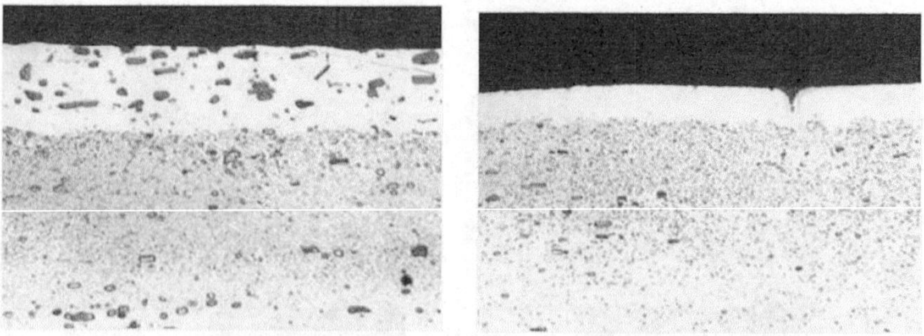

(a) (b)

10.8 Brazed sheet with Si diffusion from the clad into the core: (a) peak temperature = 570°C, dwell time = 5 min; (b) peak temperature = 620°C, dwell time = 5 min (Gao *et al.*, 2002) (courtesy of University of Kentucky, Center for Manufacturing, Brazing Laboratory).

content from 5% to 9% can increase the percentage of clad flowing to form the joint (Sontgerath *et al.*, 1996). The conventional method of predicting the available liquid filler metal flow during brazing using the lever rule (Eq. 10.1) often leads to an underestimation of the amount of residual filler (Terrill, 1966; Woods and Robinson, 1974). A more comprehensive model has been developed based on Fick's diffusion laws and non-equilibrium theory for predicting the amount of clad residue remaining after brazing (Zhao and Sekulic, 2006). There are also experimental studies that show how Si particle size can influence the flow percentage of the filler metal (Sontgerath *et al.*, 1996; Gray *et al.*, 1999). It was found that the available liquid flow for joint formation increases as the Si particle size decreases. It was suggested that a decrease in the Si particle size provides a larger total interfacial area between the Si particles and the Al matrix. Consequently, faster, more homogeneous melting of the clad and better flow would occur (Sontgerath *et al.*, 1996). However, it was also found that in CAB the effect of Si particle size is significant only when the flux load is reduced to 2 g/m^2 (Gray *et al.*, 1999) or the brazing temperature is relatively low (Doko *et al.*, 2005).

The microstructure of a multi-clad brazing sheet has been illustrated previously in Fig. 10.7(b). An interlayer such as Al or Al alloy (e.g. AA3003) can be sandwiched in between the clad and core to serve as a diffusion barrier (Kuppan, 2000). This intermediate layer not only limits the Si diffusion from the clad into the core alloy but also limits an alloying element such as Mg in the core alloy from diffusing into the clad, thereby providing better CAB brazability and corrosion resistance of the brazing sheet.

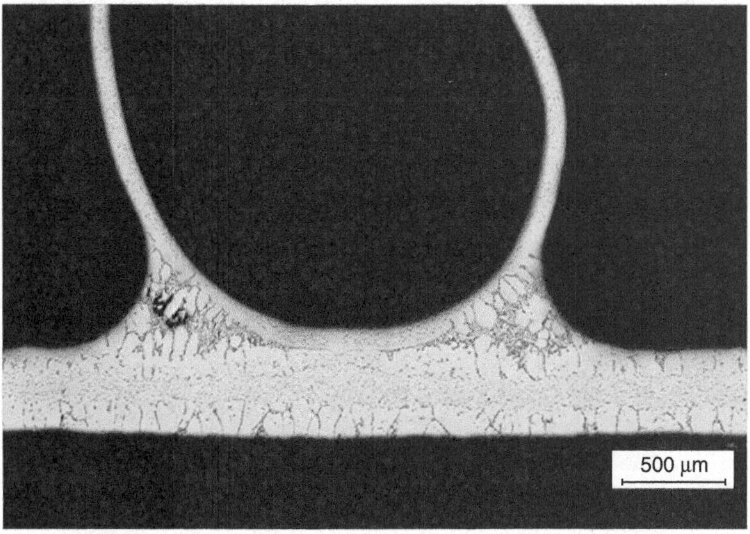

10.9 Cross-sections of a brazed fin/tube with liquid film migration (courtesy of Sapa Technology).

When clad sheet is used for brazing, penetration of liquid from the molten filler deep into the base metal is sometimes observed, as illustrated in Fig. 10.9. Such a phenomenon has been identified as liquid film migration (LFM) (Woods, 1997) and is frequently observed in O-temper materials that have been lightly worked before brazing. The phenomenon has been identified as being similar to strain induced grain boundary migration (Wittebrood, 2009). The post-braze microstructural changes due to LFM lead to reductions in brazability and mechanical strength of the brazed component. In addition, the LFM affected zone contains grain boundaries that are enriched with intermetallics, which can reduce the corrosion resistance of the heat exchanger parts (Woods, 1997). Detailed studies of the mechanisms of liquid film penetration in the CAB aluminum brazing process have been reported (Yang and Woods, 1997; Wittebrood *et al.*, 2000; Nylen *et al.*, 2000; Wittebrood, 2009).

10.4 Oxide and flux

The surface of aluminum is characterized by a tenacious oxide which re-forms rapidly after removal if the metal surface is again exposed to air or water (Jordan and Milner, 1956–7; Hatch, 1984; Shapiro, 2010). One study has shown that the thickness of the aluminum oxide film increases rapidly and reaches a value around 1 nm in the order of minutes when a fresh aluminum surface is exposed to dry oxygen (Hunter and Fowle, 1956). To illustrate the influence of the aluminum oxide on the alloy melting process, a thin disc of AA4343 filler metal (surface polished with a progressively finer series of SiC papers and polishing cloths) was heated in a hot stage microscope. An optical microscope was used to observe the changes in morphology of the alloy surface. Figure 10.10(a) shows the filler metal surface at temperatures below the Al–Si alloy solidus temperature (577°C). It appears that, upon melting of the bulk of the Al–Si alloy, a solid layer of thin, transparent oxide blankets the surface. When the sample was heated above the Al–Si alloy solidus melting temperature (see Fig. 10.10(b) and (c)), significant

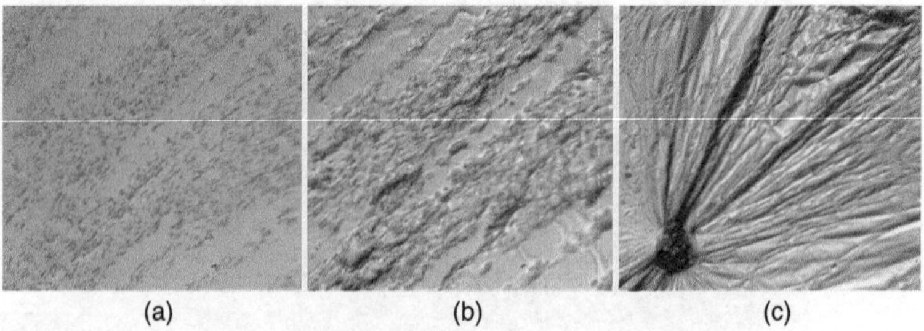

10.10 Frames (×200) showing Al–Si filler metal heated in an N_2 atmosphere (courtesy of University of Kentucky, Center for Manufacturing, Brazing Laboratory).

mobility of local spots of the blanket was observed. The observed phenomena indicate that the alloy is partially melted and the liquid phase moves underneath the oxide film, but cannot flow freely due to the restriction imposed by the oxide film on the alloy surface.

When brazing two aluminum parts, the native oxide film must be removed before melting of the filler metal to allow free flow of the molten filler into the areas to be joined and to ensure good wetting of the base metal surfaces. In the CAB process, the task of removing the oxide layer is accomplished by deposition of a fluoride based flux (potassium fluoroaluminate compounds) on the aluminum surfaces (Swidersky, 2001). The following section reviews the mechanism of oxide disruption and removal by flux at elevated temperatures and new developments in flux modification and application methods.

10.4.1 The CAB flux

Traditional brazing of aluminum utilizes chloride fluxes to remove the oxide film from the aluminum surfaces. Based on the fact that alumina (Al_2O_3) is not soluble in a flux containing mainly chlorides, such as the ones used in dip brazing, a mechanical removal theory was suggested, i.e. the molten flux penetrates along the oxide/substrate interface through cracks/defects in the oxide film (e.g. due to heating), eventually leading to the detachment of the oxide from the metal surface (Jordan and Milner, 1956–7; Terrill *et al.*, 1971). When these fluxes are used, the post-brazed parts must be thoroughly cleaned to remove the hygroscopic, corrosive flux residue. Strict environmental and safety control procedures are demanded since the brazing bath contains highly corrosive acidic chemicals.

The state-of-the-art CAB process uses a non-hygroscopic, potassium fluoroaluminate flux, which was developed during the late 1970s (Garcia *et al.*, 2010). Brazing using these fluxes is called NOCOLOK® (registered trademark of Solvay Fluor GmbH, Hannover, Germany) brazing technology (Cooke *et al.*, 1978; Garcia *et al.*, 2010). The molten flux has excellent wettability on aluminum surfaces and removes the oxide layer on both the filler and substrate surfaces and also prevents further oxidation. After brazing, the resolidified flux remains on the alloy surface as a thin film, which has very low solubility in water (Garcia and Swidersky, 2011). This glassy, residual film can stay on the alloy surface as it has very low if any corrosion effects and may provide some protection against other corrosive agents (Schwartz, 2003), although repeated thermal expansion and cooling of the heat exchanger may cause the flux to fracture and spall off. The oxide film reforms on both the base metal and joint surfaces after brazing (Swidersky, 2012).

The mechanism of oxide removal by NOCOLOK fluxes during aluminum brazing is generally considered as a dissolution process of the oxide in the molten flux, since alumina is soluble in the fluoride mixture (Zhang and Zhuang, 2008; Robert *et al.*, 1997a). Early experimental work using hot stage microscopy to study the mechanism of aluminum surface oxide removal by NOCOLOK flux suggested that the fluoride flux functions by dissolving the aluminum oxide (Field and Steward, 1987). However, recent studies using a laboratory glass tube furnace equipped with a slow-motion, high-resolution video camera suggest a similar mechanical flux removal mechanism for NOCOLOK flux as was found with chloride-based fluxes, i.e. liquid flux penetrates through the cracks in the oxide film, which are generated due to heating and the significant thermal expansion of the base aluminum. Subsequently, the entire oxide layer is lifted up and detached from the base metal. The majority of the original oxide floats on the liquid flux, while only a small amount of the oxide is dissolved by the flux (Swidersky, 2012).

Under normal conditions, NOCOLOK flux is a stable white powder. At brazing temperatures, it is a mixture of two KF–AlF_3 compounds, i.e. K_3AlF_6 and $KAlF_4$. An updated KF–AlF_3 system phase diagram was reported (Chen *et al.*, 2000) (Fig. 10.11(a)). The area expanded in Fig. 10.11(b) shows details of the phase transformation occurring in KF–AlF_3 compounds when the flux is heated to the aluminum brazing temperature. For successful brazing, it is necessary to have the flux melting slightly before the melting of the filler metal. As is shown in Fig. 10.11(b), the eutectic composition at point E_2 gives the lowest flux melting temperature at around 558°C. This requires a 44.5% mole fraction of AlF_3 in the KF–AlF_3 compound. The effectiveness of the flux is dependent upon precise control of the ratio between the two compounds $KAlF_4$ and K_3AlF_6 to form the eutectic KF–AlF_3 composition (Garcia *et al.*, 2010). The phase diagram also shows the compound compositions that have slightly higher melting temperatures but are still below the filler metal melting temperature. For example, as a pure phase, $KAlF_4$ has a melting point of 575°C, while another eutectic composition at point E_3 (51% mole fraction of AlF_3) has a melting point of 572°C (Chen *et al.*, 2000; Zhang and Zhuang, 2008).

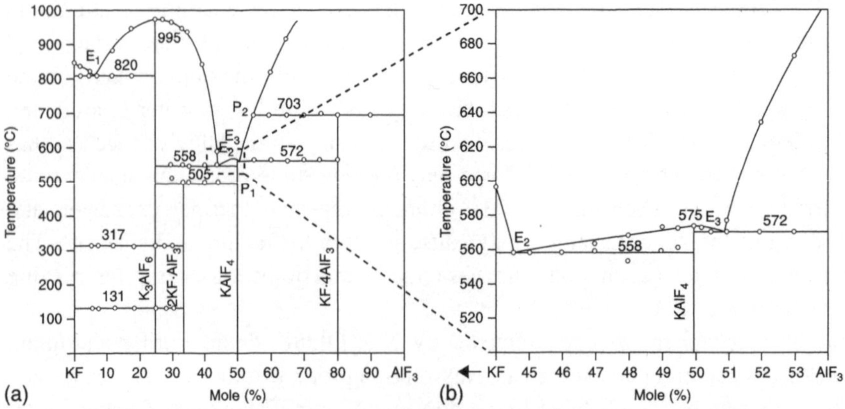

10.11 Phase diagram of the KF–AlF_3 system (Chen *et al.*, 2000).

Small quantities of HF always exist in the exhaust gases venting from CAB furnaces. One mechanism suggested for HF formation is that above 560°C $KAlF_4$ begins to vaporize (Garcia *et al.*, 2010) due to its higher vapor pressure than the other KF/AlF_3 compounds (Robert *et al.*, 1997b). Subsequently, $KAlF_4$ reacts with atmospheric moisture and generates the toxic gas HF (Garcia *et al.*, 2010). One of the suggested chemical reactions is as follows:

$$2KAlF_4 + 3H_2O \rightarrow 2KF + Al_2O_3 + 6HF$$

Such a reaction subsequently expedites the evaporation of $KAlF_4$ and leads to a shift in the flux composition away from its eutectic point (E_2 in Fig. 10.11). At too high a moisture level, the flux often develops a 'dried out' appearance and may lose its effectiveness in removing oxide (Solvay Fluor document, 2012a). Therefore, good control of the inert atmosphere (oxygen level below 100 ppm and a dew point below −40°C) in the furnace is extremely important for maintaining optimal brazing conditions. The proper treatment of the furnace exhaust gases is also necessary to prevent HF contamination of the environment (dry scrubbers filled with activated alumina: $Al(OH)_3 + 3HF \rightarrow AlF_3 + 3H_2O$).

Flux application on assembled aluminum heat exchangers is often accomplished by spraying the parts with an aqueous suspension of the flux powder, the so-called wet fluxing method. Fluxed parts must be thoroughly dried before entering the brazing furnace (Garcia *et al.*, 2010). As an alternative, dry fluxing methods such as electrostatic fluxing can be used, thereby eliminating the need for handling the flux slurry (Garcia *et al.*, 2010). Other application methods have been developed over the last few years to pre-flux aluminum parts at an earlier stage of fabrication before the heat exchanger is assembled. For example, flux is mixed with an adhesive organic carrier to be applied directly onto the heat exchanger sheet or components. There are also flux coated and flux cored Al–Si filler wires developed mainly for flame brazing. In situations where brazing components are not clad with a filler metal, a brazing paste that combines Al–Si alloy powder, flux and an organic carrier can be applied at the joint locations. This is especially effective in hard to reach or shielded locations in units which have a complex geometry. A composite of flux and an Al–Si alloy has been made in the form of rods and rings, which obviates the need to apply flux during brazing. However, the filler rod is extremely brittle due to its composite nature and has very limited use in practical applications.

10.4.2 Flux loading and Mg in aluminum alloys

Studies have shown that the initial oxide formed on aluminum surfaces is amorphous, i.e. it is an aggregate of non-ordered molecules (Kubaschewski and Hopkins, 1953; Shapiro, 2010). The oxide films are of the order of 5 nm at room temperature (Jordan and Milner, 1956–7). Exposure to elevated temperatures leads to increasing thicknesses of the oxide layer and promotes a transformation to a crystalline structure (γ-Al_2O_3) (Kubaschewski and Hopkins, 1953; Van Beek and Mittemeijer, 1984). For example, annealing brazing sheet at 500°C can generate an oxide film four times thicker than exists at room temperature (Gray *et al.*, 2011). Increasing humidity in the environment also leads to significant increases in oxide film thickness. For example, at room temperature, increasing the humidity from 50% to 100% results in a 50% thickening of the oxide film on AA3003 alloy in 9 days (Zähr *et al.*, 2010). Another study found that, over a period of 5 years, the oxide film on a pure Al surface exposed to a 100% humidity environment can be eight times thicker than after exposure in a 52% humidity environment (Godard, 1967). It has been suggested that at least two layers of oxide exist on the aluminum surface: (1) a compact barrier type oxide adjacent to the metal; and (2) a permeable outer layer, which is created by reaction between the compact oxides and environmental elements, such as moisture (Hunter and Fowle, 1956). During fabrication processes, brazing sheet materials are exposed to a series of thermal treatments that can also lead to much thicker oxides on the final products (Gray *et al.*, 2011).

From a practical standpoint, increasing the oxide thickness generally leads to greater difficulties in brazing. For production CAB, the empirical optimum value for flux loading on the surface of a heat exchanger is 5 g/m^2 (Solvay Fluor document, 2012a). Some studies have been conducted to correlate the oxide thickness and the flux loading necessary to achieve optimum brazing. It was found that the negative effect of the oxide thickness on brazability is significant when the flux loading is relatively low, i.e. below 5 g/m^2. Increasing the flux loading usually overcomes difficulties in brazing caused by a relatively thick oxide layer. For example, at a flux loading level of 2 g/m^2, an oxide thickness of over 10 nm will significantly reduce the clad flowability when compared with an oxide thickness of around 4 nm. When the flux loading was increased to 5 g/m^2 and the same series of samples used, only minor differences in clad flowability were identified (Gray *et al.*, 1999).

In addition to the oxide thickness, alloy composition also influences brazability in the CAB process. Mg is often added to Al alloys for the purpose of increasing mechanical strength. However, in the CAB aluminum brazing process, the addition of Mg to the core alloys can reduce the brazability of the system (Garcia *et al.*, 2010). When a Mg-bearing alloy is heated during the brazing cycle, Mg diffuses from the material core to the surface due to the concentration gradient generated by Mg evaporation and reactions with surface oxide and flux. Such reactions lead to the formation of complicated oxidation products that are difficult to remove. A study on oxide formation of aluminum alloys containing 2.4–8% Mg in dry air has been performed (Brouckere, 1945) and suggests that the oxide film composition mainly depends on temperature, i.e. MgO was formed as an upper layer to the Al_2O_3 at temperatures above 350°C (Kubaschewski and Hopkins, 1953). A recent real-time FTIR (Fourier Transform Infrared Spectroscopy) study of oxide formation on a Mg-bearing Al alloy during a brazing cycle also indicates that Mg transforms Al_2O_3 to the $MgAl_2O_4$ spinel and eventually forms MgO oxide

(Gray *et al.*, 2011). Mg also reacts with the flux, forming a K–Mg–F compound (Garcia *et al.*, 2010). The change of flux composition leads to a significant increase in the flux melting temperature and a subsequent loss of effectiveness in oxide removal from the Al surface. When an aluminum alloy contains a relatively small quantity of Mg (e.g. less than 0.3%), an increase in the level of flux loading may help to improve the alloy brazability (Gray *et al.*, 1999). However, with higher Mg-containing alloys, various brazing conditions must be adjusted, such as changing to faster heating rates, using significantly higher flux loadings and better control of the inert gas atmosphere in order to obtain acceptable brazing results. The slow heating rate allows excessive Mg diffusion to the surface, resulting in too much reaction between the Mg and the flux (Johansson *et al.*, 2003).

A study shows that a NOCOLOK flux with a 2% Cs content is sufficient to obtain satisfactory brazing results when an Al alloy with a relatively high Mg (0.5%) content is CAB brazed (Garcia *et al.*, 2001). Flux formulations recently developed with the addition of cesium compounds (e.g. Cs–Al–F) have proven to be effective in improving the brazability of the higher Mg-bearing alloys and are in use for commercial brazing. Figure 10.12 shows an example in which AA6061 (0.8–1.2% Mg) and AA3003/AA4045 (Mg free) brazing sheet were brazed in a CAB furnace. Standard NOCOLOK flux and NOCOLOK flux plus some cesium were applied to two samples, respectively. It was found that when regular flux was applied (Fig. 10.12(a)), the filler metal did not wet the Mg-bearing AA6061 base metal and no joint was formed. When Cs-bearing flux was applied, the joint formation was greatly improved (see Fig. 10.12(b)).

An alternative approach for high Mg applications where alloy strength is critical utilizes a multi-clad brazing sheet (see Fig. 10.7(b)). This product is expensive to produce but can be an effective device for handling the higher

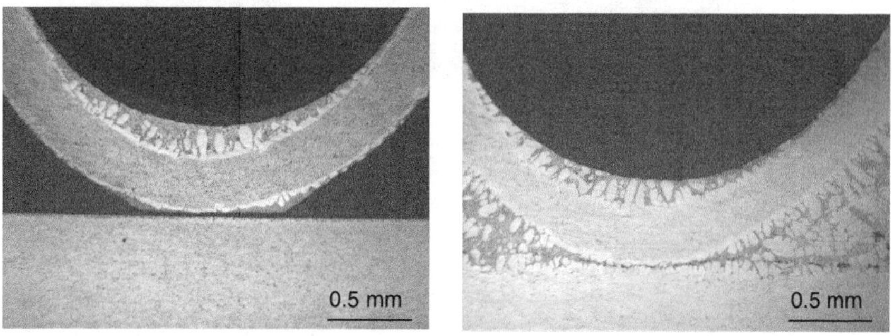

(a) (b)

10.12 Cross-sectional images of brazing results between AA6061 sheet and AA3003/AA4045 brazing sheet: (a) no joint formed using regular NOCOLOK flux; (b) brazed joint using Cs-NOCOLOK flux (courtesy of Creative Thermal Solutions, Inc.).

Mg-bearing core alloys during the CAB process. As discussed in Section 10.3.3, the interlayer serves as a diffusion barrier for reducing Mg diffusion into the clad.

10.4.3 CAB flux modification

Many studies have been devoted to flux modification to extend the application of the CAB process for Al brazing (Chen *et al.*, 1997; Garcia *et al.*, 2001). The use of cesium compounds has been mentioned above. A published phase diagram for the CsF–AlF_3 system shows that a eutectic composition that contains 42/58 mol% AlF_3/CsF can have a melting temperature as low as 471°C (Chen *et al.*, 1997) and can be very effective for brazing Mg-bearing alloys (Zhang and Zhuang, 2008). A low melting point Cs–F–Al flux is commercially available for brazing aluminum using a low melting point filler metal such as an Al–Zn alloy (www.aluminum-brazing.com). In CAB aluminum brazing, a NOCOLOK/cesium flux (Cs–K–Al–F complexes with a melting range of 545–570°C) was developed to improve brazability of Mg-containing alloys using Al–Si alloys as filler metals. It was suggested that cesium acts as a buffer for magnesium by reacting with the Mg and forming Mg–Cs–F compounds, thus suppressing the reaction between the flux and Mg and maintaining the flux activity (Garcia *et al.*, 2010). Cesium-containing fluxes are relatively expensive due to the high cost of Cs.

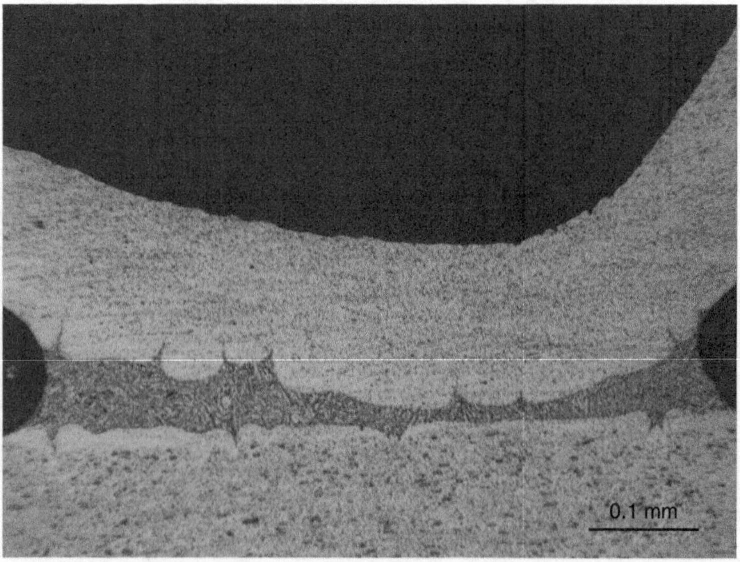

10.13 CAB brazed joint with NOCOLOK Sil-Flux (courtesy of Creative Thermal Solutions, Inc.).

The Sil-Flux is basically a mixture of NOCOLOK flux and pure Si powder. When Sil-Flux is applied at the joint locations, there is no need to supply additional filler metal. During brazing Si diffuses into the aluminum substrate and works as a melting point depressant by alloying with the Al matrix. The direct consequence of such an interaction is the formation of an Al–Si surface layer *in situ*, which melts at temperatures above 577°C (the eutectic temperature of Al–Si alloys) and flows into the joints by capillary action (Timsit and Janeway, 1993). Figure 10.13 illustrates a cross-sectional image of a brazed joint between an AA3003 substrate and fin using NOCOLOK Sil-Flux. The main application of NOCOLOK Sil-Flux in heat exchanger manufacturing is for bonding between bare fin and tube surfaces.

A NOCOLOK Zn flux has been developed to improve the corrosion performance of brazed aluminum products. It is a single-component flux of the compound $KZnF_3$. When the NOCOLOK Zn flux is applied to the Al alloy surface, the following reaction takes place in the solid state during the CAB process (Solvay Fluor document, 2012a):

$$6\ KZnF_3 + 4Al \rightarrow 6Zn + 3KAlF_4 + K_3AlF_6$$

The reaction products include potassium fluoroaluminate, which is effective in removing the aluminum oxide, and Zn, which has a high solubility in the Al matrix. The Zn diffuses into the aluminum alloy substrate surface and forms a Zn-enriched diffusion zone, which serves as a lateral sacrificial layer due to its lower galvanic corrosion potential. It was found that during brazing, a high Zn flux loading and short dwell time at the brazing temperature can help maintain a high concentration of Zn close to the alloy surface. This thin layer provides good sacrificial corrosion protection to the core (Orman *et al.*, 2008). A study has also been reported using a coating mixture of Al–12Si, Zn and flux ($KAlF_4$) powders deposited on the heat exchanger tube surfaces by the kinetic spray process. This provides a simpler pre-brazing preparation process (Yoon *et al.*, 2007).

Another modified flux product is NOCOLOK Li Flux, which is a mixture of NOCOLOK flux and lithium cryolite (Li_3AlF_6). It is claimed that this flux is as effective in the CAB process as the standard NOCOLOK flux and that the brazed aluminum surface (with the flux residue) exhibits reduced interaction with coolant fluids and enhanced corrosion resistance (Solvay Fluor document, 2012b).

10.5 Controlled atmosphere brazing (CAB) process

During CAB of compact Al heat exchangers, flow of the molten filler metal and joint formation can be influenced by many factors, such as heating and cooling cycles, furnace atmosphere, alloy compositions, surface morphologies of substrate materials, and flux loading, as well as the design of assembly fixtures. Mechanical properties of brazed parts are also significantly influenced by the brazing process conditions. This section presents some discussion of the fundamental understanding of wetting and flow behavior of molten filler metals during the CAB process, as well as a series of phenomena involving filler/base metal interactions. The influence of CAB processing conditions on material mechanical properties will also be reviewed.

10.5.1 Wetting and spreading

When brazing sheet is used in a CAB process, upon melting of the thin clad layer, capillary forces cause liquid filler metal to flow into the clearance between the parts to be bonded. The molten clad wets and interacts with the substrate aluminum alloys, forming a metallurgical bond upon solidification. As described in the previous section, in the CAB process, the flux layer will melt slightly before the filler alloy does. The molten flux removes the oxide layer from the surface and prevents re-oxidation, subsequently improving the flowability and wettability of molten filler metal (Solvay Fluor document, 2012c). The wetting and spreading of molten clad on the Al substrate surface may be affected by the presence of a liquid flux film, as well as other factors such as interactions between the liquid and solid at the interface, binary alloy compositions, etc. A sessile drop wetting/spreading test is one of the commonly used experimental techniques for studying filler metal wettability on an Al alloy surface. The classic Young's equation (Eq. 10.2) describes the relation between surface tension and equilibrium contact angle in a static situation (De Gennes, 1985), as illustrated in Fig. 10.14. It is considered to be 'mostly wetting' when the equilibrium contact angle $\theta_E \leq \pi/2$, and 'mostly non-wetting' when $\theta_E > \pi/2$ (De Gennes *et al.*, 2004). The fundamental equation is:

$$\sigma_{SV} = \sigma_{SL} + \sigma_{LV} \cdot \cos\theta_E \qquad [10.2]$$

Based on Eq. 10.2, a decrease in surface tension at the liquid/vapor interface or at the liquid/solid interface leads to a decrease in the contact angle and result in better wetting. Young's equation is based on inert wetting systems in which no interaction, such as intermetallic formation and/or substrate dissolution/erosion, occurs at the liquid and solid interfaces.

For Al brazing, the actual surface tension of the molten Al alloy is difficult to quantify, because of, among other factors, the presence of the thin but omnipresent

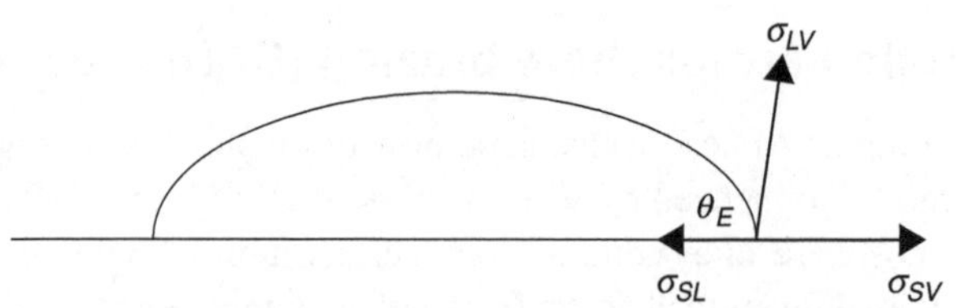

10.14 Equilibrium contact angle between a liquid and a solid.

oxide film or the liquid flux film on the surface (Hatch, 1984). Alloying additions can have different effects on the surface tension of liquid aluminum. Experiments performed in an argon atmosphere over the temperature range 700–740°C indicate that additions of Si and Cu to Al have little effect on the liquid metal surface tension. However, increasing amounts of alloying addition such as Bi, Ca, Li, Mg, Pb, Sb and Sn can lead to substantial reductions in surface tension (Hatch, 1984). Continuous efforts in alloy composition modification have been involved in developing filler alloys for fluxless brazing of aluminum under a controlled atmosphere (Schwartz, 2003; Garcia *et al.*, 2010). A detailed discussion of fluxless brazing of Al can be found in Chapter 19. Also, the surface tension of liquid Al alloys usually decreases as the temperature increases (Hatch, 1984), which is one of the reasons why increasing the brazing temperature often leads to better wetting by the filler metal on a substrate surface.

The use of hot stage microscopy to perform sessile drop wetting tests makes it possible to learn more about the dynamics of molten filler metal spreading. For example, experimental studies have captured the dynamics of the wetting process for the spreading of Al–Si filler metal on both AA3003 and AA6061 substrates under CAB conditions (peak temperature above 580°C, N_2 atmosphere). Figure 10.15(a) and (b) illustrate a sequence of images of molten AA4047 filler metal spreading on a Mg-bearing AA6061 substrate in a hot stage test with regular NOCOLOK flux and Cs-bearing flux, respectively. It is apparent that the addition of Cs to the flux significantly improves the wettability of the filler metal on a Mg-bearing alloy surface. Figure 10.16(a) and (b) illustrate respectively the

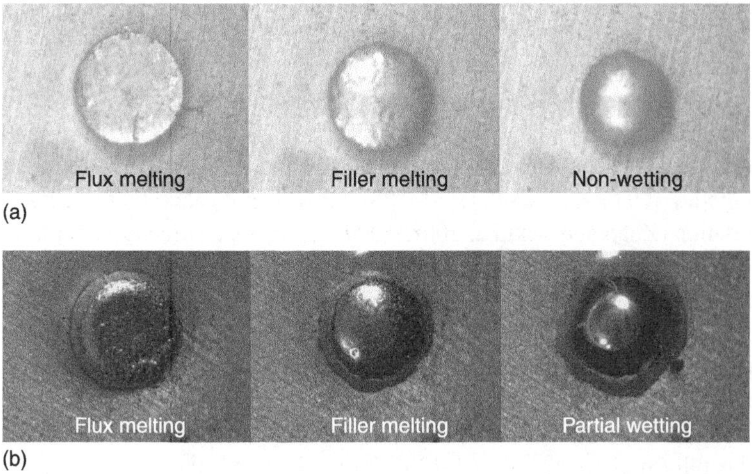

10.15 Sessile drop test of AA4047 filler metal on an AA6061 substrate with: (a) NOCOLOK flux; (b) NOCOLOK Cs flux (courtesy of Creative Thermal Solutions, Inc.).

10.16 Sessile drop test of Al–Si filler metals on an AA3003 substrate: (a) AA4343 filler; (b) AA4047 filler (courtesy of Creative Thermal Solutions, Inc.).

wetting behavior of AA4343 and AA4047 filler on a non-Mg-bearing AA3003 alloy surface.

It is well known that a power law relation $R^n \sim t$, where $n \sim 10$, is often used to describe the dynamics of droplet spreading on a non-reactive substrate surface when surface tension and viscous forces dominate (Tanner, 1978). Detailed reviews of the subject have provided confirmation of an appropriate power law in the form of $R^n \sim t$ with n varying from 2 to 10 for describing different wetting systems (Levinson *et al.*, 1988; Biance *et al.*, 2004; Mortensen *et al.*, 1997). The wettability of Al–Si filler alloys on a relatively smooth AA3003 surface have been studied. It was found that for hypoeutectic AA4343(Al–8Si) filler metal the values of n are close to the value of 10 predicted by Tanner's law. However, the spreading kinetics of a molten AA4047(Al–12Si) alloy do not generally fit such a power law relation (Zhao and Sekulic, 2008, 2009). Spreading of the Al–12Si filler metal is found to be faster and more extensive than for the hypoeutectic Al–8Si filler alloy. It has been suggested that the following factors may contribute to the differences in wetting behavior between the filler alloys: the hypoeutectic AA4343 filler is in a semi-solid state, i.e. a mixture of eutectic Al–Si liquid and solid α-aluminum solution, at the normal brazing temperature of around 600°C (Table 10.3). Therefore, the equivalent viscosity is much higher for the Al–8Si alloy than for the fully molten eutectic Al–12Si liquid. This difference may lead to a slower spreading rate. Flow dynamics are mainly controlled by viscous and capillary forces. The relatively fast spreading of the Al–12Si liquid may also indicate that another factor such as inertia is one of the dominant forces, at least in the initial stages of spreading, which control wetting kinetics (Zhao and Sekulic, 2008, 2009). The good flowability of eutectic Al–12Si filler metal is utilized when brazing parts with complicated geometries. A change of Al substrate surface morphology can also have a significant influence on the wetting behavior of molten filler metal. For example, a study shows that an increase in the Al surface roughness by coarse grinding leads to a more extensive spreading of the liquid Al–12Si alloy on an AA3003 alloy substrate (Zhao and Sekulic, 2009).

The so-called 'angle on coupon' test is very often used to evaluate the wettability of a selected filler metal or clad layer on a substrate. Figure 10.17(a) illustrates the test sample set-up. Basically, a piece of bent metal sheet is positioned on top of a flat metal coupon with the raised legs of the angle resting on a length of stainless steel wire (Bolingbroke *et al.*, 1997). This creates a joint of gradually increasing separation along which the filler metal may flow. The distance flowed is considered as a measure of the 'brazability' (Miller *et al.*, 2000). When such a test is performed using a transparent glass furnace, both the final wetting distance and the speed of flow of the liquid metal under specified brazing conditions can be measured for quantitative evaluation of the filler brazability and flowability. An example of such a test performed using a glass furnace is illustrated in Fig. 10.17(b).

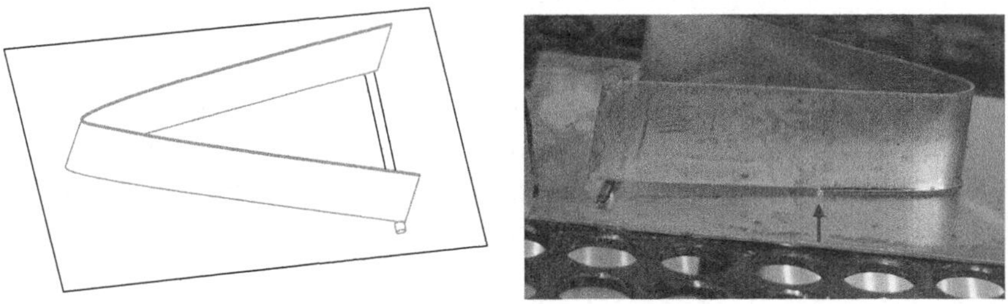

(a) (b)

10.17 In situ monitoring of liquid filler metal filling a joint of increasing clearance: (a) experimental set-up; (b) still image extracted from a real-time video of the brazing process (courtesy of Creative Thermal Solutions, Inc.).

10.5.2 Interaction of the substrate and the filler metal during brazing

In the CAB aluminum brazing process, silicon from the Al–Si filler metal diffuses into the Si-free substrate as discussed in Section 10.3.3. The extent of diffusion will depend upon the time/temperature cycle during ramp-up. When the clad melts, the interaction between the liquid filler and solid substrate becomes much more rapid and can lead to significant substrate consumption. It is considered 'dissolution' when the liquid metal stays in the joint location, i.e. the liquid is more or less 'static'. It is considered 'erosion' when the liquid metal flows and dissolves substrate material leaving 'gutters' as base metal defects where the interactions originally took place. The level of erosion can be influenced by factors such as the liquid metal flow rate, material composition, the temperature and the time available for interaction, etc. In general, dissolution leaves a full but usually large joint, which does not cause any particular problem except when exposed to very corrosive situations. However, severe substrate or tube core alloy erosion can cause problems not only during manufacture (leaking tubes) but later during the field service of the heat exchangers (initiation sites for fatigue failures). In Fig. 10.18, serious base metal consumption due to erosion is illustrated. A fatigue failure has occurred through the tube area thinned by erosion (Solvay Fluor document, 2012c).

Both a high brazing temperature and long dwell time can increase the substrate dissolution level. Studies show that brazing at a high temperature (e.g. above 610°C) has a more profound effect upon the degree of substrate dissolution than a prolonged dwell time at a moderate brazing temperature (e.g. 595°C) (Solvay Fluor document, 2012c). Figures 10.19(a) and (b) show cross-sections of CAB brazed joints produced at brazing temperatures of 600°C and 610°C, respectively. In practice, a certain level of core dissolution and/or light erosion usually exists and the industry often accepts a 10% and even greater reduction in the base metal thickness due to both phenomena (Solvay Fluor document, 2012c).

Besides the brazing conditions, alloy composition is another important factor that will influence the interfacial interactions between liquid filler metal and the solid substrate. A higher concentration of Si in the liquid filler metal will inevitably lead to a higher potential for the substrate dissolution/erosion problems. It has been discussed in a previous section that the eutectic Al–12Si alloy generally features better flowability than the hypoeutectic alloys (e.g. Al–8Si). However, the use of near eutectic Al–Si filler metal should be avoided in thin-wall brazing applications where base metal dissolution/erosion can lead to a catastrophic failure (Solvay Fluor document, 2012c).

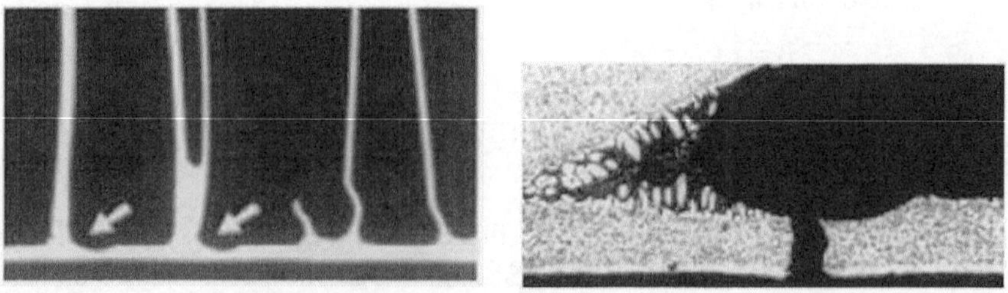

(a) (b)

10.18 Illustration of serious erosion during manufacturing of a heat exchanger: (a) erosion of tube core; (b) mechanical failure due to tube core erosion (Solvay Fluor document, 2012c).

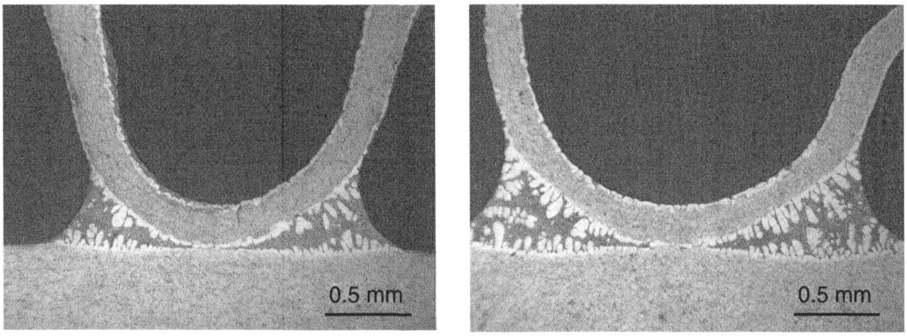

(a) (b)

10.19 Temperature effect on dissolution in a brazed joint: (a) AA3003 sheet brazed to AA3003/AA4343 fin at 600°C peak temperature; (b) AA3003 sheet brazed to AA3003/AA4343 fin at 610°C peak temperature (courtesy of Creative Thermal Solutions, Inc.).

From the equilibrium phase diagram of Al–Si alloys (Fig. 10.4), it can be seen that the lower Si content in a hypoeutectic alloy leads to a wider melting temperature range. At typical brazing temperatures of around 600°C, the hypoeutectic alloys, such as AA4343, are in the form of a liquid/solid mixture (commonly called the 'mushy zone'). A separation of the liquid and solid phases, so-called 'liquation' (Brazing Handbook, 2007; Schwartz, 2003) can happen under certain conditions. Such a phase separation often leaves a small residual conical bump on the substrate surface where liquid filler metal originally existed but has mostly flowed away. Figure 10.20 illustrates the formation of a wedge-T joint between two AA3003 plates using AA4343 as the filler metal using CAB; see Fig. 10.20(a). It is observed that capillary forces draw liquid from the mushy filler. The liquid quickly fills the gap between the two plates and leaves mainly

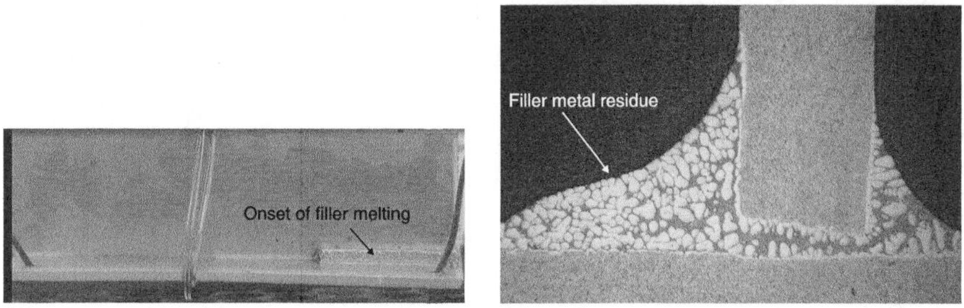

(a) (b)

10.20 Filler metal residue due to liquation during CAB: (a) onset of filler melting – frame from a movie of T joint formation using a transparent CAB furnace; (b) cross-section of a brazed joint with filler metal residue (courtesy of Creative Thermal Solutions, Inc.).

solid α-phase in place of the original alloy rod; see Fig. 10.20(b). The flowing liquid contains higher Si than the original filler metal and readily interacts with the substrate (dissolution/erosion), especially at a relatively high brazing temperature.

In practical applications, a slow heating rate and/or intricate joint design may lead to a portion of the liquid generated at a temperature just above the solidus (577°C) flowing away before the peak brazing temperature is reached. When this occurs, a large portion of the solid solution residue is left on the original surface without having contributed to joint formation and may have to be mechanically removed (Brazing Handbook, 2007).

10.5.3 Control of capillary flow in the CAB process

The manufacture of micro-channel heat exchangers involves, for example, brazing extruded tubes to headers. Should molten filler flow into the micro-channels by capillary action, then the refrigerant passages could become blocked and erosion/dissolution of the thin tube walls might take place, while seriously reducing the amount of filler metal available to form joints.

To illustrate the potential problems, CAB tests joining extruded micro-channel tubes to a header were set up in a glass brazing furnace. The header tube was cut partially open so that the joint formation process inside the header could be observed. An Al–12Si rod was used as the filler metal. A series of images capturing joint formation between the header and a micro-channel tube are presented in Plate IV in the colour section between pages 268 and 269. Upon melting, molten filler metal first fills the gap between the tube and header, as illustrated in Plate IV(a). However, extensive spreading of liquid metal into the micro-channels occurs, in this case after continuous heating for about 20 seconds as the system was approaching a peak temperature around 600°C. Once the molten filler metal reaches the entrance to a micro-channel, liquid rapidly fills the channel by capillary action. Thus, careful control of filler metal amounts, substrate surface morphology, joint geometry and brazing conditions (e.g. peak brazing temperature and dwell time) is essential to prevent issues such as micro-channel blockages.

10.5.4 Mechanical strength improvement during CAB

Some aluminum alloys, such as the Al–Cu and Al–Mg–Si alloys, for example, are heat treatable (see Table 10.1). The strength of these alloys may be enhanced by the formation of small atom clusters or intermetallic microparticles from a supersaturated Al solid solution through proper heat treatment (Callister, 2003). Such a phase transformation process is called precipitation hardening or age hardening (Callister, 2003; Hatch, 1984). When heat treatable alloys are brazed, the brazing cycle may also be utilized as a useful solution heat treatment step (Brazing Handbook, 2007). Aluminum alloys that contain only small additions of Mg (0.2–0.6%), such as AA3005 alloy, are generally considered to be non-heat treatable. However, when these Mg-bearing alloys are used as a core alloy for the brazing sheet, Si diffuses from the clad into the core during the brazing cycle, making the formation of the β-phase intermetallics Mg_2Si possible (Goodrich *et al.*, 2001), which subsequently leads to a significant effect in improving the post-brazed strength of heat exchangers.

Figure 10.21 illustrates how the brazing cycle may be used as a solution heat treatment for aluminum alloys. For an age-hardenable Al alloy, the alloying element has a relatively large solubility in the Al matrix at a high temperature T_1 compared with the solubility at room temperature. During solution heat treatment (e.g. at a brazing temperature of around 600°C), the alloying elements (e.g. Si and Mg) are dissolved in the Al matrix. After brazing a rapid quench to room temperature prevents the formation of coarse intermetallic particles (Callister, 2003; Hatch, 1984). Such rapid cooling is very difficult to accomplish in the CAB process due to the difficulties of access and the size of the brazed components. For example, the empirical optimal cooling rate for the first stage solution heat treatment of selected aluminum alloys is around 1°C/sec (Magnusson *et al.*, 1997; Stenqvist, 2001), which is slightly greater than the cooling rate that can commonly be achieved in a CAB cycle. Studies have been performed to determine the influence of the brazing cooling rate on alloy mechanical properties after aging (Kooij *et al.*, 1999). Selected heat treatable aluminum alloys were subjected to various cooling rates from 20°C/min to 90°C/min. It was found that, for all the alloys investigated, the yield strength increased with a faster cooling rate. Once efficiently quenched, the material is in a non-equilibrium state as a supersaturated solid solution (Callister, 2003). The next stage of age hardening is the precipitation heat treatment. The supersaturated material is heated to a temperature T_2 higher than room temperature (artificial aging) (see Fig. 10.21), where the diffusion rate of the alloying elements increases

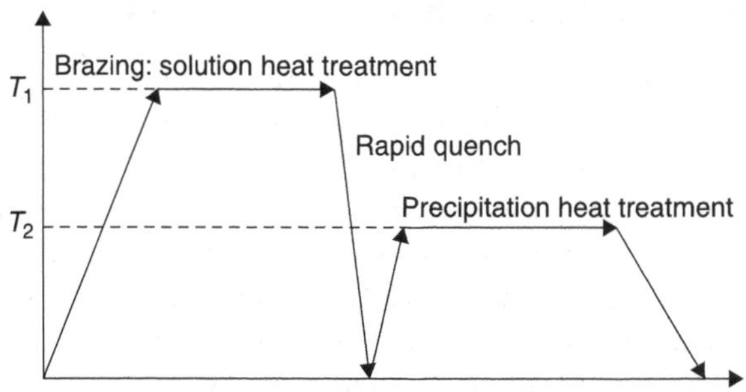

10.21 Heat treatment stages for age hardening of aluminum alloys (Callister, 2003).

significantly (Callister, 2003). For example, in Al–Mg–Si alloys, the β precipitate phase (e.g. Mg_2Si) begins to form as finely dispersed particles during the second stage of age hardening (Holmestad *et al.*, 2010). The AA6000 series alloys can be artificially aged at 160–175°C, holding for 6–20 hours (Hatch, 1984). Artificial aging is an effective strengthening process but one which adds considerable cost to the product.

When heat treatable alloys are age hardened spontaneously at ambient temperature after brazing and quenching, the process is called natural aging (Callister, 2003; Hatch, 1984). A study has shown good improvements in alloy strength by natural aging after brazing for two AA3000 series alloys (Kooij *et al.*, 1999). The compositions of these two alloys were tailored to form $MgSi_2$ and Al–Cu–Mg precipitates, respectively. Prolonged natural aging times of the order of weeks are usually required to achieve useful strengthening effects (Stenqvist, 2001). A great advantage of natural aging is the cost saving, because no additional heating at an intermediate temperature (T_2 in Fig. 10.21) is needed. It was also found that better alloy corrosion resistance can be obtained after natural aging than after artificial aging (Magnusson *et al.*, 1997).

When the heat exchanger is at the vehicle operating temperature, artificial aging can take place during service with suitable materials. For example, radiator tubes are exposed to the engine coolant at temperatures around 80–100°C during service (Stenqvist, 2001). Charge air coolers operate at even higher temperatures. The age hardening effect and corrosion behavior of an Al–Mg–Si radiator tube alloy (after brazing) during 'life cycle' tests at elevated temperatures (80°C, 100°C and 120°C) have been investigated (Kooij *et al.*, 1999). At radiator operating temperatures below 100°C, it was demonstrated that alloy mechanical strength steadily increased after up to 1400 hours in a simulated automobile radiator thermal life cycle test. The maximum strength was achieved at 120°C operating temperature after 700 hours in the life cycle test. No evidence of over-aging was identified up to 1400 hours.

10.6 Corrosion in controlled atmosphere brazing (CAB) brazed heat exchangers

It is well known that for the traditional dip-brazed aluminum parts, flux residues must be thoroughly removed because of the corrosive nature of the chloride flux (Schwartz, 2003). The chloride-based flux is highly hygroscopic and tends to absorb moisture from the environment, thus creating conditions for corrosion on the aluminum surface. When the NOCOLOK flux is used in the CAB process, a post-braze cleaning procedure is normally not required because the flux residue is hydrophobic (Schwartz, 2003). Research has shown that the flux residue does not interact with commonly used coolants, refrigerants, oils and polyalkylene glycol lubricants under normal operating conditions (Swidersky, 2001).

However, the trend towards replacing copper tubing in heat exchangers with aluminum for the HVAC&R industry, and the development of new environmentally friendly refrigerants, demands a better understanding of the corrosion impact on CAB brazed aluminum heat exchangers under different operating and environmental conditions. For example, some additives in the newly developed coolants and refrigerants may interact with flux residues; therefore, flux loading reduction becomes even more important due to the corrosion concerns in addition to the flux cost consideration (Solvay Fluor document, 2011; Garcia and Swidersky, 2011).

10.6.1 Corrosion types in CAB brazed heat exchangers

The stable oxide layer on the Al surface may impose challenges on brazing, but it also serves as a protective film that ensures good corrosion resistance of aluminum alloys under normal operating conditions. The aluminum oxide film forms rapidly when the fresh Al metal surface is exposed to air or water, and the film thickness increases at elevated temperatures or increased humidity in air (Hatch, 1984; Shapiro, 2010). Aluminum is a very active metal; without its oxide, it is less 'noble' than many other frequently used metals such as Cu and Fe. Therefore, when the oxide film on the aluminum surface is broken, the corrosion resistance is relatively low (Zhang and Zhuang, 2008; Hatch, 1984). Many factors can influence the corrosion resistance of aluminum alloys, such as alloy composition, microstructure, pH level and the type of working fluid as well as the environmental conditions (Hatch, 1984). Operating and environmental conditions impose considerable demands upon the corrosion resistance of Al heat exchangers. For example, many automotive heat exchangers, located as they usually are at the front of the vehicle, can be exposed to severely corrosive environmental conditions, such as road salts, marine saline atmospheres, impurities from exhaust condensates and cleaning agents from car washes. In the extreme, corrosion of aluminum heat exchangers leads to system leaks and the loss of coolants, requiring expensive replacements and in some cases resulting in engine damage. In other instances, leaks may not occur but the corrosion results in progressive loss of the heat transfer ability of the heat exchanger unit (e.g. debonding between fin and tube due to fillet corrosion) (Melander and Woods, 2010). Figure 10.22 illustrates a brazed aluminum condenser taken from an automobile after 6.5 years of field service; the CAB brazed joints have been corroded away and the fins have fallen out, with obvious consequences. Similarly, one of the key issues in extending the use of brazed aluminum heat exchangers in HVAC&R systems is to understand the corrosion behavior of the units, which are constantly exposed to various atmospheric and environmental conditions (Ellerbrock, 2011).

In the alloys used for CAB brazed aluminum heat exchangers, different metallurgical phases and different alloying elements are always involved. Corrosion performance depends on the electrochemical potential difference between different alloys in various heat exchanger components and also upon the

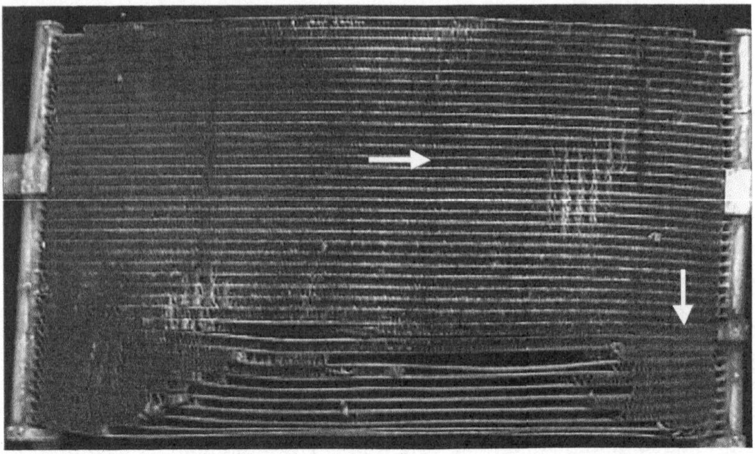

10.22 A brazed condenser after 6.5 years of service (courtesy of Sapa Technology).

micro-potential differences between various intermetallic phases in the alloys themselves (Brazing Handbook, 2007; Hatch, 1984). The corrosion potential of aluminum alloys usually decreases with the addition of alloying elements in the order Mn, Cu, Si, Mg, Zn; i.e. an Al–Mn alloy has much higher corrosion potential than an Al–Zn alloy (Meijers, 2002). Thus, in some heat exchangers Zn is added to the fin alloys to make them sacrificial to the Al–Mn or Al–Mn–Mg alloys used for the tubes and headers. Zn can also be coated onto condenser tube surfaces, where it alloys with the aluminum during brazing and forms an anodic, sacrificial surface layer. Some sacrificial alloys are often roll-clad on the inside surface of heat exchanger tube materials to provide sacrificial protection should the coolant fluids become degraded. But, in general, coolant chemistries have been improved over the years and coolant side corrosion is rarely a problem in reasonably maintained systems.

Many types of corrosion can be identified in brazed aluminum heat exchangers after service, such as: (1) galvanic corrosion; (2) pitting corrosion; (3) intergranular corrosion; (4) erosion corrosion; and (5) less commonly crevice and poultice corrosion, where corrosion is initiated at a crevice due to the formation of local differential aeration cells (Hatch, 1984) or beneath deposits. Figure 10.23 shows examples of the more common pitting and intergranular corrosion in brazed aluminum exchangers after field service.

Phase differences in the resolidified filler material after brazing can lead to galvanic corrosion of the joints under certain conditions (Brazing Handbook, 2007), as illustrated in Fig. 10.24, where Al–Si filler metal was used for joining a header and tube. The brazed joint contains some primary α-aluminum dendrites and eutectic α-aluminum alternating with almost pure Si flakes. In a corrosive environment, the α-aluminum in the eutectic mixture is preferentially corroded

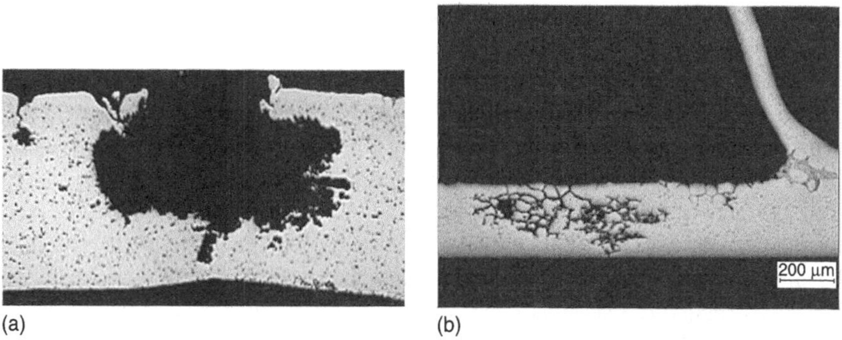

10.23 Different corrosion types identified in heat exchangers after service. (a) Pitting corrosion (courtesy of Kaiser Aluminum); (b) intergranular corrosion (Melander and Woods, 2010, © NACE International, 2010).

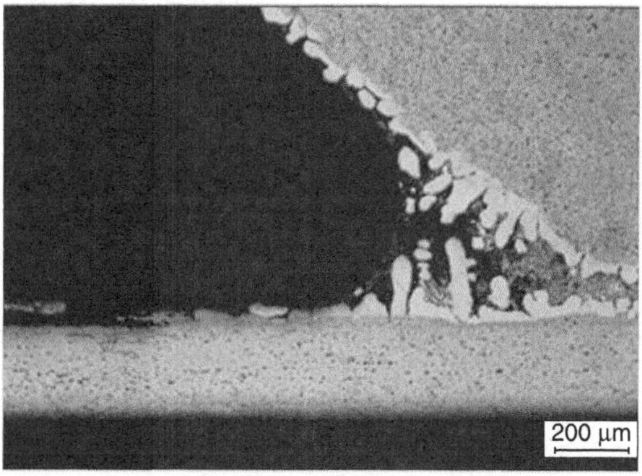

10.24 Micro-galvanic corrosion at a brazed joint (Melander and Woods, 2010, © NACE International, 2010).

because the adjacent Si flakes are highly cathodic. The primary α-aluminum dendrites are largely uncorroded because of their separation from the Si particles. Such micro-galvanic corrosion can lead to the appearance of a porous structure in the brazed joint, and in extreme cases fluid leakage will occur at a header joint, for example.

10.6.2 Corrosion control in CAB brazed heat exchangers

Air side corrosion

The switch to vacuum-brazed aluminum from copper for automobile heat exchangers in the early 1980s necessitated a significant improvement in corrosion performance. Vacuum-brazed aluminum units at that time were perforating in less than a year of 'snow belt' service. The problem was solved by the invention of a subsurface sacrificial corrosion protection mechanism whereby the pitting and intergranular modes, which resulted in rapid component perforation, were replaced by a lateral corrosion mechanism (Woods *et al.*, 1991), i.e. corrosion propagates in directions parallel to the tube surface instead of penetrating deeply into the core materials. This proved to be as effective in CAB brazed heat exchangers as it was in vacuum-brazed components. The family of materials became known as 'long-life' alloys utilizing what was termed the 'brown-band' effect, also called BDP (band of dense precipitation) (Miller *et al.*, 2000), because of the etching characteristics of this subsurface, sacrificial layer. The basic concept of the brown-band effect is the generation of a sacrificial band due to inward diffusion of silicon from the cladding during a brazing cycle and its reaction with Mn in supersaturated solid solution in the core alloy. Fine α-Al(Fe,Mn)Si dispersoids are formed in the Si diffusion zone, locally depleting the Mn content of the matrix and lowering the corrosion potential of this Mn-depleted band by approximately 30–40 mV relative to the Mn-rich core. These dispersoids are responsible for the brown-band etching effect and can be clearly identified in a metallurgical cross-section of a sheet after brazing; see Fig. 10.25(a) (Melander and Woods, 2010). The addition of some copper to the core alloy magnifies the protective effect of the brown-band zone due to the diffusion of Cu from a copper-rich core into the copper-free cladding during the brazing cycle. This sets up a copper concentration gradient below the surface, which further lowers the corrosion potential of the diffusion zone relative to the core alloy. In contrast, the diffusion of Si from the cladding into solid solution in the interfacial matrix tends to increase the potential of the diffusion zone. Thus, the cumulative effects of the decrease in Mn and the outward diffusion of Cu more than compensate for the inward diffusion of Si (Meijers, 2002). The diffusion profiles for these three elements can be calculated and the potential changes with depth below the brazed

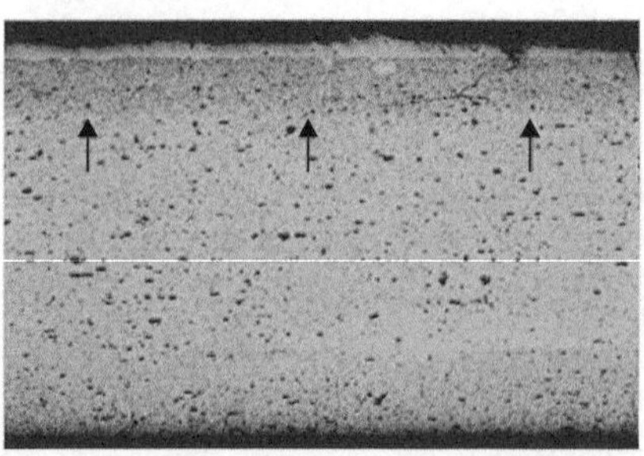
(a)

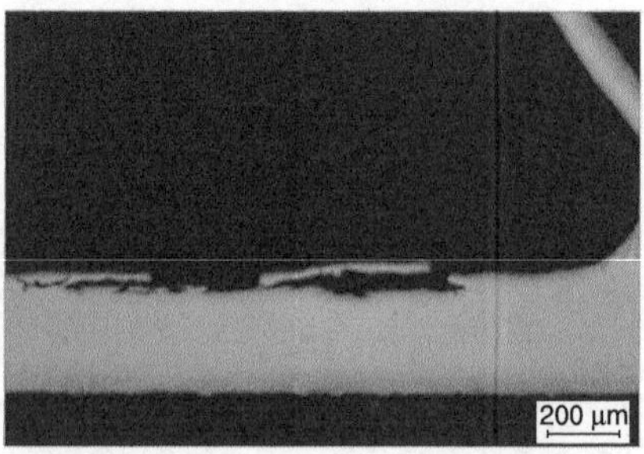

(b)

10.25 Microstructures of long-life brazed sheet: (a) the subsurface brown-band following 'mild' field service; (b) typical lateral corrosion after 21 days of SWAAT testing (Melander and Woods, 2010, © NACE International, 2010).

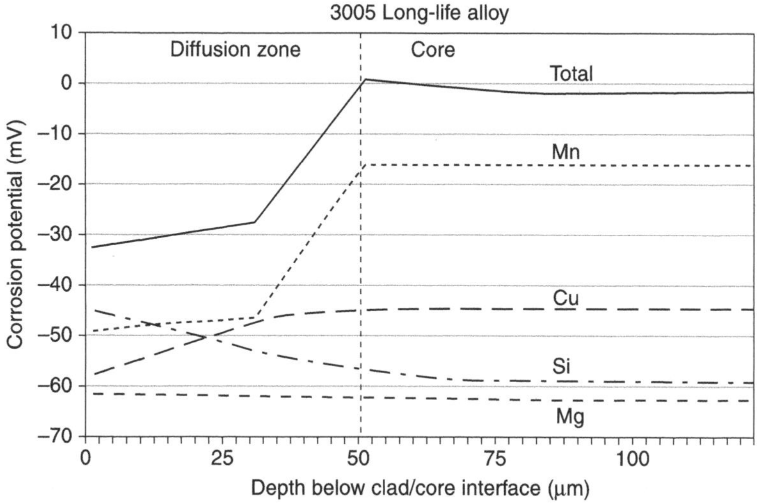

10.26 Calculated changes in the core potential profiles due to elemental diffusion gradients after a typical brazing cycle (30°C/min heating rate, 3 min dwell at 600°C) (courtesy of Sapa Technology).

sheet surface profiled as shown in Fig. 10.26. The lateralized corrosion shown in Fig. 10.25(b) is typical of the mode of corrosion found in radiators exposed to moderate to severe service environments, as well as the corrosion mode produced in the SWAAT test (Sea Water Acetic Acid Test) (ASTM G85-09, 2009). This has become the most commonly used test for evaluation and quality assurance of these long-life materials (Melander and Woods, 2010). Recent evaluations of retrieved long-service-life radiators with roll-formed tubes have confirmed that service lifetimes well in excess of 10 years can now be confidently expected (Melander and Woods, 2010). Unfortunately, this protective brown band effect has not been duplicated in extruded tubes, so the situation of premature corrosion failure in condensers still exists.

An alternative tube perforation protection mechanism was demonstrated by incorporating unusually high amounts of titanium in the cast alloy (Wade and Scott, 1986). By careful control of the casting conditions, Ti segregation to the center of the ingot core could be achieved. After rolling to sheet, a Ti-rich central zone was retained, which would become anodic to the lower-titanium-containing surface regions during corrosion exposure. Figure 10.27 shows an example of centerline corrosion of aluminum sheet due to the Ti segregation after SWAAT testing. The addition of extra Ti to the casting and the brown band effect have sometimes been combined in efforts to produce ultra-long-life alloys to be used for CAB brazed heat exchangers (Stenqvist, 2001). The concept is that, once the brown band is consumed, any further corrosive attack would again be lateralized

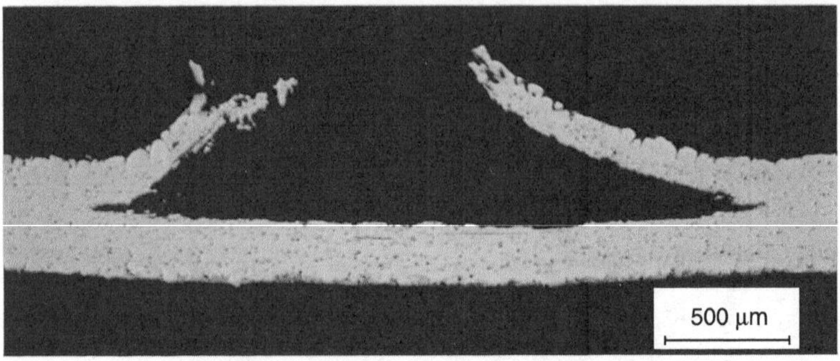

10.27 Corrosion of brazed sheet with center line Ti segregation (courtesy of Sapa Technology).

at the sheet centerline and times to perforation would be further extended. However, in practice, the necessary Ti segregation is very difficult to produce in large production-sized ingots, so the Ti effect is rarely encountered.

The use of thin walled, micro-extruded tubes in automotive condensers has greatly improved heat exchanger thermal performance. Due to the nature of the extrusion process, extruded tubes cannot be produced with a clad layer of Si-rich filler metal as are sheet products. Therefore, the brown-band corrosion resistance mechanism cannot be applied to micro-channel tubes. For joining, products using these tubes must be sprayed with filler metal deposits (e.g. Al–Si powder, Sil Flux) or brazed with claded fins. Good control of alloy microstructures and alloy compositions to improve the tube materials' corrosion resistance are essential (Brazing Handbook, 2007; Gray, 2006), especially when facing the current trends of down-gauging the tube wall thickness and the use of micro-channel heat exchangers in HVAC&R systems. Some measure of tube protection may be achieved by selecting appropriate alloy combinations for different heat exchanger components such that one component provides effective protection to the tube materials by a galvanic sacrificial mechanism. Zinc is among the few common elements that have a lower corrosion potential than Al (Hatch, 1984), so it is frequently used as an addition to fin alloys in amounts up to 2% to increase the corrosion potential difference between the fins and tubes. Such developments in fin materials make it possible to provide cathodic protection to the tube without showing excessive fin corrosion (Magnusson *et al.*, 1997). Another corrosion control method for the extruded tubes is to apply a Zn coating on the tube surface prior to assembly of the heat exchangers. Zn has a low melting point (420°C) and a high solubility in Al and diffuses easily into the tube surface during the brazing cycle. This creates a thin surface layer, which is anodic to and therefore sacrificial to the extruded tube core. Thus, the tubes are protected. However, many laboratory and field studies have shown that the Zn concentrates in the brazed metal fillets during CAB, making the joints themselves sacrificial to other components. This can lead to dramatic debonding between the fins and tubes (see Fig. 10.22) and losses in thermal efficiency as described above (Janssen *et al.*, 2011). As an alternative, the Zn coating can be applied through a Zn-containing flux. The use of Zn flux provides a lower Zn loading and a more uniform coverage of Zn on the tube surface compared with direct Zn deposition methods such as arc spraying. Consequently, post-braze Zn concentrations in the braze fillets are lower, and the selective corrosion causing joint attack and debonding can be minimized (Janssen *et al.*, 2011). It has been suggested that the furnace environment may also influence the Zn content in the fillets (Lochte and Sucke, 2011). A furnace used for condenser brazing may be contaminated with very high Zn levels due to constant exposure to Zn-coated/bearing materials. Vaporization from these accumulations subsequently leads to more Zn pick-up in the brazed joints (Lochte and Sucke, 2011).

It has been pointed out (Nordlien *et al.*, 2011) that, in addition to the efforts in alloy material development, proper design of heat exchangers and system assembly to minimize the exposure to a corrosive environment is very important. For example, corrosion failure at specific locations of a heat exchanger due to continuous exposure to wet and acidic environments may be avoided by a better design of drainage to prevent the accumulation of water and other contaminates on the heat exchanger (Nordlien *et al.*, 2011).

Coolant side corrosion

Historically, internal corrosion in automobile heat exchangers has been much less of a problem than external corrosion. In fact, over many years, improvements in coolant formulations and inhibitors have provided very good resistance to coolant corrosion when used with 'reasonably clean' make-up water. When failures have occurred, the depleted coolants or contaminated water are usually to blame. As mentioned earlier, many tube materials do incorporate an inner or waterside sacrificial clad layer of lower corrosion potential than the tube core alloy. Although many laboratory tests have demonstrated the functioning of these inner claddings, often quite dramatically, evidence for their usefulness in the field is much less convincing. Figure 10.28 illustrates light in-service pitting corrosion on an inner tube surface with an Al–Zn alloy inner cladding. In other situations where erosion corrosion occurs (see Fig. 10.29), these claddings offer no protection at all. Control of erosion corrosion has to be effected by ensuring coolants are properly maintained and by reducing the local fluid flow velocity (below 2 m/s) (Shah and Sekulic, 2003).

10.6.3 Testing methods for corrosion resistance

Due to the slow progress of heat exchanger corrosion during service, accelerated laboratory corrosion test methods have been used in alloy development programs

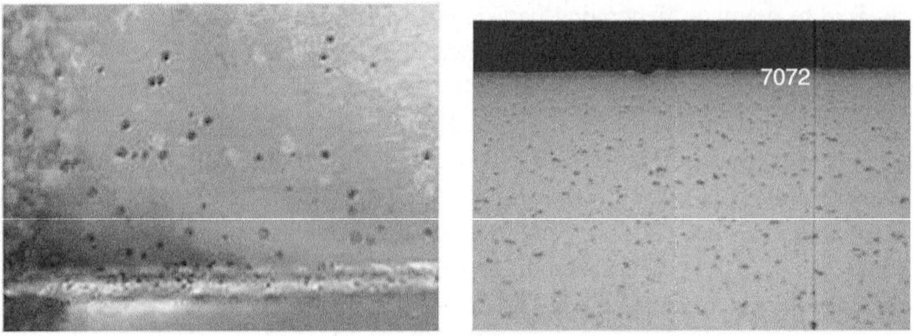

(a) (b)

10.28 Pitting, field corrosion of an AA7072(Al–Zn) inner clad radiator tube: (a) tube inner surface appearance (courtesy of Sapa Technology); (b) cross-section image (Melander and Woods, 2010, © NACE International, 2010).

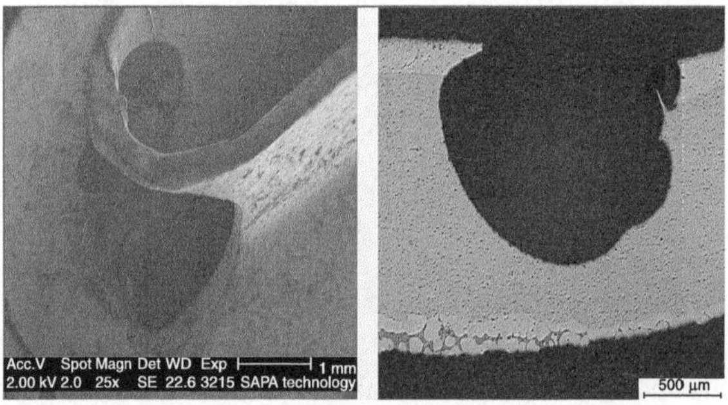

10.29 Erosion corrosion of an inner clad radiator tube (courtesy of Sapa Technology).

to replicate the mode of field service (Melander and Woods, 2010). The purpose of such accelerated tests is to predict the corrosion performance of heat exchangers after long-time usage in service. SWAAT has been widely used in the automotive industry for many years to simulate the corrosive environment that brazed heat exchangers are exposed to externally. It was found that SWAAT will produce in 2 or 3 days the intergranular or pitting attack morphologies that are also identified in radiators after a few years of field service (Scott *et al.*, 1991). Other test methods, such as the Neutral Salt Fog test (ASTM B117-09, 2009) and CASS (copper-accelerated acetic acid-salt spray testing) (ASTM B368-09, 2009), have also been used to evaluate the corrosion resistance of brazed aluminum heat exchangers. A recent comprehensive study of radiators in field service for 6–10 years has been reported (Melander and Woods, 2010). By comparing SWAAT results with the corrosion patterns found in heat exchangers used in automobiles for several years in quite severe corrosion environments, it was concluded that SWAAT is suitable for testing automotive heat exchangers exposed to harsh road conditions. However, this severe test does not fully replicate the corrosion patterns produced in automobiles operating in relatively mild, but still corrosive, field service conditions. The drawbacks of SWAAT include its relatively long test duration time and a low repeatability (Meijers, 2002), but so far no substitute accelerated corrosion test has gained wide acceptance.

The trend towards using CAB brazed aluminum micro-channel heat exchangers in stationary HVAC&R systems also raises the question of whether or not currently used accelerated test methods are suitable for these new applications. Again, SWAAT may be too severe to simulate the service conditions for heat exchangers used in stationary HVAC&R systems, even in coastal locations (Norgren and Ek, 2009). Recently, ASHRAE (American Society of Heating, Refrigerating and Air-conditioning Engineers) proposed a research project to develop a new corrosion test for CAB brazed aluminum micro-channel heat exchangers in the HVAC&R industry (Ellerbrock, 2011; Norgren and Matter, 2011), based on corrosion test criteria developed earlier for vacuum-brazed aluminum heat exchangers (Scott *et al.*, 1991).

For testing the internal corrosion resistance of CAB brazed Al heat exchangers, highly corrosive solutions such as 'OY' (Oyama River) water are often used. This test, which was developed in Japan, employs a very aggressive solution that contains NaCl, $FeCl_2$ $CuCl_2$, and $NaSO_4$ in deionized water. The tests are usually run at an elevated temperature of 88°C so that inner tube corrosion can be generated in a short time (Melander and Woods, 2010). However, it is questionable whether or not the OY test method represents any but the most severe corrosion conditions that could conceivably be encountered in field service. Therefore, different manufacturers have specified modified test procedures. For example, some companies have used diluted OY water or modified, milder acid compositions. In North America and Europe, ASTM corrosive water and even weaker solutions are sometimes used. Such variations have made the test results difficult to compare. Unfortunately, few data on internal corrosion have been reported in the literature. When results from radiators in service for 6–10 years in southern Sweden were compared with OY water test results, no resemblance in corrosion morphology was found (Melander and Woods, 2010). However, some heat exchanger failures can be encountered after very short service times due to lack of attention to inhibitor levels and/or the quality of the make-up water. So, even here, where inner claddings are used that are designed based upon OY water test results, early and sometimes catastrophic corrosive attacks may occur.

An industry-wide standardization of test procedures is needed for both internal and external corrosion testing of CAB brazed automotive aluminum heat exchangers, as well as an increased awareness of the sensitivities of CAB brazed HVAC&R heat exchangers in service.

10.7 References

Aleris document (2007), 'Aluminum heat exchanger materials', available from: http://www.aleris.com/sites/default/files/1829_AL_HeatExchanger_Br_oB_2010-09-23_RZ_WEB.pdf (accessed 10 March 2011).

ASTM B117-09 (2009), 'Standard practice for operating salt spray (Fog) apparatus', West Conshohocken, Pennsylvania, ASTM International.

ASTM B368-09 (2009), 'Standard test method for copper-accelerated acetic acid-salt spray (Fog) testing (CASS Test)', West Conshohocken, Pennsylvania, ASTM International.

ASTM G85-09 (2009), 'Standard practice for modified salt spray (Fog) testing', West Conshohocken, Pennsylvania, ASTM International.

AWS C3.7M/C3.7 (2011), 'Specification for aluminum brazing', Miami, Florida, American Welding Society.

Biance A L, Clanet C and Quere D (2004), 'First steps in the spreading of a liquid droplet', *Physical Review E*, 69, 016301-1–016301-4.

Bolingbroke R K, Gray A and Lauzon D, (1997), 'Optimisation of Nocolok™ brazing conditions for higher strength brazing sheet', SAE paper #971861.

Brazing Handbook (2007), 5th Edition, Miami, Florida, American Welding Society.

Brouckere, L De (1945), 'An electron-diffraction study of the atmospheric oxidation of aluminium, magnesium and aluminium-magnesium alloys', *Journal of the Institute of Metals*, 71, 131–147.

Callister W D (2003), *Materials Science and Engineering – An Introduction*, Massachusetts, John Wiley & Sons.

Chen R, Cao J and Zhang, Q Y (1997), 'A study of the phase diagram of AlF3-CsF system', *Thermochimica Acta*, 303, 145–150.

Chen R, Wu G H and Zhang Q Y (2000), 'Phase diagram of the system KF-AlF3', *J. Am. Ceramic Soc.*, 83, 3196–3198.

Cooke W E, Wright T E and Hirschfield J A (1978), 'Furnace brazing of aluminum with a non-corrosive flux', SAE paper #780300.

De Gennes P G (1985), 'Wetting: Statics and dynamics', *Review of Modern Physics*, 57, 827–863.

De Gennes P G, Brochard-Wyart F and Quere D (2004), *Capillarity and Wetting Phenomena – Drops, Bubbles, Pearls, Waves*, New York, Springer-Verlag.

Dehmas M, Valdés R, Lafont M C, Lacaze J and Viguier B (2006), 'Identification of intermetallic precipitates formed during re-solidification of brazed aluminum alloys', *Scripta Materialia*, 55, 191–194.

Doko T, Yanagawa Y and Tanaka S (2005), 'Change in the brazeability of brazing sheets with different Si particle size in Al-Si filler alloy layer', *Furukawa-sky review*, No.1, 27–32. Available from: http://www.furukawa-sky.co.jp/review/001/01_abst05.pdf (in Japanese, accessed 24 July 2011).

Ellerbrock D (2011), 'Considerations for HVAC accelerated corrosion testing and future directions', *2nd International Congress Aluminum Brazing Technologies for HVAC&R Aluminum-Verlag*, Düsseldorf, Germany.

Field D J and Steward N I (1987), 'Mechanistic aspects of the Nocolok flux brazing process', SAE paper #870186.

Gao F, Zhao H, Sekulic D P, Qian Y and Walker L (2002), 'Solid state Si diffusion and joint formation involving aluminum brazing sheet', *Materials Science and Engineering A*, 337, 228–235.

Garcia J, Massoulier C and Faille P (2001), 'Brazeability of aluminum alloys containing magnesium by CAB process using cesium flux', SAE paper #2001-01-1763.

Garcia J, Swidersky H-W (2011), 'Solubility characteristics of potassium fluoroaluminate flux and residues', *2nd International Congress Aluminum Brazing Technologies for HVAC&R Aluminum-Verlag*, Düsseldorf, Germany.

Garcia J, Swidersky H-W, Schwarze T and Eicher J (2010), 'Inorganic Fluoride Materials from Solvay Fluor and their Industrial Applications', in *Functionalized Inorganic Fluorides: Synthesis, Characterization & Properties of Nanostructured Solids* (ed. A. Tressaud), Chichester, UK, John Wiley & Sons Ltd.

Godard H P (1967), 'Oxide film growth over five years on some aluminum sheet alloys in air of varying humidity at room temperature', *J. Electrochem. Soc.*, 114, 354–356.

Goodrich H S, Connor Z M and Murty G S (2001), 'Aging response and elevated temperature strengthening in brazing sheet core alloys of 3xxx series aluminum', SAE paper #2001-01-1725.

Gray A, Butler C and Bosland A (2011), 'A real time FTIR study of the oxidation behavior of clad sheet during controlled atmosphere brazing', *2nd International Congress Aluminum Brazing Technologies for HVAC&R Aluminum-Verlag*, Düsseldorf, Germany.

Gray A, Flemming A J and Evans J M (1999), 'Optimizing the properties of long-life brazing sheet alloys for vacuum and NOCOLOK brazed components', *VTMS 4 Conference Proceedings*, London.

Gray A, Swidersky H W and Orman L (2006), 'Reactive Zn flux – an opportunity for controlled Zn diffusion and improved corrosion resistance', *AFC-Holcroft 11th International Invitational Aluminum Brazing Seminar*, Livonia, Michigan.

Hatch J E, editor (1984), *Aluminum: Properties and Physical Metallurgy*, Metals Park, Ohio, American Society for Metals.

Holmestad R, Marioara C D, Ehlers F J H, Torsæter M, Bjørge R, *et al.* (2010), 'Precipitation in 6xxx aluminum alloys', *Proceeding of 12th International Conference on Aluminum Alloys*, 5–9 September, Yokohama, Japan, 30–39.

Hrnjak, P (2011), 'New opportunities for Al microchannel heat exchangers', *2nd International Congress Aluminum Brazing Technologies for HVAC&R Aluminum-Verlag*, Düsseldorf, Germany.

Hunter M S and Fowle P (1956), 'Natural and thermally formed oxide film on aluminum', *J. Electrochem. Soc.*, 103, 482–485.

Janssen H, Nordlien J H, Schluter S, Insalaco J and Sicking R (2011), 'Recent alloy development for HVAC&R applications', *2nd International Congress Aluminum Brazing Technologies for HVAC&R Aluminum-Verlag*, Düsseldorf, Germany.

Johansson H, Stenqvist T and Swidersky H W (2003), 'Controlled atmosphere brazing of heat treatable alloys with cesium flux', IMechE paper C599/013/2003.

Jordan M F, Milner D R (1956–7), 'The removal of oxide from aluminum by brazing fluxes', *Journal of the Institute of Metals*, 85, 33–40.

Kaufman J G (2000), *Introduction to Aluminum Alloys and Tempers*, Materials Park, OH, ASM International.

Kooij N D A, Hurd T J, Burger A, Vieregge K and Haszler A (1999), 'High-strength heat-treatable aluminum alloys for CAB brazing', *VTMS 4 Conference Proceedings*, London.

Kubaschewski O and Hopkins B E (1953), *Oxidation of Metals and Alloys*, London, Butterworths Scientific Publications.

Kuppan T (2000), *Heat Exchanger Design Handbooks*, Marcel Dekker, New York.

Lacaze J, Tierce S, Lafont M C, Thebault Y, Pébère N, *et al.* (2005), 'Study of the microstructure resulting from brazed aluminum materials used in heat exchangers', *Materials Science and Engineering A*, 413–414, 317–321.

Levinson P, Cazabat A M, Cohen Stuart M A, Heslot F and Nicolet S (1988), 'The spreading of macroscopic droplets', *Revue Phys. Appl.*, 23, 1009–1016.

Lochte L and Sucke N W (2011), 'Zinc: Angel or devil for brazing of aluminum heat exchangers', *2nd International Congress Aluminum Brazing Technologies for HVAC&R Aluminum-Verlag*, Düsseldorf, Germany.

Magnusson A, Scholin K, Mannerskog B, Stenqvist T and Ortnas A (1997), 'Improved material combination for controlled atmosphere brazed aluminum radiators', SAE paper #971786.

Massalski T B (1986), *Binary Alloy Phase Diagrams, American Society for Metals*, Metals Park, Ohio, American Society for Metals.

Meijers S (2002), *Corrosion of aluminum brazing sheet*, PhD thesis, Technical University of Delft.

Melander M and Woods R A (2010), 'Corrosion study of brazed aluminum radiators retrieved from cars after field service', *Corrosion*, 66, 015005-1–015005-14.

Miller W S, Zhuang L, Bottema J, Wittebrood A J, De Smet P, *et al.* (2000), 'Recent development in aluminium alloys for the automotive industry', *Material Science and Engineering A*, 280, 37–49.

Mortensen A, Drevet B and Eustathopoulos N (1997), 'Kinetics of diffusion-limited spreading of sessile drops in reactive wetting', *Scripta Materialia*, 36, 645–651.

Nordlien J H, Daaland O, Lian K A and Espedal A (2011), 'Multiport extruded tube alloy development for HVAC&R applications', *2nd International Congress Aluminum Brazing Technologies for HVAC&R Aluminum-Verlag*, Düsseldorf, Germany.

Norgren S and Ek L (2009), 'Corrosion performance & testing of CAB aluminum heat-exchanger samples for HVAC&R', *1st International Congress Aluminum Brazing Technologies for HVAC&R Aluminum-Verlag*, Düsseldorf, Germany, 16–17 June 2009.

Norgren S and Matter J (2011), 'Accelerated corrosion test methods for condensers for stationary HVAC&R', *2nd International Congress Aluminum Brazing Technologies for HVAC&R Aluminum-Verlag*, Düsseldorf, Germany.

Nylen M, Gustavsson U, Hutchinson B and Örtnas A (2000), 'The mechanism of braze metal penetration by migration of liquid films in aluminum', *Material Science Forum*, 331–337, 1734–1742.

Orman L, Swidersky H W and Gray A (2008), 'Reactive zinc flux – An opportunity for controlled zinc diffusion and improved corrosion resistance – Part II', *5th International Aluminium Brazing* Congress, Düsseldorf, Germany, 6–8 May 2008.

Park C Y and Hrnjak P (2002), 'R-410A air conditioning system with microchannel condenser', *Proceedings of the 9th Purdue Refrigeration Conference*, University of Purdue.

Robert E, Olsen J E, Danek V, Tixhon E, Ostvold T, *et al.* (1997a), 'Structure and thermodynamics of alkali fluoride-aluminum fluoride-alumina melts – Vapor pressure, solubility, and Raman spectroscopic studies', *J. Phys. Chem. B*, 101, 9447–9457.

Robert E, Olsen J E, Gilbert B and Ostvold T (1997b), 'Structure and thermodynamics of potassium fluoride-aluminum fluoride melts – Raman spectroscopic and vapor pressure studies', *Acta Chemica Scandinavica*, 51, 379–386.

Schwartz M (2003), *Brazing*, 2nd edition, Materials Park, OH, ASM International.

Scott A C, Woods R A and Harris J F (1991), 'Accelerated corrosion test methods for evaluating external corrosion resistance of vacuum brazed aluminum heat exchangers', SAE paper #910590.

Shah R K and Sekulic D P (2003), *Fundamentals of Heat Exchanger Design*, New York, Wiley.

Shapiro, A E (2010), 'Brazing Q&A', *Welding Journal*, 89, 18–20.

Solvay Fluor document (2011), 'Post Braze Flux Residue and Engine Coolants'.

Solvay Fluor document (2012a), 'Brazing dictionary', available from: http://www.aluminium-brazing.com/sponsor/nocolok/Html/Dictionary/index.html (accessed 10 March 2011).

Solvay Fluor document (2012b), 'Nocolok® Li flux', available from: http://www.aluminium-brazing.com/sponsor/nocolok/Files/PDFs/31403.pdf (accessed 10 March 2012).

Solvay Fluor document (2012c), 'filler metal management in NOCOLOK® flux brazing of aluminum', available from http://www.aluminium-brazing.com/sponsor/nocolok/Files/PDFs/31352.pdf (accessed 10 March 2012).

Sontgerath J A H, Kooij N D A, Vieregge K and Haszler A (1996), 'The effect of microstructure on flow behavior of braze clad material', *Materials Science Forum*, 217–222, 1721–1726.

Stenqvist T (2001), 'A new high strength aluminum alloy for controlled atmosphere brazing', SAE paper #2001-01-1727.

Swidersky H W (2001), 'Aluminum brazing with non-corrosive flux, state of the art and trends in NOCOLOK® flux technology', *6th International Conference on Brazing, High Temperature Brazing and Diffusion Bonding*, Aachen, Germany.

Swidersky H W (2012), Personal communication.

Tanner L H (1978), 'Spreading of silicone oil drops on horizontal surface', *J. Phys. D*, 12, 1473–1484.

Terrill J R (1966), 'Diffusion of silicon in aluminum brazing sheet', *Welding Journal*, 45, 202–209.

Terrill J R, Cochran C N, Stokes J J and Haupin W E (1971), 'Understanding the mechanisms of aluminum brazing', *Welding Journal*, 50, 833–839.

Timsit R S and Janeway B J (1993), 'A novel brazing technique for aluminum and other metals', *J. Mater. Res.*, 8, 2749–2752.

Turriff D M, Corbin S F and Kozdras M (2010), 'Diffusion solidification phenomena in clad aluminum automotive braze sheet', *Acta Materialia*, 58, 1332–1341.

Van Beek H J and Mittemeijer E J (1984), 'Amorphous and crystalline oxides on aluminum', *Thin Solid Films*, 122, 131–151.

Wade K D and Scott D H (1986), 'Aluminum Alloys, Physical and Mechanical Properties', 2, 1141, Engineering Materials Advisory Services.

Wittebrood A (2009), *Microstructural changes in brazing sheet due to solid-liquid interaction*, PhD thesis, Technical University of Delft.

Wittebrood A, Kooij C J and Vieregge, K (2000), 'Grain boundary melting or liquid film migration in brazing sheet', *Materials Science Forum*, 331–337, 1743–1750.

Woods R A (1997), 'Liquid film migration during aluminum brazing', SAE paper #971848.

Woods R A and Robinson I B (1974), 'Flow of aluminum dip brazing filer metals', *Welding Journal*, 53, 440s–445s.

Woods R A, Scott A C and Harris J F (1991), 'A corrosion resistant alloy for vacuum brazed aluminum heat exchangers', SAE paper #910591.

Yang H S and Woods R A (1997), 'Mechanisms of liquid film migration (LFM) in aluminum brazing sheet', SAE paper #971849.

Yoon S, Kim H J and Lee C (2007), 'Fabrication of automotive heat exchanger using kinetic spraying process', *Surface & Coatings Technology*, 201, 9524–9532.
Zähr J, Fussel U, Ullrich H J, Turpe M, Grunenuald B, *et al.* (2010), 'Influence of the natural aluminum oxide coat on thermal joining', *DVS Berichte*, 263, 296–303.
Zhang Q Y and Zhuang H S (2008), *Brazing and Soldering Manual*, 2nd edition, Beijing, China Machine Press, in Chinese.
Zhao H and Sekulic D P (2006), 'Diffusion controlled melting and re-solidification of metal micro layers on a reactive substrate', *Heat Mass Transfer*, 42, 464–469.
Zhao H and Sekulic D P (2008), 'Wetting kinetics of a hypo-eutectic Al-Si system spreading over an aluminum substrate', *Materials Letters*, 62, 2241–2244.
Zhao H and Sekulic D P (2009), 'The influence of surface topography on wetting kinetics of solders and brazes', *4th International Brazing and Soldering Conference Proceedings*, Orlando, FL.

11
Active metal brazing of advanced ceramic composites to metallic systems

R. ASTHANA, University of Wisconsin-Stout, USA
and M. SINGH, NASA Glenn Research Center, USA

DOI: 10.1533/9780857096500.2.323

Abstract: Advanced ceramic-matrix composites (CMCs) outperform traditional ceramics in many ways and have shown potential for demanding applications. Net-shape manufacture of CMC parts is challenging, and many advanced applications demand robust and reliable integration technologies such as brazing. Brazing of CMC/metal joints is reviewed, highlighting scientific issues together with a discussion of some of the challenges that brazing of CMCs presents. Brazing practices for SiC–SiC, C–SiC, C–C, ZrB_2-based ultra-high-temperature composites, and oxide, nitride and silicate-based composites are presented. Recent research results on interface microstructure, composition and properties are discussed. Scaling effects, time–temperature–environment dependent thermomechanical properties, design guidelines and life-prediction analyses, and tools for CMC/metal joints to be used in structures constitute future research imperatives.

Key words: ceramic-matrix composites, silicon carbide/silicon carbide, carbon/carbon, ultra-high-temperature composites, interface, active brazing, shear strength, wettability, infiltration, microstructure.

11.1 Introduction

Ceramic-matrix composites (CMCs) surpass metal-matrix and polymer-matrix composites in their elevated-temperature performance and monolithic ceramics in their fracture toughness and thermal shock resistance. CMCs are designed to overcome the poor toughness of monolithic ceramics via mechanisms such as crack deflection, crack blunting, transformation toughening and frictional bonding. A ceramic-matrix composite is broadly a ceramic-based material system that is composed of a discrete monolithic ceramic constituent (reinforcement) distributed in a continuous ceramic phase (the matrix), and which derives its distinguishing characteristics from the properties of its constituents, from the geometry and architecture of the constituents, and from the properties of the boundaries (interfaces) between ceramic constituents. In a CMC, properties can be tailored, new combinations of properties are possible and the properties of the whole may be significantly different from the properties of the constituents. Unlike metal- and polymer-matrix composites, toughening rather than strengthening is the key objective in CMCs, which have a weak fiber/matrix interface so that crack deflection and frictional stresses during sliding of debonded fibers creates toughness. Thus, inherently brittle monolithic ceramics such as SiC and Al_2O_3 (fracture toughness K_{Ic} ~ 3–5 MPa.m$^{1/2}$) can be combined to yield ceramic systems whose toughness far exceeds that of the component ceramic phases. It is noteworthy that the inherent properties of the fiber and matrix constituents in a composite are largely fixed; hence, the greatest latitude in designing bulk composite properties is realized through tailoring of the interface (strictly, though, this is not true because processing conditions that lead to interface development can also modify to an extent both the fiber and matrix properties). The development of an optimum interfacial bond between the fiber and the matrix is a primary requirement for acceptable performance of a composite.

Fibers, whiskers, particulates, platelets and laminates are common reinforcement morphologies in CMCs. A number of advanced CMCs based on carbides (SiC, TiC,), oxides (Al_2O_3, ZrO_2, SiO_2), nitrides (Si_3N_4, BN, AlN), borides (TiB_2, ZrB_2, HfB_2), and glass and glass-ceramics have been developed for diverse applications during recent decades. Advanced CMCs with proven performance include SiC–SiC, SiC–Al_2O_3, C–SiC, SiC–Si_3N_4, TiC–Si_3N_4, Al_2O_3–ZrO_2, TiC–Al_2O_3 and many others. These advanced CMCs have been evaluated for applications in high-performance, high-temperature systems such as thermal structures, exhaust nozzles, turbo pump blisks, combustor liners, radiant burners, heat exchangers and a wide variety of other systems subject to severe thermal, wear, erosive and corrosive environments. For example, C–C composites containing SiC are used in lightweight brake disks, made by first fabricating the C–C composite via chemical vapor infiltration (CVI) or by polymer pyrolysis and then infiltrating the C–C composite with Si to reactively form SiC. Ceramic-matrix composites containing diamond, c-BN, B_4C and similar hard particles have been designed for cutting tools for high use temperatures and high cutting speeds. Many of these composites were originally developed for hypersonic aircraft thermal structures and advanced rocket propulsion thrust chambers. Still other composites have been developed for use at ultra-high temperatures (2173–2773 K). For example, composites based on borides of refractory metals Zr, Ti, Hf and Ta have high melting temperatures, high hardness, low volatility, and good thermal shock resistance and thermal conductivity. These materials can perform satisfactorily for short periods at temperatures in the range 2173–2773 K. These composites have good oxidation resistance, and further improvements can be achieved through the use of additives such as SiC. Comprehensive state-of-the-art reviews on processing, fabrication, properties and performance of advanced CMCs can be found in the literature (Taylor, 2000; Zweben, 2002; Lamon, 2005; DiCarlo *et al.*, 2005; Corman and Luthra, 2005; Krenkel, 2005; Lewis and Singh, 2001; Chawla, 1987).

Although net-shape manufacture of ceramic and CMC parts is preferred because of the inherent brittleness of ceramics and associated machining difficulties, many advanced technology applications of CMCs integrate geometrically simpler, discrete elements in a hierarchical fashion to create complex structural assemblies. Additionally, in real components, these CMCs must be integrated with other materials such as high-temperature metals and alloys. This demands the development or adaptation of robust and reliable joining and integration technologies. CMCs and monolithic ceramics are joined using processes such as diffusion bonding, fusion welding, adhesive bonding, active metal brazing, brazing with oxides, glasses and oxynitrides, reaction forming and several other methods (Schwartz, 1994; Locatelli *et al.*, 1995; Singh and Asthana, 2007b, 2008; Suryanarayana *et al.*, 2001).

11.2 Brazing dissimilar materials

Brazing is a relatively simple and cost-effective joining method, which is applicable to a wide range of ceramics and CMCs. Ceramic brazing utilizes either metallic or glassy filler alloys in the form of powders, pastes or rods that are melted, allowed to wet and spread in the joint region, and solidified to form joint interfaces. Ceramic brazing with metallic fillers utilizes either premetallized ceramic surfaces or braze foils, pastes or wires that usually contain a reactive filler metal such as Ti, which promotes braze wettability and flow characteristics by inducing chemical reactions with the ceramic. Titanium is one of the most commonly used active metals because it forms compounds that strongly bond to both metals and ceramics. For example, with oxide ceramics, Ti forms TiO, TiO_2 and Ti_2O_3, and with silicon carbide, silicon nitride and sialons, Ti forms titanium silicides, titanium carbides and titanium nitrides. Other reactive filler metals contain Zr, Cr, Nb and Y. Brazing is generally done in vacuum furnaces under a high-purity inert atmosphere, but noble metal brazes are used in air in order to form oxide binders (e.g. Ag–Cu_2O). Brazes must show good wetting and adherence to the substrates, be ductile and resistant to grain growth, resist creep and oxidation, have closely matched coefficients of thermal expansion (CTEs) with the joined materials, have high thermal conductivity (for thermal management), and have melting points greater than the operating temperature of the joint, but lower than the joined materials' melting temperature. A large number of metallurgical braze alloys (Table 11.1) have been designed and developed for use with ceramic-base materials, and are commercially available.

Braze fillers are available in foil, powder, paste and wire forms. Braze powders and pastes usually contain organic binders, which must be removed prior to bond formation. Braze foils and wires are used especially when there is a need to accommodate complex joint configurations. Pastes and powders may leave unwanted residue from organic binders at the joined surface, whereas foils or ribbons may be difficult to produce in alloys that contain metalloids like B, Si and P, which make the alloy inherently brittle, leading to edge cracking and difficulty in producing continuous sheets. This latter problem has been solved by rapidly solidifying braze compositions to produce ductile amorphous Ni, Co, Cu, Ti and Zr-base alloys that are readily produced in foil form.

Table 11.1 Commercial braze alloys used in CMC/CMC and CMC/metal joining

Braze	Composition (%)	T_L (K)	T_S (K)	E (GPa)	YS (MPa)	CTE ($\times 10^{-6}$ K^{-1})	K (W/m.K)	%El
Cu-ABA[a]	92.8Cu–3Si–2Al–2.25Ti	1297	1231	96	279	19.5	38	42
MBF-30[b]	Ni–4.61Si–2.8B–0.02Fe–0.02Co	1257	1327	–	–	–	–	–
MBF-20[b]	Ni–6.48Cr–3.13Fe–4.38Si–3.13B	1242	1297	–	–	–	–	–
Ticuni[a]	15Cu–15Ni–70Ti	1233	1183	144	–	20.3	–	–
Ticusil[a]	68.8Ag–26.7Cu–4.5Ti	1173	1053	85	292	18.5	219	28
Palcusil-15[a]	65Ag–20Cu–15Pd	1173	1123	–	379	–	98	23
Palcusil-10[a]	59Ag–31Cu–10Pd	1125	1097	–	327	18.5	145	18
Palcusil-5[a]	68Ag–27Cu–5Pd	1083	1080	–	333	17.2	208	11
Silcoro 75[a]	75Au–20Cu–5Ag	1168	1158	–	–	–	–	–
Cusil-ABA[a]	63Ag–35.3Cu–1.75Ti	1088	1053	83	271	18.5	180	42
Cusil[a]	72Ag–28Cu	1053	1053	83	272	19.6	371	19
Incusil-ABA[a]	59Ag–27.3Cu–12.5In–1.25Ti	988	878	76	338	18.2	70	21
Palco	65Pd–35Co	1492	1492	151.8[c]	341	14.3[d]	35	43
Palni	60Pd–40Ni	1511	1511	152.6[c]	772	15.0	42	23

[a]Morgan Advanced Ceramics; [b]Metglas, Inc.; [c]lower bound value, estimated using the rule-of-mixtures; [d]estimated using the Turner equation (Chawla, 1987). CTE, coefficient of thermal expansion; *E*, Young's modulus; *K*, thermal conductivity; T_L, liquidus temperature; T_S, solidus temperature; YS, yield strength; %El, percent elongation.

Non-metallic braze materials include glasses (or mixtures of glass with crystalline materials) such as SiO_2–MgO–Al_2O_3 and CaO–TiO_2–SiO_2 (CTS glass) compositions. The reason for using glass as a filler is that many sintered ceramics develop amorphous phases at their grain boundaries, and good wetting and bonding can be achieved between the glass filler and these grain boundary phases. Another method for joining (mostly) oxide ceramics is the molybdenum–manganese (Mo–Mn) process in which a specially formulated paint (or slurry) containing powdered Mo (or MoO_3), Mn (or MnO_2) and a glass-forming compound is applied to the ceramic surface. The painted ceramic is fired in wet hydrogen at 1500°C, which causes the glass-forming constituents from the ceramic to diffuse into the Mo layer and form a strong bond. Generally, the poor toughness, low Young's modulus and susceptibility to stress corrosion are the main challenges in using glass as a filler to join ceramics and CMCs.

Brazing of ceramics to metals also becomes feasible and joint strength improves when either the metal substrate is oxidized (resulting in an oxide/ceramic bond) or the ceramic surface is metallized (resulting in a metal/metal bond). The surface film of an active metal is deposited using sputtering, vapor deposition or thermal decomposition (e.g. Ti-coated Si_3N_4, Ti-coated partially stabilized zirconia, Cr-coated carbon and Si_3N_4 coated with Hf, Ta or Zr). Pre-coating of the ceramic with a Ti-bearing compound (e.g. TiH_2) that forms a Ti layer on the ceramic creates a wettable surface that strongly bonds the mating surfaces together upon cooling and solidification following brazing.

11.2.1 Wettability

Wettability is a major consideration in CMC/metal joining, and high-temperature wettability data can serve as a useful index for the brazability of materials. Wettability controls spreading on flat surfaces as well as capillary penetration of the open pore structure, which often characterizes CMCs. Extensive measurements of the contact angle of glasses and metallic alloys on monolithic ceramics are available in the published literature. Braze fillers containing active metals such as Ti wet oxide, carbide, nitride and carbon substrates, and, in most cases, the equilibrium contact angles approach small (e.g. near-zero) values from initially large obtuse values. Although wettability test standards are currently lacking and the measurements reveal extreme sensitivity to test conditions, the general framework for characterizing wettability invokes the concept of an angle of contact, θ (Fig. 11.1(a)), which is defined by the Young–Dupre equation:

$$\sigma_{lv}\cos\theta + \sigma_{ls} = \sigma_{sv} \qquad [11.1]$$

where the σs are the interfacial energies of the interfaces between solid (s), vapor (v) and liquid (l). For $\theta < 90°$, the solid is wet by the liquid, and for $\theta > 90°$, the solid is not wet by the liquid. The limits $\theta = 0°$ and $\theta = 180°$ define complete wetting and complete non-wetting, respectively. This exceedingly simple but

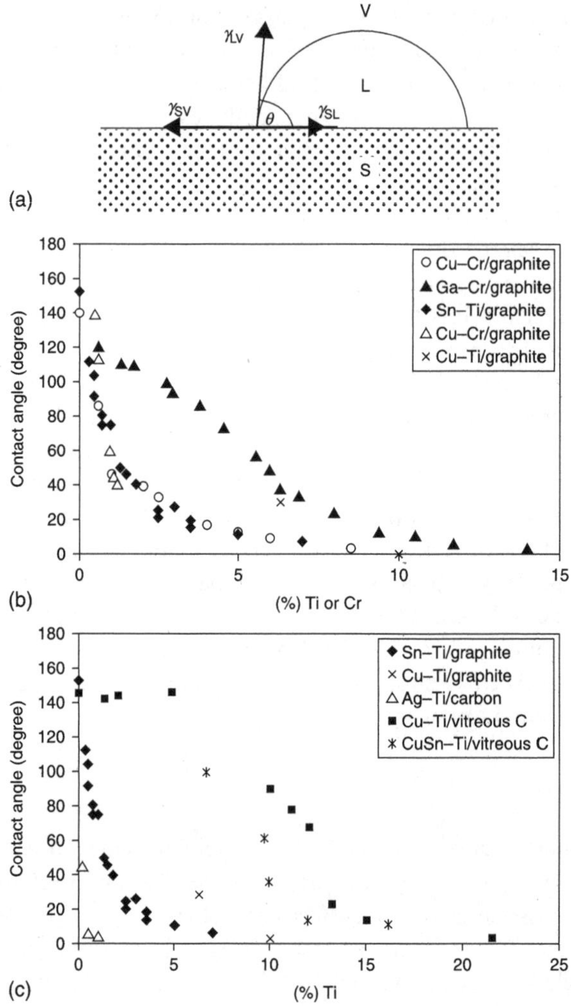

11.1 (a) Definition of contact angle; (b) and (c) effects of (b) Ti or Cr content, and (c) Ti content, on contact angle on various carbon substrates in Cu, Ag, Sn and Ga alloys.

elegant equation, derivable from thermodynamic considerations, captures the complex hydrodynamic and interfacial phenomena that govern the spreading behavior of liquids on solids. In spite of criticism from various quarters and its limitations in modeling wetting in chemically reactive systems at elevated temperatures (Sobczak *et al.*, 2005; Asthana and Sobczak, 2000), this equation is widely used as a basis of wettability measurements with braze alloys (Jacobson and Humpston, 2005). The equation has been modified to apply to reactive systems by adding a reaction free energy term, ΔG_r, which leads to a reduction in contact angle from θ_0 to θ as a result of a chemical reaction. The modified equation has the following form:

$$\cos\theta = \cos\theta_0 + \frac{(\sigma_{sl} + \sigma_{sl}^0 - \Delta G_r)}{\sigma_{lv}}$$

where θ_0 and σ_{sl}^0 are the contact angle and solid–liquid interfacial energy prior to reaction. It has also been shown that the interfacial energy term σ_{sl} is related to the Gibb's free energy of formation (ΔG) of the solid substrate that is in contact with the liquid, and the substrate's stoichiometry, temperature and atmosphere (Eustathopoulos *et al.*, 1999). ΔG is a measure of the chemical stability of the solid, and large negative ΔG values correlate with large contact angles. In other words, the wettability of a solid by liquid metals decreases as the stability of the solid increases. A few examples of wettability in high-temperature systems will be discussed below.

Both graphite and diamond are covalent, high melting point solids, which are characterized by closed stable electron configurations and strong interatomic bonds. Metals of the secondary B-subgroups in IV, V and VI periods of the periodic table (Cu, Ag, Au, Zn, Cd, In, Ge, Sn, Pb, Bi, etc.) exhibit little chemical affinity for carbon. Thus pure Ag and Cu (the base metals in a large number of brazes) do not wet graphite ($\theta \sim$ 137–140°). In contrast, transition metals show considerable interaction with carbon at high temperatures and a large work of adhesion (~20–25 kcal.mol^{-1}), where the work of adhesion, W_{ad}, is defined by $W_{ad} = \sigma_{lv}(1 + \cos\theta)$. As a result, alloying copper with Cr, V, Co, etc. reduces the contact angle of Cu on graphite (Eustathopoulos *et al.*, 1999) with Cr and V showing the greatest improvements in wettability.

Metals such as Ag, Cu and Ni are commonly used as base metals in braze fillers. The very large values of the surface tension, σ_{lv}, of these metals at their respective melting points show that these metals in a pure state will probably not wet CMCs. For example, σ_{lv} of Ni is 1796 N/m (at 1728 K), of Ag is 925 N/m (at 1233 K) and of Cu is 1330 N/m (at 1359 K), respectively (Keene, 1993). The temperature-corrected surface tension data show relatively modest decreases in σ_{lv} for these metals at temperatures exceeding their melting points, so there is only limited benefit in raising the temperature (the situation is more complex with oxidation-sensitive metals such as Al, for which raising the temperature removes the surface oxide via formation of gaseous sub-oxide, Al_2O, and oxide dissolution in the melt). The addition of small quantities of Ti or Cr to Ni improves the wetting and spreading, and leads to good bonding (Li, 1992; Grigorenko *et al.*, 1998). Figure 11.1(b) and (c) show the literature data on the effect of Ti and Cr content in some alloys on their contact angle with different types of carbon substrates. Experiments also show that, usually, short brazing times should suffice for maximum spread. For example, θ approaches near-zero contact angle values after a few minutes of contact for Cu–Ti and Cu–Sn–Ti melts in contact with vitreous C (Li, 1992; Standing and Nicholas, 1978).

Commercial brazes contain, besides active metals such as Ti, other alloying elements that are usually added to accomplish other goals (e.g. lowering the joining temperature, increasing the braze fluidity, etc.). These alloying additives could also react with the surfaces to be bonded. For example, Si and Ti in the braze Cu-ABA (Table 11.1) improve the wettability with carbon by forming titanium carbide (TiC and sub-stoichiometric carbides $TiC_{0.95}$, $TiC_{0.91}$, $TiC_{0.80}$, $TiC_{0.70}$, $TiC_{0.60}$ and $TiC_{0.48}$) and possibly also silicon carbide or a ternary $Si_xTi_yC_z$ compound. Pure Si reacts with carbon and lowers the contact angle on carbon ($\theta \sim 0°$ for Si/C (Whalen and Anderson, 1976)). The formation of TiC and SiC is thermodynamically favored at the brazing temperatures for AgCu brazes. The Gibb's free energy change (ΔG) for TiC formation via Ti + C→ TiC in the temperature range 1193–1323 K is −174 to −169 kJ and ΔG for SiC formation via Si + C→ SiC over the same temperature range is −62.4 to −61.1 kJ. Besides carbides, wettability-enhancing, metal-like interfacial oxides such as TiO may form from the reaction of Ti in the braze with the residual oxygen in the atmosphere of vacuum-brazing furnaces (TiO is a stable oxide of Ti, which forms even at very low oxygen partial pressures, $\sim 10^{-28}$ atm). Similarly, nickel in commercial brazes such as Ticuni (Table 11.1) has a higher affinity for carbon than Cu, Au and Ag, and segregates at the carbon/metal interface. Even though there are no stable carbides in the Ni–C system, the formation of metastable nickel carbide due to the interaction of liquid Ni with solid C is a possibility; this would lead to reasonably good wettability with relatively low contact angles (~68–90°) of Ni on carbon. However, as the principal constituent of Ticuni is Ti (~70%), which strongly reacts with carbon, formation of nickel carbide is less probable. Similar considerations apply to other alloying elements in brazes.

The braze metallurgy and phase transformations influence the solid–liquid reactions during brazing. In the case of the AgCuTi braze Ticusil (Table 11.1), the alloy system contains a liquid miscibility gap. The ternary Ag–Cu–Ti phase diagram shows that eutectic Ag–Cu alloys containing more than 5 at.% Ti divide the alloy into a Ti-depleted liquid and a Ti-rich liquid, the latter containing greater than 5 at.% Ti. This Ti-rich liquid reacts with carbon or SiC in the ceramic composite to form a metallurgical bond. The Ti-depleted liquid forms Ag(Cu,Ti) solid solution upon solidification, and eutectic-type phase mixtures. The Ti-rich liquid solidifies to yield Ag(Cu) precipitates and intermetallics such as AgTi, Ti_2Cu_3 and $TiCu_2$ in the braze matrix.

The chemically and structurally inhomogeneous surfaces of composites affect the wettability and brazability. Surface roughness, fiber arrangement, fiber weave pattern, surface chemical inhomogeneity and surface defects can all significantly influence the contact angle and braze flow characteristics (Sobczak *et al.*, 2005). Scant measurements of contact angle on composites currently exist (Asthana *et al.*, 2010) and this represents a research opportunity in joining science and technology.

Surface roughness can remarkably affect the wettability via contact angle hysteresis and bond strength via stress concentration in joints. As a general rule, roughness may increase the contact (bonded) area and chemical interactions in a wettable system besides promoting frictional bonding, but it may also enhance stress concentration due to notch effects in brittle ceramics. Generalizations of the roughness effects may be difficult because surface preparation (e.g. machining, grinding and polishing) not only reduces roughness but might also introduce surface and subsurface damage (grain or fiber pull-out and void formation). In addition, residual stresses may be introduced by machining and grinding. In some ceramic materials, damage induced by grinding may be healed by refiring the ground material prior to brazing. However, in the case of carbon, secondary roughening may occur during vacuum heating after polishing (Sobczak *et al.*, 2005). This is because, even though some porosity is closed by substrate grains detached during polishing, during subsequent heating under vacuum these grains are removed, along with evacuating gas, resulting in increased surface roughness.

In composites, the surface chemical and morphological inhomogeneity originating in different chemical phases and fiber architectures demands direct practical assessment of braze wettability and spreading behaviors for joining applications. Sessile-drop wettability tests with Ag and Pd-base active brazes on C–C and SiC–SiC composites reveal (Asthana *et al.*, 2010) non-uniform, anisotropic spreading, copious braze infiltration and Ti enrichment at interfaces, together with dissolution and redistribution of elements in and around the joint between the droplet and the CMC substrate, and some tendency towards interlaminar shear cracking within the composite substrates having low interlaminar shear strength. Figure 11.2 shows a solidified Ticusil braze droplet on a C–C composite where good wetting was accompanied by braze infiltration and edge cracking due to residual stresses. Contact angle data extracted from such high-temperature sessile-drop tests provide insight into braze spreading on real, heterogeneous composite surfaces. These measured angles are, however, not the equilibrium angles as they do not satisfy the Young–Dupre equation.

11.2.2 Infiltration

Brazed joints frequently exhibit braze penetration in composites that usually have a substantial degree of interconnected porosity (Fig. 11.2). The filler metal is drawn (wicked) into the porous substrate, and may leave metal-starved regions at the mating surface, which could hinder bond formation. Braze penetration in the porous substrate can occur if the critical wetting condition is satisfied, i.e. the contact angle $\theta < 90°$. This is because for spontaneous pore penetration to occur the capillary pressure, P_c (the pressure differential across the liquid front at the pore entrance) should be negative, where $P_c = -2\sigma_{lv}\cos\theta/r$, and σ_{lv} is the surface tension of the braze and r is the effective pore radius. Thus, for $\theta < 90°$, $P_c < 0$, and wicking of molten braze will occur. For $\theta > 90°$, $P_c > 0$, and external pressure will be needed to drive the liquid into the pores. Additional pressure drop will arise from forces due to fluid drag, gravity and repeated compression, decompression and changes in the direction of fluid flowing through the tortuous

(a)

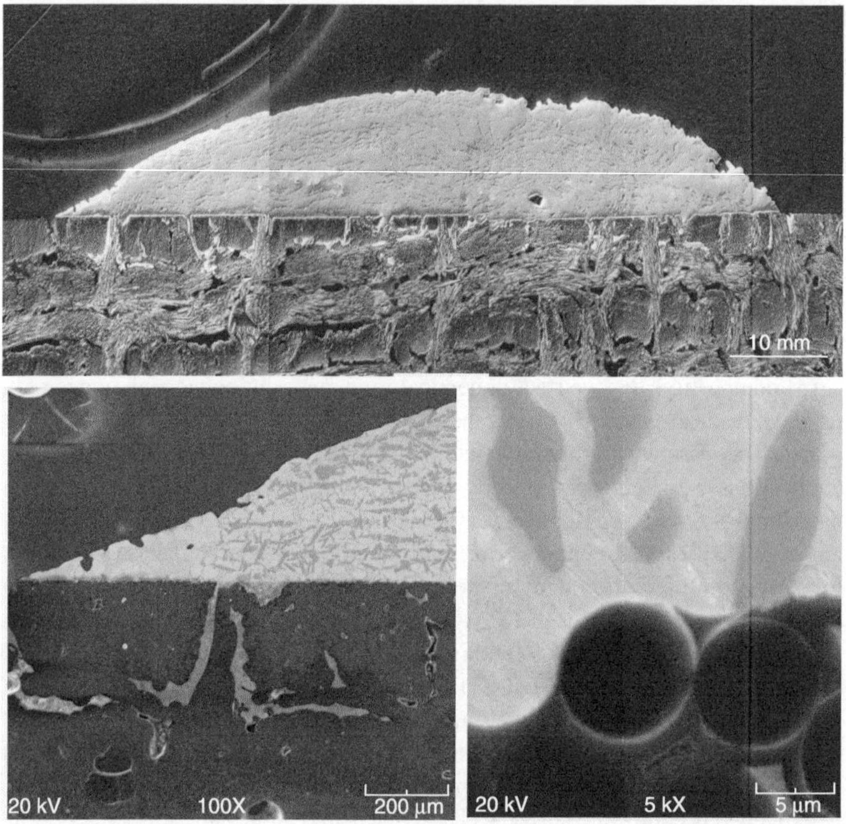

(b) (c)

11.2 (a) A sessile drop of Ag–Cu–Ti (Ticusil) braze on a C–C composite, (b) braze infiltration of porous C–C, and (c) interface between C–C and Ticusil droplet.

structure of the porous substrate. These forces govern the equation of fluid motion and the kinetics of flow in pores.

The extent of penetration of a reactive filler in porous substrates will depend also upon the volumetric changes accompanying the reaction product formation, and the porosity and continuity of the reaction layer. For example, Cu–Ti alloys both wet and impregnate the porous graphite, but Cu–Cr alloys form a reaction layer that is very dense and effectively seals the open pores in the carbon, thus blocking melt penetration. With Ti in place of Cr as an alloying element in Cu, the reaction-formed TiC layer is discontinuous, with a non-homogeneous structure, which allows pore penetration by the molten braze and continued infiltration (Sobczak *et al.*, 1997) until either the liquid supply is exhausted or the substrate becomes fully saturated with molten braze. Even in the case of a dense and impervious reaction layer, the relative rates of melt penetration and reaction layer formation will play an important role in determining the penetration kinetics; if the reaction rate is slow, then significant melt penetration could occur before the reaction products choke the liquid flow. These effects have been theoretically modeled and increase understanding of the process mechanics and kinetics (Asthana, 2002, 2000; Sangsuwan *et al.*, 2001).

11.3 Brazing ceramic-matrix composites

Both CMCs and monolithic ceramics are brazed using methods based on metallization, active brazing, use of ductile interlayers and use of composite braze fillers. The wetting behavior and reactivity of the composite constituents (e.g. fiber and matrix) and the compatibility of the brazing process with the composite fabrication technique must be addressed when joining ceramic-matrix composites to metals. Continuous fiber-reinforced composites demand a different approach to joining from particle or short fiber-reinforced composites. Because the fiber arrangement influences the component performance, then the joint design and brazing process should preserve the integrity of the fiber arrangement. Conversely, fiber orientation at the joined surface has been shown to influence the shear strength of the joint (Janczak-Rusch, 2011; Morscher *et al.*, 2006). Joints brazed parallel to the fiber direction have different strengths from those brazed in a transverse direction. Additionally, different interface morphologies and interfacial areas in fiber and particulate-reinforced CMCs will affect the wetting, braze spreading, chemical reaction and load-transfer characteristics. Figure 11.3 shows a solidified brazing front on a SiC fiber in a SiC–SiC composite in a partially wetted region of SiC–SiC, the reaction layer, and transverse fiber cracks due to local stresses. Braze fillers that produce strong joints in self-joined ceramics may

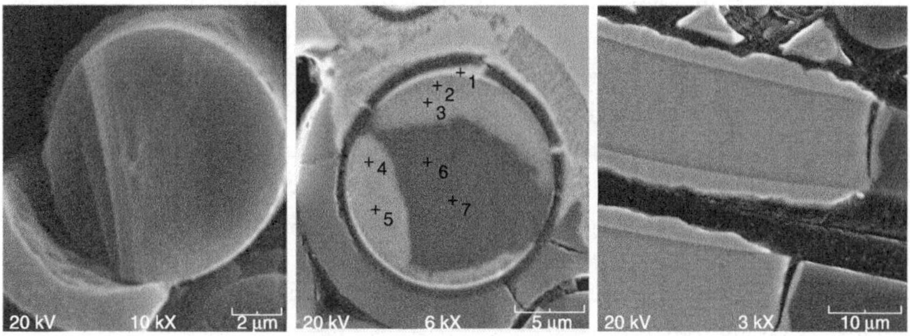

(a) (b) (c)

11.3 (a) SiC fiber in a CVI SiC–SiC composite before contact with molten braze, (b) SiC fiber partially wetted with braze Cusil-ABA, (c) partially wetted SiC fibers in a SiC–SiC composite showing surface interaction zone and transverse cracks near the wetting front.

not be best for ceramic/metal joints unless such fillers have high ductility and can accommodate residual stresses arising from the large CTE mismatch between the CMC and metals.

CMC/metal joint strength depends on a number of factors, including active metal content and the gap thickness. In the case of Ti-active fillers, relatively large (but typically less than 10%) Ti contents yield high joint strength; Ti contents greater than ~10% impair the bond strength. Alloying additions such as indium to Ag–Cu–Ti brazes increases Ti activity, thus permitting a reduction in the concentration of Ti added to braze (Nicholas, 1998). The enhanced Ti activity promotes braze reactivity and formation of titanium carbide and other phases. Theoretical models for the reaction layer growth in brazing have been attempted. For example, Torvund *et al.* (1996) modeled the coupled reaction layer growth in reactive brazing of ZrO_2-toughened Al_2O_3 with Ag–Ti filler metals, taking into account both the evolution of the titanium oxide layer at the ceramic/braze metal interface and the subsequent decomposition of the reaction products during cooling. Their model calculates individual reaction layer thicknesses and relates these directly to the concentration of the reactive element in the braze alloy.

The effect of gap thickness on CMC/metal joint strength is difficult to generalize; whereas some studies show that a small gap thickness favors high joint strength (Janczak-Rusch, 2011), others (Steffier, 2004) point out that joint strength is increased by increasing the braze layer thickness (e.g. in SiC–SiC composites brazed to 304 SS and OFHC Cu using AgCuTi braze fillers, quadrupling the braze layer thickness doubled the joint strength (Steffier, 2004)).

An important consideration in brazing CMCs is the effect of surface coatings on the brazing process. Multi-layer, multifunctional coatings are frequently applied to fibers and to CMCs to serve as a diffusion barrier to resist oxidation and corrosion, inhibit thermal degradation and promote compatibility with various matrices. For example, C–C composites are coated with SiC or Si_3N_4, and they may also contain internal oxidation inhibitors such as boron, titanium and silicon. At high operating temperatures, these elements form glassy oxide phases within the C–C composite, which seal any CTE mismatch induced cracks in the external coating. This creates a barrier to the diffusion of oxygen into the porous composite matrix and retards its oxidative degradation. An important question in joining CMCs is whether the external coating should be applied before or after brazing. This is important because most external coatings to CMCs are applied at temperatures (~1273–1673 K) that exceed the melting temperatures of common braze fillers. In addition, the effect of glass-forming inhibitors added to the composite on the wettability and adhesion of the braze must be considered.

Although a number of variants of the brazing process to join CMCs and ceramics exist, in essence, the basic process consists of placing the braze foil or paste between the metal and the composite substrates, heating the assembly to brazing temperature under vacuum (~10^{-6} torr), and isothermally holding for a predetermined time at the brazing temperature, followed by slow cooling to room temperature. Issues concerning proper fixturing, load application and the possibility of joint contamination from physical contact with a fixture need to be addressed and resolved. CMCs such as SiC–SiC, C–SiC, C–C and ZrB_2-based CMCs have been brazed to themselves and to high-temperature metals and alloys such as Ti, Cu-clad-Mo, and Ni-base superalloys (Inconel 625 and Hastelloy). Table 11.2 lists some proven CMC/metal joints.

11.3.1 Silicon carbide-silicon carbide composites

Heat- and wear-resistant SiC–SiC composites find applications in combustor liners, exhaust nozzles, re-entry thermal protection systems, hot gas filters, high-pressure heat exchangers, and fusion reactors (Lamon, 2005; DiCarlo *et al.*, 2005). Table 11.3 lists some SiC–SiC and C–SiC composites together with some of their properties.

Table 11.2 CMC/metal brazed joints

Composite	Metallic substrate	Braze
C–C[a,f]	Ti	Silcoro-75,[h] Palcusil-15[h]
C–C and SiC–SiC	Ti	Ticuni, Cu-ABA, Ticusil
C–SiC[i]	Ti and Hastelloy	MBF-20[h]
C–C[a,f]	Ti and Hastelloy	MBF-20,[h] MBF-30[h]
SiC–SiC	Hastelloy	MBF-20,[h] MBF-30[h]
C–SiC[i]	Hastelloy	MBF-30[h]
C–SiC[a,i]	Ti, Inconel 625	MBF-20,[h] MBF-30[h]
SiC–SiC	Ti	MBF-20[h]
SiC–SiC	Ti	MBF-30[h]
C–SiC[a,i]	Inconel 625	Incusil-ABA[h], Ticusil[h]
C–SiC[a,i]	Inconel 625	Cu-ABA,[h] Cusil,[h] Cusil-ABA[h]
C–SiC[a,b,i]	Ti	Cusil-ABA[g,h]
SiC–SiC[a,b,j]	Ti	Cusil-ABA[g,h]
C–SiC[b,i]	Ti, Inconel 625, Cu-clad Mo[k]	Ticusil[g]
SiC–SiC,[b,j] C–SiC,[a,i] C–C[f] (T 300)	Ti, Inconel 625, Cu-clad Mo[k]	Ticusil[g]
C–C[c,d,e]	Ti, Inconel 625, Cu-clad Mo[k]	Ticusil[g]
C–C,[c,d,e] C–SiC,[a,b,i] SiC–SiC[a,b,j]	Cu-clad Mo[k]	Cusil-ABA[g]
C–C[c,d,e]	Ti and Inconel 625	Cusil-ABA[g]
SiC–SiC[a,b,j]	Ti	Cusil-ABA[g]
C–SiC[a,b,i]	Ti	Cusil-ABA[g]
ZS, ZSS, ZSC	Ti, Inconel 625, Cu-clad-Mo	Cusil-ABA, Ticusil, MBF-20, MBF-30, Palco, Palni

[a]Polished; [b]not-polished; [c]3-D composite (CVI carbon matrix), Goodrich Corp., CA; [d]oriented fiber at the joint (3-D composite); [e]non-oriented side at the joint; [f]T 300 carbon fibers in resin-derived C matrix, C-CAT, TX; [g]braze paste; [h]braze foil; [i]T300 carbon fiber in CVI SiC matrix, GE Power System Composites, DE; [j]Sylramic SiC fiber in MI-SiC matrix, GE Power System Composites, DE; [k]H.C. Starck, Inc., MA.

Table 11.3 Examples of C–SiC and SiC–SiC composites

Composite	UTS (MPa)	*E* (GPa)	Flexural strength (MPa)	ILSS (MPa)	CTE ($\times 10^{-6}$/K)	*K* (W/m.K)
CVI C–SiC[d] (42–47% fiber)	350	90–100	500–700	35	3.0[a] 5.0[b]	14.3–20.6[a] 6.5–6.9[b]
LPI C–SiC[d]	250	65	500	10	1.16[a] 4.06[b]	11.3–12.6[a] 5.3–5.5[a]
HiPerComp SiC–SiC (22–24% fiber)[c]	–	285	–	135[c]	3.5[a] 4.07[b]	33.8[a] 24.7[b]
2D SiC–SiC (0/90 Nicalon fabric, 40% fiber)[a]	200	230	200	40	3.0,[e] 1.7[f]	19,[e] 9.5[f]
NASA 2D SiC–SiC panels (N24-C)[b]	450	210	–	<7	4.4[g]	41[h]

[a]Lamon (2005) ; [b]DiCarlo *et al.* (2005); [c]Corman and Luthra (2005); [d]Krenkel (2005); [e]in-plane value; [f]through-thickness value; [g]average value from RT to 1000°C, extracted from % expansion data in DiCarlo *et al.* (2005); [h]averaged value.
ILSS: interlaminar shear strength; UTS: ultimate tensile strength.

These composites have been brazed to themselves and to Kovar, Ti, Inconel and other high-temperature alloys with and without interlayers using Ag–Cu–Ti, Ni–Cr–B, Si–Ti, Si–Cr and Si–Ti alloys. Singh and Lara-Curzio (2001) were the first to demonstrate joining of SiC–SiC composites using Si–Ti alloys. Later, Riccardi *et al.* (2004a, 2004b) joined SiC–SiC using Si–16 at.%Ti eutectic alloy. These authors also vacuum brazed SiC–SiC composites using Si–44Cr at.% eutectic and the intermetallic $CrSi_2$ with melting temperatures of 1390°C and 1490°C, respectively. These temperatures are low enough to avoid fiber degradation in contact with Si–Cr alloys. Riccardi *et al.* noted that the joint interfaces were well bonded, defect-free and atomically sharp, and Si–Si, Cr–C and Si–Cr bond formation promoted the adhesion. The formation of atomically sharp joints suggests direct chemical bonding without interdiffusion or phase formation. The joints had high (~71 MPa) strength at room temperature and at 600°C, and failure occurred in the base material rather than in the joint, indicating that the joints were stronger than the substrates. Liu *et al.* (2010) developed a composite joining technique, using a Ni–56Si filler alloy and Kovar and Mo as interlayers. They also conducted a sessile-drop wettability test with Ni–Si alloy on SiC, which showed the non-reactive wetting characteristics with an equilibrium contact angle of 23°. The joints were fabricated by using a Ni–Si/Mo/Ni–Si structure as the interlayer. Two interfacial layers formed at the Kovar/Mo and the Mo/Ni–Si interfaces due to dissolution and interdiffusion without there being any observable reactions with the SiC.

In another study, Steffier and Tengar (2001) brazed joints of SiC–SiC composites to 304 SS and OFHC Cu using Ag–Cu–Ti brazes (Cusil-ABA and Ticusil). Using single-lap brazed joints for shear strength testing, they obtained a maximum joint strength of 25 MPa and noted that joints could be made stronger by utilizing a stepped lap joint with a series of short steps (Fig. 11.4(d)); joints could not be made stronger simply by increasing the overlap length.

Various types of SiC fibers (Nicalon, Tyranno, Sylramic, etc.) are used to fabricate SiC–SiC CMCs, and variations in their composition, morphology, surface characteristics, weave pattern, and physical and mechanical properties determine the CMC properties as well as joining response. For example, SiC fiber bonded ceramics (SA-Tyrannohex) fabricated with parallel and perpendicular Tyranno-SA™ fibers are dense, strong, tough and thermally conductive, and have been proposed for use in advanced fusion and fission reactors, hot section components of aircraft gas turbines and other demanding applications because of their excellent high-temperature stability. Recently, these CMCs have been brazed to themselves using Ag–Cu–Ti braze alloys (Matsunaga *et al.*, 2011). Brazing led to the formation of an interaction zone (~1–2 μm thick), enriched in Ti, C and Si with the Si residing between the Ag–Cu eutectic braze alloy and a Ti–C layer (presumably TiC and sub-stoichiometric carbides), which promoted wetting and bonding. Thus, the TiC phase might form in the vicinity of the SiC fibers at first, and then Si might diffuse through

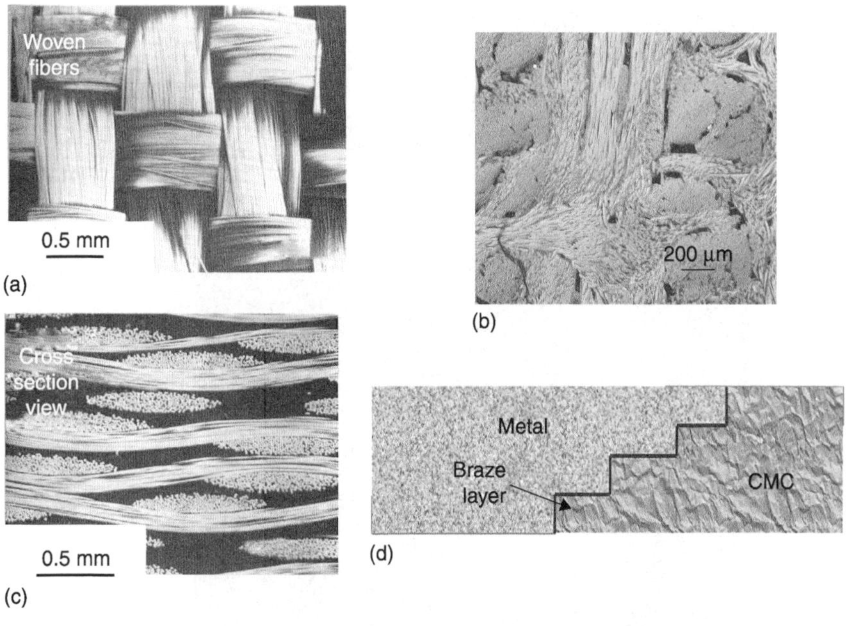

11.4 (a)–(c) Morphological inhomogeneity from fiber distribution, and (d) schematic of a stepped CMC/metal joint for improved joint strength (after Steffier and Tengar, 2001); (a) and (c) are from French and Cass, 1998.

the TiC layer and react with Ti to form titanium silicides. Microscopic examination revealed that the reacted TiC was about 0.5 μm thick (Matsunaga *et al.*, 2011).

The formation of TiC and titanium silicides ($TiSi_2$, Ti_5Si_3, $Ti_5Si_3C_x$) in the joint is thermodynamically feasible ($\Delta G < 0$); $TiSi_2$ and Ti_5Si_3 are, however, less probable than TiC (titanium silicides could form during cooling and solidification following dissolution of SiC in molten braze together with the saturation of the melt with Si). Extensive chemical reactions together with the formation of Ti–C, Ti–Si, Cu–Si and Cu–Ti phases have been reported in both brazed and diffusion bonded SiC–SiC joints in which Ti was either an ingredient of the braze or used as an interlayer (as in diffusion-bonded couples).

Composites made using various other types of SiC fibers also have been brazed. These include CVI SiC–SiC and slurry-cast, melt infiltrated (MI) SiC–SiC composites brazed to Ti, Inconel 625 and Hastelloy using Ag–Cu–Ti and amorphous Ni–Cr–Si–B brazes (Singh *et al.*, 2008a; Singh and Asthana, 2008b, 2011a). Eutectic refractory metal powders Si–B and Si–Y in conjunction with amorphous Ni-brazes were also used in selected systems to promote dissolution, chemical reaction and reinforcement of the braze matrix to achieve sound bonding. Figures 11.5–11.8 show SEM views of joint interfaces in some SiC–SiC composite systems. Microscopic examination revealed that, whereas both Si–Y (Fig. 11.7) and Si–B (Fig. 11.8) eutectic powders

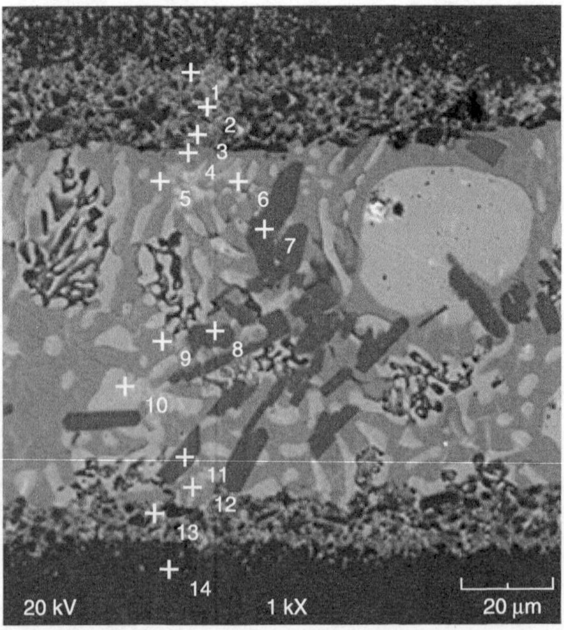

11.5 A SiC–SiC/Cusil-ABA/SiC–SiC joint showing the interaction zone and microstructure of the braze matrix. The numbers 1–14 denote point markers at which chemical composition was accessed using energy dispersive spectroscopy (EDS).

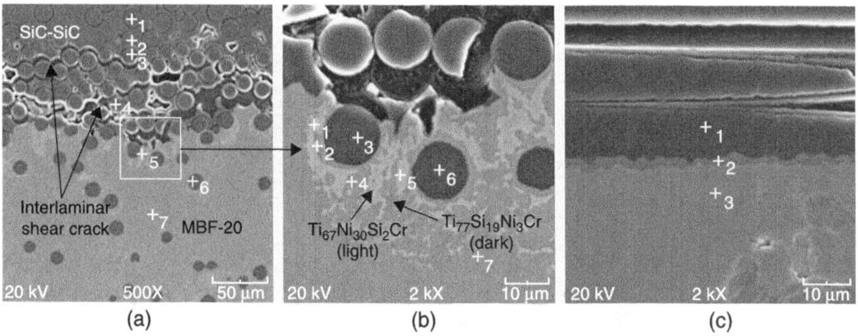

11.6 (a) and (b) SiC–SiC/MBF-20/Ti joint showing good chemical bonding but also interlaminar shear cracking in the composite, and (c) SiC–SiC/MBF-20/Si–B/Ti joint.

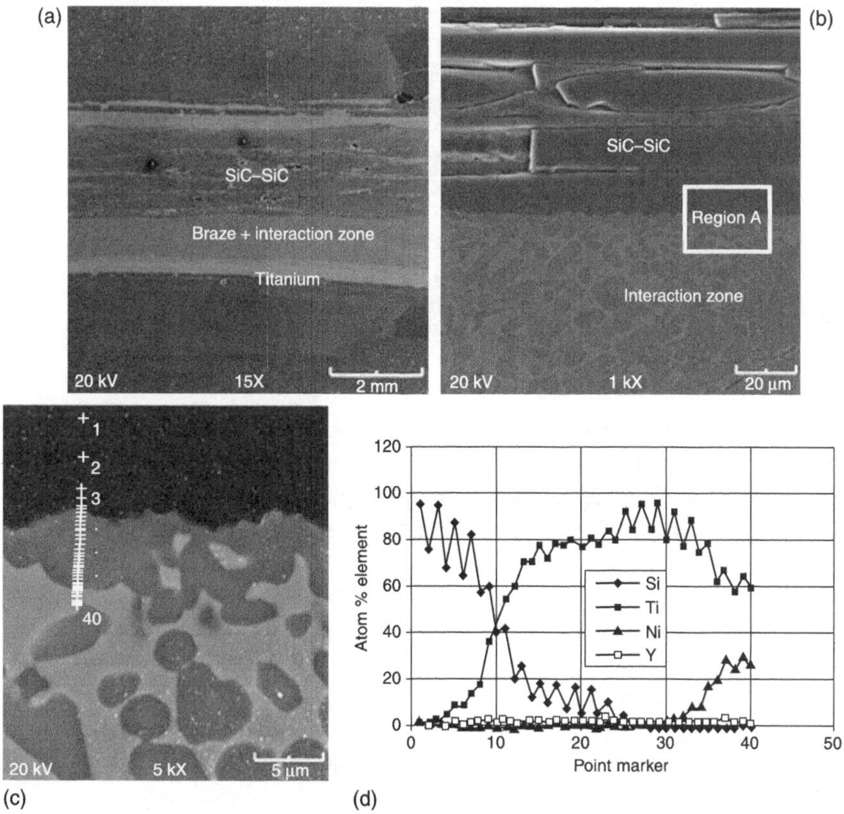

11.7 SiC–SiC/MBF-20 + Si–Y eutectic/Ti joint: (a) overall view, (b) higher magnification view of CMC/braze interface, (c) higher magnification view of region A shown in (b), and (d) elemental compositions at CMC/braze interface of (c).

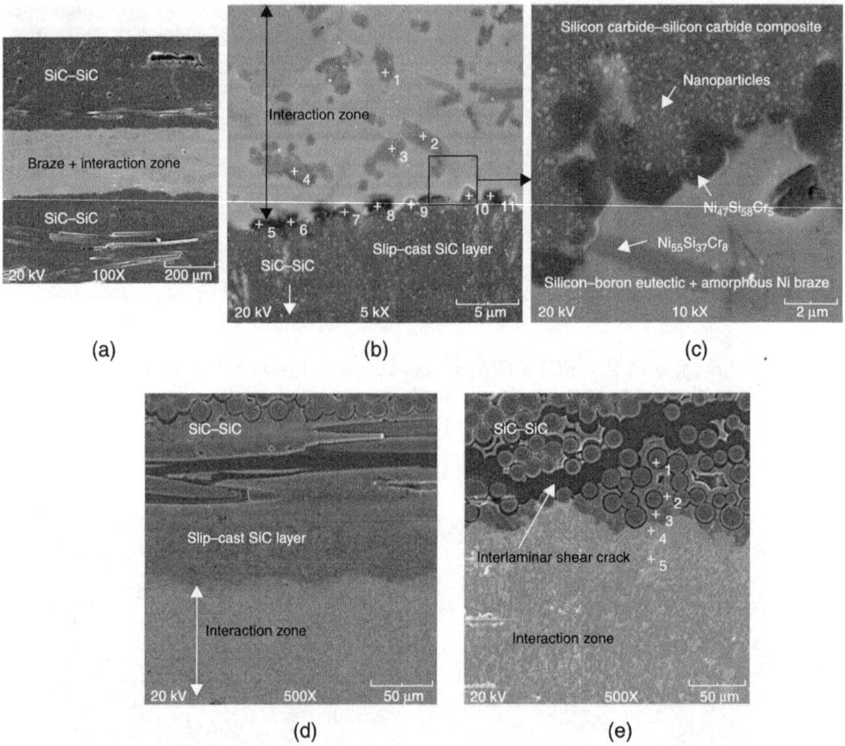

11.8 (a)–(c) SiC–SiC/MBF-20 + Si–B eutectic/SiC–SiC joint showing overall view and braze/CMC interface region; (d) and (e) show a SiC–SiC/MBF-20 + Si–B eutectic/Ti joint showing interlaminar shear cracking in the CMC.

facilitated SiC–SiC/Ti bonding, residual stresses in joints made using Si–B caused interlaminar shear (ILS) failure within the composite (Fig. 11.8(d) and (e)). CMCs with higher ILS strength and stress absorbing compliant layers have been proposed to solve the problem. The Knoop hardness distribution across SiC–SiC/Ti joints made using an Ag–Cu–Ti braze (Ticusil, Table 11.1) is shown in Fig. 11.9. Sound joints with a smooth hardness gradient were formed with this system. Preliminary test results on shear strength of SiC–SiC/Ti joints with and without polished surfaces are shown in Fig. 11.9(b) and compared with the shear strength of similar joints in C–SiC/Ti and C–C/Ti systems. A significant improvement in the shear strength of SiC–SiC/Ti joints occurred when polished CMC substrates were used for bonding; however, no beneficial effects of polishing were realized in C–SiC/Ti joints.

11.3.2 Carbon–carbon composites

Carbon–carbon composites are used in nose cones of ballistic missiles, nose cap and leading edges of the Space Shuttle, brake systems for aircraft and automobiles,

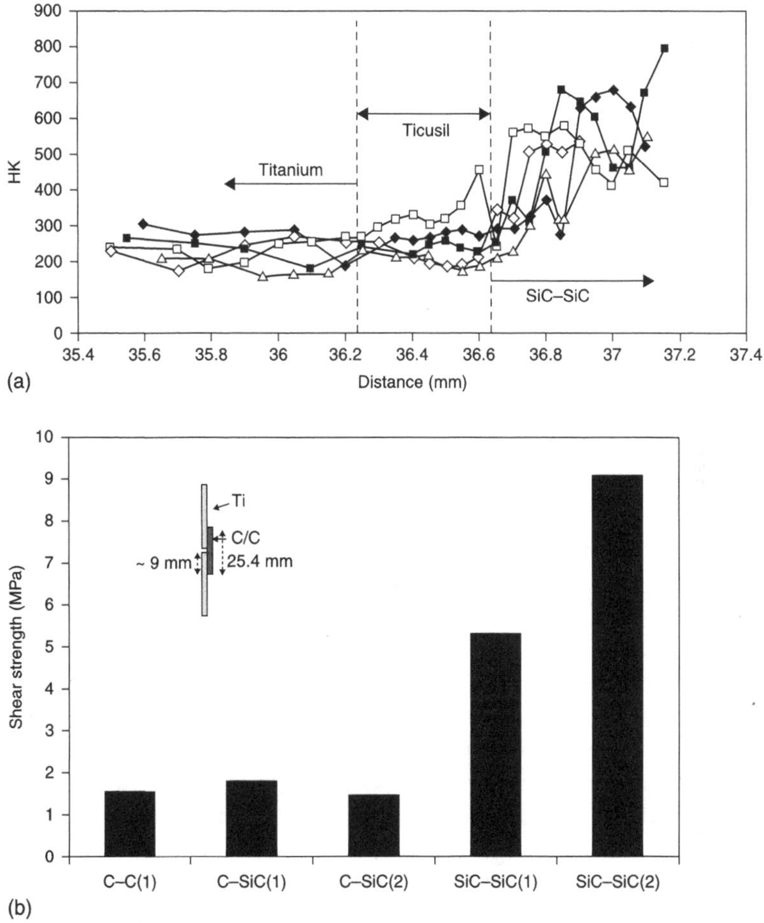

11.9 (a) Knoop hardness (HK) distribution across a SiC–SiC/Ticusil/Ti joint and (b) comparison of the shear strength of brazed SiC–SiC/Ti joints with the shear strength of similar joints between Ti and C–C and C–SiC composites. (1) As processed; (2) polished.

and high heat-flux components in thermonuclear reactors. The materials are suited for structural applications at high temperatures, or where thermal shock resistance and/or a low coefficient of thermal expansion is needed. C–C composites containing high-conductivity carbon fibers have proven to be outstanding candidates for heat dissipation and have low expansion properties for a reduced weight. The axial conductivities of high-modulus (HM) and ultra-high modulus (UHM) carbon fibers are ~120–300 W/m.K and ~500–1100 W/m.K, respectively (Taylor, 2000). The composites have been successfully brazed to a variety of metallic alloys, including Nimonic alloys (Moutis *et al.*, 2010), titanium (Singh *et al.*, 2005, 2007, 2008; Qin and Feng, 2007), titanium aluminide (Wang *et al.*, 2010), copper alloys (Salvo *et al.*, 1995; Appendino *et al.*, 2004; Centeno *et al.*,

2011), copper-clad-molybdenum (Singh *et al.*, 2007, 2008a), stainless steel (Liu *et al.*, 1994) and other alloys using a wide variety of filler metals. Collectively, these studies have examined the thermodynamics and kinetics of interfacial reactions, secondary phase precipitation, interface structure formation and mechanical properties.

With Ag–Cu–Ti active braze fillers, the joints between C–C composites and titanium form titanium carbide (TiC) and sub-stoichiometric carbides such as $TiC_{0.95}$, $TiC_{0.91}$, $TiC_{0.80}$, $TiC_{0.70}$, $TiC_{0.60}$ and $TiC_{0.48}$. Additionally, a number of Cu–Ti intermetallic compounds such as TiCu, Ti_3Cu_4, Ti_2Cu_3, Ti_2Cu, etc. also form. The spatial arrangement of these reaction layers in a joint is determined by the reaction stoichiometry; for example, it has been reported (Qin and Feng, 2007) that TiC and TiCu layers form near C–C while Ti_3Cu_4 and Ti_2Cu layers form near the Ti substrate. The reaction layer thickness increases with increasing brazing temperature and time. In active braze fillers that contain Si besides Ti (e.g. Cu-ABA), both silicon carbide and titanium carbide could form, as discussed in Section 11.2.1. The effect of joint configuration, geometry and interlayers has also been characterized. For example, fine holes drilled in the C–C composite surface strengthen C–C/metal joints made using Ag–Cu–Ti braze because of braze infiltration and solidification in the holes (Wang *et al.*, 2010).

Pd-base fillers offer higher use temperatures than Ag–Cu–Ti alloys. Commercial Pd–Co and Pd–Ni braze fillers have very good wettability with C–C (Asthana *et al.*, 2010), and the contact angles have acute values. With the addition of Si and Cr to Pd–Ni alloys, further improvements in wettability and reactivity occur; the wettability of Pd–Ni–Si–Cr braze alloy in contact with C–C composites increases with an increase in Cr content (Chen *et al.*, 2010) and is due to the formation of the chromium carbide, $Cr_{23}C_6$. Cooling and solidification of braze forms additional phases in supersaturated melt, such as the Pd–Si intermetallics, Pd_2Si and Pd_3Si. As a result of reactions and solidification, the braze alloy in the joint region becomes Ni-rich and Pd-depleted but develops sound bonding with C–C.

High heat-flux components in thermonuclear reactors require C–C composite-to-Cu alloy joints that possess excellent resistance to thermal fatigue. Such joints are usually created via brazing with or without low-CTE interlayers (Centeno *et al.*, 2011; Garcia-Rosales *et al.*, 2009; Appendino *et al.*, 2004; Casalegno *et al.*, 2009). Ferraris and coworkers investigated the joining response of C–C composites for such applications. They (Appendino *et al.*, 2004; Casalegno *et al.*, 2009) brazed Si-doped C–C composites to Cu and CuCrZr alloys using several different approaches. In one approach, Cr and Mo were deposited onto the C–C surface via slurry deposition followed by heat treatment to form micrometer-sized carbide layers. The carbide-coated surface was brought into contact with copper foil or paste, which, upon heating, led to capillary infiltration of the porous carbide layer and formed a pore- and crack-free joint. These joined samples could then be directly brazed to other alloys. In another approach (Appendino *et al.*, 2003), a Ti–Cu–Ni braze alloy was used for joining. The joints were stable to thermal fatigue to about 50 cycles over a 30°C to 450°C range, and the joint strength was comparable to the interlaminar shear strength of the C–C composite. Earlier, Ferraris and coworkers (Salvo *et al.*, 1995) had used soft and ductile rheocast Cu–Pb alloys as a thermal bond layer. Rheocast alloys have a non-dendritic, globular microstructure and excellent flow characteristics, and are formed by mechanical or electromagnetic stirring of partially solidified alloy slurries and prolonged holding in the partially solidified state to enable coalescence and globularization of primary solid dendrites. The method did not prove to be very successful because of the poor wetting characteristics (contact angle: 101–104° at 983 K under Ar) of Cu–Pb alloys on C–C, which led to very low (~1.5–3 MPa) joint strength. Wettability enhancement either by use of active fillers or by judicious surface modification could improve the joint strength.

Figure 11.10 shows joint interfaces in different C–C-to-metal brazed joints. Interface microstructures at both CMC/braze interface (Fig. 11.10(a)–(d)) and braze/metal substrate interface (Fig. 11.10(e)–(g)) show that void- and crack-free joints exhibiting intimate physical contact had formed in all cases. In joints made using Ti-bearing fillers, Ti preferentially segregated at the C–C/braze interface. Interdiffusion of alloying elements was noted for the C–C composite braze matrix and metal substrate (e.g. Ti and Cu-clad-Mo). In the case of a Cu-clad-Mo substrate joined to a C–C composite using Ag–Cu–Ti active braze alloys, the clad layer remains untransformed because the joining temperatures are below the melting point of Cu (1359 K). This provides for stress accommodation by the copper layer, aiding the stress relief provided by the ductile brazes. The braze matrix of Ag–Cu–Ti exhibits the characteristic two-phase eutectic structure with Ag-rich light-gray areas and Cu-rich dark areas (Fig. 11.10(a)). Intermetallics such as AgTi, Ti_2Cu_3 and $TiCu_2$ are also known to form in these alloys. Carbon–carbon composites have a relatively porous structure and active braze alloys can infiltrate the inter-fiber regions to distances of the order of several hundred micrometers during short (~5 min) brazing times (Fig. 11.10(c)). It has been reported that sessile drops of an active Cu–Ti braze alloy on porous graphite substrates continuously shrank in volume because of reactive infiltration of the open pores in the carbon; in fact, Cu–Ti sessile drops with high Ti content (~28% Ti) rapidly and completely disappeared into a porous graphite substrate. A porous substrate could draw the molten braze away from the interface, leaving an insufficient quantity of braze for reaction and joint formation. As a result the benefits of frictional bond formation from infiltration and solidification of the braze within surface pores could be achieved at some expense to the reactions and bond formation at the mating surface.

Carbon–carbon composites have been integrated with titanium tubular subcomponents for use in Brayton and Thermoelectric power conversion systems. Titanium is light (density: 4500 kg.m^{-3}) and has a relatively low CTE (~8.6×10^{-6}/K) that is moderately different from the CTE of C–C composites (~$0–1.0 \times 10^{-6}$/K over 20–250°C). These properties make Ti suitable for tubular components

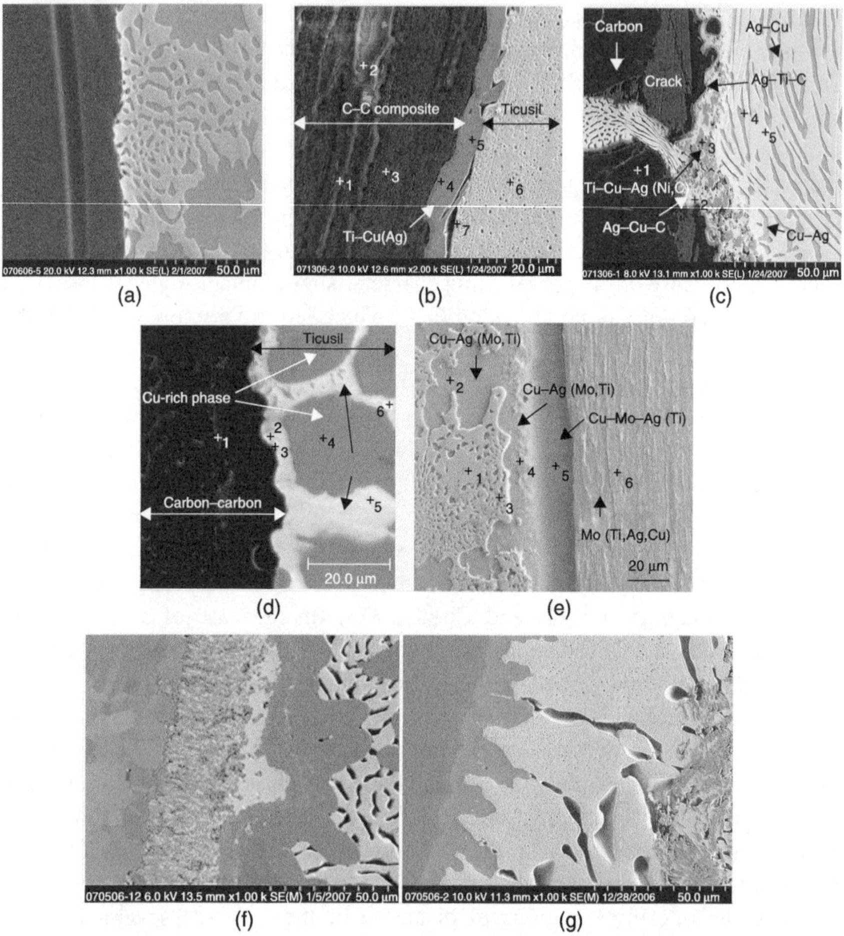

11.10 (a)–(d) C–C/braze interface in (a) C–C/Cusil-ABA/Cu-clad Mo joint, (b) C–C/Ticusil/Ti joint, (c) C–C/Ticusil/Inconel 625 joint, and (d) C–C/Ticusil/Cu-clad Mo joint; (e) Cu-clad Mo/Cusil-ABA interface, (f) Ti/Cusil-ABA interface, and (g) Inconel 625/Cusil-ABA interface.

carrying hot or cold fluid for thermal management. For a lightweight heat exchanger application, Ti tubes had to be joined to a high-conductivity C–C composite using a flexible, low-density saddle material (Poco graphite foam), which could accommodate thermo-elastic strains and prevent cracking. The foam has a small coefficient of thermal expansion, which compares favorably with the CTE of the C–C composite (the in-plane CTE of Poco foam is 1.02×10^{-6}/K and the out-of-plane CTE is -1.07×10^{-6}/K). This is beneficial in reducing residual stress and distortion. When a C–C composite is joined to Poco foam and Ti using Ag–Cu–Ti brazes (Singh *et al.*, 2008b), gaps and pores in the foam are infiltrated

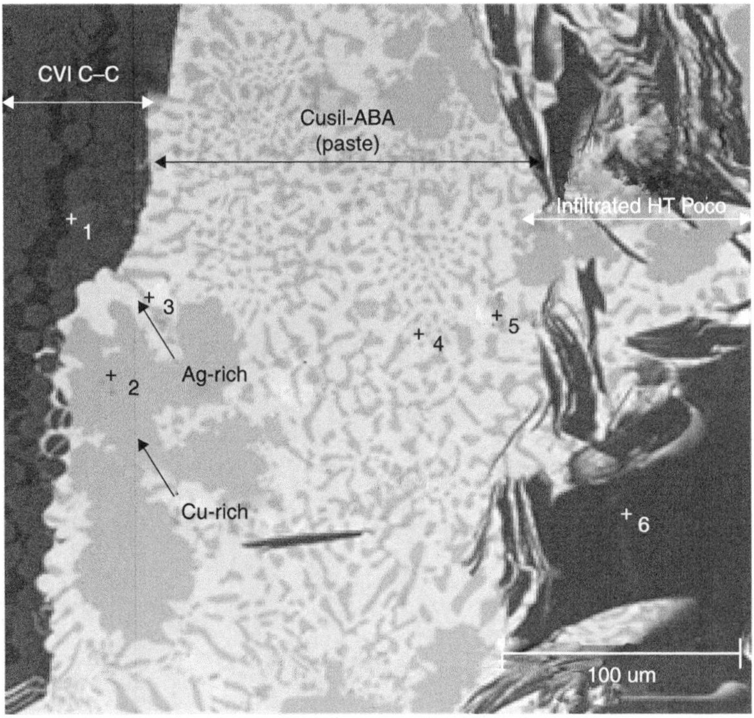

11.11 C–C composite/Cusil-ABA/HT Poco foam joint microstructure.

(braze paste exhibits better penetration characteristics than braze foil), and the joints exhibit sound bonding (Fig. 11.11).

Tensile and shear strength tests were conducted on C–C/Ti joints with and without a saddle material (graphitic foam) using different brazes and with different fiber orientations at the mating surface. The results show that the shear strength of directly bonded C–C/Ti joints, extracted from butt-strap tensile (BST) tests, is low (1.5–9.0 MPa, ±0.09–1.62 MPa) (Morscher *et al.*, 2006). Polished C–C samples yield slightly higher joint strength than unpolished samples. For C–C/Ti joints in which Ti tubes were brazed to C–C plates, the average shear strength depended not only on braze type but also on the orientation of the exterior carbon fiber tow at the joint. Joints with fiber tows perpendicular to the Ti tube axis exhibited higher failure loads than joints with parallel fibers (Morscher *et al.*, 2006). Evidently, this is due to the greater number of tows that are brazed to the Ti tube when fiber tows are oriented perpendicular to the tube axis compared with the parallel orientation. Figure 11.12(a) shows the tube-on-plate shear strength for joints made using three different active braze alloys (Ticuni, Ticusil and Cusil-ABA; Table 11.1). The standard deviation around the strength is in the range ±0.04–0.13 MPa. The load-carrying ability of the joints was affected by

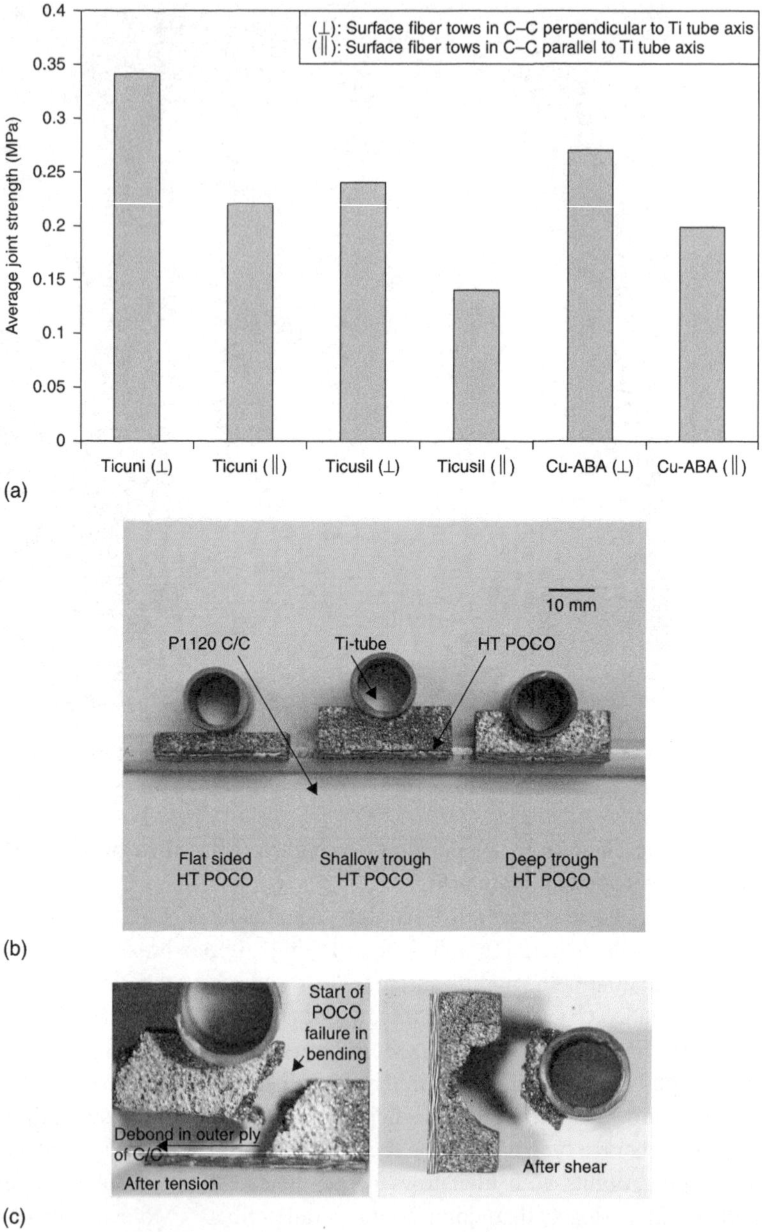

11.12 (a) Joint strength in C–C/Ti joints made using three different brazes with C fiber tows in the composite surface layer oriented either parallel or perpendicular to the titanium tube axis, (b) Ti tube/Poco foam/C–C composite joints made using flat and grooved foam, and (c) fracture surface of Ti tube/Poco foam/C–C composite joints made using Cu-ABA braze.

the bonded area and the fiber-tow orientation. For example, joints in which surface fiber tows were perpendicular to the titanium tube axis had higher joint strength than those in which surface carbon fiber tows were parallel to the tube axis (Fig. 11.12(a)).

For a Ti tube/foam/C–C plate sandwich structure, tension and shear tests were performed on flat and grooved foam surfaces (Fig. 11.12(b)) (Singh *et al.*, 2008b). It was found that the shallower the groove, the higher the stress on the tube-to-foam joint. The joint strength was greater than 12 MPa in tension or shear. Failure always occurred in the foam (Fig. 11.12(c)) regardless of the type of C–C composite used or whether the Ti tube was brazed to a curved foam plate to maximize the bonded area or to a flat foam surface to maximize stress in the joint.

Joining a C–C composite to copper-clad molybdenum can enhance thermal conduction while minimizing the residual stresses from the CTE mismatch. Theoretical projections of the effect of clad layer thickness on the effective thermal resistance of Cu-clad Mo joined to different CMCs are shown in Fig. 11.13 based on one-dimensional, steady-state heat conduction through flat joints. Joining is beneficial because the conductivity and CTE of Cu-clad Mo depend upon the clad layer thickness (Harper, 2003); for example, the thermal conductivity of Cu-clad Mo (27% Cu thickness per side of Mo substrate) is ~224 W/m.K and the CTE of Cu-clad Mo (~8.2×10^{-6}/K for 27% clad layer thickness) is about half that of pure Cu (16.5×10^{-6}/K) and only moderately high compared with the CTE of C–C (~2.0–4.0×10^{-6}/K). Additionally, the high ductility of the copper-clad layer accommodates the stresses. By controlling the clad layer thickness on molybdenum via rolling, the CTE mismatch between C–C and Cu-clad-Mo can be designed to mitigate the stresses while retaining high levels of thermal conductivity. The interface microstructure in C–C composites joined to Cu-clad Mo (Fig. 11.10) shows excellent bonding.

11.3.3 Carbon–silicon carbide composites

Carbon fiber-reinforced silicon carbide (C–SiC) composites are used in aerospace and automotive applications, and are considered to be more durable than C–C. The composites are produced by CVI, polymer infiltration and pyrolysis (PIP) and silicon MI, and have been evaluated for use in industrial gas turbine engines, combustor liner components, shrouds and expansion nozzles of rocket propulsion systems, as well as exhaust cones, engine flaps and flame holders of jet engines. Table 11.3 gives representative mechanical and thermal properties of some C–SiC composites.

C–SiC has been brazed to titanium, Inconel 625, niobium, copper-clad molybdenum and other metallic substrates (Liu *et al.*, 2011; Wang *et al.*, 2011; Lin *et al.*, 2007; Zhang *et al.*, 2002; Singh and Asthana, 2011b). In vacuum-brazed C–SiC/Nb joints made using a Ti–Ni–Nb active filler metal (Liu *et al.*, 2011), both Ti and Nb in the filler reacted with C–SiC, forming a well-bonded joint. The

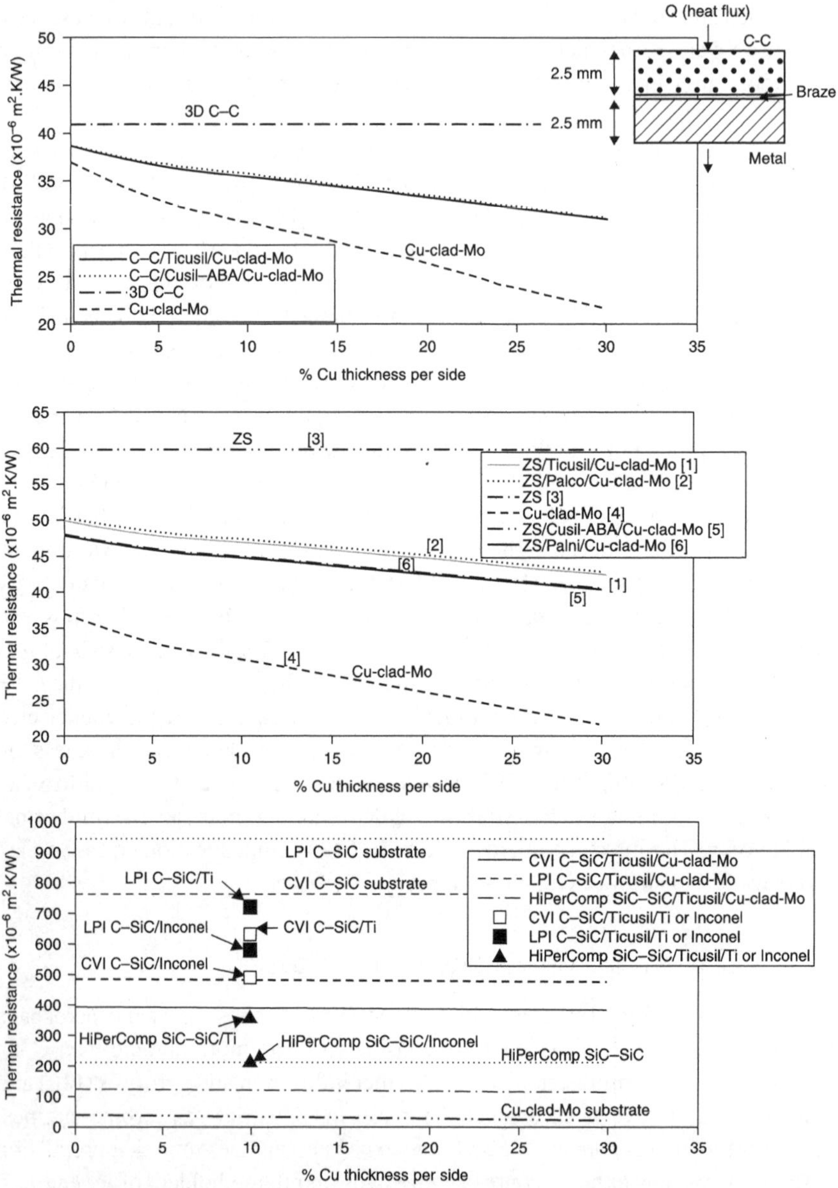

11.13 Theoretical projections of the effective thermal resistance of various CMC/Cu-clad-Mo joints shown as a function of the clad layer thickness.

ductile filler metal released the thermal stress in the joint, with the joint strength reaching values of 149, 120 and 73 MPa at 293, 873 and 1073 K, respectively.

Besides joining with conventional active filler metals, C–SiC composites have been brazed to metals using composite fillers. Composite filler metals containing dispersed fibers provide maximum joint strength, which is substantially higher than the maximum strength achieved with a fiber-free braze. The dispersed fibers mitigate the thermal stresses and reinforce the braze, both of which enhance the joint strength. For example, a (Ag–6Al) + Ti + C filler metal (as a powder mixture of Ag, Al, Ti and short carbon fibers) led to chemical reactions (TiC layer on C fibers), sound joint microstructure and high joint strength together with resistance to oxidation (Wang *et al.*, 2011). Similarly, an Ag–Cu–Ti braze containing short carbon fibers has been used to join C–SiC to Ti (Lin *et al.*, 2007); the fibers reacted with Ti, forming thin TiC_x layers on the fibers and TiC_x particles in the braze. The short fibers were uniformly distributed in the braze layer and preferentially align themselves parallel to the brazed surface.

The effect of composite surface preparation on the integrity of joints of C–SiC composites to metals was evaluated by the authors (Singh and Asthana, 2011b). The composites were fabricated using CVI and joined to Ti, Inconel 625, and Cu clad Mo using two Ag–Cu–Ti brazes (Cusil-ABA, T_L = 815°C and Ticusil, T_L 900°C) in ground and unground conditions. The effect of surface grinding on joint integrity was strongly dependent on the type of materials bonded. For example, in C–SiC/Cu-clad Mo joints made using Cusil-ABA, unground CMC substrates led to fine cracks whereas ground CMC substrates did not cause cracking. In contrast, in C–SiC/Inconel 625 joints made using Ticusil, both ground and unground CMC substrates led to cracking, whereas in C–SiC/Cu-clad Mo joints made using Ticusil neither ground nor unground CMCs led to cracking. The large differences in the CTE and yield strengths of different metallic substrates and braze interlayers lead to different thermal strains and the ability of the system to accommodate these strains. The thermal strain ($\Delta\alpha\Delta T$) is relatively low ($\sim 1.944 \times 10^{-3}$) in C–SiC/Cusil-ABA/Cu-clad Mo joints but it is significantly larger ($\Delta\alpha\Delta T \sim 7.876 \times 10^{-3}$) in C–SiC/Ticusil/Inconel 625 joints. These differences in thermal strains, in conjunction with the differences in the composite surface preparation, lead to differences in joint integrity. Additionally, surface grinding prior to joining either partially or completely removed the SiC coating applied to the C–SiC composite surface following composite fabrication, and this could also affect the wetting and bonding characteristics as well as joint integrity. The microstructural evaluation of C–SiC/metal joints also confirmed that cracking and interfacial de-cohesion are due to residual stresses during post-braze cooling rather than poor wetting of the rough (i.e. unground) composite surface by the molten braze. The C–SiC/braze interfaces showed Ti and Si enrichment, due possibly to titanium silicide formation together with the diffusion of braze constituents in the C–SiC composite. Knoop microhardness distributions across the joints did not reveal any effect due to surface preparation.

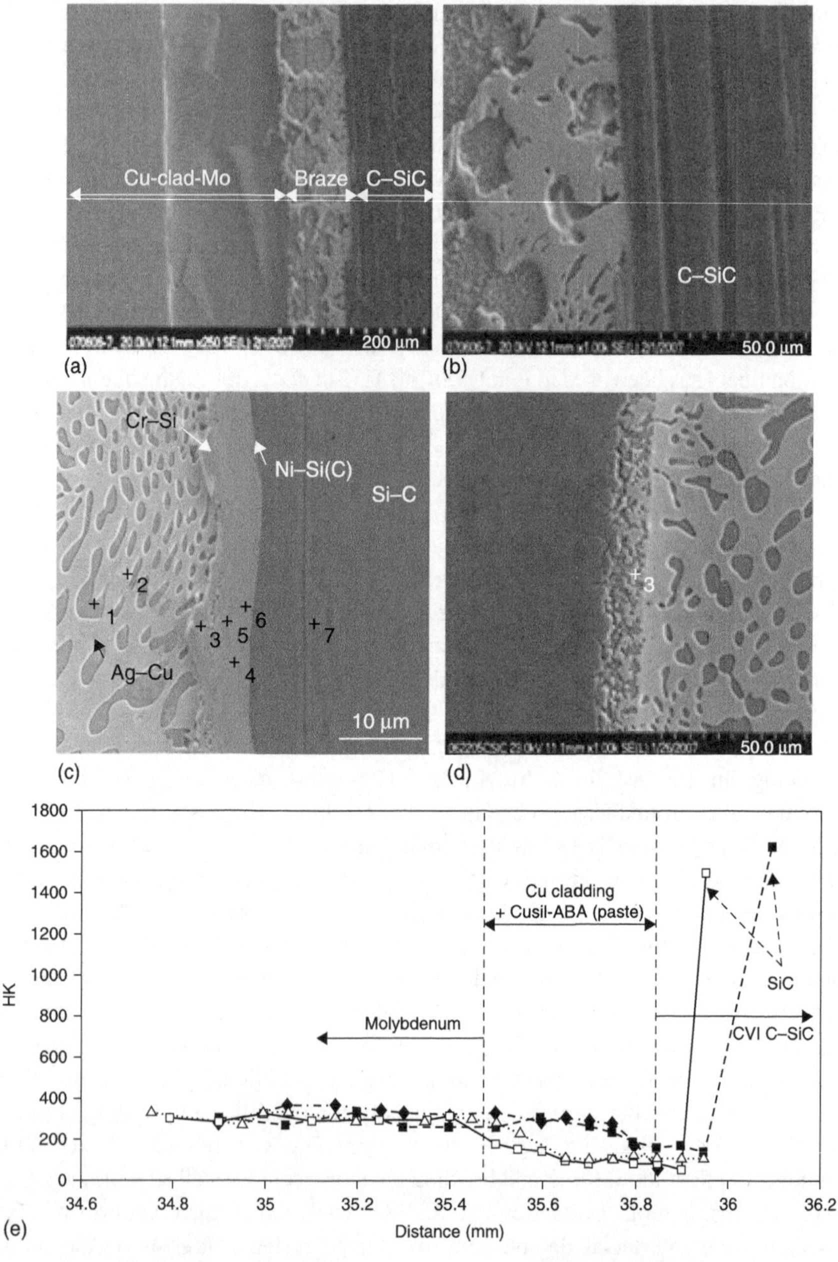

11.14 (a) and (b) C–SiC/Cu-clad Mo joint (braze: Cusil-ABA), (c) and (d) C–SiC/Inconel 625 joint (braze: Cusil-ABA), and (e) Knoop microhardness (MK) distribution in a C–SiC/Cusil-ABA/Cu-clad Mo joint.

Figure 11.14 shows the joint microstructure and Knoop hardness distribution in joints of C–SiC composite to different metals.

11.3.4 Ultra-high-temperature ceramic composites

The transition metal diborides, ZrB_2 (density: 6090 kg.m^{-3}) and HfB_2 (density: 11 200 kg.m^{-3}), with melting points greater than 3200 K have been identified as promising materials for applications in extreme environments such as the ultra-high temperatures (2173–2773 K) experienced by the sharp leading-edge components of space vehicles re-entering the Earth's atmosphere. Additives to these diborides, such as SiC and carbon, improve the thermal and mechanical properties and oxidation resistance. These diboride-based ultra-high-temperature ceramic (UHTC) composites have undergone extensive research to measure their physical, mechanical, thermal and thermomechanical properties (Gasch *et al.*, 2005; Zhu *et al.*, 2008; Bellosi and Babini, 2007; Sciti *et al.*, 2006; Monteverde *et al.*, 2002; Levine *et al.*, 2002; Tang *et al.*, 2007). Any component made from UHTC composites will need to be joined to other materials, including metals, at locations away from the hot zone. In such cases, only the UHTC will be exposed to ultra-high temperatures, while the actual joint will lie outside the hot zone and be amenable to cooling. Brazing may, therefore, be a viable approach for joining UHTC to other materials using metallic brazes.

Studies on joining diboride-based composites are relatively scarce. Muolo *et al.* (2003) utilized Ag–Zr brazes (T_L ~ 1323 K) to join a ZrB_2-based refractory to Ti–Al–V alloys and, over the last few years, the present authors utilized Ni-base metallic glass brazes (T_L ~ 1327 K) to join ZrB_2-based UHTC composites to themselves and to Ti (Singh and Asthana, 2007a), and AgCuTi and Pd-base brazes to join the UHTC composites to Cu-clad Mo (Singh and Asthana, 2009, 2010; Asthana and Singh, 2009). Palladium-based brazes (T_L ~ 1492–1511 K) have higher use temperatures than most Ag- and Ni-base brazes, besides offering excellent oxidation resistance. Other investigators (Passerone *et al.*, 2006, 2007, 2009; Voytovych *et al.*, 2007) have characterized the wettability and chemical interaction of diboride ceramics with Ag, Cu, Au and Ni. Contact angle data for Pd alloys on ZrB_2-based ceramics are, however, scarce. The contact angles of transition metals Co and Ni on ZrB_2 have been measured to be 39° and 42° at 1773 K, respectively (Passerone *et al.*, 2006, 2007). This suggests that Pd-based brazes, Palco and Palni (Table 11.1), will also probably wet ZrB_2 and ZrB_2-based composites.

Using various different braze fillers, the present authors brazed three ZrB_2-based composites to Ti, Cu-clad Mo, and Inconel 625. These composites, termed ZSS, ZSC and ZS, are: (i) ZSS (ZrB_2 + 20 v/o SiC_p + 35 v/o SiC_f), which contains SCS-9a fiber and particulate SiC, (ii) ZSC (ZrB_2 + 14 v/o SiC_p + 30 v/o C_p), which contains particulate SiC and carbon, and (iii) ZS (ZrB_2 + 20 v/o SiC_p), which contains particulate SiC. These composites were fabricated by hot pressing;

however, whereas the ZSC and ZS composites achieved full densification, the hot-pressed ZSS composites had ~30% residual porosity and micro-cracking due to residual stresses resulting from the thermal expansion mismatch between the SCS-9a fibers (CTE ~ 4.3×10^{-6}/K) and the matrix.

The three ZrB_2-base UHTC composites (ZSS, ZSC and ZS) were brazed to titanium using two boron-containing amorphous Ni-base braze alloys (MBF-20 and MBF-30). The dissolution of the composite constituents (e.g. Zr, Si) and titanium in the molten Ni-braze led to solute redistribution, interaction zone formation, and evidence of micro-cracking. The composites were also brazed to Cu-clad Mo using Cusil-ABA, Ticusil, Pd–Co and Pd–Ni brazes (Table 11.1), all of which yielded well-bonded, crack-free joints. The joints made using Pd–Co and Pd–Ni exhibited considerably thicker interaction zones than Cusil-ABA and Ticusil. Interestingly, unlike joints made using Cu-clad Mo, in which the Cu cladding acted as a stress-absorbing interlayer (ductility of Cu: 55%), in self-joined ZSC with Pd–Co braze, hairline cracks and penetration of Pd into ZSC were noted. In self-joined ZSS using Pd–Co, the pre-existing micro-cracks and pores facilitated braze penetration that led to the presence of Pd in the interaction zone away from the braze matrix and the formation of a prominent but somewhat diffuse interaction zone. Unlike Palco braze, the Palni braze led to poor bonding and extensive cracking in the UHTC joints. The lower ductility (23%) and higher yield strength (772 MPa) of Pd–Ni than Pd–Co (ductility: 43%, yield strength: 341 MPa) inhibited effective stress relief and led to unsound joints. Microstructures of some UHTC composite/metal joints are shown in Fig. 11.15(a)–(c), and multiple traces of Knoop microhardness profiles in a ZS/Palni/Ti joint are shown in Fig. 11.15(d). Within the composite matrix (Fig. 11.15(b)), the dark dispersed platelets are SiC_p and the continuous light-gray phase is the ZrB_2. The extensive interaction zone and the degradation of SCS-9a fibers in ZSS composites due to brazing are shown in Fig. 11.16.

The dissolution of Si and Zr from the ZS, ZSS and ZSC composites in molten braze could saturate the melt with solutes, resulting in the formation of intermetallic phases upon cooling and solidification, such as Pd_3Zr, Pd_2Zr, PdZr and $PdZr_2$. Similarly, in Pd–Co alloys, CoZr, Co_2Zr and $CoZr_2$ could form. In addition, silicides and complex ternary compounds could form in the interaction zone. A detailed analysis of the kinetics and thermodynamics of chemical reactions in these complex multi-component systems is difficult; however, simple thermodynamic considerations could clarify some points. For example, chemical reactions between ZrB_2 or SiC and Ni, Pd and Co in braze could form borides (Ni_2B, Ni_3B, CoB, Pd_5B), silicides ($CoSi_2$, Ni_2Si, PdSi) and carbides (Ni_3C, CoC, PdC). The free energies (ΔG) of reactions between ZrB_2 or SiC and Ni and Co, calculated as a function of temperature ($T \leq 1700$ K), show that $\Delta G < 0$ for the reaction of Ni with SiC to form Ni_2Si. In contrast, for the reaction of ZrB_2 with Ni or Co to form Ni_2B, Ni_3B, and CoB, ΔG is positive. Similarly, the formation of $CoSi_2$ and Ni_3C from the reaction of SiC with Ni or Co is unlikely. These basic

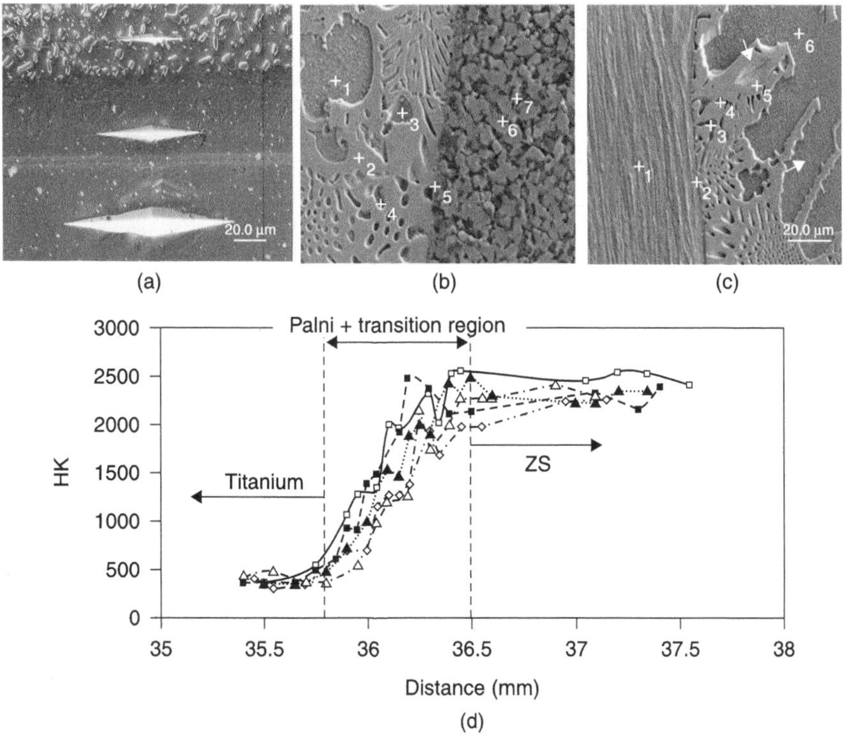

11.15 (a)–(c) Interface regions in (a) ZS/Ti joint, (b) ZSC/Ti joint, and (c) Cu-clad Mo/Cusil-ABA joint, and (d) Knoop hardness distribution in a ZS/Palni/Ti joint (symbols denote multiple traces across the same joint).

projections, based on elementary thermodynamic calculations, are only suggestive, and, in fact, certain reaction products, not feasible on simple thermodynamic grounds, have been shown to form in an actual joint. For example, nickel boride, Ni_2B, with $\Delta G > 0$ for its formation from the reaction of ZrB_2 with Ni, has been identified by electron probe microanalysis (EPMA) in an Au–Ni/ZrB_2 system under normal brazing conditions (Voytovych *et al.*, 2007). For the Au–Ni/ZrB_2 system, the chemical interaction resulted from a strong negative enthalpy of mixing of the Ni–Zr solution (Voytovych *et al.*, 2007) even though $\Delta G > 0$ for the reaction between Ni and ZrB_2 to form Ni_2B ($4Ni + ZrB_2 = 2Ni_2B + Zr$).

11.3.5 Other ceramic-matrix composites

Successful brazing of a number of other ceramic-matrix composites to metallic systems has been demonstrated in the published literature. These composites include SiC whisker-reinforced Al_2O_3, SiC and Si_3N_4 fiber-reinforced glass, quartz

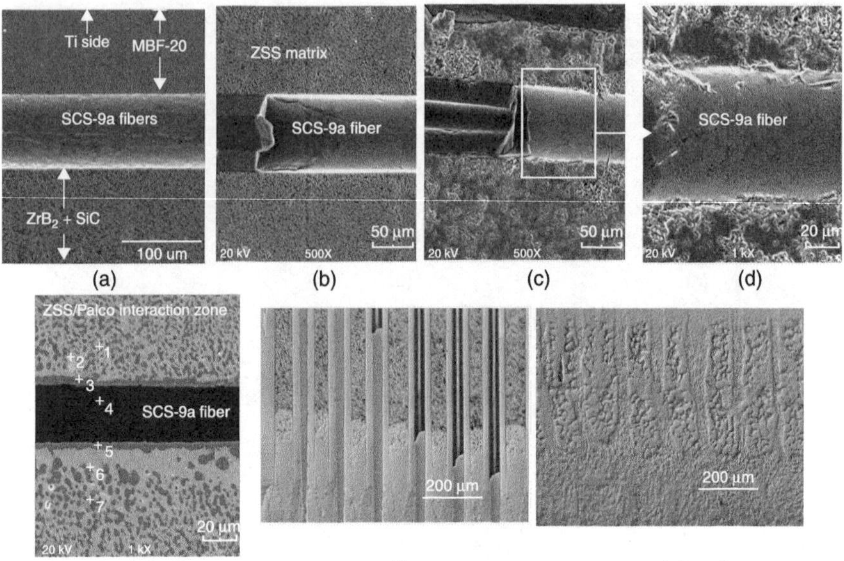

(a) (b) (c) (d)
(e) (f) (g)

11.16 Reaction and degradation of SCS-9a SiC fiber in SiC/SiC composites in contact with molten braze alloys: (a) ZSS/MBF-20/Ti joint, (b) fiber in as-fabricated ZSS composite, (c) and (d) ZSS/Palni/Inconel 625 joint, (e) ZSS/Palco/Cu-clad Mo joint, and (f) and (g) ZSS/Palni/Cu-clad Mo joint.

fiber-reinforced silica, mullite-mullite, TiN particle-reinforced Si_3N_4, AlN particle-reinforced TiB_2 and many others. Compliant layers having coefficients of thermal expansion between the CTE of the metal and the CTE of the CMC have been found to be useful for creating high strength joints (Kramer, 2010; Dixon, 1995). For example, a Kovar (CTE: 7×10^{-6}/°C) or an Invar (CTE: 11.35×10^{-6}/°C) sheet can be first welded to the metal substrate followed by brazing of the welded layer to the CMC. In many CMC systems, however, sound joints without compliant layers can form, as discussed later. Improvements in composite-to-metal joints have been achieved using composite brazing alloys (Blugan *et al.*, 2007; Janczak-Rusch, 2011). For example, Si_3N_4–TiN composites have been joined to steel using an active SiC-reinforced braze. The brazed joints showed good wetting and bonding, and uniform TiN particle distribution within the braze filler.

Oxide ceramic composites (e.g. mullite-mullite) have been brazed (Piazza *et al.*, 2003) to high-temperature alloys (e.g. Haynes-230®) using pre-metallized composite surfaces with a Ti- and Zr-activated Cu–Zn alloy followed by brazing with a Cu braze to produce excellent joints. Others have brazed ZrO_2-toughened Al_2O_3 to metals with Ag–Ti filler metals (Torvund *et al.*, 1996), tungsten carbide–nickel cermets to composite coatings (Lu and Kwon, 2002), Si–Ti–C–O fiber-reinforced CMCs to steel (Nakamura *et al.*, 1999), and SiC whisker-reinforced

Al_2O_3 to steel (Qu *et al.*, 2005). Zhao *et al.* (2011) vacuum brazed an electroless nickel-plated quartz fiber-reinforced silica composite to Invar using Ag–Cu eutectics at 1073–1163 K for 5 to 35 min. The joint strength was affected by changes in the relative amounts of the Cu–Ni eutectic and the thickness of the plated nickel; the highest shear strength was achieved at specific amounts of Cu–Ni eutectic and plating thickness. Mattia *et al.* (2005) hot pressed an AlN/TiB_2 ceramic composite and evaluated its wettability with Ag–Cu and Ag–Cu–Ti alloys using the sessile-drop method. Both the Ag–Cu alloy and pure Ag and Cu metals exhibited a non-wetting behavior whereas the active Ag–Cu–Ti alloy showed good wetting.

In the case of silicate matrix composites, Si_3N_4 fiber-reinforced cordierite glass-ceramic has been brazed to titanium and stainless steel parts (Dixon, 1995) and a SiC fiber-reinforced borosilicate glass matrix composite has been brazed to molybdenum (Janczak-Rusch *et al.*, 2005). For the SiC fiber-reinforced borosilicate glass matrix composite joined to Mo, two types of braze fillers were used: a glass braze and an active filler metal, Ag–Cu–In (Incusil-ABA, brazing temperature: 740°C). With the glass braze, the surface of the Mo substrate had to be roughened to ensure sound bonding. Higher joint strength was achieved with the active Ag–Cu–In braze than with the glass braze. Brazing was done with fibers either parallel or perpendicular to the joint. Post-braze mechanical tests revealed that in joints with parallel fibers, failure always occurred within the composite via delamination, whereas in joints with perpendicular fibers, the joints failed at the interface without delamination, indicating that the brazing zone was the weakest link. A brittle reaction layer between the borosilicate glass matrix and the Incusil-ABA brazing filler formed in the brazing zone, but no reaction between the SiC fiber and the filler metal had occurred. Sound joints formed with relatively high loading capacity and negligible thermal degradation of the glass matrix composite.

11.4 Conclusions

Advanced ceramic-matrix composites such as SiC–SiC, C–C, C–SiC, UHTC composites, and oxide, nitride and silicate-based composites are emerging as tailorable composite materials with potential to be integrated with other materials in components. These composites have been joined to metals and alloys such as Fe, Ti, Ni and Cu alloys. A number of different off-the-shelf braze compositions have been identified and successfully utilized to create sound joints with and without surface modification and stress mitigating metallic interlayers. Good wetting and bonding result from chemical interactions such as dissolution and chemical reactions of the active metal (e.g. Ti in braze) with the ceramic constituents. The microstructure, composition, durability, and mechanical, physical, and thermal properties have been measured for the fabricated CMC/joints. Issues related to optimum joint design, the influence of factors such as fiber arrangement, interlaminar shear behavior, role of fiber and composite coatings, oxidation resistance, thermal and mechanical fatigue, stress-rupture behavior, and physical, chemical and mechanical degradation with time under service conditions need to be evaluated. Life-prediction analyses for joints used in structures are also needed. Clearly, CMC/metal brazing is an area of emerging research and developmental interest.

11.5 Acknowledgment

The authors wish to thank Mr Mike Halbig, NASA Glenn Research Center, Cleveland, OH, USA, for his helpful comments.

11.6 References

Appendino P, Casalegno V, Ferraris M, Grattarola M, Merola M, *et al.* (2003), 'Joining of C–C composites to copper', *Fusion Eng. Design*, 66(8), 225–229.

Appendino P, Ferraris M, Casalegno V, Salvo M, Merola M, *et al.* (2004), 'Direct joining of CFC to copper', *J. Nuclear Mater.*, 329–333, 1563–1566.

Asthana, R (2000), 'Dissolutive capillary penetration with expanding pores and transient contact angles', *J. Colloid Interface Sci.*, 231, 398–400.

Asthana R (2002), 'Interface- and diffusion-limited rise of a reactive melt with a transient contact angle', *Metall. Mater. Trans.*, 33A, 2119–2128.

Asthana R and Singh M (2009), 'Joining of ZrB_2-based Ultra-high Temperature Ceramic Composites Using Pd-based Braze Alloys', *Scr. Mater.*, 61(3), 257–260.

Asthana R and Sobczak N (2000), 'Wettability, spreading and interfacial phenomena in high-temperature coatings', *JOM-e*, 52(1). The Minerals, Metals and Materials Society (TMS), Warrendale, PA, USA. (http://www.tms.org/pubs/journals/JOM/0001Asthana/Asthana-0001.html)

Asthana R, Singh M and Sobczak N (2010), 'Wetting behavior and interfacial microstructure of palladium- and silver-based braze alloys with C–C and SiC–SiC composites', *J. Mater. Sci.*, 45(16), 4276–4290.

Bellosi A and Babini GN (2007), 'Development and properties of ultra-high temperature ceramics – opportunities and barriers to application', in Freiman S (ed.) *Global Roadmap for Ceramics and Glass Technology*, Wiley, pp. 847–864.

Blugan G, Kübler J, Bissig V and Janczak-Rusch J (2007), 'Brazing of silicon nitride ceramic composite to steel using SiC-particle-reinforced active brazing alloy', *Ceram. Int.*, 33, 1033–1039.

Casalegno V, Salvo M, Murdaca S and Ferraris M (2009), 'One-step brazing process for CFC monoblock joints and mechanical testing', *J. Nucl. Mater.*, 393(2), 300–305.

Centeno A, Pintsuk G, Linke J, Gualco C, Blanco C, *et al.* (2011), 'Behaviour of Ti-doped CFCs under thermal fatigue tests', *Fusion Eng. Design*, 86(1), 121–125.

Chawla KK, *Composite Materials – Science and Engineering*, Springer, New York, 1987, p. 212.

Chen B, Xiong H, Mao W and Cheng Y (2010), 'Wettability and interfacial reactions of PdNi-based brazing fillers on C–C composite', *Trans. Non-Ferrous Soc. of China*, 20(2), 223–226.

Corman GS and Luthra KL (2005), 'Silicon melt infiltrated ceramic composites (HiPerComp™)', in Bansal NP (ed.), *Handbook of Ceramic Composites*, Kluwer Academic Publishers, New York, pp. 99–115.

DiCarlo JA, Yun HM, Morscher GN, Bhatt RT (2005), 'SiC-SiC composite for 1200°C and above', in Bansal NP (ed.), *Handbook of Ceramic Composites*, Kluwer Academic Publishers, New York, pp. 77–98.

Dixon DG (1995), 'Ceramic Matrix Composite-Metal brazed joints', *J. Mater. Sci.*, 30, 1539–1544.

Eustathopoulos N, Nicholas MG and Drevet B (1999), *Wettability at High Temperatures*, Pergamon, Boston, pp. 281–282.

French JD and Cass RB (1998), 'Developing Innovative Ceramic Fibers', *Ceramic Bulletin*, May, 61–65.

Garcia-Rosales C, Pintsuk G, Gualco C, Ordas N, Lopez-Galilea I, *et al.* (2009), 'Manufacturing and high heat-flux testing of brazed actively cooled mock-ups with Ti-doped graphite and CFC as plasma-facing materials', *Phys. Scr.*, T138, No. 014062.

Gasch MJ, Ellerby DT and Johnson, SM (2005), 'Ultra-high temperature ceramic composites', in Bansal NP (ed.), *Handbook of Ceramic Composites*, Kluwer Academic Publishers, New York, pp. 197–224.

Grigorenko N, Poluyanskaya V, Eustathopoulos N and Naidich Y (1998), in Bellosi A *et al.* (eds), *Interfacial Sci. of Ceram. Joining*, Kluwer Acad. Publ., Boston, pp. 69–78.

Harper CA (2003), *Electronic Materials and Processes Handbook*, McGraw-Hill, New York.

Jacobson DM and Humpston G (2005), 'Principles of Brazing', ASM International, Materials Park, OH, p. 224.

Janczak-Rusch, J (2011), 'Ceramic component integration by advanced brazing technologies', in Singh M, Ohji T, Asthana R and Mathur S (eds), *Ceramic Integration and Joining Technologies*, Wiley, Boston.

Janczak-Rusch J, Piazza D and Boccaccini AR (2005), 'Joining of SiC fibre reinforced borosilicate glass matrix composites to molybdenum by metal and silicate brazing', *J. Mater. Sci.*, 40(14), 3693–3701.

Keene BJ (1993), 'Review of data for the surface tension of pure metals', *Int. Mater. Revs.*, 38(4), 157–192.

Kramer DP (2010), 'Method of joining metals to ceramic matrix composites', US Patent 7857194.

Krenkel W (2005), 'Carbon fiber-reinforced silicon carbide composites (C-SiC, C-C-SiC)', in Bansal NP (ed.), *Handbook of Ceramic Composites*, Kluwer Academic Publishers, New York, pp. 117–148.

Lamon J (2005), 'Chemical vapor infiltrated SiC-SiC composites (CVI SiC-SiC)', in Bansal NP (ed.), *Handbook of Ceramic Composites*, Kluwer Academic Publishers, New York, pp. 55–76.

Levine SR, Opila EJ, Halbig MC, Kiser JD, Singh M, *et al.* (2002), 'Evaluation of ultra-high temperature ceramics for aeropropulsion use', *J. Euro Ceram. Soc.*, 22, 2757–2767.

Lewis D III and Singh M (2001), 'Post-processing and assembly of ceramic-matrix composites', *ASM Handbook, vol 21: Composites*, ASM International, Materials Park, OH, pp. 668–673.

Li JG (1992), 'Kinetics of wetting and spreading of Cu-Ti alloys on alumina and glassy carbon substrates', *J. Mater. Sci. Lett.*, 11, 1551–1554.

Lin G, Huang J and Zhang H (2007), 'Joints of carbon fiber-reinforced SiC composites to Ti-alloy brazed by Ag-Cu-Ti short carbon fibers', *J. Mater. Proc. Technol.*, 189(1–3), 256–261.

Liu GW, Valenza F, Muolo ML and Passerone A (2010), 'SiC/SiC and SiC/Kovar joining by Ni-Si and Mo interlayers', *J. Mater. Sci.*, 45(16), 4299–4307.

Liu JY, Chen S, and Chin BA (1994), 'Brazing of vanadium and carbon–carbon composites to stainless steel for fusion reactor applications', *J. Nuclear Mater.*, 212–215, 1590–1593.

Liu YZ, Zhang LX, Liu CB, Yang ZW, Li HW, *et al.* (2011), 'Brazing C-SiC composites and Nb with TiNiNb active filler metal', *Sci. Technol. Weld. Joining*, 16(2), 193–198.

Locatelli MR, Tomsia AP, Nakashima K, Dalgleish BJ and Glaser AM (1995), 'New strategies for joining ceramics for high-temperature applications', in *Key Engineering Materials*, vols 111–112, pp. 157–190.

Lu SP and Kwon OY (2002), 'Microstructure and bonding strength of WC reinforced Ni-base alloy brazed composite coating', *Surf. and Coat. Techn.*, 153, 40–48.

Matsunaga T, Lin H-T, Singh M, Asthana R, Kajii S, *et al.* (2011), 'Microstructure and mechanical properties of joints in sintered SiC fiber-bonded ceramics brazed with Cusil-ABA™'(unpublished).

Mattia D, Desmaison-Brut M, Tetard D and Desmaison J (2005), 'Wetting of HIPAlN-TiB_2 ceramic composites by liquid metals and alloys', *J. Eur. Ceram. Soc.*, 25(10), 1797–1803.

Monteverde F, Bellosi A and Guicciardi S (2002), 'Processing and properties of ZrB_2-based composites', *J. Eur. Ceram. Soc.*, 22(3), 279–288.

Morscher GN, Singh M, Shpargel TP and Asthana R (2006), 'A simple test to determine the effectiveness of different braze compositions for joining Ti tubes to C-C composite plates', *Mater. Sci. Eng. A*, 418(1–2), 19–24.

Moutis NV, Jimenez C, Azpiroz X, Speliotis Th, Wilhelmi C, *et al.* (2010), 'Brazing of carbon–carbon composites to Nimonic alloys', *J. Mater. Sci.*, 45, 74–81.

Muolo ML, Ferrera E, Morbelli L, Zanotti C and Passerone A (2003), 'Joining of zirconium boride based refractory ceramics to Ti6Al4V', in *Proc. of the 9th International Symposium on Materials in a Space Environment*, June 2003, Noordwijk, The Netherlands, compiled by Fletcher K, ESA SP-540, Noordwijk, Netherlands: ESA Publ. Div., pp. 467–472.

Nakamura M, Mabuchi M, Saito N, Yamada Y, Nakanishi M, *et al.* (1999), 'Joining of Si-Ti-C-O fiber reinforced ceramic composite and Fe-Cr-Ni stainless steel', *Key Eng. Mater.*, 164–161, 435–438.

Nicholas MG (1998), *Joining Processes: Introduction to Brazing and Diffusion Bonding*, Kluwer Academic.

Passerone A, Muolo ML and Passerone D (2006), 'Wetting of Group IV diborides by liquid metals', *J. Mater. Sci.*, 41, 5088.

Passerone A, Muolo ML, Novakovic R and Passerone D (2007), 'Liquid metal/ceramic interactions in the (Cu, Ag, Au)/ZrB_2 systems,' *J. Euro. Ceram. Soc.*, 27, 3277–3285.

Passerone A, Muolo ML, Valenza F, Monteverde F and Sobczk N (2009), 'Wetting and interfacial phenomena in Ni-HfB_2 systems,' *Acta Mater.*, 57, 356–364.

Piazza D, Boccaccini AR, Kaya C and Janczak-Rusch J (2003), 'Hochtemperatur-Löten von einem Mullit-Mullit Verbundwerkstoff', in *Verbundwerkstoffe*, Hrsg. Degischer, pp. 702–707.

Qin Y and Feng J (2007), 'Microstructure and mechanical properties of C-C composite/ TC4 joint using AgCuTi filler metal', *Mater. Sci. Eng. A*, 454, 322–327.

Qu S, Zou Z and Wang X (2005), 'Shear strength and fracture behavior of SiC_w–Al_2O_3 composite-carbon steel brazed joints', *Key Engineering Materials*, 297–300, 2441–2446.

Riccardi B, Nannetti CA, Petrisor T, Woltersdorf J, Pippel E, *et al.* (2004a), 'Issues of low activation brazing of SiC_f–SiC composites by using alloys without free silicon', *J. Nucl. Mater.*, 329, 562–566.

Riccardi B, Nannetti CA, Woltersdorf J, Pippel E and Petrisor T (2004b), 'Joining of SiC based ceramics and composites with Si16Ti and Si-18Cr eutectic alloys', *Int. J. Mater. Product Technol.*, 20(5–6), 440–451.

Salvo M, Lemoine P, Ferraris M, Montorsi MA and Matera R (1995), 'Cu-Pb rheocast alloy as joining materials for CFC composites,' *J. Nuclear Mater.*, 226, 67–71.

Sangsuwan P, Orejas JA, Gatica JE, Tewari SN and Singh M (2001), 'Reaction-Bonded Silicon Carbide by Reactive Infiltration', *Ind. Eng. Chem. Res.*, 40, 5191–5198.

Schwartz MM (1994), *Joining of Composite Matrix Materials*, ASM International, Materials Park, OH.

Sciti D, Guicciardi S, Bellosi A, Pezzotti G (2006), 'Properties of a pressureless sintered ZrB_2-$MoSi_2$ ceramic composite', *J. Amer. Ceram. Soc.*, 89, 2320.

Singh M and Asthana R (2007a), 'Joining of advanced ultra-high-temperature ZrB_2-based ceramic composites using metallic glass interlayers', *Materials Science & Engineering A*, 460–461, 153–162.

Singh M and Asthana R (2007b), 'Brazing of advanced ceramic composites: Issues and challenges', *Ceramic Trans.*, 198, 9–14.

Singh M and Asthana R (2008a), 'Characterization of brazed joints of carbon-carbon composites to Cu-clad-Mo', *Compos. Sci. Tech.*, 68(14), 3010–3019.

Singh M and Asthana R (2008b), 'Advanced joining and integration technologies for ceramic-matrix composite systems', in Krenkel W (ed.), *Fiber-Reinforced Ceramic Composites*, Wiley-VCH, pp. 303–325.

Singh M and Asthana R (2009), 'Joining of ZrB_2-based UHTC composites to Cu-clad-Mo for advanced aerospace applications', *Int. J. Appl. Ceram. Technol.*, 6(2), 113–133.

Singh M and Asthana R (2010), 'Evaluation of Pd-base brazes to join ZrB_2-based ultra-high temperature composites to metallic systems', *J. Mater. Sci.*, 45(16), 4308–4320.

Singh M and Asthana R (2011a), 'Initial characterization of brazed SiC–SiC composite-to-metal joints' (unpublished).

Singh M and Asthana R (2011b), 'Effect of composite surface preparation on joining response of C-SiC composites brazed to metallic systems' (unpublished).

Singh M and Lara-Curzio E (2001), 'Design, fabrication and testing of ceramic joints for high temperature SiC-SiC composites', *Trans. ASME*, 123, 288–292.

Singh M, Shpargel TP, Morscher GN and Asthana R (2005), 'Active metal brazing and characterization of brazed joints in titanium to carbon-carbon composites', *Mater. Sci. Eng. A*, 412, 123–128.

Singh M, Asthana R and Shpargel TP (2007), 'Brazing of C-C composites to Cu-clad Mo for thermal management applications', *Mater. Sci. Eng. A*, 452–453, 699–704.

Singh M, Asthana R and Shpargel TP (2008a), 'Brazing of ceramic-matrix composites to Ti and Hastelloy using Ni-base metallic glass interlayers', *Mater. Sci. Eng. A*, 498, 19–30.

Singh M, Morscher GN, Shpargel TP and Asthana R (2008b), 'Active metal brazing of titanium to high-conductivity carbon-based sandwich structures', *Mater. Sci. Eng. A*, 498(1–2), 31–36.

Sobczak N, Singh M and Asthana R (2005), 'High-temperature wettability measurements in ceramic-metal systems: Some methodological issues', *Current Opinion in Solid State & Mater. Sci.*, 9(4–5), 241–253.

Sobczak N, Sobczak J, Ksiazek M, Radziwill W and Morgiel J (1997), in *Proc. 2nd Int. Conf. on High-Temp. Capillarity*, Eustathopoulos N and Sobczak N (eds), Foundry Research Institute (Krakow), pp. 97–98.

Standing R and Nicholas M (1978), 'The wetting of alumina and vitreous carbon by Cu-Sn-Ti alloys', *J. Mater. Sci.*, 13, 1509–1514.

Steffier WS (2004), 'Brazing SiC/SiC composites to metals', NASA Tech Briefs, September 2004, p. 5.

Suryanarayana C, Moore JJ and Radtke RP (2001), 'Novel methods of brazing dissimilar materials', *Advanced Materials & Processes*, March, 29–31.

Tang S, Deng J, Wang S, Liu W and Yang K (2007), 'Ablation behavior of ultra-high temperature ceramic composites', *Mater. Sci. Eng. A*, 465, 1–7.

Taylor R (2000), 'Carbon-matrix composites', in *Comprehensive Composite Materials*, Elsevier Science Ltd., Boston, 4, pp. 387–426.

Torvund T, Grong O, Akselsen OM and Ulvensoen JH (1996), 'A model for coupled growth of reaction layers in reactive brazing of ZrO_2-toughened Al_2O_3', *Metall. Mater. Trans.*, 27(11), 3630–3638.

Voytovych R, Koltsov A, Hodaj F and Eustathopoulos N (2007), 'Reactive versus non-reactive wetting of ZrB_2 by azeotropic Au-Ni', *Acta Materialia*, 55, 6316–6321.

Wang H, Cao J and Feng J (2010), 'Brazing mechanism and infiltration strengthening of C-C composites to TiAl alloys joint', *Scr. Mater.*, 63(8), 859–862.

Wang ZP, Huang JH, Zhang H and Zhao XK (2011), 'Reactive composite brazing of C_f-SiC composites to Ti alloy with (Ag-6Al) + Ti + C composite filler materials', *Mater. Sci. Technol.*, 27(1), 49–52.

Whalen JT and Anderson AT (1976), 'Wetting of SiC, Si_3N_4, and Carbon by Si and Binary Si Alloys', *J. Amer. Ceram. Soc.*, 34(4), 378–383.

Zhang JJ, Li SJ, Duan HP and Zhang Y (2002), 'Joining of C-SiC to Ni-based superalloy with Zr/Ta composite interlayers by hot-pressing diffusion welding', *Rare Metal Mater. Eng.*, 31 Suppl 1: 393–396.

Zhao L, Zhang L, Tian X, He P and Feng J (2011), 'Interfacial microstructure and mechanical properties of joining electroless nickel plated quartz fibers reinforced silica composite to Invar', *Mater. Design*, 32(1), 382–387.

Zhu S, Fahrenholtz WG and Hilmas GE (2008), 'Enhanced densification and mechanical properties of ZrB_2-SiC processed by a preceramic polymer coating route', *Scr. Mater.*, 59, 123–126.

Zweben C (2002), 'Metal Matrix Composites, Ceramic Matrix Composites, Carbon Matrix Composites and Thermally Conductive Polymer Matrix Composites', in Harper CA (ed.), *Handbook of Plastics, Elastomers and Composites*, Chapter 5, 4th ed., McGraw-Hill, New York, pp. 321–344.

12
Brazing of metal and ceramic joints

S. HAUSNER and B. WIELAGE,
Chemnitz University of Technology, Germany

DOI: 10.1533/9780857096500.2.361

Abstract: This contribution presents the state of the art in brazing metal–ceramic compounds. General problems during brazing of metal and ceramic and different process variations with appropriate brazing materials are described. The brazing process itself and interesting examination methods for metal–ceramic compounds are explained. For a selected material combination, the microstructure and strength behaviour are described. In the final section, recent results for induction brazing of metal–ceramic compounds are introduced as an energy-efficient alternative to conventional furnace brazing processes. The process-specific basics of induction brazing and furnace brazing, as well as the microstructure and strength behaviour for both processes, are compared.

Key words: brazing metal ceramic, metal–ceramic compound, metallized ceramic, active brazing, induction brazing.

12.1 Introduction

Novel developments in technology lead to increasing demands on the materials to be used. Higher strength, lighter materials and longer lifetimes are required. Increasingly, conventional metallic materials can no longer meet these requirements. The use of ceramics, which are characterized by a number of excellent material properties, offers completely new possibilities. They have a high hardness, excellent temperature and wear resistance, high thermal and electrical insulation properties as well as good oxidation and corrosion resistance. Since ceramics also exhibit a low specific weight, this group of materials has a high application potential. In electrical engineering, ceramics are essential because of their high electrical resistance. Ceramic components such as isolators are in use in large numbers. Because of their temperature stability, ceramics are also indispensable in high-temperature technology and for industrial furnaces. The excellent wear and corrosion resistance is exploited in machinery and plant engineering, for example for guiding devices, friction bearings, rolling bearings, etc. The chemical resistance can be used for components in the chemical industry, such as pumps. In recent decades, the continuous development of production processes and manufacturing technologies for ceramics and the development of new ceramics have led to increasing application of this material group.[1–4]

Ceramics can be divided into three groups. The main distinguishing feature is the amount of glass phase: in silicate ceramics (porcelain, steatite, cordierite and mullite), the amount of glass phase is relatively high. In contrast, oxide ceramics (alumina, zirconia, etc.) and non-oxide ceramics (silicon carbide, silicon nitride, boron carbide, etc.) contain no or only a very small proportion of glass phase.[3] Most ceramics that are used in technology are oxide and non-oxide ceramics. Alumina (Al_2O_3) is the most important oxide ceramic. Because of its balanced properties, it can be regarded as a universal ceramic. Alumina is widely applied in electrical engineering, mechanical engineering, and the chemical industry and for high-temperature applications. In recent years, zirconia (ZrO_2) has become increasingly important, particularly because ZrO_2 ceramics are characterized by an atypically high fracture toughness. In addition, zirconia has a relatively high thermal expansion coefficient, which is beneficial for joining processes with metals. As zirconia exhibits polymorphism, the cubic crystal form has to be stabilized, usually with MgO or Y_2O_3. Silicon carbide (SiC), silicon nitride (Si_3N_4), boron carbide (B_4C) and boron nitride (CBN) are important non-oxide ceramics. These ceramics are based on carbides, nitrides or borides. Because of their predominantly covalent bond type, they are characterized by excellent mechanical and thermal stability. In recent years, metal–ceramic composites have become increasingly important. Ceramic matrix composites (CMCs) are usually produced by infiltration of a ceramic preform by metallic melts. Also ceramic fibres embedded in a ceramic matrix are possible. Metal matrix composites (MMCs) are made by incorporating ceramic particles or fibres into a metallic matrix. These material groups are not considered in this contribution. Rather, the brazing of monolithic ceramics for the production of metal–ceramic compounds is discussed. Brazing ceramics with metals is difficult due to their very small thermal expansion coefficient. A vast number of ceramic material variations are available. Depending on the required properties, a suitable ceramic can be selected.[1–5]

Ceramics have many advantages. However, they also have adverse properties. Ceramics exhibit a high brittleness combined with poor processing ability, and, depending on the type of ceramic, a more or less pronounced low thermal shock resistance. In addition, ceramics are very expensive, so they are only used where their specific properties are really needed. In this context, the joining of ceramics to other materials, in particular to metals, has gained great importance. By joining metals with ceramics, it is possible to produce customized components for an application. A ceramic-friendly design is necessary to reliably integrate the ceramic into the overall system. It has to be noted that ceramics have a low number of slip systems. Unlike metals, they do not possess any plastic deformation capacity for stress relief, which leads to spontaneous crack propagation. In particular, the occurrence of tensile stresses is critical for ceramics. By using a ceramic-friendly design, the tensile stresses and bending stresses should be shifted into the metal, while the ceramic can be exposed to high compressive stresses.[1–4,6]

Due to the already mentioned adverse properties of ceramics, the range of joining processes for metal–ceramic compounds is limited. Often high production costs (e.g. for grinding) have to be accepted. Some joining methods, such as riveting, cannot be used because of the impact load. Metal–ceramic compounds can be joined via force-fit, form-fit or material-fit joining mechanisms. Shrinking, clamping, screwing, casting, cementing and plugging are examples of force- and form-fitting methods. The advantage of force- and form-fitting joining is the removability of the compounds, so that these methods are employed when parts need to be exchanged or replaced. However, it is necessary to observe certain design restrictions. Compounds that are produced by these joining methods have only low temperature stability and cannot be exposed to high thermal loads. Tool holders with replaceable ceramic cutting inserts or spark plugs are examples of force- and form-fit compounds.[1–4,6]

In comparison with mechanical joints, material-fit joints have a smooth power distribution and design-effective compound shape. These joining methods include, for example, adhesive bonding, brazing and welding. Epoxy resins, polyvinyl, polymethacrylates or polyamides are used to produce metal–ceramic compounds by means of gluing. However, these compounds have only a very low temperature stability of about 200°C. Processes utilized for welding metals and ceramics include, for example, diffusion welding, friction welding and ultrasonic welding. The welding methods require a high technological effort and know-how, so they are applied comparatively rarely.[3,4,6,7]

12.2 Brazing of metal and ceramic

Due to its great flexibility and universality, the brazing of metals and ceramics has been established as a very widely used standard procedure. By means of brazing, vacuum-tight and thermally stable compounds with high strengths can be produced. The brazing of a metal and ceramic has been established in particular in electrical engineering. For electrical insulation parts, current feedthroughs, high-power electron tubes, sensors, thyristor and diode housings, the use of metal–ceramic compounds is state of the art. But also tools such as drills, cutting punches and forging dies are made of metal–ceramic compounds. The brazing of Si_3N_4 turbocharger rotors to metal shafts is a much-quoted example in engine construction. Further applications are found in aircraft and space technology as well as in fusion and nuclear engineering. Figure 12.1 shows some examples.[1–3,8–10]

The above-mentioned, highly valued properties of ceramics result from their ionic/covalent bond character with strong atomic bonding forces. This fundamentally different atomic structure in comparison to metals (with metallic bonds) poses two crucial problems when brazing ceramics to metals. The first problem results from the polar bond structure of ceramics. They have a high surface tension, so that ceramics cannot be wetted by conventional metallic melts. Furthermore, as a result of the different thermophysical properties (thermal expansion coefficient and elastic modulus), strong internal stresses are induced during brazing.[3,4]

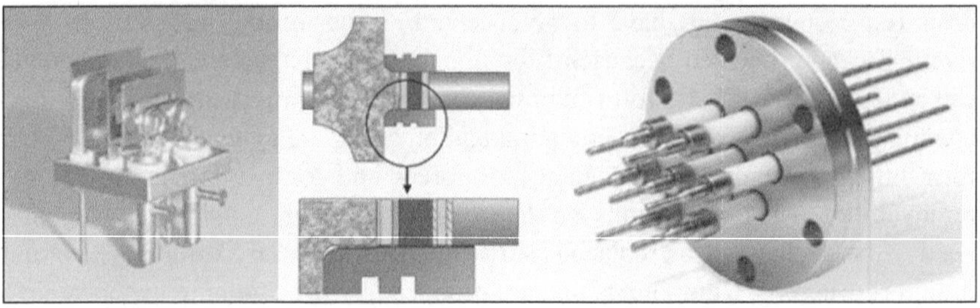

(a) (b) (c)

12.1 Examples of metal–ceramic compounds: (a) metal–ceramic compound part of an ionization manometer;[8] (b) ceramic turbocharger rotor;[1] (c) current feedthrough.[9]

The poor wettability of ceramics by conventional metallic melts is due to the different atomic structures of ceramics and metals, leading to an incompatibility of the two joint partners. In a metal bond, the atom trunks are randomly surrounded by electrons (electron gas). However, the ionic/covalent bond of ceramics is characterized by a closed, very stable electron configuration. Therefore, metallurgical reactions between the two joining parts, which are typical in metal brazing, are impossible when brazing metals to ceramics. The brazing filler metal forms drops on the ceramic without sufficient wetting (wetting angle $\theta \approx 180°$; see Fig. 12.2(a)). Only a minimal adhesion of the brazing filler metal on the ceramic

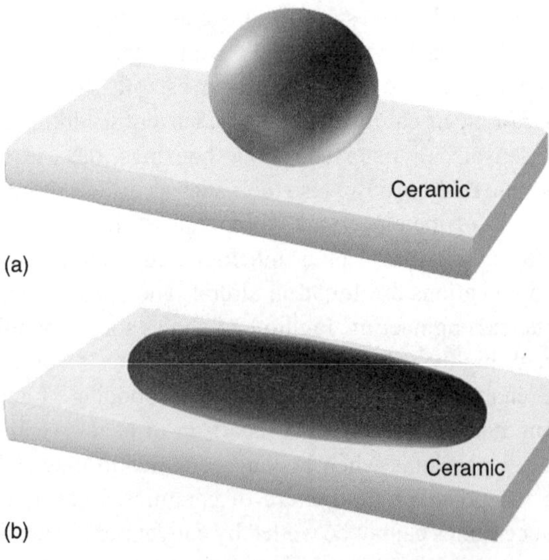

12.2 Comparison of the wetting behaviour of a ceramic with (a) a conventional metallic brazing filler metal; (b) an active-brazing filler metal.

can be achieved by epitaxial effects (structural adjustment) or through differences in potential. To allow wetting of the ceramic despite the different bond structures, two process variants for brazing metal–ceramic compounds have been established: the brazing of metallized ceramic on the one hand and active brazing on the other hand.[3,11]

12.3 Brazing of metallized ceramics

When brazing a metallized ceramic, a metallic layer is applied to the ceramic before the brazing process. This facilitates wetting at the interface between the brazing filler metal and ceramic base material. There are a number of different methods of metallization. In particular, metallization with Mo/Mn or W/Mn has been used on an industrial scale.

In this case, the ceramic to be metallized is coated with a powder (particle size 1–2 μm) of MoO_2 and Mn or MnO in an organic binder of nitrocellulose with a thickness of 10–25 μm. The oxides are reduced at temperatures between 1000°C and 1800°C in a wet H_2/N_2 atmosphere. A portion of the metals reacts with the ceramic, forming a glass phase; another part is sintered, forming the wettable metallic layer (with a thickness of about 10 μm). The glass phase, which has a very small thermal expansion coefficient ($\alpha_{\text{glass phase}} < \alpha_{\text{ceramic}}$), is located between the sintered metal layer and the ceramic. Thus, compressive stresses are induced in the glass phase during cooling. An additional Ni or Cu layer of a thickness of 2–4 μm is applied to improve the wettability. Now the metallized ceramic can be wetted by conventional vacuum-brazing alloys with an Ag or AgCu basis (see Table 12.1). In addition to the Mo/Mn-base metallization, a W/Mn-base metallization is possible in the same way. The bond strength of the metallization is of crucial importance for the quality of the compounds. It is determined by a stud pull test in accordance with ASTM F 19-64[12] and is normally between 70 and 200 MPa. The Mo/Mn or W/Mn metallization method is very often used in practice, especially in electrical engineering for mass-production components made of alumina. The metallization of ceramics by means of the Mo/Mn method for high-power electron tubes or electrical feedthroughs is one example. Even

Table 12.1 Vacuum-brazing alloys for joining metallized ceramics[3]

Filler metal	Cu (wt.%)	In (wt.%)	Pd (wt.%)	Ag (wt.%)	T_{Sol} (°C)	T_{Liq} (°C)	T_{Braze} (°C)
Ag–Cu	28	–	–	72	780	780	830
Ag–Cu–In	25	14.5	–	61.5	630	705	755
Ag–Cu–In	27	10	–	63	685	730	780
Ag–Cu–Pd	31.5	–	10	58.5	824	852	900
Ag–Cu–Pd	21	–	25	54	901	950	1000
Ag–Pd	–	–	5	95	970	1010	1050

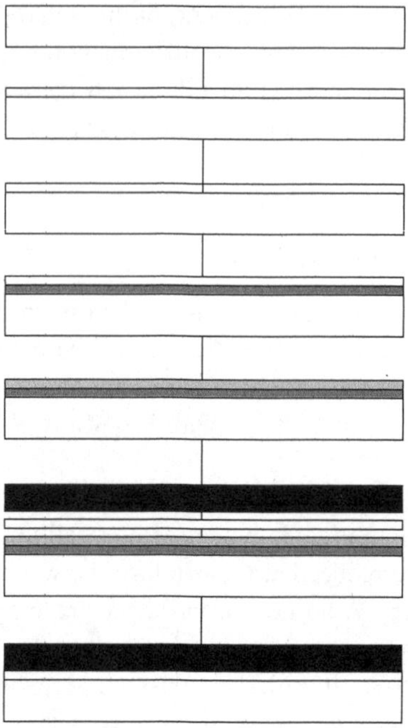

12.3 Metallization with Mo/Mn.[3,14]

though this metallization method involves very complex processes (two thermal process steps are necessary; see Fig. 12.3) that require considerable experience, this method has become widely accepted in serial production.[2–4,13–15]

In a further metallization process, the ceramic is coated with a solution of active or refractory metal salts (e.g. with lithium compounds or with salts of the IVa subgroup of the periodic table of elements). The coated ceramic is dried and subjected to thermal treatment in which the metal salts are reduced to the pure metal. This metal reacts with the ceramic and forms a bond, allowing the ceramic to be wetted by conventional brazing alloys.[2–4]

For metallizing, it is also possible to apply mixtures of metal powders and glass frits on the ceramic, using organic binders (the so-called glass frit technique[4]). The ceramic is heated so that the glass frit melts. During cooling, a mechanical bond between the metal powder, the ceramic and the glass phase is formed, which results in a wettable coating of the ceramic. Precious metals are mostly used as coating materials.[2–4]

However, only oxide ceramics can be metallized by means of these metallization methods. Recent studies deal with further methods of metallization for non-oxide ceramics. For this purpose, thermal spraying and physical vapour deposition processes such as magnetron sputtering, ion plating and vapour deposition, for example, are used. Various refractory metals, particularly titanium, are applied as coating materials. However, often insufficient bonding strength is achieved through the coating process alone. Therefore, annealing processes in vacuum are used, which result in an increased bonding strength due to the chemical interactions between the substrate and the coating material.[4]

The wettable metallization film layer, regardless of the method applied, can now be wetted by conventional metal brazing alloys. Table 12.1 shows typical alloy systems that are suitable for brazing metallized ceramics. Brazing with these filler metals is usually carried out in vacuum or in an inert gas atmosphere without the use of fluxes.[3]

Apart from these brazing alloys, it is also possible to use soft solders (soldering temperature <450°C). They are applied in particular in electrical engineering, where temperature-sensitive components have to be joined. In this case, the metallization is based on precious metals such as silver, silver/palladium, silver/platinum, gold etc.[5]

The description of the various metallization processes reveals their highly complex character, including energetic aspects. A wettable metallic coating must be generated in an upstream thermal process before each actual brazing process. Moreover, the processes require significant technical effort and experience. Only with special knowledge can the processes be optimized successfully, which hinders reproducibility between users.[4]

12.4 Active brazing of metal–ceramic compounds

Being a direct brazing method, active brazing has significant procedural and technological advantages over the aforementioned metallization processes. The wetting between the ceramic and the brazing filler metal takes place directly during the brazing process. No upstream processes are required. Even the wetting of non-oxide ceramics is possible. The so-called active-brazing filler metals are mostly based on the element Ag or the Ag–Cu system, in which the eutectic between Ag and Cu at about 28 wt.% Cu is utilized. Furthermore, active-brazing filler metals based on copper, gold and palladium are also known (Table 12.2). The brazing filler metals contain surface-active elements (usually 1 wt.% to 4 wt.% Ti, but also In, Zr, Hf or Nb), which allow a direct wetting of the ceramic due to chemical reactions (Fig. 12.2(b)).[3]

The physicochemical mechanism of wetting during the active-brazing process is very complex and will be described below. The surface-active elements added to the filler metal lead to chemical interactions with the ceramic. The active metal decomposes the ceramic at the interface between the ceramic substrate and the filler metal. Thus, reaction products with a metallic or metal-like structure are formed. This so-called reaction zone consists of metallic and/or non-metallic components. The modified chemical structure of the reaction zone causes a significant reduction of the surface tension so that the reaction zone can be wetted

Table 12.2 Commercially available active-brazing filler metals[3,16]

Active-brazing filler metal	Chemical composition (wt.%)							Melting range (°C)	Brazing temp. (°C)
	Ag	Cu	Ti	In	Ni	Al	other		
AgCu26.5Ti3	70.5	26.5	3	–	–	–	–	780–805	850–950
AgCu34.2Ti1.8	64	34.2	1.8	–	–	–	–	780–810	850–950
AgCu25.2Ti10	64.8	25.2	10	–	–	–	–	780–805	850–950
AgCu34.5Ti1.5	64	34.5	1.5	–	–	–	–	770–810	850–950
Ag–Ti	96	–	4	–	–	–	–	970	1000–1050
AgIn1Ti0.6	98.4	–	0.6	1	–	–	–	948–959	1000–1050
AgIn1Ti1	98	–	1.0	1.0	–	–	–	948–959	1000–1050
AgCu19.5In5Ti3	72.5	19.5	3	5	–	–	–	730–760	
NiTi67	–	–	67	–	33	–	–	942–980	
CuNi15Ti70	–	15	70	–	15	–	–	910–970	
AuNi3Ti0.6	–	–	0.6	–	3	–	96.4Au	1003–1030	
SnAg10Ti4	10	–	4	–	–	–	86Sn	221–300	850–950
PbIn4Ti4	–	–	4	4	–	–	92Pb	320–325	850–950

by the filler metal.[11] In the brazing of metallic materials, metallurgical effects are responsible for wetting and joint formation, whereas for the brazing of ceramics the formation of new phases at the interface is crucial. Depending on the ceramic, the reaction zone consists of nitrides, oxides, silicides or carbides and is very brittle. Therefore, the Ti content influences not only the wetting but also the mechanical properties of the compound. The formation of the reaction zone is affected by energetic and chemical boundary conditions. The process parameters during brazing, such as the brazing temperature and holding time as well as the type and concentration of the active element and the base materials, are crucial influencing factors.[2–4,11,14,16,17]

During brazing, the ceramic is dissociated by the active element. Thus, the reaction products depend on the choice of the ceramic base material. In general, the chemical reaction can be described as follows:[3]

$$\text{M-N} + \text{A} \rightarrow \text{AN} + \text{M}$$

or

$$\text{M-N} + \text{A} \rightarrow \text{AM} + \text{N}$$

where M-N is a metal–non-metal composite that is a ceramic and A is an active metal.[3]

An estimation of the reaction products, depending on the base materials used, can be carried out on the basis of a thermodynamic balancing of the possible gross reactions. However, because of the assumptions that have to be made, no definite predictions but only estimates are possible. For the most commonly used ceramics, alumina, zirconia and silicon nitride, the reaction products, which have been experimentally proven by different researchers, are briefly introduced below. With silicon nitride as the ceramic and titanium as the active element, the primarily formed reaction product is titanium nitride, followed by titanium silicide. Due to the high application potential of alumina, intensive metallurgical studies on the reaction products have been carried out. In the case of the active element titanium, TiO seems to be the major reaction product of the reaction $Ti–Al_2O_3$, but also $Ti\text{-}O_x$ phases, Ti–Al phases and more complex compounds from the involved elements can be detected. Regarding zirconia and titanium, usually a reaction layer of Ti_xO_y is formed. The titanium oxides form by reducing the oxygen content in the ZrO_2 ceramic, leading to the formation of non-stoichiometrically saturated ZrO.[3,4,18–22]

Table 12.2 shows a selection of commercially available active-brazing filler metals. The filler metals are typically available in the form of foils or strips with thicknesses from 50 to 300 μm, but also as paste or, in special cases, as wire. Due to the doping with Ti, the filler metals show no fluidity; they must be applied directly at the joining surface. The usual capillary effect cannot be exploited. It is also possible to apply the filler metal by a screen printing process. In this case, a higher active metal content is required, since the oxidation risk of the oxygen-sensitive active metal increases during the application. Depending on the alloy composition, brazing is carried out at temperatures between 700°C and 1050°C. The alloys, that contain Sn and Pb are so-called soft active solders. To achieve a wetting of the ceramic, these solders also have to be brazed at temperatures above 600°C. However, it is advantageous that these alloys remain molten down to low temperatures, and thus the thermally induced residual stresses can be kept low. But, on the other hand, these compounds exhibit only low temperature stability.[3,16]

The choice of brazing alloy, in particular the content of the active element, depends on the ceramic base materials, which affects the thermodynamic activity of the active metal. A titanium content above 2 wt.% Ti is necessary for ZrO_2, while for Al_2O_3 dispersion ceramic 1 wt.% Ti is sufficient for good wetting. For Si_3N_4, the titanium content should even be below 1 wt.%. The activity of titanium can also be influenced by alloying elements. Sn, In and Ag, for example, exhibit a wetting-enhancing effect. Table 12.3 compares the brazability of typical engineering ceramics for a few commercially available active-brazing filler metals.[3,16]

When choosing the filler metal, a compromise between good wetting and acceptable ductility must be achieved. The active metal must be available in sufficient quantity to ensure a uniform reaction zone for good adhesion and wetting of the ceramic. However, the reaction products influence not only the wetting behaviour but also the mechanical quality of the compound. Therefore, the active metal content should be limited to an amount that prevents the formation of too wide a reaction zone and thus the embrittlement of the compound.[3,16,23,24]

12.5 Influencing the mechanical properties of brazed metal–ceramic compounds

As mentioned above, the quality of an active-brazed compound is decisively influenced not only by the metallurgical aspects but also by the induced stresses. The different physical properties (thermal expansion coefficient and elastic modulus) of the metal and ceramic lead to significant residual stresses in the compound during the joining process. Figure 12.4 compares the thermal expansion coefficients of ceramics and metals. Apart from zirconia, there exist significant differences in the expansion coefficients of both material groups. These differences cause high residual stresses during the cooling process, which can significantly affect the quality or even lead to premature failure of the compound. When the temperature sinks below the solidus temperature of the filler metal, the metal shrinks more than the ceramic (schematically shown in Fig. 12.5), which can be attributed to its higher thermal expansion coefficient. This results in thermally induced stresses in the material-fit compound.[3,4,25,26]

Table 12.3 Brazability of typical engineering ceramics for a few commercially available active-brazing filler metals[3]

Filler metal	Al_2O_3	Al_2O_3–TiC	Al_2O_3–ZrO_2	ZrO_2	Si_3N_4	SiC	B_4C	AlN	PKD
AgCu19.5In5Ti3	+	+	+	+	+	+	+	+	+
AgTi	+	+	+	+	–	o	o	–	–
AgCu26.5Ti3	+	+	+	+	+	+	+	+	+
AgCu34.5Ti1.5	+	+	+	+	+	+	+	+	+
AgIn1Ti1	+	+	+	+	o	o	o	o	–
AgSn89.5Ti4	+	+	+	+	+	+	+	?	–
AuNi3Ti0.6	+	+	+	+	–	o	–	?	+

+ Suitable; o suitable with restrictions; – unsuitable.

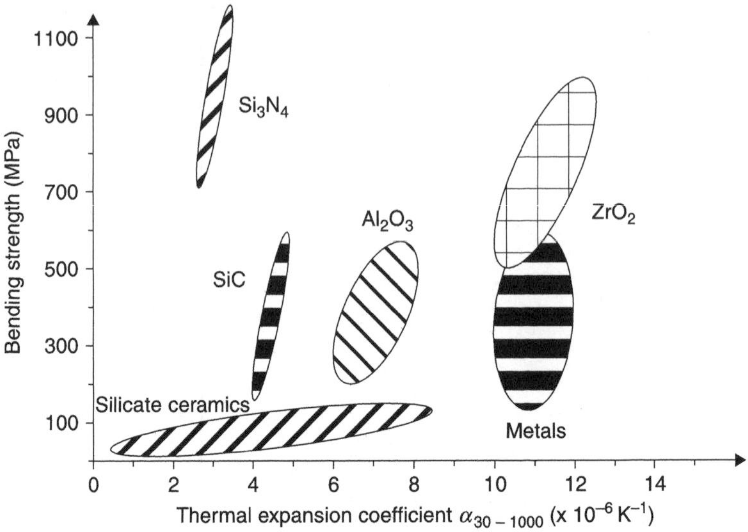

12.4 Comparison of the thermal expansion coefficient of various ceramics with those of metals.[5]

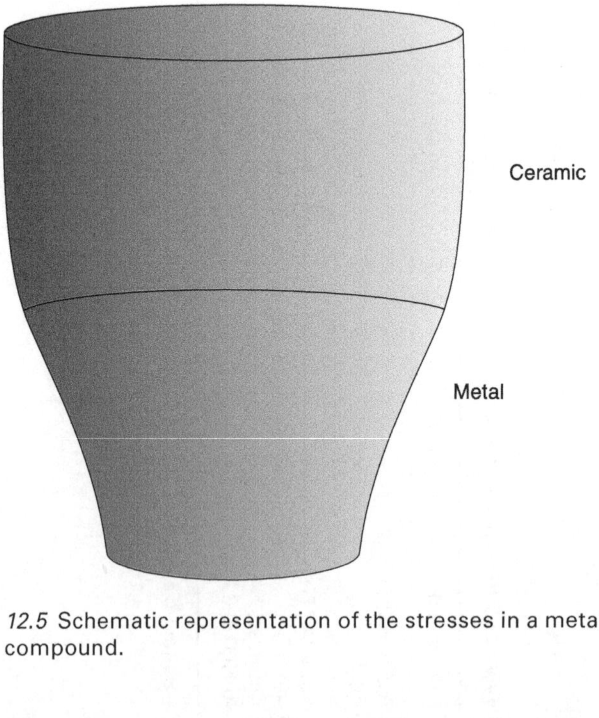

12.5 Schematic representation of the stresses in a metal–ceramic compound.

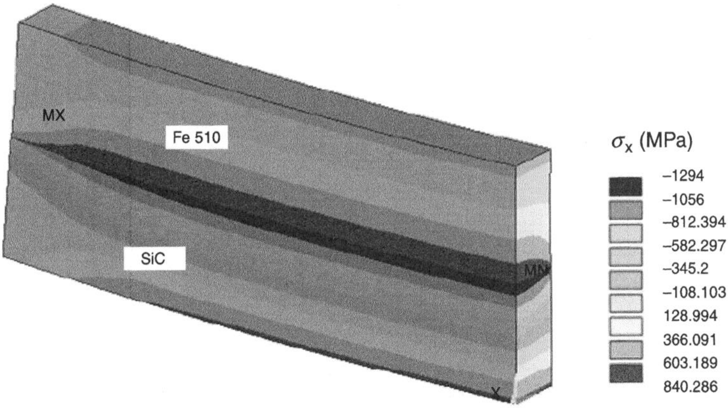

12.6 FEM simulation of thermally induced stresses of a brazed metal–ceramic compound.[27]

The intensity of the residual stresses depends not only on the thermal expansion coefficient but also on the elastic constants and the geometry of the compound. In the vicinity of the brazed seam, the tensile stresses are particularly high at the edge of the ceramic (singularity; see Fig. 12.6). Since ceramics only have a very limited tolerance for tensile stresses, such loads are critical for the ceramic base material and should be avoided by a suitable structural design where possible.[3,4,25,26]

In addition to structural design, thermally induced residual stresses can be reduced in different ways. By broadening the ductile brazing seam on the Ag or AgCu base, the stresses can be reduced through plastic deformation, and thus the strength of the compound rises. However, the increased titanium content in the widened brazing seam involves the risk of the formation of brittle phases. Thus, often soft layers (e.g. Cu or Ni) or layers with a matched thermal expansion (e.g. W, Mo, Ta) are inserted in the brazing seam. These intermediate layers do not melt during the brazing process but interact with the filler metal. The ductile soft layers allow the elimination or reduction of thermally induced stresses through plastic deformation in the interlayer material. The expansion-adapted layers are intended to relieve the ceramic by absorbing the stress. A combination of both intermediate layers is possible as well. Furthermore, the thermal expansion coefficient can be adjusted by the incorporation of particles or short fibres into the brazing seam. However, this modification of the filler metal has not yet been established in practice. Also, the use of graded interlayers is still in research and development. Hence, a graded transition between the two joining partners can be obtained. Technologically, this can be achieved by using powder metallurgical processes or thermal spraying. The residual stresses can also be reduced by the use of metals with a matched thermal expansion, such as Kovar or Invar, which have a relatively low thermal expansion coefficient. It is also advantageous to use ferritic steels

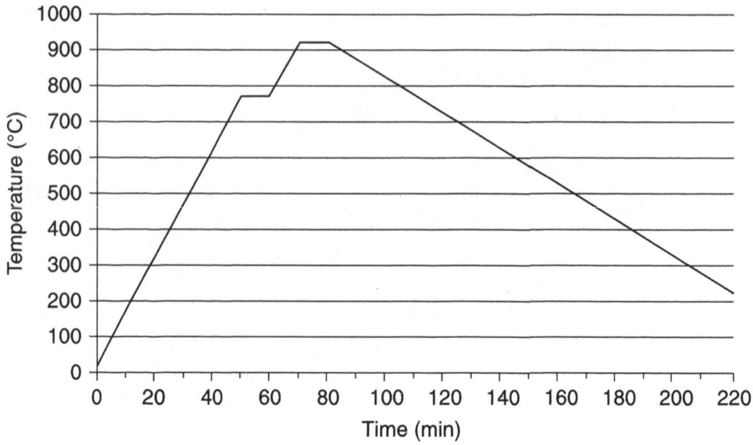

12.7 Typical temperature–time cycle.

instead of austenitic steels, because materials with a body-centred cubic structure exhibit a lower thermal expansion coefficient than materials with a face-centred cubic structure.[2–4,25,26]

The strength of the compound can also be influenced by the brazing process. With a suitable brazing temperature, holding time and heating and cooling rate, the optimum between good wettability and the minimal formation of brittle phases can be found. Figure 12.7 shows a typical temperature–time curve of a brazing process (see also Section 12.6). In particular, a cooling rate as low as possible, such as about 5 K/min, needs to be chosen to partially relax the thermally induced stresses by plastic deformation.[25]

The finite-element method (FEM) can be used to calculate the thermally induced residual stresses (see Fig. 12.6). To ensure a reliable estimate of the stresses, a significant amount of material data and in particular their temperature dependence must be known for the involved joint partners. By calculating the stresses, the structural design of a metal–ceramic compound can be optimized in terms of reliability under operating conditions. Depending on the component geometry, the calculations are very complex, particularly when plasticizing effects are taken into account. Often, the material behaviour is considered to be elastic to simplify the calculations.[3,16]

An analytical method for estimating the bond stresses in the less critical centre of the ceramic was developed by Munz and Iancu.[28] Under the simplifying asumptions that the compound consists of infinitely extended sheets, the joint partners behave elastically, the compound does not show any deflection, and the stress state is biaxial, the thermally induced residual stress state can be calculated by:[3,28]

$$\sigma_K = -(T_1 - T_0)\cdot\frac{E_K}{1-\nu_K}\cdot\left(\frac{\Sigma i\left(\frac{E_i\cdot\alpha_i\cdot h_i}{1-\nu_i}\right)}{\Sigma i\left(\frac{E_i\cdot h_i}{1-\nu_K}\right)} - \alpha_K\right) \qquad [12.1]$$

where:
E_i = elastic modulus of the involved joining parts
E_K = elastic modulus of the ceramic
ν_i = Poisson's constant of the involved joining parts
ν_K = Poisson's constant of the ceramic
α_i = thermal expansion coefficient of the involved joining parts
$(T_1 - T_0)$ = temperature difference between the solidus temperature and room temperature
h_i = thickness i of the involved materials
σ_K = thermally induced residual stresses in the ceramic[3,28]

This equation is a very simplified calculation, which gives a simple, approximate estimate of the residual stresses in the compound. The equation can be used when the geometrical influence of the compound or the influence of the intermediate layer for stress relief is to be calculated in a simple way. However, for optimizing the structural design of a joint, an FEM analysis should be selected.[3]

12.6 Preparation for and execution of the brazing process

The preparation for and execution of the brazing process will be described below. The preparation of the brazing samples is similar to that for conventional high-temperature brazing. The metallic base material and the brazing filler metal should be metallic clean and free of grease. The degreasing of the materials takes place in an ultrasonic bath followed by drying with hot air. Before degreasing, the brazing filler metal should also be roughened by a short grinding with sandpaper.[3,16,29]

The surface finish of the ceramic has a decisive influence on the compound strength. The ceramic base material should exhibit a minimum number of pores, cracks and notches. Since there is the risk that defects are inserted into the ceramic by grinding, such processing before the joining process is controversially discussed in the literature. Some sources report higher compound strengths by the grinding or polishing ceramics.[16,30]

The charging should be carried out as soon as possible after sample preparation. The joint partners should be positioned in such a way that the application of a load is possible. This weighting ensures good surface contact between the brazing filler metal and the ceramic and thus uniform wetting can be achieved. Furthermore, it is important to reiterate the fact that active-brazing materials have no fluidity, so they must be applied directly to the joining surface. When designing the brazing fixture, the differences in thermal expansion coefficients of the base materials and the high wetting ability of the active-brazing filler metals on almost all materials must be taken into consideration.[16]

The brazing is usually done in furnaces, but induction heating is also possible.[31–34] The high oxygen affinity of the active elements in the brazing filler metal requires an inert atmosphere. The contamination level of oxygen and other gases crucially affects joint formation. Too high a level of contamination leads to the risk that the active metal is not available for joint formation because it has oxidized during the brazing process. The brazing process is most often executed in high vacuum (<10^{-4} mbar). Inert shielding gases can be used, but they lead to joints of lower quality. Argon (purity >99.998) is generally used as the inert gas, but helium, neon and argon–neon are also possible.[4,16]

Figure 12.7 shows an example of the temperature–time cycle of a brazing process for a metal–ceramic compound. In general, the compound is heated to about 50–150°C below the solidus temperature of the brazing filler metal, followed by a holding time until a uniform heating of the joining parts is reached. Then the compound is heated up to brazing temperature. The holding time at the brazing temperature should be chosen such that a uniform temperature distribution can be achieved in the joining area. Normally, a holding time of between 5 and 10 minutes is used. Too long a holding time should be avoided, because the formation of brittle reaction products increases. In order to allow a relaxation of the thermally induced residual stresses, the compound should be cooled down relatively slowly. A cooling rate of 5 K/min is typical. Figure 12.8

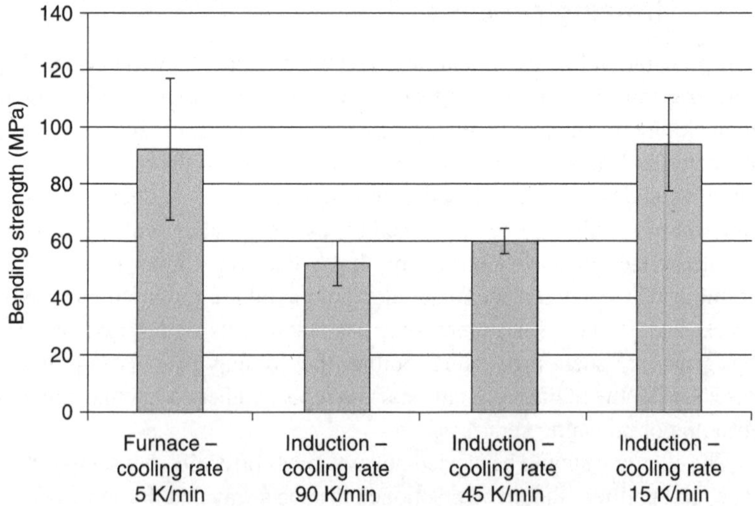

12.8 Bending strengths (four-point bending test) as a function of cooling rate (sample geometry: 9 mm × 3 mm × 50 mm; thickness of brazed seam: 50–70 μm).

shows the attainable strength values (four-point bending test) for the material combination Al_2O_3 and the iron–nickel alloy FeNi42 (Mat. No. 1.3917; ASTM F 29, F 30[35,36]) depending on the cooling rate (see also Section 12.9).[3,16,33]

Besides direct brazing with an active-brazing filler metal, metal–ceramic compounds can also be brazed directly by the use of so-called glass solders. Glass solders are special glasses with a low softening temperature, but also high-melting glass solders can be used. Glass solders have good wetting behaviour on ceramics and are suitable for soldering glasses, metals and ceramics. However, they are comparatively rarely used because of their relatively high brittleness and sensitivity to differences in expansion of the joint partners.[3,4,16]

The brazing of metal ceramic compounds is also possible with the so-called RAB (Reactive Air Brazing) process, in which mostly silver-based filler metals are used. The added oxide component CuO allows the wetting of the ceramic. This brazing process in air is being studied intensively, in particular for the brazing of fuel cells.[37]

12.7 Examination methods for brazed metal–ceramic compounds

Examination methods for brazed metal–ceramic compounds include strength tests (mostly bending tests) and optical and scanning electron microscope studies. Because of the usual high scattering of strength values in ceramic materials, bending tests of metal–ceramic compounds also lead to very high scattering values of up to ±50% compared with the mean.[16] Weibull diagrams and parameters, which are standard in ceramics research,[38] can be used to evaluate the strength values to base the measured values on a statistic. The bending test with three-point bending measures a higher strength than the four-point bending test, because the influence of the volume on the strength is of considerable importance for ceramic materials. Due to the larger stressed volume, the failure probability for the four-point bending test is higher, and lower strength values are determined. Therefore, when specifying strength values, it is crucial to ndicate whether the three- or the four-point bending test is used. Also, the geometry of the test specimens should be given, since the strength values are geometry dependent. For this reason, transferability of values is difficult. The determined data can only be used for comparative purposes.[16]

Furthermore, strength can be assessed by a shear test. Here, two butt-brazed sample plates are sheared off in a special test device. But even here the transferability of test results is complicated due to the geometry dependence.[16] The tensile test, which is widely used for testing metallic materials, is rarely applied in testing metal–ceramic compounds because the treatment of the ceramic for fixing is very problematic. In addition, ceramics are known to exhibit only a low tolerance to tensile stresses.[16]

For certain applications, e.g. in power electronics, an examination of the gas tightness of the brazed seam is of great interest. The examination is usually done on a real component using a leak detector (e.g. the helium leak test). Depending on the application, the testing of thermal shock resistance, creep behaviour and oxidation resistance are further possible examinations of metal–ceramic compounds.[16]

12.8 Example of an active-brazed metal–ceramic compound

The microstructure of an active-brazed metal–ceramic compound is described and discussed below for the example of alumina and the iron–nickel alloy FeNi42 (Mat. No. 1.3917; ASTM F 29, F 30[35,36]). The iron–nickel alloy exhibits an appropriate thermal expansion; it has a low thermal expansion coefficient below the Curie temperature, so that the induced bond stresses are comparatively low. The alumina ceramic used has a purity of 99.7%. The brazing metal is a commercially available active-brazing filler metal based on Ag with 26.5 wt.% Cu and 3 wt.% Ti (see Table 12.2). Figure 12.7 shows the temperature–time curve, which is typical for brazing metal–ceramic compounds. The brazing was carried out in a vacuum furnace with a heating rate of 15 K/min. The heating increases to 150 K below the solidus temperature of the filler metal, followed by an intermediate dwell for sufficient warming of the samples. Then the compound is heated up to the brazing temperature at 900°C with a holding time of 5 min. In order to relax the thermally induced stresses by plastic deformation, the cooling rate is slow (5 K/min).

Figure 12.9 shows an optical microscope image of a metal–ceramic compound. The formation of the reaction layer is clearly visible at the interface with the

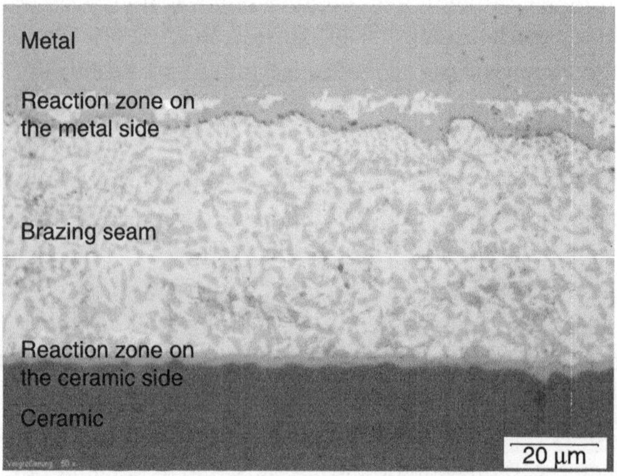

12.9 Optical microscope image of a metal ceramic joint (Al_2O_3 + FeNi42).

ceramic. The reaction zone is formed due to chemical reactions with the ceramic through the active metal (the formation of titanium oxides). The titanium can form a variety of sub-oxides and oxides with thermodynamic binding parameters very close to each other. A very clear presentation of the possible interactions and reaction products of titanium and alumina can be found in Bang and Liu.[18] The brazing filler metal also interacts with the metal. A closed, meandering belt can be seen on the metal side, which is characteristic of brazing compounds with thermal-expansion-matched metals. The main alloy constituents, Fe and Ni, diffuse into the AgCu matrix, where they interact with the active metal titanium, and form intermetallic phases. The brazing seam is composed of the Ag–Cu eutectic with primarily separated Cu dendrites.

12.9 Induction brazing of metal–ceramic compounds

In recent years, energy efficiency has increasingly become the focus of research due to rising energy prices, dwindling resources and increasing CO_2 emissions. In this context, it is crucial to adapt production processes to become more energy efficient. Therefore, in a sub-project of the Cluster of Excellence the 'Energy-Efficient Product and Process Innovation in Production Engineering' (eniPROD®), the brazing of metals and ceramics by induction is being examined as an energy-efficient alternative to conventional furnace brazing processes.[31–34] Partial heating during induction brazing results in a high energy-saving potential, in particular when brazing large components as well as individual parts and small batches. Heating in a furnace is an indirect heating method in which the heat must be transferred by radiation and/or convection to the component. This method of heating is characterized by high losses and long processing times. In contrast, induction heating is a direct heating method, in which the heat is generated directly in the workpiece. The electrical energy can be transformed into heat almost without losses, resulting in a very high efficiency. Induction heating is based on the principle of magnetic induction and is one of the electrothermal processes that can be used to heat all electrically conductive materials. Alternating current flows through an inductor and generates an alternating electromagnetic field. When conductive material is introduced into the alternating field, eddy currents arise in this material, which counteract the resistivity of the material and thus lead to a warming of the material. The producible heat depends especially on the conductivity and permeability of the material as well as on the frequency of the alternating field. Inductive brazing of metals and ceramics poses two basic problems:

- Since ceramics have no or only a low electrical conductivity, they cannot be heated by induction. However, to ensure a wetting of the ceramic, the ceramic must have a minimum temperature (see Section 12.4).
- The specific advantages of the induction brazing process lead to a completely changed process control in comparison with furnace brazing, characterized by faster heating rates, shorter holding times and faster cooling rates (see Fig. 12.10).

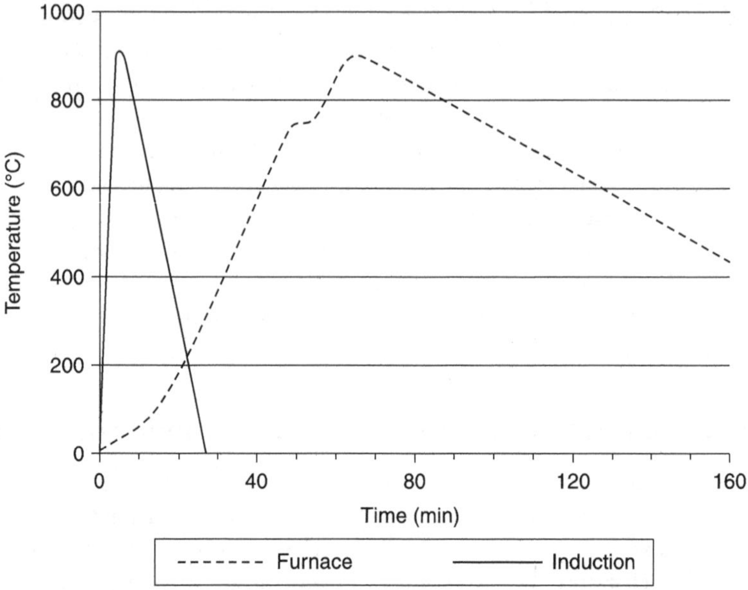

12.10 Comparison of the temperature–time cycle of furnace and induction brazing.

In addition to this energy saving potential, the shorter process times are advantageous for brazing of temperature-sensitive components. However, the influence of the different process control on the quality of the compounds must be investigated. In particular, the holding time, which influences the wetting behaviour of the ceramic, and the cooling rate, which influences the strength of the compound, are of crucial importance.

It has been shown[31–34] that metal–ceramic compounds can be produced successfully by induction heating even with non-conductive ceramics. Furthermore, the joints show equivalent qualitative properties in comparison to furnace brazing. The microstructure and the strength behaviour (four-point bending test) were compared for both processes to investigate the influence of the differing process control on the properties of the compounds. As ceramic base materials, alumina (purity: 99.7%) and MgO-stabilized zirconia were used. As metallic base materials, the iron nickel alloy FeNi42 (Mat. No. 1.3917; ASTM F 29, F 30[35,36]), the iron nickel cobalt alloy FeNiCo29 18 (Mat. No. 1.3981, ASTM F 15[39]) and the stainless steel CrNi18 10 (Mat. No. 1.4301, AISI 304) were used. As brazing material, a commercially available active-brazing filler metal (AgCu26.5Ti3) foil was used. For the microstructure investigation of the joint surfaces, the samples were 10 mm × 15 mm, with a height of 9 mm for the ceramic and 3 mm for the metal (Fig. 12.11(a)). The geometry of the samples for the four-point bending test was based on the sample geometry described in DVS Merkblatt

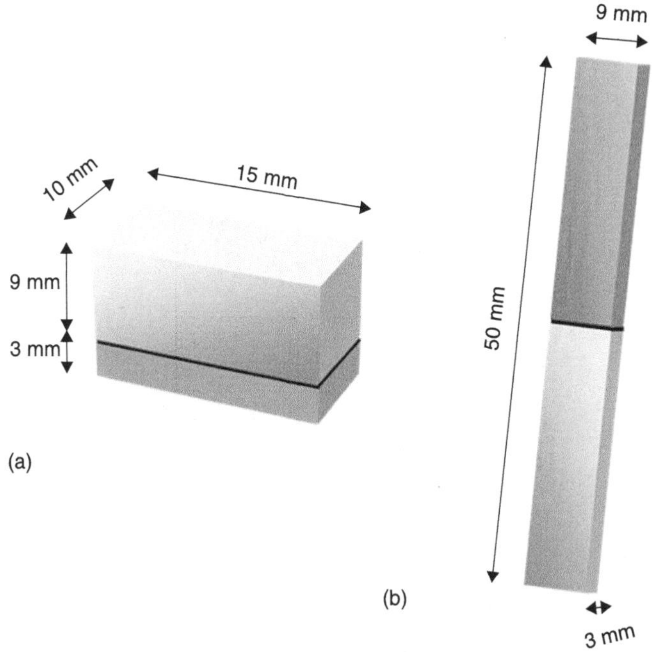

12.11 Sample geometry for (a) microstructure investigation and (b) four-point bending test.

3102.[16] The total length of the bending samples was 50 mm with a joint area of 9 mm × 3 mm (Fig. 12.11(b)).

The inductive brazing tests (frequency range of the induction system used: 10–25 kHz) were carried out in an evacuatable quartz tube. Different temperature–time cycles were used (Table 12.4, brazing cycles 2 and 3). For comparison, conventionally brazed samples were brazed in a high-temperature vacuum furnace with a typical temperature–time curve for metal–ceramic compounds (Table 12.4, brazing cycle 1). For the two processes, the brazing was performed under vacuum (10^{-4} mbar range) with the manufacturer's recommended brazing temperature of 900°C.

Table 12.4 Temperature–time cycles

Brazing cycle	Heating rate	Holding time	Cooling rate
1 (Furnace)	15 K/min	5 min	5 K/min
2 (Induction)	350 K/min	1 min	90 K/min
3 (Induction)	150 K/min	2 min	45 K/min

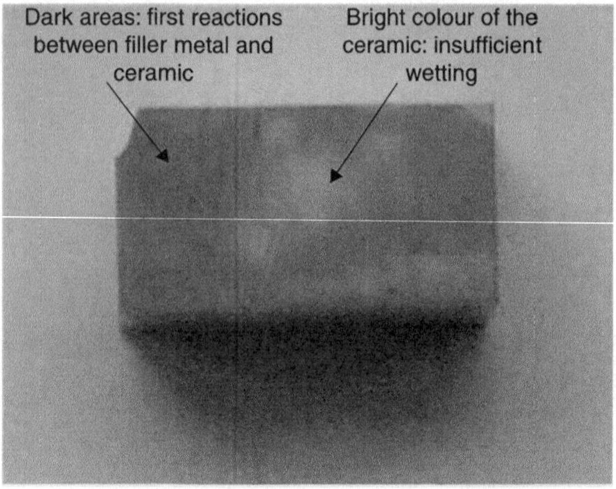

12.12 Joining surface after too fast a heating rate.

In order to ensure wetting of the ceramic, the ceramic must be heated by thermal conduction through the filler metal or the metal. Thus, an adjustment of the heating rate is necessary. The ceramic cannot be sufficiently heated and wetted with too fast a heating rate. Figure 12.12 shows the joint surface of a ceramic with such a heating rate (500 K/min) whereby no binding could be achieved. In certain areas, a reaction has indeed already taken place with the ceramic (dark areas), but the bright colour of the ceramic is still recognizable in some places. Insufficient wetting was achieved. For the specimen geometries used, adequate heating and thus wetting of the ceramic can be achieved with optimal heating rates of 100–200 K/min. These values cannot be assumed as generally valid, because the required heating rate strongly depends on the material used and component geometry. The heating rate must be adjusted according to the component to be brazed.

Figures 12.13–12.15 show optical microscope images for the material combination CrNi18 10 and Al_2O_3, and Fig. 12.16–12.18 present scanning electron microscope images of inductively brazed specimens (brazing cycles 2 and 3, Table 12.4) as well as a furnace-brazed sample (brazing cycle 1, Table 12.4). A comparison of the elemental mapping images for induction brazing and furnace brazing is illustrated in Table 12.5. All three joints show the typical three-zone structure of metal–ceramic compounds with the reaction zone between metal and filler metal (1), the brazing seam (2) and the reaction zone between filler metal and ceramic (3); see Fig. 12.13–12.15. For the various brazing cycles, marked differences in the microstructure of the individual zones are evident. For induction brazing with brazing cycle 2 (Fig. 12.13), corresponding to the fast heating rate and short holding time, the compound exhibits a relatively thin

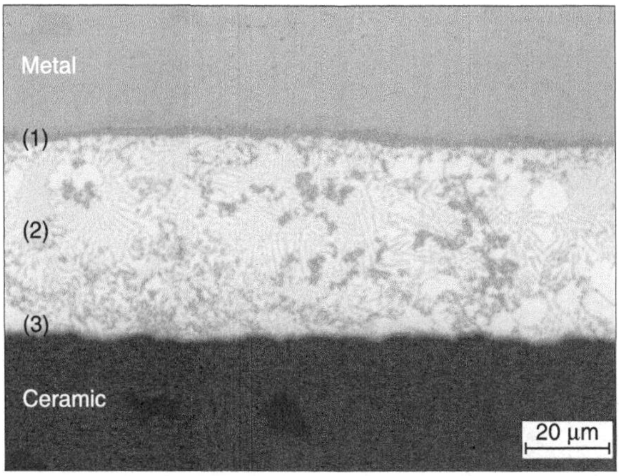

12.13 Induction-brazed joint (brazing cycle 2, Table 12.4).

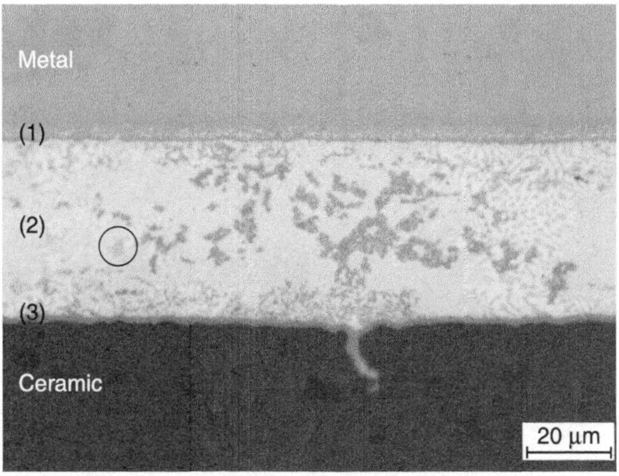

12.14 Induction-brazed joint (brazing cycle 3, Table 12.4).

reaction zone (≈ 2 μm; see Fig. 12.16(a)) between the filler metal and the metallic base material. Energy-dispersive X-ray spectroscopy (EDX) analysis shows that the reaction layer consists mainly of titanium and copper from the filler metal as well as iron and nickel from the metal (see Table 12.5). In comparison to the other two brazing processes, the metallic base material is hardly affected due to the short holding time (Fig. 12.16(a)). This is confirmed by an EDX analysis of the brazed seam (see Fig. 12.19). In comparison to furnace brazing, the elements iron, nickel and chromium from the steel are present only in very small quantities.

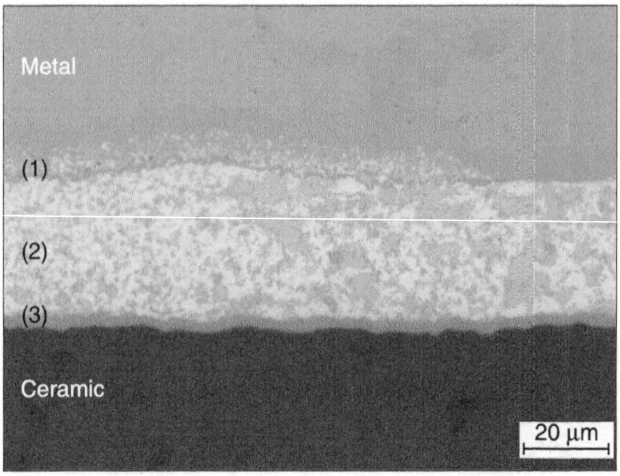

12.15 Furnace-brazed joint (brazing cycle 1, Table 12.4).

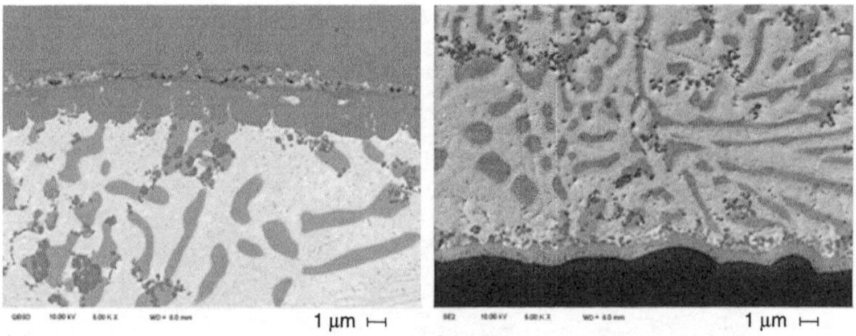

12.16 (a) Metal side and (b) ceramic side of induction-brazed joint (brazing cycle 2, Table 12.4).

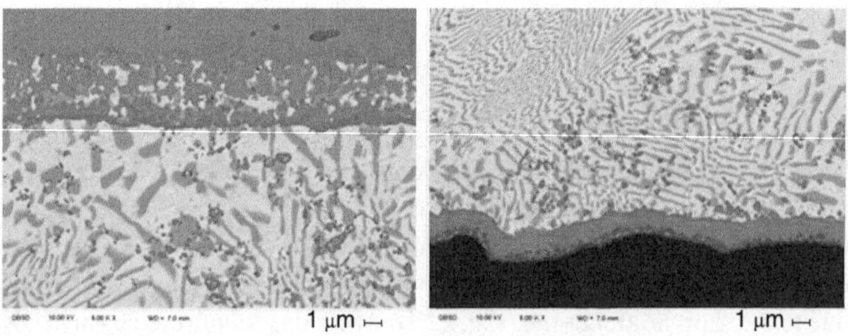

12.17 (a) Metal side and (b) ceramic side of induction-brazed joint (brazing cycle 3, Table 12.4).

Table 12.5 Elemental mapping images

	Induction – brazing cycle 2	Furnace – brazing cycle 1
SEM	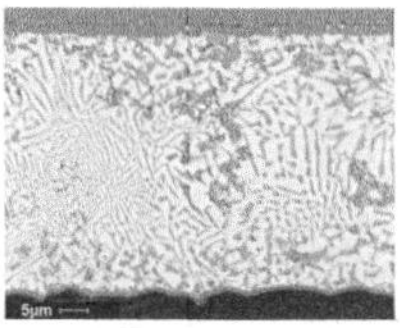	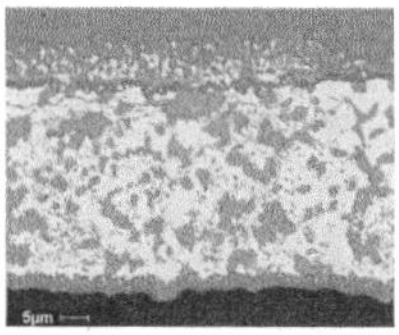
Ag		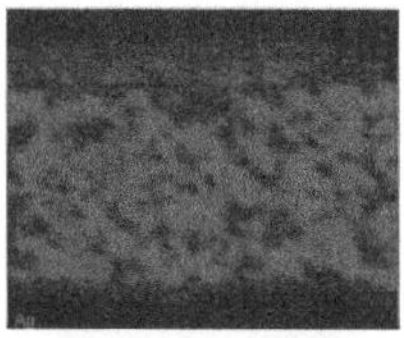
Cu	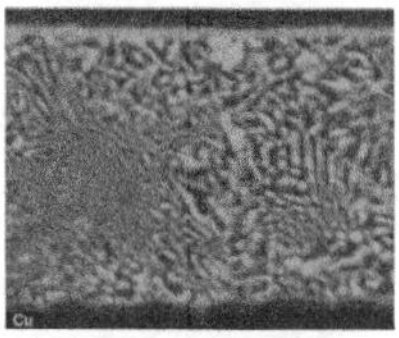	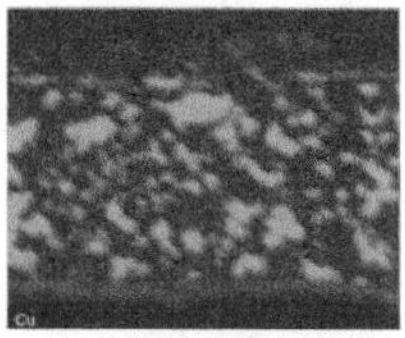
Ti	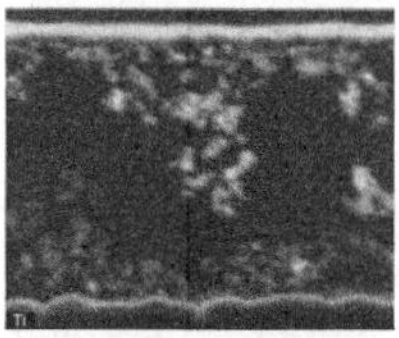	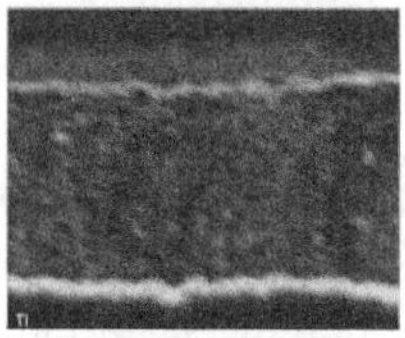
Fe		
Cr		
Ni		

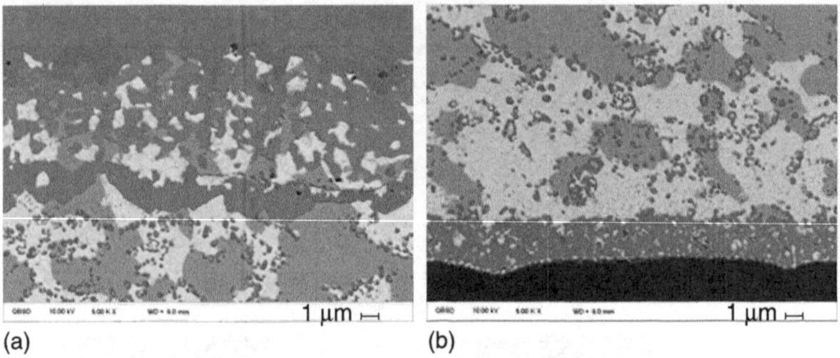

12.18 (a) Metal side and (b) ceramic side of furnace-brazed joint (brazing cycle 1, Table 12.4).

Due to the fast heating rate and the short holding time, the reaction zone between the filler metal and the ceramic has a small thickness (≈ 0.8 μm) (Fig. 12.16(b)). There was no intensive interaction with the ceramic. The EDX analysis shows that the reaction zone mostly contains the elements titanium, copper, silver and aluminium. The elemental mapping images (Table 12.5) also show that the titanium content is higher on the metal side than on the ceramic side.

The brazed seam has a fine eutectic structure (Fig. 12.13), which consists of the main components of the brazing filler metal, silver and copper. The silver matrix contains finely dispersed, dark intermetallic precipitates. Within the eutectic structure, larger dark precipitates can also be seen. The presence of titanium in the brazing seam indicates the formation of intermetallic phases such as Cu_3Ti.[40]

For brazing cycle 3 (Table 12.4) of induction brazing, the reaction zone between the filler metal and the metallic base material has a thickness of ≈ 3.5 μm (Fig. 12.17(a)). The EDX analysis of the reaction zone and the brazing seam (see Fig. 12.19) shows a slightly higher proportion of the elements iron, chromium and nickel in the two zones, since the metal is more intensely affected by the brazing material through the longer holding time.

The reaction zone between the filler metal and the ceramic is also more pronounced (thickness ≈ 1.3 μm; see Fig. 12.17(b)). The EDX analysis of the brazed seam (Fig. 12.19) shows a higher aluminium content due to a stronger reaction with the ceramic. In addition, the elements iron and nickel from the steel have been measured in the reaction zone on the ceramic side. These elements diffuse through the brazing seam to the ceramic side. So, the stronger reaction with the metal can be confirmed.

Also in this brazed seam, a eutectic structure of silver and copper with CuTi precipitates is recognizable (Fig. 12.14). Furthermore, the first large mainly copper crystals can be detected, which are indicated by the circle in Fig. 12.14.

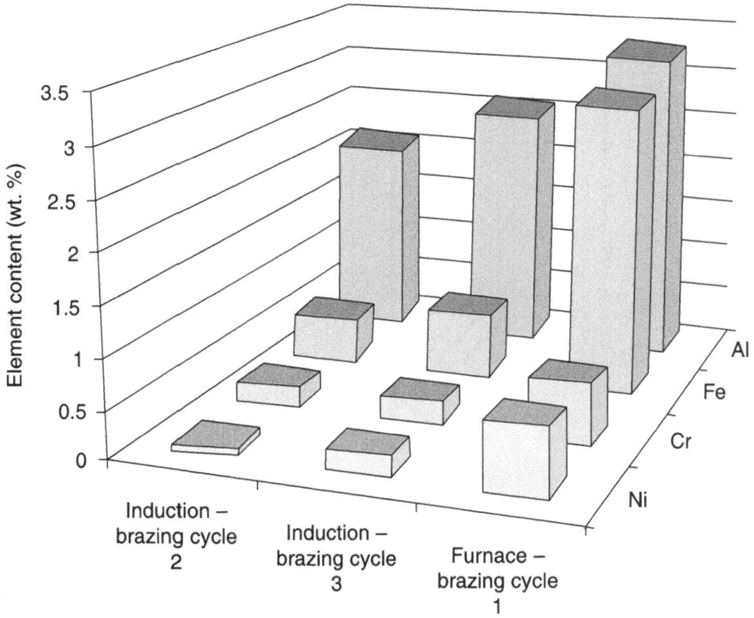

12.19 EDX analysis of the brazed seam without reaction zones.

The conventional furnace-brazed compound (brazing cycle 1, Table 12.4) has the thickest reaction zone on the metal side (≈ 8 μm; see Fig. 12.18(a)), as well as on the ceramic side (≈ 2.2 μm; see Fig. 12.18(b)). SEM images of both reaction zones show that strong interactions between the filler metal and the base materials took place. EDX analysis detects the presence of the elements iron, nickel and chromium from the metal inside the brazing seam (see Fig. 12.19), as well as on the ceramic side (see Table 12.5). The elemental mapping images also show that, in contrast to induction brazing, the titanium content on the ceramic side is higher than on the metal side.

Instead of the eutectic structure, large mainly copper crystals can be observed (Fig. 12.15). In comparison to induction brazing, the dark CuTi precipitates are much more finely distributed (see Table 12.5).

Due to the strong geometry dependence of the strength, the influence of the different microstructures must be assessed for each component. It is possible that the more finely dispersed CuTi precipitates in the furnace-brazed sample have a more beneficial effect on the strength than the precipitates appearing in clusters in the induction-brazed samples. A generalized statement about the influence of the reaction zone thickness on the strength is not possible. On the ceramic as well as on the metal side, the induction-brazed reaction zones are thinner than the furnace-brazed ones (see Fig. 12.20). A thicker reaction zone does not necessarily require a higher strength. Due to the resulting reaction products (oxides in the case of the

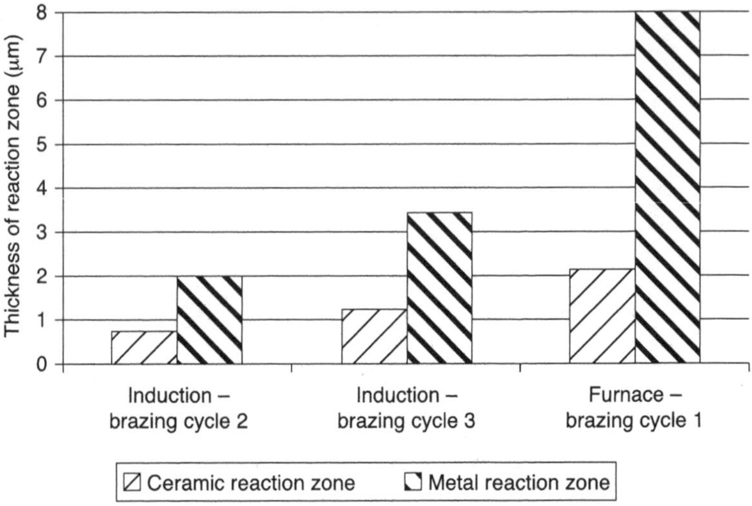

12.20 Comparison of thickness of reaction zones.

present ceramic), the reaction zone contributes not only to the wetting but also to the embrittlement of the compound. The optimum thickness of the reaction zone depends on the material combination and has therefore to be determined for real components.

The strength tests show that, when alumina is used, the crack pattern corresponds to a typical concave crack in a ceramic, which is caused by the different thermal expansion coefficients (Fig. 12.21(a)). Thus, the microstructure of the brazed

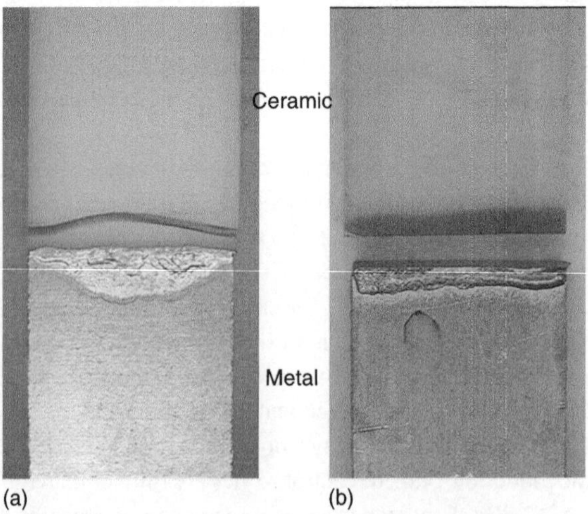

12.21 Crack pattern of (a) alumina and (b) zirconia.

seam has no influence on the strength. However, it must be noted, again, that the strength depends strongly on the sample's geometry. Hence, with a different component geometry, the strength can behave completely differently. In addition, the concave crack pattern in the ceramic can also be attributed to the relatively thin brazing seam (ca. 50–75 μm). Since the influence of different process operations should be studied individually, only one brazing foil was applied. Broadening of the brazed seam or the insertion of ductile interlayers may mean that the brazed joint is the weak spot and so the crack occurs in the brazed seam. In this case, the microstructure of the brazed seam is crucial for the strength.

When alumina is used, the crack always has the crack pattern shown in Fig. 12.21(a), and therefore the amount of thermally induced residual stress significantly affects the strength and can be influenced by the cooling rate. For the material combination FeNi42 and Al_2O_3, Fig. 12.8 shows the attainable bending strength values for different cooling rates of induction brazing in comparison to conventional furnace brazing. With decreasing cooling rate, the bending strength increases due to the reduction of residual stresses by plastic deformation. The strength of furnace-brazed samples (bending strength: 93 MPa) can be achieved by induction heating with a cooling rate of 15 K/min (bending strength: 94 MPa). Furthermore, the smaller standard deviations of the bending strength of induction-brazed samples, which can also be observed in other material combinations, have a positive impact.

When zirconia is used, the crack occurs at the interface between the brazing seam and the ceramic (Fig. 12.21(b)). Therefore, the holding time, and hence the thickness of the reaction zone and thus the wetting, is crucial to the strength. With a suitable holding time in induction brazing, it is possible to achieve an equivalent strength to furnace-brazed samples.

In consequence, depending on the ceramic used, induction-brazed compounds can exhibit qualitatively comparable properties to furnace-brazed compounds, if the respective process parameters are adjusted properly.

When induction heating is used, it must also be observed that the heat flow, which heats the ceramic by thermal conduction from the metallic joining partner, results in a temperature gradient in the ceramic. This leads to additional stresses, which may superpose the thermally induced stresses. How this affects the mechanical properties of the compound must be examined for each respective component.

12.10 Conclusion

Ceramics are characterized by a number of excellent material properties (see Section 12.1). Since demands on materials are steadily increasing, ceramics are a very interesting and important alternative to conventional materials. In this context, the joining of ceramics and metals plays an important role. Due to its considerable flexibility and universality, brazing has established itself as a widely used standard process for joining ceramics and metals (see Section 12.2). By means of brazing, vacuum-tight and thermally stable compounds with high strength can be produced. However, ceramics cannot be wetted by conventional metallic melts because of the different atomic structure of ceramics (ionic/covalent bonds) in comparison to metals (metallic bonds). To allow wetting of the ceramic, two process variants for brazing metal–ceramic compounds have been established: brazing of metallized ceramics (see Section 12.3) and active brazing (see Section 12.4). In brazing metallized ceramics, a metallic layer is applied to the ceramic before the brazing process. This facilitates wetting at the interface between the brazing filler metal and ceramic base material. The metallized ceramic can be wetted by conventional vacuum-brazing alloys with an Ag or AgCu basis. Since metallization is a very complex process, another method that causes direct wetting during brazing has been developed: active brazing. Commercially available active-brazing filler metals are mostly based on the element Ag or the Ag–Cu system. These brazing filler metals contain surface-active elements (usually 1 wt.% to 4 wt.% Ti, but also In, Zr, Hf or Nb), which directly wet the ceramic due to chemical reactions. Depending on the alloy composition, brazing is carried out at temperatures between 700°C and 1050°C. Due to the high oxygen affinity of the active elements, brazing is executed in vacuum or under an inert gas. Section 12.6 deals with the preparation for and execution of the brazing process.

The different physical properties (thermal expansion coefficient and elastic modulus) of metals and ceramics lead to significant residual stresses during the cooling process, which can significantly affect the quality or even lead to premature failure of the compound (see Section 12.5). These thermally induced residual stresses can be reduced in several ways. In addition to a suitable structural design, slow heating and cooling rates should be used. The residual stresses can also be reduced by using metals with a matched thermal expansion, such as Kovar or Invar, which have a relatively low thermal expansion coefficient. It is also possible to minimize the stresses by broadening the ductile brazing seam or inserting soft layers.

The quality of metal–ceramic compounds is assessed in addition to microstructure studies on strength tests (see Section 12.7). Mostly, the four-point bending test is used. Further examination methods for metal–ceramic compounds are, for example, testing gas tightness, thermal shock resistance, creep resistance and oxidation resistance.

Section 12.8 describes the microstructure and strength behaviour of an active-brazed metal–ceramic compound for a selected example. Furthermore, induction brazing is introduced as an energy-efficient alternative to conventional furnace brazing processes (see Section 12.9). The process-specific basics of induction brazing and furnace brazing, as well as the microstructure and strength for both processes, are compared. The presented results show that metal–ceramic compounds can be produced successfully by induction brazing, even with non-conductive ceramics. Induction brazing can be easily integrated into the manufacturing process, especially in serial production, and thus represents a very interesting and energy-efficient alternative to conventional furnace-brazing processes. To ensure wetting the ceramic must be heated by thermal conduction through the filler metal or the metal. Therefore, it is necessary to adjust the heating rate. Depending on the ceramic used, induction-brazed compounds can exhibit qualitatively comparable properties to furnace-brazed compounds if the process parameters (holding time and cooling rate) are optimal.

12.11 Acknowledgements

The Cluster of Excellence 'Energy-Efficient Product and Process Innovation in Production Engineering' (eniPROD®) is funded by the European Union (European Regional Development Fund) and the Free State of Saxony.

European Union
Europe funds Saxony.
EFRE
European Regional Development Fund

12.12 References

1. Nicholas M G (1990), *Joining of Ceramics*. London: Chapman and Hall, ISBN 0 412 36750 5.
2. Schwartz M (1990), *Ceramic Joining*. ASM International, Materials Park, Ohio 44073.
3. Boretius M, Lugscheider E, Tillmann W (1995), *Fügen von Hochleistungskeramik. Verfahren – Auslegung – Prüfung – Anwendung*. Düsseldorf: VDI-Verlag.
4. Kollenberg W (2004), *Technische Keramik. Grundlagen, Werkstoffe, Verfahrenstechnik*. Essen: Vulkan-Verlag.
5. Verband der Keramischen Industrie (2003), *Technische Keramik*. Lauf: Fahner, ISBN: 3-9241578-77-0.
6. Tietz H D (1994), *Technische Keramik*. Düsseldorf: VDI Verlag.
7. Salmang H, Scholze H (2007), *Keramik*. Berlin, Heidelberg, New York: Springer.
8. IPT-Albrecht GmbH (2010), *Hochleistungstechnologie für Forschung und Produktentwicklung*. Available from: http://www.ipt-albrecht.de [Accessed 9 June 2010].
9. Pfeiffer Vacuum (2010), Available from: http://www.pfeiffer-vacuum.com [Accessed 9 June 2010].
10. Schmoor H (2003), Aktivlöten – aktuelle und potenzielle Anwendungen. In: H P Degischer: *Verbundwerkstoffe. 14. Symposium Verbundwerkstoffe und Werkstoffverbunde*. pp. 691–696. Weinheim: Wiley-VCH Verlag.

11. Naidich J V (1981), The wettability of solids by liquid metals. In: *Progress in surface and membrane science*, edited by: D A Cadenhead and J F Danielli, Vol. 14, New York: Academic Press.
12. ASTM F 19-64 (2005), *Standard Test Method for Tension and Vacuum Testing Metallized Ceramic Seals*. Philadelphia: American Society for Testing and Materials.
13. Zigerlig B (1981), *Metall-Keramik-Technologie im Elektronenröhrenbau bei BBC und Prüfung metallisierter Aluminiumoxidkeramik*. Referat am 24.09.1981 im Rahmen des DVS/DKG-Gemeinschaftsausschusses W3.
14. Mizuhara H, Oyama T(1991), Ceramic/Metal Seals. *ASM Handbook, Vol. 4: Ceramic and Glasses*, ASM International, Materials Park, Ohio 44073.
15. Mayer H, Heinicke E (1999), Elektrische Isolation mit Aluminiumoxid. Oxidkeramik-Metall-Verbundbauteile. *Vakuum in Forschung und Praxis* 11, 2, pp. 83–85.
16. DVS (2005), Merkblatt 3102 *Herstellen von Keramik-Keramik- und Keramik-Metall-Verbindungen durch Aktivlöten*. Düsseldorf: DVS.
17. Tillmann W, Buschke I, Lugscheider E (1995), Herstellungsmöglichkeiten von hochtemperaturbeständigen Verbunden nichtoxidischer Ingenieurskeramiken mittels verschiedener Lotkonzepte. *DVS-Berichte* 166, pp. 110–114.
18. Bang K S, Liu S (1994), Interfacial reaction between alumina and Cu-Ti filler metal during reactive metal brazing. *Welding Journal* 73, 3, pp. 54–60.
19. Suganuma K (1990), Joining silicon nitride to metals and to itself. In: *Joining of ceramics*, edited by M G Nicholas, London: Chapman and Hall.
20. Klomp J T (1987), Interface chemistry and structure of metal-ceramic interfaces. In: *Fundamentals of diffusion bonding*, edited by Y Ishida, Amsterdam, Oxford, New York, Tokyo: Elsevier Science Publishers, p. 43.
21. Nicholas M G, Crispin R (1985), The role of titanium in active and activated brazing of alumina. *Fortschrittsberichte der deutschen keramischen Gesellschaft*, Bd. 1, p. 3.
22. Hanson W B, Ironside K I, Fernie J A (2000), Active metal brazing of zirconia. *Acta mater.* 48, pp. 4673–4676.
23. Pak J J, Santella M L, Fruehan R J (1990), Thermodynamics of Ti in AgCu Alloys. *Metallurgical Transactions Part B* 21, 2, pp. 349–355.
24. Nicholas M G (1986), Active Metal Brazing. *Br. Ceram. Trans. J.* 85, pp. 144–146.
25. Lugscheider E, Tillmann W (1992), Herstellung von Keramik-Metall-Verbunden – konstruktive und fügetechnische Aspekte. *VDI Berichte* 965.1, pp. 227–240.
26. Maier H R, Magin M, Fischer S (1995), Abbau und Steuerung von Verbundspannungen in Keramik-Metall-Verbunden – Fortschritte bei experimentellen und theoretischen Analysen. *DVS-Berichte* 166, pp. 155–160.
27. Schüler H (2001), *Simulation von Lötprozessen beim Metall-Keramik-Löten*. Dissertation. TU Chemnitz.
28. Munz D, Iancu O T (1988), Spannungszustände in der Verbindung und ihre mathematische Erfassung. Unterlagen zum Seminar 'Fügen von Hochleistungskeramik', Seminar der Technischen Akademie Wuppertal.
29. Sugunama K (1990), Recent advances in joining technology of ceramics to metals. *ISIJ International* 30, 12.
30. Frey T, Haubenreich A, Girmscheid R, Metzler P E (2003), *Vergleichende Festigkeitsuntersuchungen an Mg-PSZ und Aluminiumoxid*, Schriftenreihe der Georg-Simon-Ohm-Fachhochschule Nürnberg 20, ISSN:1616-0762.
31. Wielage B, Hoyer I, Hausner S (2010), Niederenergetische Verbindungsverfahren – Metall-Keramik-Aktivlöten. In: *Energieeffiziente Produkt- und Prozessinnovationen in der Produktionstechnik*, ISBN 978-3-942267-00-7, Chemnitz, pp. 307–324.

32. Wielage B, Hoyer I, Hausner S (2010), Beitrag zum Induktionslöten von Metall-Keramik. In: *DVS-Bericht* Bd. 263: Hart- und Hochtemperaturlöten und Diffusionsschweissen, ISBN 978-3-87155-589-3, Düsseldorf, pp. 331–336.
33. Wielage B, Hoyer I, Hausner S (2011), Induktionslöten der Fe-Ni-Legierung 1.3917 mit Al_2O_3 unter Verwendung eines Ag-Cu-Ti-Aktivlotes. In: *Tagungsband zum 18. Symposium Verbundwerkstoffe und Werkstoffverbunde*, ISBN 978-3-00-033801-4, Chemnitz, pp. 387–392.
34. Wielage B, Hoyer I, Hausner S (2011), Induktionslöten von ZrO_2 und der Kovar-Legierung FeNiCo29 18. *Keramische Zeitschrift* 63, 5, pp. 344–348.
35. ASTM F 29-97 (2009), *Standard Specification for Dumet Wire for Glass-to-Metal Seal Applications*. Philadelphia: American Society for Testing and Materials.
36. ASTM F 30-96 (2009), *Standard Specification for Iron-Nickel Sealing Alloys*. Philadelphia: American Society for Testing and Materials.
37. Dabbarh S, Pfaff E, Ziombra A, Bezold A (2010), 'Reactive air brazing' sauerstoffleitender Keramik mit CrNi-Stahl. In: *DVS Berichte* 263, pp. 338–343.
38. Batdorf S B, Sines G (1980), Combining data for improved Weibull parameter estimation. *J. Am. Ceram. Soc.* 63, pp. 214–218.
39. ASTM F 15-04 (2009), Standard Specification for Iron-Nickel-Cobalt Sealing Alloy. Philadelphia: American Society for Testing and Materials.
40. Kar A, Mandal S (2007), Characterization of interface of Al_2O_3-304 stainless steel braze joint. *Materials Characterization* 58, 6, pp. 555–562.

13

Brazing of carbon–carbon (C/C) composites to metals

T.-T. IKESHOJI, Tokyo Institute of Technology, Japan

DOI: 10.1533/9780857096500.2.394

Abstract: The brazing of carbon–carbon (C/C) composites and metals is a necessary manufacturing technique for some high-performance applications. In this chapter, the fundamentals of brazing C/C composites to metals are described including the wettability of the brazing filler on C/C composites and the coefficient of thermal expansion (CTE) difference between C/C composites and metals. Commercially available brazing fillers for the brazing of C/C composites and various metals are listed. The effect of fiber orientation on the joint strength of C/C composites and titanium is described. The manufacturing method for overcoming the mismatch in CTE between C/C composites and metals is also described.

Key words: aerospace, high-speed train, ITER, active brazing filler, super alloy, anisotropic material, residual stress, fracture dynamics.

13.1 Introduction

The application of carbon–carbon (C/C) composites is starting to expand from fields such as the aerospace industry to new areas such as high-speed trains. These applications take advantage of the excellent properties offered by C/C composites, including high-temperature resistance, ultra-low weight, high Young's modulus and high strength. However, C/C composites are difficult to join to metals and ceramics. Currently, mechanical joining techniques are primarily used, as common arc welding will fuse and destroy the structure of C/C composites, leading to the loss of key properties. Brazing is considered to be the only promising welding technique for joining C/C composites. In this chapter, the brazing techniques developed for joining C/C composites to metals are described.

In the first part of this chapter, the properties of C/C composites are briefly described, with an analysis of the phenomena that occur during the brazing process. Brazing filler alloys for the brazing of C/C composite's to metals are then described. As far as the author is aware, at the time of writing the brazing filler alloys recommended for C/C composite brazing are not supplied commercially. The required specification of brazing filler alloys for brazing C/C composites is therefore discussed, before the brazing filler alloys reported as feasible for use are introduced. Finally, the other key factors which must be considered during the brazing of C/C composites to metals are discussed.

13.2 Carbon–carbon composites

Carbon–carbon composites (C/C composites) are composite materials consisting of a matrix of graphite reinforced by carbon fibers. C/C composites offer excellent features, including high heat resistance, ultra-low weight and almost negligible thermal deformation. Since their original development for military aerospace use, it has been recognized that C/C composites provide solutions for applications requiring low weight at high temperatures.

C/C composites are now expected to be the material of choice for the next generation of rocket engines. Current rocket engines operate at approximately 1600°C; an operating temperature which is expected to be raised to 2000°C. In addition, requests have been made for a lighter weight engine capable of facilitating the longer lifetime required by a reusable system. C/C composites meet these specification requirements. The European Aeronautic Defence and Space Company (EADS) has already developed such a C/C composite rocket engine, which is currently undergoing evaluation (Knoche *et al.*, 2006). For the next generation rocket engine, plans have been made to manufacture both the nozzle skirt and throat using C/C composites (Vignoles *et al.*, 2010; Li *et al.*, 2011), while research into their possible application in conventional rocket nose cone is also being conducted (Yamamoto *et al.*, 1996). At the 2011 Paris Airshow, EADS announced the R&D for its supersonic cruiser project. The planned supersonic cruiser will use C/C composites for both the engine's hot sections and its air frame (Steelant, 2010). Japan Aerospace Exploration Agency (JAXA) explored a similar supersonic cruiser project in the middle of the 1990s, leading to the development of its ATREX (Air Turbo Ramjet EXpander cycle) engine, featuring C/C composite turbine rotator disks (Goto *et al.*, 2003).

The newest application of C/C composites is the ITER divertor. The ITER (International Thermonuclear Experimental Reactor) is the world's largest tokamak-type nuclear fusion reactor, and is currently under construction in Cadarache, France. The reactor features a huge donut-shaped vacuum vessel, with the high-temperature plasma inside confined by a magnetic field. Figure 13.1 shows a cross-section of the reactor. Its bottom section is called the divertor, from which the heat and the helium ash are extracted. Facing the high-temperature plasma, the divertor's walls are protected by armor made from C/C composites. Inside the C/C composite armor, copper alloy tubes are inserted for cooling. The gap between a tube wall and armor hole is filled using a brazing technique, to ensure heat removal. Similar research into the manufacture of fusion reactor armor is being conducted by the Joint European Torus project in Oxfordshire, UK (Hirai *et al.*, 2007).

Another potential area for C/C composite application, high-speed rail, has gained the attention of governments around the world due to its large capacity and perceived environmentally friendly systems. In the future Japanese high-speed trains may be equipped with a C/C composite disk brake system (Okezawa *et al.*,

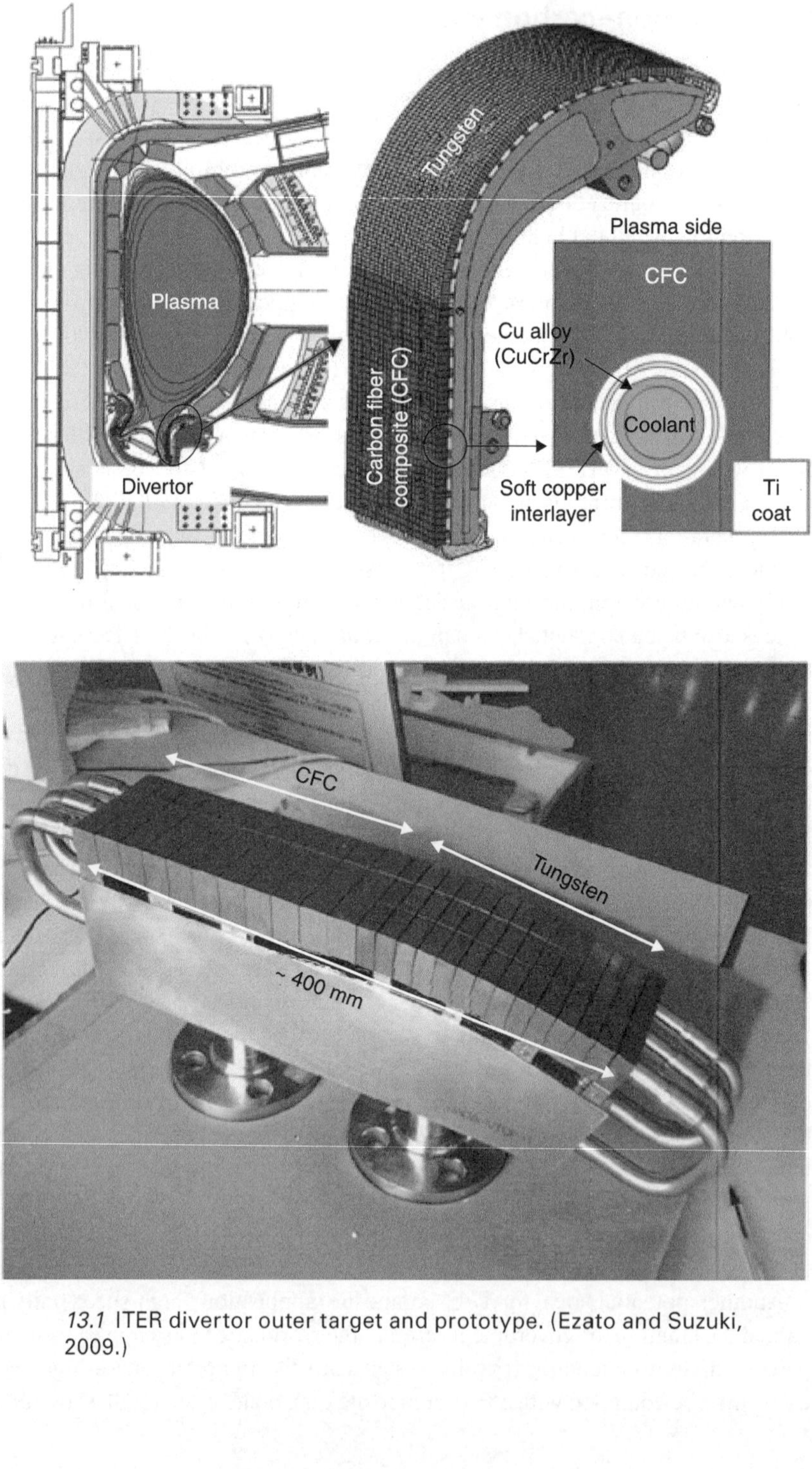

(a)

(b)

13.1 ITER divertor outer target and prototype. (Ezato and Suzuki, 2009.)

1997; Takemoto *et al.*, 1998). The brake disk is expected to experience temperatures of over 1000°C when decelerating from 350 km/h, making high heat resistivity an essential requirement. The current wrought steel brake disks could not support such a load at these high temperatures, whereas the heat resistivity and high specific strength of C/C composites meet the specifications.

The application of C/C composites has been expanding steadily in various fields. However, at this stage C/C composites are still tailor-made and relatively expensive materials. If techniques for joining C/C composites with other materials can be developed, a standardized and mass-produced material may be combined with metals or ceramics. If this situation becomes feasible, C/C composite prices will be reduced, and many more applications will appear.

13.2.1 Characteristics of C/C composites

Structures of C/C composites

Carbon–carbon composites consist of carbon fibers in a graphite matrix. The carbon fibers used have high modulus, and exhibit relatively high tensile strength (Mallick, 2007). Carbon fibers are manufactured by graphitization of precursors such as a textile or pitch. Textile precursors are the filament used in the textile; rayon filaments, phenol filaments and acryl filaments, for example. The most commonly used textile precursors are acryl filaments or polyacrylonitrile (PAN) filaments, while the pitch precursor is developed from petroleum products.

Another fabrication method commonly used is the vapor-phase growth process. Vapor-grown carbon fiber (VGCF) is manufactured by heat decomposition of hydrocarbon gas at high temperature in the presence of a transition metal catalyst. The process is similar to the fabrication of carbon nanotubes. However, the VGCF method can only produce short fibers, just centimeters in length (Jayasankar *et al.*, 1995; Mishra and Ting, 2008; Mukai *et al.*, 2000).

A comparison of the physical properties of pitch, PAN and VGCF fibers are shown in Table 13.1. This differences for pitch, PAN and VGCF fibers originates from the degree of graphite crystallization, with the degree of graphite crystallization further affecting the anti-corrosion properties of the fibers. In pitch

Table 13.1 Physical properties of carbon fibers

	Mesophase pitch*	PAN†	VGCF‡
Diameter, μm	10	7	5–8
Density, g/cm^3	1.98	1.75–1.93	2.20
Tensile strength, MPa	3500	3800–5900	5000–8000
Tensile modulus, GPa	200	290–590	500

*Nippon Graphite Fiber Co. †TORAY Industries Inc. ‡Inagaki, 2009.

carbon fiber, the graphite crystals are arranged randomly about the fiber axis. In contrast, the graphite crystals in VGCF fiber are arranged in parallel to the axis, like the annual growth rings of wood. This ordered arrangement produces a density as high as graphite, along with robust mechanical properties. PAN carbon fiber combines elements of both of these arrangements, featuring randomly arranged graphite crystals in the middle of the fiber, hemmed by the 'annual growth ring'-like arrangement of graphite crystals. Hence, PAN carbon fiber exhibits the intermediate properties of both pitch and VGCF carbon fiber. The degree of graphite crystallization affects the anti-corrosion properties of fibers. As described later, during the brazing process the fibers in C/C composites could be corroded by the active element in the brazing filler. VGCF has high corrosion resistance, PAN carbon fiber has intermediate, and pitch carbon fiber has low.

Once the carbon fibers have been produced, they are used to weave carbon cloth. The structure of a cloth can be one-dimensional (1D), two-dimensional (2D) or three-dimensional (3D). In a one-dimensional structure, all fibers are oriented in one direction. A two-dimensional structure features two or more fiber directions, with the variations in weaving just the same as those seen in ordinary textiles. A three-dimensional structure can be created by stitching a stack of woven fabric. Regardless of the dimensions of the cloth structure, bulk C/C composites are made by laying up these thin cloths to the required thickness and pasting them together using plastic or pitch binder. The lay-up is then heated to graphitize the binders, producing graphite-reinforced carbon fiber. However, this cloth has numerous voids inside. By repeating the infiltration of pitch and graphitizing at high temperature, the fraction of voids is reduced, and fine voids can further be filled using the chemical vapor deposition technique.

Mechanical properties

C/C composites with PAN or pitch carbon fiber may exhibit lower density than graphite, due to the lower density of these carbon fibers in comparison to graphite. The average density of graphite is 2.2 g/cm^3. However, if PAN or pitch carbon fibers are used in C/C composites, this density could fall into the range of 1.7–2.2 g/cm^3. In practice, the density of commercially supplied C/C composites ranges from 1.4 to 2.0 g/cm^3 as shown in Table. 13.2, which is somewhat lower than expected. This is due to the inner porosity of the composite, which is introduced during the carbonization process. The gases decomposed from the precursors during the process are insufficiently extruded, and remain in the composite as pores or voids. By repeating the carbonization process, these pores and voids can be filled. However, total densification requires time and expense, and is practically impossible, meaning C/C composites are largely porous materials.

C/C composites have strong anisotropic properties, and, with such adaptive potential, they can be seen as tailor-made materials. By altering the combination of carbon fibers and the matrix structure, the characteristics of C/C composites can be

Table 13.2 Examples of commercially supplied C/C composites

	Tokarec, CC26NF*	Tokarec, CC27MFP*	Across, AC300†	Across, FC500†
Density, g/cm^3	1.40	1.62	1.7	1.5
Flexural strength, MPa	100	170	250	98
Tensile strength, MPa	100	150	250	170
CTE, $\mu m \cdot m^{-1} \cdot K^{-1}$	0.8	0.8	0–1	0–1
Thermal conductivity, W/Km	2.6	8		
Typical applications	Pipe	Crucible	Structure	Fastener

*Tokai Carbon Co., Ltd. †Across, Inc.

adapted in accordance with a designer's needs. While homogeneous characteristics can be developed through the laminate surface direction, the characteristics of the direction of lamination are usually different from those in the laminate surface.

One of the mechanical properties in which C/C composites have strong anisotropy is the coefficient of thermal expansion (CTE). A 2D cross-ply laminate C/C composite has a CTE of $\pm$ 0.1 $\mu m \cdot m^{-1} \cdot K^{-1}$ in the laminate plane (Ishikawa *et al.*, 1991). However, its cross-sectional CTE is almost 10 $\mu m \cdot m^{-1} \cdot K^{-1}$, which is a value comparable to pure titanium (8.6 $\mu m \cdot m^{-1} \cdot K^{-1}$), pure iron (11.8 $\mu m \cdot m^{-1} \cdot K^{-1}$), 0.30%C steel (11.5 $\mu m \cdot m^{-1} \cdot K^{-1}$) or martensitic stainless steel (SUS410, 10.4 $\mu m \cdot m^{-1} \cdot K^{-1}$). In the lamination direction, the CTE value could be considered comparable with graphite. The CTE of graphite varies with the degree of crystallization from 2 to 26 $\mu m \cdot m^{-1} \cdot K^{-1}$ (Martin and Entwisle, 1963), while the degree of crystallization of a C/C composite matrix is intermediate, so its CTE value falls around 10 $\mu m \cdot m^{-1} \cdot K^{-1}$.

13.3 Brazing filler alloys for brazing of carbon–carbon composites and metals

13.3.1 Wetting characteristics of C/C composites

The wetting behavior of molten filler alloys on C/C composites is similar to that on graphite, as is expected from the fact that C/C composites consist of both graphite fibers and a graphite matrix. On graphite, brazing filler alloys are required to contain carbide-forming elements, such as Cr, Mn, Mo, Si, Ti, V and Zr. These alloys with the carbide-forming elements are called active brazing filler alloys (Donnelly and Slaughter, 1962). The presence of these active elements in the brazing filler alloys is considered necessary for the wetting of C/C composites.

The wetting behavior of Ti-containing brazing filler alloys on C/C composites and graphite are compared in Fig. 13.2 (Ikeshoji *et al.*, 2010). The C/C composite

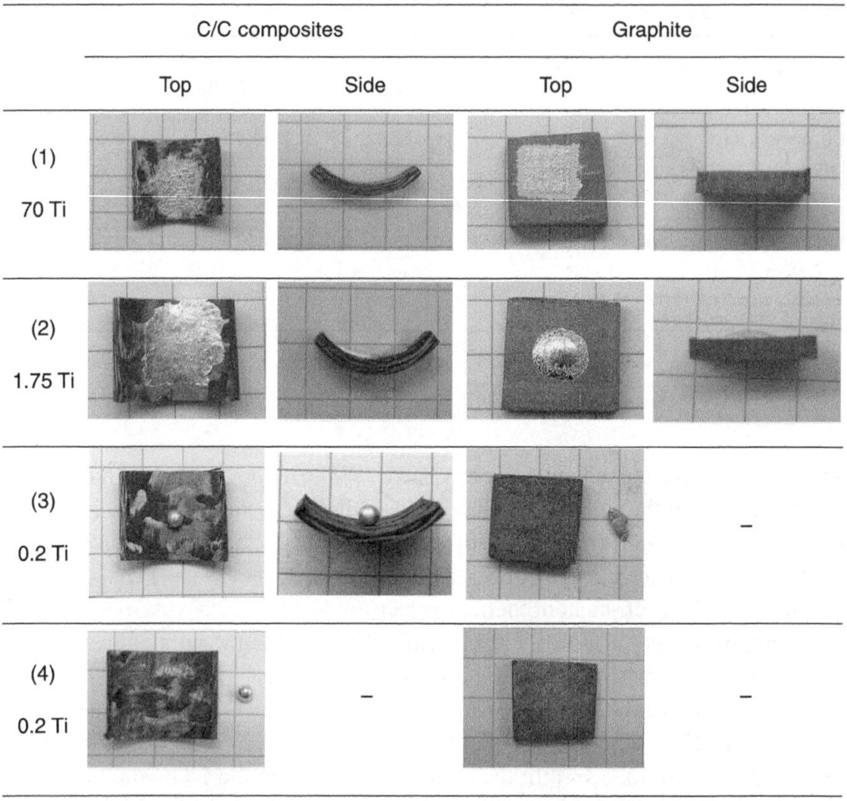

13.2 Macroscopic appearance of results of wetting test specimens. Brazing filler alloys are: (1) 70Ti–15Cu–15Ni, (2) Ag–35.25Cu–1.75Ti, (3) Al–0.8Fe–5.3Si–0.2Ti and (4) Al–6.3Cu–0.3Mn–0.2Si–0.2Ti. The spacing of the background grid is 6.35 mm.

substrate is the wall of a two-layer angle-interlock woven pipe. The brazing filler alloys contain various titanium fractions from 0.2 to 70 mass%. The wetting tests were conducted in a vacuum atmosphere, with the specimens first baked at 250°C for 1800 s and heated up to 20–30° below the brazing filler metal's solidus. For 70 mass% Ti alloy, the solidus was 960°C, making the dwelling temperature 930°C. The solidus of 1.75 mass% Ti alloy was 830°C, and that of 0.2 mass% Ti alloy was 650°C. The specimens were then rapidly heated up above their liquidus. For a brazing filler with 70%Ti the maximum temperature was 1000°C, and for 1.75% Ti and 0.2% Ti the maximum temperature was 900°C. After heating ceased, the specimens were cooled down in the vacuum furnace.

The brazing filler alloys with 1.75 mass% Ti wet very well both C/C composites and graphite. The 0.2 mass% Ti seemed unable to wet either type of substrate. On the C/C composites, the impulsive force between the alloy and substrate makes the brazing filler alloy into a perfect sphere. However, 70 mass% Ti proved an excessive amount for wetting, and the molten brazing filler alloy infiltrated both the C/C composites and graphite. The side views show a flat surface with almost no residual alloy on either substrate. These results suggest there must be a suitable fraction of the carbide-forming element in the active brazing alloys. It should be noticed that molten active brazing filler alloys can infiltrate into C/C composites. In addition, because similar wetting behavior was observed for the wetting tests on graphite, the suitable fraction required to wet C/C composites could be estimated from the wetting tests of active brazing alloys on graphite.

The contact angle of metal–Ti binary alloys on graphite has been measured, and decreases with Ti fraction (Fig. 13.3) (Singh *et al.*, 2005b). The copper alloys require at least 10 mass% Ti to wet graphite. When the substrate is vitreous graphite, 15 mass% Ti is necessary for wetting. In contrast, Ag–Ti alloys wet very well with just 2 mass% Ti.

The wetting of Ag–Ti binary alloys on graphite is similar on C/C composites (Fig. 13.4) (Okamura *et al.*, 1996). Pure silver does not wet C/C composite substrates, but the addition of just 1 mass% Ti improves the contact angle by several degrees. The minimum contact angle is achieved with the addition of 2–3 mass% Ti in silver. These experimental results suggest that the most suitable fraction of Ti in alloys is around 2 mass%.

In the brazing process, silver is usually utilized in the form of an Ag–Cu eutectic alloy to lower the melting point. Ti can dissolve in both Ag and Cu, so an increase in the fraction of Ti in Ag–Cu alloys to more than 2% might be more efficient. In addition, Cu and Ti can easily form an intermetallic compound, meaning the effective fraction of Ti will be smaller than the nominal fraction. Hence,

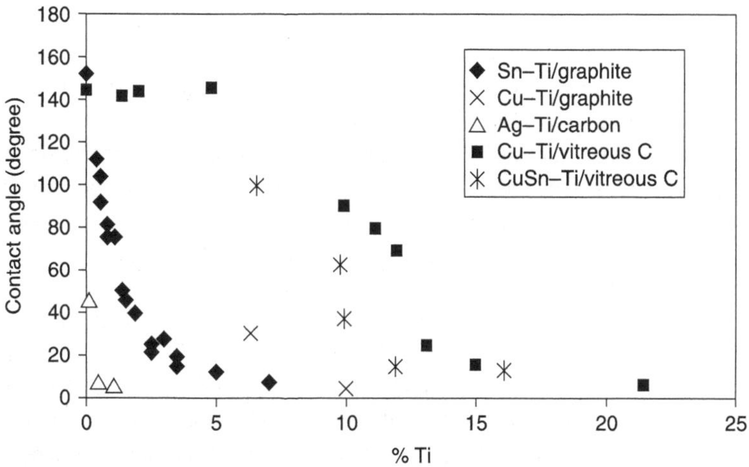

13.3 Contact angle of M–Ti alloys on graphite (Singh *et al.*, 2005b). (Data from studies cited in Standing and Nicholas, 1978; Li, 1992; Grigorenko *et al.*, 1998; Humenik and Kingery, 1954; Naidich, 1981.)

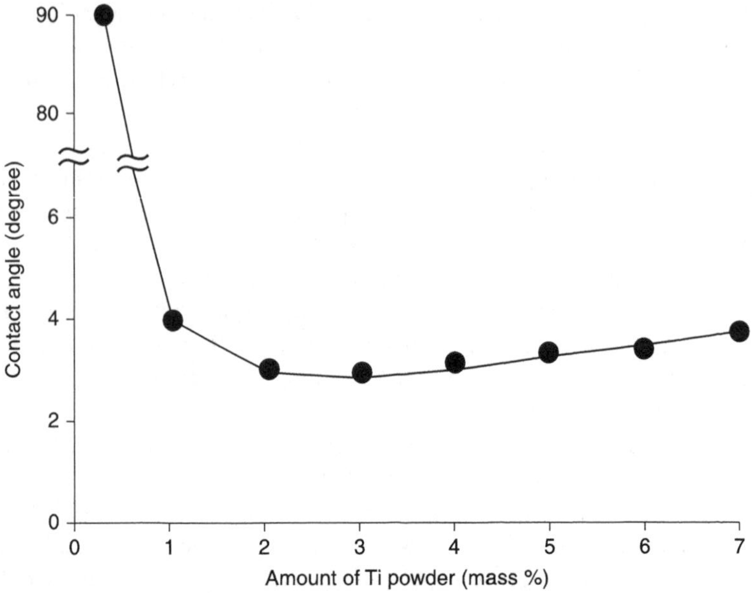

13.4 Effect of Ti fraction in Ag on the contact angle on C/C composites; temperature, 1123 K. (Okamura *et al.*, 1996)

Ag–Cu–Ti alloys should have around 3–4 mass% Ti. This can, however, be altered in practice. Cusil-ABA®, which is the most promising brazing filler for C/C composites and metal brazing, is an Ag–Cu alloy foil containing flaked Ti. This structure suppresses the solid solution of Ti in Cu and the formation of the Cu–Ti intermetallic compound. The flake-Ti-dispersion structure facilitates the effective use of the greatest amount of Ti as the active element in the brazing process. As such, Cusil-ABA contains only 1.75% Ti in the Ag–Cu alloy.

Other than the alloys containing Ti, brazing filler alloys containing Cr can wet C/C composites (Chen *et al.*, 2010). Ni–Cr–Pd alloys have better wettability on increasing the Cr fraction. With 12% Cr, the contact angle decreases drastically by several degrees. Chromium works as an active element and the formation of a $Cr_{23}C_6$ layer seems to encourage wetting. Therefore Pd–(35.2–30)Ni–(12–25)Cr, Ni–33Cr–24Pd–4Si and Ni–14Cr–10.5P–0.1Si (BNi-7) wet well the C/C composite substrate (Chen *et al.*, 2010; Liu *et al.*, 1994).

The contact angle of various brazing filler alloys on C/C composites is shown in Fig. 13.5. Cusil-ABA, Ag–Cu–1.8Ti, Ag–Cu–2Ti and Ni–72Ti contain titanium as the active element. BNi-7, BNi-2, 39Ni–33Cr–24Pd–4Si and 30Ni–40Cr–30Ge contain chromium. 39Ni–33Cr–24Pd–4Si alloy has the smallest contact angle of 3.5° (Okamura *et al.*, 1996). It is thought that the collaboration between the formation of Cr_xC_y and the solid solution of Pd dissolving graphite produce this low contact angle (Okamura *et al.*, 1996). The final three types of brazing filler alloys, 48Ni–31Mn–21Pd, Palni® and Palco™ (Pd–35Co) utilize palladium.

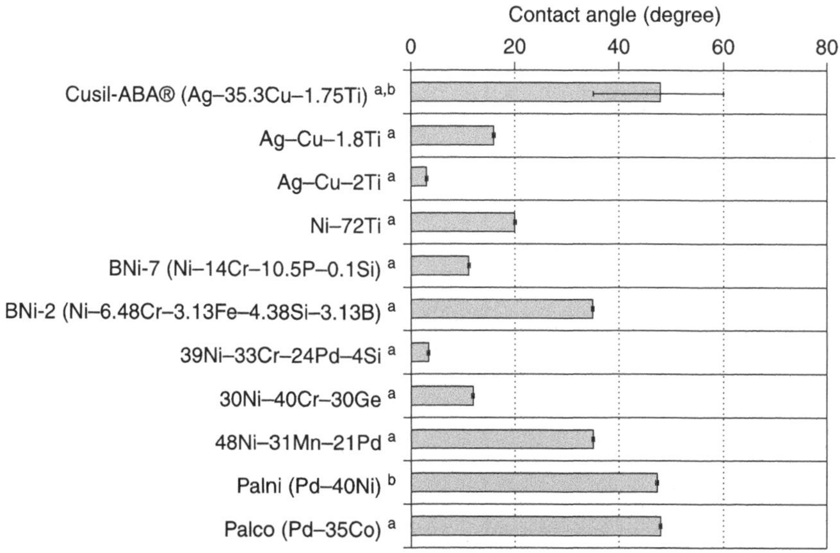

13.5 Contact angle of brazing filler alloys on C/C composites. (Data from (a) Okamura *et al.*, 1996 and (b) Asthana *et al.*, 2010.)

According to the C–Pd binary phase diagram, palladium dissolves several mass percent of graphite at room temperature.

13.3.2 Brazing filler alloys for C/C composites

There is no brazing filler alloy exclusively recommended for commercial C/C composites/metal brazing. As previously mentioned, C/C composites consist of carbon fiber and a graphite matrix, which have similar properties to graphite. Therefore, fundamentally, the brazing filler alloys for graphite can be applied to C/C composites. The commercially supplied brazing alloys capable of facilitating the brazing of graphite are summarized in Table 13.3. Most of these brazing filler alloys contain titanium as the active element to form carbides.

The possible combinations of metal substrates and brazing fillers are listed in Table 13.4, which was originally created by Singh and Asthana (2007). Their results are noted as (a). Notes (b)–(d) were added by the author. Singh and Asthana's work evaluated joint quality, qualitatively based on the metallurgical soundness of the joint microstructures. Further details of joint microstructures and strength are described in their works (Singh *et al.*, 2005a, 2005b, 2006; Singh and Asthana, 2007; Morscher *et al.*, 2006; Asthana *et al.*, 2006). As previously mentioned, no brazing filler alloy is currently commercially supplied for exclusive use with C/C composites, but some commercially supplied brazing filler alloys have been reported as feasible for joining C/C composites and metals (Table 13.5).

Table 13.3 Brazing alloys recommended for the brazing of graphite

Manufacturer	Title	Nominal composition, %	Temperature range, °C
Wesgo Metals	Copper-ABA®	Cu-3Si-2Al-2.25Ti	1025–1050
	Nioro®-ABA™	Au-15.5Ni-1.75V-0.75Mo	970–1020
	Ticuni®	Ti-15Cu-15Ni	960–1100
	Ticuni-60®	Ti-15Cu-25Ni	920–980
	Silver-ABA®	Ag-5Cu-1Al-1.25Ti	920–950
	Ticusil®	Ag-26.7Cu-4.5Ti	920–960
	Cusil-ABA®	Ag-35.2Cu-1.75Ti	830–850
	Cusin-1-ABA®	Ag-34.2Cu-1Sn-1.75Ti	810–860
	Incusil®-ABA®	Ag-27.2Cu-12.5In-1.25Ti	720–750
BrazeTec	BrazeTec®-CB1	Ag-19.5Cu-5In-3Ti	850–950
	BrazeTec®-CB4	Ag-26.5Cu-3Ti	850–950
	BrazeTec®-CB10	Ag-25.2Cu-10Ti	850–950
Titanium Brazing	TiBraze® 375	Ti-37.5-Zr-15Cu-10Ni	850–900
	TiBrazeAl® 462	Al-5.3-Si-0.8-Fe-0.3-Cu-0.2Ti	650–680
	TiBrazeAl® 655	Al-6.3-Cu-0.3-Mn0.2-Si-0.2-Ti-0.2Zr	650–670
Lucas-Milhaupt	Hi-Temp 095	Cu-9.5Ni-38Mn	950–1090
Tokyo Braze	TB-608T	Ag-28Cu-2Ti	780–800
	TB-629T	Ag-24Cu-14In-2Ti	620–720

Temperature, °C (axis: 500, 700, 900, 1100)

Table 13.4 Brazability of C/C composites and metals

Metal	Brazing filler alloy	Bonding	Remark
Ti	Ticuni[a]	Good	σ_{max}: 0.21–0.32 MPa*
	Ticusil[a]	Good	σ_{max}: 0.14–0.24 MPa*
	Cu–ABA[a]	Good	σ_{max}: 0.19–0.26 MPa*
	Cusil–ABA[a,b]	Good	τ_{max}: 10–35 MPa
	MBF–20[a]	Good	
	MBF–30[a]	Good	
Hastelloy (Ni–Cr–Mo alloys)	MBF–20[a]	Fair	
	MBF–30[a]	Fair	
Inconel 625 (Ni–Cr–Fe–Mo alloys)	Ticusil[a]	Good	
	Cusil–ABA[a]	Good	
SUS316	BCu–1[c]	Good	
	BNi–7[c]	Good	Poor ductility
Cu	Ag–Cu–1.8Ti[d]	Good	τ_{max}: 12–18 MPa
	Ag–28Cu–2Ti[d]	Good	τ_{max}: 12–18 MPa
	Cu–30Ti[d]	Good	τ_{max}: 3–10 MPa
	39Ni–33Cr–24Pd–4Si[d]	Good	τ_{max}: 1–8 MPa
	BNi–7[d]	Good	τ_{max}: 1–10 MPa
	Cu–50Pb (Rheocast)[e]	Good	τ_{max}: 1.5 MPa
Cu–1Cr	Ag–28Cu–2Ti[d]	Good	τ_{max}: 14–18 MPa
30Cu–70W sintered	Ag–28Cu–2Ti[d]	Good	τ_{max}: 12–18 MPa
W	Cu–Cr[f]	Good	

σ_{max}, tensile strength; τ_{max}, shear strength.

[a]Singh and Asthana, 2007; [b]Okamura *et al.*, 1996; Li *et al.*, 2011; [c]Liu *et al.*, 1994; [d]Ikeshoji *et al.*, 2011; [e]Salvo *et al.*, 1995; [f]Koppitz *et al.*, 2007.

*The specimen configuration was a pipe/plate joint (Morscher *et al.*, 2006).

Source: Based on Singh and Asthana's work (Singh and Asthana, 2007); other published works have been added in the same format.

For the combination of C/C composites with titanium, Ag–Cu–2Ti alloys are currently the first candidates for the brazing of C/C composites and metals. Cusil-ABA consists of titanium flakes dispersed in the middle of the Ag–Cu eutectic foil. The alloyed Ag–Cu–Ti requires a titanium fraction of approximately 4 mass%. Cu and Ti form intermetallic compounds, and the amount of titanium, that is effective as the active element is smaller than the nominal fraction.

For the brazing of C/C composites to Ni-based refractory alloys, the Ni-based brazing filler alloys are considered to be a reasonable choice. However, for Hastelloy/C/C composite brazing, a sound joint could not be obtained with MBF-20 or MBF-30. This appears to be due to a mismatch in the CTE between the substrates. Hastelloy's CTE varies among the minor compositions, but its range is usually from 11 to 16 $\mu m \cdot m^{-1} \cdot K^{-1}$, whereas that of C/C composites in the laminate surface is 0–1 $\mu m \cdot m^{-1} \cdot K^{-1}$. Besides the problem of the CTE mismatch, an

Table 13.5 Commercially supplied brazing filler alloys reported as feasible for joining C/C composites and metals

Name	Composition	Remark
Ticuni[a]	15Cu–15Ni–70Ti	
Ticusil[a]	Ag–26.7Cu–4.5Ti	
Cusil-ABA[a]	Ag–35.3Cu–1.75Ti	Cladded foil
TB-608T[c]	Ag–28Cu–2Ti	
BCu-1	99.9Cu	
Cu-ABA[a]	Cu–3Si–2Al–2.25Ti	
MBF-20[b]	Ni–6.48Cr–3.13Fe–4.38Si–3.13B	Amorphous of BNi-2
MBF-30[b]	Ni–4.61Si–2.8B–0.02Fe–0.02Co	Amorphous of BNi-3
BNi-7	Ni–14Cr–10.5P–0.1Si	
BCu-1	99.9Cu	
Gemco[a]	Cu–12Ge–0.25Ni	

[a]Wesgo, Morgan Advanced Ceramics, Hayward, CA, USA.

[b]Metglas, Inc., Conway, SC, USA.

[c]Tokyo Braze Co. Ltd, Tokyo, Japan.

enhancement of the Cr fraction might improve the wetting and brazability of C/C composites. By using Ag–Cu–Ti alloys, Ni-based refractory alloys can be brazed to C/C composites, but this could lead to the heat resistivity being restricted to lower than 800°C due to the Ag–Cu eutectic layer contained in the microstructure of the brazing interface.

Stainless steel can also be brazed to C/C composites with Ni-based brazing filler alloys (Liu *et al.*, 1994). Using BNi-7 (Ni–14Cr–10.5P–0.1Si), the 316 stainless steel can form a brazed joint with C/C composites. However, this joint has poor ductility, and observation of the microstructure reveals that the Cr is depleted by carbide formation at the C/C composite interface, and that a Ni-rich layer is formed in the middle of the brazing interface. This could make the joint relatively brittle.

Copper alloys can be brazed to C/C composites with both Ag–Cu-based brazing filler alloys and Ni-based brazing filler alloys. The CTE of copper is 16.5 $\mu m \cdot m^{-1} \cdot K^{-1}$. While this is large in comparison to C/C composites, it has a relatively low Young's modulus. The Young's modulus of copper is approximately 110–128 GPa, while Hastelloy's Young's modulus is approximately 190–210 GPa. Inconel alloys have values comparable to Hastelloy. Due to its elastic deformation capabilities, copper could relieve the thermal residual stress induced during the brazing process.

Direct brazing is achieved using active brazing filler alloys containing Ti, Cr and Si. Sufficient fractions of those elements enhance the wetting of brazing filler alloys on C/C composites, and make it possible to form sound joints. However, the mismatch in CTE between C/C composites and metals still presents a major obstacle. The techniques to relieve thermal residual stress will be described in later sections.

13.3.3 Infiltration of molten brazing filler

C/C composites are porous materials, with voids distributed between the yarns and at the interface of plies (Fig. 13.6(a)). Some pores open to the surface of the C/C composite. These voids are introduced during the fabrication process. In addition to the voids, C/C composites have relatively weak sections in the microstructure. In the interface of fibers, yarns and plies, some graphite phase remains insufficiently crystalized, and at the atomic scale these parts are porous, like the grain boundary of metals.

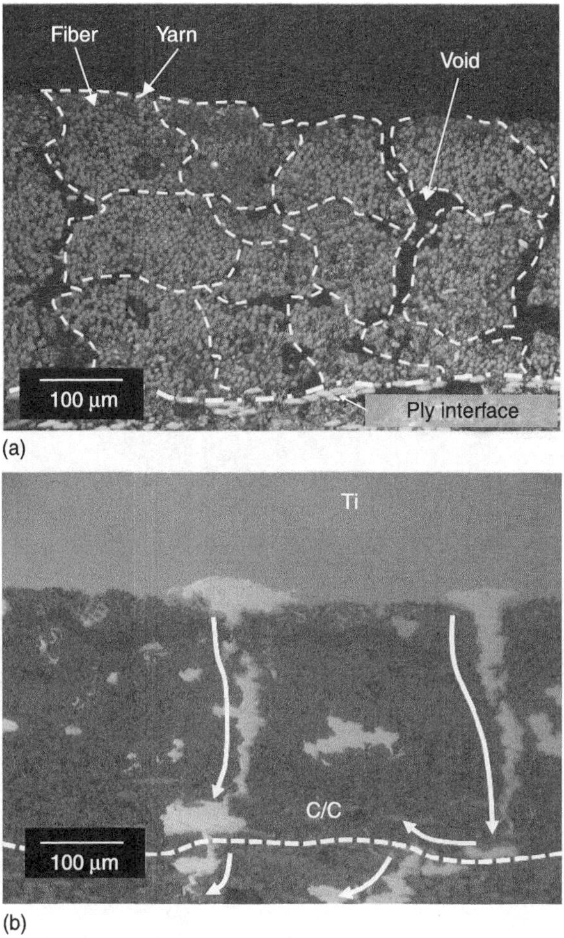

13.6 Voids inside C/C composites and the infiltration of brazing filler alloy. (a) Cross-section of an as-received C/C composite. (b) Infiltration of brazing filler alloy when a C/C composite is brazed to pure titanium with Ag–35.25Cu–1.75Ti (Cusil-ABA).

When C/C composites are brazed using active brazing filler alloys, the molten brazing filler alloys smear into the pores open to the brazing interface surface. Figure 13.6(b) shows a cross-section of the brazed interface of a C/C composite and pure titanium, brazed using Ag–35.25Cu–1.75Ti (Cusil-ABA). The molten alloy infiltrates the voids between the yarns and reaches to the ply interface. Along the ply interface, the molten alloy spreads and fills voids, before beginning to penetrate the voids leading to the inside of further plies. Precise observation at the vicinity of the brazing interface shows that the molten alloy smears into the interface between yarns and even penetrates between fibers. This means the graphite matrix is corroded by the active brazing filler alloy. The infiltration of the active brazing filler becomes more severe with any increase in the Ti fraction. Figure 13.7 shows the cross-section of a sessile drop on a C/C composite (Asthana *et al.*, 2010). The sessile drop is the active brazing filler alloy with 4.5% Ti (Ag–26.7Cu–4.5Ti; Ticusil). After 300 s of keeping time, the active brazing filler alloy infiltrates deeply inside the C/C composite substrate and spreads along the ply interfaces. In the center of Fig. 13.7(c), it can be seen that the bright white metal encircles the dark carbon fibers. This indicates corrosion of the graphite matrix around the fibers. It is also suggested that copious braze infiltration assists cracking of the substrate near the droplet periphery.

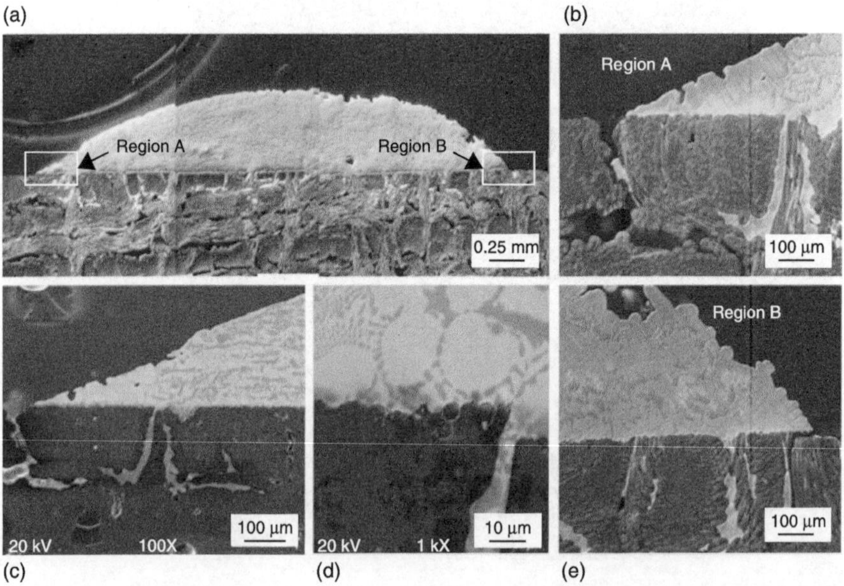

13.7 Sessile drop test of Ag–26.7Cu–4.5Ti (Ticusil) on C/C composite at 800°C after 300 s. (a) Overall view of the drop; (b) and (c) infiltration in region A in (a); (d) infiltration and corrosion around a fiber are observed at the brazed interface near the droplet center; (e) infiltration at region B in (a) (Asthana *et al.*, 2010).

As C/C composites are porous materials by nature, infiltration of the active brazing filler alloys is considered unavoidable. However, the infiltration is considered to cause cracking of C/C composite substrates and degradation of the substrate by corrosion. Protection against infiltration might therefore be required during the brazing process. Effective protection methods have not yet been proposed. But, if the pits on the surface of C/C composites could be filled, the infiltration of the molten brazing filler might be blocked. For example, carbon coating via physical vapor deposition (PVD) might be a plausible method for filling in the pits on a C/C composite surface.

13.3.4 Brazing conditions for C/C composites

The brazing conditions for C/C composites are almost the same as those for graphite, yet the composite structure of C/C composites leads to some differences in the conditions.

The porous structure of C/C composites means they easily adsorb water and vapor. Before being heated for brazing, C/C composites must be dried. The drying conditions typically involve keeping the C/C composites above 150–200°C for more than 1 hour. It is preferred that drying is conducted in a vacuum or inert gas atmosphere. If 'wet' C/C composites are heated to more than 800°C, the H_2O decomposes on the surface, forming CH_4, CO and CO_2. These reactions can lead to the degradation of the C/C composite.

As well as requiring a vacuum or inert gas atmosphere, the brazing of C/C composites and metals requires the partial pressure of oxygen to be lowered as far as possible to prevent the oxidation of the C/C composite, the metal and the brazing filler alloy.

The brazing temperature depends on the brazing filler alloy, and is not considered to be excessively high. One reason for this is the mismatch in the CTE, as mentioned above. Another reason is the infiltration of the molten brazing filler into the C/C composites described in the previous section. The active brazing filler alloys wet graphite well, and easily smear into the pores of the C/C composite. Therefore, it is better that the brazing time is not excessively long.

13.4 Anisotropy of carbon–carbon composites and their brazing with metals

13.4.1 The brazability of 2D laminates and their orientation to the brazing surface

As previously mentioned, C/C composites have strong anisotropy in the CTE. The 2D cross-ply laminate of C/C composites has a CTE of 0–1 $\mu m \cdot m^{-1} \cdot K^{-1}$ in the laminate plane, but in its cross-section surface the CTE is almost 10 $\mu m \cdot m^{-1} \cdot K^{-1}$.

This mismatch in CTE always causes problems for the brazability, suggesting that the brazability of C/C composites to metals must be affected by the anisotropy in CTE.

The brazability of 2D cross-ply laminate C/C composites to pure titanium has been investigated experimentally (Fig. 13.8) (Ikeshoji *et al.*, 2009). Three types of 2D cross-ply laminate C/C composites were studied. The felt laminate consists of plies with non-woven short length fibers. The others are typical $[0°/90°]_S$ cross-ply laminates, but with a different laminate orientation to the brazing interface. The felt-type substrate cannot form firmly brazed joints. When the 2D cross-ply substrate is brazed on its laminate surface, the brazed joint can be obtained to a fairly good standard. In the aforementioned experiment, a sound joint was formed when the 2D cross-ply was brazed at its cross section. The fracture styles are schematically illustrated in Fig. 13.9. A brazed joint hardly forms between the felt-type 2D laminated C/C composite and the metal, because the felt surface cannot withstand the compression stress induced by the mismatch in CTE. On the surface of the 2D cross-ply laminates, the metal can be brazed, but its joint strength is not strong. The infiltration of active brazing filler inside the C/C

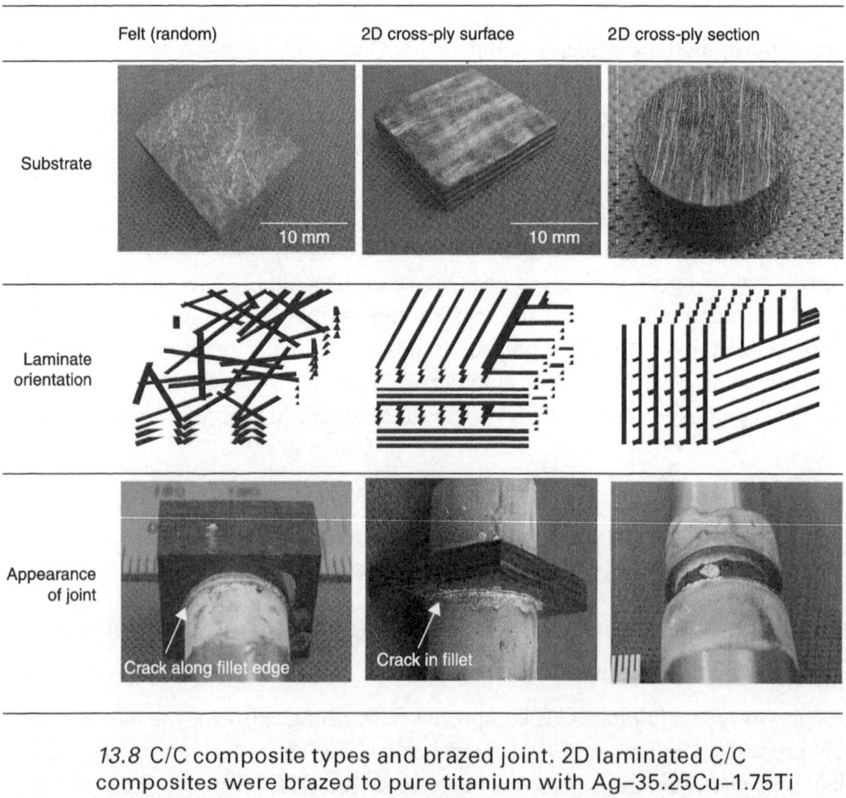

13.8 C/C composite types and brazed joint. 2D laminated C/C composites were brazed to pure titanium with Ag–35.25Cu–1.75Ti (Cusil-ABA®).

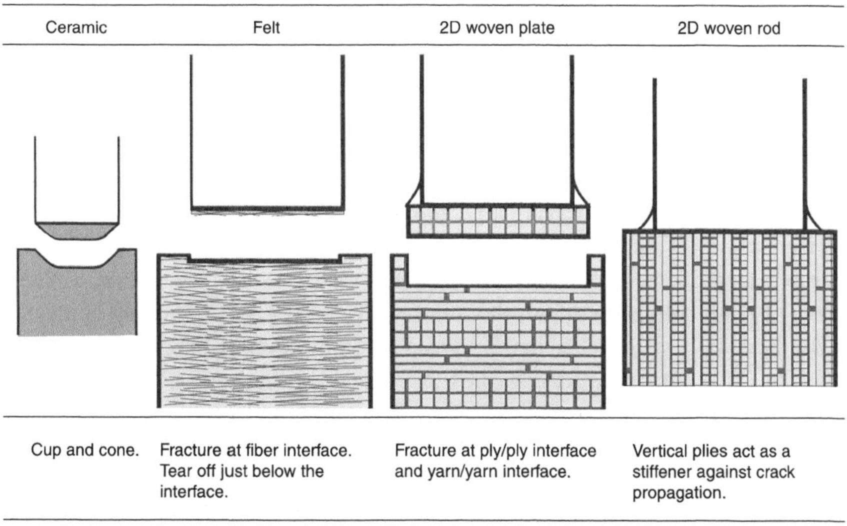

13.9 Effect of lamination orientation on the strength of a brazed joint.

composite weakens the substrate and cracks are introduced, again as a result of the compression stress caused by the mismatch in CTE. When the cross-section of the laminates is brazed to metals, a fairly strong joint can be obtained. This effectiveness is considered to be due to the relatively large CTE in this direction.

13.4.2 The effect of fiber arrangement on brazing interface and brazed joint strength

Even when the cross-section of lamination is brazed to the metal, the anisotropy in the laminates affects the brazed joint strength (Ikeshoji *et al.*, 2011). As shown in Fig. 13.10, three angle directions of 2D cross-ply laminated C/C composites were studied, and their top surface was brazed to pure titanium using Cusil-ABA.

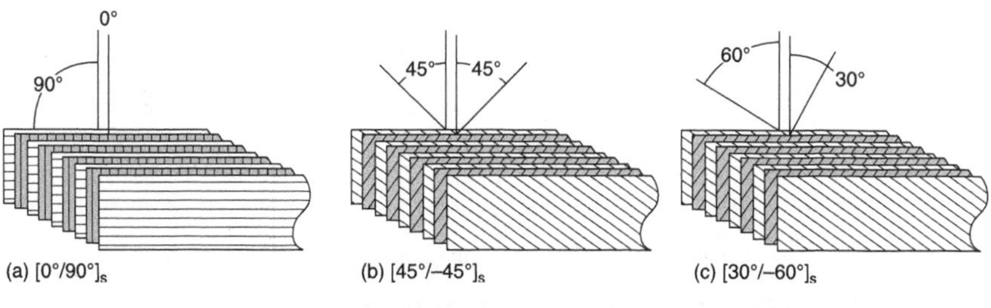

13.10 Angles of 2D cross-ply C/C composites. The top surface is the cross-section of the lamination to be brazed with metal.

The shear strength of brazed joints varies according to the cross-ply angle. The joint with $[0°/90°]_S$ cross-ply angle has the strongest shear strength value. The joints with $[45°/–45°]_S$ and $[30°/–60°]_S$ have a comparable shear strength, but their value is almost half that of $[0°/90°]_S$ (Fig. 13.11). Most joints rupture in the C/C composite, but some of the joints with $[45°/–45°]_S$ and $[30°/–60°]_S$ cross-ply laminates break at the brazing interface. The microstructure at the brazing interface reveals that the brazing filler alloys infiltrate along the fibers, and that the crack initiates at the tip of the infiltrated brazing filler alloy (Fig. 13.12). The stress placed on the open crack at the tip is mildest in the case of the $[0°/90°]_S$ cross-ply laminates. Therefore, the joints with the $[0°/90°]_S$ cross-ply laminates have the strongest shear strength. The severe infiltration of brazing filler alloys into the C/C composites leaves voids in the brazing filler layer at the interface. The $[45°/–45°]_S$ and $[30°/–60°]_S$ cross-ply laminates can be infiltrated by brazing filler alloys in two directions. In contrast, $[0°/90°]_S$ cross-ply laminates have one

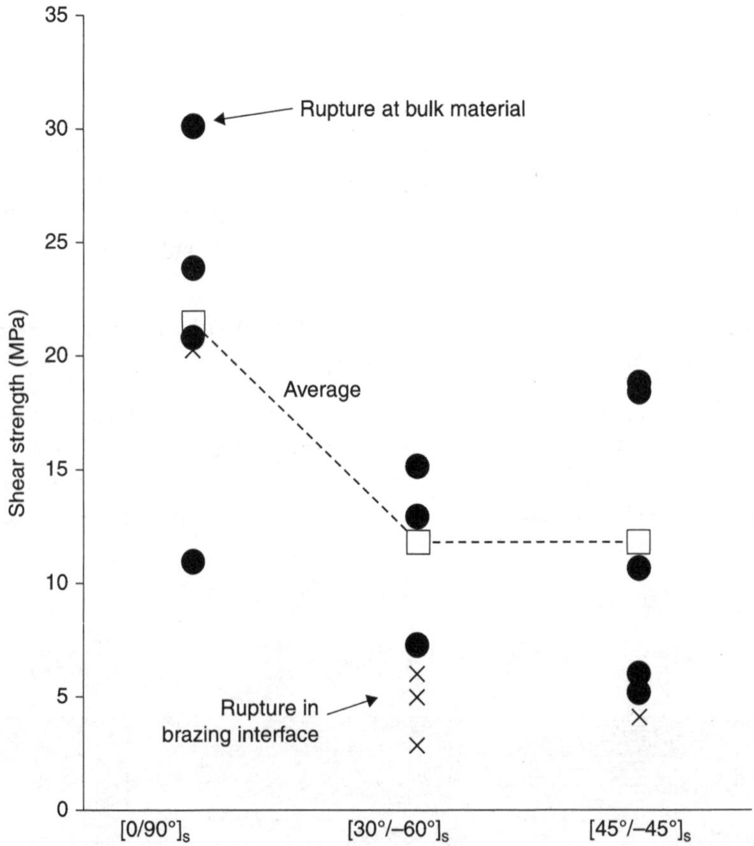

13.11 Shear strength of brazed joint of titanium and C/C composites with various cross-ply angles.

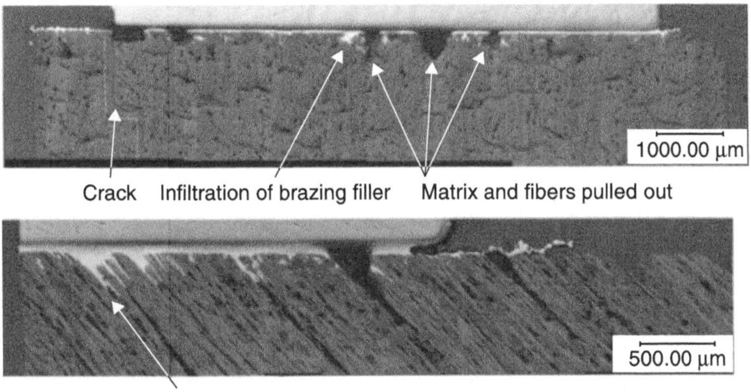

13.12 Cross-section of shear ruptured brazed joint [45°/–45°]. (a) Side by side cross-section. (b) Longitudinal direction cross-section.

direction. Therefore a shortage of brazing filler alloy occurs in the former two types of cross-ply angle laminates, and their joints break at the brazing filler interface.

13.5 Indirect methods for brazing carbon–carbon composites to metals

Active brazing fillers make it possible to braze C/C composites to metals directly. Their use, however, restricts the selection of brazing fillers and possible combinations of substrates. The major obstacles are the wetting of brazing filler and mismatches in the CTE. These obstacles can be removed via surface modification techniques and by the insertion of interlayers.

13.5.1 The surface modification technique for enhancing wetting on C/C composites

As mentioned in the previous sections, C/C composites are difficult to wet, and can be considered as porous materials. The difficulty in wetting them can be overcome by the addition of a carbide-forming element, i.e. utilization of active brazing filler alloys. Once wetting by the molten brazing filler alloys has been achieved, their infiltration into the pores and voids becomes a problem. However, surface modification techniques can be used to solve the problem of wetting and infiltration.

In the fabrication of the ITER divertor, chromium-modified C/C composites were brazed to a CuCrZr alloy (Salvo *et al.*, 2008). On C/C composite surfaces, high-purity chromium was deposited via the slurry technique. At a high temperature of 1300°C, the C/C composites were heat treated for 1 h to allow the chromium carbide-forming reaction to occur. As shown in Fig. 13.13, the chromium carbide layer was formed tightly along the C/C composite surface, filling in the surface pores. The chromium carbide layer enhanced the wetting of the molten brazing filler alloy, but simultaneously prevented the liquid metal's infiltration.

The conventional Mo–Mn method is considered applicable for metalizing C/C composite surfaces, because both molybdenum and manganese easily form carbides. The ordinary Mo–Mn method uses a mixed slurry of molybdenum, molybdenum oxides, manganese and manganese oxides. To reduce those oxides, the ceramic substrate pasting slurry is heated up to 1500°C in a moist atmosphere. However, in the case of C/C composites, the material could be degraded in the moist atmosphere at high temperature. These oxides should not, therefore, be added to the slurry, and the heat treatment should be conducted in vacuum of an inert gas atmosphere. This treatment is also applicable to the sintering process of the nickel layer. After heat treatment of the nickel layer, the surface modified C/C composite substrate is ready to braze. The Mo–Mn method is not a fuss-free process, but its metalized surface can be wetted by most ordinary brazing fillers without the necessity for active elements.

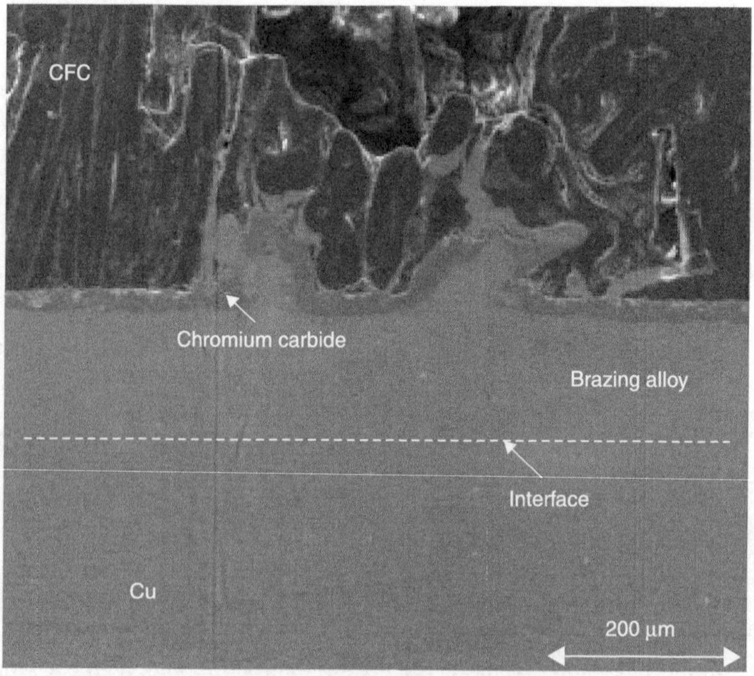

13.13 Scanning electron micrograph of the cross-section of a C/C composites/Cu/CuCrZr brazed by Gemco(R) alloy. CFC in the figure means C/C composites. (Salvo *et al.*, 2008.)

Another method to prevent the severe infiltration of active brazing filler alloys into C/C composite substrates is the insertion of a thin foil between the brazing interfaces. In the case of brazing C/C composites with pure titanium, insertion of pure copper foil with a thickness of 200 μm meant that, while the average tensile strength of the brazed joint was comparable to that of the brazed joint without foil, the variability became narrower (Ikeshoji *et al.*, 2010).

13.5.2 Insertion of an interlayer

C/C composites have a much lower CTE than most metals (Table 13.6). The first approximation of the residual thermal stress is $\frac{(\alpha_1 - \alpha_2)E_1E_2}{(E_1 + E_2)\Delta T}$, where α, E are the CTE and the elastic modulus and ΔT is the difference between the brazing temperature and room temperature. When the surface of a C/C composite with 2D laminates (α_1 = 0 μm·m^{-1}·K^{-1} and E_1 = 110 GPa) and pure titanium (α_1 = 8.6 μm·m^{-1}·K^{-1} and E_1 = 105 GPa) are brazed at 850°C, the residual thermal stress in the C/C composite will be approximately 381 MPa in compression. The compressive stress exceeds the C/C composite's compressive tensile strength of 150 MPa and the substrate will rupture. In contrast, when a cross-section of a C/C composite with 2D laminates (α_1 = 8 μm·m^{-1}·K^{-1} and E_1 = 5 GPa) is brazed to pure titanium, the residual compressive thermal stress becomes 2.3 MPa. This example is merely a first approximation and exaggerates the residual stress in the substrates, but does demonstrate that differences in CTE and brazing temperature are dominant in the problem of residual thermal stress. A precise calculation of thermal stress is conducted using finite element analysis (Liu *et al.*, 1994).

Table 13.6 Coefficient of thermal expansion

Substrate		CTE, μm·m^{-1}·K^{-1}	Young's modulus, GPa
C/C composite	Laminate surface (xy)	0–1	50–300
	Cross section (z)	1–12	3–100
Cu		16.5	110–128
CuCrZr*		15.7	128
Nb		7.3	105
W		4.5	411
Ti (Grade-7)		8.6	105
Ti-6Al-4V (Grade-5, annealed)		8.6	113
SUS316		16	196
SUS304		17.3	193–200

*Sha *et al.*, 2011.

Table 13.7 Effect of interlayer for brazing C/C composites to metals

Substrate	Interlayer	Brazing filler	Remarks
Cu–1Cr[a]	Cu (0–4mm)	Ag–Cu–2Ti	at 2 mm, τ_{max}: 14–18 MPa
30Cu–70W sintered[a]	Cu (0–4mm)	Ag–Cu–2Ti	at 2 mm, τ_{max}: 12–18 MPa
SUS316[b]	Nb (0–5mm)	BNi-7*	at 2 mm, τ_{max}: 18.5 MPa
SUS316[b]	W (0–5mm)	BNi-7*	at 2 mm, τ_{max}: 13.2 MPa

*Ni–14Cr–105P–0.1Si.

[a]Okamura *et al.*, 1996; [b]Liu *et al.*, 1994.

A reduction in the residual thermal stress can be achieved by inserting an interlayer between the C/C composite and the metal substrate (Table 13.7). Okamura *et al.* showed the effectiveness of the insertion of an OFHC copper plate for the brazing of C/C composites with Cu–Cr alloys, and C/C composites with Cu–W alloys. The shear strength of the brazed joint was measured for different plate thicknesses. It was suggested that a plate thickness of 2 mm was sufficient to improve the shear strength for both metal substrates. A similar experiment was conducted for the combination of C/C composites with 316 stainless steel (Liu *et al.*, 1994). A niobium plate and tungsten plate were inserted as interlayers. In both cases, a 2 mm plate thickness was sufficient to improve the joint shear strength. The benefit of the interlayer is not only the reduction of the residual thermal stress, but also improvement of fatigue stress. The ITER divertor consists of C/C composite blocks pierced by a CuCrZr cooling pipe (Casalegno *et al.*, 2008). Its fabrication process can be broadly divided into three steps. First, the pure copper is cast inside the hole of a C/C composite block, whose wall surface is modified with chromium. The CuCrZr cooling pipe is brazed to the pure copper pipe using the Gemco brazing filler alloy. Finally, the C/C composite/Cu block is brazed to the Cu/CuCrZr pipe. The thickness of the pure copper interlayer is 2–3 mm in total (Fig. 13.14). The interlayer in this configuration demonstrated

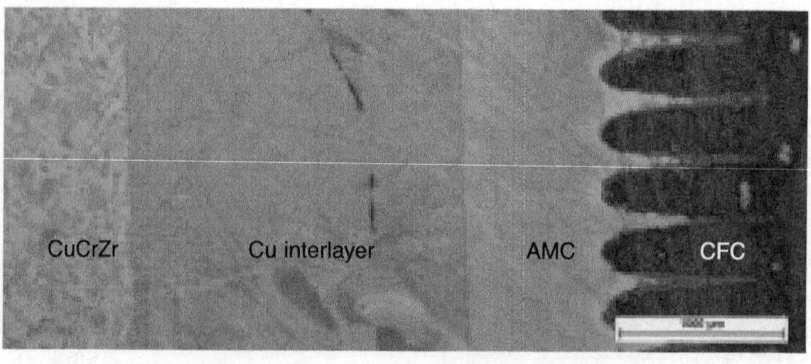

13.14 Cross-section of CFC/Cu bonding of bi-layer type. The image clearly shows the good joining quality of the Cu interlayer. (Greuner *et al.*, 2009.)

sufficient fatigue strength under the operating conditions for the ITER (Greuner *et al.*, 2009).

13.6 Conclusion

The brazing of carbon–carbon composites and metals is restricted by three main problems: the wettability of brazing fillers on C/C composites, the infiltration of the molten brazing filler into the C/C composites, and the mismatch in the coefficient of thermal expansion of C/C composites and metals. The problem of the wettability of brazing filler on the C/C composites can be solved using active brazing filler alloys, which contain a carbide former, Ti or Cr for example, leading to the formation of a reaction layer on the surface of the C/C composite, which improves wettability. Infiltration of molten brazing filler metal originates from the porous nature of the C/C composites. Current results show that deep infiltration should be avoided, because the infiltrated alloys act as crack nucleation sites. The mismatch in CTEs could be overcome by the weaving of fibers in C/C composites, or by altering the orientations of fibers to the joining interface. The insertion of an interlayer into the brazing surface will relieve the severe residual stress.

These difficulties could be overcome by using currently available brazing fillers and established methods. However, the development of a brazing filler specifically designed for C/C composites will provide a much better solution. Furthermore, in resolving problems the suitable methods for each situation should be considered individually. Most applications of C/C composites are still confined to the aerospace industry and high-end science, but, as techniques for the joining of metals develop, these applications will begin to be explored across a wide range of fields.

13.7 References

Asthana, R., Singh, M. and Shpargel, T. *Brazing of ceramic-matrix composites to Ti using metallic glass interlayers*. 30th International Conference on Advanced Ceramics and Composite, 2006, Cocoa Beach, Florida, USA.

Asthana, R., Singh, M. and Sobczak, N. 2010. Wetting behavior and interfacial microstructure of palladium- and silver-based braze alloys with C-C and SiC-SiC composites. *Journal of Materials Science*, 45, 4276–4290.

Casalegno, V., Salvo, M., Ferraris, M., Smeacetto, F., Merola, M. *et al.* 2008. Non-destructive characterization of carbon fiber composite/Cu joints for nuclear fusion applications. *Fusion Engineering and Design*, 83, 702–712.

Chen, B., Xiong, H. P., Mao, W. and Cheng, Y. Y. 2010. Wettability and interfacial reactions of PdNi-based brazing fillers on C-C composite. *Transactions of Nonferrous Metals Society of China*, 20, 223–226.

Donnelly, R. G. and Slaughter, G. M. 1962. The Brazing of Graphite. *Welding J.*, 41, 461–469.

Ezato, K. and Suzuki, S. 2009. ITER Divertor Prototypes Pass High Heat Flux Test 'Prototypes Fabricated by JAEA Withstand High Heat Flux'. *JAEA R&D Review*.

Goto, K., Hatta, H., Kogo, Y., Fukuda, H., Sato, T. *et al.* 2003. Carbon-carbon composite turbine disk for the air turbo ramjet engine (ATREX). *Advanced Composite Materials*, 12, 205–222.

Greuner, H., Swirth, B., Boscary, J., Chaudhuri, P., Schlosser, J., *et al.* 2009. Cyclic heat load testing of improved CFC/Cu bonding for the W 7-X divertor targets. *Journal of Nuclear Materials*, 386–388, 772–775.

Grigorenko, N., Poluyanskaya, V., Eustathopoulos, N. and Naidich, Y. 1998. Kinetics of spreading of some metal melts over covalent ceramic surfaces. In: Bellosi, A., Kosmac, T. and Tomsia, A. (eds) *Interfacial Science of Ceramics Joining*. Boston: Kluwer Academic Publishers.

Hirai, T., Bondarchuk, E., Borovkov, A. I., Koppitz, T., Linke, J., *et al.* 2007. Development and testing of a bulk tungsten tile for the JET divertor. *Physica Scripta*, T128, 144–149.

Humenik, M. and Kingery, W. D. 1954. Metal-Ceramic Interactions: III, Surface Tension and Wettability of Metal-Ceramic Systems. *Journal of the American Ceramic Society*, 37, 18–23.

Ikeshoji, T.-T., Kunika, N., Suzumura, A. and Yamazaki, T. *Brazing of Copper-coated Carbon/Carbon Composites to Titanium Alloys*. 4th International Brazing and Soldering Conference, 2009, Orlando, Florida, USA. American Welding Society.

Ikeshoji, T. T., Suzumura, A. and Yamazaki, T. *Brazing of C/C composites and titanium alloys with inserting OFHC copper foil*. Brazing, High Temperature Brazing and Diffusion Bonding LÖT2010, 2010, Aachen, Germany. DVS, 23–28.

Ikeshoji, T.-T., Amanuma, T., Suzumura, A. and Yamazaki, T. *Shear strength of brazed joint between titanium and C/C composites with various cross-ply angles*. JSME/ASME 2011 International Conference on Materials and Processing ICMP2011, 2011, Corvallis, Oregon, USA. JSME.

Inagaki, M. 2009. *Carbon, Old but New Materials*. Tokyo: Morikita Publishing Co., Ltd.

Ishikawa, T., Hayashi, Y., Fukunaga, H., Matsushima, M. and Noguchi, T. 1991. Experimental examination of theory of CTE (coefficient of thermal-expansion) control technology. *JSME International Journal Series I – Solid Mechanics Strength of Materials*, 34, 178–186.

Jayasankar, M., Chand, R., Gupta, S. K. and Kunzru, D. 1995. Vapor-grown carbon-fibers from benzene pyrolysis. *Carbon*, 33, 253–258.

Knoche, R., Koch, D., Tushtev, K., Horvath, J., Grathwohl, G., *et al*: Interlaminar properties of 2D and 3D C/C composites obtained via rapid-CVI for propulsion systems. *Proceedings 5th European Workshop on Thermal Protection Systems and Hot Structures*, 17–19 May 2006, Noordwijk, The Netherlands.

Koppitz, T., Pintsuk, G., Reisgen, U., Remmel, J., Hirai, T., *et al.* 2007. High-temperature brazing for reliable tungsten-CFC joints. *Physica Scripta*, T128, 175–181.

Li, J. G. 1992. Kinetics of wetting and spreading of Cu-Ti alloys on alumina and glassy carbon substrates. *Journal of Materials Science Letters*, 11, 1551–1554.

Li, K. Z., Shen, X. T., Li, H. J., Zhang, S. Y., Feng, T. and Zhang, L. L. 2011. Ablation of the carbon/carbon composite nozzle-throats in a small solid rocket motor. *Carbon*, 49, 1208–1215.

Liu, J. Y., Chen, S. and Chin, B. A. 1994. Brazing of vanadium and carbon-carbon composites to stainless-steel for fusion-reactor applications. *Journal of Nuclear Materials*, 212, 1590–1593.

Mallick, P. K. 2007. *Fiber-Reinforced Composites: Materials, Manufacturing, and Design*, 3rd Ed., Boca Raton: CRC Press.

Martin, W. H. and Entwisle, F. 1963. Thermal expansion of graphite over different temperature ranges. *Journal of Nuclear Materials*, 10, 1–7.

Mishra, D. K. and Ting, J.-M. 2008. Multi-centimeter length vapor grown carbon fibers for industrial applications. *Diamond and Related Materials*, 17, 598–601.

Morscher, G. N., Singh, M., Shpargel, T. and Asthana, R. 2006. A simple test to determine the effectiveness of different braze compositions for joining Ti-tubes to C/C composite plates. *Materials Science and Engineering A-Structural Materials Properties Microstructure and Processing*, 418, 19–24.

Mukai, S. R., Masuda, T., Hashimoto, K. and Iwanaga, H. 2000. Physical properties of rapidly grown vapor-grown carbon fibers. *Carbon*, 38, 491–494.

Naidich, Y. V. 1981. The wettability of solids by liquid metals. In: Cadenhead, D. A. and Danielli, J. F. (eds) *Progress in Surface and Membrane Science.* New York: Academic Press.

Okamura, H., Kajiura, S. and Akiba, M. 1996. Bonding between carbon fiber/carbon composite and copper alloy. *Quarterly Journal of Japan Welding Society*, 14, 39–46.

Okezawa, K., Takahashi, K., Matsui, A., Morimoto, T., Seto, T. *et al.* 1997. Development of C/C Composites Disk Brake of High Speed Shinkansen. *Mitsubishi Heavy Industry Technical Review*, 34, 406–409.

Salvo, M., Lemoine, P., Ferraris, M., Appendino Montorsi, M. and Matera, R. 1995. Cu–Pb rheocast alloy as joining material for CFC composites. *Journal of Nuclear Materials*, 226, 67–71.

Salvo, M., Casalegno, V., Rizzo, S., Smeacetto, F., Ferraris, M. *et al.* 2008. One-step brazing process to join CFC composites to copper and copper alloy. *Journal of Nuclear Materials*, 374, 69–74.

Sha, J. J., Hao, X. N., Wang, J. and Gao, X. W. *Numerical analysis of thermally-induced residual stresses in plasma facing components for fusion reactor.* 2nd International Conference on Manufacturing Science and Engineering, ICMSE 2011, 9 April 2011–11 April 2011, Guilin, China. Trans Tech Publications, 1614–1620.

Singh, M. and Asthana, R. 2007. Brazing of Advanced Ceramic Composites: Issues and Challenges. In: Ewsuk, K. (ed.) *Characterization and Control of Interfaces for High Quality Advanced Materials II: Ceramic Transactions.* Wiley-American Ceramic Society.

Singh, M., Shpargel, T. P., Morscher, G. and Asthana, R. 2005a. Active metal brazing of carbon-carbon composites to titanium. In: Singh, M., Kerans, R. J., Laracurzio, E. and Naslain, R. (eds) *High Temperature Ceramic Matrix Composites 5.* Westerville: American Ceramic Society.

Singh, M., Shpargel, T. P., Morscher, G. N. and Asthana, R. 2005b. Active metal brazing and characterization of brazed joints in titanium to carbon-carbon composites. *Materials Science and Engineering A – Structural Materials Properties Microstructure and Processing*, 412, 123–128.

Singh, M., Shpargel, T., Asthana, R. and Morscher, G. N. 2006. Effect of composite substrate properties on the mechanical behavior of brazed joints in metal-composite system. In: Stephens, J. J. and Weil, K. S. (eds) *Brazing and Soldering.* AMS International and AWS.

Standing, R. and Nicholas, M. 1978. The wetting of alumina and vitreous carbon by copper-tin-titanium alloys. *Journal of Materials Science*, 13, 1509–1514.

Steelant, J. 2010. Hypersonic Technology Developments with EU Co-Funded Projects *AVT-185 RTO AVT/VKI Lecture Series*. von Karman Institute, Rhode St Genèse, Belgium.

Takemoto, T., Sakagami, S., Yamashita, Y. and Nakajima, S. 1998. Friction characteristics evaluation of carbon/carbon composite produced by the carbon powder sintering method. *SEI Technical Review*, 152–157.

Vignoles, G. L., Aspa, Y. and Quintard, M. 2010. Modelling of carbon-carbon composite ablation in rocket nozzles. *Composites Science and Technology*, 70, 1303–1311.

Yamamoto, M., Atsumi, M., Yamashita, M., Imazu, I. and Morita, S. 1996. Development of C/C composites for OREX (orbital reentry experimental vehicle) nose cap. *Advanced Composite Materials*, 5, 241–247.

Part III
Applications of brazing and brazed materials

14
Brazing of cutting materials

W. TILLMANN, A. ELREFAEY and L. WOJARSKI,
Technical University of Dortmund, Germany

DOI: 10.1533/9780857096500.3.423

Abstract: An important brazing application is the production of cutting and machining tools. This chapter provides an overview of brazing alloys for the joining of cutting materials. Process-related problems and solutions are presented and discussed. High strength values can be achieved with the right filler alloy and joint design.

Most cemented carbides consist of a hard material such as tungsten carbide and a metallic binder. Due to the high carbide content, wetting and bonding reactions have to be monitored. Addition of active elements can help to induce a wetting reaction. Brazing ceramic cutting materials requires the use of active filler alloys to achieve a wetting reaction. The chapter reviews the alloys required.

A joint can be weakened by interfacial effects or mismatches in physical properties. Employing the right filler metal with an adapted brazing process can significantly improve joint quality. Stress calculations based on the finite element method contribute to better joint design with reduced stress levels.

Key words: brazing, cutting tools, ceramic, carbide, diamond.

14.1 Introduction

Of all the manufacturing processes that can be applied to shape and form raw materials into useful products, the machining process has always been one of the most important operations. Cutting tools play a vital role in the efficiency and reliability of this operation. Consequently, much attention has been directed to the selection and manufacturing of cutting tools. Cutting, machining, sawing or drilling tools are often provided with removable inserts comprising conventional materials such as cemented carbides and ceramics, as depicted in Fig. 14.1. However, super abrasive materials containing diamond and/or cubic boron nitride provide enhanced machining performance over conventional materials and are widely used as tool inserts. Due to the high material cost of super abrasives, fabrication techniques have been developed and optimized to reduce the amount of super abrasives in the insert. One such technique is the manufacture of tipped inserts. A tipped insert consists of an insert body and an abrasive tip of super abrasive material, with the insert body being usually made of cemented tungsten carbide. The super abrasive tip is attached to a corner or edge of the insert body by a brazing process. Brazing provides a sufficient binding force to withstand the cutting forces and heat and is a convenient method to attach small abrasive tips (Webb and Eiler, 2010). However, the brazing process is not an easy task.

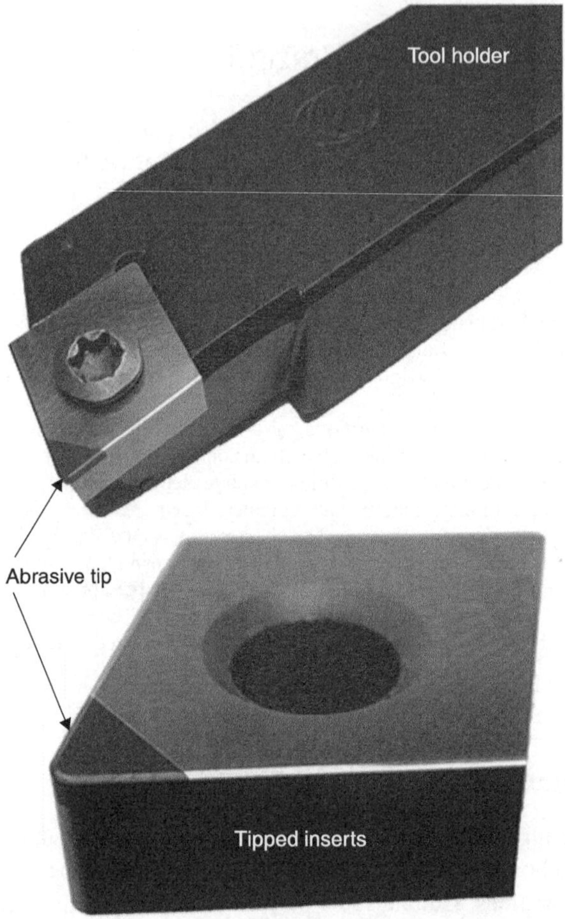

14.1 Cutting tool with removable inserts.

Difficulties which arise from the dissimilarity in material composition, physical characteristics and wetting are some of the challenges. Furthermore, the resulting bond strength is a function of various factors, including the clearance between the parts, the brazing material used, the joint interface, the brazing process and the brazing conditions.

The challenge in brazing technology is to find solutions that meet today's multiple business priorities and, at the same time, create bonds that fulfil performance needs in the most cost-effective and environmentally friendly way. This chapter is an attempt to explore the most common factors that control the quality of brazing cutting materials, such as the wetting of parent metals, processes employed, the selection of brazing filler metals, and metallurgical and mechanical challenges. More specifically, the chapter highlights the novel innovations and new trends in brazing cutting materials. Furthermore, case studies representing brazing of the most common cutting materials will be presented and discussed to explore the ideas in more detail. The aim of this chapter is to provide technicians, designers and engineering students with basic knowledge and to suggest how the potential of the process can be used.

14.2 Cutting materials

With the progress of the industrial world, there has been a need to continuously develop and improve cutting tool materials and their geometry. Cutting tools are often of interest when manufacturers look for improvements in overall productivity. Technologies such as high-speed machining and dry machining place extreme demands on cutting tools. Since different machining applications require different cutting tool materials, many types of tool materials, ranging from high carbon steel to ceramics and diamonds, are used as cutting tools in the current metalworking industry. It is important to be aware of the fact that differences do exist among tool materials, as schematically presented in Fig. 14.2 (Kalpakjian and Schmid, 2006). The following section will briefly discuss the most important types of cutting materials and their different properties and applications.

14.2.1 Cemented carbides

Cemented (or sintered) carbides, known collectively in most parts of the world as hard metals, are a range of very hard, refractory, wear-resistant alloys made by

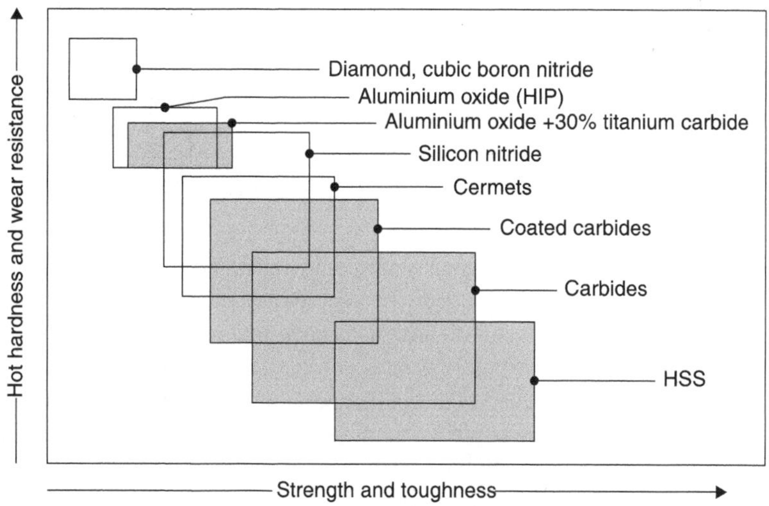

14.2 Schematic representation of the difference in properties of tool materials. HIP, hot isostatic pressing; HSS, high strength steel.

powder metallurgy techniques. The minute carbide or nitride particles are cemented by a binder metal that is liquid at the sintering temperature. Early pieces of cemented carbide were too brittle for an industrial use, but it was soon discovered that mixing tungsten carbide powder with up to 10% of metals such as iron, nickel or cobalt allowed pressed compacts to be sintered at about 1500°C to gain a product with low porosity, very high hardness and considerable strength (Jones *et al.*, 2004). Tungsten carbide (WC) is the most common hard phase, and cobalt (Co) alloy is the most common binder phase. These two materials form the basis for cemented carbide structures and grades. Other types of cemented carbide have been developed. In addition to these simple WC–Co compositions, cemented carbides may contain varying proportions of titanium carbide (TiC), tantalum carbide (TaC) or niobium carbide (NbC) and others. Cemented carbides, which have the cobalt binder phase alloyed with, or completely replaced by, other metals such as nickel, chromium, iron, molybdenum or alloys of these elements, are also produced. This combination of materials makes them ideal for use as metal cutting tools (Sandvik, n.d.).

In many cases, for tools for finish machining (especially the turning of special alloys), usually a TiC-based cermet with Ni and Ni (Mo) alloys or a steel binder is used. In spite of Ni and Ni alloys being the most widely used binder materials for TiC cermets, Fe–Ni iron alloys are attractive. Cermets based on TiC are often used because of properties such as low density, relatively high mechanical strength, high oxidation resistance and a low (close to steel) thermal expansion coefficient. Titanium carbide cermets with a steel binder are used as materials for high-cycle fatigue-stamping and as blanking or die materials. A key feature of this material is the potential to vary its composition so that the resulting physical and chemical properties ensure maximum resistance to wear, deformation, fracture, corrosion and oxidation. As this material is expensive and has low rupture strengths, it is normally used in the form of tips, which are brazed or clamped to a steel shank. The clamped tips are generally used as throw away inserts (Klaasen *et al.*, 2003; Laansoo *et al.* 2004).

14.2.2 Coated carbides

Carbide inserts coated with wear-resistant compounds perform well and have a long tool life and are the fastest growing segment of the cutting tool materials spectrum. The use of coated carbide inserts has permitted a significant increase in machining rates in comparison to the rates possible with uncoated carbide tools. The first coated insert consisted of a thin titanium carbide layer on a conventional WC substrate and was made using a chemical vapour deposition (CVD) process. Ever since, various single and multiple coatings of carbides and nitrides of titanium, hafnium, zirconium and coatings of oxides of aluminium and zirconium have been developed to increase the range of applications for coated carbide inserts (Cubberly and Bakerjian, 1989). The TiC–Al_2O_3 and TiC–Al_2O_3–

TiN-coated tools are generally employed at higher speeds than the TiC–TiCN–TiN-coated tools. When machining steels at higher speeds, the alumina layers provide excellent crater resistance. These grades work well on both irons and steels but are not recommended for materials such as aluminium or titanium (Davis, 1995).

The physical vapour deposition (PVD) coating process, commonly used on high-speed steel tools, was tested on carbide and consistently produced a damage-free interface. Subsequent cutting tests with PVD coated tools confirmed that an increase of the tool life could be observed when compared with CVD coated tools. A special group of PVD coated carbides are recommended for heavy or interrupted cutting applications (Whitney, 1994). These inserts are available in a wide variety of chip control geometries in all standard insert configurations. There is little doubt that the coated carbides come closest to an all-purpose cutting tool material for ferrous alloys. Their success is the reason why coated carbides account for approximately 60% of all sales of indexable inserts.

14.2.3 Ceramics

Ceramic tools are far superior to sintered carbides in respect of hot hardness, chemical stability and resistance to heat and wear, but are lacking in fracture toughness and strength. They are well suited for machining cast iron, hard steels and superalloys. Two types of ceramic cutting tools are available: alumina-based and silicon nitride-based ceramics. The alumina-based ceramics are used for high-speed semi- and final finishing of ferrous and some non-ferrous materials. The silicon nitride-based ceramics are generally used for rougher and heavier machining of cast iron and superalloys. In the last few years, remarkable improvements in strength and toughness, and hence the overall performance of ceramic tools, have become possible by several means, which include (Whitney, 1994):

- Sinterability, microstructure, strength and toughness of Al_2O_3 ceramics were improved to some extent by adding TiO_2 and MgO.
- Transformation toughening by adding an appropriate amount of partially or fully stabilized zirconia to Al_2O_3 powder.
- Isostatic and hot isostatic pressing (HIP).
- Introducing nitride ceramic (Si_3N_4) with a proper sintering technique – this material is very tough but prone to built-up-edge formation when machining steels.
- Adding a carbide like TiC (5–15%) to Al_2O_3 powder – to impart toughness and thermal conductivity.
- Reinforcing oxide or nitride ceramics with SiC whiskers, which enhance the strength, toughness and life of the tool and thus spectacularly increase productivity.
- Toughening Al_2O_3 ceramic by adding a suitable metal like silver, which also imparts thermal conductivity and self-lubricating properties; this novel and inexpensive tool is still in an experimental stage.

14.2.4 Diamond

Cutting tools made from industrial grade and mined single-crystal diamonds have been used for many years. However, many applications use polycrystalline diamond tools. Polycrystalline diamond (PCD) tools consist of fine diamond crystals compacted and bonded together under high pressure and temperature. Usually the diamond is bonded to a standard carbide insert as a single tip on the insert. The insert is then ground (and perhaps polished) to give a very smooth finish on the diamond. A PCD tool is an excellent general purpose tool for machining non-ferrous and non-metallic, abrasive materials (Rufe, 2002). When polycrystalline diamond blanks are bonded to a tungsten or tungsten carbide substrate, cutting tools are produced that are not only high in hardness and abrasion resistance but also greater in strength and shock resistance (Cubberly and Bakerjian, 1989).

The most common applications for PCD tools include copper and aluminium alloys, machined at high cutting speeds. It is a standard practice for aluminium automobile wheels. Another common application for PCD tools is the machining of non-metallics such as hard reinforced plastics and materials such as granite and marble. Because PCD is much more abrasion resistant than carbide, it can be used at higher cutting speeds and/or has a longer tool life than carbides for the same machining operation.

14.2.5 Cubic boron nitride

Next to diamond, cubic boron nitride (CBN) is the hardest material presently available. Tools are available as both tipped inserts (like PCD) and also in solid CBN inserts. The solid inserts cost about three times as much as the tipped inserts, but offer multiple cutting edges and a much tougher cutting material. The extreme hardness, toughness, chemical and thermal stability, and wear resistance of CBN are the advantages that make it suitable for high material removal rates as well as precision machining, imparting an excellent surface integrity to the products. Such unique tools are effectively and beneficially used in machining a wide range of work materials, covering high carbon and alloy steels, non-ferrous metals and alloys, exotic metals like Niard, Inconel Nimonic, and many non-metallic materials, which are difficult to machine with conventional tools. It is firmly stable at temperatures up to 1400°C.

CBN must compete with lower-priced ceramic tools for cutting steels in the 55–63 Rockwell Hardness (HRC) range. A fine alumina/TiC ceramic is about 10% the price of a single-tipped CBN insert and can have four to eight cutting edges. It is wise to consider the entire economic situation of any machining operation in order to justify high-performance and high-price cutting tool materials such as CBN.

A machining operation where solid CBN inserts are better than other tool materials is face milling high-strength steels (more than 60 HRC). Solid

CBN inserts have an incredible toughness and, when correctly applied, can withstand the most severe interruptions without chipping and breaking (Whitney, 1994).

14.3 The main factors controlling the quality of joints

A detailed literature search was conducted to identify the factors that affect the brazing process. The parameters that influence the quality and strength of joints vary to a great extent. In this section, the main factors, identified as wetting, processes and equipment, filler metals, and metallurgical and mechanical features of the joint, are presented and discussed. These factors have major effects on the geometries and microstructures of braze joints, and eventually determine the properties of the joints.

14.3.1 Wetting

The key physical phenomenon that makes brazing work is wetting. The liquid must strongly adhere to the base component surfaces to flow readily over those surfaces and fill the narrow gaps between the components. Wetting should be good enough to fill the gaps completely. Any voids left at the interface would significantly degrade the mechanical properties of the joint. Good wetting is the first important criterion when choosing a braze material for use in particular brazing applications (Dudiy, 2002). Detailed information on the wetting phenomenon is presented in Chapter 1 of this book. However, it can be concluded that wetting is the ability of the molten brazing filler metal to adhere to the surface of a metal in a solid state and, when cooled below its solidus temperature, to make a strong bond with that metal. Wetting is a function, not only of the nature of the brazing filler metal, but also of the degree of interaction between the materials (ceramics, and so on) to be joined. There is considerable evidence that, in order to wet well, a molten metal must be capable of dissolving, or alloying with, some of the metal on which it flows (Olson *et al.*, 1993).

Wetting of tool material surfaces can be obtained by two general methods: applying coatings that promote wetting to the surfaces prior to brazing (metallization) and alloying braze filler metals with elements that activate wetting (active brazing). In active brazing processes, an active metal such as Ti, Cr, Zr, etc. is used to react and improve the wettability of the braze alloy with the tool material. The application of brazing alloys, which contain reactive metals, requires that joining is performed in a vacuum with a very low oxygen potential, or in a dry inert-gas atmosphere with a low dew point to prevent the active elements from reacting with the atmosphere. If these requirements cannot be met, the material surfaces should be metallized prior to brazing. Several techniques are available, such as thermal spraying, electrolytic coating, electron-beam and laser coating, vapour-phase deposition (chemical or physical) and hot isostatic pressing. The techniques most commonly used are, however, the moly-manganese (Liu *et al.*, 2007; Akelsen, 1992) and metal hybrid powder processes (Kohl, 1967). Joining of metallized ceramics can subsequently be performed with conventional brazing alloys (typically BCu, BAg and BAu alloys) and the reduction of metal oxides is essential to provide the required wetting and spreading.

The effect of an active filler metal composition on its wettability to tool materials has been reported in several investigations (Tillmann *et al.*, 2009a, 2010a). The wetting properties of commercial active brazing alloys and tailor-made fillers on CVD diamond thick film and cemented carbide substrates were studied. Moreover, the effect of titanium and chromium on the wettability of the filler metals was investigated. It was found that the wettability of the investigated filler metals was better on CVD diamonds than cemented carbides. Table 14.1 lists the active filler metals used in this study. Figure 14.3(a) shows wetting angle measurements on CVD diamonds in comparison to cemented carbides, which are shown in Fig. 14.3(b). Most of the investigated active brazing alloys showed good wetting qualities on CVD diamonds. The AgCu-based and tailor-made CuSn-base fillers containing Ti or Cr as active elements had contact angles between 7° and 14° on diamond. It is also clear from the result that the effect of the brazing temperature on the wetting angle was not obvious and the holding time at the brazing temperature showed insignificant effects as well. On the other hand, Ti as an active element was much better in achieving wetting on diamond than Cr in the tailor-made filler metal. Figure 14.4 shows CVD diamond substrates wetted by CuSnTi1.5 and CuSnCr3 brazing alloys and the measured values of the contact angles.

In another study, the wetting behaviour of Cu–Cr, Cu–Ni–Cr, Cu–Si–Cr and Cu–Ni–Si–Cr alloys on sintered SiC was investigated. The addition of Cr to Cu base alloys significantly improves their wettability on SiC. A duplex layer, containing Cr_3C_2 and a mixture of Cu and C phases, stems from the reaction between the Cu–Cr or Cu–Ni–Cr alloys and SiC. Cu–Si–Cr alloys suppressed the

Table 14.1 Active filler metals used to braze CVD diamond and cemented carbide

Filler metal	Composition (wt.%)	Concentration of active elements (wt.%)
CB4	70.5 Ag; 26.5 Cu	3 Ti
Incusil ABA	59 Ag; 27.25 Cu; 12.5 In	1.25 Ti
NiBSi-M	Ni-basis with 3.2 B	4.5 Si
L-Ni_2	Ni-basis with 3 Fe; 3 B	4.5 Si; 7 Cr
L-Ni_3	Ni-basis with 3 B; 0.5 Fe	4.5 Si
L-Ni_6	Ni-basis with 11 P	Modified with 3 Cr
Tailor-made	Cu-basis with approx. 18 Sn	1.5 Ti or 3 Cr

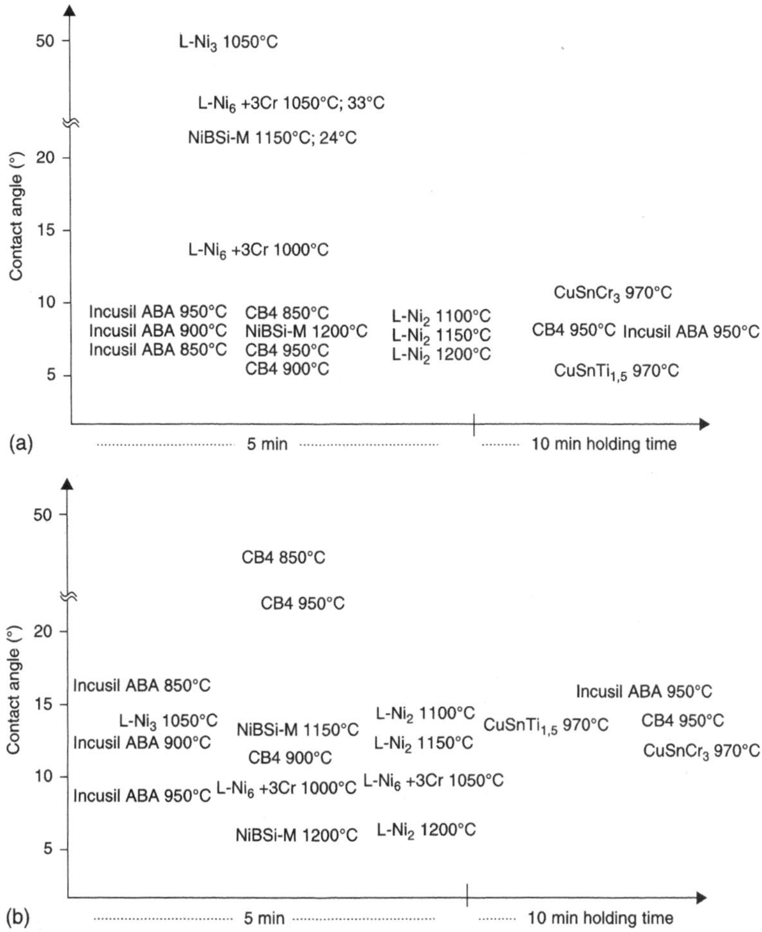

14.3 Wetting angle measurements on CVD diamonds (a) in comparison to cemented carbides (b).

formation of this duplex layer above some critical Si content and still achieved good wetting, while Cu–Ni–Si–Cr alloys with low levels of Cr suppressed the formation of the duplex layer, but this alloy with high levels of Cr induced the formation of the duplex layer. The lowest contact angles always formed in the presence of an interfacial reaction (Xiao and Derby, 1998).

An interaction layer formed by active braze fillers has significant influence on the wetting of materials by brazing fillers. Droplets of CuAg alloys without Ti do not wet alumina. They form contact angles much higher than 90°, which are typical of noble metal/ionocovalent oxide systems. Additions of 3 and 8 at.% Ti lead to a dramatic decrease in the steady contact angle to about 10°, a value which is typical of metal/metal systems, not of metal/ceramic ones. This is attributed

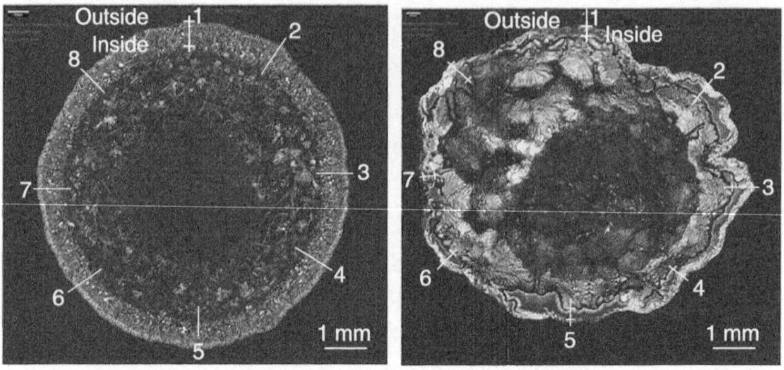

Position	Outside	Inside
1	5.8°	11.9°
2	6.2°	21.9°
3	8.5°	14.4°
4	8.6°	16.0°
5	9.1°	10.3°
6	6.7°	11.4°
7	8.1°	19.0°
8	6.7°	20.1°
Average	7.4°	15.6°

Position	Outside	Inside
1	4.9°	26.0°
2	17.0°	19.2°
3	15.2°	36.2°
4	19.6°	39.3°
5	9.6°	15.9°
6	14.7°	37.2°
7	15.6°	41.3°
8	18.2°	16.4°
Average	14.3°	28.9°

14.4 Wetting of CVD diamond substrate by CuSnTi1.5 and CuSnCr3 brazed filler metals.

to the formation of continuous layers of metal-like M6X-type Ti_3Cu_3O and Ti_4Cu_2O compounds at the interface, through reactions between the Ti-containing alloy and alumina. When the Ti content in the alloy is reduced to 0.7 at.%, M6X-type compounds are no longer formed and wetting is controlled by a $Ti_{1.75}O$ layer. This change in the interface chemistry is responsible for a significant increase in the contact angle from about 10° to 60–65° (Voytovych *et al.*, 2006).

In general, wetting is a complex physical phenomenon, involving many delicate processes at different length and time scales. In the first approximation, it is controlled by the strength of the adhesion between the liquid and the substrate, in comparison with the liquid surface tension. However, even at this simplest level, the microscopic mechanisms behind the wetting of materials by metals have, for a long time, lacked understanding, being a challenge to both experiment and theory.

14.3.2 Processes and equipment

This section is an attempt to present the most important items that are relevant to the process of manufacturing brazed joints of cutting materials. It is worth noting that, due to space limitations in the chapter, brief overviews of the techniques are presented with some applications to clarify the processes as far as possible.

Heating method

In general, several different sources of heat are used in typical brazing processes, e.g. flame (torch), furnace, infrared, induction, dip, laser, and resistance. Torch brazing and induction brazing are commonly used with BAg, BCu, copper–manganese and copper–manganese–nickel filler metals for brazing carbide tools to steel. These filler metals are also used in controlled atmosphere furnaces, vacuum, or, less frequently, in an induction brazing apparatus arranged for use in a protective atmosphere. On the other hand, many ceramics, particularly oxides like Al_2O_3 and ZrO_2, have low thermal conductivities compared with metals. The result of this characteristic is that ceramics take longer to heat to a certain temperature than most metals and alloys, and they are difficult to heat evenly by localized heating techniques. Also, most ceramics are electrical insulators, especially at low temperatures, and therefore cannot be directly heated by induction. Ceramics can be induction heated using a metal or graphite susceptor, but this approach is difficult to control and not recommended. Because of their unique electrical and thermal properties, ceramics are best suited to furnace brazing. For most applications, vacuum or controlled atmosphere brazing in inert gas is the process most frequently used for brazing ceramics (AWS, 1991).

Since ceramics do not heat as rapidly as metals, heating rates should be controlled during brazing so that both materials are maintained at the same temperature. Once melted, braze filler metals will be drawn to regions where the temperature is highest. If a metal component of a ceramic-to-metal joint reaches the brazing temperature faster than the ceramic component, then the filler metal may be drawn away from the joint, which results in a defective joint or no joint at all. One technique for minimizing this problem is to equilibrate the temperature of an assembly to just below the solidus temperature of the filler metal, followed by slow controlled heating to the final brazing temperature. The rapid cooling of ceramic-to-metal joints should also be avoided. If the cooling rate is too fast, the metal component will cool more rapidly than the ceramic and residual stress problems will be magnified. In addition, slow controlled cooling will permit some plastic deformation of ductile braze filler metals under stress due to thermal expansion coefficient mismatches, and thereby reduce the residual stress level retained in the joint at room temperature (AWS, 1991).

The brazing of diamond can be carried out with the same equipment as for brazing of metals; in particular electric resistance furnaces, induction heating equipment, and gas torches or plasma heat sources can be used as well. When using active brazing alloys, containing chemically active elements – titanium, chromium, vanadium and other elements – the brazing of diamonds should be carried out in high vacuum. This is the most efficient method of securing the diamond (Naidich *et al.*, 2007).

By far the most important factor to consider when brazing diamonds is the graphitization of the diamond. Once the diamond has undergone this transformation, it loses its desired mechanical properties. Graphitization occurs as a function of the type of diamond, oxygen level and temperature. For example, for sintered PCD, graphitization can be enhanced above 700°C when it contains up to 10vol.% metallic phase. The brazing temperatures for this material are limited to a maximum of 1000°C with slow heating and cooling rates between 700° C and 1000°C (AWS, 1991).

While a complete graphitization of the diamond will have detrimental effects on the mechanical properties, graphitization is used in the brazing process. Metals do not diffuse into and are not dissolved in the diamond. However, diamonds form carbides relatively easily. A very careful control of the temperature during the brazing cycle is not only important for effective wetting, but also has a direct effect on the strength of the joint. From this point of view, vacuum brazing is the joining technology which is used frequently for the production of diamond tools. However, extensive reactions which occur during brazing with such a prolonged heating period may have an adverse effect on the base metal and the diamond grits. Additionally, a reduction of vacuum also decreases the temperature of the start of graphitization of the diamond. In air, graphite can be detected on a diamond surface at 500–900°C (Naidich *et al.*, 2007).

To avoid these problems, induction heating has been used for brazing diamond tools under the protection of an inert gas. The benefits of induction brazing include the ability to provide localized heating and precise control of both the heating rate and heating temperature, which can decrease or minimize the damage to the diamond (Chattopadhyay *et al.*, 1991; Xu *et al.*, 2006). There is a drawback when using the induction heating process, especially when different shapes of tools are brazed, since the induction loops have to be redesigned and reproduced and the process becomes more complex. An alternative solution to this problem is to apply heat to activate the brazing process in a narrowly selected area and carry out the brazing process continuously to fully heat the whole brazed area. Laser beams with high energy density and low energy input have the properties necessary for this type of brazing. These attributes can achieve the goals of short heat treatment duration, small heat-affected zone and consequently minimal deformation of the substrate, and high efficiency in the production of diamond tools (Zhao *et al.*, 2009; Jiang and Zhang, 2010). Other processes, such as high-vacuum infrared systems and electric resistance furnaces, are also used for brazing diamond tools (Radtke, 2009). In general, there are various considerations to take into account when selecting the brazing process for a particular application. Some of these considerations are listed in Table 14.2 (Schwartz, 2003).

It is also worth noting that the holding time at the brazing temperature plays an essential role in controlling the strength of the joints. Figure 14.5 shows the effect

Table 14.2 Characteristics of heating methods

Method	Characteristics					
	Capital cost	Usage cost	Flux required	Production rate	Complex geometry	Feasible sizes
Torch	Large	Medium/high	Yes	Medium	Low	Large
Furnace (atmosphere)	Medium/high	Medium/high	Yes/no	High/low	High	Large
Furnace (vacuum)	High	Low	No	High/low	High	Large
Induction	Medium/high	Medium/high	Yes/no	Low	Medium	Medium
Infrared	Medium	Low	Yes/no	Low	High	Medium

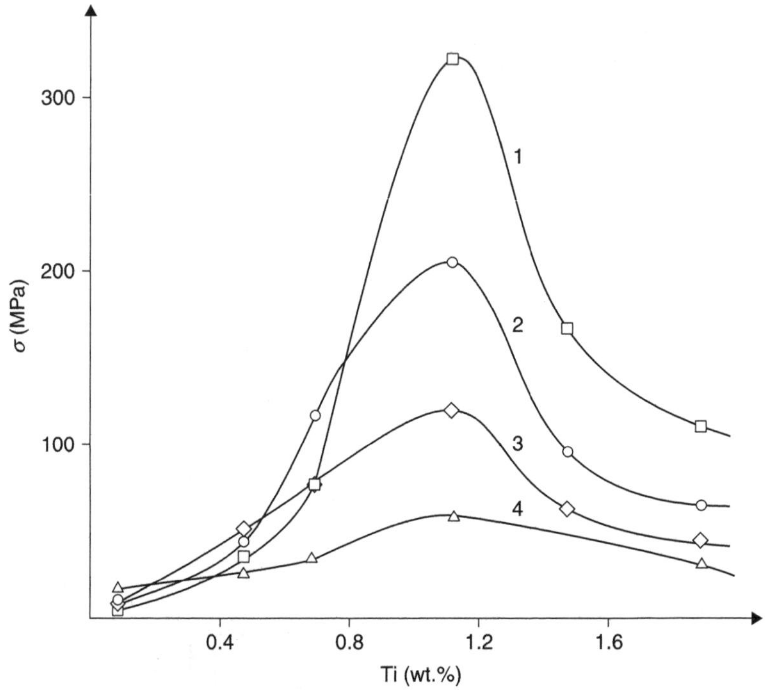

14.5 Dependence of the bonding strength of Cu–Ga–Ti alloys with diamond on the titanium concentration at a brazing temperature of 1000°C and time: (1) 0.5; (2) 1; (3) 2; (4) 5 min.

of the holding time on the bonding strength of a diamond–metal contact. Approximately 327 MPa was recorded for the (Cu–Ga)–1.1% Ti alloy at a relatively short holding time of 0.5 min. Increasing the contact time between the melt and the diamond, like an increase in the titanium concentration of the alloy to 1.5–1.8%, greatly decreases the contact strength. The decrease of the strength with increasing temperature and time, displacement of the diamond–metal interface and of the fracture surface zone to the depth of the diamond, and the increase of the thickness of the carbide interlayer make it possible to assume that diffusion processes are the main factor in softening, and the following interpretation of the softening phenomenon can be proposed. Because of the differences in the diffusion coefficients of the carbon atoms into metal $D1$ and of the metal atoms into diamond $D2$ at the ratio $D1 \gg D2$ in the subsurface regions of the diamond crystal, there are regions with increased vacancy concentration, which can coalesce into micro- and macropores. The vacancies, micropores and dislocation disruptions in the diamond crystal, which are formed as a result of differences in the diffusion flows C → Me and Me → C, 'loosen up' and soften the diamond crystal.

Joint design

When two dissimilar materials are bonded by the fusion of a braze filler material, the difference in the coefficient of thermal expansion (CTE) for each material is important. Materials used in cutting tools (carbides, ceramics, diamonds, steel, etc.) have significantly different CTEs. Therefore, as the bonded materials cool after brazing, the contraction of the materials is different and a strain is set up in the joint, the severity of which depends on the type of materials and the cooling curve. Strength data from direct brazing of Si_3N_4 to metals or oxide ceramics (Al_2O_3, ZrO_2 and MgO) with 50 at.% Cu–50 at.% Ti fillers (vacuum, 1000°C, 5 min) are presented graphically in Fig. 14.6 versus the difference in coefficient of thermal expansion. In spite of the large scatter observed, joint strength tends to decrease with increasing mismatch (Naka and Okamoto, 1985). A similar trend has been found for an aluminium brazing alloy (Naka *et al.*, 1989). Therefore, thermal stresses originate and sometimes exceed the elastic stress of the materials, and cracks appear. This factor must be considered in the design of joints.

The size and shape of the brazed joint must be considered when selecting the filler metal form. A simple braze of well-proportioned parts, which are not excessively long (less than 12.7 mm) or of a large surface area, presents no serious issues. As a rule, normal strains are observed in the assembly, because generated stresses are easily absorbed or relieved by the braze material. However, on longer joints (more than 12.7 mm long) the strain is proportionately greater. In many cases, cored shims, made of copper or nickel, are used to absorb the contracting

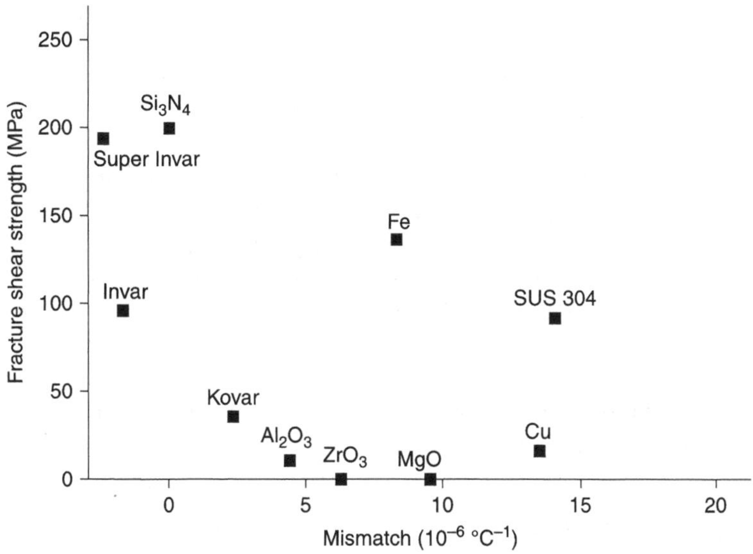

14.6 Effect of CTE mismatch on Si_3N_4–metal and Si_3N_4–oxide ceramic joint strength; direct brazing with 50 at.% Cu–50 at.% Ti filler in vacuum at 1000°C for 5 min.

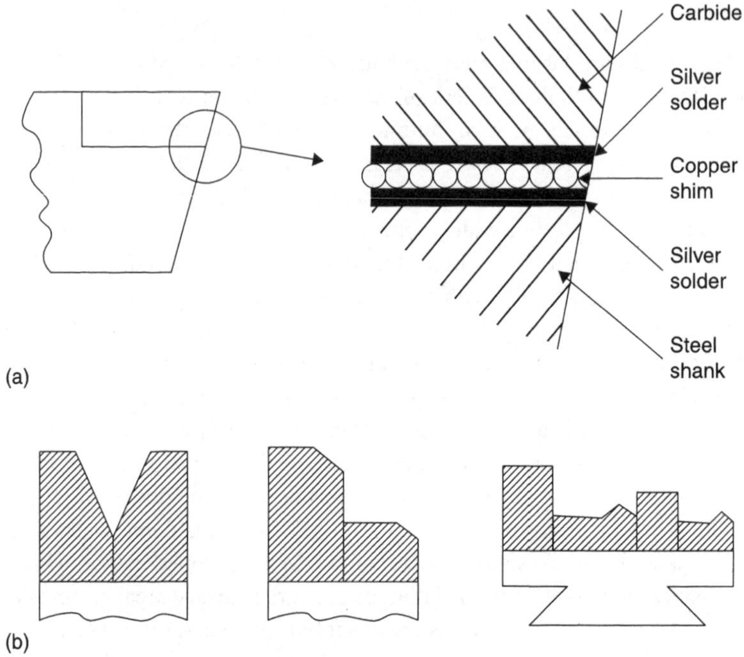

14.7 Joint design to accommodate stresses: (a) sandwich braze assembly; (b) use of multiple inserts.

stresses (see Fig. 14.7(a)). With pieces longer than 76.2 mm, it may be necessary to divide the brazed segments into small segments, as shown in Fig. 14.7(b). The filler metal joining the ends of the segments will serve to relieve the excessive stresses that would occur in a single length (AWS, 1991).

Stress may often be relieved or reduced by brazing only on one surface. Additionally, by pre-stressing the assembly beyond normal, it is often possible to have the part come out almost straight when the clamp is removed and the part cools. One way of doing this is to clamp the assembly in a jig.

Another factor, which has a significant effect on the joint, is joint clearance. Small joint clearances are preferable for brazing. Since brazing relies on capillary attraction, the design of the joint must provide an unobstructed and unbroken capillary path to enable a flux to escape as well as allow the filler metal to flow into the joint. If the brazed joint is free of defects, its strength will then depend on the joint thickness (Schwartz, 1995).

Surface preparation and cleaning

Cleaning all surfaces that are involved in the formation of the desired brazed joint is necessary to achieve successful and repeatable brazed joints. All obstructions to the wetting, flow and diffusivity of the molten filler metal must be removed from both surfaces to be brazed prior to assembly. Ultrasonic cleaning or scrubbing with a mild detergent is effective for removing loose particles and some organic oils. Metallic contaminants can be removed by immersion in dilute acids followed by rinsing in a neutralizing solution.

Carbide surfaces should be grit blasted or ground on a silicon carbide or diamond wheel to remove any surface carbon enrichment, in particular the sintering skin. It may be desirable to polish ground ceramic surfaces, or to heat treat them at a very high temperature, prior to the brazing operation, since grinding operations, which are often used to shape ceramic components, typically produce a high degree of near-surface micro-cracking. This surface damage can increase the probability that residual stresses will produce cracks during the cooling of ceramic-to-metal joints (AWS, 1991).

Temperature and time

The temperature of the brazing filler metal has a direct effect on the wettability of the joining surfaces. The selection of the brazing temperature requires an understanding of the wetting and flow characteristics of the filler metal. Clearly, the temperature must be above the solidus temperature of the filler metal and below the melting point of the parent metal. Braze filler metals can be complex alloys with melting occurring over a range of temperatures. Within this range a temperature is selected that will give the best braze quality. At excessive temperatures or even excessive brazing times, alloying between the filler metal and the base materials can occur. This happens in all brazings, and must be controlled in some cases where joint degradation can occur. Typical brazing temperature ranges and methods of brazing for various filler metals are schematically shown in Fig. 14.8.

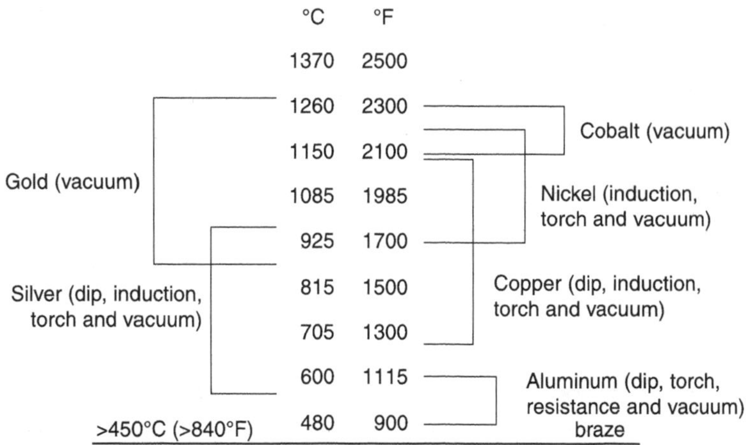

14.8 Typical brazing temperature ranges and methods of brazing for various filler metals.

Studies with an Ag–Cu–Ni braze material have shown that, for the same brazing temperature, braze durations ranging from 1 min to 10 min had a significant impact on joint strength. It was determined that the diffusion of the filler metal and base material into one another was excessive, resulting in weak joints (Etter *et al.*, 1978). Sometimes even the range of temperatures between liquid and solid is too high. In experiments brazing Si_3N_4 and SiC with Ni–Cr–Si alloys, it was found that within the range of melting temperatures there was excessive ingress of Si from the parent material. This caused excessively brittle joints containing cracks and voids (Ceccone *et al.*, 1995; Knott and Ssemakula, 2000).

Brazing temperature and holding time were proven to be controlling factors in many investigations, especially in regard to joint strength. When brazing a SiC whisker reinforced Al_2O_3 ceramic composite (SiCw/Al_2O_3) to carbon steel in a tube resistance furnace using Ag–Cu–Ti filler metals, the joining temperature and holding time were strongly affected by interfacial reactions. When the brazing temperature was low or a short holding time was used, the interfacial reaction between the composite and filler metal was insufficient and both the interfacial bonding and reaction-layer strength were low; therefore, the joint strength was also low. With increasing brazing temperature or holding time, the interfacial reaction became sufficient, and both the interfacial bonding and reaction-layer strength were increased, as was the joint strength. However, when the brazing temperature was too high or the holding time too long, the reaction layer became too thick and brittle. Hence, the thermal stress close to the interface increased remarkably. Therefore, joint strength decreased. Figure 14.9 shows the effect of the brazing temperature and holding time on the joint strength (Qu *et al.*, 2005). Almost the same trend was reported by Chiu *et al.* (2008) while brazing WC–Co hardalloy and carbon steel using Cu and bronze alloy as insert filler metals in a vacuum furnace. With an increased brazing temperature and bonding time, the thickness of the Fe–Co–Cu interaction layer increases as well, which has a positive effect on the joint strength. This phenomenon was also reported

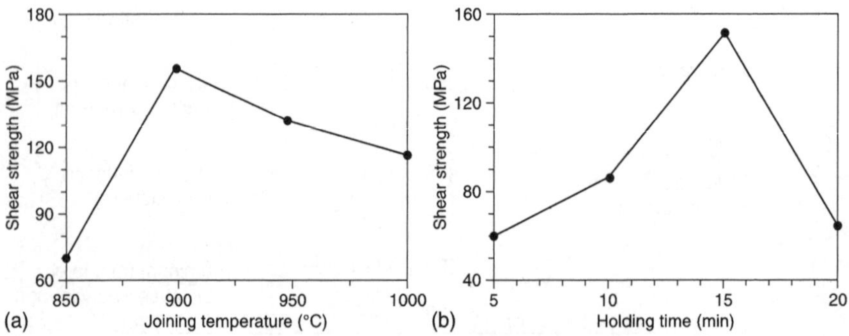

14.9 Effect of brazing temperature (a) and holding time (b) on SiCw/Al_2O_3 to carbon steel joint strength.

by Tillmann *et al.* (2009b) while investigating a cermet to steel joint under ambient atmosphere conditions using induction heating.

It was reported that when diamond is brazed to steel or a hard metal (tungsten carbide) using a silver–copper–titanium filler alloy the adherence depends on the brazing temperature and time. In this case, the kinetics, which controls the diffusion and reactivity of the brazing filler with the diamond and with the tungsten carbide, strongly depends on these parameters, resulting in TiC formation at the interface. It is possible to obtain good adherence at low brazing temperatures (for example 850°C) but the brazing time has to increase. At higher temperatures (for example 920°C), the adhesion strength decreases, probably due to the increasing thickness of the carbide at the interface (Santos *et al.*, 2004; Buhl *et al.*, 2010).

14.4 Brazing filler metals

Brazing filler metals must possess four basic characteristics to yield quality joints. First, filler materials must have mechanical and physical properties to meet the intended service conditions. These include ductility, toughness, corrosion resistance, electrical and thermal conductivity, as well as temperature and fatigue resistance. The CTE of the filler metal should closely match that of the materials being joined or, where severe gradients exist, bridge the difference in CTEs. Second, the melting point of the filler metal should be below the solidus of the base material but as high as necessary to meet the operating temperature requirement. Third, the composition must be sufficiently stable that a separation of constituents or liquidation does not occur. Additionally, the filler metal must be chemically compatible with the mating materials to avoid adverse reactions or subsequent corrosion. Finally, the brazing filler metal must have the ability to wet the surfaces to be joined to form a continuous quality joint (Steven and Mukasa, 2000).

14.4.1 Filler metals used for carbide tools

Silver-base filler metals, copper–zinc filler metals and copper are often used on carbide tools. Silver-base fillers that contain nickel are generally recommended, because nickel improves the wettability. The RBCuZn-D and BCu filler metals also have been used, particularly where a post-braze heat treatment is required. The BCu filler metal retains practically all of its strength up to a temperature of 540°C; however, it requires a hydrogen atmosphere furnace to obtain the best brazing results. 85Ag–l5Mn and 85Cul5Mn filler metals are used where the brazed joint is subjected to elevated temperatures in service and also for wetting the titanium-base or chromium-base carbides. Additionally, the BNi filler metals with boron and a 60Pd–40Ni filler metal have been used successfully to braze nickel- and cobalt-bonded cermets of tungsten carbide, titanium carbide and niobium carbide. Filler metals can also provide a cushioning or shock absorber effect that actually helps the carbide in use to survive. This is generally accomplished in one of three ways. The first is the use of a soft filler metal, such as 50% Ag with cadmium. This is a pure cushioning effect. However, Cd should not be used because of its toxic effects and the legal situation in most industrialized countries. The second is through the use of manganese. Manganese is a hard material, but it has an absorbing impact, more like a shock absorber. The third method is the use of trimetal sandwiches. These are typically filler metal/copper/filler metal. During the brazing process, the copper anneals to a dead soft condition (Schwartz, 2003).

Brazing carbides for use at elevated temperatures usually employ Au–18Ni and copper-based Cu–10Co–5Mn alloys. The joints brazed with these fillers suffer from a rather high melting temperature, which increases the possibility of high thermal stresses. In addition the Cu-based filler has low corrosion resistance because of its high manganese content. It was shown that nickel–palladium-based MBF-1002 exhibits high-temperature strength, good corrosion resistance and good erosive resistance (Bose *et al.*, 1986). However, a brittle fracture can occur in the cemented carbide parts in the vicinity of the joint at room temperature. Furthermore, a metallographic analysis of joints clearly indicates the formation of a detrimental microporosity at the interface of the cemented carbide parts. The porosity is formed due to leaching of cobalt out of the cemented carbide. The higher the brazing temperature, the more intensive the leaching. In search of lower melting point brazing alloys, Rabinkin and Pounds (1989) investigated a number of amorphous brazing filler alloys. In order to reduce the cobalt concentration in the brazement, they developed a nickel–palladium-based alloy, MBF-1011, which contained cobalt and molybdenum. The role of the molybdenum was to preserve the low melting temperature of the alloy while decreasing the overall mobility of its atomic species. The alloy formed a ductile, amorphous foil after rapid solidification. The interface was devoid of porosity with moderate leaching of cobalt. The eutectic mixture of the braze formed upon solidification had a favourable morphology: rounded particles of intermetallic $(Ni, Pd, Co, Mo)_3Si$ phase evenly distributed in the ductile β-(Ni, Pd, Co, Mo) solid solution matrix. As a result, the shear strength of the samples brazed using amorphous MBF-1011 was found to be the highest among the competing alloys.

14.4.2 Filler metals used for ceramics

As mentioned earlier, when a ceramic is pre-metallized to facilitate brazing, copper, silver–copper and gold–nickel filler metals can be used. Titanium or zirconium hydride can also be decomposed at the ceramic–metal interface to form an intimate bond. Non-metallized ceramics have usually been brazed with Cu–X or Ag–Cu–X brazes, where X is typically a group of elements such as Ti, Zr and Hf, and Ti–Zr alloys (often with addition of Be).

Attempts have been made to develop aluminium-based alloys. The results indicate that the contact angle on alumina is relatively high compared with that of titanium-containing alloys. Considering the low surface free energy of pure liquid aluminium at the melting temperature, relatively little adhesion may occur. This observation implies that brazing with aluminium fillers may require high temperatures (minimum 1050°C). As a consequence, the use of such fillers may be critical. On the other hand, wetting of alumina can be enhanced by adding copper to the aluminium filler. In this situation, the contact angle approaches 50° for copper amounts between 4 and 30 wt.% (Naka *et al.*, 1984).

In many cases, there is no need for a brazed joint that can resist high temperatures. According to Kapoor and Eagar (1989a, 1989b), the development of tin-based brazes with a lower melting temperature than Ag- and Cu-based fillers would minimize residual stresses. Although their results from wetting angle measurements show that a brazing temperature of at least 900°C is required, the lower solidification temperature may reduce the residual stress, provided that the solidification range is sufficiently small. It may be tentatively suggested that titanium additions beyond 2 at.% should be avoided due to the substantial increase in the solidification range between 4 and 40 at.% Ti (from about 180 to about 650°C) (ASM, 1973; Akelsen, 1992).

14.4.3 Filler metals used for diamond

A variety of filler materials have been used for bonding diamond or other carbon-containing abrasives by brazing or soldering. The brazing alloys used for brazing can be divided into two groups. The first group includes metals or alloys that do not interact chemically with carbon. To ensure chemical bonding with diamond, when using these brazing alloys, the latter is coated with a metallic film – an interlayer which, forming a strong bond with the diamond, has metallic properties. This film efficiently wets and bonds with the metallic brazing alloy. The second group includes the brazing alloys with a sufficiently high chemical affinity for carbon, including carbide-former elements, which form strong adhesion bonds with diamond. The content of the chemically active elements is usually not high, 1–10% (Naidich *et al.*, 2007).

As mentioned earlier, one of the most important characteristics of the brazing alloys for brazing diamonds is the melting point, which should not exceed the temperature of rapid graphitization of the diamond. Silver and copper enter into eutectic equilibrium at 779°C when 28% of copper is added to pure silver. This alloy system forms the basis of the widely used ternary and quaternary silver-base brazes for diamond brazing. The addition of metals such as zinc, cadmium and tin decreases the melting point of the brazing alloy, which has a positive effect on the performance of diamond tools. Adding a few per cent of titanium drastically improves the wetting of braze on diamond. With the increase in titanium, the strength of the joints decreases due to the increase of the TiC layer thickness and the amount of intermetallic. Chen *et al.* (2009) have reported that the Ti content must be under 4 wt.%.

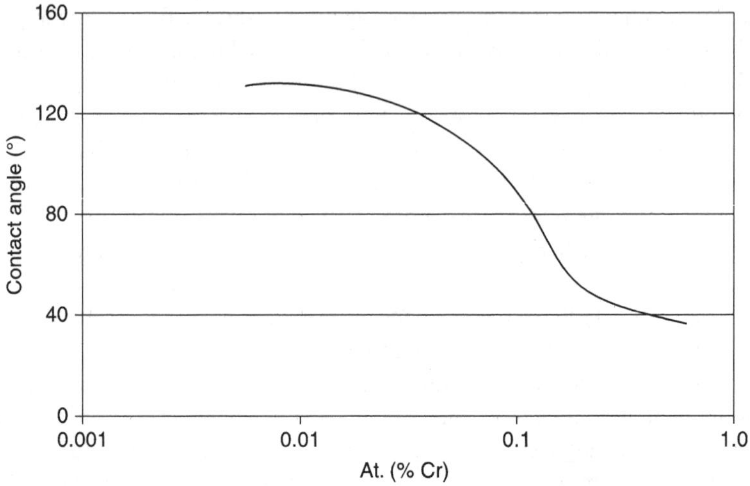

14.10 Effect of Cr addition to braze alloys on the wetting of a diamond surface.

A high-temperature brazing alloy based on nickel and containing elements such as chromium, boron, silicon, titanium and iron was proposed by several authors to braze diamond and diamond/metal cutting tools. A thin reaction layer has been detected, indicating that the surface of the diamond has been partly decomposed under the formation of carbides (Cr, Si). It was proved in wetting experiments that Cr additions promote a wetting reaction on a diamond surface, as is shown in Fig. 14.10 (Tillmann, 2003a).

Alloys on a Cu–Sn base with additions of titanium have been described and investigated quite extensively. According to Kizikov and Lavrinenko (1973), the optimum ratio of the component from the viewpoint of the efficiency of the diamond tools produced is the composition containing 20% Sn, 10% Ti and Cu, at a brazing temperature of 960°C. Cu–Sn–Ti filler metals probably show the best combination of properties, especially due to their relatively high strength and erosion resistance compared with Ag–Cu alloys and their low melting point compared with Ni-based ones (Klotz *et al.*, 2008).

14.5 Induced stresses in brazed joints

The production of joints by brazing involves the application of brazing cycles in order to melt the inserted filler metal, which bonds the joining partners after solidification due to interfacial reactions at the surfaces of the applied materials. The temperature change during the brazing cycle leads to an elongation of the joining partners during heating and to shrinkage during the cooling process. If no bond is realized, the materials remain stress free at the brazing temperature as well as at room temperature after the brazing cycle, as shown in Fig. 14.11 (Iancu, 1989).

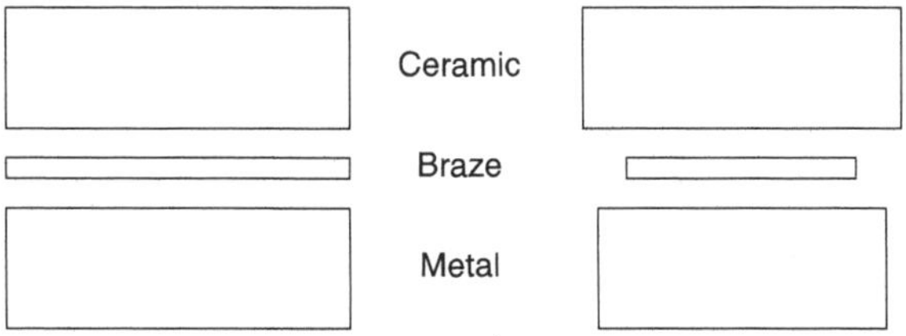

14.11 Free shrinkage of different materials.

The change in material closure, as well as the mismatch in the CTE and different material transformations of the joining partners in combination with different elastic–plastic material properties, makes the generation of residual stresses inevitable. Generally, the material with the lower CTE experiences compressive stresses while the other is under tension, as illustrated in Fig. 14.12 (Iancu, 1989).

Especially in the case of joints consisting of ceramic materials and metals, the generation of residual stresses has to be taken into account because of high differences in thermal expansion and a negligible ductility and high Young's modulus of the ceramics (Fig. 14.13), which leads to a lack of relaxation of the stresses as a result of plastic deformation (Tillmann *et al.*, 2010b). The high induced thermal stress level leads to a degradation of the resistance of the joints against external loads and can even cause premature failures during the cooling process after brazing.

Thus, it is of major importance to consider the thermally induced stress state within metal-to-ceramic material joints in order to mitigate premature damage by selecting the most suitable design. In this regard, much effort has been devoted in recent years to give a better understanding of the factors influencing the stress level, which has led to strategies to reduce the stresses. For this purpose both analytical and numerical models have been introduced to evaluate the stress distribution within brazed joints. However, calculations can only provide sound

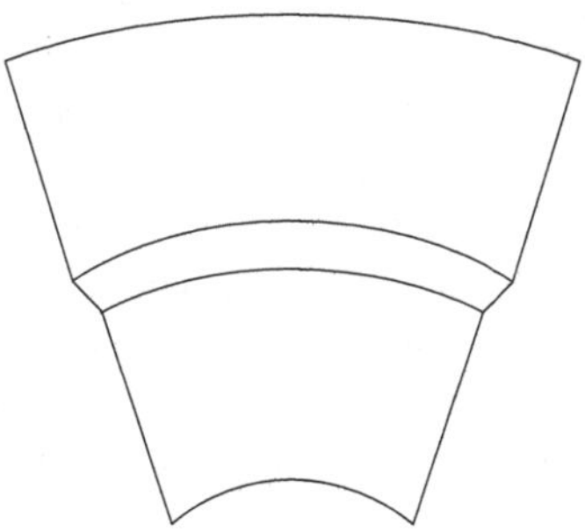

14.12 Deformation of joint consisting of materials with different CTEs. *T* = room temperature.

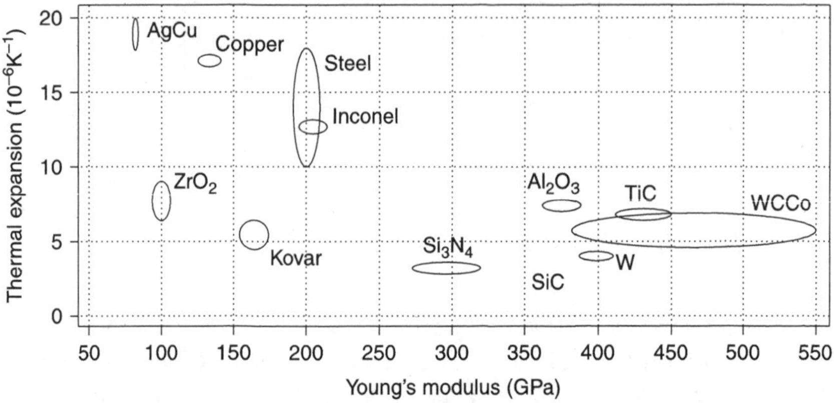

14.13 Thermomechanical properties of different materials.

results if they are accompanied by appropriate measuring methods (Wang *et al.*, 1996).

The applicable measurement techniques can be classified into non-destructive and destructive procedures. The non-destructive methods make use of the diffraction of radiation or neutrons due to the lattice structure of the materials, according to Bragg's law, which is distorted by residual stresses. Hence, the shift of the diffraction angle compared with the angle in the stress-free state correlates with the degree of the residual stress level and can be calculated taking Young's modulus into account. Using X-rays, only residual stresses close to the external surfaces can be detected. In order to determine stresses in deeper layers of the material, a radiation with higher intensity, such as synchrotron radiation, is necessary. Another possibility is to make use of neutron diffraction, because of the high penetration depth of neutrons into materials. But the resolution of neutron diffraction is very poor and results in a gauge volume of the order of 1 mm^3. Thus, a choice must be made between high resolution and high penetration depth (Pintschovius *et al.*, 1997; Galli, 2007).

The destructive methods are based on the relief of residual stresses due to the removal of material. Stress relief is a result of specimen deformation due to the stiffness decrease caused by the removal. Because of the hardness and wear resistance of ceramic materials, methods like layer peeling and hole drilling are accompanied by difficulties in grinding and drilling and have to be carried out very carefully in order to prevent stress generation by the machining process. Furthermore, these methods only allow the determination of an integral stress distribution along the machining direction. Hence, the application of these techniques interdicts the localization of stress concentrations within the joint (Galli, 2007).

Therefore, non-destructive diffraction techniques are most frequently used for the verification or adjustment of calculated residual stress distributions at the surface in brazed joints. In this regard, Iancu formulated a two-dimensional analytical model for the estimation of residual stresses and compared the results with X-ray measurements (Iancu and Munz, 1989). Due to the strong hypotheses used as a basis for the model's assumptions, restrictions concerning the valid geometry and material behaviour had to be met. The model is only applicable to laminate plane joints with an infinite width and linear elastic material behaviour. In this context, neglecting plastic deformation leads to an overestimation of the residual stress level. In order to minimize the mismatch between calculated and measured stresses, Iancu recommended substituting the solidus temperature of the braze, which is commonly selected as a starting temperature for the generation of residual stresses, by a normalized temperature. To determine this normalized temperature, he suggested decreasing the starting temperature in order to lower the calculated stresses according to the measured stress level at a random superficial spot.

Using this approach, Iancu investigated ZrO_2 to steel joints by measuring the stresses at the outer side of the ceramic and steel side (Iancu and Munz, 1989). After adjusting of the starting temperature in order to match the measured stress level at the outer ceramic side, in some cases the error between the measured and calculated stresses in the steel side was in the range of only a few per cent, but significantly below 50% of the measured values. Furthermore, he proved that the model is able to reproduce the stress change caused by the variation of the thickness of the joining couple due to the influence of stiffness.

A useful version of the model for multi-layer systems was formulated by Malzbender and Steinbrech (2004). Instead of the assumption of temperature-independent CTEs, which are fixed at $(T_{\text{brazing}} - T_{\text{room temperature}})/2$ in Iancu's model, the following formula allows for the temperature dependence:

$$\sigma_{res,i,y} = \left(\frac{1}{r} - \frac{1}{r_0}\right)\frac{E_i}{1-v_i}(y - y_{na}) + \left(\frac{E_i}{1-v_i}\sum_{j=1}^{n}\frac{E_j}{1-v_j}t_j\int_{T_0}^{T}(a_j - a_i)\,dT\right)\left(\sum_{j=1}^{n}\frac{E_j}{1-v_j}t_j\right)^{-1}$$

with

$$\frac{1}{r} - \frac{1}{r_0} = \frac{1}{D^*}\sum_{i=1}^{n}\frac{E_i t_i\left(2\sum_{j=1}^{i-1} t_j + t_i\right)\sum_{j=1}^{n}\frac{E_j}{1-v_j}t_j\int_{T_0}^{T}(\alpha_j - \alpha_i)\,dT}{2\sum_{i=1}^{n} E_i w_i t_i}$$

where σ = stress; r = radius of curvature; E = Young's modulus; v/x = Poisson's ratio; $y - y_{na}$ = distance to neutral axis; α/γ = coefficient of thermal expansion; T = temperature; t = thickness of the layer; D^* = flexural stiffness; w = width of the layer; n = number of layers. Subscripts 'res', '0', 'i' and 'j' represent resulting, initial, regarded layer and runnability parameter, respectively.

The capability of the analytical model has been exploited within Schüler's investigations of the influence of tungsten intermediate layers in SiC to steel joints with an AgCu-based braze (Schüler, 2001). The studies revealed that one can significantly reduce the residual stress level with the right design. However, a poor design might even lead to an increase of tensile stresses within the ceramic, which is particularly critical (see Fig. 14.14).

Hence, the dominating factors for the resulting stresses are the CTE mismatch and stiffness of the joining partners. Therefore, the model shows good applicability for materials with a low formability and distinct elastic behaviour. On the other hand, the model cannot be used to explain the decrease of the stress level when substituting the tungsten layer with copper. The reason for the decrease in stress despite the higher CTE of copper is due to the high number of glide planes, which lead to high formability and hence the ability to relieve stresses (Schüler, 2001).

In order to get sophisticated stress distributions for non-linear material behaviour, devoid of restrictions regarding the joint geometry, numerical approaches are increasingly used and steadily refined. They have high reproduction accuracy of real stress situations. In this regard, the most common approach is the finite element method, which is based on the decomposition of

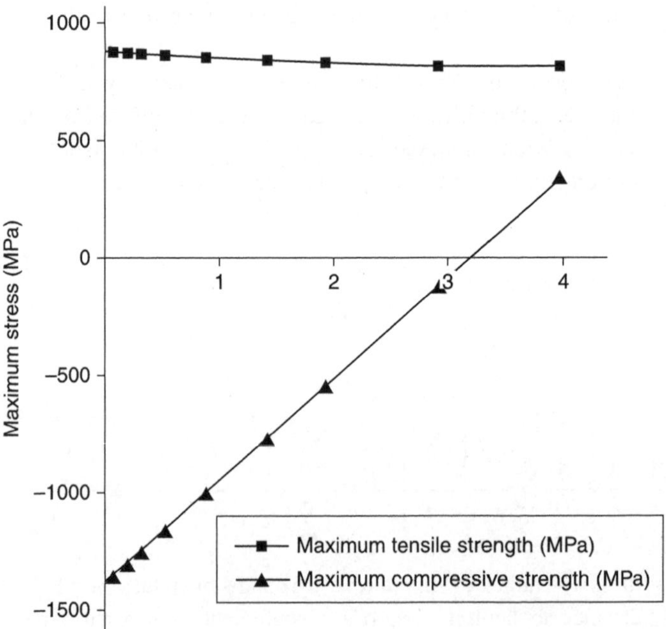

14.14 Influence of intermediate tungsten layer on the residual stress state from an analytical model.

the body of interest into simply shaped parts. The mathematical problem is solved for each individual part in order to determine the response of the complete component.

As a consequence of the rapid advancement of computing capacities, calculations are more accurate and significantly quicker. Furthermore, the complexity of the material models in finite element analyses is steadily growing, with the objective of minutely reproducing material behaviour. In the 1990s merely linear elastic simulations of brazed joints were carried out, with very few advantages compared with an analytical model. The most valuable benefit of linear elastic numerical calculations is the omission of the restrictions regarding the shape and complexity of the joint geometry and temperature-dependent material behaviour. Nowadays, elastic–plastic models, which allow for the frequently seen distinct plasticity of the filler metal, are state of the art. To implement these models, it is very important to have appropriate material databases for the filler metals as well as the base materials. In this context, there are three different approaches for modelling the joint area (Mackerle, 2003).

Within the simplest procedure, the joint area is assigned the properties of the filler metal. Hence, the reaction products are disregarded. Nevertheless, this model is a powerful tool for designing the joint geometry. The capability of this model has been exploited within Schnee's investigations on cutting wheels (Schnee *et al.*, 2010). His aim was to optimize the assembly of wheels within an industrial production line. He found that the braze foil width between the steel wheel and the cemented carbide inserts is a significant parameter for the stress level and suggested an ideal width. Furthermore, the location of the stress maximum in the calculations correlated with the real fracture path.

Moreover, the model was frequently used in investigations in the development of different strategies for stress reduction in metal-to-ceramic material joints. These investigations can be divided into five different procedures:

- selection of suitable materials
- stiffness decrease
- application interlayer in the joint area
- reinforcement of the filler metal
- heat treatment of the joint after brazing

The choice of materials in the majority of cases is very restricted. Each field application usually requires special properties, which narrow down the number of applicable materials. However, it is favourable to minimize the mismatch of CTE by selecting ceramics with a high CTE and metals with a low CTE, like Kovar. Furthermore, the filler metal should have the highest CTE, whereas a braze with an intermediate value creates joints with the most disadvantageous stress state (Galli, 2007).

Iancu's investigations showed that the thickness, respectively stiffness, of the joining partners has a significant influence on the resulting stress level. Besides variations of the thicknesses, which are commonly not applicable due to geometrical restrictions in industrial applications, it is possible to pattern the braze. Eager and Park used different patterned Ni foils for transient liquid phase brazing of Si_3N_4 to Inconel 718 (see Fig. 14.15) (Park *et al.*, 2005). According to finite element calculations as well as diffraction measurements, the residual stress level could be decreased significantly due to appropriate patterning of the interlayer in the joint area. However, the patterning reduces the brazed area, which decreases the resistivity of the joint against external loads. It was found that a certain degree of porosity enhances the joint strength in spite of the smaller joint area. However, oftentimes the patterning is not applicable because of wetting and flowing of the braze.

The use of interlayers aims either to enhance plastification within the joint area with ductile interlayers or to shift the stress maxima from the ceramic material into the intermediate layer with low CTE interlayers. Copper is commonly used as a material for ductile interlayers. It has a low yield stress and high plastic elongation. A low yield stress is particularly favourable for the reduction of residual stresses within joints with a high as well as a low CTE mismatch (Galli, 2007). Nevertheless, their application is limited due to poor corrosion resistance at high temperatures and an oftentimes chemically aggressive environment. On the other hand, interlayers with a low CTE, such as molybdenum or tungsten,

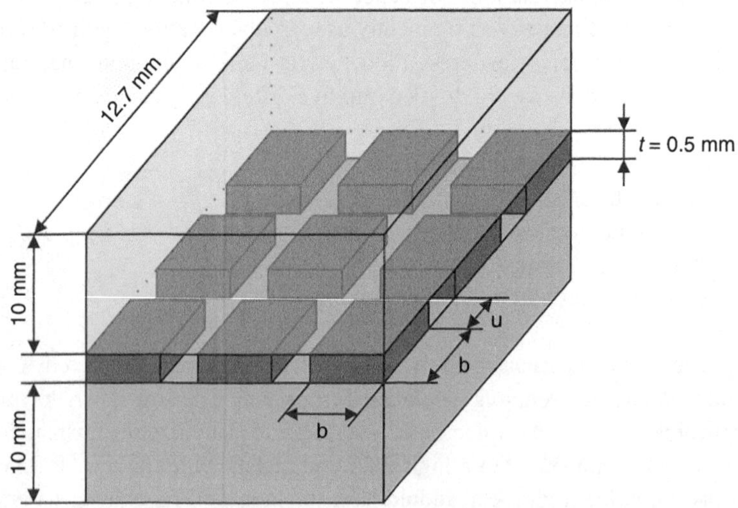

14.15 Assembly with patterned filler metal to reduce residual stresses.

have been analysed in numerous practical and theoretical investigations. Schüler's theoretical prediction of the stress decrease in Si_3N_4/ stainless steel joints due to a thick tungsten interlayer, using the analytical model, was confirmed by Tae-Woo's measurements of residual stresses (Tae-Woo *et al.*, 2001). Tae-Woo observed a significant decrease of residual stresses within the ceramic when increasing the thickness of the interlayer. After exceeding a certain value for the thickness, the fracture path shifts from the ceramic into the interlayer as desired. The critical thickness depends on the joint design. A compromise between both approaches is to apply multi-layers, which exploit the full potential of the high ductility as well as the low CTE, resulting in high strength (Park *et al.*, 2002; Bissig *et al.*, 2007).

Another very promising strategy is to lower the CTE of commonly used filler alloys by reinforcement with particles or fibres consisting of low CTE materials. In order to impregnate the joint area with the reinforcement, the braze is usually applied as a powder, which is mixed with the low CTE materials. Alternatively, Elsener and Klotz presented a novel braze alloy paste based on the CuSnTi system, which formed homogeneously dispersed TiC nanoparticles during the brazing process in the joint area. The TiC particles were a result of the reaction between the organic binder and the Ti in the braze (Elsener and Klotz, 2007). Possible reinforcement materials are SiC, TiC or ZrO_2 powders as well as carbon fibres. Such reinforced active brazing alloys showed a significant increase in joint strength compared with conventional brazements. Investigations of Si_3N_4–steel joints using Incusil ABA with 30 vol.% SiC compared with joints without reinforcement revealed an average strength increase of over 20% (Blugan *et al.*, 2007). Nevertheless, when applying reinforced brazes, challenges such as overcoming the lack of chemical compatibility of the reinforcement as well as preventing the agglomeration or even dissociation of particles have to be faced (Galli, 2007).

Furthermore, a rather practical approach to the decrease of residual stresses is heat treatment. It is commonly used for stress release in metallic materials and is often used on production lines. A stress decrease due to thermal cycles has been observed in brazed joints as well. The reasons for the stress release were found to be similar, such as annealing, and resulted in improved joint strength (Blugan *et al.*, 2007). This can be explained by the advances in finite element models, which take the visco-elastic material behaviour of the metallic materials as well as creep characteristics into account (Qiao *et al.*, 2003; Wielage *et al.*, 2009). Therefore, predictions of joint properties by finite element analyses have become more reliable and precise, and provide a better understanding of the stress situation as well as stress influencing factors. Nowadays the models are used not only for the calculation of tensile stresses but increasingly to record stress development during use, thus assisting in component optimization (Cazajus *et al.*, 2001; Tillmann *et al.*, 2010c).

14.6 Case studies

Some important applications of brazing of cutting materials will be highlighted in this section. The focus is to explore the relation between the various brazing parameters used in brazing processes, metallurgical characterization and mechanical features of the joints.

14.6.1 Brazing of carbide tools

In order to take advantage of the unique properties of cemented carbides, such as their outstanding wear resistance, high level of compressive strength, hardness and rigidity, and at the same time to minimize the consequences of their inherent brittleness, it is advisable and more economical to join the carbide to a tougher material. Steel, as well as other non-ferrous alloys, has been widely utilized. However, a problem stems from the fact that quality joints between hard metals such as WC–Co and structural steel cannot be easily achieved due to the formation of cracks and other defects during the brazing and/or cooling stages, which arise from the high residual stresses created by thermal contraction and expansion. Post-braze heat treatment to reduce the residual stress could be used as an effective treatment of this problem. During this process, the joint samples are held at a certain temperature during cooling after brazing. This post-braze heat treatment leads to a local plastic deformation in the area where residual stress was concentrated, and therefore the residual stress can be partly reduced. In order to reduce the residual stress sufficiently, the holding temperature should be 450°C or above. Too high a holding temperature will cause recrystallization of the cemented carbide and reduce its strength. Besides, too high a holding temperature will lead to an increase in the evaporation of some elements in the brazing foil, such as zinc in copper-based filler metals, which influence the performance of the brazed joint. The optimum heat treatment process cools the specimen slowly from the brazing temperature to 450°C after brazing, then maintains it at a constant temperature of 250°C for 1 hour, and then cools the specimen to room temperature for about 4–5 hours, as reported by Zhang *et al.* (2007). Figure 14.16 shows the difference in the strength of brazed joint before and after heat treatment. The strength of the joint using the filler alloy CuZnMn reached 263.81 MPa, an increase of 11.8% after the heat treatment. The strength of the joint using the filler alloy CuZnMnCo reached 285.08 MPa after the heat treatment.

Another problem is the substantial coarsening of the WC grains with longer brazing times. One solution to this problem, reported by Lee *et al.* (2006), is to add VC, Cr_3C_2, TaC, TiC or Mo_2C to inhibit grain growth. Each of these additions gives certain benefits and also certain disadvantages. Cr_3C_2 is quite useful for its corrosion resistance, and this was tested in different concentrations in WC–Co alloys, which were brazed with carbon steel under various conditions. It was

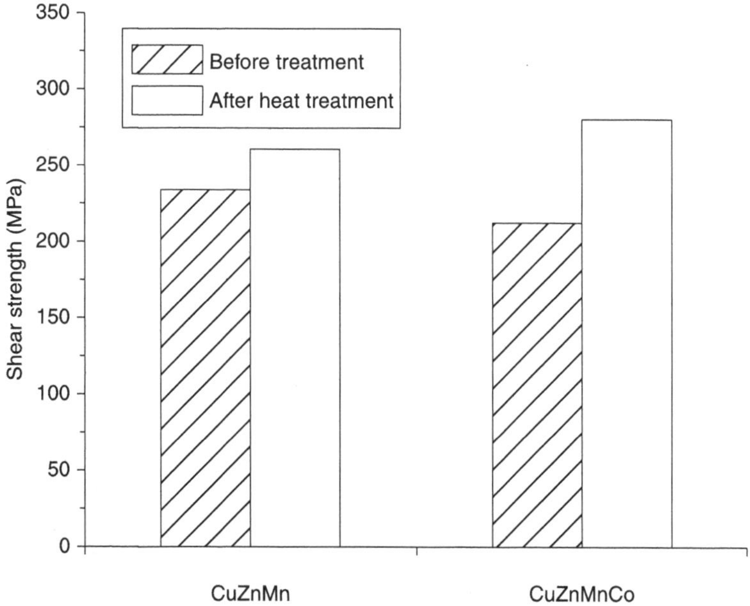

14.16 The strength difference between a brazed WC–Co and structural steel joint before and after heat treatment.

found that increasing amounts of Cr_3C_2 did in fact prevent the formation of Co_3W_3C and inhibited the coarsening of WC grains. At and above 1% of Cr_3C_2, the shear strength deteriorated due to the formation of exceptionally brittle Cr_7C_3 in the insert layer, which counteracted the original purpose of adding Cr_3C_2. Brazing was carried out at 1050°C for 10 min. A composite brazing filler metal was used. Cu–1Zn–0.7Si foil, with a thickness of 100 μm, was placed on the carbon steel part, and amorphous Ni–15Cr–3.7B foil, with a thickness of 40 μm, was placed on the WC–Co part in order to minimize the formation of intermetallics at the interface. The recommended amount of Cr_3C_2 was 0.5%, which produced the highest shear strength of 370 MPa for the brazed joints. Furthermore, due to the carbon potential between steel and cemented carbide, the WC can be decarburized, which leads to brittle η-carbides ($Me_6C/Me_{12}C$). The embrittlement of the joint can significantly weaken joint quality. Appropriate brazing alloys, the use of interlayers or moderate joining conditions can influence the formation of the unwanted carbides. Figure 14.17 shows a joint between cemented carbide and steel. Within the joint, a Co–Fe phase has formed, which acts as a bridge for carbon migration, leading to the formation of brittle carbides. The mechanical properties are degraded, and the overall soundness of the joint has changed (Tillmann, 2003b).

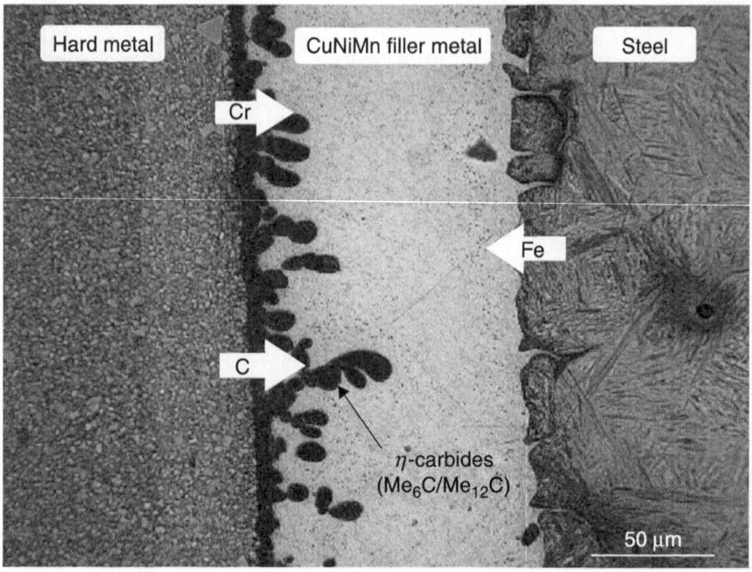

14.17 Microstructure of a cemented carbide to steel brazed joint.

The most common defects which originate during brazing of cemented carbides to steel are presented in Fig. 14.18. An irregular base metal surface, due to manufacturing defects, could be a reason for insufficient braze metal or a poorly wetted area at the interface, which results in cracks within the filler metal, as clearly shown in Fig. 14.18(a). It has also been observed in several cases that a difference in CTEs between the brazed materials usually induces cracks at the interfacial region (see Fig. 14.18(b)). Additionally, η-carbides are also considered a detrimental defect to joint quality, as described earlier (Tillmann, 2009). It is worth noting that there is a well-known defect that usually occurs when interdiffusion between two metals is uneven, where atoms of one metal have a higher mobility and diffuse across the junction more rapidly than atoms of the other metal; this imbalance results in vacancies or voids at the interface between the two metals, commonly referred to as Kirkendall voids or Kirkendall porosity. This phenomenon has been observed in thin film couples of Al–Au, Au–Sn, Au–Pb, Au–Sn/Pb, Cu–Au, Cu–Pt and Pt–Ir. Kirkendall porosity can be avoided by careful choice of the diffusing species. Employing the right filler metal together with an adapted brazing process can significantly improve joint quality. Annealing, possibly combined with a hot isostatic press, has been used to suppress Kirkendall porosity formation (Dini, 1993).

Brazing filler metals containing nickel and/or manganese have been shown to have improved wetting characteristics on cemented carbides that are difficult to

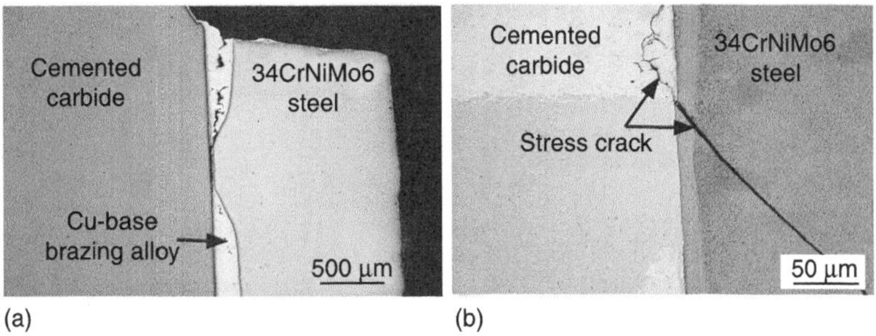

14.18 The most common defects which originate during brazing of cemented carbides to steel. (a) Defect caused by irregular base metal surface; (b) defect caused by the difference in CTEs between the brazed materials.

wet. Some of the specialized brazing filler metals for this type of component are listed in Table 14.3 (Saru Silver, n.d.).

14.6.2 Brazing of ceramic tools

As discussed earlier in this chapter, there are two main problems when brazing ceramics. The first is the poor wettability of ceramics by most metals and alloys and the second is the difference in physical properties between ceramics and metals. There are two ways to overcome the wetting problem. The first approach includes pre-metallizing the surface to be joined and subsequently brazing in vacuum or in shielding gas, or by using the so-called active filler metals in a one-step brazing process. The latter contain small amounts of very reactive metals, which are able to induce a wetting reaction by forming an interfacial reaction layer. Table 14.4 summarizes the chemical composition along with the solidus and liquidus temperatures of widely used commercial filler alloys for brazing ceramics (do Nascimento *et al.*, 2003). Furthermore, Table 14.5 gives some examples of ceramic/ceramic or ceramic/metal brazed joints and the suitable brazing alloys (Mizuhara, 1986).

On the other hand, in practice, there are several ways to reduce residual stresses in ceramic-to-metal joints to acceptable levels. The first is by selecting a ceramic and metal with the same or similar thermal expansion coefficients. This approach is effective, but is often not practical because the components of a joint are usually selected for other reasons, such as strength or corrosion or oxidation resistance. Another method for reducing residual stresses is by using compliant interlayers when making the joints. In this case, a thin sheet of a highly ductile material or one with a thermal expansion coefficient close to that of the ceramic is inserted between the ceramic and metal parts making the desired joint. A wide variety of methods using different interlayers have been developed for ceramic/metal systems, for example soft metal interlayers (Nicholas and

Table 14.3 Filler metals for brazing tungsten carbides

Nominal composition (wt.%)					Melting range	Operating temp.	International standards		Main application
Ag	Cu	Zn	Ni	Others	(°C)	(°C)	IS 2927	AWS A5.8	
50	15.5	15.5	3	16Cd	645–690	660	BACuAg12	BAg-3	A general purpose alloy with excellent mechanical properties. The nickel improves wetting on carbides. A less free-flowing alloy giving a thicker braze joint, better able to accommodate higher stresses.
49	16	23	4.5	7.5Mn	625–705	690	–	–	The addition of manganese enhances wetting of difficult carbides containing titanium and tantalum, which are difficult to wet under normal brazing conditions.
49	27.5	20.5	0.5	2.5Mn	670–690	690	–	–	For brazing tungsten carbide and substances that are difficult to wet.
40	30	28	2	–	671–779	780	BACuAg18	BAg-4	An economical alloy with good wetting characteristics. Used extensively for carbide tip brazing.
40	30	25	5	–	660–860	860	–	–	For tungsten carbide and stainless steels.
27	38	20	5.5	9.5Mn	680–830	840	SC9	–	Economical alloy with good wetting characteristics used for brazing hard metals to steel mountings. Addition of nickels and manganese improves wettability on tungsten and molybdenum materials.
–	50	39.7	10	0.3Si	890–920	910	–	RBCuZn-D	Used for brazing tungsten carbide. It is also used with steel, cast iron and malleable iron, nickel and nickel alloys.
1	50	39.7	9	0.3Si	890–920	910	–	–	Nibro with 1% silver to improve joint strength.
–	97	–	3	0.03B	1080–1100	–	–	–	Used for furnace brazing without flux. Used to fill gaps up to 0.5 mm.

–	86.5	–	2.5	11Mn	965–995	–	–	–	Excellent gap-bridging properties. Inclusion of manganese content requires it to be brazed in inert atmosphere. If required it can also be brazed in air atmosphere using flux. Equivalent to J&M C Bronze.
–	86	10	–	4Co	980–1030	–	–	–	Used in the brazing of rock drills having good wetting properties and strength. Equivalent to J&M D Bronze
–	57	39	–	2Mn2Co	890–930	930	–	–	Suitable for brazing cemented carbide for rock drills and similar percussively loaded joints. Equivalent to F Bronze of J&M.
–	96.6	–	2.7	0.7Si	1090–1100	1100	–	–	Most suitable for brazing carbide to rock drills.
–	55	35	6	4Mn	880–920	910	–	–	Low melting improved Nibro for carbides, tool steels, stainless steels and nickel alloys.

Table 14.4 Commercial alloys (wt.%) for brazing ceramics

Commercial name	Ag	Cu	Ti	Au	Sn	Others	$T_{solidus}$ (°C)	$T_{liquidus}$ (°C)
Cusil ABA*	63.00	35.25	1.75	–	–	–	780	815
Cusil 1 ABA*	63.00	34.25	1.75	–	1.00	–	775	806
Silver ABA*	92.75	5.00	1.25	–	–	1.00Al	860	912
Incusil ABA*	59.00	27.25	1.25	–	–	12.50In	605	715
Ticusil*	68.80	26.70	4.50	–	–	–	830	850
Gold ABA*	–	–	0.60	96.40	–	Ni–3.00	1003	1030
Nioro ABA*	–	–	–	82.00	–	15.60Ni–0.75Mo–1.75V	940	960
CB1†	72.50	19.50	3.00	–	–	5.00In	730	760
CB2†	96.00	–	4.00	–	–	–	–	970
CB4†	70.50	26.50	3.00	–	–	–	780	805
CB5†	64.00	34.50	1.50	–	–	–	770	810
CB6†	98.00	–	1.00	–	–	1.00In	948	959
CS1†	10.00	–	4.00	–	86.00	–	221	300

*Wesgo, Inc. Brazing Alloys, Belmont, CA, EUA. †Degussa, Germany.

Table 14.5 Examples of ceramic/ceramic or ceramic/metal brazed joints

Ceramic substrate	Member	Brazing alloy	Temp. °C
PSZ	PSZ	70% Ag, 27% Cu, 3% Ti	850
PSZ	Ductile iron	70% Ag, 27% Cu, 3% Ti	860
PSZ	410 Stainless steel	81% Au, 18% Ni, 1% Ti	1000
PSZ	1010 Steel	70% Ag, 27% Cu, 3% Ti	850
PSZ	Copper	70% Ag, 27% Cu, 3% Ti	850
PSZ	Alloy of Ni,Fe,Co	60.5% Ag, 24 %Cu, 14% In, 1.5% Ti	770
2% Al_2O_3,13% Y_2O_3,balance Si_3N_4	2%Al_2O_3,13% Y_2O_3,balance Si_3N_4	97% Cu, 3%Zr	1125
2% Al_3O_3,13% Y_2O_3,balance Si_3N_4	Ductile iron	70% Ag, 27% Cu, 3% Ti	850
3,5% Al_2O_3,13% Y_2O_3,balance Si_3N_4	Alloy of Ni,Fe,Co	60.5% Ag, 24% Cu, 14% In, 1.5% Ti	770
3.5% Al_2O_3,13% Y_2O_3,balance Si_3N_4	1010 Steel	70% Ag, 27% Cu, 3% Ti	850
2% Al_2O_3, 13% Y_2O_3,balance Si_3N_4	410 Stainless	60.5% Ag, 24% Cu, 14% In, 1.5% Ti	770
99% SiC, 1% B_2O_3	Copper	60.5% Ag, 24% Cu, 14% In, 1.5% Ti	770
99% SiC, 1% B_2O_3	1010 Steel	60.5% Ag, 24% Cu, 14% In, 1.5% Ti	770

Crispin, 1982), composite interlayers (Kawasaki and Watanabe, 1987), laminate interlayers (soft metal/low expansion and hard metal) (Koizumi *et al.* 1987; Suganuma *et al.* 1987). The interlayer material accommodates the residual stresses by self-deformation, or by isolating the ceramic from the most intense stresses. Post-brazing heat treatment and proper joint design are sometimes used to relieve stresses as well (AWS, 1991).

An interesting study was conducted to evaluate of the interlayer while brazing Si_3N_4/TiN ceramic composite to steel, using two Ti-activated brazes, CB6 (Ag 98.4, In 1, Ti 0.6, wt.%) and a Cu74.5, Sn14, Ti10, Zr 1.5, wt.% alloy. In addition, the properties of a two-layered brazing filler system with and without WC particle reinforcement were investigated (Blugan *et al.*, 2004). Six different joints were investigated and are illustrated graphically in Fig. 14.19. The use of a two-layered brazing system has successfully led to an increase in the flexural joint strength over the commercially available braze CB6. The two layer (CuSnTiZr +Incusil 15) braze system gave an average strength of approximately 400 MPa and a maximum strength of 474 MPa for Si_3N_4/TiN to steel joints. This compares with an average strength of 311 MPa and a maximum strength of 375 MPa when CB6 was used. A metallographic SEM examination of braze systems is shown in Fig. 14.20. A two-layer braze system has been used to reduce thermal stresses produced upon cooling, which led to increasing flexural strengths of the joints. Furthermore, the chemistry in the braze zone was altered by preventing detrimental reactions (i.e. the formation of the brittle Laves phases; see Fig. 14.20(a) and (b)). On the other hand, particle reinforcement by WC can theoretically lower the CTE of the brazing alloy. However, the average strength of the joint was approximately 40% lower than the joint without particle reinforcement. Particles not only affect the coefficient of thermal expansion of the braze zone, but can also alter the chemical reactions occurring in multi-layer brazings (see Fig. 14.20(c)). Optical microscopy and SEM of the fracture surfaces confirmed the presence of pores in the braze and there was also evidence of fracture/tearing/cracking near the pores, indicating that the porosity is very likely due to the WC particles added to this braze alloy. The results show that particle selection cannot be based solely on the coefficient of thermal expansion and Young's modulus of the reinforcing material. In further research by the same author, 10 and 30% vol. SiC particles were used to braze the same materials with Incusil ABA. SiC achieved better results than WC. An average four-point flexural strength of nearly 400 MPa was achieved at room temperature when using Incusil ABA + 30 vol.% SiC (sandwich foil system) compared with 330 MPa with only Incusil ABA. At a test temperature of 250°C, relaxation of residual stresses in the joints results in an average flexural strength of approximately 520 MPa when using Incusil ABA + 10 vol.% SiC (Blugan *et al.*, 2004).

A recent study to braze Si_3N_4 ceramic to 42CrMo steel using Ag–Cu–Ti + Mo composite filler (He *et al.*, 2010) has been reported. The effect of Mo particle content on the microstructure and mechanical properties of the joints was investigated.

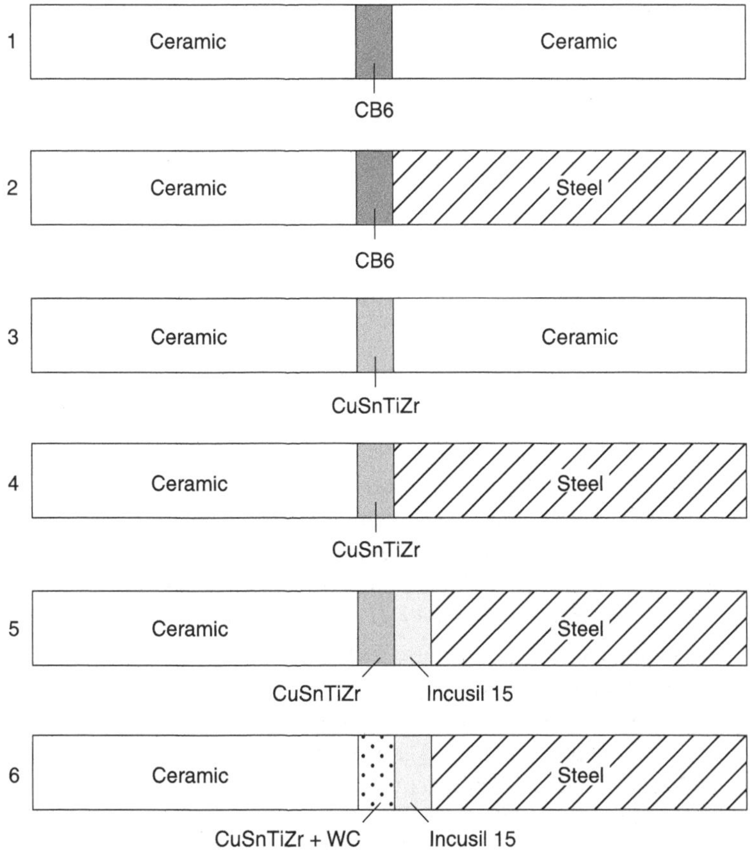

14.19 Schematic view of different joint designs for brazing a ceramic/steel joint.

Defect-free joints were created when the Si_3N_4/42CrMo steel joints were brazed with an Ag–Cu–Ti + Mo composite filler. The maximal bending strength reached 587.3 MPa with 10 vol.% Mo particles in the joint, at which the joint strength was 414.3% higher than the average strength without Mo particle addition.

14.6.3 Brazing of diamond tools

The brazing processes of diamond materials are either single stage, using active alloys (in melting, these alloys wet the diamond surface and form strong interphase bonds with carbon atoms) or two stage, in which the diamond is coated in advance with a metallic film (the diamond is metallized) and the brazing process can then be carried out using a conventional brazing alloy used for joining metals.

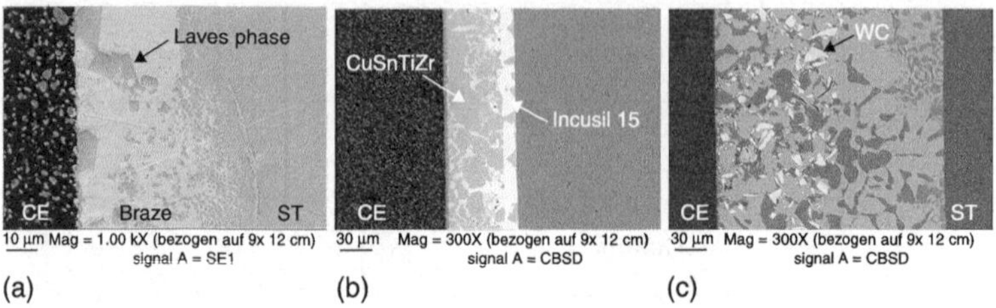

14.20 SEM micrographs of brazing zone in joint with CuSnTiZr single brazing layer (a); joint brazed with double layer CuSnTiZr/Incusil 15 braze (b); joint brazed with CuSnTiZr+35% vol. WC/Incusil 15 braze (c). CE, ceramic; ST, steel.

The interfacial carbide reaction between diamond and the brazing alloy is the key factor in controlling joint performance. In contrast to the active brazing of oxide or monoxide ceramics, diamond does not require extremely reactive filler agents to promote the interfacial interaction. Whereas high-performance ceramics require refractory metals such as titanium, zirconium or hafnium to induce a wetting reaction, diamonds can be wetted successfully by moderately reactive metals such as chromium or silicon (Tillmann and Osmanda, 2005).

A study on brazing diamond grits onto a steel substrate using a Ni-based filler alloy (Ni–Cr–B–Si) used a laser beam in an argon atmosphere. The Ni–Cr alloy in the laser brazing process showed good wetting of the diamond grains and the steel substrates. In the interfacial reaction region of the diamond grains and the filler metal, the chromium in the Ni-based alloy was concentrated at the surface of the diamond grain, and two kinds of compounds (Cr_7C_3 and Cr_3C_2) were formed, through which good joining between the filler metal and diamond grains was accomplished. Figure 14.21 shows an SEM image of the diamond brazed using the Ni-based filler alloy. A good joint was also achieved between the steel substrate and the filler metal. Hence, with the bridge effect of the Ni–Cr filler as an active brazing alloy, the diamond abrasive grains and the steel substrate can form a hard joint to meet the requirements of heavy load grinding (Yang *et al.*, 2008).

Joints between CVD diamond and hard metal are frequently investigated because of the wide range of applications. Active filler metals such as silver-based or copper-based alloys are usually used to braze this type of joint. It has been proved that filler metals with a lower concentration of the active element, such as 1.25 wt.% Ti, produce reliable interaction layers between diamond and cemented carbide. Figure 14.22 shows good wetting and limited interaction layers both the diamond and hard metal to the AgCuInTi1.25 brazing alloy. Meanwhile, higher concentrations of active elements, such as 3 wt.% Ti, showed a thick interaction layer on the diamond and cemented carbide, which has a deteriorating effect on the mechanical properties of the joint, especially if the brazing temperature is not as low as possible (Tillmann *et al.*, 2009a).

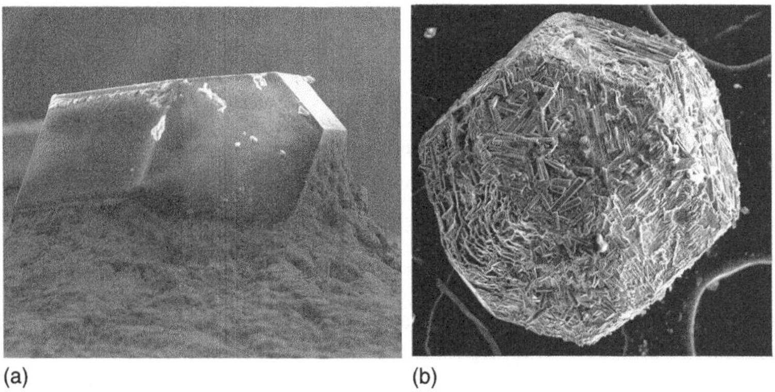

14.21 SEM micrograph of diamond brazed with a Ni-based filler alloy (a); carbides formed at the diamond surface (b).

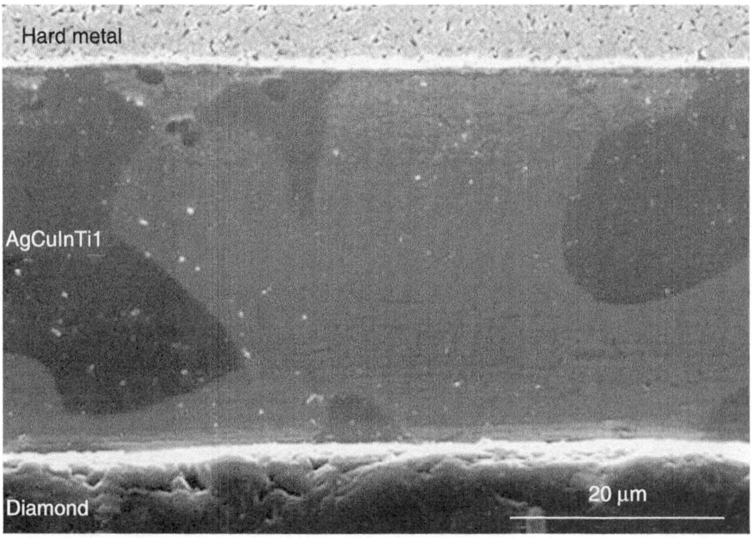

14.22 SEM image of diamond/cemented carbide joint using a AgCuInTi1.25 brazing alloy.

A joint between diamond segments and a steel blank has been used for many years to manufacture diamond blades in either wetting or drying applications. There are many types of diamond blades, and they are used in many industries, for example, in the construction industry to cut stone, concrete, asphalt, bricks, glass and ceramics, in the information technology industry to cut semiconductor materials, and in the gem industry to cut gemstones, including diamonds. Synthetic diamond particles are welded to the outside edge of a circular saw blade in a vacuum-brazing furnace. All of the diamond particles are on the exterior cutting edge of the blade. Depending on the manufacturer's recommended blade application, vacuum-brazed blades will cut a wide variety of materials, ranging from concrete to masonry, stone and brick, to steel, various irons, even plastics, tile, wood and glass. Smaller, finer synthetic diamond grits will produce a smoother finish with no chipping on tile or burring on steel. Larger diamond grits have a faster cutting speed, but will likely cause chipping, burring or cracking of the material (Litchfield, 2005).

It is worth mentioning that the presence of metals of the iron group in the interphase region results in the graphitization of the diamond while heating the composition and will greatly reduce the strength of the brazed joint. In this case, it is better to use refractory metals – molybdenum, tungsten, niobium, tantalum, etc. – but the use of these metals as structural materials is limited. When using steel dies or holders with diamond tools, it is necessary to take measures to prevent the penetration of the iron, nickel and manganese atoms into the interphase region. This is achieved by increasing the thickness of the metallizing coating, using two-stage brazing technology.

14.7 Conclusion and future trends

Developments and innovation in brazing technology mainly focus on the two main branches of brazing: brazing filler metals and brazing techniques. During the last few years, new alloys, with compositions which are modifications of previously known ones, have appeared on the world market. The motivation was either to improve the characteristic properties of the brazed joint, such as strength, wetting and corrosion resistance, especially at high temperatures, or to reduce the cost of the materials. Thus, the selection of an active alloy or coating material and the optimization of the contact and brazing conditions play a vital role in controlling the strength of the cutting tool.

A promising technology, which is expected to influence the science of brazing in the near future, is nanotechnology. In many industrial sectors, nanotechnology is considered to be a driving force of innovation. An interesting approach in brazing is based on the combination of size effects with exothermic reactions. All chemical reactions involve energy transformations and here energy released in the form of heat in the exothermic reaction is used for joining.

With modern thin-film deposition techniques, fully dense multi-layer materials can be fabricated with similar exothermic reactions. Such reactive multi-layer foils consist of hundreds or thousands of nanometre scale alternating layers of two or more materials, known as reactants. When heat is applied locally to the foil, it undergoes self-propagating high-temperature synthesis (SHS) (Qiu, 2007). So far, self-propagating reactions have been reported in Ti/B, Ni/Si, Zr/Si, Rh/Si, Ni/Al, Monel (7Ni:3Cu)/Al, Ti/Al, Pd/Al, Pt/Al and CuO_x/Al multi-layer materials.

Reactive multi-layer materials are most commonly fabricated using PVD methods such as magnetron sputtering and electron-beam evaporation. The PVD methods involve creating a vapour of a material, known as the source material, and then depositing the vapour onto a substrate. The rate at which the vapour is deposited is controllable, allowing the growth of layers with thickness ranging from nanometres to micrometres. Alternatively, multi-layer foils can also be made by cold rolling. In use, the nanofoils are combined with additional brazing foils placed between the two components to be joined. A small applied pressure allows the braze to flow and wet all surfaces. The reaction can be triggered electrically or thermally. A 9 V battery is sufficient to create a short circuit current for ignition. By varying the thickness and composition of the nanofoils, the maximum temperature, the speed and the absolute heat input of the joining process can be controlled. Figure 14.23 shows a schematic drawing of a self-propagating reaction in a multi-layer foil (Kim *et al.*, 2008).

Reactive multi-layer foils provide a unique opportunity to dramatically improve conventional soldering and brazing technologies by using the foils as local heat sources to melt solder or braze layers and thereby join components. The only limitation of this process up till now is its suitability to be used as a heating source and as a brazing filler metal. Additional applications are being developed, such as a multi-layer system that is optimized for use as filler metal, as well as higher energy foils. The chemistry of this foil will produce more heat and energy to melt

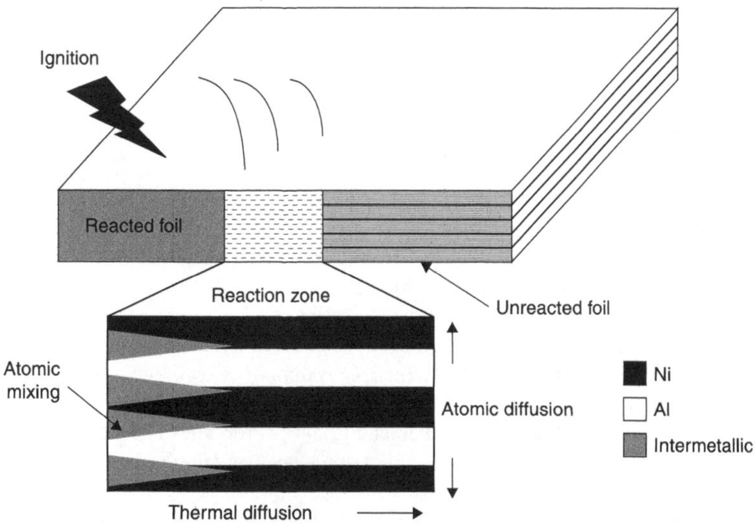

14.23 A self-propagating reaction in a multi-layer foil.

braze layers with higher melting temperatures. The essential advantage when simultaneously using the size effects and exothermic reactions for joining is the fact that the heat is released directly in the joining area. Thereby, the heat input is reduced significantly in comparison to conventional joining methods. The joint thus does not experience thermally induced mechanical stresses. Moreover, the joining process does not take much time because of the high velocity of propagation.

Despite the widespread use of the vacuum-brazing process in joining a variety of cutting materials, it is not suitable when the joining partners are made from materials that tend to be damaged by reducing atmospheres at high temperatures, for example high-temperature fuel cells or gas separation membranes for CO_2-free power plants. A process known as Reactive Air Brazing (RAB) has become increasingly important in recent years and is an alternative to vacuum brazing. RAB has proven to be advantageous, since materials that are sensitive to reducing atmospheres or vacuum can be joined. A RAB braze is typically made of a precious metal (e.g. silver) matrix and a reactive component, in most cases CuO, which improves wetting on ceramic and oxidized metal surfaces. Therefore brazing can be processed in air. The additional costs of the precious metal-based RAB brazes compared with nickel-based vacuum brazes can potentially be compensated by lower processing and investment costs compared with vacuum-brazing facilities. Research is in progress to develop the process and improve joint characteristics. In this regard, small percentages of Pd, up to 5 mol.%, were employed successfully to extend the temperature used, because the high-temperature application of CuO–Ag brazes is ultimately limited by the eutectic transformation present in this system (Darsella and Weil, 2007). Furthermore, the addition of 0.5 mol.% titania (TiO_2) to the Ag–CuO brazes was found to increase the wettability during RAB on alumina surfaces (Weil *et al.*, 2006).

14.8 References

Akelsen O M (1992), 'Review: Advances in brazing of ceramics', *J Mater Sci*, 27, 1989–2000.

ASM International (1973), *Metals Handbook, Metallography, Structures and Phase Diagrams*, American Society for Metals.

AWS (1991), *Brazing Handbook*, Miami, Florida, USA.

Bissig V, Janczak-Rusch J and Galli M (2007), 'Selection and design of brazing fillers for metal-ceramic joints', *THERMEC International Conference on Processing and Manufacturing of Advanced Materials*, 5, 539–543.

Blugan G, Janczak-Rusch J and Kuebler J (2004), 'Properties and fractography of Si_3N_4/TiN ceramic joined to steel with active single layer and double layer braze filler alloys', *Acta Mater*, 52, 4579–4588.

Blugan G, Kuebler J, Bissig V and Janczak-Rusch J (2007), 'Brazing of silicon nitride ceramic composite to steel using SiC-particle-reinforced active brazing alloy', *Ceram Int*, 33, 1033–1039.

Bose D, Datta A, Rabinkin A and DeCristofaro N H (1986), 'High strength nickel-palladium-chromium brazing alloys', *Weld J*, 65, 23s–29s.

Buhl S, Leinenbach C, Spolenak R, and Wegener K (2010), 'Influence of the brazing parameters on microstructure, residual stresses and shear strength of diamond–metal joints', *J Mater Sci*, 45, 4358–4368.

Cazajus V, Lorrain B, Welemane H and Karama M (2001), 'Residual Stresses in Ceramic Metal Assembly after Brazing Process', *Advances in Science and Technology*, 45, 1543–1550.

Ceccone G, Nicholasa M G, Peteves S D, Kodentsov A A, Kivilahti J K, *et al.* (1995), 'The Brazing of Si_3N_4 with Ni-Cr-Si Alloys', *J Eur Ceram Soc*, 15, 562–563.

Chattopadhyay K, Chollet L and Hintermann H E (1991), 'Induction brazing of diamond with Ni-Cr hardfacing alloy under argon atmosphere', *Surf Coat Technol*, 45, 293–298.

Chen Y, Hong J X, Fu Y C and Su H H (2009), 'Effect of Ti addition on shear strength of brazing diamond and Ag based filler alloy', *Key Eng Mater*, 416, 264–268.

Chiu L H, Wang H F, Huang C P, Hsu C T and Chen T C (2008), 'Effect of brazing temperature on the microstructure and property of vacuum brazed WC-Co and carbon steel joint', *Adv Mater Res*, 47–50, 682–685.

Cubberly W and Bakerjian R (1989), *Tool and Manufacturing Engineers Handbook*, Society of Manufacturing Engineers.

Darsella J T and Weil K S (2007), 'The effect of palladium additions on the solidus/liquidus temperatures and wetting properties of Ag–CuO based air brazes', *J Alloys and Compd*, 433, 184–192.

Davis J R (1995), *Tool Materials*, ASM International, American Society for Metals.

Dini J W (1993), *Electrodeposition – The Materials Science of Coatings and Substrates*, William Andrew Publishing/Noyes, Westwood, New Jersey, USA.

do Nascimento R M, Martinelli A E and Buschinelli A J A (2003), 'Review article: Recent advances in metal-ceramic brazing', *Cerâmica*, 49, 178–198.

Dudiy S (2002), *Microscopic Theory of Wetting and Adhesion in Metal-Carbonitride Systems*, PhD, Chalmers University of Technology, Göteborg University, Göteborg, Sweden.

Elsener H R and Klotz U E (2007), 'Modifizierte Aktivlote – entstanden durch Zersetzung von organischen Bindern und in-situ gebildeten TiC Nanopartikeln', *8. Internationales Kolloquium Hart- und Hochtemperaturlöten und Diffusionsschweissen*, 243, 27–31.

Etter D E, Egleston E E and Jaeger RR (1978), *Migration of iron, nickel, cobalt and chromium associated with silver brazing during ceramic-to-metal joining*, Technical Report No. MLM-2526, Monsanto Research Corp, Miamisburg, OH, USA.

Galli M (2007), *The constitutive response of brazing alloys and residual stresses in ceramic-metal joints*, Ecole Polytechnique Federale de Lausanne, Lausanne.

He Y M, Zhang J, Sun Y and Liu C F (2010), 'Microstructure and mechanical properties of the Si_3N_4/42CrMo steel joints brazed with Ag-Cu-Ti plus Mo composite filler', *J Eur Ceram Soc*, 30, 3245–3251.

Iancu O T (1989), *Berechnung von thermischen Eigenspannungsfeldern in Keramik/Metall-Verbunden*, progress report VDI, Universität Karlsruhe.

Iancu O T and Munz D (1989), 'Residual thermal stress in a ceramic/solder/metal multilayered plate', *3rd International Conference on Joining Ceramics, Glass and Metal, Oberursel*, 257–264.

Jiang Z and Zhang C (2010), 'Analysis on microstructure of laser brazing diamond grits with a Ni-based filler alloy', *Adv Mater Res*, 97–101, 3879–3883.

Jones F D, Ryffel H H, Oberg E, McCauley C J and Heald R M (2004), *Machinery's Handbook*, Industrial Press.

Kalpakjian S and Schmid S R (2006), *Manufacturing Engineering and Technology*, Pearson Education, Upper Saddle River, NJ.

Kapoor R R and Eagar T W (1989a), 'Oxidation Behavior of Silver and Copper Based Brazing Filler Metals for Silicon Nitride/Metal Joints', *J of the American Ceramic Society*, 72(3), 448–454.

Kapoor R R and Eagar T W (1989b), 'Tin-based reactive solders for ceramic/metal joints', *Metall Trans*, 20B, 919–924.

Kawasaki A and Watanabe R (1987), 'Finite element analysis of thermal stress of the metal/ceramic multi-layer composites with compositional gradients', *J Jpn Inst Met*, 51, 525–529.

Kim J S, LaGrange T, Reed B W, Taheri M L, Armstrong M R, *et al.* (2008), 'Imaging of transient structures using nanosecond in situ TEM', *Sci*, 321, 1472–1475.

Kizikov E D and Lavrinenko I A (1973), 'A study of adhesion and contact interaction between copper–tin–titanium alloys and diamond', *SintetAlmazy*, 6, 21–25.

Klaasen H, Kübarsepp J, Laansoo A and Preis I (2003), 'Dual compounds "cemented carbide-steel" produced by diffusion bonding', *Proc of Eur Powder Metall Conf*, 20–22 October, Valencia, 155–159.

Klotz U E, Liu C, Khalid F A and Elsen H R (2008), 'Influence of brazing parameters and alloy composition on interface morphology of brazed diamond', *Mater Sci Eng A*, 495(1–2), 265–270.

Knott S K and Ssemakula M E (2000), 'An experimental investigation into factors affecting the braze strength of CBN cutting tools', *Scientific Literature Digital Library and Search Engine (CiteSeerX)*.

Kohl W H (1967), *Handbook of Materials and Techniques for Vacuum Devices*, Reinhold Publishing Corp, New York.

Koizumi M, Takagi T, Suganuma K, Miyamoto Y and Okamoto T (1987), *High Tech Ceramics*, Elsevier Amsterdam.

Laansoo A, Kübarsepp J and Klaasen H (2004), 'Diffusion welding and brazing of tungsten free hardmetals', in *Proc. of 7th Intl Conf on Brazing, High Temperature Brazing and Diffusion Welding, DVS*, Aachen, 27–30.

Lee W B, Kwon B D and Jung S B (2006), 'Effects of Cr_3C_2 on the microstructure and mechanical properties of the brazed joints between WC-Co and carbon steel', *Int J Refract Met Hard Mater*, 24, 215–221.

Litchfield M W (2005), *Renovation: Completely Revised and Updated*, Taunton Press, Newtown, Connecticut.

Liu G, Qiao G, Wang H and Jin Z (2007), 'Microstructure and Strength of Alumina-Metal Joint Brazed by Activated Molybdenum–Manganese Method', *Key Eng Mater*, 353–358, 2049–2052.

Mackerle J (2003), 'Finite element analysis and simulation of machining: an addendum. A bibliography (1996–2002)', *International Journal of Machine Tools and Manufacture*, 43(1), 103–114.

Malzbender J and Steinbrech R W (2004), 'Mechanical properties of coated materials and multi-layered composites determined using bending methods', *Surface and Coatings Technology*, 176(2), 165–172.

Mizuhara H (1986), *Method of brazing ceramics using active brazing alloys*, US Patent No. 4591535, 27 May 1986.

Naidich Y V, Umanskii V P and Lavrinenko L A (2007), *Strength of the Diamond–Metal Interface and Brazing of Diamonds*, Cambridge International Science Publishing Ltd, Cambridge UK.

Naka M, Hirono Y and Okamoto I (1984), 'Wetting of alumina by molten alumina and alumina-copper alloys', *Trans JWRI*, 13, 201–206.

Naka M, Kubo M and Okamoto I (1989), 'Brazing of Si_3N_4 to metals with Al filler (Report II): Si_3N_4/Fe, Ni, Cu, Al or SUS304 joint', *Trans JWRI*, 18, 195–198.

Naka M and Okamoto I (1985), 'Wetting of Silicon Nitride by Copper-Titanium or Copper-Zirconium Alloys', *Trans JWRI*, 14, 19–34.

Nicholas M G and Crispin R M (1982), 'Diffusion bonding stainless steel to alumina using aluminum interlayers', *J Mater Sci*, 17, 3347–3360.

Olson D L, Siewert T A, Liu S and Edwards G R (1993), *ASM Handbook: Welding, Brazing, and Soldering*, ASM International.

Park J W, Mendez P F and Eagar T W (2002), 'Strain energy distribution in ceramic-to-metal joints', *Acta Materialia*, 50(5), 883–899.

Park J W, Mendez P F and Eagar T W (2005), 'Strain energy release in ceramic-to-metal joints by ductile metal interlayers', *Scripta Materialia*, 53(7), 857–861.

Pintschovius L, Schreieck B and Eigenmann B (1997), 'Neutron, X-ray, and finite-element stress analysis on brazed components of steel and cemented carbide', *MAT-TEC – Analysis of Residual Stresses from Materials to Bio-Materials*, Gournay-Sur-Marne, 307–312.

Qiao G J, Zhang C G and Jin Z H (2003), 'Thermal cyclic test of alumina/Kovar joint brazed by Ni-Ti active filler', *Ceramics International*, 29(1), 7–11.

Qiu X (2007), *Reactive Multilayer Foils and Their Applications in Joining*, MSc thesis, Tsinghua University.

Qu S, Zou Z and Wang X (2005), 'Shear strength and fracture behavior of SiCw/Al_2O_3 composite-carbon steel brazed joints', *Key Eng Mater*, 297–300, 2441–2446.

Rabinkin A and Pounds S (1989), 'Brazing cemented carbides: Specifics, braze optimization, and custom-designed Metglas[R] brazing filler metals', *Int J of Refract Met Hard Mater*, 8, 224–231.

Radtke R P (2009), *Thermally Stable Diamond Brazing*, US Patent No. 7487849 B2, 10 Feb 2009.

Rufe P D (2002), *Fundamentals of Manufacturing*, Society of Manufacturing Engineers, USA.

Sandvik Hard Materials, *Cemented carbide, Sandvik new developments and applications*, catalogue, [online], available at: http://www.sandvik.com/sandvik/0130/HI/SE03411.nsf/88c2e87d81e31fe5c1256ae80035acba/651f6e334db04c46c125707600562c88/$FILE/Cemented%20Carbide.pdf [accessed 07 April 2011].

Santos S I, Casanova C A, Teixeira C R, Balzaretti N M and Jornada J A (2004), 'Evaluation of the adhesion strength of diamond films brazed on K-10 type hard metal', *Mater Res*, 7, 293–297.

Saru silver, *Brazing solutions*, brochure, Sardhana Road, UP, India. Available from: http://www.sarusilver.com/catalogue.pdf [accessed 1 December 2011].

Schnee D, Zenk C, Hafner T, Meyer U, Magin M, *et al.* (2010), 'Einfluss der Bandbreite von Schichtloten auf die Festigkeit der Verbindung von Hartmetall und Stahl bei Sägeblättern', *9th International Conference Brazing, High Temperature Brazing and Diffusion Welding*, 263, 226–229.

Schüler H (2001), *Simulation von LötprozssenbeimMetall-Keramik-Löten*, TU Chemnitz.

Schwartz M M (1995), *Brazing for the Engineering Technologist*, Chapman and Hall, London.

Schwartz M M (2003), *Brazing*, ASM International, Materials Park, Ohio.

Steven K K and Mukasa E S (2000), 'An experimental investigation into factors affecting the braze strength of CBN cutting tools', *Scientific Literature Digital Library and Search Engine (CiteSeer X)*, http://www.faim2000.isr.umd.edu/faim/export/Faim20 (accessed 22 May 2011), http://citeseerx.ist.psu.edu/viewdoc/summary?doi=10.1.1.41.5433.

Suganuma K, Okamoto T, Koizumi M and Shimada M (1987), 'Joining of silicon nitride to silicon nitride and to Invar alloy using an aluminium interlayer', *J Mater Sci*, 22, 359–364.

Tae-Woo K, Hwi-Souck C and Sang-Whan P (2001), 'Re-distribution of thermal stress in a brazed Si_3N_4 /stainless steel joint using laminated interlayers', *J Mat Sci Let*, 20, 973–976.

Tillmann W (2003a), 'Concepts for innovative cutting tools based on modern joining technologies', *International Conference on Welding and Joining of Materials*, 20–27 October 2003, Lima, Peru.

Tillmann W (2003b), 'Einsatz von Diamantwerkzeugen in der Gesteinsbearbeitung', *Werkstoffwissenschaftliche Schriftenreihe*, 61, 73–79.

Tillmann W (2009), 'Löten artungleicher Verbunde', *Hartlöten und Hochtemperaturlöten seminar*, Technische Akademie Esslingen (TAE), Ostfildern, Germany.

Tillmann W and Osmanda A M (2005), 'Production of diamond tools by brazing', *Mater Sci Forum*, 502, 425–430.

Tillmann W, Osmanda A M and Yurchenko S (2009a), 'Investigations of contact angles of active brazing fillers on diamond-layers by optical 3D-microscopy', *Proc 4th Int Brazing & Soldering Conf*, 26–29 April 2009, Florida, USA.

Tillmann W, Osmanda A M, Yurchenko S, Magin M and Useldinger R (2009b), 'Strength Properties of Induction Brazed Cermets', *Proc Int Conf High Perform P/M Mater*, 25–29 May, Reutte, Austria.

Tillmann W, Osmanda A M, and Yurchenko S (2010a), 'Untersuchung der Einflüsse unterschiedlicher Aktivlote und Lötprozessparameter auf die Fügezonenausbildung und die Eigenschaften von CVD-Diamantdickschicht-Hartmetall-Lötverbunden', *Proc LÖT 2010*, 15–17 June 2010, Aachen.

Tillmann W, Wojarski L and Lehmert B (2010b), 'Investigation of mechanical and metallurgical properties of brazed ceramic and cemented carbide joints', *IIW 2010 – International Conference on Advances in Welding Science & Technology for Construction, Energy & Transportation*, Istanbul, 337–341.

Tillmann W, Wojarski L and Lehmert B (2010c), 'The Quality of Brazed Ceramic and Cemented Carbide Joints. A Mechanical and Metallurgical Assessment', *Advances in Science and Technology*, 108(64), 108–114.

Voytovych R, Robaut F and Eustathopoulos N (2006), 'The relation between wetting and interfacial chemistry in the Cu Ag Ti/alumina system', *Acta Mater*, 54, 2205–2214.

Wang X L, Hubbard C R, Spooner S, David S A, Rabin B H, *et al.* (1996), 'Mapping of the residual stress distribution on a brazed zirconia-iron joint', *Mater Sci Eng A*, 211(1–2), 45–53.

Webb S W and Eiler J T (2010), *Cutting tool inserts and methods to manufacture*, US Patent No. 7824134 B2, 2 Nov 2010.

Weil K S, Kim J Y, Hardy J S and Darsell J T (2006), 'Improved Wetting Characteristics in TiO_2-Modified Ag-CuO Air Braze Filler Metals', *Proc 3rd Int Conf on Brazing and Soldering*, 120–124, ASM International, Materials Park, OH, USA.

Whitney E D (1994), *Ceramics Cutting Tools – Materials, Development and Performance*, William Andrew Inc, Park Ridge, New Jersey, Noyes Publications.

Wielage B, Hoyer I and Weis S (2009), 'Contribution to creep behavior of Al_2O_3-particle-reinforced SnAg solders', *4th International Brazing and Soldering Conference*, Orlando, 49–54.

Xiao P and Derby B (1998), 'Wetting of silicon carbide by chromium containing alloys', *Acta Mater*, 46, 3491–3499.

Xu Z, Xu H, Fu Y, Xiao B and Xu J (2006), 'Induction brazing diamond grinding wheel with Ni-Cr filler alloy', *Mater Sci Forum*, 532–533, 377–380.

Yang Z, Xu J, Fu Y and Xu H (2008), 'Laser brazing of diamond grits with a Ni-based brazing alloy', *Key Eng Mater*, 359–360, 43–47.

Zhang J, Jin LY, Xu JC and Liu XQ (2007), 'Microstructure and properties of brazing joint between YG8 cemented carbide and A3 steel', *Solid State Phenomena*, 127, 265–270.

Zhao B, Xu X, Cai G and Kang R (2009), 'Experimental research on laser brazing of diamond grits with a Ni-based filler alloy', *Key Eng Mater*, 416, 396–400.

15
Coating techniques using brazing

H. KRAPPITZ, Innobraze, Germany

DOI: 10.1533/9780857096500.3.472

Abstract: Although brazing is commonly regarded as a joining technique, the following chapter will introduce possibilities for using brazing as a coating technology, which offers new and unique solutions for applying functional coatings to surfaces. Details are presented of some fundamental aspects of this coating technology with regard to processes, metallurgical considerations and mechanical properties, followed by the various methods for producing coatings by brazing. Examples of applications are explained with regard to the various coating techniques. The chapter gives an overview of the major technologies of coating by brazing and demonstrates that this relatively unknown coating technique offers complementary possibilities to other well-established coating processes. It is shown that the specific possibilities of coating by brazing create a niche where this coating technique can offer superior solutions.

Key words: coating by brazing, wear protection, repair brazing, abrasive coatings, turbine repair.

15.1 Introduction

Brazing is commonly regarded as a joining technique. Due to its specific capabilities, however, brazing offers new and unique solutions to apply functional coatings to surfaces of components, working under severe conditions of mechanical, thermal or chemical load or a combination thereof.

By forming a metallurgical bond, brazing offers the possibility of producing joints of high strength. Applied as a coating technique, this results in high bond strength for the coating. In contrast to most welding processes, brazing also allows materials that belong to very different groups to be combined and joined. By using powder metallurgical methods, brazing can produce coatings for many different combinations of materials, tailored to specific applications. This chapter will introduce solutions with blends of different metals, as well as between metals and carbides or even – using special filler metals – between metals and ceramic materials including diamond and cubic boron nitride.

Following some fundamentals of brazing, with special regard to coating technology, there will be an introduction to the different methods for using brazing as a coating technique. This starts from the well-known cladding of surfaces with solid segments of cemented carbides, using filler metal and flux. It will also cover methods for manufacturing and applying abrasive surfaces, consisting of hard and sharp-edged particles in a filler metal matrix. The application of pre-sintered sheets of well-defined composition to surfaces, as well as the restoration of worn surfaces with such compositions using polymer bond green tapes, will be explained. With regard to wear protection, a method will be introduced to produce coatings of high carbide concentration, using suspensions of carbides and filler metals. Another promising technique uses mats of carbides and filler metals, to form coatings of well-defined geometry.

The application of such coatings will be explained for surface restoration and wear protection and for protection against corrosion and oxidation. In plant engineering, coatings of very specific properties may be required, e.g. self-lubricating coatings, coatings of defined porosity or coatings with characteristic physical or chemical properties. Examples of such applications will be introduced.

15.2 Fundamentals of brazed coatings

Almost all of the coatings produced by brazing consist of two or even more constituents, typically applied in the form of powders or products manufactured from powders. At least one of these constituents remains solid during the brazing process, while at least one of the other constituents forms a liquid phase at the brazing temperature. The free spaces between the non-melting powder particles form capillary gaps for the melting constituents. The melting constituent is regarded as the brazing filler metal and may be heat treated above its liquidus temperature or just between the solidus and liquidus. At the brazing temperature the filler metal melts and wets the non-melting constituent, to form a dense layer of coating material. At the same time, some excess of filler metal wets the substrate to be coated and forms a strong bond between the coating and the surface of the substrate. As a result, the applied material forms a uniform surface layer of constant thickness and well-defined contours (Fig. 15.1).

15.2.1 Metallurgical considerations

In the molten stage, the filler metal may dissolve considerable amounts of the non-melting material, depending on the solubility of this material in the filler metal. Also, diffusion of elements of the filler metal into the non-melting powder and into the substrate occurs. The intensity of such reactions is controlled by the metallurgical composition of the system, the specific surface area of the non-melting powder and the process parameters of the brazing process. It is increased by a large contact area between filler metal and the non-melting powder surface. Also, narrow gaps between the solid powder particles form short diffusion paths and thereby increase the diffusion flux according to Fick's first law of diffusion. This contributes to a more intense reaction as well. The effect may be detrimental with regard to the required coating properties. An excess of liquid phase by molten material and subsequent deformation or even flow off of the coating, for example, may lead to scrap. The ratio of solid phase and liquid phase, therefore, has to be well balanced, to allow

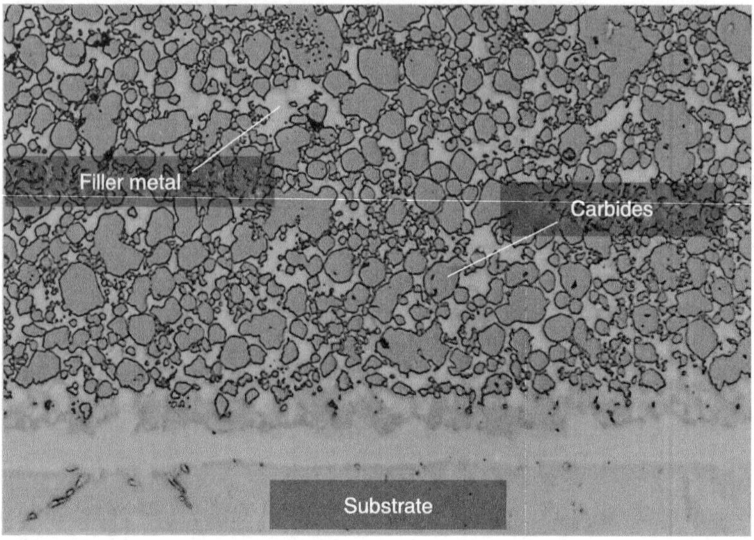

15.1 Structure of brazed coating, non-melting particles (carbides) in a nickel filler metal matrix.

wetting and densification of the applied material, but to have sufficient solid material available to keep the contours as sharp and precise as required.

Also, metallurgical incompatibilities could result in brittle phase formation and could lead to unacceptable cracks in the surface. For most applications, however, diffusion processes between filler metal and the non-melting constituent are an important prerequisite for obtaining the required properties of the coatings. An example of such an application will be given in Section 15.3.3 with regard to repair brazing of turbine components.

Besides the two constituents of a brazed coating as mentioned above, further additives may be applied to a coating system, either to obtain specific properties of the surface or to enhance wetting of the filler metal with the non-melting constituents. As an example, the reactive element Ti may be added, to allow wetting of ceramic materials like cubic boron nitride (CBN) or diamond grit.

15.2.2 Porosity and packing density

When applying a coating to a surface, it is an objective for almost any application to obtain coatings of low porosity. Any volume of powder, however, is a packing of particles and consists of solid particles and of free spaces between these particles. The ratio of the volume of particles versus the total volume of the bulk material is called the packing density. The value of the packing density depends, among other things, very much on the packing order of the particles, the shape of the particles and their grain size distribution. To obtain dense brazed coatings, the packing density of the applied coating material should already be relatively high in the 'green stage' before brazing. Modelling of such systems is difficult, because in a random packed structure of particles of different grain size there is no analytical method available to calculate the packing density. There are models available, however, to calculate the packing density of powders with a defined packing structure (Steinhaus, 1999). For monosized spheres the highest possible packing density is obtained when the spheres form a face centred cubic array (Fig. 15.2). Under such conditions, the theoretical maximum packing density is about 74%. Under real conditions, particles will not find the optimum position in a face centred cubic array. Also, such particles will not have a constant diameter and perfect spherical shape. The author's own experiments with gas atomised spherical metal powders of NiCrBSi with grain sizes between 10 μm and 106 μm showed a lower packing density of about 63 to 67%, depending on the method of densification. With irregularly shaped particles, the packing density is further reduced, depending on the morphology of the particles.

On the other hand, a carefully selected grain size distribution can contribute to increased packing density. Adding a second fraction of non-melting particles may increase packing density considerably. The maximum size of such a second fraction is obtained when the particles of the fine fraction fit into the free spaces of the first fraction. Under the condition of spherical-shaped particles, the maximum diameter of a second fraction of powder added to the first fraction can be calculated as given in Fig. 15.3. This calculation shows that the finer fraction should be smaller by a factor of six to fit into the free spaces between the coarser particles without decreasing the packing density of this fraction. Using a blend of fine carbide powders of 2–4 μm in a fraction of coarser carbide powders of about 20 μm, for example, was successful in increasing the packing density to 70% in experiments with optimised powder blends.

The brazing process applied as a coating technique aims to transform a porous packing of particles into a dense layer of compound material with minimum porosity.

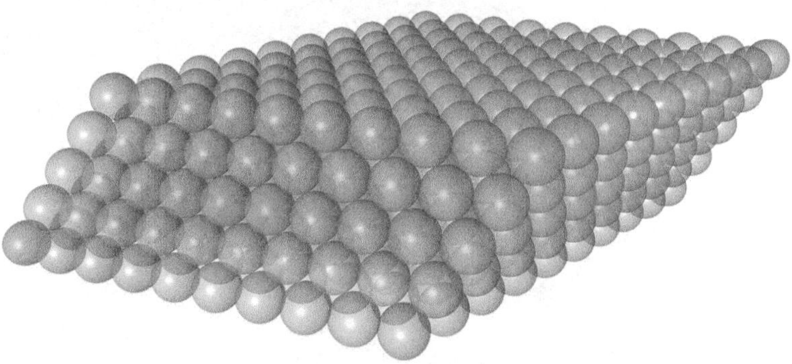

15.2 Highest possible packing density of spheres, obtained by a face centred cubic array.

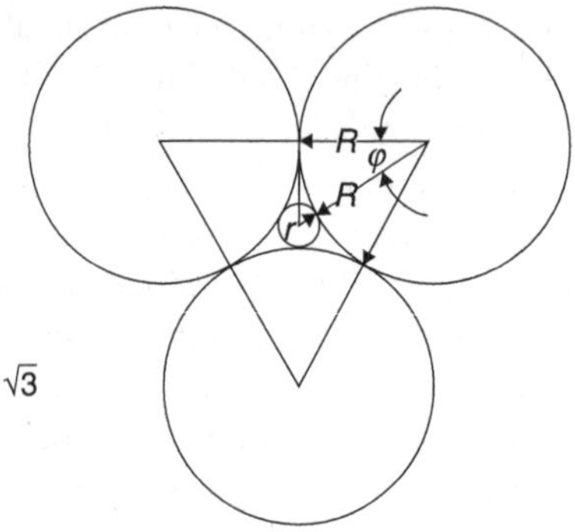

$$\cos \varphi = \frac{R}{R+r}$$

$$\varphi = 30° \Rightarrow \cos \varphi = \frac{1}{2}\sqrt{3}$$

$$r_{max} = R(\frac{2}{\sqrt{3}} - 1)$$

$$r_{max} \approx 0.1547 * R$$

15.3 Optimum grain size of second fraction between spheres of a first fraction.

In the case of porous-free coatings, this process should lead to a structure in which the free spaces between the non-melting particles of the coating are filled completely with brazing filler metal. This defines the minimum quantity of filler metal required to produce dense coatings. With a packing density of e.g. 64% for the non-melting powder, a minimum volume of 36% of filler metal is required to fill the free spaces between the non-melting particles. Some additional filler metal is required, to form a joint between the coating and the substrate. In repair brazing of turbine components, which is one of the important applications for coating by brazing, a filler metal ratio of 40% to 60% is therefore required in the relevant specifications (Pratt and Whitney, 2010; Budinger *et al.*, 1993; Wayne and Budinger, 2002).

If the coating material is a homogeneous mix of non-melting material and filler metal (Fig. 15.4), the filler metal particles themselves contribute to the packing

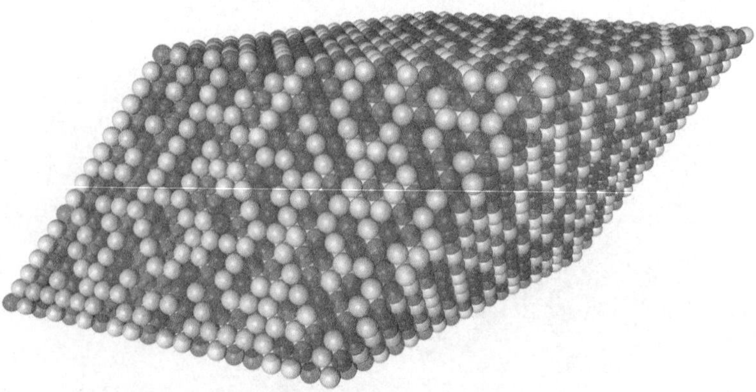

15.4 Homogeneous mix of non-melting material (dark spheres) and filler metal (light spheres).

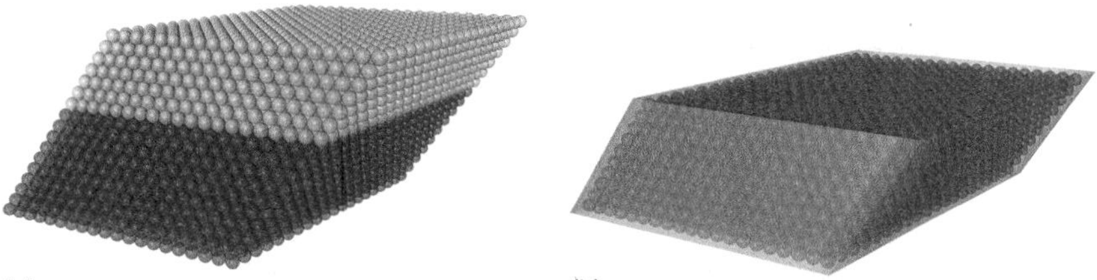

(a) (b)

15.5 (a) Layer of filler metal (light spheres) on top of non-melting material layer (dark spheres). (b) Layer of non-melting material, infiltrated by filler metal.

density of the system. With 40 vol.% of filler metal, therefore, the non-melting material contributes just 60% to the maximum packing density of e.g. 64%. Under such conditions, the packing density of non-melting particles would be just 0.6 × 64% = 38.4%. As a result, the layer would have to shrink by 36% during the brazing process to form a dense coating without pores. This is different if the filler metal is added as a second layer to the coating (Fig. 15.5(a)). In this case, the non-melting particles can form a high packing density, which is infiltrated by the melting filler metal during the brazing process (Fig. 15.5(b)). The geometry of the non-melting layer, therefore, does not have to shrink to obtain full density. This allows the application of coatings that have the same geometry in the brazed state as in the green state before brazing.

15.2.3 Strength properties

Due to the many different material combinations that can be produced by brazing as a coating process, it is not possible to give general values of strength, as might be the case for other coating processes. There are, however, some general considerations with regard to bond strength between coating and substrate. Coating by brazing forms a bond between materials at the interface. The bond strength is comparable to the strength of brazed joints in conventional brazing. If the coating material and base material of the substrate have similar physical and mechanical properties, the strength of the joint can be close to the strength of the base material. With dissimilar materials the discontinuities at the interface have a major influence on the strength values. In particular different coefficients of thermal expansion (CTEs) can cause high internal stresses in the coating. In the worst case, such stresses may be so high that the coating may chip off completely due to crack propagation parallel to the interface zone. Figure 15.6 explains the mechanism that leads to thermal mismatch stress. Two materials of different CTE are heated up to the brazing temperature and will expand according to their coefficient of thermal expansion. The filler metal wets the substrate, and after solidification during the cooling cycle a strong bond is formed. During further cooling it is not possible for both elements of the joint to contract according to their material constants, as each

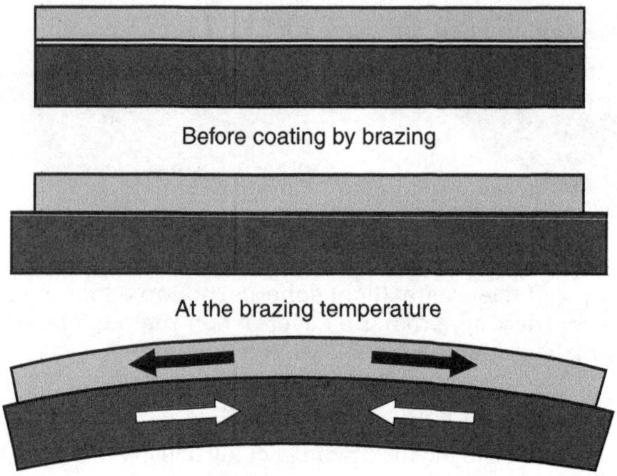

15.6 Thermal mismatch stress, caused by different coefficients of thermal expansion.

element obstructs the contraction of the other. This leads to CTE mismatch stresses, which, if they reach sufficiently high levels, may cause complete failure of the joint. With coatings of dissimilar materials, such stresses typically have the most important influence on the bond strength of the coating. The magnitude of mismatch stresses, however, not only depends on the combination of materials, but also on the geometry and design of the coated components. In Section 15.3.1 possibilities will be discussed for minimising the level of CTE mismatch stresses.

15.3 Classification of brazed coatings

This section will explain the different variations of coating techniques available to produce coatings by brazing.

15.3.1 Cladding with segments

This coating technique can be regarded as a joining technique as well as a coating technique and thereby forms the link between both. Segments, typically cemented carbide plates or other wear-resistant materials, are brazed to the surface of a component. Applications are found in the tools industry, mining industry, wood and paper processing and minerals processing. Brazing provides the means of joining with the highest strength and thermal stability. Figure 15.7 shows milling cutter tools with brazed hard metal inserts. These tools were brazed at 1080°C in a conveyor-belt furnace under a reducing atmosphere of pure hydrogen. A foil of 0.2 mm thickness with composition CuMn12Ni3 was used as a filler metal. Carbide segments can be applied as a single piece with well-defined geometry or,

15.7 Milling cutter tools with brazed hard metal preforms.

in the case of larger surface areas, can be assembled to form a mosaic-like surface. As well as furnace brazing, torch brazing and induction heating in air are very common heating methods. When brazing in air, however, fluxes have to be used to remove the surface oxide layers during the brazing process. Wetting of cemented carbides is generally difficult, however, especially when the cemented carbides contain low amounts of cobalt binder phase and carbides of Ti, Ta or Nb besides the commonly known tungsten carbide. Fluxes for cemented carbide brazing are therefore usually more aggressive than standard brazing fluxes, in order to remove also of higher chemical stability. To obtain joints of high strength, special filler metal compositions are used. The most common compositions are silver or copper based and additionally contain manganese and nickel or cobalt, to improve wetting (Table 15.1).

Table 15.1 Common filler metals used for cemented carbide brazing (ISO 17672, 2010)

Designation	Composition	Melting range
Ag 449	Ag: 49.0, Zn: 23.0, Cu: 16.0, Mn: 7.5, Ni: 4.5	680–705°C
Cu 470a	Cu: 60.0, Zn: 39.7, Si: 0.3	875–895°C
Cu 773	Cu: 48.0, Zn: 41.8, Ni: 10.0, Si: 0.2	890–920°C
Cu 595	Cu: 84.2, Mn: 12.5, Ni: 3.3	965–1000°C
Cu 141	Cu: 99.9, P: 0.03	1085°C

Joints produced by brazing with special filler metals and fluxes can obtain shear strength values well above 300 N/mm^2. The coefficient of thermal expansion ratio between steel and cemented carbides, however, is in the range of 2:1. As explained in Section 15.2.3, this may lead to high internal stresses in a joint. Strategies to minimise such stresses include the design of the parts as well as appropriate filler metal selection. The design should ensure a rigid backing for the carbide coating by the steel part, to avoid tensile stresses in the carbide cladding caused by bending.

Selection of low-melting filler metals results in a small temperature interval between the brazing temperature and room temperature and thereby limits the total thermal contraction as a result of the temperature difference. For this reason, low-melting silver-based filler metals are frequently used to braze cemented carbide plates. An additional advantage of this group of filler metals is their high ductility and relatively low yield strength. The CTE stresses can be limited very effectively by plastic deformation of the filler metal layer between substrate and coating. The best results for stress reduction are obtained if the thickness of the ductile filler metal layer is as high as possible. In capillary brazing, however, the brazing gap is limited to 0.2–0.3 mm, to keep the molten filler metal in the capillary gap. To overcome this limitation and to apply much thicker layers of filler metal, special foils are used, which consist of a ductile non-melting copper foil cladded with low-melting silver filler metal on both sides. These so-called 'trifoils' allow high plastic deformation and are frequently used for carbide brazing of critical components. For higher-melting filler metals, trifoils are also available with a Ni mesh as a centre layer (Bronny and Schimpfermann, 2011).

The brazing process can also be modified to lower the level of CTE mismatch stresses. Low cooling rates down from the brazing temperature leave sufficient time for the steel substrate to undergo plastic deformation at elevated temperatures. Such creep effects contribute to a decrease of stress levels in the joint. For some critical applications with components of larger surface area, it has even been known to stretch the steel part during cooling or in a following heat-treating process by mechanical straightening. Such straightening induces tensile stresses in the steel part, which lead to plastic deformation of the steel, while the cemented carbide part is under compressive stresses. As cemented carbides can withstand compressive stresses much better than tensile stresses, at the end of such a straightening process the internal stresses are effectively lowered. Large surface areas, such as carbide-plated industrial paper knives of length 1000 mm and more, can be coated by this technique.

For application under extreme conditions of wear, segments of polycrystalline diamond (PCD), chemical vapour deposited diamond or boron nitride with a backing of cemented carbides are selected. Such applications can be found in fibreboard and metal matrix composite machining and in oil and gas drilling. For oil and gas drilling the highest reliability and strength are required, as replacement of the cutter head during drilling may incur immense costs. Besides the commonly known silver- and copper-based filler metals, special gold containing brazes,

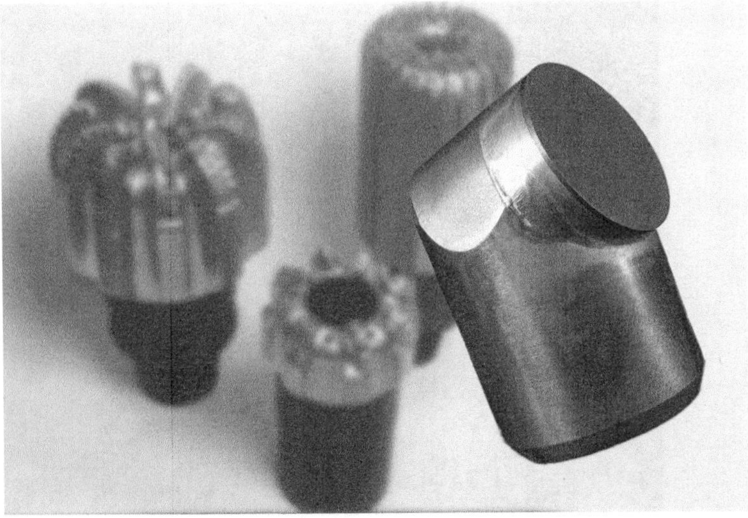

15.8 Brazed PCD cutter insert for oil and gas drilling © Background Graphics: WEG (2008).

which combine the highest strength and ductility, are used for such applications. Using gold–copper–manganese filler metals (Tank, 1997), joints can be obtained with shear strength values well above 800 MPa. Figure 15.8 shows a PCD cutter bit on a cemented carbide backing, brazed to a steel shank. In a subsequent step, the tool is brazed onto a cutter head for oil and gas drilling.

15.3.2 Abrasive surfaces

When surfaces with high friction or for abrasive tools are required, brazing offers a superior solution. Hard particles produced from crushed cemented carbides or from ceramic material chips are embedded in a metal matrix and bonded to the surface of a component by brazing. The metal matrix is formed by a brazing filler metal. To produce such coatings, the surface of the tool or component is initially coated with an adhesive layer. This could be a special glue as well as a suitable brazing paste, which offers sufficient adhesion. The carbide particles are then sprinkled onto the adhesive surface. Some of the applied particles stick to the surface, while an excess of material, which has not bonded, can be separated from the part by gravity. If not applied with the paste, a well-adjusted quantity of filler metal is then added to the surface. Filler metals are selected from copper-based, copper–tin, nickel or silver–copper alloys.

In a brazing process under protective atmosphere or in a vacuum, the filler metal melts, wets the carbide particles and forms a bond with the substrate. As the joint between the materials is a strong material bond, the particles do not need to

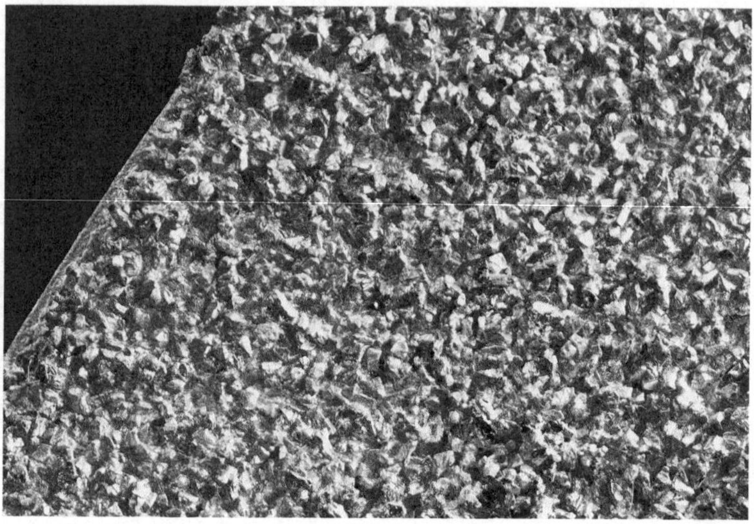

15.9 Surface of an abrasive coating, applied by protective atmosphere brazing. Carbide particles of crushed tungsten carbide, brazed with CuSn6 filler metal to a carbon steel substrate.

be embedded completely in the metal matrix, as would be the case for sintered or galvanic coatings. With a properly adjusted quantity of filler metal, the particles show excellent performance because of the large number of free cutting edges. This improves efficiency in comparison to sintered cutting segments. Figure 15.9 shows the surface of an abrasive coating, applied by brazing. Coarse crushed tungsten carbide particles were brazed with filler metal CuSn6 to a carbon steel substrate in a conveyor-belt furnace under a protective atmosphere of hydrogen.

Applications of abrasive surfaces are found in the tooling industry. Such tools are used for grinding or cutting operations. Figure 15.10 shows a segment of a saw blade. Tungsten carbide was brazed to the surface of a cutting disc with a copper-based filler metal. This disc is used to cut porous concrete and similar construction materials.

Besides tungsten carbide, other hard particles may be used for abrasive tools. If so-called active elements like Ti or Zr are added to the filler metals, these filler metals will even wet ceramic materials under suitable conditions of brazing atmosphere and temperature. Cubic boron nitride as well as diamond grit can be used to produce tools for grinding extremely hard materials.

15.3.3 Surface restoration by brazing

Gas turbine components require high reliability under service conditions of an elevated temperature, corrosive attack and high mechanical load. Blades and vanes of gas turbines for stationary use as well as for aircraft engine applications,

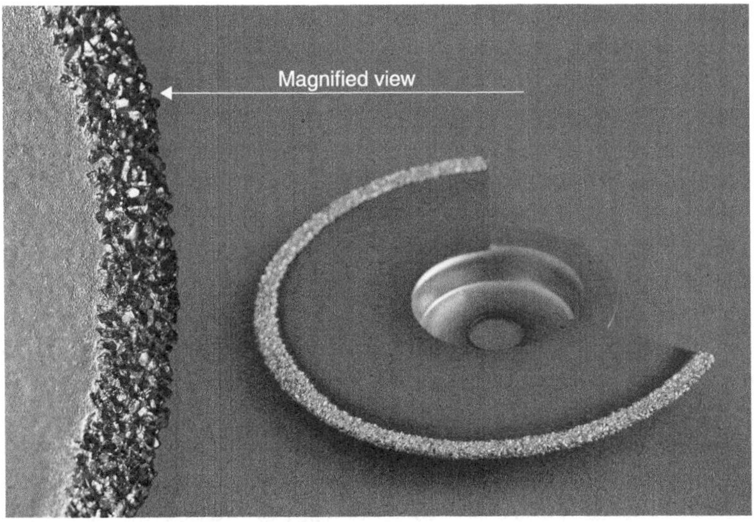

15.10 Brazed tungsten carbide coated segment of a saw blade.

therefore, are commonly produced from high-performance materials. Special nickel and cobalt-based superalloys with improved heat resistance and high strength at elevated temperature are specified for this application. New developments comprise single-crystal blades as well as intermetallic compound materials, e.g. titanium aluminides.

After a certain time under service conditions, these components show damage such as cracks, pitting and material loss. Due to high costs, reliable repair methods are required so that the worn components can be reused. The restored part should perform as reliably as a new component. As most welding techniques would damage the structure of these high-performance materials, the best solution to repair worn blades and vanes is offered by brazing techniques. More detailed information is given by Lugscheider *et al.* (1995) and Wayne and Budinger (2002).

Pre-sintered sheets

A proven solution for repairing worn gas turbine components is the application of pre-sintered preforms. Mixtures of powders with a similar composition as the base metal (the metal matrix powder) and a second powder, a so-called diffusion braze, are processed in combination with organic binders, to bring the mixed powders into the form of a sheet or tape of defined thickness. Uniaxial pressure is applied to the green tape, to obtain a high green density. Sintering under a controlled atmosphere will form a metal sheet material from these tapes. The sintering process is performed as liquid phase sintering at temperatures close to the liquidus temperature of the filler metal. Typically diffusion brazes contain boron, which acts as a melting depressant for the filler metal. While the filler metal forms some amount of molten phase at the sintering temperature, the base metal powder has a solidus temperature well above the sintering temperature. Due to the presence of the liquid phase, the sinter process leads to a material of low porosity within relatively short sinter cycle times. The formation of the liquid phase is very sensitive to the process parameters, time and temperature of the sintering process.

In a second step, the resulting metal sheets can be machined into the required shape. Figure 15.11 shows the geometry and microstructure of pre-sintered preforms, produced from NiCrCo base metal powder (In939) and NiCrCoB filler metal (DF3), used for repair brazing of gas turbines. The bright areas in the microstructure are the high melting particles of In939, embedded in the filler metal matrix. The preforms are attached to the turbine component by tack welding and brazed to the surface. Some additional filler metal, applied as a paste, forms the bond between the substrate and coating. Special cleaning processes are performed on the components before brazing, to remove any impurities and oxides from the surface and to ensure proper wetting of the coating on the substrate. The boron in the filler metal portion of the sheet can diffuse very quickly in nickel and cobalt alloys during the sintering and brazing cycle, decreasing the boron concentration in the filler metal. Due to this effect, the remelting temperature of the coating increases considerably, thus allowing high-temperature applications for such coatings.

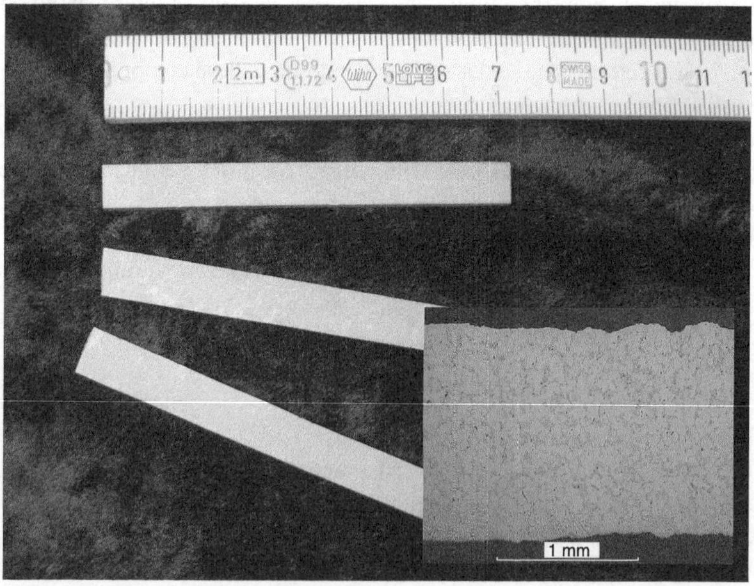

15.11 Pre-sintered preform strips for repair brazing of gas turbine blades. Particles of In939 in a filler metal matrix of DF3. Geometry (left) and microstructure (right) of pre-sintered strips before coating.

Pre-sintered preforms allow a very clean controlled atmosphere during the brazing cycle, as no organic binder is left in the sheet material when it is brazed to the substrate. As the pre-sintered sheets have a low porosity due to the sintering heat treatment, there is little further shrinkage in the coating operation by brazing. For simple geometries, the pre-sintered sheets offer a simple and reliable solution; however, it is difficult to obtain a proper fit to the substrate when coating complex three-dimensional shapes.

Polymer bond tapes

Besides pre-sintered sheets, another technique is applied successfully to restore worn surfaces of turbine components. A blend of base metal powder and filler metal powder is processed together with suitable polymers, to form a flexible mat material. There are many different compositions of base metal and filler metal in use; an overview of common compositions and combinations can be found in Sulzer (2011). In the green state, polymer bond mats can be formed easily into any desired shape, using conventional cutting or stamping operations. These flexible mat preforms can then be easily applied to a surface, using a glue to keep the green tape in position. The main requirement for this glue is that it keeps the mat in position not only at ambient temperature, but also at elevated temperatures during the subsequent brazing cycle. At the same time, after brazing it should not leave any harmful residues at the interface between the base metal and coating, which would weaken the joint and thereby would decrease the bond strength.

The brazing cycle comprises several steps. In the first step, the binder is carefully decomposed by thermal debindering at temperatures up to approximately 600°C under protective atmosphere or in a vacuum. The heating rate should be relatively slow, to avoid blisters caused by the vapour of the decomposing binder. In the next step, under a clean vacuum atmosphere, the temperature is slowly raised to the liquid phase sintering temperature. This temperature is typically slightly above the liquidus temperature of the filler metal. The temperature is maintained for a period of time, which is longer than the conventional holding time for brazing processes, to allow intense diffusion between the filler metal and base metal powder. There is no general recommendation for the brazing temperatures and holding times, as the values depend very much on the composition of base metal and filler metal powder, their ratio in the composition and the metallurgical interaction of both powders, as well as on the requirements of the turbine component.

During liquid phase sintering, the density of the coating increases to almost full density. This requires shrinkage of the randomly packed powder structure by about 40%. Shrinkage is a three-dimensional process and would lead to dimensional changes of the preformed tape. For coating purposes, however, it is preferable to allow shrinkage only in the direction perpendicular to the surface, while the other dimensions should remain stable. This can be realised to a major

15.12 Coating on a turbine tip of brazed CBN particles in a heat resistant MCrAlY metal matrix.

degree, if the glue obstructs the shrinkage in the two other directions by locking the coating material to the surface of the turbine component during the sintering process. Figure 15.12 shows a blade tip coated with hard particles. A flexible mat of matrix powder, filler metal powder and embedded particles of cubic boron nitride has been mounted and brazed to protect the tip against wear. The resulting coating is part of an abradable seal, which helps to increase the efficiency of gas turbines by closing the gap between the rotating blade and the casing (shroud).

15.3.4 Wear protection with polymer bond tapes

Wear protection is a major application of coating processes by brazing. Surfaces with very well-defined properties and precise contours close to the final dimensions can be obtained. This reduces the requirement for expensive machining of difficult-to-machine wear-resistant coating materials.

Unlike the surface restoration of turbine components, the non-melting phase is usually a type of carbide of metallic nature. While tungsten carbides WC and W2C are most common, other carbides are also used, such as chromium carbide, titanium carbide or vanadium carbide. Filler metals are mainly nickel based, but also a group of low-melting iron-based alloys or even copper-based filler metals of high ductility can be used, depending on the requirements of the application (Koschlig, 1994; Schwarz, 2003). Figure 15.13 shows the structure of a high carbide-containing coating of chromium carbide in a metal matrix of NiCrBSi. It

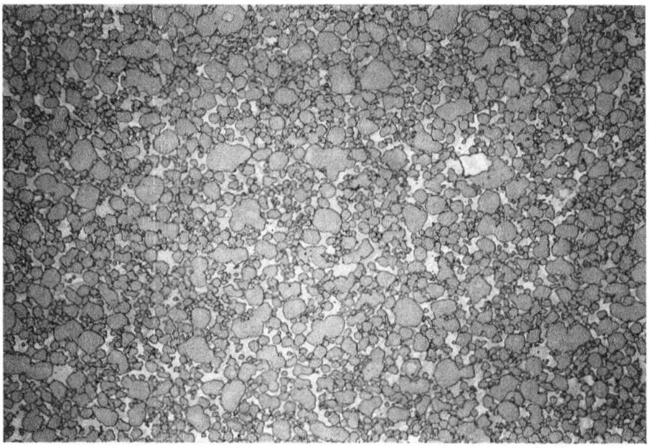

15.13 Structure of a high carbide-containing coating of chromium carbide in a metal matrix of NiCrBSi.

has been shown that extremely high carbide concentrations can be obtained with this coating technique.

Except for the wetting process, there should not be too much diffusion or alloying between the carbides and the metal matrix, as such effects would lead to the formation of brittle phases, e.g. eta carbides. The conditions for a brazing process with carbide-containing systems therefore have to be modified. Brazing temperatures should be as low as possible while still allowing proper wetting and flow of the filler metal, and the time at the brazing temperature should be short in order to avoid excessive alloying. These metallurgical requirements are in contrast with the requirement to obtain low porosity in liquid phase sintering. To overcome this problem, a process with two or more layers has been developed.

Multi-layer polymer bond tapes

With multi-layer polymer bond tapes, the filler metal constituent is separated from the carbides in the tape. This leads to a system in which a pure carbide layer mat of high packing density is formed. By simple cutting or stamping operations, this carbide mat can be easily brought into the desired geometry. The thickness of such a layer varies between 0.5 mm and 3 mm. Due to the high flexibility of such polymer bond mats, it is possible to align these mats with the curved surfaces of wear parts (Fig. 15.14). Detailed descriptions of the process are given in Gill (2006) and Innobraze (2012).

The carbide mat is attached to the surface that requires protection, using a glue of adapted composition. On top of this first layer of the pure carbide mat a second congruent layer is applied, which consists of only filler metal powder. The quantity of the filler metal supply is controlled by the thickness of this second layer in relation to the thickness of the carbide mat.

15.14 Flexible mats, produced from polymer bond powders.

A subsequent furnace process at approximately 1100°C creates the final coating under a hydrogen atmosphere or in a vacuum. The molten filler metal infiltrates the applied carbide layer and at the same time brazes it to the substrate material. Figure 15.15 demonstrates the coating process in a conveyor-belt furnace. Technical data of such coatings are given in Table 15.2. The technical data of the coating material may differ considerably from the properties of steel, which is commonly used as a base material. In particular the mean coefficient of thermal expansion (CTE) plays an important role with regard to mismatch stresses, as

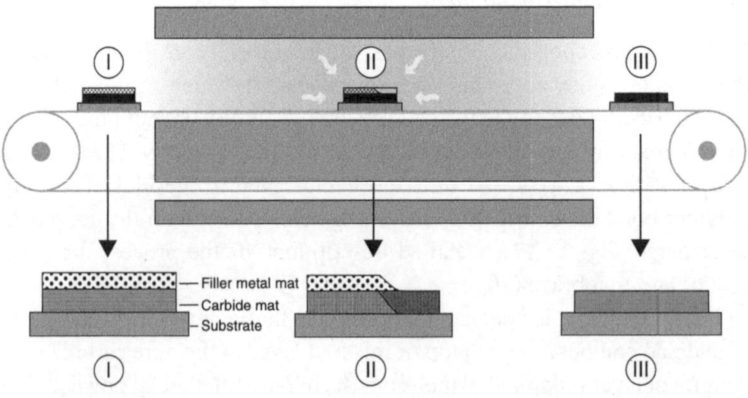

15.15 Coating process by brazing in a conveyor-belt furnace (schematic).

Table 15.2 Properties of brazed coatings

Coating type	Hardness (HV10/HRA)	Density (g/cm^3)	α-thermal ($10^{-6}/K$)
WC/NiCrBSi	1240/88	13.0	8.1
Cr_3C_2/NiCrBSi	1150/86	7.0	11.4
WC/CuMnNi	–/64	13.1	9.6

already explained in Section 15.2.3. A coating of WC/NiCrBSi has a low CTE and causes high mismatch stresses in combination with steel parts of larger surfaces. Cr_3C_2/NiCrBSi coatings, however, show thermal expansion behaviour very similar to that of constructional steel, yielding considerably reduced stresses in the coated parts.

As the packing density of the carbides is already at the maximum of up to 70 vol.% in the green tape, the carbide tape does not have to shrink to obtain fully dense coatings. Instead the molten filler metal infiltrates the carbide mat and fills the free spaces between the carbide particles. As a result, it is possible to have fully dense coatings of extremely high carbide concentration. Due to the lack of shrinkage, the finished coating has the same size and contour as the carbide cloth in the green state, when it was easy to machine.

As an example of a coating close to the final contour, Fig. 15.16 shows a debarker tip, used to remove the bark from the trunks of trees. At the cutting

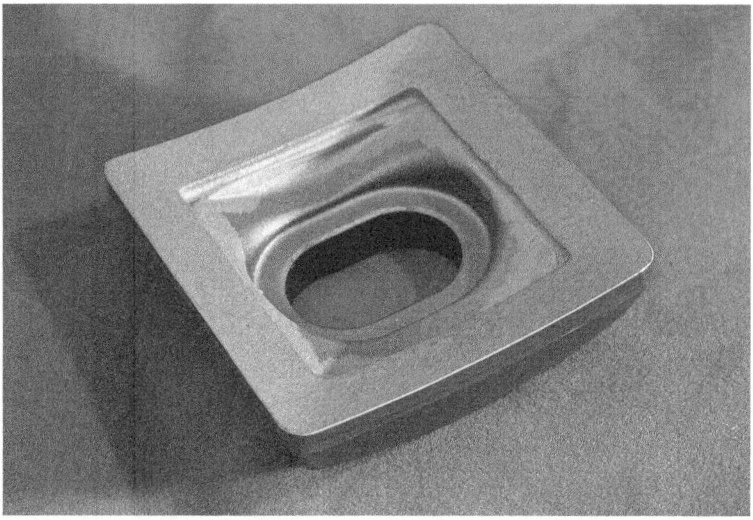

15.16 Debarker tip, coated with a flexible carbide mat by infiltration.

edges, a coating in the form of a frame can be seen. This frame consists of tungsten carbide in a matrix of NiCrBSi, brazed to the substrate. The hardness of this material is well above 85 Rockwell A and makes the tip wear resistant under the conditions of severe abrasion in service. Any machining of such hard and wear-resistant materials would be difficult and expensive. In the green state, however, the carbide mat can be easily brought into the required shape by simple cutting or stamping operations. During the infiltration brazing process, the shape remains as it was in the green state. The expensive diamond grinding process can be limited to just sharpening the cutting edges. This saves production costs and provides superior properties to the part with regard to cutting life cycle and reliability.

Due to its ability to produce coatings of precise contours, coating by brazing may even be used as a forming process. Figure 15.17 shows a test sample of a grinding track. Again the carbide profiles can be produced by simple cutting operations from the soft mat material in the green state. The green tungsten carbide profiles are then laid onto a flat steel plate, using a template for precise positioning. After addition of a suitable amount of filler metal to the carbide profiles, an infiltration brazing process under hydrogen forms the surface structure and results in a strong joint between the carbide profiles and steel plate. Using this technique, refiner discs for paper pulp production can be produced very successfully. The lifetime of such refiner discs for processing pulp filled with abrasive additives, was many times longer than conventional refiner discs, manufactured from chilled cast iron.

Multi-layer tapes also allow the production of coatings of different layers with specific properties. A buffer layer of high ductility and low coefficient of thermal

15.17 Grinding track made from carbide strips formed by infiltration brazing.

expansion, for example, can be inserted between a surface layer of high carbide concentration and a steel substrate, to reduce CTE stresses. With an increasing number of layers, it is even possible to produce a quasi-graded structure in the coating, allowing a smooth transition of material properties from the base metal to the surface of the coating.

15.3.5 Dip application of slurries

The application of base metal and filler metal layers using tapes offers very precise application of coating materials from the aspect of quantity and geometry as well as creating high-performance coatings with regard to material properties. For certain applications, however, this is not required. A simple but reliable process may be required instead, which mass produces wear parts at low cost. Again, coating by brazing offers an economic solution. The protective layers can be applied by dipping the wear parts into slurry, composed of filler metal and carbides in a liquid binder. Such slurries typically consist of approximately 20 wt.% binder, 20–50% of non-melting particles (carbides) and 30–60% of filler metal. Some slurries, however, do not contain any non-melting constituents at all, but are brazed below the solidus and liquidus temperatures instead, to leave some of the filler metal in the non-molten stage in order to avoid excessive flow.

The viscous binder of the slurry is intended to keep the metal and carbide particles in a homogeneous distribution so they stick to the surface of the wear part after application. Such binders consist of a solvent, a resin or polymer as a thickener and rheological additives. The additives may improve the anti-settling behaviour of the solid particles, avoid sagging of the applied slurry coating or support wetting of the liquid slurry on the surface of the wear part. After drying, additional additives may contribute to enhanced adhesion of the applied coatings on the surface and to improved handling properties in the green state. All ingredients of such binders have to be selected in such a way as not to disturb the brazing process or to have any detrimental effect on the furnace from evaporating decomposition products.

In the coating process, the part is dipped into the slurry. Depending on the viscous properties of the slurry and its wetting properties on the steel surface, a layer of a certain thickness is deposited onto the part. In the next step, the solvent is dried, which is mainly done at elevated temperature in an air oven. Finally, the dried parts are placed in a brazing furnace under a protective atmosphere, where the polymers of the binder decompose. This degradation of the binder polymers happens at temperatures up to approximately 500°C. On further heating, the filler metal melts and wets the non-melting particles and the surface of the wear part, to form a dense coating.

Applications of such coatings are found in agricultural products or equipment for mechanical processing equipment. Figure 15.18 shows an example of a pin mill. The pins are coated with a slurry of tungsten carbide and the eutectic nickel

15.18 Coated pin of a pin mill after deposition of slurry coating with WC and filler metal NiP11 by dipping. Pins brazed in a vacuum furnace at 1020°C.

phosphorus filler metal NiP11 in a matrix of organic binder. One hundred pins are inserted into a cradle and coated simultaneously by dipping into the slurry. After dipping, the adhering slurry is dried and the pins are brazed at 1020°C in a vacuum furnace. This application demonstrates a very cost-effective solution for obtaining coatings of high hardness.

15.3.6 Spray application of suspensions

Filler metals and carbides in an organic binder can also be deposited onto the surface of wear parts by spray processes. Suitable slurries with homogeneous and finely dispersed particles with a narrow grain size distribution are also called suspensions. These suspensions for spray application are modified to have low viscosity under high shear rates but at the same time a high viscosity and yield point at low shear rates. Such rheological behaviour is called structural viscosity. It allows the slurries to be deposited in a spray process with high shear rates using a conventional spray gun and at ambient temperature (Krappitz, 2006). After spray deposition the suspension is no longer sheared and the viscosity increases, which prevents the suspension from running off.

This technology produces coatings of 0.05 to 0.3 mm thickness. Thin layers of approximately 0.1 mm final thickness can be sprayed using this method in a single pass, while thicker coatings can be produced from multiple layers with intermediate drying under hot air of each sprayed layer.

The spray process is very similar to a paint spray process with regard to equipment and methods of application and thus offers similar possibilities. This process allows coating of complex geometries like grinding tracks, slotted screens or, using special spray equipment, the inner surfaces of tubes and elbows. At the same time it is possible to produce a very smooth layer of controlled thickness using automated and computer-controlled spray processes.

After deposition of the suspension layer and after drying, the coated parts are brazed under protective atmosphere or in a vacuum furnace. The resulting layers show very low porosity (<1%) and high carbide concentration. High contents of hard phases, well above 60% by volume, lead to high hardness values of approximately 65 HRC. Investigations of wear behaviour under standardised conditions, as well as parts under service conditions, revealed notably increased wear resistance compared with nitrided, borided or thermally sprayed coatings.

Figure 15.19 shows a segment of a twin screw extruder, coated with chromium carbide in a corrosion-resistant matrix of NiCr14P10 filler metal (BNi-7). This machine is used to process ceramic bodies and the lifetime of this segment could be increased very significantly. The example demonstrates that it is possible, using the suspension spray process, to coat even very complex geometries with a smooth layer of constant thickness.

Another specific advantage of a spray coated and brazed layer is shown in Fig. 15.20. The slotted screen is used to mechanically separate the liquid phase from the solid phase of sludge. The screen bars form a flat surface or are arranged to create a basket with narrow gaps. In the separation process, the sludge is passed

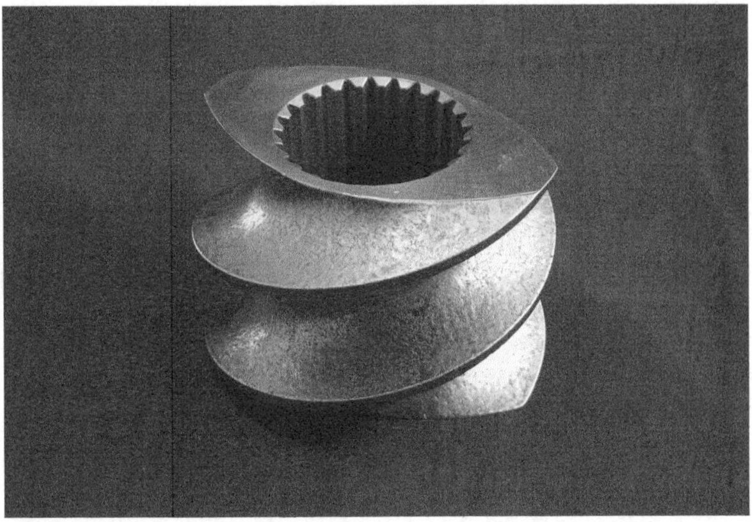

15.19 Segment of a twin screw extruder, coated with chromium carbide in a corrosion resistant filler metal matrix of NiCr14P10 and brazed in a vacuum furnace at 1020°C.

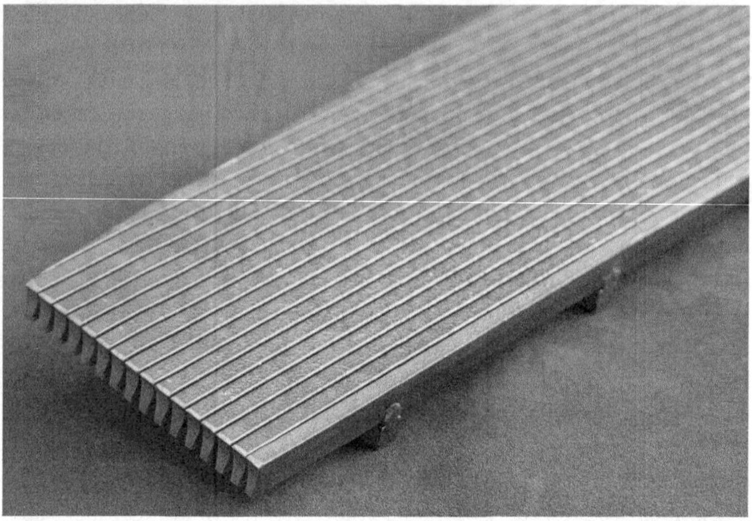

15.20 Segment of a slotted screen, coated with tungsten carbide using a suspension spray process.

over the surface of the sieve under compressive load, which may cause considerable wear of these components. A wear-protective coating is formed applying a smooth layer of hard material to the working surface of the screen bar, without bridging the narrow gaps of the screen. This can be achieved by spray coating with a carbide and filler metal suspension. A very uniform layer of carbide-containing material can be deposited by spraying, leaving the narrow gaps open after brazing. Other possible means of applying carbide coatings are very limited.

Sprayed coatings of higher thickness, up to 2.5 mm, can be obtained with slurries of higher viscosity, as described by Lugbauer (2009). Wear plates produced by this method show a smooth and homogeneous surface and can be processed by cold forming after brazing.

Successful investigations have demonstrated that suspension coatings can also be brazed using induction (Kortenbruck, 1998) as well as high-density infrared heating (Krappitz *et al.*, 2001). These methods can apply coatings without using a furnace. Large parts can be coated selectively, without heating the whole component. Up to now, however, there has been very limited application of these heating sources.

15.4 Functional coatings

Besides wear and corrosion protection, other requirements may be met by functional coatings. High friction surfaces improve safety by preventing slips and are useful for improving the grip of handling equipment. For such applications, surfaces can be applied as shown in Figure 15.9. Sharp-edged particles of crushed metals or carbides are brazed to the surface with copper- or nickel-based filler metals. Applications include emergency brake pads in elevators or grippers in the paper industry.

Unlike high-friction surfaces, coatings with embedded solid lubricants are capable of optimising the tribological behaviour of bearing surfaces and contributing to low friction and wear. Self-lubricating surfaces can be produced, using polymer bond mats with graphite inclusions. Figure 15.21 shows a microsection of a copper–tin layer with embedded graphite particles (dark phase), added to improve emergency running characteristics.

Nuclear fusion research requires an ultra-high vacuum (UHV) of less than 10^{-7} mbar. Residual deuterium–tritium and helium exhaust reaction products from the plasma have to be pumped inside the reactor. In order to generate such an extreme vacuum, materials with extremely high specific surface area, like zeolites or activated charcoal, are used to absorb gas molecules and thereby bring the pressure down to UHV conditions. For the fusion experiment, a cryosorption panel had to be produced which would have an extremely high specific surface area and at the same time an optimised heat flow, in order to cool the surface down to the cryogenic operating temperature. This was solved using pellets of activated charcoal, brazed to the surface of a copper plate with a eutectic silver–copper–titanium filler metal. In the experiment, this solution showed superior properties in comparison with high-temperature ceramic glues. Figure 15.22(a) shows the panel as brazed, while Fig. 15.22(b) gives a magnified view.

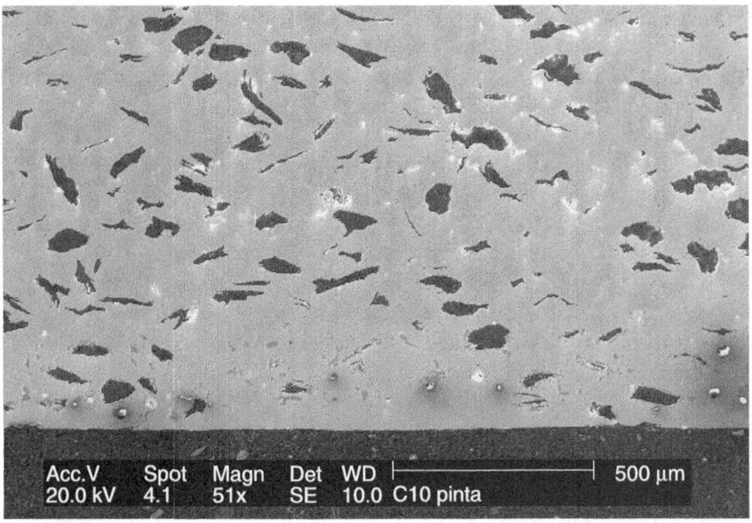

15.21 Microsection of a copper–tin layer with embedded graphite particles.

(a) (b)

15.22 Cryosorption panel, using activated charcoal vacum brazed at 950°C to the surface of a copper plate with filler metal AgCu27Ti3. (a) As-brazed, (b) magnified view.

15.5 Conclusion

The above application examples prove that brazing offers complementary possibilities to other well-established coating processes. The specific possibilities of coating by brazing create a niche where this coating technique can offer superior solutions.

15.6 References

Bronny, M. and Schimpfermann, M. (2011), 'Hartlöten von HW-Schneidwerkzeugen', Umicore AG & Co. KG; Hanau, available from: http://www.technicalmaterials.umicore.com/de/bt/BT_HartloetenHW.pdf [accessed August 2011].

Budinger D., Ferrigno S. and Murphy W. (General Electric Company) (1993), 'Alloy powder mixture for brazing of superalloy articles', United States Patent 5240491, 31 August 1993.

Gill, C (2006), 'Tungsten carbide for extruder wear parts', *Feed International*, February, Watt Publishing, Rockford, IL, USA.

Innobraze (2012), 'BRAZECOAT® – the convincing solution against wear', available from: http://www.innobraze.com [accessed January 2012].

ISO 17627 (2010), *Brazing — Filler metals*, Beuth-Verlag, Berlin.

Kortenbruck, G. (1998), *Einsatzmöglichkeiten der Löttechnologie zur Herstellung verschleißfester Funktionsschichten*, Thesis, Aachen University of Technology, Shaker Verlag, Aachen, Germany.

Koschlig, M. and Krappitz, H. (1994) *The BrazeCoat process, wear resistant coatings close to final contour built up by using organic-binder-bonded carbide and brazing preforms*, Degussa AG, Hanau, Germany; available from Innobraze GmbH, Esslingen, Germany.

Krappitz, H. (2006), 'Build-up Brazed Wear-protection Coatings', in Bach, F., Laarman, A., Möhwald, K. and Wenz, T., *Modern Surface Technology*, Wiley-VCH Verlag, Weinheim, Germany, 239–252.

Krappitz, H., Smith, R. and Blue, C. (2001), 'Braze infiltrated wear coatings from powder suspensions', *Proceedings of the 6th international conference Brazing, High Temperature Brazing and Diffusion Welding*, Aachen, Germany, May 2001, DVS-Verlag.

Lugbauer, M. (2009), 'Creating a new technology', Paper presented in Herzogenburg, Austria, May 2009, available from: http://www.eriknetwork.net/erikaction/presentations/Day2-Busatis-Innovation-Assistant.pdf [accessed August 2011].

Lugscheider, E., Schmoor, H. and Eritt, U. (1995), 'Optimization of repair-brazing processes for gas turbine blades', *Brazing, High Temperature Brazing and Diffusion Welding, conference proceedings*, Aachen, Germany, DVS-Verlag, 259–262.

Pratt & Whitney (2010), Specification PWA 1185, 'Cobalt Alloy Filler Metal'.

Schwarz, M. (2003), *Brazing*, 2nd ed., ASM International, Materials Park, Ohio, USA, p. 229.

Steinhaus, H. (1999), *Mathematical Snapshots*, 3rd ed., New York: Dover Publications, pp. 202–203.

Sulzer (2011), *Braze Materials Guide*, Sulzer Metco, Wohlen, Switzerland, April 2011.

Tank, K. (De Beers Industrial Diamond Division) (1997), *Method of bonding two bodies together by brazing*. European Patent EP 0636445 B1, 1 October 1997.

Wayne, C. and Budinger, D. (General Electric Company) (2002), *Turbine engine component having wear coating and method for coating a turbine engine component*, United States Patent 6451454, 17 September 2002.

WEG Wirtschaftsverband Erdöl- und Erdgasgewinnung e.V (2008), *Erdgas Erdöl – Entstehung, Suche, Förderung*, Wirtschaftsverband Erdöl- und Erdgasgewinnung e.V, p. 17.

16

Metal–nonmetal brazing for electrical, packaging and structural applications

C. A. WALKER, Sandia National Laboratories, USA

DOI: 10.1533/9780857096500.3.498

Abstract: Metal–nonmetal brazing is an established joining method used to fabricate products such as hermetic electronic packages, insulators for power generation and turbo-machinery components. Brazing presents opportunities for the materials engineer seeking to utilize recently engineered materials in advanced applications and extreme environments. Three commonly used brazing methods used for joining metals to nonmetals will be discussed: conventional brazing methods that use metallization coatings on the nonmetal surface to be brazed; active brazing methods that eliminate the need for metallization coatings; and direct brazing methods utilizing conventional brazing filler metals to join and seal packages without prior metallization.

Key words: metal–nonmetal brazing, active brazing alloys, brazed tensile strength, metal–ceramic brazing, niobium–sapphire brazing.

16.1 Introduction

Compared with other permanent joining methods, the ability to bond vastly different materials effectively is perhaps the greatest strength of brazing technology. Specialized welding processes have been developed that allow joining of different metal types, such as ferrous metals to non ferrous metals, or nickel-based superalloys to copper alloys; however, brazing methods can be used to join not only different metal families, but also metals to nonmetals. Only brazing has demonstrated the ability to routinely, reliably and permanently join metals to nonmetals for commercial applications such as high-voltage insulators, optical sensors and vacuum breaks, etc. While soldering methods can be used to join dissimilar materials, they are limited to low-temperature environments and are not generally recognized as creating high-strength permanent joints for structural applications. Although metal–nonmetal brazing technology is a time-tested and proven joining method used for fabrication of high-voltage, high-current devices and insulators for the electrical power industry, it can offer a host of opportunities for today's materials engineer or designer seeking to utilize the unique material properties of recently engineered materials for advanced applications. For example, highly desirable properties of sapphire and oxide ceramics, such as chemical resistance and high-voltage insulation, can be effectively married to readily machined metals or metal alloys. Oxide and nitride ceramics can be joined to heat-resistant metal alloys for use in hot exhaust gas environments such as turbo-machinery applications. Metallization of the nonmetal surface to promote proper braze filler metal wetting and flow, a costly and time-consuming requirement, can be eliminated in many cases when modern brazing methods are utilized.

16.1.1 Aim of the chapter

The intent of this metal–nonmetal brazing chapter is to familiarize designers and engineers with the concepts necessary to properly design and specify brazements for metal–nonmetal brazed assemblies from the perspective of the applications. This focus includes a discussion of the three most commonly used methods of joining metals to nonmetals: brazing methods that use metallization coatings on the nonmetal surface to be joined; brazing methods that utilize so-called 'active' metal additions to brazing filler metals, which allow the direct brazing of metals to nonmetals; and specific metal–nonmetal systems that allow the use of conventional brazing filler metals to join and hermetically seal packages and high-reliability devices. Additionally, test sample geometries, typical thermal profiles and expected tensile strengths for the various brazing methods will be detailed.

16.2 Designing and specifying a brazement

Prior to attempting a metal to ceramic braze, take a good, long look at Fig. 16.1, a variant of a figure reported by Nascimento *et al.* (2003). The post-joining stresses displayed in Fig. 16.1 show the simplified direction of the resultant stresses in a two-dimensional depiction of a metal–ceramic brazement, where the metal substrate has a higher coefficient of thermal expansion than the ceramic material ($\alpha_M > \alpha_C$). In limited cases, such as when the refractory materials molybdenum, tungsten and other selected alloys are brazed to alumina ceramics, $\alpha_M < \alpha_C$ and the resultant stresses will be in the opposite directions. When joining dissimilar materials, especially metals to nonmetals, it is vital to understand that residual stresses will most likely remain in the joined assembly. It is even more important to realize that, for any given set of materials, the higher the joining temperature and the larger the brazement the higher the resultant residual stresses. Many failed metal–ceramic brazes occur when this important rule is overlooked or forgotten.

16.2.1 Brazement functional requirements

For all brazed assemblies the concept of looking at the entire package is used. This process includes: analysis of the joint function; selection of a proper materials set; design and configuration of the braze joint; and requirements for braze fixturing. This 'package concept', first documented by Budde (1963), has undergone continuous improvements and upgrades as metal–ceramic joining technology improvements have been realized, and will be considered initially to ensure that the brazing engineer or designer is made aware of several of the many factors that must be considered when designing a brazed component or assembly.

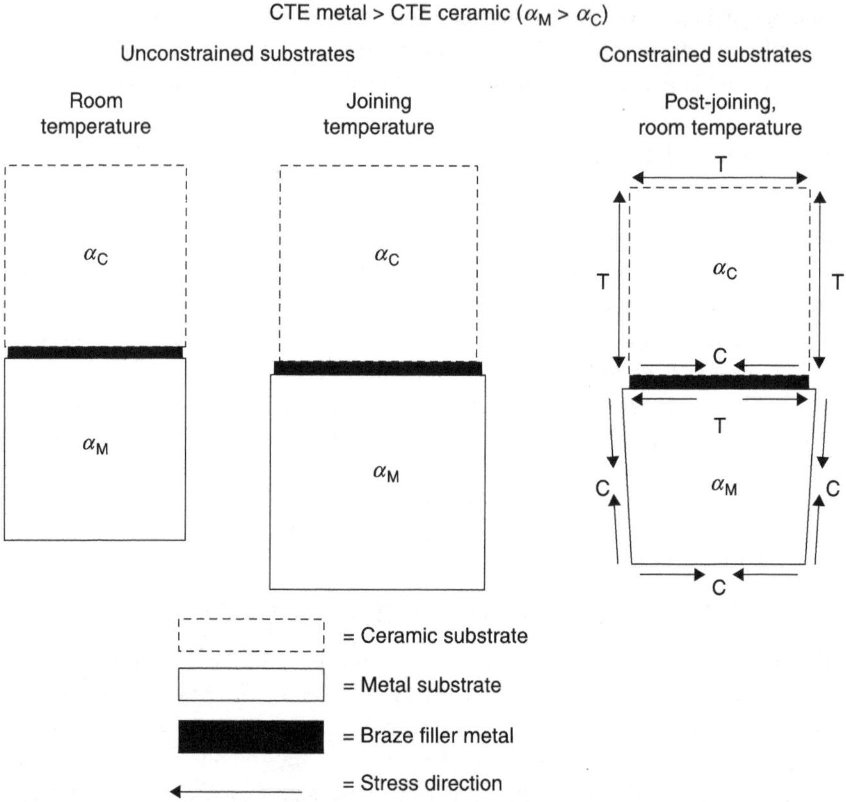

16.1 Post-joining stresses, showing the simplified direction of the resultant stresses in a two-dimensional depiction of a metal–ceramic brazement, where the metal substrate has a higher coefficient of thermal expansion than the ceramic material ($\alpha_M > \alpha_C$). T, tension; C, compression.

When analyzing the joint function, many factors, including service temperature, repetitive thermal cycling, mechanical loading or pressurization, high-temperature oxidizing or corrosive environments, electrical requirements and subsequent joining requirements, must be considered. Additional brazing package design resources are Chapter 2 of the *AWS Handbook*, 5th Edition (AWS, 2007) and Chapter 1 of *Principles of Brazing*, by Jacobson and Humpston (2005).

16.2.2 Brazement materials selection

During the second phase, materials selection, candidate materials can be eliminated from further consideration after comparing material properties data with requirements previously listed in the analysis of the joint function. Displayed in Table 16.1 are various material properties for selected metals and nonmetals

Table 16.1 Physical and mechanical properties of select metals and nonmetals

Material	CTE, linear (μm/m-°C), (20–1000°C)	Yield strength (MPa)	Flexural strength (MPa)	*Elongation or fracture toughness
Metals				
SS316/SS304	19.8	210	–	58%
Kovar	11.3	345	–	30%
Niobium	8.5	207	–	30%
Molybdenum	6.5	415	–	2–17%
Tungsten	5.2	750	–	1–4%
Nonmetals				
94–96% Al_2O_3	7.2–8.5		275–350	4.0–5.0 MPa-$m^{0.5}$
98–99% Al_2O_3	8.2–9.8		330–400	4.0–5.4 MPa-$m^{0.5}$
Sapphire[†]	8.3–9.0		480–895	6.0 MPa-$m^{0.5}$
Silicon nitride	3.1–3.3		650–900	2.2–8.5 MPa-$m^{0.5}$
Silicon carbide	4.3–4.6		400–440	4.6 MPa-$m^{0.5}$

Notes: Properties shown for metals represent annealed conditions. CTE values are averages for the temperature range 0–1000°C.

* Elongation at break percentage is given for metals, fracture toughness for nonmetals in K_{Ic}.

[†] Sapphire CTE varies with crystallographic orientation.

used typically to produce brazed metal–nonmetal structures. During the materials selection phase two key issues arise that must be addressed at the outset: (1) differences in the coefficients of thermal expansion (CTEs) of metals and nonmetals, and (2) poor wettability of the nonmetal surface by the liquid braze filler metal. Techniques used to remedy the inability of brazing filler metals to wet ceramics and other nonmetals adequately will be addressed when selecting a metal–ceramic brazing method later in the chapter. In the following sections, steps to assist the brazing designer through the materials selection decision-making process will be reviewed in greater detail. A typical metal–nonmetal brazed component will be used to aid in the discussion and materials selection process. For this example, the designer has been given the task of fabricating an optically transparent window assembly for use in high-vacuum vessels. The brazed assembly will be joined to a flange in a subsequent process, either by welding or by using threaded fasteners. To aid in keeping the materials selection process simple, service conditions and operating environments of the devices will be considered to have minimal effects.

16.2.3 Metal substrate selection

The differences in CTE or CTE mismatch must take into account the residual stresses that the brazed assembly will encounter during the brazing temperature cycle and any environmental extremes that the brazed assembly will encounter in service (Rosebury, 1965). Because nonmetals generally have strengths and fracture toughness values far lower than their metal counterparts, two methods can be employed to create acceptable low to moderate residual compressive stresses in the ceramic members while avoiding the creation of undesirable residual tensile stresses: utilizing metals or metal alloys with similar CTEs to the ceramic materials; or incorporating a braze joint design that leaves the nonmetal member of the brazement in a state of compression. For most brazements, controlled expansion alloys in the iron–nickel–cobalt or iron–nickel families will be required. There are instances when the joint function will require refractory metals such as molybdenum, tungsten and niobium. As braze joint sizes increase, the CTE mismatch between the ceramic and metal member must decrease to ensure sound brazements. Selection of the metal component to be brazed requires diligence to ensure that the joint function is not compromised. In choosing a metal substrate for our example brazed window assembly for vacuum service, the designer will want to select a low expansion metal such as niobium, molybdenum or Kovar. As can be seen in Table 16.1, all of these metals have coefficients of thermal expansion (CTE) similar to alumina-based ceramics or sapphire. If, however, the designer chooses to weld the metal substrate to typical stainless steel vacuum flange material, Kovar would be a logical choice. Although the austenitic stainless steels such as SS304 and SS316 have substantially higher CTE values than the other materials shown in Table 16.1, they may be utilized when the windows to be brazed are small. Residual stresses from CTE mismatch in brazed assemblies are directly proportional to substrate size.

16.2.4 Nonmetal substrate selection

Aluminum oxide ceramic, or alumina, is the preferred nonmetal choice for many structural ceramic components and high-voltage insulators. Alumina ceramic is relatively inexpensive, readily obtained and has metallization schemes compatible with the user-friendly 94% to 97% compositions that can be commercially obtained. Higher-purity alumina compositions (>99%) are more difficult to metallize and are not as easily brazed using active filler metal brazing processes as are the lower-purity compositions of alumina. When more than one material alternative is acceptable, consider availability, ease of use and cost prior to making a final decision. For the simple brazed vacuum viewport under consideration the material of choice from those listed in Table 16.1 would be sapphire, as optical clarity is a requirement. Many transparent high-temperature materials are suitable for vacuum vessel viewports depending upon individual end-user requirements such as transmission within the visible, infra-red (IR) or ultra-violet (UV) light spectrum. Commercial high-vacuum equipment suppliers include viewport material selection guides in their literature, which will greatly assist in final viewport material selection. It is the responsibility of the brazing engineer to determine the most practical method for brazing the viewport material to the metal housing.

16.2.5 Brazing filler metal selection

The third material to be considered is the brazing filler metal. Functional requirements will drive the selection of which filler metal family is selected. High-stress, low-ductility filler metals are to be avoided. Successful candidate filler metals are generally copper, silver or gold based. Active braze filler metals that promote wetting of nonmetal materials through the formation of metal oxides, nitrides or carbides are available for processing in temperature ranges similar to the conventional braze filler metals. Helpful guides and literature to assist in filler metal selection can be found on commercial brazing filler metal manufacturers' websites, such as Wesgo Metals (2009) and Lucas-Milhaupt (2011). In addition to selection guidance, many manufacturers provide mechanical and physical properties for the desired brazing filler metal alloys. For the Kovar body-sapphire window assembly under consideration, the designer has many options for a suitable brazing filler metal. To simplify the process and eliminate the need to metallize the sapphire window, an active braze filler metal is selected for this application. Because filler metals having lower tensile strengths and greater elongation are preferable to harder, low-ductility filler metals when making metal–nonmetal brazements, a high silver content filler metal such as 97Ag–1Cu–2Zr or 98Ag–2Zr will be utilized. As experience is gained, the brazing engineer will discover that several brazing filler metals can work equally well for a given application. Because of this, the ultimate choice of filler metal might be due to availability, cost or other factors such as compatibility with the desired brazing method and equipment requirements.

16.2.6 Braze joint design

An important piece of the design package is the brazement design or configuration. A seemingly innocuous change to a braze joint design can have a cascading effect. For example: a designer makes the decision to use a pin and sleeve-type joint versus a lap-shear joint for an electrical feed-through. This one seemingly small decision effectively prohibits the use of an active braze filler metal. Additional resources are now required to apply a proper metallization scheme to the inner diameter of the ceramic component, which in turn uses additional resources. Figure 16.2 graphically depicts the additional process requirements necessary when standard metallization schemes are employed. As there are many factors to be considered in finalizing the brazement joint design, the braze joint designer is encouraged to take advantage of resources and design guides available to assist in this process, such as the *AWS Brazing Handbook* (2007). Some tips to remember:

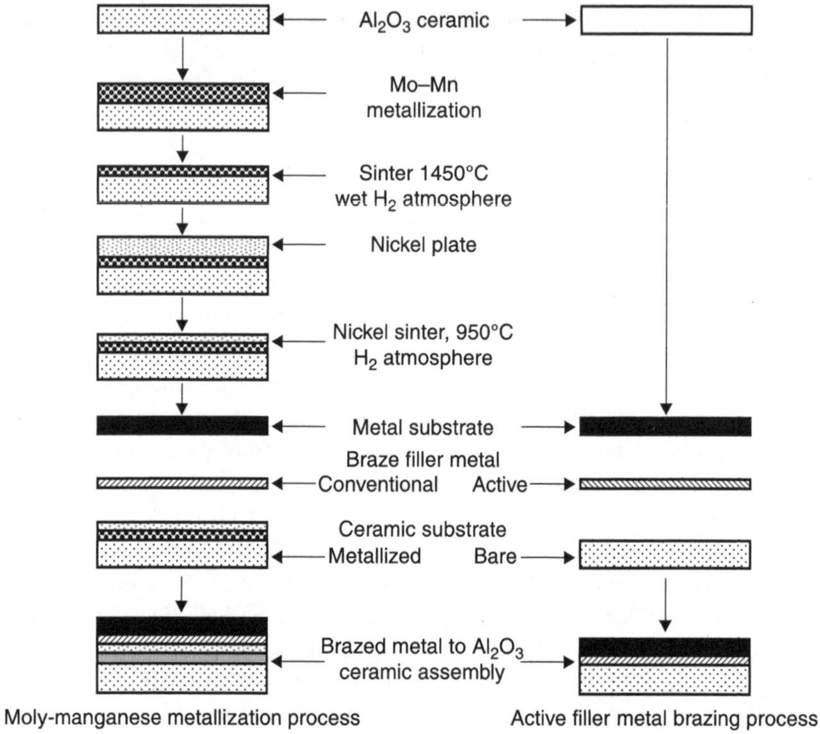

16.2 The additional process requirements necessary when standard metallization schemes are employed.

a properly designed braze joint will maintain a proper braze joint clearance at the brazing temperature; excessively long braze joints, those exceeding 2.5 mm, should be avoided as they require too much filler metal to be adequately filled; applying too much brazing filler metal leads to oversized braze fillets and high levels of undesirable stress. For large assemblies, it may be necessary to create a graded CTE joint by joining metals with progressively lower CTEs in order to obtain a joint with the least amount of stress. An example would be brazing a SS316 substrate to a niobium substrate, followed by brazing to a sapphire substrate. Reductions in joint geometry give corresponding reductions in the CTE mismatch and the resulting residual stresses within the brazement. Likewise, selecting the lowest-melting brazing filler metal that meets the design and service requirements will usually result in lower residual stresses in the brazed assembly. For the sample active-brazed Kovar-sapphire viewport under consideration, a simple lap-shear joint where the round sapphire window is placed on a counter-bored Kovar surface will provide a stable, minimal stress assembly. As the window diameter approaches sizes greater than 75 mm, the designer might want to use the graded approach mentioned previously to minimize residual stresses in the viewport assembly.

16.2.7 Brazement fixturing

The final concept to be considered when designing and specifying a braze joint is the braze fixturing. Ideally, metal–ceramic assemblies are self-fixturing: each component maintains the components and filler metal preforms it touches in proper alignment with one another. In reality it is necessary to use fixtures that are non wettable by the liquid braze filler metal to maintain the subassembly components in their proper positions throughout the furnace-brazing thermal cycle. To accomplish this, the thermal expansion of the brazing fixture must be similar to that of the brazing components. Brazing fixtures should have low thermal mass and be compatible with the selected brazing process. High scrap rates, low product yields and inferior products are oftentimes the consequences of too little time spent on brazing fixturing details. Schwartz (2003) emphasizes the relationship between proper fixture design and the efficiency or productivity of the brazing system. An attractive feature of the lap-shear joint design chosen for the sample active-brazed Kovar-sapphire viewport is that the required fixturing is minimal. A brazing preform stamped or cut approximately the same size as the faying surfaces is placed between the Kovar body and the sapphire window. The only additional fixturing requirement is to preload the window assembly. Calculate the braze joint area and then apply a force of 30–50 g/cm^2 centrally over the sapphire window. If brazing multiple assemblies at one time, maintain adequate spacing between assemblies to ensure all of the assemblies can be uniformly heated.

16.2.8 Brazement design and specification shortcut

For those already familiar with metal–ceramic brazing and brazing processes, the following set of substrate and brazing filler metal materials is suggested as a shortcut through the brazed component specification and design process: 94% alumina for the ceramic material; ASTM-F15 (UNS: K94610) iron–nickel–cobalt sealing alloy, commonly referred to as Kovar, for the metal substrate; silver-based titanium-containing material, such as Cusil-ABA, for the brazing filler metal. Experienced brazed assembly designers or fabricators will recognize that many metal–ceramic assemblies are currently fabricated using these three materials that fully comply with all of the joint functional requirements with no prior metallization of the ceramic substrate. If a joint design specific to active brazing filler metals and processes cannot be utilized in the component, then metallization must be used and a conventional filler metal substituted for the active braze filler metal. For any joint functional requirements not met by this shortcut material set, the experienced designer can replace or modify the non-complying substrate as needed.

16.3 Metallization schemes

After choosing the material set, the brazing method, type of metallization and method of application can be selected. The metallization to be used is determined by many factors, including cost, availability, joint geometry, brazing temperature and service temperature. As referred to previously, the nonmetal surface, typically alumina ceramic, will comprise aluminum oxides and various glass oxides, although with certain nonmetals there may be nitrides or carbides present. The low surface energies of oxides and other nonmetals prohibit the wetting and spreading of conventional brazing filler metal alloy (Kapoor and Eagan, 1988). The metallization process mechanically and chemically bonds a metal coating or scheme of coatings to the low surface energy substrate, resulting in a highly wettable high surface energy metallic surface. Although there are many methods (Nascimento *et al.*, 2003) by which this metallization is applied, two will be considered at this time: molybdenum–manganese metallization followed by nickel plating (Mo–Mn/Ni) and thin-film deposition.

16.3.1 Molybdenum–manganese metallization

The Mo–Mn/Ni process has been around for many decades and is a historically proven, capable method of metallizing oxide ceramics such as alumina (Phillips, 1969). In addition to the process details depicted in Fig. 16.2, steps to ensure thickness uniformity and dimensional control must be implemented as part of the process development phase. The Mo–Mn coating can be applied using a brush or with automated equipment such as screen-printers or robotically controlled sprayers. No matter which application method is employed, it is critical that the metallization slurry mixture be stirred constantly to prevent segregation of the components, which leads to non-uniform thicknesses. After drying in air or under heat lamps, the metallized component is placed in a controlled atmosphere furnace and fired at a temperature of approximately 1500°C for 30 to 45 min in an atmosphere of wet hydrogen (dew point ~ 30°C) or a combination of wet hydrogen and nitrogen and slowly cooled. Upon cooling, the metallization coatings are inspected for adequate and uniform thickness. Upon passing inspection, the parts are prepared for nickel plating. The component is prepared for the plating bath and a soft nickel coating is plated over the previously metallized surface. After the nickel plating steps are complete, the component is again inspected and prepared for the nickel densification process. To accomplish the nickel densification step, the component is again placed within the hydrogen furnace and fired in the same wet gas mixture as the Mo–Mn coating, but this time at a temperature of 900°C for 30 min. Upon cool-down the component is reinspected to ensure all the necessary areas are adequately coated and plated. A total metallization thickness of 0.025 mm is desirable, including a densified nickel plating thickness ranging from 0.0075 to 0.010 mm. Moly–manganese/nickel plating processes are attractive and cost effective in high volume quantities, but the additional processing requirements for these types of metallization coatings make them unattractive for small lot production runs and one-of-a-kind applications. An easily performed literature search will yield many valuable information resources for these types of coatings, as well as sources that provide commercial metallization services.

16.3.2 Thin-film metallization

Like Mo–Mn/Ni-plated metallization, thin-film metallization coatings also can be utilized to successfully increase the surface energy of nonmetal faying surfaces to levels where they behave like a metallic surface. Early references such as Reed and McRae (1965) describe the use of thin-film evaporation techniques to deposit metals onto ceramic surfaces similar to the electron-beam evaporation or sputtering methods used currently by the thin-film industry to deposit a two or three-layer scheme consisting of a reactive metal, followed by a noble metal (Walker and Hodges, 2008). For three-layer systems, a barrier layer is used to inhibit diffusion of the noble layer into the reactive layer. The reactive layer is typically one of the periodic table IVA elements, such as titanium, zirconium or hafnium. Readily obtainable and capable of forming the stable oxides necessary for sufficient chemical bonding, titanium is the most commonly used reactive layer. Reactive layer thicknesses of 250–500 nm are generally sufficient for successful brazing. If an intermediate layer is desired to limit or mitigate diffusion of the reactive layer into the molten filler metal, a thin layer of a refractory metal approximately 100–200 nm thick can be used. The top, exposed layer is generally a noble metal such as gold, palladium or platinum with a thickness ranging from 250 to 500 nm. It is critical when selecting a material system that applicable phase diagrams are consulted to identify brittle intermetallic phases that may form, potentially degrading joint strength, or hindering the formation of hermetic brazements if required. For example, replacing the top layer of gold in a thin-film metallization system with platinum will prevent the formation of Au–In intermetallic phases when the use of lower melting Ag–Cu–In filler metals is desired. An informative treatise on the use and application of phase diagrams for the brazing process engineer can be found in Jacobson and Humpston's book, *Principles of Brazing* (2005). Because thin films can be dissolved rapidly by the molten braze filler metal, proper characterization and analysis of the brazements is recommended during the development process.

16.4 Brazing method selection

The three most commonly used brazing methods for assembly of metal–nonmetal components are: conventional brazing, active filler metal brazing and direct brazing. While gaining in usage and popularity, the active brazing method still comes in a distant second in overall popularity as a brazing method for metal–nonmetal, and in particular metal–ceramic, brazing. Direct brazing remains the least popular method of the three metal–nonmetal brazing methods. The most popular metal–ceramic joining method, conventional brazing, will be discussed first. Considerations to assist the brazing designer in selecting the most appropriate brazing method will follow the brazing method descriptions.

16.4.1 Conventional brazing method

The standard or conventional brazing method is widely used today and will continue in this role in the future. Conventional brazes are performed using conventional or traditional filler metals, primarily from the silver-based family for metal–ceramic brazing. There is a great deal of historical data available to the brazing designer or process engineer (Rosebury, 1965; Kohl, 1967) with regard to filler metal selection and recommended thermal cycles to fabricate hermetic and strong metal–ceramic brazements. Also, conventional filler metals are less expensive than similar active brazing filler metals. Performing the braze is trivial as long as the metallization coatings remain uniform and are repeatable, run to run, lot to lot, and so on. The downside to conventional brazes is that a metallization coating is required in order to modify the nonmetal faying surface so that it can be adequately wetted by the liquid filler metal. In some applications and geometries, the required metallization process is elaborate, expensive and time-consuming. The desire to expose metal–ceramic joints to greater service temperatures requires brazing at temperatures that rapidly dissolve metallization coatings into the liquid braze filler metal. In cases when Mo–Mn is used with nickel plating, the nickel layers completely dissolve into the filler metal, leaving a non hermetic, low-strength brazement. As discussed previously, metallization processes can be complicated, costly and in some cases difficult to obtain. For large quantities and single geometries, however, conventional brazing methods might be the most cost effective.

16.4.2 Active brazing method

Active brazing processes are surging in popularity as metal–ceramic brazing applications become more widely used. The ability to braze a metal to a nonmetal without the need for metallization coatings not only saves valuable product development funds, but can also shave days or weeks from the required time to get a product from the conceptual or design prototype stage to final production. As designers and engineers are becoming familiar with active braze joint design requirements, required brazement design features are incorporated in the initial phases. It is important for designers to understand that braze joints having the design features necessary to be successfully brazed using active brazing methods can also be brazed using conventional brazing methods following the proper application of the metallization coating. The opposite, however, is not true. In fact, many metal–ceramic braze joints designed for conventional brazing are unsuitable for brazing using active brazing methods. Active brazing processes require the use of a special filler metal alloy that contains a stable oxide, carbide or nitride former, such as titanium, zirconium or hafnium (Kapoor and Eagan, 1989). Although other elements such as chromium, vanadium and molybdenum have been successfully utilized as active filler metal alloy additions (Hosking *et al.*, 2000), group IVA elements are the most commercially successful. Fully alloyed active braze filler metal compositions are currently available and should be utilized whenever possible versus filler metals fabricated by cladding active filler metal layers to conventional braze filler metals. As reported by Vianco *et al.* (2003), only small quantities of the active element within the braze filler metal are necessary for hermetic, completely bonded brazements.

Once the functional requirements, set of materials, joint design and fixturing requirements have been satisfied, the active braze process may be undertaken. As with all fluxless furnace-brazing processes, the components should be cleaned thoroughly and handled using contaminant-free gloves or finger cots. The joint designs used for active brazing allow the filler metal to be preplaced in the correct position between the metal and nonmetal faying surfaces. Standard machine finish roughness values of 1.5 μm are adequate for the metal faying surfaces while values not exceeding 0.8 μm are recommended for the nonmetal surface to be brazed. Care should be taken to ensure that any fixturing used will not come into contact with the liquid active braze filler metal. While inert gases, high-vacuum and reducing atmospheres may all be utilized for active brazing processes, high-vacuum environments of less than 1.3×10^{-3} Pa are recommended for consistent, high-quality results.

16.4.3 Direct brazing method

Schematically, direct brazing processes are similar to active brazing processes: the brazing filler metal is preplaced between the metal and nonmetal faying surfaces of the assembly, heated until complete melting has occurred, and cooled. Both types of brazing are employed to manufacture hermetically sealed packages and high-voltage switches (Walker *et al.*, 2010). Direct brazing, however, accomplishes this feat, brazing directly to bare nonmetal substrates, without the use of specialized active filler metal brazing alloys. Differences in these joining methods are more readily understood when the bonding schemes and material interactions of active and conventional brazing filler metals are compared. As discussed previously for the active brazing process, a Group IVA element such as titanium alloyed in the active brazing filler metal migrates through the molten filler metal, combines with nonmetal surface oxides, nitrides or carbides and forms strong, stable chemical bonds. In the case of direct brazing, however, there are no reactive elements or stable oxide-forming elements present in the brazing filler metal. In fact, the braze filler metals commonly used in direct brazing applications are gold- or silver-based conventional brazing filler metals, materials with Gibbs free energy levels unable to form oxides at typical brazing temperatures (Coughlin, 1954). The key to direct brazing, then, is the interaction between the metal and nonmetal substrates, which occurs in the presence of the molten filler metal. Direct brazing is possible when a metal substrate has limited solubility in a molten filler metal and migrates to the nonmetal substrate, forming a relatively stable chemical bond. Direct brazing processes consistently produce hermetic assemblies; however, tensile strengths approximately 50% less than those observed in active or conventionally brazed metal–ceramic test specimens are observed (Stephens *et al.*, 2005). For material sets well matched with respect to CTE, such as niobium metal joined to alumina ceramic or single-crystal sapphire, direct brazing processes can be utilized and meet or exceed the joint functional requirements. Unlike active brazing, the surface finish of the nonmetal faying surface should be ≤ 0.2 μm for consistent joining. While not always practical to put into place, studies have shown that joint strength improvements may be obtained when braze joint loads are increased (Marks *et al.*, 2001). Because no metallization coatings are required and conventional braze filler metals may be used, direct brazing is the least expensive method used to join metals to nonmetals. For many applications, low brazement strengths and a limited material set to choose from eliminate direct brazing during the down-selection process. For specific material sets such as niobium and sapphire, however, direct brazing remains a versatile, though not well-known brazing method.

16.4.4 Selecting the appropriate brazing method

Selecting the proper brazing method for a specific metal–ceramic brazing assembly can be compared to selecting the proper hand tool to perform a specific task. Oftentimes, the toolbox contains several different tools that can perform a similar function; however, selecting the correct tool will make the task less difficult and yield better, more consistent results. Knowing the capabilities of the equipment available is a critical step in selecting a brazing method. Certain types of equipment, such as high-vacuum furnaces, can be utilized for conventional, active and direct brazing processes. Atmospheric furnaces that utilize inert gases or reducing environments may be used for conventional and direct brazing processes as long as contaminant gases such as oxygen and water vapor are maintained at low levels, typically in the single-digit parts-per-million (ppm) range. Reducing environments can be used for many applications but can ruin assemblies if the metal substrates (i.e. tantalum, titanium, titanium alloys) are susceptible to hydrogen embrittlement. Again using the toolbox analogy, not all tools exist in everyone's toolbox. In many cases, the designer must work within the constraints of the furnace-brazing equipment on hand to perform the given task, in which case certain options for the brazing process will not be available. A second consideration in selecting a brazing method is the brazement design. Conventional, active and direct brazing methods will all work acceptably well if the braze filler metal can be preplaced in the braze joint proper. If, however, a specific braze joint geometry requires the filler metal to be drawn or wicked into the brazement volume using capillary attraction, conventional brazing methods must be utilized. As mentioned previously, when the conventional brazing method is selected, the designer is also resigned to utilizing a metallization coating on the nonmetal substrate faying surface. The direct brazing method works quite well for limited material systems and filler metal schemes, but does require flat, smooth faying surfaces. Of the three methods discussed, direct brazing has the most limited application for most designers. Designers must also consider the cost associated with selecting a brazing method. Conventional brazing methods require less stringent surface finishes of the metal and nonmetal substrates, which results in lower substrate fabrication costs. The requirement for a metallization coating adds additional costs, which vary greatly depending on the method of application and quantities. Although the brazing filler metal costs are typically marginal compared with the costs of procuring materials, fabricating samples and fixturing, the designer should be aware that active brazing filler metals required for active brazing processes cost approximately five times more than their conventional counterparts, which will cut into the cost savings realized by omitting metallization. To summarize, the brazing method selected will vary depending upon equipment requirements, joint design and the costs associated with the nonmetal surface preparation. The informed design engineer, knowing upfront which design features drive specific brazing methods, can avoid the unfortunate experience of being forced to use a specific brazing process, especially if that method does not exist in their toolbox.

16.5 Performing the brazing operation

Brazing furnace cycles are very similar for all of the discussed metal–ceramic brazing methods. A typical brazing thermal profile begins with a ramp rate of 10–15°C per min to a temperature of 25–40°C below the filler metal solidus temperature and includes a uniform temperature soak for a period of minutes to allow the components to come into thermal equilibrium. The temperature is then increased to a point 20–30°C above the filler metal liquidus temperature and soaked for a period of one to several minutes. Standard cooling rates are 10–25°C per min. When adequately cooled, the assemblies may be removed from the furnace, again utilizing contaminant-free gloves or finger cots.

16.5.1 Braze processing tips

Required vacuum pump-down times will be significantly decreased if the vacuum furnace hot-zone temperature is approximately 20–30°C above ambient room conditions any time the furnace chamber is exposed to air. This is especially important when high humidity conditions are present. For active brazed assemblies where shallow braze filler metal wetting-angle fillets on the nonmetal substrates are desired or are beneficial to the brazement, longer duration soak periods at the peak temperature may be employed. While in most cases a peak soak duration of 2 to 5 min will suffice, Walker *et al.* (2010) have reported positive results when using peak soak times as great as 240 min. Uncontrolled furnace cooling can often be utilized without causing damage to small assemblies. Large assemblies, as well as those made from materials having excessive differences in coefficients of thermal expansion, will benefit from multiple extended soak periods at elevated temperatures during the cooling cycle or slower cooling rates to allow stress relaxation to occur. Switching from a high-vacuum atmosphere to a partial pressure atmosphere during the brazing thermal cycle is a method used to eliminate evaporation of high vapor pressure brazing filler materials throughout the furnace hot-zone and chamber. When utilizing this process it is recommended that the furnace be maintained at an elevated temperature, dependent upon the material set, until an adequate vacuum base-pressure is reached by the furnace chamber. After a reasonable period of time has elapsed, the atmosphere can be switched from high vacuum to an inert gas partial pressure of 100–250 Pa for the duration of the thermal cycle. For example, a silver-based filler metal has a vapor pressure of 2.5×10^{-2} Pa at the brazing temperature of 850°C. When performing the brazing cycle, the furnace can be allowed to equilibrate and soak at a temperature of 600°C, where the vapor pressure of silver is 8.0×10^{-6} Pa. After a sufficient period, determined during the process development stage, the furnace can be switched to partial pressure and the brazing process completed with minimal silver evaporation.

16.6 Testing the brazements

Brazed test specimens should be fabricated using the selected set of substrate or component materials and brazing filler metals. When several candidate materials can be used, screening experiments are often performed as part of the final down-selection process. Tensile and/or shear test specimens can be used to duplicate the geometry and joint loading during service of the desired component. Certain brazing test samples, such as those shown in Fig. 16.3, and designated as ASTM F19 tensile buttons (ASTM, 2005), can be used for two separate tests: helium leak testing and tensile testing brazement. ASTM F19, *Standard test method for tension and vacuum testing metallized ceramic seals*, describes fully the geometry and dimensions of the tensile specimens, as well as detailed schematics for the gripping apparatus required for proper completion of the test procedure. For critical applications it is important to properly test and analyze brazements to fully qualify the brazing process selected, prior to proceeding into production of components. This point should be stressed to

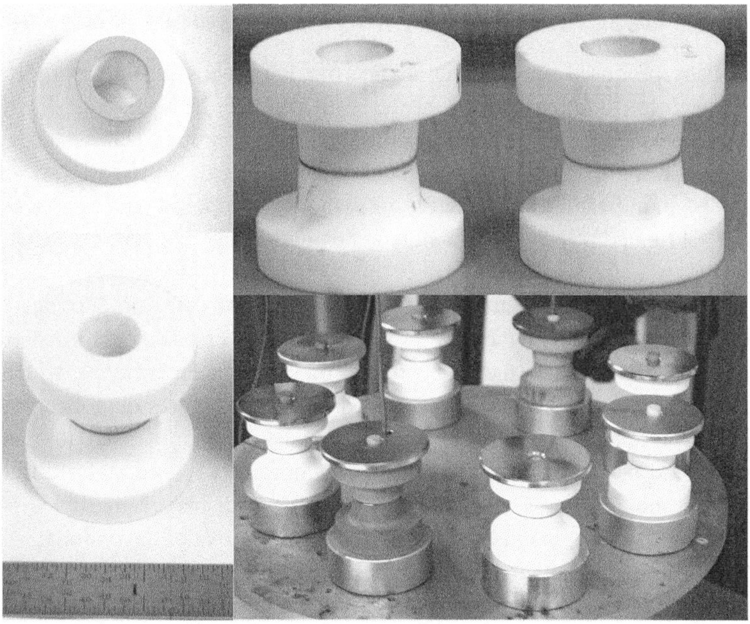

16.3 Brazing test samples that can be used for two separate tests: helium leak testing and tensile testing brazement.

all personnel involved when fabricating brazed assemblies consisting of nonmetal components susceptible to brittle fracture and failure mechanisms. Sound engineering judgment and due diligence during the testing and qualification phases are required, especially if the brazements have been previously identified as being either Class A or Class B, as defined by the American Welding Society (AWS, 2008c). As stated in AWS C3.6M/C3.6:2008 *Specification for furnace brazing*: 'Class A joints are those joints subject to high stresses, cyclic stresses, or both, the failure of which could result in significant risk to persons or property, or could result in a significant operational failure.' The specification defines Class B joints as 'those joints subjected to low or moderate stresses, cyclic stresses, or both, the failure of which could result in significant risk to persons or property, or in significant operational failure.' The product engineer is encouraged to become familiar with the AWS C3.2 *Standard method for evaluating the strength of brazed joints* (AWS, 2008a) and AWS C3.3 *Recommended practices for the design, manufacture and examination of critical brazed components* (AWS, 2008b), as well as AWS C3.6 *Specification for furnace brazing*. These specifications can be invaluable for determination of proper sampling plans and other quality assurance provisions.

16.7 Test results and analysis for select material sets

For any proposed set of materials and brazing processes, the testing results and sample cross-section analysis are used to validate the successful down-selected candidates. Brazed ceramic tensile button assemblies, fabricated using the proposed material set and manufactured according to ASTM F19 specifications, can be an indicator of expected results when the proposed packages or assemblies are fabricated. It should be noted, however, that the ceramic tensile button geometry, effectively a ceramic–metal–ceramic brazement, is a 'balanced' assembly that can at times show positive results that the product design cannot duplicate, especially when the proposed product or package consists of brazed joints made with thick metal substrates and large brazement areas, which result in large residual stresses within the brazed assembly. Test results for ceramic tensile button assemblies fabricated with materials commonly used to manufacture electronic packages or high-voltage insulators and brazed using different filler metals, metallizations and brazing methods are displayed in Tables 16.2–16.5. Table 16.2 shows results for the FeNiCo alloy, Kovar, conventionally brazed to 94% alumina ceramic using two types of metallization processes: molybdenum–manganese with nickel plating, and a sputtered thin-film of titanium and gold. As displayed in Table 16.2, gold- and silver-based conventional filler metals produce high tensile strength brazements that are hermetically sealed, within the detectable limits of the helium mass-spectrometer leak detection equipment used. A representative metallographic cross-section of a conventionally brazed tensile button, having a standard metallization coating (Mo–Mn/Ni plate), is displayed in Fig. 16.4. Desirable characteristics observed in conventionally brazed metal–ceramic cross-section images are discernible reaction layers between the Mo–Mn metallization and the ceramic substrates, clearly defined Ni plating layers, minimal erosion of the Kovar or other base metal materials and well-bonded void-free filler metal fillet areas. Figure 16.5 shows a thin-film metallized ceramic substrate that has been conventionally brazed to a Kovar substrate. Vastly differing reaction layer interfaces result from using these two metallization methods, as can be readily observed in Fig. 16.4 and 16.5. The Mo–Mn layer has reacted with the glassy phase of the ceramic and diffused well beyond the alumina substrate surface (Fig. 16.4), whereas the well-bonded titanium layer of the thin-film metallization remains on the surface of the ceramic substrate (Fig 16.5). Note in Fig. 16.5 that the protective gold thin-film layer has diffused into the braze joint fillet, and is no longer visible along the titanium thin-film layer between the braze filler metal and the ceramic substrate, consisting primarily of titanium as the protective gold layer has diffused into the braze joint fillet.

Test results for the same set of materials, 94% alumina ceramic and Kovar but joined using active brazing and direct methods, are displayed in Table 16.3. The average tensile strength values displayed in Tables 16.2 and 16.3 are very similar for the same materials joined using conventional and active brazing processes. Also

Table 16.2 Average tensile strengths of conventionally brazed metallized tensile buttons

Braze filler material	Metallization	Tensile button material	Metal interlayer material	Peak brazing temperature (°C)/time (minutes)	Furnace atmosphere	Av. tensile strength (MPa)
Mo–Mn metallization						
Au–62Cu–3Ni	Mo–Mn/Ni	94% Al_2O_3	Niobium	1059/1	UHV	90
Au–62Cu–3Ni	Mo–Mn/Ni	94% Al_2O_3	Niobium	1059/2	UHV	71
Au–62Cu–3Ni	Mo–Mn/Ni	94% Al_2O_3	Niobium	1059/5	UHV	57
50Au–50Cu	Mo–Mn/Ni	94% Al_2O_3	Kovar	1000/3	Hydrogen	118
35Au–65Cu	Mo–Mn/Ni	94% Al_2O_3	Kovar	1040/3	Hydrogen	100
72Ag–28Cu	Mo–Mn/Ni	94% Al_2O_3	Kovar	810/3	Hydrogen	99
Au–13Ag–10Ge	Mo–Mn/Ni	94% Al_2O_3	Kovar	495/3	Hydrogen	108
Au–13Ag–10Ge	Mo–Mn/Ni	94% Al_2O_3	Kovar	455/5	Hydrogen	111
Thin-film metallization						
50Au–50Cu	Ti–Au	94% Al_2O_3	Kovar	1000/3	Hydrogen	102
50Au–50Cu	Ti–Au	94% Al_2O_3	Kovar	1020/10	Hydrogen	89
50Au–50Cu	Ti–Pd–Au	94% Al_2O_3	Kovar	1000/3	Hydrogen	111
50Au–50Cu	Ti–Pd–Au	94% Al_2O_3	Kovar	1020/10	Hydrogen	81
72Ag–28Cu	Ti–Au	94% Al_2O_3	Kovar	810/3	Hydrogen	90
Ag–27Cu–10In	Ti–Pt	94% Al_2O_3	Kovar	755/2	Argon	91
Ag–27Cu–10In	Ti–Pt	Kovar	LTCC	755/2	Argon	57
Ag–27Cu–10In	Ti–Au	Kovar	LTCC	755/2	Argon	45
Ag–27Cu–10In	Ti–Pd	Kovar	LTCC	755/2	Argon	26

Notes: Results are for ASTM F19 tensile button assemblies. Braze filler metal compositions are shown in weight percents. UHV: vacuum atmospheres $\leq 1.3\times10^{-4}$ Pa at peak brazing temperature. LTCC: low temperature co-fired ceramic.

Table 16.3 Brazed tensile button strengths using active filler metal and direct brazing methods

Braze filler material	Tensile button material	Metal interlayer material	Peak brazing temperature (°C)/time (minutes)	Furnace atmosphere	Av. tensile strength (MPa)
Active brazed					
35Au–62Cu–2Ti–1Ni	94% Al_2O_3	Kovar	1006,1026/6,8	UHV/Argon	76, 97
35Au–62Cu–2Ti–1Ni	94% Al_2O_3	Niobium	1059/1–5	UHV	65, 66
97Ag–1Cu–2Zr	94% Al_2O_3	Kovar	990/5	UHV/ Hydrogen	106
97Ag–1Cu–2Zr	94% Al_2O_3	Kovar	965, 985, 3, 5	Argon	140, 170
97Ag–1Cu–2Zr	94% Al_2O_3	SS304L®	985/5	UHV	114
Cusil-ABA	94% Al_2O_3	Kovar	1040/2	Hydrogen	100
Cusil-ABA	94% Al_2O_3	Kovar	825–1040/2–10	Argon	97, 76
Cusil-ABA	94% Al_2O_3	Kovar	825–1040/2–10	UHV	110, 76
Cusil-ABA	94% Al_2O_3	Copper	855/5	UHV	84
Cusil-ABA	94% Al_2O_3	Niobium	855/5	UHV	115
Incusil-ABA	94% Al_2O_3	Kovar	755/5	UHV	99
Incusil-ABA	94% Al_2O_3	Niobium	755/5	UHV	80
Direct brazed					
50Au–50Cu	94% Al_2O_3	Niobium	1000/3	UHV	61
35Au–62Cu–3Ni	94% Al_2O_3	Niobium	1060/1, 3	UHV	61, 74
35Au–62Cu–3Ni	94% Al_2O_3	Niobium	1060/240	UHV	96
35Au–62Cu–3Ni	94% Al_2O_3	Niobium	1060/120	UHV	39
92Au–8Pd	94% Al_2O_3	Niobium	1270/4	UHV	88

Notes: Results are for ASTM F19 tensile button assemblies. Braze filler metal compositions are shown in weight percents. UHV: vacuum atmospheres $\leq 1.3\times10^{-4}$ Pa at peak brazing temperature. Cusil-ABA nominal composition: 63Ag–35.25Cu–1.75Ti; Incusil-ABA nominal composition: 59Ag–27.25Cu–12.5In–1.25Ti.

shown is the lower average strength obtained using the direct brazing method. Similar hermeticity test results were obtained as the higher tensile strength methods. Energy dispersive x-ray spectroscopy (EDS) can be used to show the migration of the active element through the molten braze filler metal. In Fig. 16.6, a cross-section of an active-brazed sample shows how most of the active element, Zr in this case, has moved through the joint fillet and reacted with the ceramic substrate. A benefit of using the active brazing method can be seen in Table 16.4, where the effects of an expanded brazing process window, time and temperature in this case, are detrimental to conventionally metallized and brazed assemblies, while the active brazed and thin-film metallized samples remain largely unaffected. This is due to the nickel plating being consumed by the liquid brazing filler metal, leaving a glassy Mo–Mn

Table 16.4 The effects of temperature, time and metallization type on average tensile strengths

Braze filler material	Metallization	Nonmetal substrate	Metal substrate	Peak brazing temperature (°C)/ time (minutes)	Furnace atmosphere	Av. tensile strength (MPa)
Thin-film metallizations						
50Au–50Cu	Ti/Au	94% Al_2O_3	Kovar	1000/3	Hydrogen	102
50Au–50Cu	Ti/Au	94% Al_2O_3	Kovar	1020/10	Hydrogen	89
50Au–50Cu	Ti/Pd/Au	94% Al_2O_3	Kovar	1000/3	Hydrogen	111
50Au–50Cu	Ti/Pd/Au	94% Al_2O_3	Kovar	1020/10	Hydrogen	81
Mo–Mn metallizations						
Au–62Cu–3Ni	Mo–Mn/Ni	94% Al_2O_3	Niobium	1059/1*	UHV	90
Au–62Cu–3Ni	Mo–Mn/Ni	94% Al_2O_3	Niobium	1059/5*	UHV	57
50Au–50Cu	Mo–Mn/Ni	94% Al_2O_3	Kovar	1028/23.5†	Hydrogen	92
50Au–50Cu	Mo–Mn/Ni	94% Al_2O_3	Kovar	1028/23.5‡	Hydrogen	119
Active brazed						
Cusil-ABA	n/a	94% Al_2O_3	Kovar	845/5	Hydrogen	99
Cusil-ABA	n/a	94% Al_2O_3	Kovar	1040/2	Hydrogen	100

Notes: Results are for ASTM F19 tensile button assemblies. Braze filler metal compositions are shown in weight percents. UHV: vacuum atmospheres $\leq 1.3\times10^{-4}$ Pa at peak brazing temperature. Cusil-ABA nominal composition: 63Ag–35.25Cu–1.75Ti.

* Nickel plating thickness 10.2 μm.
† Total time above liquidus, nickel plating thickness 5.1 μm.
‡ Total time above liquidus, nickel plating thickness 10.2 μm.

Table 16.5 Active and direct brazed niobium–alumina and niobium–sapphire tensile strengths

Braze filler material	Tensile button material	Metal interlayer material	Peak brazing temperature (°C)/time (minutes)	Furnace atmosphere	Av. tensile strength (MPa)
35Au–62Cu–3Ni	Niobium	Sapphire	1060/2	UHV	35
35Au–62Cu–2Ti–1Ni	Niobium	Sapphire	1060/2	UHV	22
97Ag–1Cu–2Zr	Niobium	Sapphire	985/5	UHV	38
35Au–65Cu	Niobium	Sapphire	1060/240	UHV	6
Cusil-ABA	94% Al_2O_3	Niobium	855/5	UHV	102
35Au–65Cu	94% Al_2O_3	Niobium	1060/2	UHV	55
35Au–65Cu	94% Al_2O_3	Niobium	1060/240	UHV	96
35Au–65Cu	94% Al_2O_3	Niobium	1260/120	UHV	39

Notes: Results are for ASTM F19 tensile button assemblies. Braze filler metal compositions are shown in weight percents. UHV: vacuum atmospheres $\leq 1.3 \times 10^{-4}$ Pa at peak brazing temperature. Cusil-ABA nominal composition: 63Ag–35.25Cu–1.75Ti.

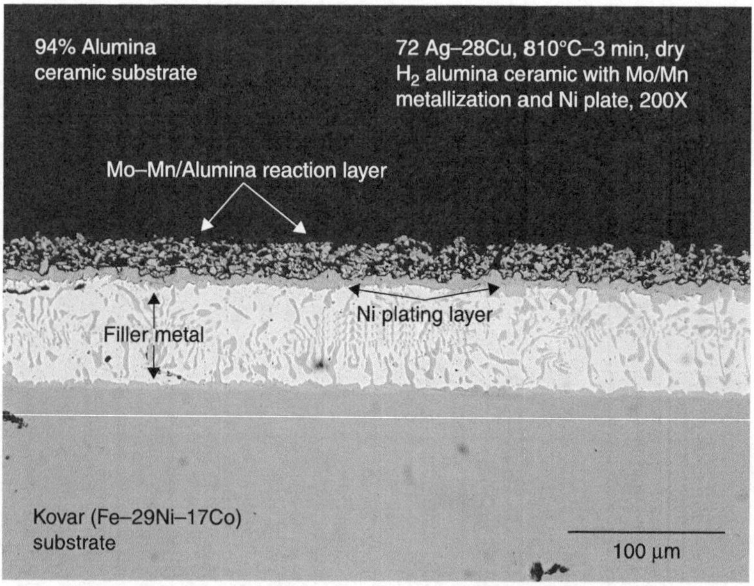

16.4 Representative metallographic cross-section of a conventionally brazed tensile button, having a standard metallization coating (Mo–Mn/Ni plate).

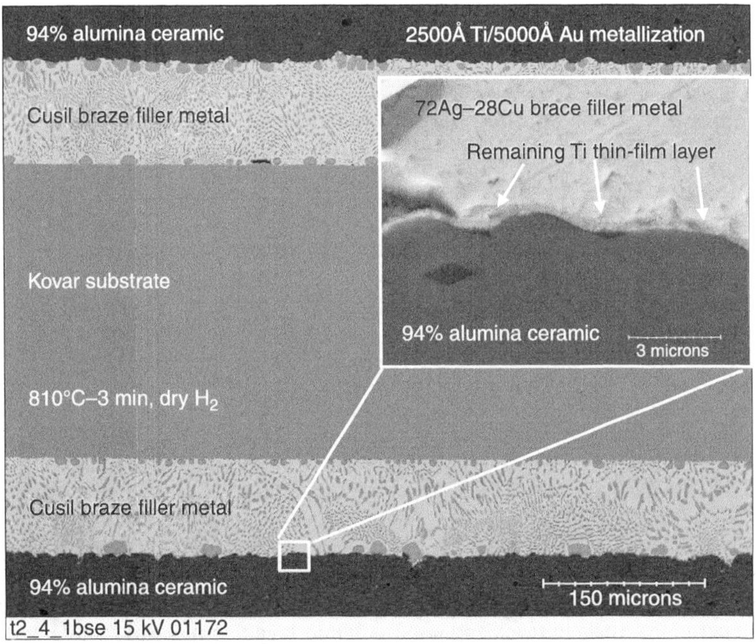

16.5 Thin-film metallized ceramic substrate that has been conventionally brazed to a Kovar substrate.

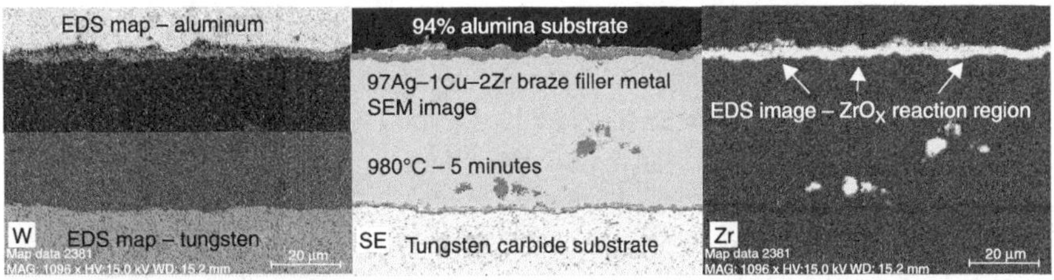

16.6 Cross-section of an active-brazed sample shows how most of the active element, Zr in this case, has moved through the joint fillet and reacted with the ceramic substrate.

surface that does not bond well with conventional filler metals. The titanium thin-film metallization layer remains intact, as does the reaction zone of the active brazed samples, which can be important considerations if long thermal cycles or rework brazing cycles are important to the process or product.

Table 16.5 shows sample tensile button data when niobium is brazed to alumina ceramic. Niobium has a CTE which closely matches that of alumina or single-crystal sapphire, making it an ideal substrate in many cases for metal–ceramic

Table 16.6 Average tensile strengths of selected non standard tensile button assemblies

Braze filler material	Tensile button material	Interlayer material	Peak brazing temperature (°C)/time (minutes)	Furnace atmosphere	Av. tensile strength (MPa)
63Ag–27Cu–10In	Kovar	LTCC	755/3	Argon	32
Incusil-ABA	Kovar	LTCC	755/5	UHV	55
Incusil-ABA	Kovar	LTCC	755/5	Hydrogen	6
Incusil-ABA	Kovar	LTCC	755/5	Argon	3
Incusil-ABA	Kovar	94% Al_2O_3	755/5	Hydrogen	61
Incusil-ABA	Kovar	94% Al_2O_3	755/5	Argon	48
35Au–62Cu–3Ni	Niobium	Sapphire	1060/240	UHV	6
35Au–62Cu–3Ni	Niobium	Sapphire	1060/2	UHV	35
35Au–62Cu–3Ni	Niobium	Sapphire	1060/2	UHV	35
97Ag–1Cu–2Zr	Niobium	Sapphire	985/5	UHV	38
82Au–18Ni	94% Al_2O_3	n/a	1000/5	Hydrogen	93
Nioro-ABA	94% Al_2O_3	n/a	1000, 1020/5	Hydrogen	103, 99
Nioro-ABA	98% Al_2O_3	n/a	1000, 1020/5	Hydrogen	35, 25

Notes: Non-standard tensile button assemblies use metal tensile buttons in place of ceramic tensile buttons and nonmetal interlayers in place of metal interlayers. LTCC: low temperature cofired ceramic; UHV: chamber pressure at brazing temperature ≤1.3×10^{-4} Pa. Incusil-ABA nominal composition: 59Ag–27.25Cu–12.5In–1.25Ti. Nioro-ABA nominal composition: 82Au–15.5Ni–1.75V–0.75Mo.

joining. Also included in Table 16.5 are data for non standard samples where the tensile button geometries have been inverted to show the effect of a metal–nonmetal–metal sample versus the standard nonmetal–metal–nonmetal (ASTM F19). Performing such tests for material sets having large CTE differences will have dramatically different results with samples fracturing during the cooling portion of the brazing thermal cycle.

Tensile strengths and leak test results for a variety of material sets are displayed in Table 16.6. Low-temperature co-fired ceramic (LTCC) is a glass-ceramic material commonly used in electronic packaging applications. While LTCC materials, such as DuPont 951 Green Tape, are typically joined using soldering methods, applications not originally considered by material engineers or designers require higher joining methods, i.e. brazing (Walker *et al.*, 2006).

16.8 Future trends

The future of metal–nonmetal brazing continues to appear positive. As materials are pushed to greater operating extremes, commercially available joining methods, with service temperatures several hundred degrees higher than those currently used, will be required. For low-stress applications, present brazing technology requires that service temperatures remain within the 750°C–800°C range. This limitation is imposed by both the commercially available brazing filler metals, commercial metallization coatings and the controlled-expansion metal alloy substrate materials to which the nonmetals are being joined. Turbine engines and turbo-machinery applications with ever increasing efficiency requirements will be among the many products that rely on metal–nonmetal assemblies in order to take advantage of the intrinsic hot exhaust-gas corrosion resistance of oxide and nitride ceramics. However, lower than desired fracture toughness of the ceramic materials, especially for critical rotating mechanisms, will require mating to suitable metal components with sufficiently high strength and corrosion resistance for the desired application.

Because current commercially available controlled-expansion metal alloys are unable to meet the service requirements, there is a need for metallurgical research efforts to develop new controlled-expansion metal alloys capable of being joined at higher temperatures. The importance of the co-development of high service temperature filler metals and controlled-expansion alloys cannot be overemphasized. A third factor not to be overlooked in manufacturing and obtaining high-temperature assemblies is the nonmetal member. As the joining and service temperatures increase it becomes even more critical that the entire brazement, metal substrate and nonmetal substrate have coefficients of thermal expansion matched as closely as possible in order to avoid excessive residual stresses in the brazed assemblies. This important but often overlooked detail is made more difficult due to the crystalline (low CTE) nature of structural ceramics and the amorphous (high CTE) nature of the metal alloy substrates. When these materials are joined using high CTE brazing filler metals with low ductility, the resulting brazements are predictably disastrous.

The nonmetals family, which today mainly comprises 92%–96% aluminum oxide materials, will need to be augmented with additional nonmetal materials having chemical and oxidation resistance and electrical isolation properties with a CTE more closely matched to controlled-expansion metal alloys. To continue to move metal–nonmetal brazing forward in the future it will be important, therefore, to focus development efforts on the main issues: high service temperature filler metals; high-temperature metallization schemes; high service temperature controlled-expansion alloys; and the commercialization of new nonmetal materials with coefficients of thermal expansion closely matching those of the metal alloys.

16.9 Sources of further information and advice

Metal–nonmetal brazing encompasses a variety of brazing materials and methods. There are many resources available to those interested in gaining more knowledge in this evolving joining field. The following books and publications listed are but a few of the many references for the designer or process engineer: *Brazing Handbook*, 5th Edition (AWS, 2007), is a comprehensive and up-to-date general brazing handbook complete with filler metal compositional information as well as chapters on metal–nonmetal brazing; Schwartz's (1990) *Ceramic Joining*, though published over 20 years ago, remains a sound and informative text covering nearly all aspects of metal–ceramic joining; *Principles of Brazing*, by Jacobson and Humpston (2005), contains more processing information than the previously mentioned texts. Especially useful are the chapter detailing the assessment of joint quality and the previously mentioned chapter explaining the application of phase diagrams to brazing. Two publications by Mizuhara, a pioneer of active brazing processing and technology, *Ceramic-to-Metal Joining: Past, Current and Future* (2000)and *High-reliability joining of Ceramic to Metal* (Mizuhara *et al.*, 1989), provide a unique view of the role that active brazing filler metals fill from the perspective of a filler metal researcher. A highly detailed publication by Stephens *et al.* (2006), *The evolution of a ternary Ag-Cu-Zr active braze filler metal for Kovar®/alumina braze joints*, describes efforts to develop a silver-based zirconium containing active braze filler metal. The 97Ag–1Cu–2Zr (weight percents) active braze filler metal described by Stephens in the conference proceedings paper has been made commercially available and capable of producing metal–ceramic brazements of the highest tensile strengths. Although information regarding direct brazing applications is limited, the journal publication *Ceramic joining IV. Effects of processing conditions on the properties of alumina joined via Cu/Nb/Cu interlayers*, by Marks *et al.* (2001), describes in detail experimental process parameters used in a laboratory setting to fabricate alumina joints and the resulting strengths obtained using direct brazing methods. Likewise, Song *et al.* (1997) have authored a very informative article describing the bonding mechanisms for niobium and sapphire titled *Extraordinary Adhesion of Niobium on Sapphire Substrates*, for those desirous of more literature on this least used and understood brazing method. To further understand the complexities in forming the required metal–nonmetal reaction layer and thorough microstructural analysis, readers are encouraged to examine publications by Dezellus and Eustathopoulos (2010) and Hosking *et al.* (2000). The brazement designer or engineer will find the American Welding Society Standard, *AWS C3.2M/3.2:Standard Method for Evaluating the Strength of Brazed Joints* (AWS, 2008a), extremely valuable for information regarding the selection, fabrication and testing of metal–ceramic brazements.

16.10 References

ASTM (2005) *Standard test method for tension and vacuum testing metallized ceramic seals*, West Conshohocken, Pennsylvania, American Society for Testing Materials.

AWS (2007) *Brazing handbook*, 5th Edition, Miami, Florida, USA, American Welding Society.

AWS (2008a) *AWS C3.2M/3.2:2008: Standard method for evaluating the strength of brazed joints*, Miami, Florida, American Welding Society.

AWS (2008b) *AWS C3.3M/3.3:2008: Recommended practices for the design, manufacture, and examination of critical brazed components*, Miami, Florida, American Welding Society.

AWS (2008c) *AWS C3.6M/3.6:2008: Specification for furnace brazing*, Miami, Florida, American Welding Society.
Budde F (1963) 'The package concept of ceramic-to-metal seals', *Advances in electron tube techniques, Proceedings of the sixth national conference*, New York, New York, The Macmillan Company.
Coughlin J (1954) *Bulletin 542, Contributions to the data on theoretical metallurgy: XII heats and free energies of formation of inorganic oxides*, Washington DC, Bureau of Mines.
Cusil-ABA® and Incusil-ABA® active brazing filler metals are registered trademarks of the Morgan Crucible Company plc, 1985.
Dezellus O and Eustathopoulos N (2010) 'Fundamental issues of reactive wetting by liquid metals', *Journal of Material Science*, 45, 16, 4256–4264.
Hosking F M, Cadden C H, Yang N Y C, Glass S J, Stephens J J, *et al.* (2000) 'Microstructural and mechanical characterization of actively brazed alumina tensile specimens', *Welding Journal*, 79, 8, 222s–230s.
Jacobson DM and Humpston G (2005) *Principles of brazing*, Metals Park, Ohio, ASM International.
Kapoor R K and Eagar T E (1989) 'Brazing alloy design for metal/ceramic joints', *Ceramic Engineering and Science Proceedings*, 10, 11–12, 1613–1639.
Kapoor R K, Podszus E S and Eagar T E (1988) 'Wettability of silver based reactive metal brazing alloys on alumina', *Wettability of brazing alloys*, 22, 8, 1277–1279.
Kohl, W H (1967) *Handbook of materials and techniques for vacuum devices*, New York, New York, Reinhold publishing company.
Kovar® low expansion metal alloy is a registered trademark of CRS Holdings, a subsidiary of Carpenter Technology Corporation, 1935.
Lucas-Milhaupt (2011) VTG/High Purity Silver, Gold, Palladium. https://www.lucasmilhaupt.com/en-US/products/fillermetals/vtghighpuritysilvergoldpalladium/12/ [Accessed 24 February 2012]
Marks R A, Suga J D and Glaeser A M (2001) 'Ceramic joining IV. effects of processing conditions on the properties of alumina joined via Cu/Nb/Cu interlayers', *Journal of Materials Science*, 36, 5609–5624.
Mizuhara H (2000) 'Ceramic-to-metal joining: past, current and future', *Advanced brazing and soldering technologies*, Metals Park, Ohio, ASM International.
Mizuhara H, Huebel E and Oyama T (1989) 'High-reliability joining of ceramic to metal', *Ceramic Bulletin*, 68, 9, 1591–1599.
Nascimento R M, Martinelli A E and Buschinelli A J A (2003) 'Review article: recent advances in metal-ceramic brazing', *Cerâmica*, 49, 310, 178–198.
Phillips W M (1969) *'TR32-1420:Metal-to-ceramic seals for thermionic converters, a literature study'*, NASA Technical Reports, Pasadena, California, Jet Propulsion Laboratory.
Reed L and McRae R C (1965) 'Evaporated Metallizing on Ceramics', *Bulletin of the American Ceramic Society*, 44, 1, 12–13.
Rosebury F (1965) *Handbook of electron tube and vacuum techniques*, Reading, Massachusetts, Addison-Wesley publishing company, Inc.
Schwartz M M (1990) *Ceramic joining*, Metals Park, Ohio, ASM International.
Schwartz M M (2003) *Brazing*, 2nd edition, Metals Park, Ohio, ASM International.
Song G, Remhof A, Theis-Bröhl K and Zabel H (1997) 'Extraordinary adhesion of niobium on sapphire substrates', *Physical Review Letters*, 79, 25.
Stephens J J, Hlava P F, Walker C A, Headley T A and Boettcher G E (2005) *Direct brazing process to alumina ceramic using conventional braze alloys and a niobium interlayer*, Annual TMS Meeting, San Francisco, California, USA.

Stephens J J, Hosking F M, Yost F G, Walker C A and Dudley E (2006) 'The evolution of a ternary Ag-Cu-Zr active braze filler metal for Kovar®/alumina braze joints', *Proceedings of the 3rd International Brazing and Soldering Conference*, Metals Park, Ohio, ASM International.

Vianco P T, Stephens J J, Hlava P F and Walker C A (2003) 'A barrier layer approach to limit Ti scavenging in FeNiCo/Ag-Cu-Ti/Al2O3 active braze joints', *Welding Journal*, 82, 9, 252s–262s.

Walker C A and Hodges V C (2008) 'Comparison of metal-ceramic brazing methods', *Welding Journal*, 87, 10, 43–50.

Walker C A, Trowbridge F R, Wagner A R (2010) 'Direct brazing of sapphire to niobium', *Welding Journal*, 89, 3, 50–55.

Walker C A, Uribe F, Monroe S L, Stephens J J, Goeke R S and Hodges, V C (2006) 'High-temperature joining of low-temperature co-fired ceramics', *Proceedings of the 3rd International Brazing and Soldering Conference*, Metals Park, Ohio, ASM International.

Wesgo Metals (2009), *Alloy Selection Chart*. Available from http://www.wesgometals.com/resources/alloy-selection/ [Accessed 24 February 2012]

17

Glasses and glass-ceramics as brazing materials for high-temperature applications

M. SALVO, V. CASALEGNO, S. RIZZO, F. SMEACETTO, A. VENTRELLA and M. FERRARIS, Politecnico di Torino, Italy

DOI: 10.1533/9780857096500.3.525

Abstract: Glass-ceramics can be versatile joining materials with tailorable thermal and mechanical properties; they are not affected by oxidation and can be used as pressureless joining materials above their glass softening point. The glassy phase can be minimized by a suitable thermal treatment in order to improve the glass-ceramic creep behaviour. The use of glasses and glass-ceramics as sealants in solid oxide fuel cells and as joining materials for SiC-based materials will be discussed.

Key words: glass and glass-ceramic joining materials, high-temperature applications, solid oxide fuel cell, SiC-based materials, mechanical shear tests.

17.1 Introduction

Glass-ceramics can be versatile joining material with tailorable thermal and mechanical properties; they are not affected by oxidation and can be used as pressureless joining materials above their glass softening point. The glassy phase can be minimized by a suitable thermal treatment in order to improve the glass-ceramic creep behaviour. The use of glasses and glass-ceramics[1] as (1) sealants in solid oxide fuel cells and (2) joining materials for SiC-based materials will be discussed.

1. One important challenge in the fabrication and commercialization of planar solid oxide fuel cells (SOFCs) is the development of sealant materials that are thermomechanically and thermochemically stable in both oxidizing and wet-reducing environments at 800 °C for long-term exposures. A critical point in the fabrication of a planar SOFC is the sealing of the ceramic electrolyte with the metallic interconnect, in order to obtain a hermetic (gas-tight) joint. Among several possible joining (sealant) materials, glasses, glass-ceramics and glass-based composites seem to offer the most suitable solutions. In this chapter some glass-ceramic compositions are reviewed and discussed as well as their performance in SOFC-operating conditions.
2. SiC-based materials have a brittle nature and consequent poor workability which limit their extensive use. The joining of SiC-based materials is one of the

[1] According to ASTM standard C162, glass is an inorganic product of fusion that has cooled to a rigid condition without crystallizing; a glass-ceramic is a solid material, partly crystalline and partly glassy, formed by the controlled crystallization of a glass.

most challenging technological issues for materials in nuclear and high-temperature applications since the extreme environment, i.e. cyclic and transient thermal loads, steep thermal gradients and high fluxes of energetic neutrons, put high demands on the joining materials. This chapter will focus on the available scientific literature with special emphasis on glass-ceramics as joining materials. The advantages of using glass-ceramics as joining materials will be discussed.

17.2 Glass and glass-ceramic sealants for solid oxide fuel cells

Solid oxide fuel cells (SOFCs) are highly efficient energy conversion devices, which produce electricity by the electrochemical reaction between a fuel and an oxidant (Steele and Heinzel, 2001). Among the different SOFCs, the planar type, which is expected to be cost effective and mechanically reliable (Singh, 2007), offers an attractive potential for increased power densities compared with other concepts. A fuel cell device consists of an anode electrode (exposed to the fuel), an electrolyte, and a cathode electrode (exposed to the oxidant). The repeating unit of a planar solid oxide fuel cell is formed by anode–electrolyte–cathode and interconnects. The interconnect links the anode of one cell to the cathode of the neighbouring cell. In most SOFC stack designs, the interconnect is sealed to the ceramic cell components (Fig. 17.1). A critical point in the fabrication of a planar SOFC is the sealing of the ceramic electrolyte with the metallic interconnect, in order to obtain a hermetic (gas-tight) joint (Jordan, 2008). Currently, five main approaches are being studied for sealing SOFCs: brazing, compressive seals, and glass, glass-ceramic and glass matrix composite seals (Fergus, 2005; Singh *et al.*, 2007; Le *et al.*, 2007; Bansal and Gamble, 2005; Chou *et al.*, 2007b; Chou *et al.*, 2007c; Gross *et al.*, 2010; Wang *et al.*, 2011; Smeacetto *et al.*, 2008a). Among several possible joining (sealant) materials, glasses, glass-ceramics and glass-based composites seem to offer the most suitable solutions. Four important thermal properties must be considered when designing a SOFC seal glass: glass transition temperature (*Tg*), glass softening temperature (*Ts*), coefficient of thermal expansion (CTE) and thermal stability. A sealant must be stable for thousands of hours at 800 °C in oxidizing and reducing atmospheres and withstand

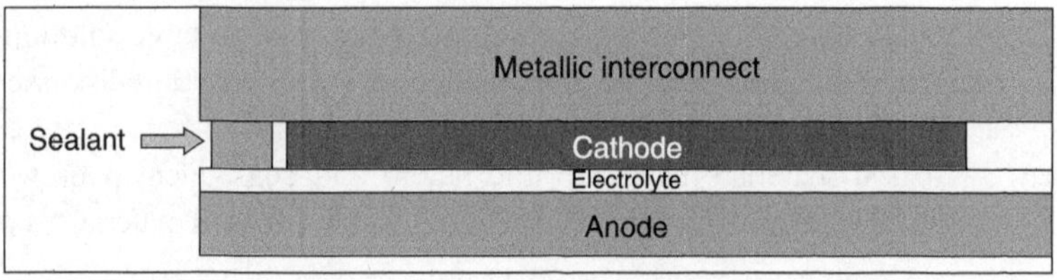

17.1 Schematic of the assembly of a solid oxide fuel cell (SOFC). Not to scale.

thermal cycles between room temperature (RT) and the cell operating temperature. Many glass-forming systems have been considered as SOFC sealants, including those based on phosphates, borates and silicates (Mahapatra and Lu, 2010a). Glass-ceramics, which are prepared by crystallization of glasses, are considered advantageous for SOFC sealant applications because, with careful choice of the glass composition, glasses and glass-ceramics meet, in principle, most of the requirements of an ideal sealant. Glass-ceramics, which can be prepared by controlled sintering and crystallization of glasses, are typically stronger than the parent glass. Moreover, it is possible to tailor the properties of the resulting glass-ceramic sealant (for example its thermal expansion) by controlling the kinetics of crystallization and the crystalline phases that are formed. To develop a suitable glass-ceramic sealant, it is therefore necessary to understand crystallization kinetics, sealing properties and chemical interaction when in contact with other components of the cell. Finally, glass-ceramic sealants have good hermeticity and are usually thermally and environmentally stable.

The CTEs of SOFC components are generally $10.5 \times 10^{-6}\,°C^{-1}$ for the ceramic electrolyte (yttria-stabilized cubic zirconia, YSZ), $10–14.0 \times 10^{-6}\,°C^{-1}$ for the anode (for example NiO–YSZ composite), and $11.0–15.0 \times 10^{-6}\,°C^{-1}$ for the metallic interconnect (austenitic and ferritic stainless steels) (Yang, 2008). Generally a CTE in the range $10.0–12.0 \times 10^{-6}\,°C^{-1}$ is required for a glass-ceramic sealant. Concerning silica-based glass-ceramic systems, several compositions have been proposed in the literature in the last five years. Most of the compositions proposed as sealants for SOFCs can be classified into:

- BaO-containing borosilicate glasses (Pascual *et al.*, 2006; Meinhardt *et al.*, 2008);
- $SrO–Al_2O_3–La_2O_3–B_2O_3–SiO_2$ seal glass systems (Mahapatra *et al.*, 2009; Chou *et al.*, 2007a; Goel *et al.*, 2010);
- CaO- and MgO-containing borosilicate glasses (Pascual *et al.*, 2007; Loehman *et al.*, 2002; Zhang *et al.*, 2008);
- NaO-containing silicate glass systems (Smeacetto *et al.*, 2009, 2008a, 2008b; Nielsen *et al.*, 2007; Chou *et al.*, 2011).

Complete and interesting reviews of the different glass and glass-ceramic systems employed as joining and sealant materials for SOFCs, as well as possible interfacial reactions between glasses and metals, are given by Donald *et al.* (2011) and Fernie *et al.* (2009).

As in any other joining process, good bonding and sealing between either the glasses or glass-ceramics and the metallic interconnect can be achieved if both the interface chemistry and wetting between the dissimilar materials are controlled. Reactions at the sealant/metal interface are not a serious problem during seal formation (except in the case of large differences in thermal expansion coefficients of the substrates and the reaction products), but can become excessively critical and dangerous for SOFC durability during cell operation.

Most of the authors focus their research on the glass and glass-ceramic to metallic interconnect bonding. Anyway, it must be kept in mind that the critical point in the fabrication of planar SOFCs is the joining of the ceramic electrolyte to the metallic interconnect. Few authors, in their morphological characterizations, show the SEM cross-section of the joined 'sandwich', i.e. metallic interconnect/sealant/ceramic electrolyte (Smeacetto *et al.*, 2009, 2008a, 2008b, 2008c; Goel *et al.*, 2010; Lin *et al.*, 2011). This is an extremely important issue, since one of the most important challenges in development of SOFC seals is the chemical and thermomechanical compatibility of sealant with both the metallic interconnect and the ceramic electrolyte.

As an example of a complete glass-ceramic sealant design and characterization for an SOFC, one Na-containing silica-based glass-ceramic will be considered in this chapter. The designed sealant is a barium-free and boron-oxide-free silica-based glass containing sodium. The heat-resistant metal alloys used as the interconnect for this study were Crofer22APU (manufactured by ThyssenKrupp, Germany containing 22 wt.% chromium and exhibiting a thermal expansion coefficient of $11.5 \times 10^{-6}\,°C^{-1}$ from RT to 700 °C) and AISI 430 (thermal expansion coefficient of $11.9 \times 10^{-6}\,°C^{-1}$ from RT to 500 °C). The metal interconnects was pre-oxidized at 950 °C in air. The yttria 8 mol.% zirconia wafer (with a thermal expansion coefficient of $10.5 \times 10^{-6}\,°C^{-1}$, RT to 800 °C) had a thickness of 200 μm. The sealants were produced as glasses by melting the appropriate oxide and carbonate raw materials in different proportions and by heating at 1500 °C for 1–2 h in a platinum crucible; the melts were cast on a brass plate. Differential thermal analysis (DTA), heating stage microscopy (HSM) and dilatometry are currently used for the complete thermal and thermomechanical characterization of the glasses. HSM has also been used for studies of wettability of glasses on Crofer22APU and YSZ substrates. Table 17.1 summarizes characteristic temperatures and thermomechanical properties for the glass system characterized and discussed in the following sections (composition, mol.%: 53–58% SiO_2, 6–8% Al_2O_3, 24–26% CaO, 10–12% Na_2O).

The joined samples were obtained by placing the pre-oxidized Crofer22APU plates on a YSZ surface with a glass slurry (a mixture of glass powder 38 μm < size <75 μm, dispersed in ethanol with solid content 40 wt.%) sandwiched in between,

Table 17.1 Characteristic temperatures and thermomechanical properties for the SiO_2-Al_2O_3-CaO-Na_2O system

α_{glass} (10^{-6}/°C) (150–450 °C)	$\alpha_{glass\text{-}ceramic}$ (10^{-6}/°C) (150–450 °C)	T_g (°C)	Tx (°C)	T_p (°C)	T at η 10^4 dPas (°C)	T_{FS} (first shrinkage) (°C)	T_{MS} (maximum shrinkage) (°C)	T_{HSF} (hemisphere) (°C)	T_F (flowing) (°C)
9.6	10.7	670	881	904	1165	725	840	1163	1180

then by heating at 900 °C for 30 min with an optimized thermal process determined by heating stage microscopy and DTA. Crystalline phase compositions of the sealant material after the joining heat treatment were determined by X-ray diffraction (XRD) analysis. Cross-sections of joined samples were morphologically characterized by scanning electron microscope (SEM) observations and energy-dispersive X-ray spectroscopy (EDS) analysis. Joined Crofer22APU/sealants/YSZ samples were exposed to a H_2 – 3 vol.% H_2O atmosphere (800 °C for 300–500 h) and to thermal cycling and thermal ageing in air, as well as in SOFC relevant conditions (i.e. dual atmosphere at 800 °C up to 500 h).

17.2.1 Joined ceramic electrolyte/metallic interconnect sample characterization

Figure 17.2 shows an SEM micrograph of a Crofer22APU/sealant/anode-supported YSZ cross-section. The joint region has an average thickness of 150–200 μm. A very good adhesion between the glass-ceramic and both Crofer22APU and YSZ can be observed. The glass-ceramic thermal expansion coefficient was found to be $10.7 \times 10^{-6}\,°C^{-1}$, an intermediate value between Crofer22APU ($11.2–11.5 \times 10^{-6}\,°C^{-1}$) and YSZ ($10.5 \times 10^{-6}\,°C^{-1}$). No cracks are observed near the interface of the glass-ceramic with either Crofer22APU or YSZ. Moreover, a Vickers indentation was performed at the Crofer22APU/glass-ceramic sealant interface; it showed that the joining material is in a favourable compressive stress state: cracks propagate preferentially parallel to the interface (Fig. 17.3). Figure 17.4 is a magnification (rectangular area in Fig. 17.2) of the interface between the glass-

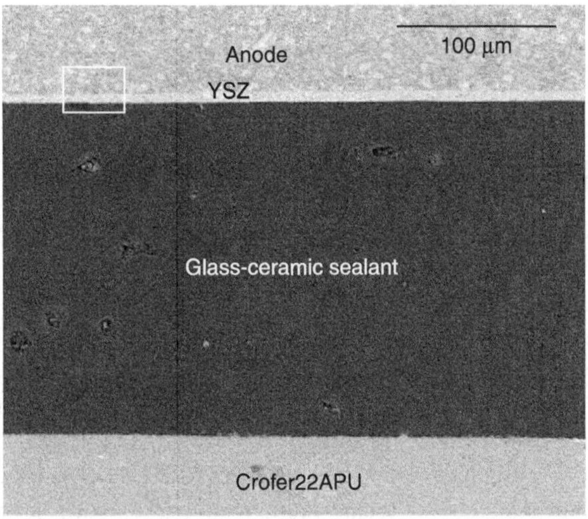

17.2 Cross-section of a Crofer22APU/glass-ceramic sealant/anode-supported YSZ electrolyte from Smeacetto (2008c).

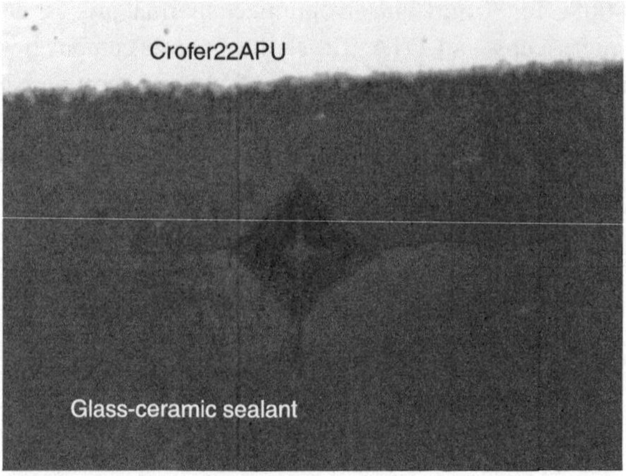

17.3 Vickers indentation at the Crofer22APU/glass-ceramic sealant interface: cracks propagate preferentially parallel to the interface (500x).

17.4 Cross-section of the interface between the glass-ceramic sealant (left) and the anode-supported YSZ electrolyte (right).

ceramic and the YSZ: the interface is continuous without the presence of pores or cracks. The YSZ electrolyte is supported on the porous NiO–YSZ anode.

Concerning the bonding at the interface between Crofer22APU and the glass-ceramic sealant, the glass-ceramic was successfully sealed to the Crofer22APU as received as well as pre-oxidized (950 °C, 2 h air atmosphere). The oxidation of Crofer22APU in air resulted in a chromium–manganese oxide layer at the alloy surface. Figure 17.5 shows an SEM cross-section of the interface between pre-oxidized Crofer22APU/SACN (silicon oxide-aluminum oxide-calcium

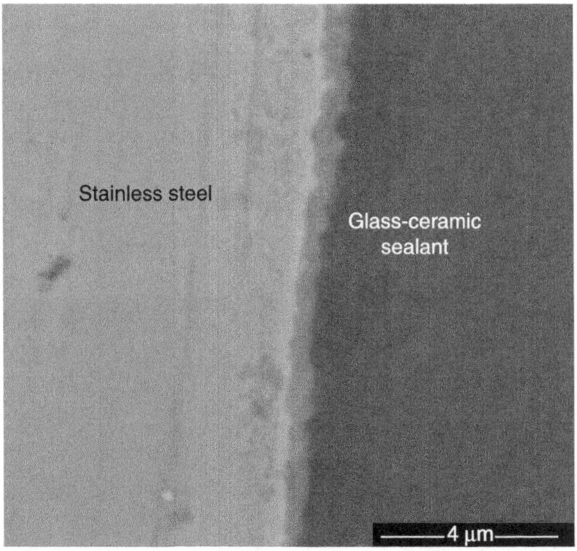

17.5 Cross-section of the interface between pre-oxidized Crofer22APU/SACN glass-ceramic sealant.

oxide-sodium oxide glass-ceramic) sealant. Pre-oxidation of the metallic interconnect prior to the sealing process was also found to be beneficial in avoiding diffusion of Cr, Mn and Fe into the glass-ceramic; this effect (during cell operation) will be discussed in the following paragraph.

In order to measure the bonding strength of the glass-ceramic joined Crofer22APU, square Crofer22APU samples were joined by the glass-ceramic to obtain a sandwich-like structure. A modification of ASTM C633-0139 was designed and performed according to Smeacetto *et al.* (2008c), where the bond strength of the joint between the Crofer22APU and the glass-ceramic was investigated by carrying out uniaxial tensile tests at room temperature. The average tensile strength value was about 6 MPa. It is noteworthy that the fracture always occurred through the glass-ceramic and never at the metallic interconnect/glass-ceramic interface. This evaluation suggested that the bonding strength at the sealant/metal interface was high with very good adhesion.

Various examples of protective coatings applied to the metallic interconnect (in order to avoid Cr evaporation with cathode poisoning, reactions between seal and metal, formation of undesired products with swelling of the metal and failure at the interface) can be found in the literature (Mahapatra *et al.*, 2010b; Chou *et al.*, 2008).

Very good results were obtained with AISI 430 as the interconnect material using the same glass-ceramic material. It must be highlighted that in order to obtain good bonding between AISI 430 and the glass-ceramic sealant the pre-oxidation heat treatment (900 °C, 2 h, air) was essential, since the glass-ceramic layer was completely detached from the as-received AISI 430 after joining.

17.2.2 Joined sample performance and tests

In this section some tests exposing Crofer22APU/sealant/YSZ joined samples to an H_2–3 vol.% H_2O atmosphere, plus thermal cycles in air (from RT to 800 °C) and exposure to SOFC relevant conditions (i.e. dual atmosphere) are reported and discussed.

Figure 17.6 shows the Crofer22APU/glass-ceramic sealant interface and relative EDS line scans after 500 h of exposure to the H_2–3% H_2O atmosphere; it

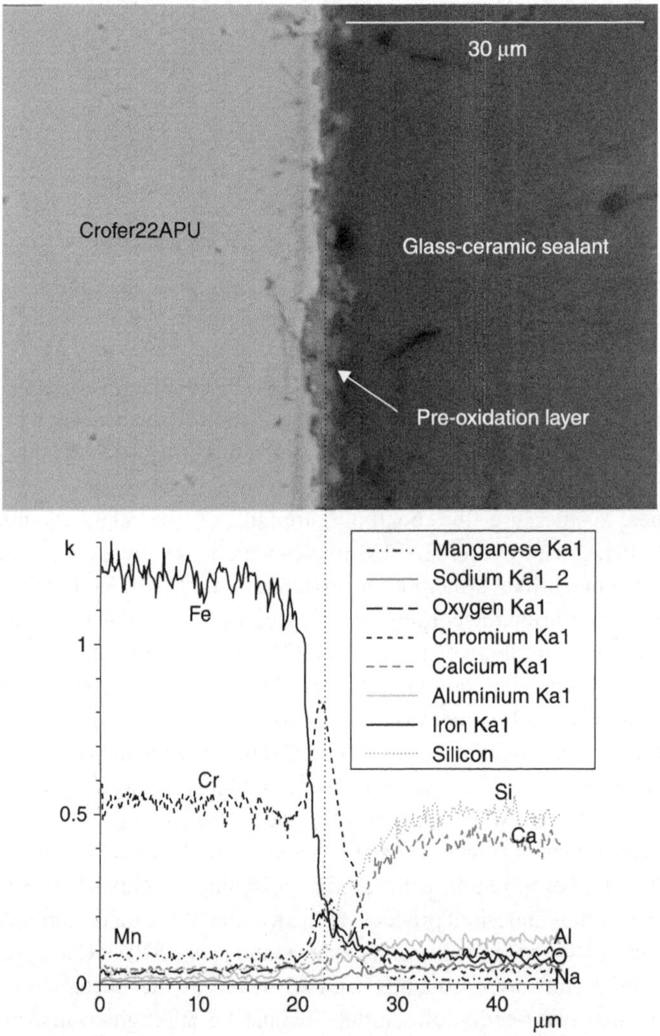

17.6 Crofer22APU/glass-ceramic sealant interface and relative EDS line scans after 500 h of exposure to the H_2–3% H_2O atmosphere (Smeacetto, 2009).

is evident that no reaction between the chromium–manganese oxide scale and the glass-ceramic components took place and consequently no new interfacial products were found at the interface (Smeacetto *et al.*, 2009).

The problem for SOFC components with dissimilar thermomechanical properties sealed together becomes even more challenging as there is also a requirement for thermal cycle stability of the joined structures. In view of this, the sealants must survive for several hundreds of thermal cycles (from RT to 800 °C) during SOFC operations. Any crack which forms within the sealant or poor adhesion at the interfaces with both metallic interconnect and ceramic electrolyte can cause leakage, which leads to mixing of the fuel and oxidant with consequent lower cell performance and efficiency.

Joined Crofer22APU/glass-ceramic sealant/anode-supported YSZ samples submitted to thermal cycles and thermal ageing at 800 °C for 500 h in air showed no evidence of interfacial failure and cracks; slight migration of both Cr and Mn into the glass-ceramic was observed (Smeacetto *et al.*, 2011a).

17.2.3 Testing of joined Crofer22APU/glass-ceramic sealant/ anode supported cell in SOFC relevant conditions

In this test Crofer22APU plates (pre-oxidized at 95 °C for 2 h in air) and an anode-supported cell (ASC) substrate based on NiO–YSZ (8 mol.% Y_2O_3 stabilized ZrO_2), with a thin YSZ electrolyte and a cathode made of a LSM (lanthanum strontium manganite)/YSZ functional layer and a LSM current collecting layer were used. A joined single repeating unit (SRU) (Crofer22APU plate/glass-ceramic sealant/ASC) was obtained by placing the cell on the top of the slurry coated Crofer22APU. Figure 17.7 shows the scheme of the single repeating unit and the test facility.

In this experiment, the electrical behaviour of the SRU in ageing conditions after 460 h at 800 °C at open circuit voltage (OCV), was evaluated. The OCV was found to be stable at a value of 1075 mV, very close to the theoretical reversible voltage of 1080 mV.

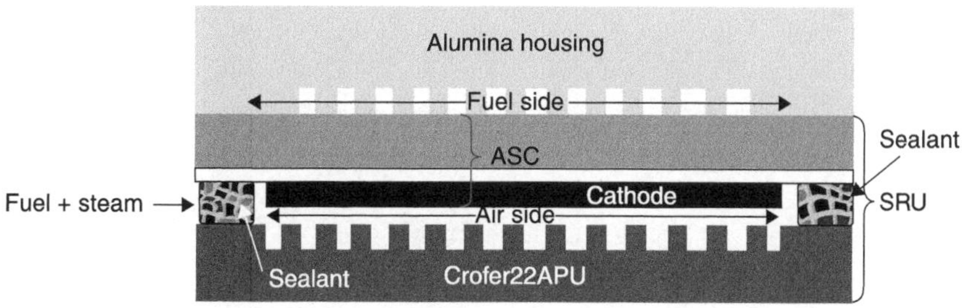

17.7 Scheme of the single repeating unit and the test facility. ASC: anode supported cell; SRU: single repeating unit.

This result was the evidence of an effective gas stream separation between the anode and cathode side provided by the glass-ceramic sealant. In order to validate this assumption, SRU polished cross-sections of the joined structures after test were submitted to post-mortem SEM and EDS analyses.

An SEM cross-section of the SRU after thermal ageing experiments in a dual atmosphere for 460 h is presented in Fig. 17.8. It can be observed that, despite the stresses induced during cutting operations, no cracks are detected in the whole length of the glass-ceramic sealant, as observed on several cross-sections obtained by polishing; moreover, the adhesion between the glass-ceramic, Crofer22APU and YSZ is still sound. The absence of cracks near the interface of the glass-ceramic with Crofer22APU and the anode-supported electrolyte demonstrates its good thermomechanical compatibility with both substrates.

17.2.4 Future trends

Sealants are still one of the main technical barriers in the commercialization of advanced planar SOFCs; a critical point in the fabrication of planar SOFCs is still the sealing of the electrolyte with the metallic interconnect, in order to obtain a reliable, long-lasting, hermetic (gas-tight) joint able to survive working conditions such as on/off operation mode, blackouts, etc. Repair issues also remain to be solved, with some advances in the field of laser sealing showing promising results.

SOFC researchers and developers are also working on materials that can provide high ionic conductivity at low temperatures, such as 650 °C. Lowering of the SOFC operating temperature from 800 °C to 650 °C, and even to a lower

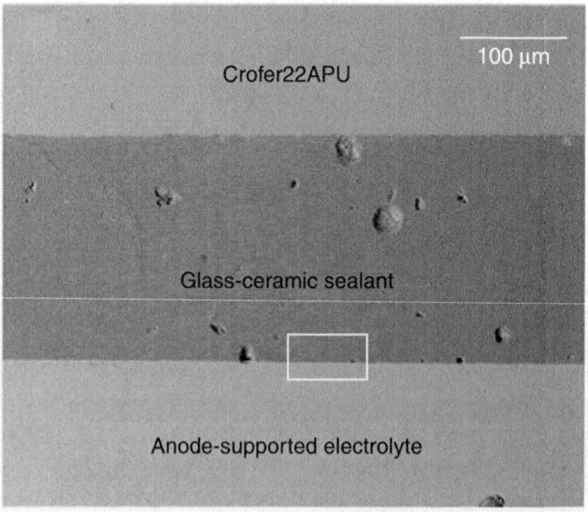

17.8 Cross-section of the joined single repeating unit after thermal ageing experiments in dual atmosphere for 460 h (Smeacetto, 2011b).

value, would be a fundamental step towards the commercialization of SOFCs. This will significantly reduce the thermomechanical constraints for all the involved materials, as well as the overall cost, and it will simplify maintenance.

17.3 Glass and glass-ceramic joining for SiC-based materials

The use of SiC-based materials (SiC, SiC/SiC, C/SiC) for structural and functional applications is considered in several high-temperature applications, i.e. advanced fission and fusion systems, space propulsion systems, etc. (Snead *et al.*, 2011). The potential of SiC-based materials is mainly due to their unique thermomechanical properties in harsh environments, and low-activation characteristics for nuclear and fission applications. In most cases, however, the use of SiC-based materials depends on joining them to themselves or to other materials in order to manufacture the final components.

There are many possible techniques for joining SiC-based materials to themselves and to dissimilar materials: diffusion bonding using various active fillers (Cockeram, 2005), transient eutectic phase methods such as nano-infiltration and transient eutectic phase (NITE) (Katoh *et al.*, 2004; Hinoki *et al.*, 2005; Jung *et al.*, 2009), laser joining (Lippmann *et al.*, 2004), selected area chemical vapor deposition (Harrison and Marcus, 1999), glass-ceramic joining (Katoh *et al.*, 2000; Ferraris *et al.*, 2008, 2011), solid state displacement reactions (Henager *et al.*, 2007), pre-ceramic polymer routes (Lewinsohn *et al.*, 2001), reaction forming (Singh, 1999), and brazing (Riccardi *et al.*, 2004; Singh *et al.*, 2008).

Transient liquid phase (TLP) bonding uses a thin liquid interlayer between the two surfaces to be joined. It has been applied to many metallic systems and can produce a bond at a lower temperature than the one finally used. TLP has been tailored for a number of ceramics using interlayers based on glasses (such as oxynitrides) (Glass *et al.*, 1998).

TLP bonding has been adapted in some joining techniques for SiC-based materials for nuclear applications: the NITE (nano-infiltration and transient eutectic phase) process was developed by employing a transient liquid eutectic phase method utilizing a mixture of β-SiC nano-powder below 30 nm and SiO_2, Al_2O_3, Y_2O_3 as joining material. The NITE joining process is performed under an argon atmosphere by applying a pressure of 20–40 MPa at a temperature ranging between 1400 and 1900 °C. Complete descriptions of this method are reported in Katoh *et al.* (2004) and Hinoki *et al.* (2005). To the best of the authors' knowledge, no results have been published on neutron-irradiated NITE joints, but a TEM analysis of a NITE SiC/SiC composite using dual-ion irradiation gave promising results (Kishimoto *et al.*, 2007).

Lippmann *et al.* (2004) joined SiC by a laser joining process based on the local melting of a glass joining material (based on Al_2O_3, Y_2O_3 and SiO_2) by laser radiation; gas-tight binding between shaped bodies made of these ceramic materials was obtained. The joints obtained were temperature resistant at 1600 °C and above. At the moment, this joining technique is available for joining ceramic parts, but it has not been tested on SiC/SiC composites. The main advantage of this process is the local heating, thus avoiding undesirable effects of heating the whole component.

Esposito *et al.* (1999) proposed a pressureless joining for SiC by using an Al–Y–Si–O glass interlayer and heating at a slightly higher temperature than the melting of the glass. An alternative to using a glass interlayer is to apply pressure onto two liquid phase sintered silicon carbide pieces placed in close contact at a temperature above the grain boundary softening point, thus producing a direct joining (Esposito *et al.*, 1999). At the bonding temperature, the glass boundary phase flows through the interface driven by the applied pressure and the capillary pressure that builds up between the two contacting ceramic surfaces, promoting their adhesion. The need for pressure during the joining cycle limits the use of this technique to samples with simple geometries. However, very high-strength joints can be obtained, comparable to the ceramic to be joined.

Perham *et al.* (1999) reported the manufacture of SiC joints using a cordierite glass-ceramic deposited by a tape casting method. Cordierite, with glass-ceramic processing, remains amorphous and wets the SiC substrate; subsequent heat treatment crystallizes a multiphase interlayer with a CTE close to the SiC one. The SiC joined to the cordierite glass-ceramic showed flexural strengths higher than 500 MPa at room temperature, with the fracture propagating into the bulk ceramic.

Lee *et al.* (1998) joined sintered SiC using a mixture of MgO, Al_2O_3 and SiO_2 powders (MAS filler) and a heat treatment at 1500 °C for 30 min in an inert atmosphere. An oxycarbide glass joint interlayer was produced by the reaction of the MAS filler with the SiC matrix. The flexural strength for these joints was 342–380 MPa, which is close to that of the as-received SiC specimen up to 800 °C.

Coon and Neilson (1989) developed a process based on the MgO–Li_2O–Al_2O_3–SiO_2 glass system for SiC/SiC joining where the joining temperature was kept below 1200 °C in order to avoid degradation of this composite.

17.3.1 Glass and glass-ceramic joining of SiC-based materials for nuclear applications

In the field of nuclear energy production (fusion and fission), requirements for SiC-based joints are extremely severe: a joining material must be compatible with a neutron environment and the joining technique must comply with the fusion nuclear reactor design where SiC/SiC components several metres long and 3 mm thick must be joined in a reliable and feasible way (Giancarli *et al.*, 2002; Riccardi *et al.*, 2000). The very few available solutions for SiC/SiC joining for nuclear applications are reported or reviewed in Ehrlich *et al.* (2001), Riccardi *et al.*

(2000), Tavassoli (2002), Colombo *et al.* (2000), Katoh *et al.* (2000), Singh (1999), Gasse *et al.* (1997), Lewinsohn *et al.* (2002) and Hinoki *et al.* (2005); brazing is one of the most promising joining techniques.

Glasses and glass-ceramics have been proposed as joining materials for SiC-based materials because of their unique properties: hermeticity, tailorable properties (i.e. characteristic temperature, coefficient of thermal expansion and wettability suitable for an SiC substrate), LAM (low-activation material), probable neutron irradiation stability, thermomechanical stability at working temperatures (higher than 1000 °C), easy joining technique (such as pressureless joining), not affected by oxidation and eventually self-sealant.

Most silicate glasses have been reported to exhibit good wettability on SiC substrates (Yurkov *et al.*, 1991). Several glassy systems used for joining SiC-based components are based on the $MgO–Al_2O_3–SiO_2$ system, mainly for three reasons: 1) it has a small wetting angle on the SiC substrate (less than 10 °); 2) the thermal expansion coefficient of these systems can be custom-made to be close to that of SiC itself; 3) the reaction at high temperatures of these glassy systems with SiC leads to oxycarbide glass, which exhibits high mechanical resistance at high temperatures.

As discussed above, the refractoriness of the glassy joining material can be enhanced by controlled crystallization, the so-called glass-ceramic method, in order to obtain a chemically and thermomechanically compatible ceramic or glass-ceramic material. These joining materials are mechanically reliable up to the softening point of the residual amorphous phase, which can be minimized by suitable thermal treatment (Hinoki *et al.*, 2005).

Irradiation-induced creep and void swelling are two major material issues in nuclear reactors. They affect material properties because of the change in dimensions with increasing irradiation dose and the increasing stress measured at a constant temperature and dose (Baldev *et al.*, 2008).

In addition to the general benefits reported above, glass-ceramic joining materials are advantageous for joining large structures (for example, as required in a nuclear reactor): no pressure is required during the joining process since it is performed above the glass softening point. For nuclear-related applications, a joining material should be a low-activation material (LAM). Surveys of the safety-relevant activation properties of joining materials (including glass-ceramics based on $SiO_2–Al_2O_3–BaO–B_2O_3$, $SiO_2–Al_2O_3–MgO$ and $CaO–Al_2O_3$) for SiC-based components are given in Zucchetti *et al.* (2001) and Ferraris *et al.* (2008).

In Ferraris *et al.* (1998, 2011b) and Katoh *et al.* (2000) and the joining of SiC/SiC composites and SiC was obtained using a calcia–alumina (CA) glass-ceramic, which does not contain boron oxide (incompatible with fusion applications) and has high characteristic temperatures (softening point at about 1300 °C).

Figure 17.9 shows an SEM cross-section of a $C_aO–Ap_2O_3$) glass-ceramic joined SiC sample. The average apparent shear strength of the joined structures was 28 MPa at room temperature for joined SiC/SiC composites and 36 ± 8 MPa for

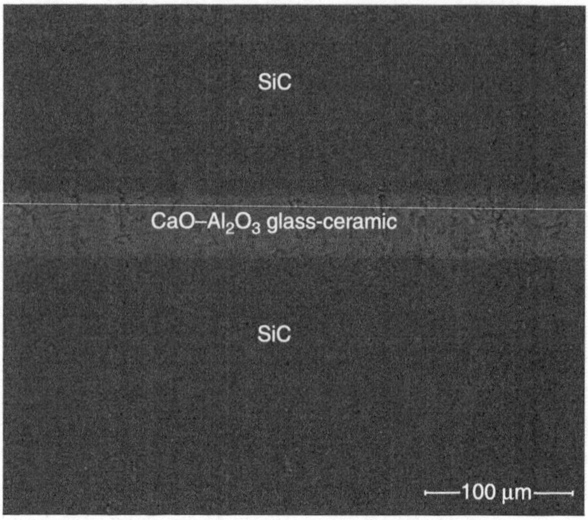

17.9 $CaO–Al_2O_3$ (CA) glass-ceramic joined SiC: the thickness of the joining materials is approximately 50 μm.

joined SiC, measured by single lap test in compression. A pure shear strength of 104 ± 25 MPa was measured by torsion tests of CA joined SiC.

In Ferraris *et al.* (2008) SiC/SiC (chemical vapour deposited SiC-coated 2D composites) were joined by a glass ceramic based on $SiO_2–Al_2O_3–Y_2O_3$ (SAY) designed to be a low-activation joining material (Fig. 17.10). Three different

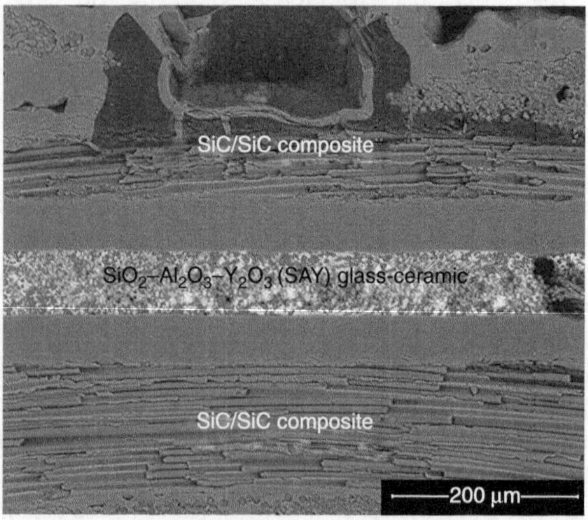

17.10 Back-scattered SEM micrograph of the cross-section of a SiC/SiC composite joined by $SiO_2–Al_2O_3–Y_2O_3$ (SAY) glass-ceramic.

kinds of joined samples were manufactured to couple the reliability of the machined joint (e.g. mortise and tenon like) with the sealant properties of the glass-ceramic joining material. The bending strength results are very promising: a shear strength higher than 120 MPa at room temperature was obtained.

Irradiation test results were performed on these joined samples: SAY joined SiC/SiC was still joined after irradiation at 600 °C, 16.3×10^{24} n/m^2 and at 820 °C, 31–32 $\times 10^{24}$ n/m^2. The samples were submitted to bending tests; within the statistical limitations of this study, an encouraging 118 MPa was obtained for type 2 SAY joined SiC/SiC, comparable to 122 ± 10 MPa measured on the non-irradiated joined sample (Ferraris *et al.*, 2012).

17.3.2 Mechanical shear tests for joined ceramic samples

The increasing application of joining and integration technologies in aerospace as well in ground-based systems requires more emphasis on analytical models to predict failure and consequently for the safety standards. As soon as analytical models and calculation codes become available, they must be validated by experimental testing. A widely accepted standard to test the mechanical strength of joined ceramics and composites is still unavailable and measuring the shear strength of the same joining material with different test methods can produce different results (Serizawa *et al.*, 2007; Ferraris *et al.*, 2010b; Ventrella *et al.*, 2010).

A homogeneous pure shear stress state is not obtainable with most of the currently used single or double lap tests, which give the apparent and not the pure shear strength of the joined samples; in addition, the presence of several different apparent shear strength tests in the literature makes comparison quite impossible.

The asymmetrical four-point bending test (ASTM C1469-10, 2010) is the standard test used the ASTM to test joined ceramics and composites, but notches have to be cut in the joined area when the joining material strength is high. When the shear strength of the joining material approaches that of the substrate to be joined, ASTM C1469-10 is not suitable.

With torsion tests (ASTM F734-95, 2006; ASTM F1362-97, 2003) the pure shear loading strength can be measured without using notches (Ferraris *et al.*, 2010a). Some examples of tested configurations are sketched in Fig. 17.11. In the first row samples are loaded in torsion; for all other tests, shear is obtained by loading the specimens in compression.

17.3.3 Future trends

Glass and glass-ceramic joining for SiC-based materials is a promising, pressureless, user-friendly way to widen the use of SiC-based components and their maintenance. The availability of an almost infinite range of glass and glass-ceramic compositions can give solutions for joining SiC components to themselves and to dissimilar materials.

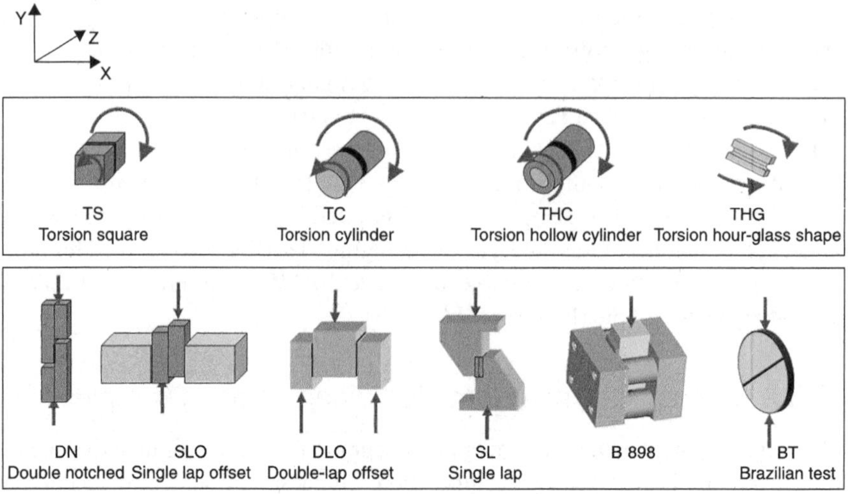

17.11 Shear and apparent shear tests for joined ceramics.

A critical point is still the softening of the amorphous phase, which must be minimized by suitable thermal treatment: further studies are sought on this topic. Repair issues also remain to be solved, with some advances in the field of laser sealing showing promising results.

The lack of published studies of high-temperature mechanical tests on glass-ceramic joined SiC components should be filled by the scientific community to make these solutions reliable and widely accepted.

Also, the implementation of a suitable mechanical test to measure the effectiveness of the joined components should be a topic of more intensive investigation.

Finally, glass-ceramic joined SiC-based components, which gave promising but not statistically relevant results, should be further tested in a nuclear environment.

17.4 References

ASTM C1469-10 (2010), *Standard test method for shear strength of joints of advanced ceramics at ambient temperature.*

ASTM F1362-97 (2003), *Standard test method for shear strength and shear modulus of aerospace glazing interlayer materials.*

ASTM F734-95 (2006), *Standard test method for shear strength of fusion bonded polycarbonate aerospace glazing material.*

Baldev R, Vijayalakshmi M, Rao V P R and Rao S K B (2008), 'Challenges in materials research for sustainable nuclear energy', *Harnessing Materials for Energy MRS Bulletin*, 33(4), 327–340.

Bansal N P and Gamble E A (2005), 'Crystallization kinetics of a solid oxide fuel cell seal glass by differential thermal analysis', *J Power Sources*, 47, 107–115.

Chou Y-S, Stevenson J and Gow R N (2007b), 'Novel alkaline earth silicate sealing glass for SOFC. Part I. The effect of nickel oxide on the thermal and mechanical properties', *J Power Sources*, 168, 426–433.

Chou Y-S, Stevenson J and Gow R N (2007c), 'Novel alkaline earth silicate sealing glass for SOFC. Part II. Sealing and interfacial microstructure', *J Power Sources*, 170, 395–400.

Chou Y-S, Stevenson J W and Singh P (2007a), 'Novel refractory alkaline earth silicate sealing glasses for planar solid oxide fuel cells', *J Electrochem Soc*, 154, B644–B651.

Chou Y-S, Stevenson J W and Singh P (2008), 'Effect of aluminizing of Cr-containing ferritic alloys on the seal strength of a novel high-temperature solid oxide fuel cell sealing glass', *J Power Sources*, 185, 1001–1008.

Chou Y-S, Thomsen E C, Williams R T, Choi J-P, Canfield N L, *et al.* (2011), 'Compliant alkali silicate sealing glass for solid oxide fuel cell applications: Thermal cycle stability and chemical compatibility', *J Power Sources*, 196, 2709–2716.

Cockeram B V (2005), 'Flexural strength and shear strength of silicon carbide to silicon carbide joints fabricated by a molybdenum diffusion bonding technique', *J Am Ceram Soc*, 88, 1892–1899.

Colombo P, Riccardi B, Donato A and Scarinci G (2000), 'Joining of SiC/SiC_f ceramic matrix composites for fusion reactor blanket applications', *J Nucl Mater*, 278, 127–135.

Coon D N and Neilson Jr RM (1989), 'Wetting of silicon carbide surfaces by MgO-Li_2O-Al_2O_3-SiO_2 glasses', *Ceram Eng Sci Proc*, 10, 1735–1744.

Donald I W, Mallinson P M, Metcalfe B L, Gerrard L A and Fernie J A (2011), 'Recent developments in the preparation, characterization and applications of glass- and glass–ceramic-to-metal seals and coatings', *J Mater Sci*, 46, 1975–2000.

Ehrlich K, Gasparotto M, Giancarli L, Le Marois G, Malang S and Van der Schaaf B (2001), *European Material Assessment Meeting*, June 2001 EFDA-T-RE-2.0.

Esposito L, Bellosi A and Landi S (1999), 'Interfacial forces in Si_3N_4- and SiC-based systems and their influence on the joining process', *J Am Ceram Soc*, 82, 3597–3604.

Fergus J W, 'Sealants for solid oxide fuel cells', *J Power Sources*, 2005, 147, 46–57.

Fernie J A, Drew R A L and Knowles K M (2009), 'Joining of engineering ceramics', *Int Mater Rev*, 54, 283–331.

Ferraris M, Casalegno V, Rizzo S, Salvo M, Van Staveren T O and Matejicek J (2012), 'Effects of neutron irradiation behavior on glass ceramics as pressure-less joining materials for SiC based components for nuclear applications', submitted to *Journal of Nuclear Materials*.

Ferraris M, Salvo M, Casalegno V, Ciampichetti A, Smeacetto F and Zucchetti M (2008), 'Joining of machined SiC/SiC composites for thermonuclear fusion reactors', *J Nucl Mater*, 375, 410–415.

Ferraris M, Salvo M, Casalegno V, Han S, Katoh Y, *et al.* (2011b), 'Joining of SiC-based materials for nuclear energy applications', *J Nucl Mater*, Vol., 417, 2011, 379–382.

Ferraris M, Salvo M, Casalegno V, Rizzo S and Ventrella A (2010a), 'Joining and integration issues of ceramic matrix composites for nuclear applications', *Ceram Trans*, 220, 173–186.

Ferraris M, Salvo M, Isola C, Appendino Montorsi M and Kohyama A (1998), 'Glass-ceramic joining and coating of SiC/SiC for fusion applications', *J Nucl Mater*, 258–263, 1546–1550.

Ferraris M, Ventrella A, Salvo M, Avalle M, Pavia F and Martin E (2010b), 'Comparison of shear strength tests on AV119 epoxy -joined carbon/carbon composites', *Composites: Part B*, 41, 182–191.

Gasse A, Saint Antonin F and Coing Boyat G (1997), 'Specific non-reactive BraSiC alloys for SiC–SiC joining', *Report CEA-Grenoble*. DEM n. DR 25.

Giancarli L, Golfier H, Nishio S, Raffray A R, Wong C P C and Yamada R (2002), 'Progress in blanket designs using SiC_f/SiC composites', *Fus Eng Des*, 61–62, 307–318.

Glass J, Mahoney F M, Quillan B, Pollinger J P and Loehman R E (1998), 'Refractory oxynitride joints in silicon nitride', *Acta Mater*, 46, 2393–2399.

Goel A, Pascual M J and Ferreira J M F (2010), 'Stable glass-ceramic sealants for solid oxide fuel cell', *Int J Hydrogen Energy*, 35, 6911–6923.

Gross S M, Federmann D, Remmel J and Pap M (2011), 'Reinforced Composite Sealants for SOFC applications', *J Power Sources*, 196(17), 5, doi:10.1016/j.jpowsour.2011.02.002.

Harrison S and Marcus H L (1999), 'Gas-phase selective area laser deposition (SALD) joining of SiC', *Mater Des*, 20, 147–152.

Henager C H, Shin Y Jr, Blum Y, Giannuzzi L A, Kempshall B W and Schwarz S M (2007), 'Coatings and joining for SiC and SiC-composites for nuclear energy systems', *J Nucl Mater*, 367–370, 1139–1143.

Hinoki T, Eiza N, Son S, Shimoda K, Lee J and Kohyama A (2005), 'Development of joining and coating technique for SiC and SiC/SiC composites utilizing NITE processing', *Ceram Eng Sci Proc*, 26, 399–405.

Jordan R (2008), 'Is the future of SOFCs sealed in a glass?', *Amer Ceram Soc Bull*, 87, 26–29.

Jung H-C, Park Y-H, Park J-S, Hinoki T and Kohyama A (2009), 'R&D of joining technology for SiC components with channel', *J Nucl Mater*, 386–388, 847–851.

Katoh Y, Kohyama A, Nozawa T and Sato M (2004), 'SiC/SiC composites through transient eutectic-phase route for fusion applications. Part 1', *J Nucl Mater*, 329–333, 587–591.

Katoh Y, Kotani M, Kohyama A, Montorsi M, Salvo M and Ferraris M (2000), 'Microstructure and mechanical properties of low activation glass-ceramic joining and coating for SiC/SiC composites', *J Nucl Mater*, 283–287, 1262–1266.

Kishimoto H, Ozawa K, Hashitomi O and Kohyama A (2007), 'Microstructural evolution analysis of NITE SiC/SiC composite using TEM examination and dual-ion irradiation', *J Nucl Mater*, 367–370, 748–752.

Le S, Sun K, Zhang N, Shao Y, An M, *et al.* (2007), 'Comparison of infiltrated ceramic fiber paper and mica base compressive seals for planar solid oxide fuel cells', *J Power Sources*, 168, 447–452.

Lee H L, Nam S W, Hahn B S, Park B H and Han D (1998), 'Joining of silicon carbide using $MgO-Al_2O_3-SiO_2$ filler', *J Mater Sci*, 33, 5007–5014.

Lewinsohn C A, Colombo P and Reimanis I (2001), 'Stresses occurring during joining of ceramics using preceramic polymers', *J Am Ceram Soc*, 84, 2240–2244.

Lewinsohn C A, Jones R H, Colombo P and Riccardi B (2002), 'Silicon carbide-based materials for joining silicon carbide composites for fusion energy applications', *J Nucl Mater*, 307–311, 1232–1236.

Lin S-E, Cheng Y-R and Wei W C J (2011), 'Synthesis and long-term test of borosilicate-based sealing glass for solid oxide fuel cells', *J Eur Ceram Soc*, 31, 1975–1985.

Lippmann W, Knorr J, Wolf R, Rasper R, Exner H, *et al.* (2004), 'Laser joining of silicon carbide – a new technology for ultra-high temperature resistant joints', *Nucl Eng Des*, 231, 151–161.

Loehman R E, Dumm H P and Hofer H (2002), 'Evaluation of sealing glasses for solid oxide fuel cells', *Ceram Eng Sci Proc*, 23, 699–705.

Mahapatra M K and Lu K (2010a), 'Seal glass for solid oxide fuel cells', *J Power Sources*, 195, 7129–7139.

Mahapatra M K, Lu K and Bodnar R J (2009), 'Network structure and thermal property relation of a novel high temperature seal glass', *Appl Phys A*, 95, 493–500.

Mahapatra M K, Lu K, Liu X and Wu J (2010b), 'Compatibility of a seal glass with $(Mn,Co)_3O_4$ coated interconnects: effect of atmosphere', *Int J Hydrogen Energy*, 35, 7945–7956.

Meinhardt K D, Kim D S, Chou Y S and Weil KS (2008), 'Synthesis and properties of a barium aluminosilicate solid oxide fuel cell glass-ceramic sealant', *J Power Sources*, 182, 188–196.

Nielsen K A, Solvang M, Nielsen S B L, Dinesen A R and Larsen P H (2007), 'Glass composite seals for SOFC application', *J Eur Ceram Soc*, 27, 1817–1822.

Pascual M J, Guillet A and Durán A (2007), 'Optimization of glass–ceramic sealant compositions in the system MgO–BaO–SiO_2 for solid oxide fuel cells (SOFC)', *J Power Sources*, 169, 40–46.

Pascual M J, Lara C and Durán A (2006), 'Non-isothermal crystallisation kinetics of devitrifying RO-BaO-SiO_2 (R = Mg, Zn) glasses', *Eur J Glass Sci Tech B*, 47, 572–581.

Perham T J, de Jonghe L C and Moberly Chan W J (1999), 'Joining of silicon carbide with a cordierite glass-ceramic', *J Am Ceram Soc*, 82, 297–305.

Riccardi B, Fenici P, Frias Rebelo A, Fiancarli L, Le Marois G and Philippe E (2000), 'Status of the European R&D activities on SiCf/SiC composites for fusion reactors', *Fus Eng Des*, 51–52 , 11–22.

Riccardi B, Nannetti C, Petrisor T, Woltersdorf J, Pippel E, *et al.* (2004), 'Issues of low activation brazing of SiCf/SiC composites by using alloys without free silicon. Part 1', *J Nucl Mater*, 329–333, 562–566.

Serizawa H, Fujita D, Lewinsohn C A, Singh M and Murakawa H (2007), 'Numerical analysis of mechanical testing for evaluating shear strength of SiC/SiC composite joints', *J Nucl Mat*, 367–370, 1223–1227.

Singh M (1999), 'Design, Fabrication and Characterization of High Temperature Joints in Ceramic Composites', *Key Engineering Materials*, 415, 164–165.

Singh M, Asthana R and Shpargel T (2008), 'Brazing of ceramic-matrix composites to Ti and Hastelloy using Ni-base metallic glass interlayers', *Mater Sci Eng A*, 498, 19–30.

Singh M, Shpargel T P and Asthana R (2007), 'Brazing of stainless steel to yttria-stabilized zirconia using gold-based brazes for solid oxide fuel cell applications', *Int J Appl Ceram Technol*, 4, 119–133.

Singh R N (2007), 'Sealing technology for solid oxide fuel cells (SOFC)', *Int J Appl Ceram Technol*, 4, 134–144.

Smeacetto F, Chrysanthou A, Salvo M, Moskalewicz T, D'herin Bytner F, *et al.* (2011a), 'Thermal cycling and ageing of a glass-ceramic sealant for planar SOFCs', *Int J Hydrogen Energy*, 36, 11 895–11 903.

Smeacetto F, Chrysanthou A, Salvo M, Zhang Z and Ferraris M (2009), 'Performance and testing of glass-ceramic sealant used to join anode-supported-electrolyte to Crofer 22 APU in planar SOFCs', *J Power Sources*, 190, 402–407.

Smeacetto F, Salvo M, Ferraris M, Casalegno V and Asinari P (2008a), 'Glass and composite seals for the joining of YSZ to metallic interconnect in solid oxide fuel cells', *J Eur Ceram Soc*, 28, 611–616.

Smeacetto F, Salvo M, Ferraris M, Casalegno V, Asinari P and Chrysanthou A (2008c), 'Characterization and performance of glass-ceramic sealant to join metallic interconnects to YSZ and anode-supported-electrolyte in planar SOFCs', *J Eur Ceram Soc*, 28, 2521– 2527.

Smeacetto F, Salvo M, Ferraris M, Cho J and Boccaccini A R (2008b), 'Glass–ceramic seal to join Crofer 22 APU alloy to YSZ ceramic in planar SOFCs', *J Eur Ceram Soc*, 28, 61–68.

Smeacetto F, Salvo M, Leone P, Santarelli M and Ferraris M (2011b), 'Performance and testing of joined Crofer22APU-glass-ceramic sealant-anode supported cell in SOFC relevant conditions', *Mater Letters*, 65, 1048–1052.

Snead L L, Nozawa T, Ferraris M, Katoh Y, Shinavski R and Sawan M (2011), 'Silicon carbide composites as fusion power reactor structural materials', *J Nucl Mater*, 417, 330–339.

Steele B C H and Heinzel A (2001), 'Materials for fuel-cell technologies', *Nature*, 414, 345–352.

Tavassoli A A F (2002), 'Present limits and improvement of structural materials for fusion reactors – a review', *J Nucl Mater*, 302, 73–88.

Ventrella A, Salvo M, Avalle M and Ferraris M (2010), 'Comparison of shear strength tests on AV119 epoxy-joined ceramics', *J Mater Sci*, 45, 4401–4405.

Wang S F, Chiu T W, Lu C M, Wu Y C and Yang Y C (2011), 'La_2O_3-Al_2O_3-B_2O_3-SiO_2 glasses for solid oxide fuel cell applications', *Int J Hydrogen Energ*, 36, 3666–3672.

Yang Z (2008), 'Recent advances in metallic interconnects for solid oxide fuel cells', *Int Mater Rev*, 53, 39–54.

Yurkov A L, Polyak B I and Murahver T V (1991), 'Interaction between silicon carbide and melt of aluminoborosilicate glass', *J Mater Sci Letters*, 10, 1342–1343.

Zhang T, Brow R K, Reis S T and Ray C S (2008), 'Isothermal crystallization of a solid oxide fuel cell sealing glass by differential thermal analysis', *J Am Ceram Soc*, 91, 3235–3239.

Zucchetti M, Ferraris M and Salvo M (2001), 'Safety aspects of joints and coatings for plasma facing components with composite structures', *Fus Eng Des*, 58–59, 939–943.

18

Brazing of nickel-based filler metals for pipes and other components in contact with drinking water

I. HOYER and B. WIELAGE, Chemnitz University of Technology, Germany

DOI: 10.1533/9780857096500.3.545

Abstract: The amendment of the Drinking Water Directive has reduced the permissible nickel ion content in drinking water from 50 μg/l to 20 μg/l. The development of draft standards has begun on a European level so that national standards can be adjusted and transformed to meet European and international standards. For example, DIN EN 15664-1, valid since June 2008, regulates the influence of metallic materials on water intended for human consumption. Since heat exchangers joined with nickel-based filler metals are applied in the equipment for drinking water systems, there is a possible risk of exceeding the limits for nickel ions. So far, components joined with nickel-based filler metals have been applied according to the 5% rule defined in DIN 50930-6. Under the present circumstances, the ban on nickel-based filler metals as a result of the amendment of the Drinking Water Directive will cause severe economic losses for the suppliers of solders as well as for customers. Nickel-based filler metals have not yet been included in the list of harmless materials. Investigations on the release of nickel ions into drinking water carried out according to DIN EN 15664-1 are presented. These investigations are one step forward for the inclusion of nickel-based filler metals in the list of harmless materials. Further measures, which will also be reported in this contribution, have to be taken.

Key words: nickel-based brazing fillers, drinking water, corrosion test, nickel leaching, nickel ion content in drinking water, heat exchangers.

18.1 Introduction: brazing filler metals for corrosion-resistant applications

Nickel-based alloys are typical choices for corrosion-resistant joints in large-scale applications. They are alloyed with the elements boron, silicon and phosphorus, which lower the melting point. One advantage of these metalloids is that they improve the wettability of the base materials. However, they can result in the formation of brittle intermetallic phases. In particular, boron binds with chrome, which is alloyed to increase corrosion resistance, to form brittle chromium boride. In the worst case these phases develop as a continuous band in the middle of the gap (see Fig. 18.1) and can result in a drastic degradation of strength.[1]

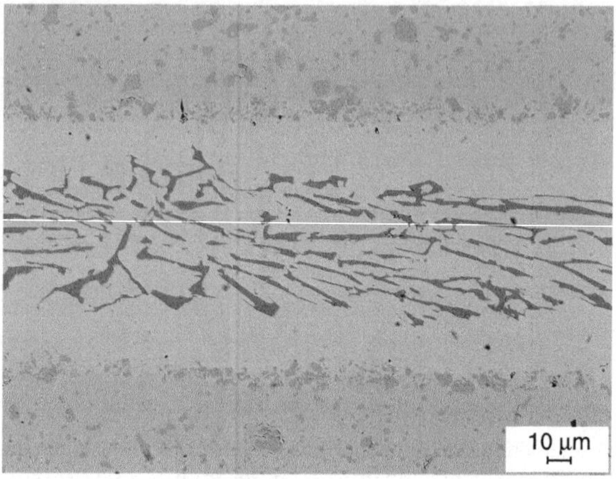

18.1 Continuous band of brittle phases in the middle of the gap; filler, NI102.[1]

The typical dependence of the gap properties of nickel-based fillers on the gap width can be explained as follows: the band of brittle phases can be avoided with short diffusion paths in strict accordance with a critical gap width. Depending on the filler metal, brazing temperature and holding time, the critical gap width can be adjusted between 50 and 200 μm. As a result of this gap width adjustment, the metalloids diffuse into the base metal without forming bands of brittle phases in the middle of the gap. Above the critical gap width such a band of brittle phases is inevitable, as shown in Fig. 18.1. However, creating these very narrow gaps results in high manufacturing costs. These high costs can be reduced with appropriate process parameters, such as extended holding times or a suitable heat treatment in order to dissolve the formed hard phases or to reduce their amount. It is, however, a time-consuming and costly process, which can result in damage to the base metal. In addition, complete dissolution of the brittle phases cannot be achieved in phosphorus-containing nickel-based fillers due to the low diffusion rate of phosphorus. Furthermore, boron-containing filler metals diffuse at the edge of the soldered seam or into the base material because of boron's high rate of diffusion. The high affinity between boron and chromium can result in the reduction of chromium in the base material in combination with the formation of chromium borides. If the base-material content close to the gap decreases below the critical value of about 13 wt.% there is insufficient corrosion resistance.[1,2]

Because nickel-based alloys are characterized by excellent flow properties, good gap-filling assets and very good mechanical bond strength, they have a wide range of applications despite the risk of brittle phases. Furthermore, they have been the subject of a significant amount of research for a long time. The research includes both process technology and metallurgical approaches.[2–9]

18.2 Materials and components in drinking water installations

The quality and the composition of drinking water are subject to strict guidelines. Detailed conditions for the chemical composition are imposed in addition to microbiological requirements. It is very important to observe the maximum permissible limits on metallic elements for drinking water installation systems through a selection of appropriate materials.

The usability of a material in drinking water installation systems is limited by technical considerations (the tolerated mass transfer between the material and drinking water), health and safety, and the impact on the smell and taste of the drinking water.[10] The feasibility of a material as a component of a drinking water installation system is determined by its compliance with a specified parameter for drinking water regulation (PW). This parameter correlates to the average weekly water consumption by consumers. The evaluation of a material is based on Eq. 18.1 using data determined by DIN EN 15664-1 for pipes with small diameters and with an installation parameter B:[11]

$$M(T_n) \le W(15) = \mathrm{PW} \cdot \left(\frac{A}{B}\right) \qquad [18.1]$$

where:

M = arithmetic average of the measured values for the concentration C
T_n = last measurement phase
PW = parameter value
A = ratio of the area of water surface in contact with the material to be evaluated to the entire area of water surface in contact with the test track according to DIN 50930-6
B = installation parameter from DIN 50930-6

A material can be used in drinking water installation systems if the condition in the above equation is satisfied. Evaluating this is difficult at present, because there are several valid standards (EN 15664-1 and DIN 50930-6), which must be considered.

Table 18.1 shows the determined parameter values (PW) of the materials that are commonly used for components of drinking water installation systems, such as pipes, fittings and valves.

The standard DIN 50930-6 includes additional limits on the pH, the contained total organic carbon (TOC), and acids and bases as well as the maximum capacity for the unavoidable mass of accompanying elements (e.g. arsenic, sulfur, phosphorus) for the applicability of the materials. It regulates the use conditions for

- carbon and zinc-plated steels
- copper, tin-plated copper and copper alloys
- stainless steel
- lead

Table 18.1 Determined parameter values for materials used in drinking water installation systems according to DIN 50930-6

Installation component	Material	Parameter	Parameter value (PW)
Pipes and fittings	Copper	Cu	2 g/m^3
		Pb	10 mg/m^3
		Cd	5 mg/m^3
	Zinc-plated steel	Nitrite	0.5 g/m^3
		Ammonium	0.5 g/m^3
	Lead	Pb	10 mg/m^3
	Carbon steel	Fe	200 mg/m^3
Valves and connecters	Cu–Sn–Zn alloys	Cu	2 g/m^3
		Pb	10 mg/m^3
		Ni	20 mg/m^3
	Cu–Zn alloys	Cu	2 g/m^3
		Pb	10 mg/m^3
		As	10 mg/m^3
		Cu	2 g/m^3
	Cu–Ni alloys	Ni	20 mg/m^3
		Cu	2 g/m^3
Water heater		Ni	20 mg/m^3
		Nitrite	0.5 g/m^3
		Ammonium	0.5 g/m^3

Nickel and nickel alloys are not listed explicitly in this standard. However, nickel-based materials are applied as brazing fillers in plate heat exchangers and valves in the layered systems in crucial areas, such as in domestic installations and the chemical and food industry.

Valves for drinking water installation systems are coated with chromium to protect against mechanical abrasion and for decorative reasons. During manufacture a nickel layer is applied first with a thickness of 0.5 to 25 μm by galvanic deposition in order to achieve good adhesion of the chrome layer. Openings in the components are left open during the application of nickel layers, so that interior surfaces will also be partially coated. In the subsequent application of the chrome layer these areas are only partly covered and can release nickel into drinking water. The valves may only be used if 'no more than 20% of the total water-wetted inner surface of the fitting' is covered with the nickel layer.[12]

The investigations in Werner and Klinger[12] for heavy metal ion release were performed according to DIN EN 15664-1. Flexible connection hoses with nickel-plated connectors were connected to a test system in accordance with DIN EN 15664-1. The samples were tested in their original condition and after removal of the nickel layer. The test water was a medium-hard groundwater with a pH of 7.1 to 7.4. After predetermined periods of stagnation the test samples were removed and the arithmetic mean of the metal-ion concentration determined.

The results showed that the threshold value of the nickel-ion concentration (20 μg/l) was sometimes clearly exceeded. Depending on the type of connection tube, the concentration did not increase until a few weeks later. However, the nickel concentration then decreased in the course of the experiments to below the determined limit values. This time-dependent nickel ion leaching is not to be neglected. Because of this, requirements for nickel-coated components of drinking water installation systems must be redefined.[12]

In general, these values are indicative of the unsuitability of nickel materials in general, including for a plate heat exchanger as explained above. They are a low-cost alternative with relatively low investment costs and operating expenses in comparison to the industrially used pipe bundle heat exchangers and spiral heat exchangers. Furthermore, plate heat exchangers have high heat transference capacity and a compact design suitable for small spaces. The steel X5CrNiMo17-12-2 (1.4401) is almost exclusively used for heat exchangers in building technology. These plate heat exchangers are constructed of profiled metal sheets, which are layered into parcels and joined by either screwing or brazing them together. The plate-type geometry forms a parallel or superimposed system of narrow ducts. The construction of a plate heat exchanger can be seen in Fig. 18.2. Drinking and heating water flows in counter-currents separated by plates through the ducts. The coined shape of the plates causes a highly turbulent percolation, which, in combination with the plate's low thickness of approximately 150 μm, results in effective heat transfer from the primary to the secondary side.[13,14]

Heat exchangers are joined by high-temperature brazing with integrated heat treatment, and utilize copper or nickel-based fillers. The process parameters are a

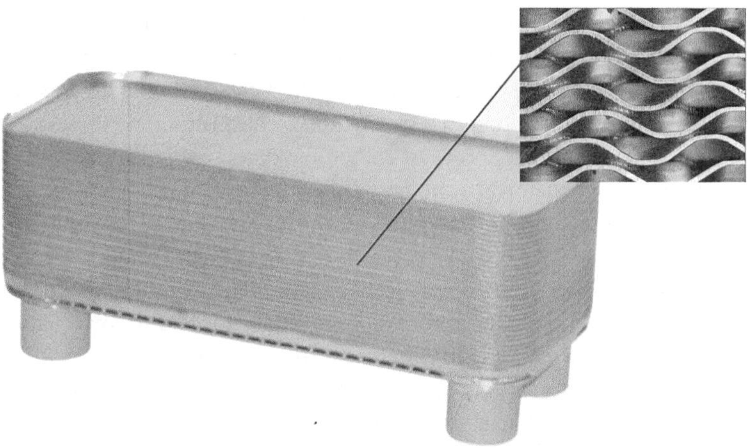

18.2 Architecture of a plate heat exchanger.[13]

vacuum of approximately 5×10^{-4} mbar and a heat treatment temperature of 1040°C. This temperature is held steady until the components are homogeneously heated. Only then is the temperature increased to a brazing temperature of 1100°C. Rapid cooling with nitrogen occurs after the brazing process. This ensures solution annealing for the non-stabilized austenitic steel 1.4401. During the heat treatment the secondary phases (carbide removals and intermetallic phases) dissolve and the quick cooling avoids further release. The generated austenite is supersaturated with alloy additions and corresponds to the equilibrium solubility at the annealing temperature.[15–17] A similar procedure is also used for tool steels, which are joined and tempered in a combined brazing–heat-treatment process.

Plate heat exchangers are exposed to thermal, mechanical and corrosive loads during use and therefore require a suitable choice of material, including filler materials. The maximum allowed temperature for drinking water is 60°C, while heating water temperatures can reach up to 90°C. A high thermodynamic demand must be expected because the temperature of the drinking water at the entry of a plate heat exchanger is between 10°C and 15°C. If there is an irregular workload the alternating between cooling and reheating results in marked thermal cycling, which can lead to cracking. The mechanical stress results from the pressure inside the drinking water pipelines of approximately 5 bar. However, pressure peaks up to 20 bar are possible, especially with quickly closing armatures (magnet valves). This is where brazed heat exchangers offer a decisive advantage, because seals are not required, which leads to a higher pressure loading capacity than with screwed heat exchangers. Nevertheless, the focus of a heat exchanger's use must be corrosive stress, which will be considered below.[18–20]

In the past, serious corrosion damage was often found in copper-brazed plate heat exchangers used for warming drinking water. Plate heat exchangers show no or only limited corrosion on the side for heating water, but strong corrosion on the side for drinking water (see Fig. 18.3). This is due to the different amounts of dissolved oxygen in the diverse water cycles. Fresh drinking water is highly oxygenated, while in a closed heating water cycle, oxygen barely exists, because it is used up by corrosion processes within a short time. The presence of oxygen always leads to the possibility of cathodic oxygen reduction. This is the driving force for electrochemical (anodic) dissolution of metal. Hence, an enhanced danger of corrosion can be found in the circulation of oxygenated drinking water.[21,22]

Although copper and stainless steel are corrosion-resistant materials with high electrochemical potential, a corrosion cell can arise in a combination of both materials. Copper is a slightly more non-precious metal than stainless steel, so local corrosion can appear and copper can dissolve anodically. The reaction proceeds with high intensity in conductive water (200–1000 µS), because in a plate heat exchanger the surface difference between the steel and copper-based filler is very large. The dissolution of the copper-based filler results in leakage on

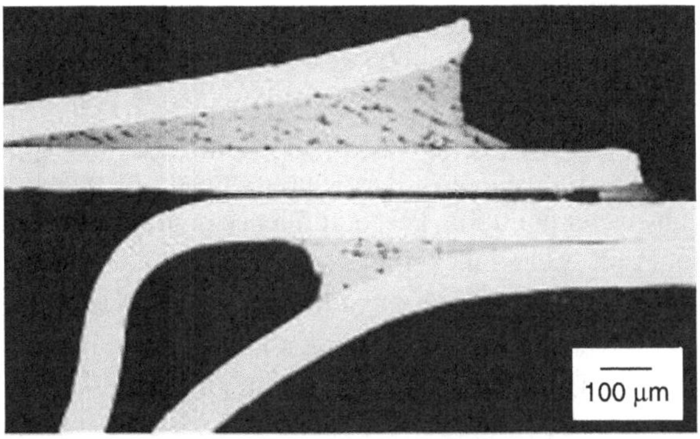

18.3 Cross-section of the edge of a plate heat exchanger: completely dissolved filler on the side for drinking water, no corrosion on the side for heating water.[20]

the one hand and increased copper-ion concentration in drinking water on the other. The heating water and the drinking water can mix or leak out to the exterior surface. Especially in warm water systems, this leads to the limit value for copper ions being exceeded and, in combination with zinc-coated steel pipes, to heavy corrosion problems. In order to prevent pitting corrosion components of copper alloys in assembly, lines should not use zinc-coated steel pipes.[23] That is why, according to the appropriate standard of the German Institute for Standardization DIN 1988-7 (Technical rules for drinking water installations. Part 7: Avoidance of corrosion damages and stone education[23]), the combination of copper and zinc-coated steel is not allowed. Only plate heat exchangers used for heating water are brazed using copper-based fillers. These corrosion phenomena also depend on the chemical composition of the water (salt, sulfate and chloride content). For the warming of drinking water, nickel-based fillers are used because of their higher corrosion resistance.[13,14,20,23–26]

18.3 Current drinking water regulations and standards

The Drinking Water Directive (DWD), amended on 21 May 2001, came into effect on 1 January 2003. It requires the implementation of the EC guideline 98/83 on 'the quality of water for human use' into national law. The directive gives limit values for pathogens (microbiological requirements) and indication parameters, as well as for chemical substances, to protect the health of the consumers in the European Union by making sure that water is safe to drink.[27] The allowed concentration of leached nickel ions was reduced from 50 µg/l to 20 µg/l in drinking water by an amendment at the beginning of 2003. At the same time, norm drafts began to be developed to adapt or transfer national standards to the European level. In installations of drinking water systems, nickel-based fillers are often used for joining heat exchangers. Therefore, there is a danger of exceeding the limit value for leached nickel ions. In order to measure the amount of leached nickel ions, a procedure is required to test a representative water sample for average weekly parameters, according to the Drinking Water Directive. Until then, DIN 50930-6 (Corrosion of metallic materials inside of pipes, tanks and apparatus corrosion by water pollution, Part 6: Influence of drinking water quality) is valid for the amount of leached nickel ions. As of June 2008, the standard for testing is DIN EN 15664-1 (Influence of metallic materials on water intended for human consumption – Dynamic rig test for assessment of metal release – Part 1: Design and operation[28]). Until now the application of nickel-based fillers to component joints has followed the 5% rule found in DIN 50930-6: the 'use of nickel-based fillers in components, e.g., for electric pipe heaters, is allowed if the water-contacting surface is not greater than 5% of the whole water-contacting surface of the component. In all other cases an individual verification is required.' Until now, this verification had followed the German Institute for Standardization 50931-1.[29] Since DIN EN 15664-1 came into effect, there is uncertainty with regard to the test rules of DIN 50930-6,[11] which is not completely aligned with DIN EN 50930-6. Until July 2009 all standards mentioned above were valid, which caused confusion among consumers. In the meantime DIN 50931-1[29] has been withdrawn.

The use of nickel-based fillers has been questioned by various people, such as by Pajonk:[13] 'With the implementation of the Drinking Water Directive nickel-based fillers are no longer permitted because of their ability to leach nickel ions into the drinking water.' For participating industries the search for different fillers in conjunction with approved corrosion tests began. Restrictions were imposed on individual companies on the use of nickel-based fillers, which did not have proof of appropriate testing, in order to comply with the Drinking Water Directive before DIN EN 15664-1 came into effect. This is especially true for foil-brazed plate heat exchangers, for which it is assumed that the limit values cannot be observed due to the large brazing-coated surface in contact with drinking water.

Appropriate tests so that nickel-based fillers can be included in the list of acceptable substances (CPDW: Construction Products in contact with Drinking Water), have not been performed so far. The investigations are very complex and expensive and thus, in particular for SMEs, extremely difficult to implement.

Due to the decrease of the maximum permissible leached nickel ions in drinking water, efforts have been made to forgo nickel-based fillers in drinking water installations in the future. Iron-based fillers have great potential for use in such installation systems and a high economic advantage. To ensure sufficient corrosion resistance, the composition of iron-based fillers should be based on the widely used austenitic stainless steels. These have relatively high nickel content, depending on the type of steel up to 31 wt.%. Schwenk *et al.* have already shown that even the high nickel content in austenitic steels (and nickel-based fillers to be developed) is not expected to increase nickel ion leaching.[30] It can also be shown that in liquids with low pH values, even in the presence of strong complexing agents, nickel ion leaching does not increase. Based on these results, the use of stainless steels in drinking water installations in association with DIN 50930-6[11] has been allowed, regardless of the composition. The leaching of iron or chromium ions into drinking water is not expected, because Fe(III) and Cr(III) oxides are nearly insoluble in water in comparison with Ni(II) oxides. When using an iron-based filler with a similar chemical composition to stainless steel the limits for metallic elements in the Drinking Water Directive will not be exceeded.

18.4 Test rig and samples

Currently it is not known whether brazing filler metals have been tested according to DIN EN 15664-1 and DIN 50931-1. A test rig was designed in Chemnitz in accordance with DIN EN 15664-1, Figure A.2, for fittings. The prescribed form pieces (Fig. 18.5) with a length of 100 mm and an inner diameter of 17 ± 0.3 mm have to be arranged per test section (a total of three per sample) with five pieces each (Fig. 18.4).

This results in a ratio of the water-contacting surface of the sample to the total surface area of 0.9 per test line. The test rig was designed for three fillers tested in parallel. Stainless steel 1.4404 (X2CrNiMo17-12-2) was used as reference material. This differs from the reference material in EN 15664-1 (1.4401 X5CrNiMo17-12-2) due to the required welds (see Fig. 18.5).

The inside of the samples was brazed or coated with the particular filler metal. This was achieved through uniform application of brazing filler paste on the inside diameter and subsequent brazing of the fittings per charge in a vacuum-brazing process at a suitable working temperature. The brazing filler metals (standard

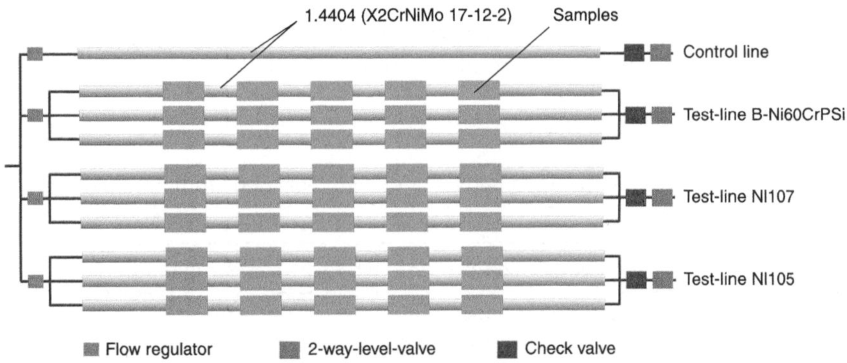

18.4 Schematic configuration of a test rig according to DIN EN 15664-1.

18.5 Sample with brazing coat inside.

nickel-based fillers NI105, NI107 and B-Ni60CrPSi) were chosen in collaboration with a consortium of industries (Table 18.2).

The test rig uses a programmable logic controller (PLC) to monitor and regulate tests in order to guarantee continuous operation in accordance with the 24 h cycle defined in DIN EN 15664-1. This simulates the drinking water flow rate in an average household with stagnation periods between 0.5 and 8 h. Sampling is performed according to Table 18.3. Because of the validity of both standards at the beginning of the tests, the removal cycles were combined to allow flexible analysis.

In addition to the regular 24 h cycle, longer cycle times of stagnation can occur, for example during the holiday season, so water analyses were carried

Table 18.2 Chemical composition and recommended brazing temperature of the tested nickel-based filler metals

Designation	Chemical composition (wt.%)	Brazing temperature (°C)
NI105	71 Ni, 19 Cr, 10 Si	1190
NI107	76 Ni, 14 Cr, 10 P	980
B-Ni60CrPSi	60 Ni, 30 Cr, 6 P, 4 Si	1090

Table 18.3 Comparison of the removal cycles

Standard	Planned removal week										
	1	2	3	4	6	8	12	16	18	21	26
DIN 50931-1	x	x	x		x		x		x		x
DIN EN 15664-1				x		x	x	x		x	x
removal	x	x	x	x	x	x	x	x	x	x	x

out after stagnation times of 48 h and 72 h in weeks 4 and 12 according to DIN 50931-1.

The connection of the test rig to the urban drinking water supply in Chemnitz was a standard drinking water installation (supply pipe: copper). The temperature in the test room was approximately 20°C. The water temperature differs dependent upon the season. Analysis of the test water, as prescribed in DIN EN 15664-1, was performed in a standardized manner, and the results are shown in Table 18.4. The exceeding of the test water limit value for copper (maximum 77 μg/l) and partially

Table 18.4 Chemical composition of the test water in Chemnitz

Parameter	Measured range	Unit
pH value	8.4–8.6	
Electrical conductance at 20°C	234–245	μS/cm
Dissolved oxygen	9.47–11.4	mg/l
Acid capacity, at pH value of 4.3	0.81–0.86	mmol/l
Total organic carbon (TOC)	1.6–1.7	mg/l
Calcium	27.7–29.1	mg/l
Magnesium	5.7–6.2	mg/l
Potassium	1.3–2.3	mg/l
Sodium	6.4–9.9	mg/l
Phosphate–phosphorus	0.04	mg/l
Silicic acid	5.22–7.61	mg/l
Chloride	19.9–21.2	mg/l
Nitrate	15.5–16.6	mg/l
Sulfate	35.5–39.8	mg/l
Copper (limit value: 10 μg/l)*	**35–77**	μg/l
Iron (limit value: 50 μg/l)*	20–30	μg/l
Manganese (limit value: 5 μg/l)*	< 20[†]	μg/l
Zinc (limit value: 50 μg/l)*	10–**60**	μg/l
Nickel	2	μg/l

*Values determined by the DIN EN 15664-1. The exceedances of concentrations in the test water are marked in bold.

[†]The detection limit of the analysis method used by Südsachsen Wasser GmbH is 20 μg/l.

for zinc (maximum 60 μg/l) for technical reasons is accepted according to the Drinking Water Directive (limit value for copper is 2 mg/l). All samples were analyzed by the certified laboratory of Südsachsen Wasser GmbH. The test volume was approximately 50 ml per sampling. The measurement error for the nickel ion analysis is 15.9%.

18.5 Test results

18.5.1 Determination of parameters *A* and *B* (according to DIN 50930-6)

Parameter *A* is the ratio of the area of the specimen surface in contact with water (A_p) to the total surface area (A_g). The calculations for these areas are listed in Table 18.5.

The ratio of the areas for the samples gives parameter *A* equal to 0.0732. Parameter *B* can be found in the standard DIN 50930-6 in table 3 ($B = 1.0$ for DN 10 to DN 25).

With the parameter value for nickel (PW = 20 μg/l) as well as parameters *A* and *B*, *W*(15) can be determined from Eq. 18.1 to be 1.46 μg/l. At the end of the test period the average value *M*(*T*) must be less than or equal to *W*(15). Table 18.6 lists the maximum concentration and the average values for the measured phases T_{21} and T_{26}. The measurements in Table 18.6 show that at the end of the test period the average values of nickel-ion concentration were below the limiting value *W*(15). As a result the given nickel-based fillers can be regarded as suitable for use in drinking water installations.

Table 18.5 Calculation of sample and total surface area to determine parameters *A* and *B* according to DIN 50930-6

Dimensions	Calculation
Sample surface:	
n_P = 15	$A_P = n_P \cdot d_{iP} \cdot \pi \cdot l_P$;
d_{iP} = 17 mm	
l_P = 100 mm	$A_P = 80\ 111\ mm^2$;
Total surface:	
n_R = 4	$A_G = A_R - A_P = n_R \cdot d_{iR} \cdot \pi \cdot l_R - A_P$;
d_{iR} = 17 mm	
l_R = 5500 mm	$A_G = 1\ 094\ 845\ mm^2$;

A_G (mm²), total test surface; A_P (mm²), sample surface; $n_{P/R}$ number of samples/reference samples; $d_{iP/R}$ (mm), inner diameter of sample/control line; $l_{P/R}$ (mm), length of sample/control line.

Table 18.6 Average and maximum nickel-ion concentration for test periods T_{21} and T_{26}

Filler metal	$C_{max}(T_{21})$ (μg/l)	$C_{max}(T_{26})$ (μg/l)	$M(T_{21})$ (μg/l)	$M(T_{26})$ (μg/l)	$M(T_{26}) \leq W(15)$ ($W(15)$ = 1.46μg/l)
NI105	1.4	1.4	1.2	1.2	fulfilled
NI107	1.5	1.4	1.3	1.1	fulfilled
Ni60CrPSi	2.4	1.1	1.5	1.1	fulfilled

C (μg/l), ion concentration; C_{max} (μg/l), maximum ion concentration; M (μg/l), average of concentration C; T_n (h), test period.

18.5.2 Metal-ion concentration according to periods of stagnation and measuring phases

Concentration is plotted against the stagnation period to create stagnation curves for each measurement phase. The time course of metal ion release is presented as the mean concentration against the duration of the experiment. To create the stagnation curves for every test line (n), the metal-ion concentration (c^*_n) for each stagnation (t) and operating time (T) must be determined according to Eq. 18.2. The metal-ion concentration of individual test lines c_n (T, t) is reduced by the blank value of the control lines (c_{CL}).

$$c^*_n (T, t) = c_n (T,t) - c_{CL} (T,t) \qquad [18.2]$$

Negative results for metal-ion concentration must be specified as values below the detection limit. If the measured concentration of metal in the control line is below the limit, it is: $c^*_n (T,t) = c_n (T,t)$.[28]

18.5.3 Nickel-based filler metals NI105, NI107, Ni60CrPSi

The concentration of nickel ions in drinking water was determined for the three nickel-based fillers Ni60CrPSi, NI107 and NI105. In none of the tested brazing filler metals did the measured concentrations exceed the established limit of the drinking water regulations. The concentration of the measured metal in the test line should be corrected for the blank value (concentration of metal measured in the control line) in the creation of stagnation curves according to Eq. 18.2. This results in some negative values for c^*_n, which should be interpreted as values below the detection limit. Nevertheless, in order to illustrate the metal concentration in the test line, the correction of c^*n was waived and the stagnation curves for the

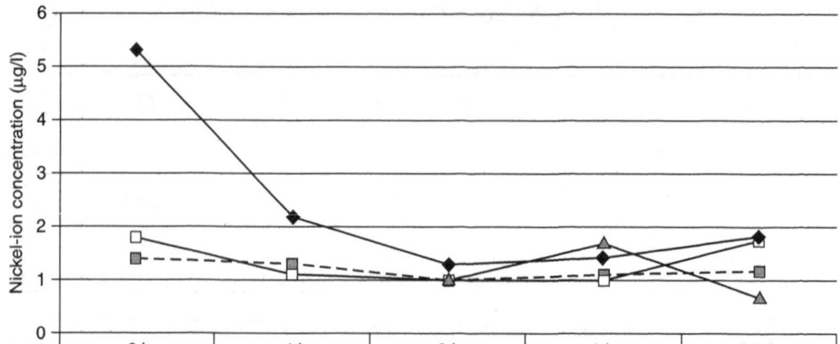

	8 h	4 h	2 h	1 h	0.5 h
Week 1	5.3	2.2	1.3	1.45	1.8
Week 2	1.4	1.3	1	1.1	1.15
Week 3		1.1	1.0	1.7	0.7
Week 4	1.8	1.1	1	1.0	1.8

18.6 Graphs of stagnation for weeks 1 to 4 for test-line NI105.

test line (NI105) and the control line (1.4404) are juxtaposed for comparison (see Fig. 18.6–18.11).

The measured concentrations of nickel ions in the control group and the test line showed similar values. The reason is the relatively high proportion of the reference surface material 1.4404 in the test leads. The nickel alloys comprise only about 7% of the total surface area in contact with water. Under these test conditions, i.e. for area ratios of approximately 1:14 (brazing filler to the base material), the nickel-ion concentration was determined by the reference material. Nevertheless, the results are extremely positive considering that the brazing joints often make up a lower proportion than 7% of the total component.

1st to 4th week

In the first weeks of the experiment, strong interactions between the materials and the drinking water took place, because the formation of the protective passivation layer had not yet occurred, so that relatively high concentrations of nickel ions were measured. For the first week values determined after 0.5 and 8 h for c^*_n with the help of Eq. 18.2 were not below the detection limit:

$$c^*_n(1, 0.5) = 0.5\,\mu g/l$$

$$c^*_n(1, 8) = 2.4\,\mu g/l$$

These results show that most nickel ions from nickel-based fillers dissolved after a stagnation period of 8 h. Reactions between the contact partners are

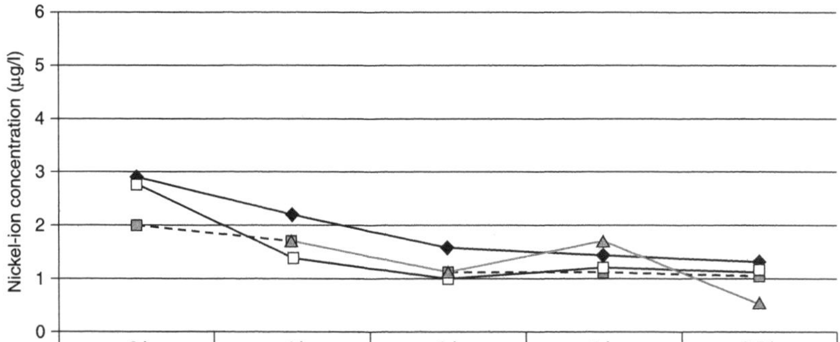

	8 h	4 h	2 h	1 h	0.5 h
Week 1	2.9	2.2	1.6	1.45	1.3
Week 2	2	1.7	1.1	1.1	1.05
Week 3		1.7	1.1	1.7	0.6
Week 4	2.8	1.4	1	1.2	1.1

18.7 Graphs of stagnation for weeks 1 to 4 for control line 1.4404.

increasingly inhibited with the formation of the passivation layer. The passivation of the material is reflected in the decreasing values of the nickel-ion concentration.

6th to 16th week

At week 10 the formation of the passivation layer was considered complete. The concentration of nickel ions remained largely constant even at high

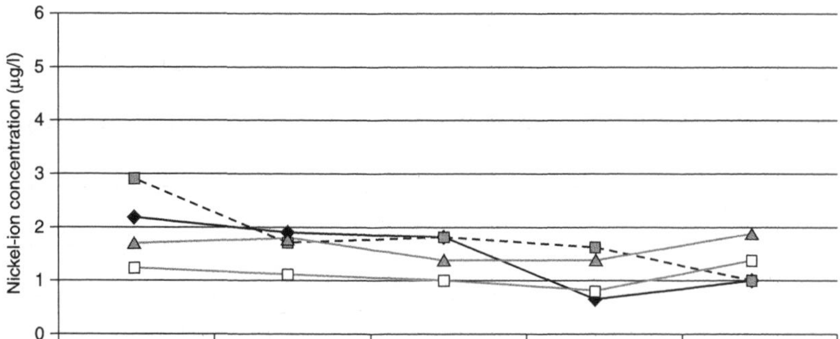

	8 h	4 h	2 h	1 h	0.5 h
Week 1	2.2	1.9	1.8	0.7	1.0
Week 2	2.9	1.7	1.8	1.6	1.0
Week 3	1.7	1.8	1.4	1.4	1.9
Week 4	1.2	1.1	1	0.8	1.3

18.8 Graphs of stagnation for weeks 6 to 16 for test line NI105.

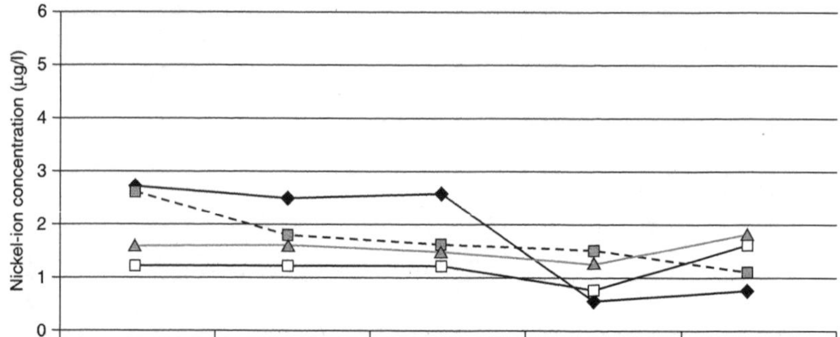

	8 h	4 h	2 h	1 h	0.5 h
Week 6	2.7	2.5	2.6	0.6	0.8
Week 8	2.6	1.8	1.6	1.5	1.1
Week 2	1.6	1.6	1.5	1.3	1.8
Week 16	1.2	1.2	1.2	0.8	1.6

18.9 Graphs of stagnation for weeks 6 to 16 for control line 1.4404.

stagnation periods (4 h or 8 h). The measured values for the test line were 1.0 to 1.2 µg/l.

18th to 26th week

The reactions that result in the passivation of the material surface prevented contact between the partners, so that there was no further increase in nickel-ion

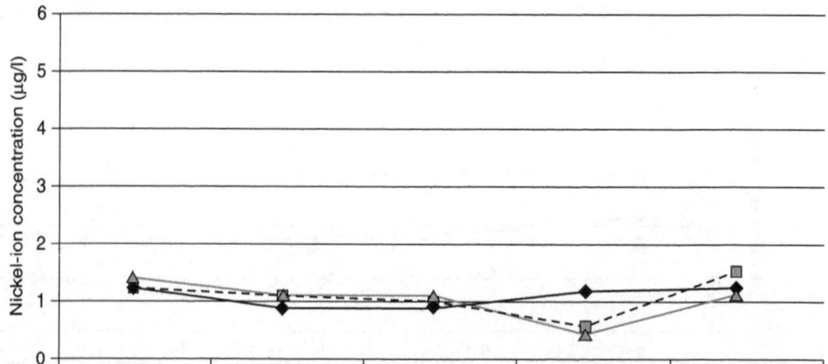

	8 h	4 h	2 h	1 h	0.5 h
Week 18	1.2	0.9	0.9	1.2	1.2
Week 21	1.2	1.1	1	0.6	1.5
Week 26	1.4	1.1	1.1	0.5	1.1

18.10 Graphs of stagnation for weeks 18 to 26 for test line NI105.

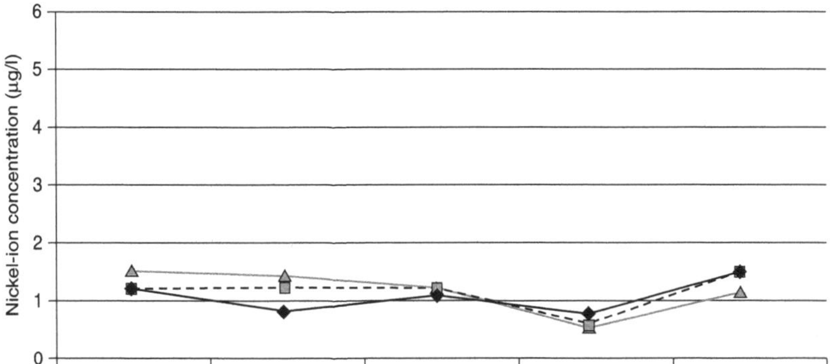

	8 h	4 h	2 h	1 h	0.5 h
Week 18	1.2	0.8	1.1	0.8	1.5
Week 21	1.2	1.2	1.2	0.6	1.4
Week 26	1.5	1.4	1.2	0.5	1.1

18.11 Graphs of stagnation for weeks 18 to 26 for control line 1.4404.

concentration. Accordingly, in the last few test weeks almost constant values from 1.1 to 1.4 µg/l were measured.

Figure 18.12 shows the average metal-ion concentration of the tested nickel-based fillers (NI105, NI107 and Ni60CrPSi) and the reference material (1.4404) over the entire measurement period of 26 weeks. Regardless of the material, significantly different values were observed until week 7. A measurement phase with higher concentration values (e.g. measured phases 1 and 3) was generally followed by lower values (e.g. measured phases 2 and 4). The different values are due to the relatively high margin of error of about 16% and the natural fluctuations in water composition. Between weeks 6 and 18 the metal-ion concentrations decreased significantly because of the formation of the passivation layer. From week 19 on hardly any changes occurred in the metal-ion concentration. The measured values for all tested materials were within the range of 0.9 to 1.5 µg/l.

The different stagnation times simulate normal use by the consumer. Longer consumer absences, such as holidays, are taken into account by removing samples after 48 h (Week 3) and 72 h (Week 13). Figure 18.13 lists all averaged metal-ion concentrations for each stagnation time. It turned out that the values of the tested materials for stagnation times of 0.5 to 8 h are lower than the prescribed limit values. After 48 h the stagnation values for the metal-ion concentration are slightly higher than after 72 h. The reason for this was that the passivation layer had not yet fully developed at the time of removal at week 3 (48 h). At week 13 (72 h), due to the lower concentrations it was assumed that reactions between the materials and the drinking water were inhibited through the passivation layers. In the test line NI107 higher concentrations were measured after 72 h than after 48 h of stagnation. This contradiction can be explained only by analyzing

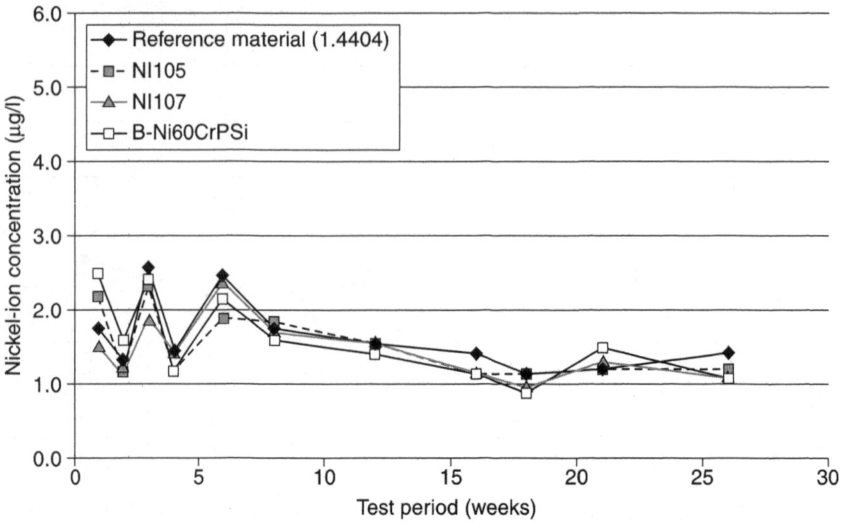

18.12 Average metal-ion concentration of the tested nickel-based fillers measured over the entire experimental period of 26 weeks.

previous inaccuracies. The measured maximum concentration of 7.9 μg/l (1.4404 at 48 h) shows that even after a long stagnation the limit value of 20 μg/l is not exceeded.

The review of the nickel-based alloys Ni60CrPSi, NI105 and NI107 by means of the dynamic test stand trials according to DIN EN 15664-1 showed that at no

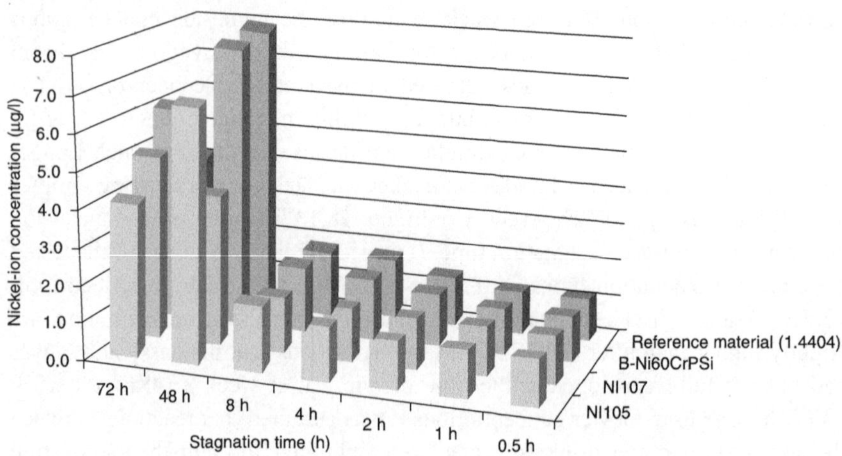

18.13 Average metal-ion concentrations for the individual periods of stagnation.

time, even after stagnation periods of up to 72 h, was the prescribed limit for the nickel-ion concentration exceeded. The concentrations fell significantly during the experiment due to the passivation of the materials and remained almost constant during the final test section.

18.6 Conclusion

In the Drinking Water Directive (DWD), amended on 21 May 2001, the allowable nickel-ion content in drinking water was reduced from 50 μg/l to 20 μg/l. At the same time the development of norm drafts started to adapt or transfer national standards to meet the European level. Thus, for example, DIN EN 15664-1 regulates the amount of leached metallic ions in drinking water and has been valid since June 2008. In installations of drinking water systems nickel-based fillers are often used to join heat exchangers. Hence the danger exists that the limit for leached nickel ions could be exceeded. Until now the use of nickel-based fillers to component joints has been regulated by the 5% rule, which can be found in DIN 50930-6. A prohibition of nickel-based fillers would mean severe economic losses for different parts of the relevant industries, such as the producers of fillers, users and so on. Nickel-based fillers are not yet included in the list of acceptable materials. Investigations have shown that the limit value for the nickel-ion content in drinking water, as required in the Drinking Water Directive, is not exceeded when using various nickel-based fillers. An important comparison has been developed with the reference material (1.4404). There are no significant differences for all investigated nickel-based fillers. In some analyses the nickel-based alloys have even lower values. The tests are valid only for soft, alkaline water, such as exists in the area of Chemnitz.

18.7 References

1. Hoyer, I.: *Beitrag zur Entwicklung von Hochtemperaturloten auf Eisenbasis*. Dissertation. TU Chemnitz 2005.
2. Wielage, B.; Hoyer, I.: *Modifizierte Ni-Basis-Standardlote zum Hochtemperaturlöten von hochlegierten, stark korrosiv beanspruchten Stählen*. AiF Abschlussbericht. AiF-Nr.: 14.439 B. Chemnitz 2007.
3. Steffens, H.-D.; Dammer, R.: Thermodynamisch-metallkundliche Betrachtung während des Hochtemperaturlötens durch Zwangspositionierung der Fügepartner. In: *DVS-Berichte*, Band 92. Hochtemperaturlöten, S. 19–23. Düsseldorf: DVS-Verlag 1984.
4. Steffens, H.-D.; Wielage, B.; Dammer, R.; Lebküchner-Neugebauer, J.: Fertigungstechnische Beeinflussung der Erstarrungsvorgänge beim Hochtemperaturlöten unter Druckaufbringung. In: *Materialwissenschaft und Werkstofftechnik*, Bd. 22, Heft 6, S. 197–202, 1991.
5. Steffens, H.-D.; Wielage, B.; Weiss, B.-Z.: Beitrag zur Phasenanalyse in hochtemperaturgelöteten Verbindungen. In: *DVS-Berichte*, Band 92. Hochtemperaturlöten, S. 77–87. Düsseldorf: DVS-Verlag 1984.

6. Lugscheider, E.; Cosack, T.: Erosionsverhalten von Nickelbasisloten beim Hochtemperaturlöten. In: *Schweissen und Schneiden*, Bd. 41, Heft 7, S. 432–345, 1989.
7. Lugscheider, E.; Aulerich, M.: Innovative Fügetechniken für Mikrobauteile. In: *DVS-Berichte*, Band 192. Hart- und Hochtemperaturlöten und Diffusionsschweissen, S. 265–266. Düsseldorf: DVS-Verlag 1998.
8. Kim, D.S.: *Entwicklung niedrigschmelzender Hochtemperaturlote auf Silizidbasis zum Fügen von Stählen*. Dissertation. Technische Hochschule Aachen 1990.
9. Schittny, T.: *Untersuchungen zum Hochtemperaturlöten mit nicht kapillarem Lötspalt*. VDI.
10. Liessfeld, R.: Die neue Trinkwasserverordnung – wichtige Pflichten der Wasserversorger. *Energie Wasser Praxis*, Heft 1, S. 10–15, 2003.
11. N.N.: DIN 50930-6: Korrosion metallischer Werkstoffe im Innern von Rohrleitungen, Behältern und Apparaten bei Korrosionsbelastung durch Wasser. Teil 6: Beeinflussung der Trinkwasserbeschaffenheit. Berlin: Beuth-Verlag 2001.
12. Werner, W.; Klinger, J: Untersuchungen zur Nickelabgabe von flexiblen Anschlussschläuchen mit vernickelten Verbindern für Sanitärarmaturen. *Energie Wasser Praxis*, Heft 3, S. 67–71, 2009.
13. Pajonk, G.: Korrosionsschäden an gelöteten Plattenwärmetauschern. In: *Sonderbände der praktischen Metallographie*, Bd. 36, S. 371–378, 2004.
14. Stichel, W.: Korrosionsprobleme mit gelöteten Plattenwärmetauschern aus nichtrostendem Stahl. In: *Materials and Corrosion*, Bd. 49, Heft 3, S. 189–194, 1998.
15. Bargel, H.-J.; Schulze, G.: *Werkstoffkunde*. Berlin, Heidelberg: Springer Verlag 2005.
16. Kawe, W.; Martinez, L.: Hochtemperaturlöten von Plattenwärmetauschern. In: *VDI-Berichte*, Nr. 1072, S. 187–192, 1993.
17. Schneemann, K.: Schädigungen an austenitischen Stählen im Kraftwerk. In: *Nichtrostende Stähle im Kraftwerks- und Anlagenbau. Tagungsband zur 6. Jahrestagung Schadensanalyse 1994 in Würzburg*, S. 207–236. Düsseldorf: VDI-Verlag 1994.
18. N.N.: Gelötete Plattenwärmetauscher. Informationsbroschüre der Firma GEA Ecoflex.
19. Klümper, T.H. u.a.: Wasserverwendung – Trinkwasser – Installation. München, Wien: Oldenburg Industrieverlag 2000.
20. Köhler, S.: Schäden an gelöteten Plattenwärmetauschern bei der Erwärmung von Trinkwasser. In: *Materials and Corrosion*, Bd. 50, S. 227–232, 1999.
21. N.N.: DIN EN 14868: Korrosionsschutz metallischer Werkstoffe – Leitfaden für die Ermittlung der Korrosionswahrscheinlichkeit in geschlossenen Wasser-Zirkulationssystemen. Berlin: Beuth Verlag 2005.
22. N.N.: DIN EN 12502-1: Korrosionsschutz metallischer Werkstoffe – Hinweise zur Abschätzung der Korrosionswahrscheinlichkeit in Wasserverteilungs- und speichersystemen. Teil 1: Allgemeines. Berlin: Beuth Verlag 2005.
23. N.N.: DIN 1988-7: Technische Regeln für Trinkwasser-Installationen. Teil 7: Vermeidung von Korrosionsschäden und Steinbildung. Berlin: Beuth Verlag 2004.
24. Lugscheider, E.; Minarski, P.: Untersuchungen zur Korrosionsbeständigkeit hochtemperaturgelöteter Werkstoffe in Trinkwasser. In: *Schweissen und Schneiden*, Bd. 41, Heft 11, S. 590–595, 1989.
25. N.N.: DIN EN 806-4: Technische Regeln für Installationen innerhalb von Gebäuden für Trinkwasser für den menschlichen Gebrauch. Teil 4: Installation. Berlin: Beuth Verlag 2007.

26. Erning, J.W.; Klein, U.: Korrosion von kupfergelöteten Plattenwärmeübertragern – die Folgen. In: *VDI-Berichte*, Nr. 1765. Korrosionsschäden in der Industrie, S. 85–95. Düsseldorf: VDI-Verlag 2003.
27. N.N.: Verordnung zur Novellierung der Trinkwasserverordnung. DVGW Deutsche Vereinigung des Gas- und Wasserfaches 2001.
28. N.N.: DIN EN 15664-1: Einfluss metallischer Werkstoffe auf Wasser für den menschlichen Gebrauch – Dynamischer Prüfstandversuch für die Beurteilung der Abgabe von Metallen; Teil 1: Auslegung und Betrieb Berlin: Beuth-Verlag 2001.
29. N.N.: DIN 50931-1: Korrosionsversuche mit Trinkwässern. Teil 1: Prüfung der Veränderung der Trinkwasserbeschaffenheit. Berlin: Beuth Verlag 1999.
30. Schwenk, W.: Untersuchungen über die Nickelabgabe nichtrostender Chrom-Nickel-Stähle in Trinkwasser. In: *gwf. Das Gas- und Wasserfach. Ausgabe Wasser, Abwasser*, Bd. 133, Nr. 6, S. 281–286, 1992.

19

Fluxless brazing of aluminium

D. K. HAWKSWORTH, Sapa Group, Canada

DOI: 10.1533/9780857096500.3.566

Abstract: The creation of a thermally induced metallic bond between aluminium components in the absence of powdered brazing fluxes has been achieved using a number of different techniques. Oxide disruption to promote the wetting and flow of filler metal on the surface is necessary for brazed joint formation. This can be achieved by making modifications to the brazing sheet surface, filler metal alloy and the use of composite alloy–flux materials. Different inert atmosphere brazing techniques and associated brazing mechanisms are described, including nickel brazing, filler metal alloying, electrochemical surface treatment and composite filler material developments.

Key words: fluxless brazing, self-fluxing, flux-free, surface-modified brazing sheet, filler alloy modified brazing sheet, spray formed composite brazing sheet.

19.1 Introduction

The aluminium brazing industry has since the late 1970s been seeking a method by which products can be brazed without the need for the complex and expensive application of brazing fluxes to components and assembled units prior to brazing. Processes that are less susceptible to fluctuations in oxygen and moisture levels in the brazing atmosphere have also been sought, but are much more difficult to achieve. Careful control and application of flux is a critical step in the brazing of aluminium under an inert atmosphere and is a significant factor in determining quality and recovery. Brazing of aluminium is particularly difficult because an oxide film forms instantaneously on the surface when exposed to air and the barrier action of existing and freshly formed aluminium oxide hinders wetting, inhibits capillary flow and prevents uniform brazed joint formation. To enable intimate contact between the molten filler metal and the base metal it is necessary to disrupt the oxide through mechanical fracture, the use of an inorganic salt that acts as a flux, or through reactions between elements deposited on the surface and the aluminium oxide and underlying metal. To prevent re-oxidation of the molten filler metal an inert brazing atmosphere free from oxygen and water vapour is necessary.

In general, the less control there is of the process, which includes surface cleanliness, component fit-up and brazing furnace atmosphere, the more flux is required to compensate in order to achieve satisfactory brazed joints (Fig. 19.1). However, increasing flux loading often compromises the overall quality (appearance, clogged fin packs, poor thermal performance, coolant degradation and performance) of the brazed product.

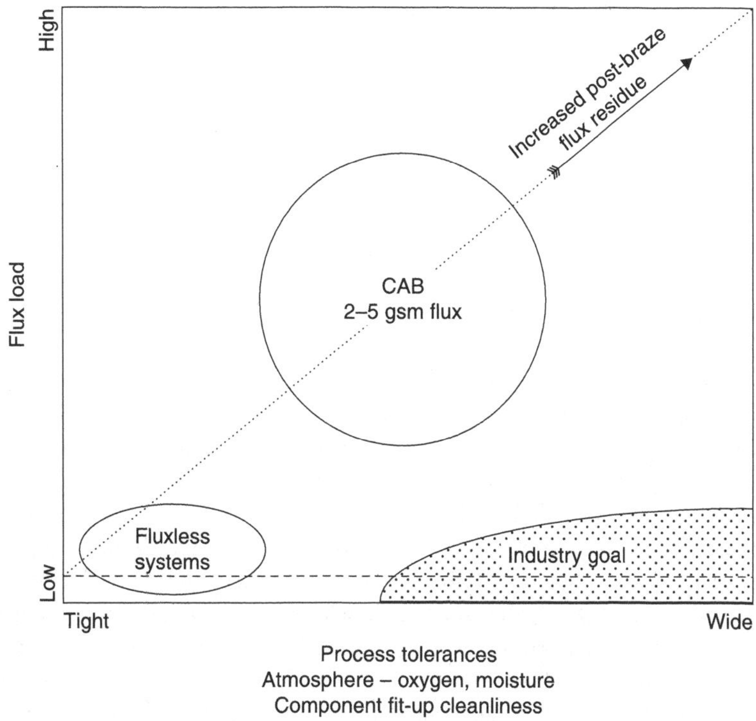

19.1 The influence of flux loading on brazing process control. CAB, controlled atmosphere brazing.

The development of fluxless brazing processes has been driven by the need to simplify manufacturing processes, the evolution of complex heat exchanger designs, the need to minimize residual post-braze flux entrapped inside a brazed product, and growing environmental and health and safety concerns.

19.2 Definition of fluxless brazing

Fluxless brazing refers to the brazing of an assembly of aluminium components, without the need to apply particulate flux to individual components or assembled components prior to brazing, under an inert atmosphere, usually nitrogen. There are two categories, namely:

- Completely flux-free systems (similar to vacuum brazing)
- Self-fluxing material systems.

19.3 Controlled atmosphere brazing process limitations

The use of potassium alumino-fluoride fluxes (based on the KF-AlF_3 system, registered as NOCOLOK Flux and developed by Alcan International during the

1980s) for high-volume mass-production of brazed aluminium automotive heat exchangers is a well-established industrial process. Heat exchangers such as radiators, condensers and evaporators are made up of assembled components with faying surfaces that are predominantly external, exposed to the brazing atmosphere with relatively easy access for flux application. Methods for flux application have included dipping assemblies into flux slurry solutions, spraying of slurry solutions and electrostatic spraying of dry flux onto heat exchangers prior to brazing. Brush or spray bottle application of concentrated slurry in localized sensitive, difficult to braze areas is also sometimes necessary. Binders for applying paint-type flux coatings are also used for some components.

As controlled atmosphere brazing techniques have improved, flux loadings have been reduced from a typical load of 15 to 25 gm^{-2} down to less than 2 gm^{-2} depending on the assembly being brazed and the furnace conditions. Post-braze residues of recrystallized brazing flux and flux oxide reaction products are always left on the surface after brazing. These salt crystals can be in the form of large globules collected on the underside of the brazed product where excess amounts of flux have been added and flowed under gravity, or as a fine layer of crystals (characteristic of the brazing conditions and oxide surface) uniformly covering the aluminium surface. Excessive post-braze flux residues can compromise surface uniformity and dimensions and can also contribute to flux dust generation. Flux particulates on internal brazed surfaces can break free, clogging tube microports, and have also been reported to accelerate degradation of some coolant solutions to form gels (Gershun and Woyciesjes, 2006).

Successful brazing relies on the proper control of oxygen concentration and dew point in the furnace atmosphere to prevent the uncontrolled oxidation of freshly exposed molten filler metal and highly reactive brazing flux. For typical controlled atmosphere brazing using a flux loading <3 gm^{-2}, an oxygen concentration below 100 ppm and a moisture content <128 ppm (dew point lower than −40°C) are generally regarded as the standard (Claesson *et al.*, 1995). Many modern industrial controlled atmosphere brazing furnaces are capable of running with oxygen levels <50 ppm.

Heat exchanger design has also evolved to a point at which flux application becomes a constraint. Examples include compact heat exchangers with high density fin packs and closely spaced tubes with small internal diameters, the use of folded B-type instead of welded tubes or welded tubes with stuffed convoluted tabulators, and complex stacked plate designs. Heat exchangers that have a large internal braze surface area with intricately designed components that have narrow channels and restricted ports are difficult to flux using conventional techniques. Large enclosed internal volumes cannot be easily evacuated and purged with inert gas prior to brazing. Binder decomposition products, where paint flux applications have been used, also become entrapped and these pyrolysates react with both the flux and aluminium surface during brazing, inhibiting brazed joint formation.

Also associated with heat exchanger design is the need for high-strength lower-gauge brazing sheet. These alloys often contain levels of magnesium that are incompatible with the conventional potassium alumino-fluoride brazing flux. As heat exchanger design and manufacturing process technologies have evolved, so has the need to simplify the manufacturing process. Concerns for health and safety in dealing with fine particulate fluoride salts and environmental implications of effluent disposal have also prompted moves to develop brazing technologies that eliminate (or minimize) flux application steps and associated flux handling.

The development and evolution of modern continuous and semi-continuous brazing furnaces with precision thermal profile control and low oxygen and moisture-containing atmospheres has made it possible to braze with very low flux loadings, and in some cases no flux at all. The evolution of aluminium brazing technology and associated process changes are illustrated in Fig. 19.2.

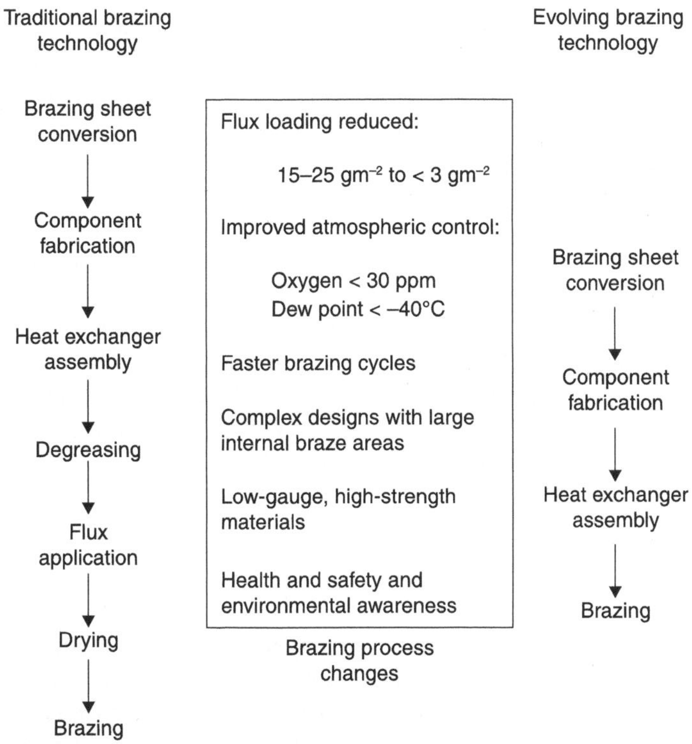

19.2 Brazing technology process evolution.

19.4 Background chemistry and metallurgy influencing fluxless brazing

19.4.1 The metallic bond

To create a thermally induced metallic bond between aluminium components, changes in the adjacent surface phases are necessary. Metallurgical and chemical processes take place both on the surface and within the materials during brazing. Changes in the adjacent surface phases are necessary to allow the atoms of the metal or alloy to contribute electrons to the formation of an 'electron cloud', which is shared by the entire solid, and thus the formation of a metallic (interatomic) bond. A metallic bond is characterized by the conduction of electricity and heat, which is a result of the free movement of electrons through the matrix of positive metal ions.

During brazing, molten filler metal must wet and flow across the surface of a solid base core alloy of the brazing sheet in order to form a good braze joint or brazing fillet. Spreading and wetting will be influenced by the thickness, morphology and composition of the reaction layer. The contact angle is a function of temperature and time and can be attributed to the effects of alloying (base metal dissolution) and composition. These factors all change interfacial energies and the tendency to approach equilibrium. Wetting is a function of the braze filler metal, and the nature of the surface to be joined. In order to wet well, a molten metal must be capable of dissolving or alloying with some of the metal on which it flows, to form a thin reaction layer. Joint strength is influenced by the reaction layer thickness and morphology.

Surface cleanliness affects wetting. Dust particulates, residual grease and machine oils, and surface oxides can impede the uniform wetting and spreading of a molten filler metal. The adjacent surfaces of the parts that are to be brazed must therefore be kept free of such contaminants. The presence of adsorbed molecules on a metal surface markedly reduces the solid/vapour surface energy. This increases the contact angle and retards wetting; hence the importance of clean brazing surfaces. The presence of an oxide layer and small amounts of impurities on the surface can affect the liquid/vapour surface energy. The function of a flux is to remove the oxide layer in the area to be brazed, expose clean base metal and prevent re-oxidation.

19.4.2 Aluminium surface chemistry

Aluminium oxide chemistry and transformations that can occur during component preparation, under the influence of oxygen and water in the furnace atmosphere, and which are influenced by the thermal history of the components during processing have a significant role to play in the brazed joint formation mechanism of all brazing systems, both fluxed and fluxless.

Aluminium is a highly reactive element, deriving stability as a hard, strong white metal from a 4 nm thick oxide film that forms rapidly in 100 ps on exposure to oxygen (Campbell *et al.*, 1999; Emsley, 1991). The conditions of the oxidizing atmosphere have a significant effect on the surface film growth rate, composition, surface topography and crystallography (Field, 1989). The characteristics of the oxide and the transformations that occur during the brazing process and how these influence filler metal behaviour are key to understanding fluxless brazing systems.

During prolonged heating at 500–600°C the oxide grows to 100–400 nm (Jordan, 1956; Olefjord *et al.*, 1994; Wittebrood *et al.*, 1997). At low temperatures ≤300°C the oxide can be thought of as a continuous amorphous film, which thickens as a function of temperature. At high temperatures duplex oxide films develop comprising both amorphous and crystalline oxides. The amorphous inner layer has a thickness of 1–1.5 nm. The growth of the oxide film occurs by the outward diffusion of cations through the oxide film at low temperatures, and the inward migration of oxygen at high temperatures (Eq. 19.1). The rate of film growth or thickening is temperature dependent and ultimate thickness is reached in about 20 h.

$$4Al^{3+} + 12e^{-} + 3O_2 \rightarrow 2Al_2O_3 \qquad [19.1]$$

The oxide of aluminium is amphoteric, acting primarily as a base (Srivatrava and Farber, 1978). It also ionizes to a weak acid. Al_2O_3 appears as various polymorphs and hydrated species depending on the method of formation. There are two anhydrous forms, α-Al_2O_3 and γ-Al_2O_3, and hydrated forms corresponding to the stoichiometries AlO.OH and $Al(OH)_3$. Under moist oxidizing environments at temperatures >360°C up to a limiting temperature of about 550–590°C, boehmite forms according to Eq. 19.2. The dehydration of boehmite occurs at a temperature <450°C. Poorly recrystallized boehmite dehydrates at a temperature as low as 350°C to form γ-Al_2O_3 (Eq. 19.3). At high temperatures aluminium also reacts directly with water to form crystalline γ-Al_2O_3 (Eq. 19.4). Water may come from the environment, or from the dehydration of boehmite (Eq. 19.3).

$$2Al + 4H_2O \rightarrow 2AlO.OH + 3H_2O \qquad [19.2]$$

$$2AlO.OH \rightarrow \gamma\text{-}Al_2O_3 + H_2O \qquad [19.3]$$

$$2Al + 3H_2O \rightarrow \gamma\text{-}Al_2O_3 + 3H_2 \qquad [19.4]$$

19.4.3 Oxide film modification and transformation during brazing

Modification of brazing sheet surfaces using fluoride-ion-based conversion coatings has been attempted with some success. The fluoride ion is thought to disrupt the oxide growth mechanism by virtue of its relative size and its ability to enter the oxide and occupy potential migration sites for oxygen (Prigmore, 1993). The outward diffusion of cations (Al^{3+}) is further disrupted by the formation of complex salts between the cation and the fluoride ion. Aluminium forms a series of complex salts with fluoride ions that range from AlF^{2+} to AlF_6^{3-}, a situation which may prevail in the oxide film as well (Drazic *et al.*, 1983; Lorking and Mayne, 1996). Fluoride-stabilized oxides are assumed to promote the wetting of filler metal during brazing.

Reactive alloying elements such as lithium and magnesium form thick, friable crystalline oxide films in the presence of trace amounts of oxygen in the brazing atmosphere. Magnesium-containing alloys form a mixture of oxides as the braze cycle develops (Gray *et al.*, 2011). Diffusion of Mg from the core transforms the Al_2O_3 to the $MgAl_2O_4$ spinel, and as this phase saturates MgO continues to form. The different spatial volumes occupied by these new complexes within the oxide cause fracture of the oxide, thus allowing the exudation and wetting of molten filler metal. The formation of low-melting Mg_2Si (553°C) within the aluminium–silicon filler alloy also contributes to increased filler metal activity and brazed joint formation.

19.5 Fluxless brazing processes

Brazing under an inert atmosphere with either a self-fluxing or completely flux-free brazing sheet has been achieved with varying levels of success. The manufacture and preparation of these materials add additional levels of complexity, but simplify the downstream brazing process. Three distinct brazing systems have been developed:

- Surface-modified brazing sheet
  - Chemical conversion
  - Mechanical imbedding of flux
- Filler metal alloy modifications
- Composite filler metal–flux materials.

The mechanisms by which brazed joints are formed differ in each system. Surface-modified systems braze by a mechanism similar to conventional systems in which flux is applied to the assembled components. Oxide disruption is initiated from the surface and subsequent melt activity determined by the action of the applied flux or coating. In conventional controlled atmosphere brazing the flux operates at the brazing temperatures by melting, spreading and dissolving the oxide film (Field and Steward, 1987; Jensen, 1969). Surface-initiated brazing mechanisms are sensitive to the condition of the surface (cleanliness), flux loading and quality of chemical modification and brazing atmosphere.

Filler metal alloy modified systems are extremely sensitive to the brazing atmosphere and condition of the oxide. Oxygen and water (dew point) levels in

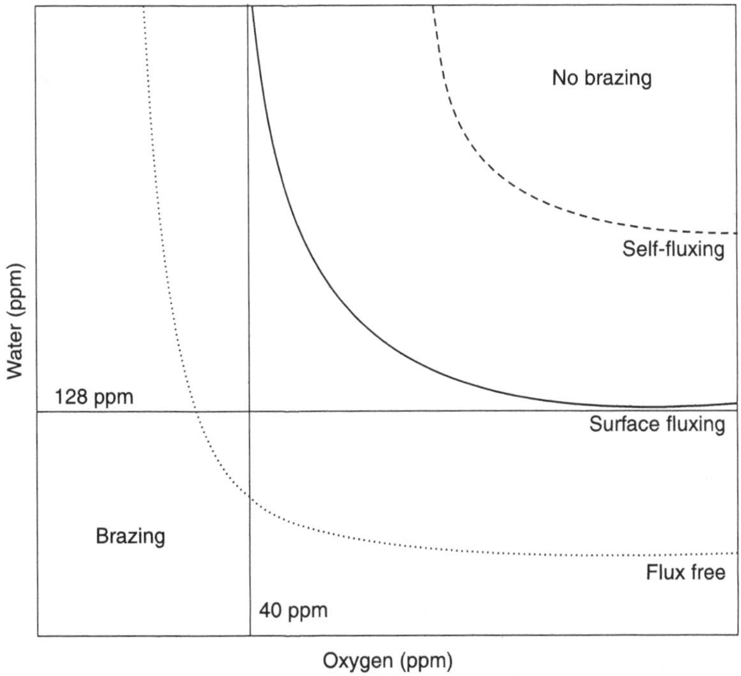

19.3 Relative sensitivity of brazing systems to oxygen and water in the brazing atmosphere. Flux-free systems require very low oxygen and moisture levels (<30 ppm; dew point −50°C). Typical limits for traditional flux-coated systems are 40 ppm oxygen and 128 ppm water (dew point −40°C) whereas self-fluxing composite systems are more tolerant to high levels of oxygen (>2000 ppm) and higher dew points (−7°C).

the brazing atmosphere are critical and have to be held under tight control for effective brazing (Fig. 19.3). Brazing is initiated from below the surface and depends on the natural fracturing of the oxide, which occurs with increased temperature resulting from chemical transformations within the oxide and differences in expansion coefficient. Molten filler metal does not wet an oxide naturally, resulting in relatively slow brazed joint formation, often characterized by the appearance of irregularly shaped brazing fillets and joint porosity.

Composite self-fluxing systems are the least sensitive to atmospheric and surface conditions. During brazing, active flux is liberated from within the filler metal and acts from beneath the surface oxide along the metal–oxide interface, thus lifting the oxide (and any surface debris) and exposing a very active molten filler metal layer that rapidly wets, spreads and flows into the gaps between components to form uniform brazed joints characterized by a smooth fillet meniscus.

19.5.1 Surface-modified systems

Various binderless flux-coating systems have been developed that include the use of plasma, cold spraying and kinetic energy metallization. Pre-fluxed strips consisting of flux that is mechanically embedded in the strip during rolling were evaluated by the Ford Motor Company (Evans, 2000). Alcoa experimented with a high-speed spraying method to embed flux into the brazing sheet surface (Zhao *et al.*, 2007). These techniques, although proven on a small scale and used to produce prototype brazed units, were not suitable for high-volume brazing sheet conversion.

Chemical modifications to the brazing sheet surface involve deposition of a metal such as nickel or the formation of a fluoride-based conversion coating that changes the behaviour of the surface oxide during brazing. Nickel brazing technology is the most successful of the fluxless brazing systems developed during the late 1970s and early 1980s (Cheadle and Dockus, 1988). This nickel-promoted fluxless aluminium brazing technology is owned and used by Dana Canada Corporation (Long Manufacturing). Their patented process for electrodeposition and chemical deposition of nickel (and other metals) onto the surface of clad brazing sheet has been used for commercial mass production of evaporators, oil coolers and other automotive heat exchangers. A thermite reaction between the nickel and the aluminium–silicon alloy causes sufficient disruption of the surface oxide to promote filler metal interaction with adjacent surfaces, creating a brazed joint (Fig. 19.4). Corus have also more recently developed a similar patented technology for nickel brazing (Mooij *et al.*, 2000), but have not supplied any commercial volumes of plated brazing sheet.

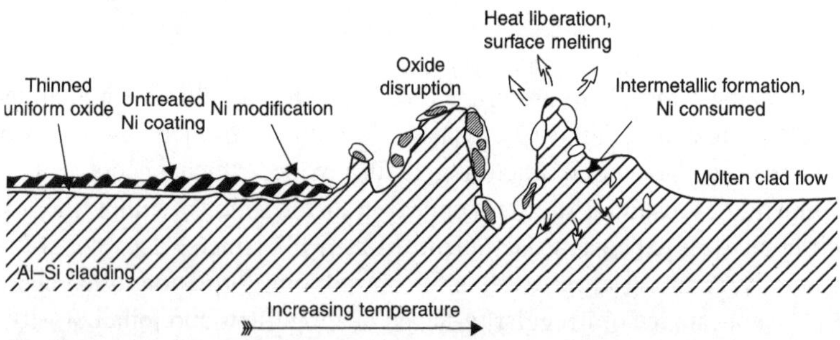

19.4 Schematic representation of the mechanism for nickel-promoted fluxless brazing. The nickel reacts exothermically with the aluminium surface forming a ternary Al–Si–Ni eutectic, which effectively fractures the oxide, enabling flow of the molten filler metal. Filler metal flow is aided by the liberation of heat from the reaction. An intermetallic phase of Al_3Ni is formed, which is dispersed uniformly in the molten cladding. (Illustration provided by Dana Canada Corporation)

Fluoride-ion conversion coatings on aluminium brazing sheet have been used with limited success. These techniques have frequently been used in combination with filler metal alloy modifications (e.g. Bi). Fluoride-ion conversion coatings are assumed to inhibit oxide film growth. During brazing the fluoride complex is active at the interface between oxide and metal, lifting the oxide and thereby promoting filler metal flow.

Childree (1999) reported that treatment of clad aluminium components in dilute aqueous NaF or KF solutions improved brazed quality without the addition of the flux. The formation of an oxyfluoroaluminate was assumed to promote wetting of the molten filler metal. This technique, when used in combination with calcium additions (0.01–0.1%) to the filler, improved brazing of Mg-containing alloys. Safrany *et al.* (2007) created a conversion coating by pretreating the clad brazing sheet surface with an alkaline solution, and thereafter a sulfuric acid electrolyte containing K^+ and F^- ions added in the form of a $KCl+NH_5F_2$ mixture. The low pH of the electrolyte, less than 3, was necessary to activate the surface to promote stable fluoride complex formation.

19.5.2 Filler metal alloy modified systems

Alloying of aluminium–silicon filler metals with elements such as magnesium, lithium, bismuth and strontium is a common approach that has been used to modify the properties of molten filler metal (Schultze and Schoer, 1973; Schmatz and Winterbottom, 1983; Childree, 1999). Surface active elements such as lithium are reported to reduce surface tension of the molten filler metal, thereby promoting wetting. Elements such as strontium can influence the viscosity of molten filler and hence capillarity in a brazed joint. A number of specialized filler metal alloy variants have been used, without the need for flux application, for niche products brazed under carefully controlled atmospheres, and in some cases in combination with an additional outer clad layer. This multiclad brazing sheet consists of a thin unalloyed outer cladding, which dissolves in the inner eutectic filler layer during brazing, thus disrupting the residual oxide. Provided the atmosphere is completely inert so that re-oxidation cannot recur, the molten filler metal can flow into the gaps between components to produce brazed joints.

Furukawa has developed a multiclad fluxless inert-atmosphere brazing process referred to as exudation brazing (Takeno and Suzuki, 2005). A five-layer brazing sheet is used with an external clad layer of high-purity aluminium alloy. The technique relies on the exclusion of external oxygen by the formation of close-contact overlap joints between components. During brazing the thin outer clad layer dissolves in the molten filler metal, resulting in the fragmentation of the surface oxide. Exudation through the fragmented oxide allows wetting of the molten filler and the formation of a brazed joint without the addition of a brazing flux. This process is dependent on exudation of molten filler metal through the oxide skin; wetting of the surfaces to be joined before oxidation of the molten

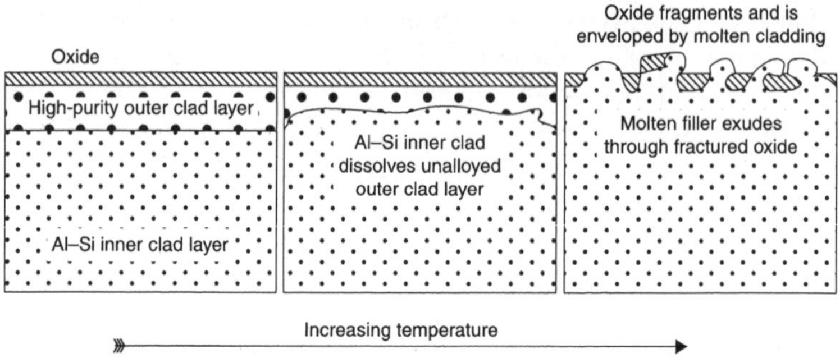

19.5 Schematic representation of the mechanism for multiclad fluxless exudation brazing. On heating, the eutectic Al–Si inner cladding melts and dissolves the high-purity outer cladding, thereby fragmenting the surface oxide. Molten filler metal then exudes through the disrupted oxide, enveloping it in the process. Provided re-oxidation of the exuded filler does not occur, brazed joints can be formed.

filler can occur (Fig. 19.5). This fluxless brazing process appears to have limited application and cannot be used for low-melting-point copper-containing alloys, casting alloys and alloys with magnesium levels exceeding 1%.

Wittebrood *et al.* (2010) and Kirkham *et al.* (2010) demonstrated that it was possible to braze modified aluminium–silicon alloys clad to brazing sheet under conditions of low oxygen content. They assessed the brazing performance using an angle-on-coupon brazing test of various flux-free alloy variants under brazing atmospheres of different oxygen concentrations. Where the oxygen levels exceeded 30 ppm incomplete brazed joints were formed, and in some cases less than 10 ppm oxygen was required to produce a satisfactory brazed joint. Brazed joint formation was very slow, with gradual wicking of the filler metal along the interface between faying surfaces as molten filler exuded through cracks in the residual oxide.

Janssen (2010) demonstrated that suitable brazed joints could be achieved by taking advantage of a local micro-atmosphere effect by 'brazing in a box'. Aluminium–silicon filler metal alloyed with magnesium, indium and bismuth was used. The indium and bismuth were added to reduce surface tension of the molten filler metal, and the magnesium acted as an oxygen scavenger or getter. To minimize interaction of molten filler metal and the core alloy it was recommended that the brazing cycle time be reduced when brazing without flux under an inert atmosphere.

19.5.3 Composite filler metal–flux materials

Self-fluxing liner materials consisting of a eutectic filler alloy with finely dispersed flux particles within the alloy matrix have been produced by two different methods. The first involves mixing a filler alloy powder with an appropriate ratio of potassium fluoro-aluminate flux and compressing into a solid as described by Iwai (1992) and Dromard (2004). The second is for the production of a spray-formed composite that can be further fabricated into a liner plate for cladding onto brazing sheet (Ogilvy *et al.*, 2008).

The Showa patent of 1994 describes a mixture of flux and filler metal powder compressed to form a brazing skin, which could be laminated to a core alloy (Koichiro, 1995). This material has been used to produce brazed joints without the need for additional applied flux. Flux and metal powders with defined particle size achieved by pulverizing to ensure homogeneity were compacted by hot pressing to form a rigid integral piece. Having been produced from pulverized powder, this material contains considerable amounts of internal oxide and interconnected porosity and therefore requires increased relative amounts of flux. This renders the compressed material less malleable and prone to cracking during rolling and forming. This technology is not known to be in commercial use.

Iwai (1992) describes a 'brazing agent' that forms a brazing sheet cladding. The material is made up of a mixture of blended powders consisting of an aluminium powder of 99.5% purity, a silicon powder, a zinc powder, a tin powder, an indium powder and a fluoride flux powder of aneutectic composition of KF and AlF_3. Blended powders are placed in an oven and vacuumed at 500°C, after which the batch is heated to 480°C and subjected to a hot pressing process. The resulting block is then hot extruded at 500°C. Good braze test results were achieved using either a 5 or a 10% silicon powder and a flux content of 5, 8 or 10%. Assembled test pieces heated under nitrogen to a brazing temperature of 600 to 620°C for 10 min produced brazed joints. Brazed joints were also reported to have been produced in the absence of nitrogen. A similar hot isostatic-pressing technique used to produce a solid bar that can be rolled into a form suitable for brazing is described by Dromard (2004). In this case a powdered Zn–Al alloy with a particle size of 50 to 300 μm was mixed with 10% cryolite flux with a particle size of between 1 μm and 10 μm.

Rheocasting or thixoforming has been used to produce a brazing composite as described in Koichiro's patent (1995). A molten aluminium silicon alloy was allowed to solidify to a semi-molten state, upon which flux powder was added and the semi-molten mixture stirred and subsequently allowed to cool and solidify. It was claimed that this composite material could be used to produce a brazed joint without further application of flux to the test pieces.

None of the three processes described above are known to be in commercial use as far as can be ascertained.

The use of pressed solid alloy powders and thixotropic mixing to make a self-fluxing material can result in the inclusion of excessive oxides, which can have a detrimental impact on brazability. Minimizing or even eliminating oxide inclusion can be overcome by spray forming a metal–salt composite under an inert atmosphere using cospray technology developed by Sandvik Osprey. Working in collaboration, Sapa AB and Sandvik Osprey have invented a metal matrix composite comprising a eutectic aluminium–silicon alloy within which small particles of potassium alumino-fluoride salt are dispersed (Ogilvy *et al.*, 2008). The use of this composite material for brazing has been registered as Trillium™ Technology.

The cospray technique used to produce a solid spray-formed composite material involves consolidating an atomized stream of molten metal alloy into a spray of droplets into which salt particles are introduced. These coincident streams are collected and cooled to produce a composite solid that can be further processed by either one or a number of possible conversion processes. Conversion processes may include direct collection, forging, extrusion and rolling to manufacture either a composite liner-clad brazing sheet or a composite brazing sheet, wire or shim. The spray-forming technique results in very little oxidation of the materials and the inherently low oxygen content of the composite does not interfere with the effectiveness of the salt, which forms a flux on the surfaces to be joined during brazing.

A spray-formed composite material is characterized by non-macro-segregated, rapidly solidified phases of silicon and aluminium. Primary silicon particles have an average size less than 1 μm and a maximum size typically less than 5 μm (Fig. 19.6). The Si particle sizes in spray-formed composite materials are much smaller than those made using casting, including direct chill casting or rheocasting processes. Small Si particles melt rapidly during brazing, enhancing

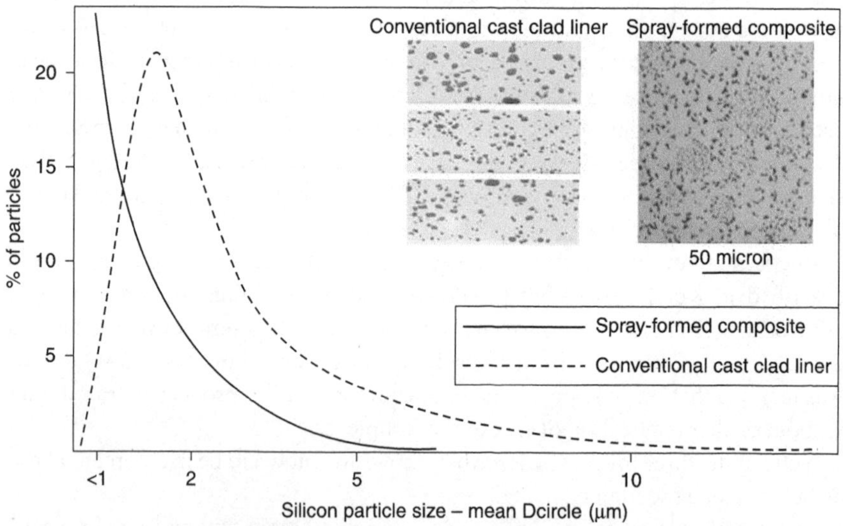

19.6 Relative silicon particle size in composite spray-formed filler material (light particles) and conventional cast liner cladding. Micrographs of the spray-formed composite show salt particles (dark), which are of the same size as the silicon particles in the cast liner material.

filler metal flow and brazed joint formation. These small uniformly distributed silicon particles are generated without the addition of elements such as strontium, which create nucleation sites for precipitation. Excess levels of strontium are known to have a detrimental effect on braze joint formation. The small finely distributed silicon particles are also advantageous when extremely thin clad layers are required, making it possible to create a continuous film of molten metal on the surface rather than discontinuous localized pools of molten metal, which can interact with and dissolve underlying core material, resulting in non-homogeneous post-braze metallurgical properties.

The particle size of the inorganic salt phase distributed within the metal matrix is bi-modal in character. Extremely fine salt particles less than 1 μm in size are observed within the spray-formed matrix. Some particles as small as 50 nm have been observed using transmission electron microscopy and are generally located at grain triple points (Fig. 19.7). A micro-segregated phase of larger particles and

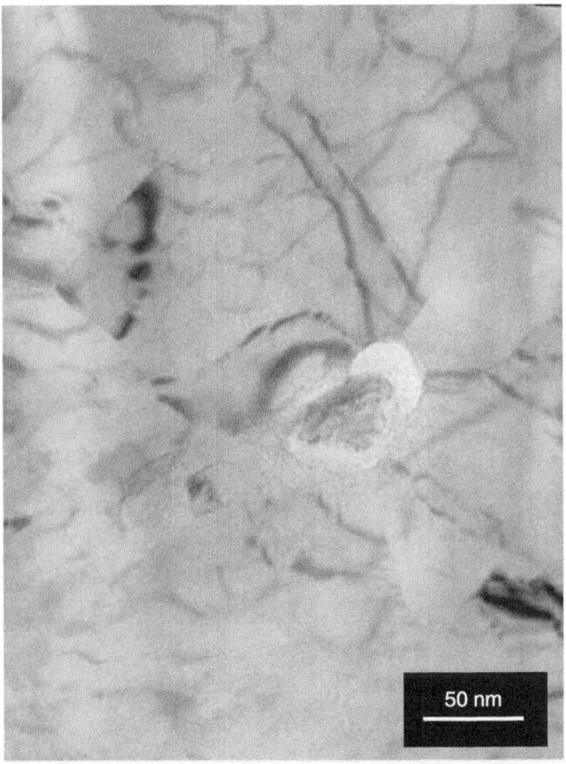

19.7 A transmission electron micrograph showing a grain boundary triple point at which a nanoparticle of salt is located within a spray-formed composite of aluminium–silicon alloy and potassium fluoro aluminate salt. (Innoval Technology, Sapa Heat Transfer)

agglomerates, ranging from 5 up to 200 μm in size respectively, corresponds to the last parts of the composite to solidify. Differential scanning calorimetry displays a small endotherm around 550°C (sometimes followed by a second melting peak at 563°C) corresponding to the onset of fluxing activity. A sharp eutectic endotherm at 577°C corresponds to the melting of the aluminium–silicon alloy.

Composite filler material has been evaluated under a wide range of brazing conditions in both the laboratory and industrial controlled atmosphere brazing furnaces. The material has proven to be extremely robust with improved fluxing activity and shows less vulnerability and sensitivity to conditions that compromise braze joint formation using conventional fluxing and especially other fluxless systems.

A study of brazing under severe background atmospheric alterations by Yu *et al.* (2012) found that spray-formed composite (Trillium)-clad brazing sheet showed much greater resilience to increased oxygen levels and humidity content than traditional clad brazing sheet coated with brazing flux. The limits for traditional brazing sheet with a flux loading of about 10 gm^{-2} were found to be about 200 ppm oxygen and a dew point temperature of −18°C. More resilient Trillium-clad brazing sheet produced consistent uniform brazed joints under a brazing atmosphere with oxygen content in excess of 2000 ppm and a dew point temperature less than −7°C (Fig. 19.8).

The wetting and flow of Trillium composite filler in a brazed joint observed in a laboratory glass furnace can be extremely rapid, with the formation of a uniform, evenly distributed residuum on the surface of the brazing sheet and smooth symmetrical brazing fillets. This very active molten filler material can fill larger gaps than is generally accepted, forming smooth uniform brazing fillets between

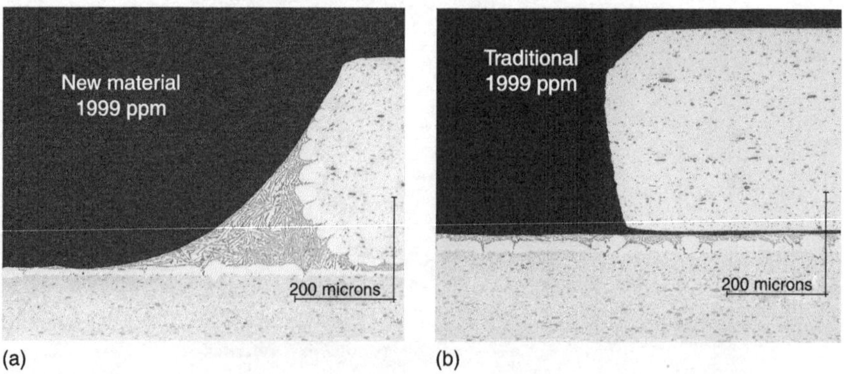

19.8 Cross-sections of comparative joint zone of Trillium brazing sheet (left) and standard Al–Si clad brazing sheet coated with 10 gm^{-2} of flux (right) brazed in an atmosphere with an oxygen content of 1999 ppm. (University of Kentucky, Sapa Heat Transfer)

components without excessive interaction with the solid core alloy (Fig. 19.9 and 19.10). Lubricant residues have little influence on brazing and do not impart the same levels of post-braze discoloration as observed with alternative brazing systems. Products have a very clean shiny post-braze appearance (Fig. 19.11), the throwing power of the flux released from the composite acting on the unclad surface as well.

Differences in brazing mechanisms between traditional fluxed brazing sheet and composite-clad brazing sheet can be explained in terms of the location of the flux and the fluxing action during brazing (Fig. 19.12).

Flux applied to the surface of conventional brazing sheet sits on top of the oxide and any other surface debris. During the heating cycle this flux is exposed

19.9 Surface topography and brazed joint appearance of a composite spray-formed liner-clad radiator tube and an unfluxed header plate. (Sapa Heat Transfer)

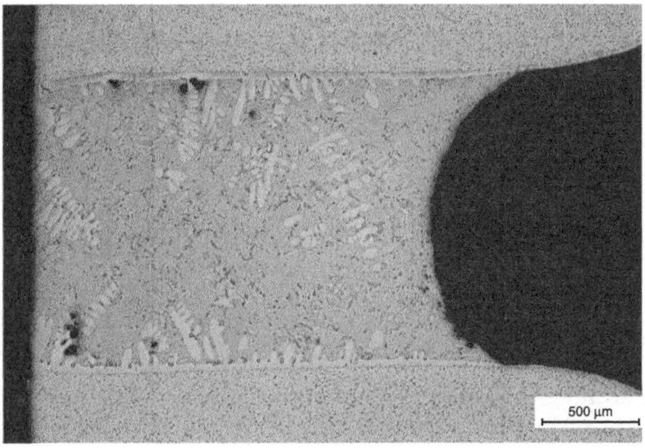

19.10 Microsection of a plate-to-plate brazed joint showing the excellent gap filling properties of spray-formed composite filler metal. (Sapa Heat Transfer)

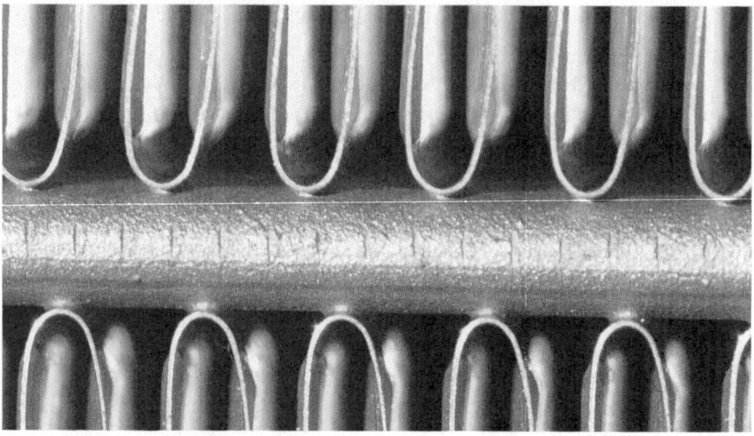

19.11 The general appearance of a spray-formed composite clad brazed radiator. (Sapa Heat Transfer)

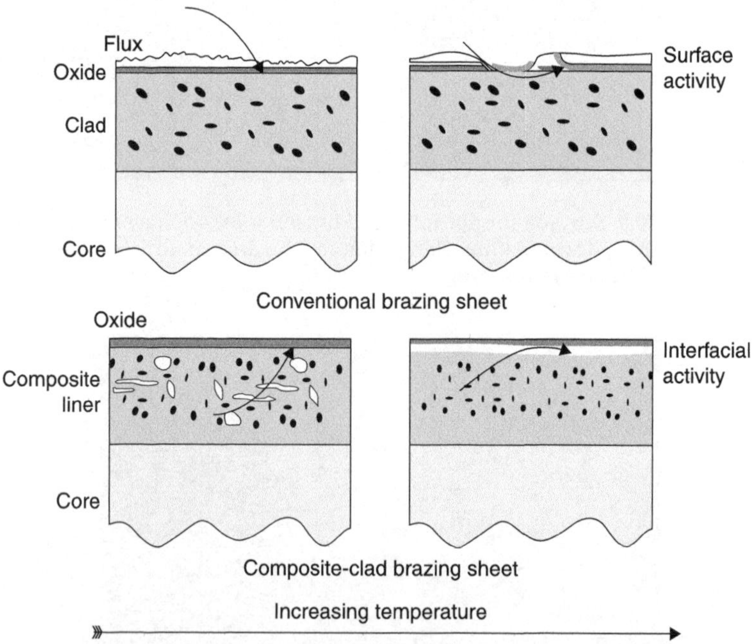

19.12 Schematic representation of comparable mechanisms for composite-clad brazing sheet and conventional flux-coated brazing sheet. The fluxing action of conventional brazing sheet is activated from the surface of the oxide. The fluxing action of composite and clad brazing sheet is initiated from within the clad layer and acts on the interface between the clad and the oxide.

to the furnace atmosphere and reacts with any residual oxygen and moisture that might be present. This has the initial effect of cleaning the furnace atmosphere when very low levels of oxygen and humidity are present. However, there is a threshold above which joint formation rapidly deteriorates. Flux reaction products include HF and aluminium oxide, which, together with the growth in thickness of surface oxide, especially in the presence of high moisture levels, consume the applied flux. The melting point of the reacted flux also rapidly increases, often causing the salt to solidify. These combined effects prevent the disruption of the surface oxide, thus inhibiting filler metal flow and brazed joint formation.

The fluxing action of composite brazing sheet is initiated from within the filler metal-clad layer and beneath the residual surface oxide. This fluxing activity can only commence when the filler metal starts to melt and the immiscible molten salt segregates towards the surface. Prior to melting, the oxide prevents the flux from reacting with atmospheric and other surface contaminants. Once molten, the immiscible molten flux acts on the interface between the metal and the oxide, detaching the oxide to allow molten filler metal to wet, spread and flow into closely spaced gaps between components to form very good brazed joints on cooling.

19.6 Conclusion: a summary of fluxless brazing processes

In conclusion, a number of fluxless brazing processes, often application dependent, have been developed and implemented with varying levels of success. These processes are summarized as follows.

19.6.1 Surface-modified systems

Nickel-plated brazing sheet

- Successful proprietary industrial process with limited use
- Conventional controlled atmosphere brazing conditions required
- Thermite reaction between nickel and filler metal disrupts surface oxide, thereby promoting wetting and brazed joint formation

Fluoride-ion conversion coatings

- Fluoride-ion complex inhibits oxide film growth, contributes to filler metal flow and joint formation
- Successful laboratory results reported
- Good quality brazing atmosphere required: O_2 <40 ppm
- Often used in combination with filler metal alloy modifications

Mechanical binderless flux coatings

- Similar properties to conventional flux-applied systems

19.6.2 Filler metal modified systems

- Surface and alloy active element additions promote wetting and oxide fracture
- Oxide fracture is essential to allow filler metal interaction, wetting and brazed joint formation
- Very clean brazing atmosphere with oxygen levels less than 30 ppm is required to prevent re-oxidation of molten filler metal
- Localized atmosphere effects where magnesium acts as an oxygen scavenger can promote brazed joint formation

19.6.3 Composite filler metal systems

- Brazing flux is liberated from within the filler when it melts
- Less sensitive to furnace atmosphere and surface contaminants than other systems
- Filler metal is very active, melts rapidly, forms stable melt layer and uniform brazed joints

19.7 References

Campbell T, Kalia R, Nakano A, Vashishta P, Ogata S and Rodgers S, 1999, Dynamics of Oxidation of Aluminum Nanoclusters using Variable Charge Molecular-Dynamics Simulations on Parallel Computers, *Physical Review Letters* 82, 4866.

Cheadle B and Dockus K, 1988, *Inert Atmosphere Fluxless Brazing of Aluminum Heat Exchangers*, SAE paper 880446.

Childree D, 1999, *flux and fluxless brazing of Al-Si filler metals containing low levels of calcium*, IMechE C543/045, 493–514.

Claesson E, Engstrom H, Holm T and Scholin K, 1995, *Nitrogen Flow Optimization System in CAB (NOCOLOK) Furnaces*, VTMS Conference, London, England.

Drazic D M, Zecevic S K, Atanasoski R T and Despic A R, 1983, The effect of Anions on the Electrochemical Behaviour of Aluminium, *Electrochim. Acta* 28, 751.

Dromard J, 2004, *Fabrication of a brazing product by isostatic compression of a mixture of powders of scouring flux and filler metal to form a homogeneous solid block*, Patent Application FR 2855085.

Emsley J, 1991, *The Elements*, 2nd Ed., Oxford, Oxford University Press.

Evans, T, 2000, *Method for making brazing sheet*, US Patent 6120848.

Field D, 1989, Oxidation of Aluminium and its Alloys, *Treatise on Material Science and Technology*, Vol. 31, Academic Press, New York, pp. 523–537.

Field D and Steward N, 1987, *Mechanistic aspects of the Nocolok flux brazing process*, SAE Technical Paper Series, 870186, International Congress and Exposition, Detroit, Michigan, 23–27 February 1987.

Gershun A and Woyciesjes P, 2006, Engine coolants and coolant system corrosion, *ASM Handbook, Corrosion: Environments and Industries*, Vol. 13C, pp. 531–537.

Gray A, Butler C and Bosland A, 2011, *A real time FTIR study of the oxidation behaviour of clad sheet during controlled atmosphere brazing*, 2nd International Congress HVAC&R, Dusseldorf.

Iwai I, 1992, *A brazing agent and a brazing sheet both comprising an aluminum alloy containing a flux*, EP 0 552 567.

Janssen H, 2010, *Materials for Fluxfree Brazing*, 6th International Congress Aluminium Brazing, Dusseldorf, Germany, 5 May 2010.

Jensen B, 1969, *Phase and structure determination for new complex alkali aluminium fluorides*, Institute of Inorganic Chemistry, Norwegian Technical University, Trondheim.

Jordan M F, 1956–7, *Journal of the Institute of Metals*, 85, 33–35.

Kirkham S, Wittebrood A and Jacoby B, 2010, *The evolution of flux-free brazing of aluminium*, 6th International Congress Aluminium Brazing, Dusseldorf.

Koichiro F, 1995, *Production of flux-containing aluminium alloy brazing filler metal*, Japanese Patent JP7001185.

Lorking K F and Mayne J E O, 1996, *Brit. Corros. J.*, 1, 181–182.

Mooij J, Wittebrood A and Joseph J, 2000, *Brazing sheet product and method of its manufacture*, US Patent 6379818.

Ogilvy A, Hawksworth D and Abom E, 2008, *A brazing piece comprising a composite material including an inorganic flux*, UK Patent Application GB2447486.

Olefjord I, Stenqvist T and Karlsson Å, 1994, *Surface oxides of braze clad aluminium*, ICAA-4, Atlanta.

Prigmore R M, 1993, personal communication.

Safrany S, Mediouni M, Henry S and Dulac S, 2007, *Method for forming a conversion layer on an aluminium alloy product prior to fluxless brazing*, US Patent 2007/0204935.

Schmatz D and Winterbottom W, 1983, A fluxless process for brazing aluminium heat exchangers in inert gas, *Welding Journal*, 31–38.

Schultze W and Schoer H, 1973, Fluxless brazing of aluminum, *Welding Journal*, October, 644.

Srivatrava R D and Farber M, 1978, Thermodynamic Properties of GrpIII Oxides, *Chem. Rev.*, 78, 627.

Takeno S and Suzuki Y, 2005, *Development of Thermal Products Using the EB Method*, Furukawa Review No. 27.

Wittebrood A, Jacoby B and Kirkham S, 2010, *The Evolution of Flux-free Brazing of Aluminium*, 14th International Invitational Aluminium Brazing Seminar, Novi, MI, USA, 28 October 2010.

Wittebrood A, van der Veldt T, Vieregge K and Haszler A, 1997, VTMS 3, SAE971858.

Yu C-N, Hawksworth D, Lui W and Sekulic D, 2012, *Al brazing under severe alterations of background atmosphere: A new vs. traditional brazing sheet*, 5th International Brazing and Soldering Conference, Las Vegas, Nevada, USA, April.

Zhao Z, Gillispie B, Irish M and Rosen J, 2007, *Kinetic spray deposition of flux and braze alloy composite particles*, US Patent 00710029370.

Index